# Chronic Allograft Failure

## Natural History, Pathogenesis, Diagnosis and Management

Nasimul Ahsan, MD, FACP, MBA

Department of Medicine
Mayo Clinic – College of Medicine
Mayo Clinic Transplant Center
Jacksonville, Florida, USA

LANDES BIOSCIENCE
AUSTIN, TEXAS
USA

# CHRONIC ALLOGRAFT FAILURE:
## NATURAL HISTORY, PATHOGENESIS, DIAGNOSIS AND MANAGEMENT

Landes Bioscience

Please address all inquiries to the Publisher:
Landes Bioscience, 1002 West Avenue, Austin, Texas, USA 78701
Phone: 512.637.6050; FAX: 512.637.6079
www.landesbioscience.com

Cover Art: Chimera. Apulian red-figure dish, ca. 350-340 BC. Musée du Louvre, Paris, France. Campana Collection, 1861.

ISBN: 978-1-58706-153-0

**Library of Congress Cataloging-in-Publication Data**

Chronic allograft failure : natural history, pathogenesis, diagnosis, and management / [edited by] Nasimul Ahsan.
   p. ; cm.
 Includes bibliographical references and index.
 ISBN 978-1-58706-153-0
 1. Homografts. 2. Transplantation of organs, tissues, etc. 3. Graft rejection. I. Ahsan, Nasimul.
 [DNLM: 1. Transplantation, Homologous--adverse effects. 2. Graft Rejection. 3. Replantation. WO 660 C557 2008]
 RD120.78.C47 2008
 617.9'54--dc22

         2008012648

# ABOUT THE EDITOR

NASIMUL AHSAN is a Professor of Medicine of Mayo Clinic – College of Medicine and a Consultant of Mayo Clinic Transplant Center, Jacksonville, USA. His main research interests include transplant immunology and immunosuppressive management of transplant patients. He has authored numerous peer-reviewed articles and previously edited a book entitled *Polyomaviruses and Human Diseases*. He is a member of several national and international scientific organizations. Nasimul Ahsan's academic degrees include Doctor of Medicine from Chittagong Medical College (Chittagong, Bangladesh) and Master in Business Administration from Kellogg School of Management, Northwestern University (Evanston, Illinois, USA). He has obtained training in Pathology (University of Maryland, Baltimore, Maryland, USA and Penn State University – College of Medicine, Hershey, Pennsylvania, USA), Family Practice and Internal Medicine (Eastern Virginia Graduate School of Medicine, Norfolk, Virginia, USA) and Clinical Nephrology (University of Texas, Southwestern Medical School, Dallas, Texas, USA).

# DEDICATION

To an outstanding clinician and teacher, Ali A. Choudhury, MD, FRCP (London), who inspired me to obtain training in Nephrology and Transplantation at the prestigious University of Texas Southwestern Medical Center in Dallas, Texas, USA.

To my parents for their everlasting love. To my family: Arzumand (wife), Shaon (daughter), and Naveed (son) for their constant support and dedication.

# CONTENTS

# EDITOR

## Nasimul Ahsan

Department of Medicine
Mayo Clinic – College of Medicine
Mayo Clinic Transplant Center
Jacksonville, Florida, USA
Email: ahsan.nasimul@mayo.edu
*Chapter 37*

# CONTRIBUTORS

Rodolfo Alejandro
Clinical Islet Transplant Program
Diabetes Research Institute
University of Miami Miller School of Medicine
Miami, Florida, USA
Email: ralejand@med.miami.edu
*Chapter 30*

Mohammad Ali
Department of Hepato Biliary Pancreatic Surgery and Liver
BIRDEM Hospital
Shahbag, Bangladesh
Email: hbbirdem@yahoo.com
*Chapter 17*

Francisco G. Alvarez
Department of Transplant Medicine
Mayo Clinic
Jacksonville, Florida, USA
Email: alvarez.francisco@mayo.edu
*Chapter 15*

John A. Belperio
Department of Medicine
Division of Pulmonary and Critical Care Medicine
    and Pediatrics
David Geffen-School of Medicine
University of California – Los Angeles
Los Angeles, California, USA
Email: jbelperio@mednet.ucla.edu
*Chapter 13*

Roy D. Bloom
Renal-Electrolyte and Hypertension Division
University of Pennsylvania Medical Center
Philadelphia, Pennsylvania, USA
Email: rdbloom@mail.med.upenn.edu
*Chapters 3, 39*

Hugo Bonatti
Department of Surgery
University of Virginia Health System
Charlottesville, Virginia USA
Email: hbe3@hscmail.mcc.virginia.edu
*Chapter 35*

Jon Carthy
The James Hogg iCAPTURE Centre for Cardiovascular
    and Pulmonary Research
St. Paul's Hospital
Vancouver, British Columbia, Canada
Email: jcarthy@mrl.ubc.ca
*Chapter 10*

Jeremy R. Chapman
Department of Renal Medicine
Westmead Hospital
Westmead, Australia
Email: jeremy_chapman@wsahs.nsw.gov.au
*Chapter 25*

Pauline Chen
Division of Transplant Surgery
The Mount Sinai Medical Center
New York, New York, USA
Email: pwchen@post.harvard.edu
*Chapter 18*

Jeffrey S. Crippin
Division of Gastroenterology
Washington University School of Medicine
St. Louis, Missouri, USA
Email: jcrippin@wustl.edu
*Chapter 20*

Patrick G. Dean
Division of Transplantation Surgery
Mayo Clinic College of Medicine
Rochester, Minnesota, USA
Email: dean.patrick2@mayo.edu
*Chapter 31*

Misha Denham
Clinical Islet Transplant Program
Diabetes Research Institute
University of Miami Miller School of Medicine
Miami, Florida, USA
Email: mdenham@med.miami.edu
*Chapter 30*

David M. Dickinson
Scientific Registry of Transplant Recipients
*and*
Arbor Research Collaborative for Health
Ann Arbor, Michigan, USA
Email: david.dickinson@arborresearch.org
*Chapter 4*

B. Dominguez-Gil
Organizacion Nacional de Transplantes
Hospital 12 de Octubre
Madrid, Spain
Email: bdominguez@msc.es
*Chapter 36*

Cinthia B. Drachenberg
Department of Pathology
University of Maryland School of Medicine
Baltimore, Maryland, USA
Email: cdrac001@umaryland.edu
*Chapter 33*

David B. Erasmus
Lung Transplant Program
Department of Transplant Medicine
Mayo Clinic
Jacksonville, Florida, USA
Email: erasmus.david@mayo.edu
*Chapter 14*

Raquel N. Faradji
Clinical Islet Transplant Program
Diabetes Research Institute
University of Miami Miller School of Medicine
Miami, Florida, USA
Email: rfaradji@med.miami.edu
*Chapter 30*

Thomas W. Faust
Gastroenterology Division
University of Pennsylvania Medical Center
Philadelphia, Pennsylvania, USA
Email: thomas.faust@uphs.upenn.edu
*Chapter 3*

Ryan Fields
Department of Surgery
Washington University School of Medicine
St. Louis, Missouri, USA
Email: fieldsr@wustl.edu
*Chapter 12*

Maria Tercsa Gandolfo
Departments of Pathology
Johns Hopkins University School of Medicine
Baltimore, Maryland, USA
Email: mgandol1@jhu.edu
*Chapter 6*

Xochiquetzal J. Geiger
Department of Laboratory Medicine and Pathology
Mayo Clinic
Jacksonville, Florida USA
Email: geiger.xochiquetzal@mayo.edu
*Chapter 26*

Matthias Glanemann
Department of General, Visceral and Transplantation Surgery
Charité, Campus Virchow-Klinikum
Universitätsmedizin Berlin
Berlin, Germany
Email: matthias.glanemann@charite.de
*Chapter 19*

Trudie Goers
Department of Surgery
Washington University School of Medicine
St. Louis, Missouri, USA
Email: goerst@wustl.edu
*Chapter 12*

Atoussa Goldar-Najafi
Lahey Clinic Medical Center
Burlington, Massachusetts, USA
Email: atoussa.goldar-najafi@lahey.org
*Chapter 21*

Lee R. Goldberg
Cardiovascular Disease Division
University of Pennsylvania Medical Center
Philadelphia, Pennsylvania, USA
Email: lee.goldberg@uphs.upenn.edu
*Chapter 3*

Brigette Gomperts
Department of Medicine
Division of Pulmonary and Critical Care Medicine
    and Pediatrics
David Geffen-School of Medicine
University of California – Los Angeles
Los Angeles, California, USA
Email: bgomperts@mednet.ucla.edu
*Chapter 13*

Elizabeth K. Gross
Department of Surgery
University of Minnesota
Minneapolis, Minnesota USA
Email: gross374@umn.edu
*Chapter 32*

Rainer W.G. Gruessner
Department of Surgery
University of Arizona
Tucson, Arizona USA
Email: rgruessner@surgery.arizona.edu
*Chapter 32*

Amanda Jabin Gustafsson
Department of Clinical Research and Education
Karolinska Institutet
Stockholm South General Hospital
Stockholm, Sweden
Email: amajab@ki.se
*Chapter 28*

Philip F. Halloran
Alberta Transplant Applied Genomics Centre
*and*
Division of Nephrology
University of Alberta
Edmonton, Alberta, Canada
Email: phil.halloran@ualberta.ca
*Foreword*

Heather Heine
The James Hogg iCAPTURE Centre for Cardiovascular
    and Pulmonary Research
St. Paul's Hospital
Vancouver, BC, Canada
Email: hheine@mrl.ubc.ca
*Chapter 10*

Walter C. Hellinger
Division of Infectious Diseases
Mayo Clinic
Jacksonville, Florida, USA
Email: hellinger.walter@mayo.edu
*Chapter 35*

Atul Humar
Departments of Medicine and Transplant Infectious Diseases
University of Alberta
Edmonton, Alberta, Canada
Email: ahumar@ualberta.ca
*Chapter 34*

Anikphe Imoagene-Oyedeji
Hospital of the University of Pennsylvania
Philadelphia, Pennsylvania, USA
Email: aimoagene@hotmail.com
*Chapter 39*

Md. Shahidul Islam
Department of Clinical Research and Education
Karolinska Institutet
Stockholm South General Hospital
Stockholm, Sweden
Email: shaisl@ki.se
*Chapter 28*

Bela Ivanyi
Department of Pathology
University of Szeged
Szeged, Hungary
Email: ivanyi@patho.szote.u-szeged.hu
*Chapter 27*

Kishore R. Iyer
Intestinal Transplant and Rehabilitation
    Recanati/Miller Transplantation Institute
The Mount Sinai Medical Center
New York, New York USA
Email: kishore.Iyer@msnyuhealth.org
*Chapter 22*

Basit Javaid
Kidney and Pancreas Transplant Program
Stanford Hospital and Clinics
*and*
Division of Nephrology
Department of Medicine
Stanford University School of Medicine
Palo Alto, California USA
Email: bjavaid@stanford.edu
*Chapter 2*

Tiffany E. Kaiser
Division of Pharmacy
University of Cincinnati Medical Center
Cincinnati, Ohio, USA
Email: tiffany.kaiser@uc.edu
*Chapter 16*

Michael P. Keane
Department of Medicine
Division of Pulmonary and Critical Care Medicine
    and Pediatrics
David Geffen-School of Medicine
University of California – Los Angeles
Los Angeles, California, USA
Email: mpkeane@mednet.ucla.edu
*Chapter 13*

Cesar A. Keller
Lung Transplant Program
Department of Transplant Medicine
Mayo Clinic
Jacksonville, Florida, USA
Email: keller.cesar@mayo.edu
*Chapters 14, 15*

Urmila Khettry
Department of Pathology
Tufts University School of Medicine
Boston, Massachusetts, USA
*and*
Lahey Clinic Medical Center
Burlington, Massachusetts, USA
Email: urmila.khettry@lahey.org
*Chapter 21*

Andras Khoor
Department of Pathology and Laboratory Medicine
Mayo Clinic
Jacksonville, Florida, USA
Email: khoor.andras@mayo.edu
*Chapter 14*

Jon A. Kobashigawa
Heart Transplant Program
David Geffen School of Medicine
University of California – Los Angeles
Los Angeles, California, USA
Email: jonk@mednet.ucla.edu
*Chapter 9*

Robert M. Kotloff
Pulmonary, Allergy and Critical Care Division
University of Pennsylvania Medical Center
Philadelphia, Pennsylvania, USA
Email: robert.kotloff@uphs.upenn.edu
*Chapter 3*

Yogish Kudva
Division of Transplantation Surgery
Mayo Clinic College of Medicine
Rochester, Minnesota, USA
Email: kudva.yogish@mayo.edu
*Chapter 31*

Deepali Kumar
Division of Infectious Diseases
University of Alberta
Alberta, Canada
Email: deepali.kumar@ualberta.ca
*Chapter 34*

Susan Lerner
Recanati/Miller Transplantation Institute
The Mount Sinai Medical Center
New York, New York, USA
Email: susan.lerner@mountsinai.org
*Chapter 18*

Gregory N. Levine
Arbor Research Collaborative for Health
Ann Arbor, Michigan, USA
Email: greg.levine@arborresearch.org
*Chapter 4*

Martin L. Mai
Department of Transplantation Medicine
Mayo Clinic
Jacksonville, Florida USA
Email: mai.martin@mayo.edu
*Chapter 26*

Roslyn B. Mannon
Transplantation Branch
National Institutes of Diabetes, and Digestive
    and Kidney Diseases
National Institutes of Health
Bethesda, Maryland, USA
Email: rozm@intra.niddk.nih.gov
*Chapter 5*

Paul Martin
Division of Liver Diseases
Recanati/Miller Transplantation Institute
The Mount Sinai Medical Center
New York, New York, USA
Email: paul.martin@mountsinai.org
*Chapter 18*

Bruce McManus
The James Hogg iCAPTURE Centre for Cardiovascular
    and Pulmonary Research
St. Paul's Hospital
Vancouver, British Columbia, Canada
Email: bmcmanus@mrl.ubc.ca
*Chapter 10*

Herwig-Ulf Meier-Kriesche
Kidney and Pancreas Transplant Program
University of Florida
Gainesville, Florida, USA
Email: meierhu@medicine.ufl.edu
*Chapter 23*

Andrea Meinhardt
Departments of Cardiology and Cardiovascular Surgery
University Hospital Center
*and*
Lausanne and Geneva-Lausanne Transplant Network
University of Lausanne
Lausanne, Switzerland
Email: andrea.meinhardt@chuv.ch
*Chapter 8*

Thalachallour Mohanakumar
Department of Surgery, Pathology and Immunology
Washington University School of Medicine
St. Louis, Missouri, USA
Email: kumart@msnotes.wustl.edu
*Chapter 12*

Kathy Monroy
Clinical Islet Transplant Program
Diabetes Research Institute
University of Miami Miller School of Medicine
Miami, Florida, USA
Email: kmonroy@med.miami.edu
*Chapter 30*

José M. Morales
Renal Transplant Unit
Nephrology Department
Hospital 12 de Octubre
Madrid, Spain
Email: jmorales@h12o.es
Chapter 36

Alice Mui
St. Paul's Hospital
Jack Bell Research Centre
University of British Columbia
Vancouver, British Columbia, Canada
Email: amui@interchange.ubc.ca
Chapter 10

Guy W. Neff
Division of Hepatology
Department of Transplantation
University of Cincinnati Medical Center
Cincinnati, Ohio, USA
Email: guy.neff@uc.edu
Chapter 16

Volker Nickeleit
Nephropathology Laboratory
Department of Pathology
The University of North Carolina at Chapel Hill
Chapel Hill, North Carolina, USA
Email: volker_nickeleit@med.unc.edu
Chapter 38

Natascha C. Nüssler
Department of General, Visceral
    and Transplantation Surgery
Charité, Campus Virchow-Klinikum
Universitätsmedizin Berlin
Berlin, Germany
Email: natascha.nuessler@charite.de
Chapter 19

John C. Papadimitriou
Department of Pathology
University of Maryland School of Medicine
Baltimore, Maryland, USA
Email: jpapa001@umaryland.edu
Chapter 33

Manuel Pascual
Departments of Cardiology and Cardiovascular Surgery
University Hospital Center
and
Lausanne and Geneva-Lausanne Transplant Network
University of Lausanne
Lausanne, Switzerland
Email: manuel.pascual@chuv.ch
Chapter 11

Jignesh K. Patel
Heart Transplant Program
University of California – Los Angeles
Los Angeles, California, USA
Email: jpatel@mednet.ucla.edu
Chapter 9

Breay W. Paty
Departments of Medicine
Clinical Islet Transplant Program
University of Alberta
Edmonton, Alberta, Canada
Email: breay.paty@ualberta.ca
Chapter 29

Hamid Rabb
Department of Medicine
Johns Hopkins University School of Medicine
Baltimore, Maryland, USA
Email: hrabb1@jhmi.edu
Chapter 6

Raymund R. Razonable
Division of Infectious Diseases and Internal Medicine
and
William J von Leibig Transplant Center
Mayo Clinic
Rochester, Minnesota, USA
Email: razonable.raymond@mayo.edu
Chapter 35

Mohammed Shawkat Razzaque
Department of Pathology
Nagasaki University Graduate School of Biomedical Sciences
Nagasaki, Japan
and
Department of Developmental Biology
Harvard School of Dental Medicine
Boston, Massachusetts, USA
Email: mrazzaque@hms.harvard.edu
Chapter 7

Camillo Ricordi
Clinical Islet Transplant Program
Diabetes Research Institute
University of Miami Miller School of Medicine
Miami, Florida, USA
Email: cricordi@med.miami.edu
Chapter 30

Gonzalo P. Rodriguez-Laiz
Intestinal Transplant and Rehabilitation
Recanati/Miller Transplantation Institute
The Mount Sinai Medical Center
New York, New York USA
Email: gonzalo.rodriguez-laiz@mountsinai.org
Chapter 22

Paola Romagnani
Department of Clinical Pathophysiology
Excellence Center for Research
Transfer and High Education DENOTHE
University of Florence
Florence, Italy
Email: p.romagnani@dfc.unifi.it
*Chapter 24*

John D. Scandling
Kidney and Pancreas Transplantation
Stanford Hospital and Clinics
University School of Medicine
Palo Alto, California USA
Email: jscand@stanford.edu
*Chapter 2*

Douglas E. Schaubel
University of Michigan Kidney Epidemiology
    and Cost Center
Ann Arbor, Michigan, USA
Email: deschau@umich.edu
*Chapter 4*

Ludwig K. von Segesser
Departments of Cardiology and Cardiovascular Surgery
University Hospital Center
*and*
Lausanne and Geneva-Lausanne Transplant Network
University of Lausanne
Lausanne, Switzerland
Email: ludwig.von-segesser@chuv.ch
*Chapter 11*

Charles Seydoux
Departments of Cardiology and Cardiovascular Surgery
University Hospital Center
*and*
Lausanne and Geneva-Lausanne Transplant Network
University of Lausanne
Lausanne, Switzerland
Email: charles.seydoux@chuv.ch
*Chapter 11*

A.M. James Shapiro
Department of Surgery
University of Alberta
Edmonton, Alberta, Canada
Email:  amjs@islet.ca
*Chapter 29*

Harsharan K. Singh
Nephropathology Laboratory
Department of Pathology
The University of North Carolina at Chapel Hill
Chapel Hill, North Carolina, USA
Email: harsharan_singh@med.unc.edu
*Chapter 38*

Titte R. Srinivas
Glickman Urologic and Kidney Institute
Cleveland Clinic
Cleveland, Ohio, USA
Email: srinivt@ccf.org
*Chapter 23*

Barbara Stange
Department of General, Visceral
    and Transplantation Surgery
Charité, Campus Virchow-Klinikum
Universitätsmedizin Berlin
Berlin, Germany
Email: barbara.stange@charite.de
*Chapter 19*

Mark D. Stegall
Division of Transplantation Surgery
Mayo Clinic College of Medicine
Rochester, Minnesota, USA
Email: stegall.mark@mayo.edu
*Chapter 31*

Takashi Taguchi
Department of Pathology
Nagasaki University Graduate School
    of Biomedical Sciences
Nagasaki, Japan
Email: taguchi@nagasaki-u.ac.jp
*Chapter 7*

Raphael Thuillier
Transplantation Branch
National Institutes of Diabetes, and Digestive
    and Kidney Diseases
National Institutes of Health
Bethesda, Maryland, USA
Email: thullierr@niddk.nih.gov
*Chapter 5*

Jennifer Trofe
Department of Pharmacy
Hospital of the University of Pennsylvania
Philadelphia, Pennsylvania, USA
Email: jennifer.trofe@uphs.upenn.edu
*Chapter 39*

Giuseppe Vassalli
Departments of Cardiology and Cardiovascular Surgery
University Hospital Center
*and*
Lausanne and Geneva-Lausanne Transplant Network
University of Lausanne
Lausanne, Switzerland
Email: giuseppe.vassalli@chuv.ch
*Chapters 8, 11*

Pierre Vogt
Departments of Cardiology and Cardiovascular Surgery
University Hospital Center
*and*
Lausanne and Geneva-Lausanne Transplant Network
University of Lausanne
Lausanne, Switzerland
Email: pierre.vogt@chuv.ch
*Chapter 11*

Hani M. Wadei
Department of Transplantation Medicine
Mayo Clinic
Jacksonville, Florida, USA
Email: wadei.hani@mayo.edu
*Chapter 26*

Andrew Y. Wang
Gastroenterology Division
University of Pennsylvania Medical Center
Philadelphia, Pennsylvania, USA
Email: andrew.wang@uphs.upenn.edu
*Chapter 3*

Samuel Weigt
Department of Medicine
Division of Pulmonary and Critical Care Medicine
    and Pediatrics
David Geffen-School of Medicine
University of California – Los Angeles
Los Angeles, California, USA
Email: sweigt@mednet.ucla.edu
*Chapter 13*

Robert A. Wolfe
Arbor Research Collaborative for Health
Ann Arbor, Michigan, USA
Email: robert.wolfe@arborresearch.org
*Chapter 4*

E. Steve Woodle
Division of Transplant Surgery
University of Cincinnati Medical Center
Cincinnati, Ohio, USA
Email: woodlees@uc.edu
*Chapter 16*

Harold C. Yang
Central Pennsylvania Transplant Associates
Pinnacle Health System
Harrisburg, Pennsylvania, USA
Email: hyang@pinnaclehealth.org
*Chapter 1*

# FOREWORD

This book addresses one of the largest unmet needs in transplantation, the need to reduce late allograft loss. Each year a percentage of the existing organ transplant patients will lose their grafts, many of them apparently due to immunologic causes. In the case of kidney transplantation, we believe that around 5,000 people in the USA per year will lose their kidney transplants and have to return to dialysis. This is in addition to the problem of death with function, but not completely unrelated to it.

Recent years have seen the emergence of a new concentration on the phenotypes and of late kidney transplant failure and their pathogenesis. The principal phenotypes seem to be related to antibody-mediated injury, including the transplant glomerulopathy and related pathologies; a rather non-specific group that has interstitial fibrosis and tubular atrophy without an obvious cause; and recurrent disease. The role of calcineurin inhibitor toxicity also has to be considered although its contribution as a distinct phenotype may be less than its contribution to the pathogenesis of other conditions. Finally, a number of people present with a rather severe T cell-mediated rejection or antibody-mediated rejection, often in the context of a deficiency of their immunosuppressive prescription. In some cases this is non-compliance and in other cases it may be related to inappropriate "minimization".

This book pulls together the science in this area and should serve as a source of information and as a catalyst for further research. It is particularly remarkable that no good case counting is done on the phenotypes of late kidney allograft failure in the national databases of any country. This is one area where progress needs to be made. Once we identify the phenotypes of the late failing allograft, we are in a better position to develop evidence about how to extend the survival of these grafts and prevent these phenotypes or at least slow their deterioration.

As one tries to piece together the conditions that contribute to late allograft loss, one should always remember the issues of health care delivery. We are never sure that patients are being adequately followed, either because their caregivers are not adequately trained, or because they themselves no longer adhere to the follow-up program. Thus the problem of late allograft failure has a component of biology and pathogenesis, a component of drug management, and a component of the science of delivering care to the hundreds of thousands of persons in the world who now live with kidney transplants. To that end, I believe this book is going to make a useful contribution to the literature.

*Philip F. Halloran, MD, PhD, OC, FRSC*
*Professor of Medicine*
*Director, Alberta Transplant Applied Genomics Centre*
*Canada Research Chair in Transplant Immunology*
*Editor-in-Chief, American Journal of Transplantation*
*Division of Nephrology*
*University of Alberta*
*Edmonton, Alberta, Canada*

# PREFACE

*"In our daily patients we witness human nature in the raw—despair, courage, understanding, hope, resignation, heroism. If alert, we can detect new problems to solve, and new paths to investigate."*

—Joseph E. Murray, MD
Nobel Laureate, Medicine, 1990

In 1902, the first successful experimental kidney transplant was performed in Vienna Medical School, in Austria. Fifty years later, human solid organ transplantation became a reality with the performance of the first successful kidney transplant in Boston by Dr. Joseph E. Murray. Since then an explosion of modern immunosuppression, improvement of surgical techniques, expansion in transplant immunology and better peri-transplant care have allowed successful kidney, heart, lung, liver, pancreas, islet cell, and small bowel transplants. These experiences are successful when compared to the alternatives but there is a constant struggle to get even 50% of the grafts from deceased donors to survive more than a year. However, science continued to advance knowledge of the immune response. With this came more and increasingly powerful tools for the clinician. Suddenly, success rates of 80-90% at one year were attainable. In the current era, it is reasonable to expect that most allografts will serve their recipients through their life span and death with preserved graft function is probably the ultimate goal for all transplant recipients. However, long term allograft survival has not paralleled improvements made in the decades in short term survival. For example, almost half of all deceased donor renal allografts will be lost within 10 years post-transplantation from chronic allograft nephropathy and more and faster rate in other organs due to: cardiac allograft vasculopathy, bronchiolitis obliterans, vanishing bile duct syndrome, islet cell exhaustion, and chronic enteropathy. Then there are other causes such as recurrent diseases and infectious complications compromising allograft function.

So, what prompted this book was a seeming imbalance between advancements made in science and what appears to be known generally in understanding the science of allograft loss. Many monographs and even volumes of admirable papers have been published on this subject. An explosion of information in this field mandates both a far reaching scope of coverage and in depth analysis to present the complete and current picture. To meet these objectives, the book has been arranged to contain ample outline of mechanism and management of chronic allograft loss. The use of multiple authors from around the globe was essential to ensure the all-inclusive nature of this text. Many of these authors are pioneers in the field, and all have extensive experiences studying and treating organ transplant.

The book is divided into several sections covering the entirety of chronic allograft loss from basic science consideration to clinical implications. We start with the historic perspective of organ transplant which is followed by an analysis of the data from the Scientific Registry of Transplant Research/Organ Procurement Transplant Network. A general knowledge of transplant immunology and basic mechanism of allograft rejection are discussed in four chapters. In the next 25 chapters, the book then describes the patho-physiology and management of chronic allograft loss involving individual organs. Five chapters discussed the roles played by infectious agents, particularly those by cytomegaloviruses and emerging viruses. The final chapter presents the novel therapy and available pharmacotherapeutic options to manage chronic allograft loss in organ transplant.

From the description just given, it is evident that the book has no higher goal than that of a compilation, with the addition of whatever information the authors may have from some of their own work. Because it is comprehensive, this book has broad applications for a variety of readers, including medical students, immunologists, pathologists, and transplant specialists, as well as patients. A few who read our book may be attracted to study transplant science. Altogether it is earnestly hoped that the information contained in this book may be found useful, facilitating future research in this field.

We wish, as well, to recognize and honor careers of several friends and colleagues who contributed unstintingly to research in the field of transplantation. To that end, the authors and editor have done what they can do, and tried to present the views and experiments of everyone, as best as possible. If any mistakes have occurred, and in a work like the present it is very possible, I shall thankfully receive notification of such errors and shall take the earliest opportunity to correct them. Our book will be an agent of change and betterment. If our work benefits patients and draws investigators into our field, we are satisfied.

*Nasimul Ahsan, MD, FACP, MBA*
*Professor of Medicine*
*Mayo Clinic – College of Medicine*
*Mayo Clinic Transplant Center*
*Jacksonville, Florida, USA*

# ACKNOWLEDGEMENTS

The editor is deeply indebted to each contributor to this first edition of "Chronic Allograft Loss". The painstaking revision and responses to suggestions for appropriate changes and willingness of the participants to adhere to a standard format in order to achieve a uniform style is gratefully acknowledged.

In collaboration with Landes Bioscience, this is the second book edited by me in less than two years. Again, high tribute goes to those members of the Landes Bioscience staff responsible for the publication. Ronald G. Landes has provided strong support and advice in the preparation of all aspects of this work and his encouragement and enthusiasm have been very functional in making it a reality. The contribution of Cynthia Conomos, who has been involved in all aspects of this book with an impressive commitment to detail, is gratefully acknowledged. Appreciation is also expressed to Celeste Carlton and the rest of the dedicated staff of Landes Bioscience, who skillfully processed the illustrations and prepared the thorough index.

Special recognition must be given to my colleagues at the Mayo Clinic Transplant Center, Jacksonville, Florida, USA and members of my MBA class at the Kellogg School of Management, Evanston, Illinois, USA for their guidance in every aspect including selection of the contributors and lay outs of the topics. To many of my international and US colleagues, my gratitude can not be overstated.

Finally, a thoroughly dedicated colleague and my loving wife Arzumand Ahsan has brought her many talents in editorial preparation to the compilation of the book. Her critical review of every chapter has been extraordinary, and her commitment and excitement in developing this book has been prime stimulus deserving of the highest praise.

# CHAPTER 1

# Introducing Chronic Graft Failure

Harold C. Yang*

Over the past 50 years the short-term improvement in one year graft survival in solid organ transplantation has improved dramatically. The latest statistics from the Scientific Registry of Transplant Recipients (SRTR) for 2006 demonstrate graft survivals in kidney, liver, lung and heart transplant patients at the first year that approach or exceed 90%.[1] These advances have occurred despite the fact that recipients are older with more comorbid illnesses and cadaver donors are less ideal.[2-4] Progress in surgical techniques and preservation has contributed to this progress in one year graft survival, particularly for liver and lung transplantation.[5-7] The biggest improvement has been in our understanding of immunosuppression.[8-10] Pretransplant work-ups allow the identification of specific alloantibody to be avoided in the donor.[11-14] Flow crossmatches at the time of transplant allow faster and more sensitive detection of antibody that may be missed by conventional techniques.[15-17] After transplantation, there are a greater variety of potent pharmacological agents to choose from and there is a better rationale of how to administer them. Drug regimens can be individualized for groups, such as African Americans or diabetics, but also for specific patients.[18-21] Monitoring is no longer limited to measuring trough levels at every clinic visit. Immune cell function and short term area under the curve (AUC) calculations have become commonplace.[22-27] If a patient has an adverse reaction to a particular class of drugs, that drug class can be minimized or eliminated after transplantation with equivalent immunologic results. Long-term single antigen assays can now detect the new onset of antibody against donor.[28,29] When it is detected and confirmed by biopsy there are techniques to mitigate their presence and improve allograft function.[30-32]

The majority of renal recipients are at highest risk of death from coronary vascular disease, much like the general population.[33,34] While that observation is laudable chronic graft failure continues to complicate long-term morbidity and survival in all solid organs. This book attempts to identify the factors that contribute to this failure of long-term function.

The first article by Javaid and Scandling provides an overview of chronic allograft failure in the past and present. These insights give us a glimpse of what the future might hold. Bloom et al discuss the indications and contraindications for heart, liver, kidney, pancreas and lung transplantation. Dickinson et al provide us with the methodology and statistics for graft and patient survivals in these solid organs used by the Scientific Registry of Transplant Recipients (SRTR).

The immunology of chronic allograft injury is outlined by Thuillier and Mannon. Because chronic allograft dysfunction is not solely an immunologic process Gandolfo and Rabb provide the basis for ischemia reperfusion injury. Taguchi and Razzaque go on to further describe the end product of immunologic and non-immunologic fibroproliferative injury and its relationship to increase expression of heat shock protein 47 (HSP 47).

The pathology of heart allograft rejection is addressed by Carthy, et al while Patel and Kobashigawa describe the macroscopic lesion and the clinical implications of cardiac vasculopathy. Vassalli et al provide some insight into experimental ex vivo gene therapy of heart transplantation. Goers T et al review the basic science of lung allograft failure. Belperio et al describe the cascade of cytokine responses in the development of lung allograft dysfunction while Erasmus et al outline the diagnosis of chronic lung failure. Alvarez and Keller provide us with current treatment options for those with bronchiolitis obliterans.

Kaiser and her co-authors give us an excellent overview of liver transplantation. Ali provides us with the pathogenesis, diagnosis and management of patients with chronic live failure. Lerner et al define chronic liver dysfunction while Stange et al summarize graft loss due to vascular complications. Crippin describes the clinical characteristics of late allograft failure while the pathologic basis is provided by Khettry and Goldar-Najafi. For small bowel transplantation Rodriguez-Laiz and Iyer delineate chronic enteropathy.

For renal allografts, Meier-Kriesche provides the epidemiology associated with survival while Romagnani outlines predictive parameters for failure. Champman and Ivanyi describe the classical pathology associated with chronic allograft nephropathy and dysfunction. Wadei et al outline risk factors and management of recurrent glomerular disease while Morales and Dominguez-Gil discuss hepatitis C as a singular risk factor for graft loss after renal transplantation. Ahsan provides us with the natural history and management of polyoma BK viral infection while Nickeleit and Singh discuss the pathologic manifestations of clinical disease.

The management of diabetes is separated into islet and whole pancreas transplantation. Gustafsson and Islam describe the cellular structure and physiology of islets of Langerhans while Paty and Shapiro outline clinical islet transplantation. Faradji et al provides us with metabolic indicators of islet cell dysfunction. Dean et al discuss pancreas and islet allograft failure while Gross and Gruessner specifically address chronic pancreas allograft failure. Papadimitrious and Drachenberg provide insight into the pathological aspects of pancreas allograft failure.

Non-HLA causes of chronic graft failure include CMV, which is discussed by Bonatti et al. Kumar and Humar address other emerging viruses in transplantation. Meinhardt and Vassali address immune tolerance as it relates to dendritic cells. Finally, Trofe et al discuss the pharmacotherapeutic options in solid organ transplantation.

The goal of this book is to provide the reader with an overview of long term problems in solid organ transplantation. This overview not limited to the diagnosis of immunologic and non-immunologic causes of chronic graft loss but current management as well as novel therapies the may help mitigate this problem.

*Harold C. Yang—Central Pennsylvania Transplant Associates, Pinnacle Health System, 205 Front Street, Harrisburg, PA 17105, USA. Email: hyang@PINNACLEHEALTH.org

*Chronic Allograft Failure: Natural History, Pathogenesis, Diagnosis and Management,* edited by Nasimul Ahsan. ©2008 Landes Bioscience.

## References

1. Scientific Registry of Transplant Recipients. Data as of January 1, 2007. Available at: www.ustransplant.org.
2. US Renal Data System: Chapter 4, Treatment modalities, in USRDS 2004 Annual Report. Bethesda: The National Institutes of Health, National Institute of Diabetes and Digestive and Kidney Diseases, 2004. Available at: http://www.usrds.org/atlas.htm.
3. 2004 Annual report of the US Organ Procurement and Transplantation Network and the Scientific Registry of Transplant Recipients: Transplant Data 1994-2003. Department of Health and Human Services, Health Resources and Services Administration, Healthcare Systems Bureau, Division of Transplantation, Rockville, MD; United Network for Organ Sharing, Richmond, VA; University Renal Research and Education Association, Ann Arbor, MI 2004, Chapter I, Trends and Results for Organ Donation and Transplantation in the United States, 2004:I-1-I-8.
4. 2004 Annual report of the US Organ Procurement and Transplantation Network and the Scientific Registry of Transplant Recipients: Transplant Data 1994-2003. Department of Health and Human Services, Health Resources and Services Administration, Healthcare Systems Bureau, Division of Transplantation, Rockville, MD; United Network for Organ Sharing, Richmond, VA; University Renal Research and Education Association, Ann Arbor, MI 2004, Chapter VI, Current Status of Kidney and Pancreas Transplantation in the United States, 1994-2003, VI-1-VI-13.
5. Lucey MR, Merion RM, Beresford TP. Liver Transplantation and the Alcoholic Patient: Medical, Surgical and Psychosocial Issues. Cambridge University Press, 1994:81-83.
6. Harringer W, Haverich A. Heart and Heart-Lung Transplantation: Standards and Improvements. World J Surg 2002; 26(2):218-25.
7. Ginns LC, Wain JC. Lung transplantation. In: Ginns LC, Cosimi AB, Morris PJ, eds. Transplantation. Blackwell Science, 1999:490-550.
8. Morris PJ. Transplantation-A medical miracle of the 20th century. NEJM 2004; 351:2678-80.
9. Curran MP, Keating GM. Mycophenolate sodium delayed release:Prevention of renal transplant rejection. Drugs 2005; 65:799-805.
10. Shapiro R, Young JB, Milford EL et al. Immunosuppression: Evolution in practice and trends, 1993-2003. Am J Transplant 2005; 5:874-886.
11. Bielmann D, Honger G, Lutz D et al. Pretransplant risk assessment in renal allograft recipients using virtual crossmatching. Am J Transplant 2007; 7(3):626-632.
12. Zeevi A, Girnita A, Duquesnoy R. HLA antibody analysis: sensistivity, specificity, and clinical significance in solid organ transplantation. Immunol Res 2006; 36(1-3):255-264.
13. Saidman SL. Histocompatibility testing for highly sensitized transplant candidates. Transplant Proc 2007; 39(3):673-75.
14. Kerman RH. Understanding the sensitized patient. Heart Fail Clin 2007; 3(1):1-97.
15. Gebel HM, Bray RA. Sensitization and sensitivity:defining the unsensitized patient. Transplantation 2000; 69(7):1370-74.
16. Vaidya S, Cooper TY, Avandasalehi J et al. Improved flow cytometric detection of HLA alloantibodies using pronase: potential implications in renal transplantation. Transplantation 2001; 71(3):422-28.
17. Pelletier RP, Admas PW, Hennessy PK, Orosz CG. Comparison of crossmatch results obtained by ELISA, flow cytometry, and conventional methodologies. Hum Immunol 1999; 60(9):855-61.
18. Srinivas TR, Meier-Kriesche HU, Kaplan B. Pharmacokinetic principles of immunosuppressive drugs. Am J Transplant 2005; 5(2):207-17.
19. Dirks NL, Huth B, Yates CR, Meibohm B. Pharmacokinetics of immuno-suppressants: a perspective on ethnic differences. J Clin Pharmacol Ther 2004; 42(12):701-18.
20. First MR. An update on new immunosuppressive drugs undergoing pre-clinical and clinical trials: potential applications in organ transplantation. Am J Kidney Dis 1997; 29(2):303-17.
21. Klupp J, Holt DW, van Gelder T. How pharmokinetic and pharmacody-namic drug monitoring can improve outcome in solid organ transplant recipients. Transpl Immunol 2002; 9(2-4):211-14.
22. Hooper E, Hawkins DM, Kowalski RJ et al. Clin Transplant 2005; 19(6):834-39.
23. Jin Y, Hernandez A, Fuller L et al. A novel approach to detect donor/recipient immune responses between HLA-identical pairs. Hum Immunol 2007; 68(5):350-61.
24. Sottong PR, Rosebrock JA, Britz JA, Kramer TR. Measurement of T-lymphocyte responses in whole-blood cultures using newly synthesized DNA and ATP. ClinDiagn Lab Immunol 2000; 7(2):307-11.
25. LeMeur Y, Buchler M, Thierry A et al. Individualized mycophenolate mofetil dosing bsed on drug exposure significantly improves patient outcomes after renal transplantation. Am J Transplant 2007; 7(11):2496-503.
26. Kuypers DR. Immunosuppressive drug monitoring-what to use in clinical practice today to improve renal graft outcome. Transpl Int 2005; 18(2):140-50.
27. Pawinski T, Hale M, Korecka M et al. Limited sampling strategy for estimation of mycophenolic acid area under the curve in adult renal transplant patients treated with concomitant tacrolimus. Clin Chem 2002; 48(9):1497-504.
28. Leffell MS, Montgomery RA, Zachary AA. The changing role of antibody testing in transplantation. Clin Transpl 2005; 19(2):259-71.
29. Vaidya S, Partlow D, Sussking B et al. Prediction of crossmatch outcome of highly sensitized patients by single and/or multiple antigen bead luminex assay. Transplantation 2006; 82(11):1524-28.
30. Akalin E, Watschinger B. Antibody-mediated rejection. Semin Nephrol 2007; 27(4):393-407.
31. Kobashigawa JA. Contemporary concepts in noncellular rejection. Heart Fail Clin 2007; 3(1):11-15.
32. Locke JE, Zachary AA, Haas M et al. The utility of splenectomy as recuse treatment for sever acute antibody mediated rejection. Am J Transplant 2007; 7(4):842-46.
33. Veneta JP, Pascual M. New treatments for acute humoral rejection of kidney allografts. Expert Opin Investig Drugs 2007; 16(5):625-33.
34. Wolfe RA, Ashby VB, Milford EL et al. Comparison of mortality in all patients on dialysis, patients on dialysis awaiting transplantation, and recipients of a first cadaveric transplant. NEJM 1999; 341:1725-30.
35. Port FK, Wolfe RA, Mauger EA et al. Comparison of survival probabilities for dialysis patients vs cadaveric renal transplant recipients. JAMA 1993; 270:1339-1343.

# CHAPTER 2

# Chronic Allograft Failure:
## Past, Present and Future

Basit Javaid and John D. Scandling*

## Introduction

Transplantation is the treatment of choice for irreversible organ damage. According to United Network for Organ Sharing (UNOS) data, by the end of June 2007, 400,291 solid organ transplants had been performed in the United States since 1988.[1] In 2006, a total of 28,933 solid organ transplants including kidney, liver, pancreas, heart, lung and intestine, either individually or in combination with other organs were performed. About 97,000 patients were on the wait list for an organ transplant by mid 2007. The number of solid organ transplants performed in the United States each year has nearly doubled over the past two decades, which speaks to the wider acceptability and success of organ transplantation.

Ever since the initial reports of successful transplantation of kidneys procured from related and un-related healthy human volunteers, or of liver transplants obtained from deceased donors in late 1960s,[2] short-term patient survival rates for kidney and liver transplants have continually improved.[3,4] One year patient survival rates for recipients of kidney transplants in the United States now exceed 95%,[1] in sharp contrast to the one-year patient survival rate of 67% for recipients of living-related, and 33% for recipients of living-unrelated, donor kidney transplants in the early 1960s.[2] The short term survival rate for liver transplants has also improved significantly. In the 1960s survival in the first recipients of liver transplants was limited to only a few months.[2] In comparison, nowadays over 86% percent of recipients live beyond the first year.[1] Similar improvement is also evident in thoracic organ transplantation. Survival rates for heart and lung transplant recipients are now over 87% and 83% respectively at one year after transplantation.[1] In the first decade of heart transplantation survival at one year was less than 50%;[2] in 1987 the one-year survival rate for lung transplantation was 35%.[5]

It is encouraging to observe that short-term graft survival for solid organ transplants continues to improve. According to United States Renal Data System (USRDS) data, 90-day graft survival following deceased donor kidney transplantation improved from 86% in 1990 to 94% in 2004 and one-year graft survival improved from 78% in 1990 to 89% in 2003. Unfortunately, improvement in short-term graft survival rates has not consistently transpired into improved long-term graft outcome and the overall gain in unadjusted long-term graft survival rate for kidneys from deceased donors has been inconsequential.[6-8] The cumulative increase in long-term graft survival for first kidney transplants performed between 1988 and 1995 is less than six months (Fig. 1).[9] Similar trends have also been observed in other solid organ transplants.[10-12] The five-year graft survival rates between 1997 and 2004 for kidney, liver and heart transplants, which constitute over 90% of all solid organ transplants performed in the United States, are shown in Table 1.[1] The lack of improvement in long-term graft life and the similarities in the trend for long-term graft outcomes for the major transplanted organs suggest commonality of potential barriers to prolonged survival.

## Chronic Allograft Loss: Current Survival Outcomes

Despite the heterogeneity of disease mechanisms that require organ transplantation, the major causes of chronic graft loss among all major organ transplant recipients are relatively similar. Recipient death with a functioning graft, chronic rejection and recurrence of the primary disease in the transplanted organ are the leading causes of graft loss over time in solid organ transplant recipients (Table 2).

In the case of heart or lung transplant recipients, aside from the few exceptions wherein retransplantation occurs, graft survival equates to patient survival. On the other hand, recipients of kidney and liver transplants often outlive their transplant organs. In the U.S., retransplantation accounted for 11.4% of all kidney and 8.2% of all liver transplants in 2006.[1] In the same year, the retransplantation rate for heart and lung transplant patients was 3%.[1] Multiple retransplants are not uncommon in kidney and liver transplant recipients. In the U.S. in 2005, each day about 77 individuals received a solid organ transplant—kidney, liver, heart, lung, pancreas, or intestine. Every eleventh transplant that year, much like the preceding ten years, was a repeat transplant necessitated by the loss of a previously functioning transplant organ. In the vast majority of patients loss of the transplant organ was the result of chronic allograft dysfunction.[1]

*Table 1. Graft survival rates*

| Transplant Organ | Graft Survival Rates | | |
|---|---|---|---|
| | 1 year | 3 year | 5 year |
| Kidney | 91.6% | 81.9% | 71.4% |
| Liver | 81.9% | 71.9% | 65.3% |
| Heart | 86.4% | 77.3% | 69.9% |
| Lung | 82.4% | 61.7% | 45.8% |

Graft survival rates for transplants performed in the United States based on UNOS data.[1] 1 year survival based on 2002-2004 transplants; 3 year survival rates based on 1999-2002 transplants; 5 year survival rates based on 1997-2000 transplants.

*Corresponding Author: John D. Scandling—Medical Director, Kidney and Pancreas Transplantation Stanford Hospital and Clinics, Professor of Medicine, Stanford University School of Medicine, 750 Welch Road, Suite 200, Palo Alto, CA 94304-1509, USA. Email: jscand@stanford.edu

*Chronic Allograft Failure: Natural History, Pathogenesis, Diagnosis and Management*, edited by Nasimul Ahsan. ©2008 Landes Bioscience.

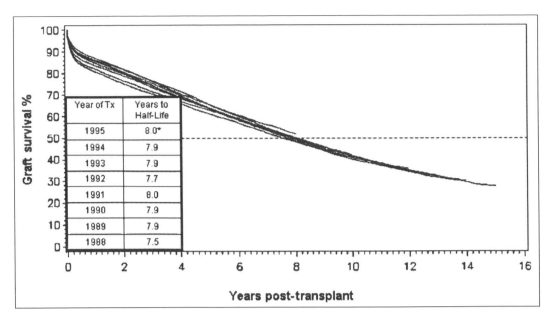

Figure 1. Kidney transplant survival rates. Overall graft survival by year of transplantation. Reprinted from: Meier-Kriesche HU et al. Am J Transplant 2004; 4(8):1289-95;[9] ©2004 with permission from Blackwell Publishing.

Acute rejection rates in U.S. adult first kidney transplant recipients during the first and second year after transplantation declined between 1995 and 2000 without a significant improvement in overall graft survival, suggesting that reducing the rate of acute rejection may not lead to improved long-term graft survival.[3] Death with a functioning graft is one of the most common causes of transplant kidney loss, accounting for the loss of up to half of all functioning grafts.[13-15] Chronic allograft nephropathy, a consequence of immunologically mediated injury or chronic rejection, accounts for 30 to 40% of graft failure and is the primary cause of chronic allograft dysfunction.[15] Chronic toxic effects of the calcineurin inhibitors, late acute rejection, recurrent or a new primary kidney disease and the relatively recently recognized problem of polyoma virus nephropathy are other contributors to chronic renal allograft loss.[14-17]

Late liver allograft dysfunction may result from a variety of causes and varies significantly with the underlying cause of liver failure. Most late causes of liver allograft dysfunction present simply as abnormalities of routinely monitored serum chemistry tests. It can be difficult to distinguish among the potential causes of dysfunction due to overlapping of underlying factors. Recurrence of the primary disease is the single largest cause of late allograft loss in this population. Infections (viral hepatitis); autoimmune diseases (autoimmune hepatitis, primary biliary cirrhosis, primary sclerosing cholangitis); malignancy; hepatotoxic factors (alcoholism); metabolic disorders (nonalcoholic steatohepatitis); and other disease conditions, such as idiopathic granulomatous hepatitis, can recur in the transplant liver.[18] Recurrent infections and autoimmune processes are responsible for the vast majority of late graft failures in liver transplant recipients.[4] Recurrence of hepatitis C liver disease accounts for up to 60% of all graft failures over time.[4] Less common is graft loss to chronic allograft rejection, primarily manifest as biliary epithelial senescence, with or without bile duct loss and foam cell obliterative arteriopathy.[18] Similar to the situation in kidney transplantation, a decrease in the incidence of early graft loss in liver transplantation has not resulted in improved long term graft outcome.[4] Comparison of two successive 5-year periods (1997-2002 vs. 1992-1996) among 35,186 U.S. deceased donor adult liver transplant recipients showed identical 5-year graft survival, 67.5 vs. 67.4% respectively, although one-year graft survival had improved from 81.0 to 83.5% (Fig. 2). The absence of correlation between short- and long-term survival reflects a lack of understanding of the factors leading to chronic allograft dysfunction and the absence of effective therapy which could prolong graft life. Aside from hepatitis C and primary sclerosing cholangitis, long-term liver allograft survival was independent of the primary cause of native liver failure and did not change over the decade from 1992 to 2002.[4]

Early outcomes after heart transplantation have improved significantly over the past decade. Deaths in the first posttransplant year have steadily decreased from 179 deaths per 1000 patient-years at risk in 1995 to 131 deaths in 2003.[12] Nonetheless, five-year survival, based on cohorts of recipients transplanted between 1995 and 1999, remained relatively unchanged, with patient and graft survival rates of approximately 70% (Fig. 3). This has been corroborated recently in the report of the International Society for Heart and Lung Transplantation (ISHLT).[19] One-year survival has improved but the long term attrition rate has remained relatively unchanged since 1982. Graft loss or death after the fifth posttransplant year was attributable mainly to cardiac allograft vasculopathy (CAV) in up to 30% of the cases, a prevalence

---

*Table 2. Causes of chronic allograft loss*

**I. Death with functioning allograft**

**II. Chronic rejection**
- *Kidney*—Chronic allograft nephropathy (CAN)
- *Liver*—Biliary epithelial senescence and foam cell obliterative arteriopathy
- *Heart*—Chronic allograft vasculopathy (CAV)
- *Lung*—Brochiolitis obliterans syndrome (BOS)

**III. Recurrent diseases**
- *Kidney*—Glomerulonephritides
- *Liver*—Viral hepatitides; autoimmune diseases; malignancy; hepatotoxic factors; metabolic disorders; idiopathic granulomatous hepatitis
- *Heart*—Amyloidosis; Chagas' disease
- *Lung*—Sarcoidosis; Pulmonary lymphangioleiomyomatosis

**IV. Miscellaneous**
- *Kidney*—Chronic calcineurin-inhibitor nephrotoxicity (CIN); late acute rejection; chronic hypertension; chronic obstruction; chronic bacterial pyelonephritis; polyoma virus nephropathy
- *Liver*—Thrombosis; primary graft failure
- *Heart*—Viral infections
- *Lung*—Brochoalveolar carcinoma

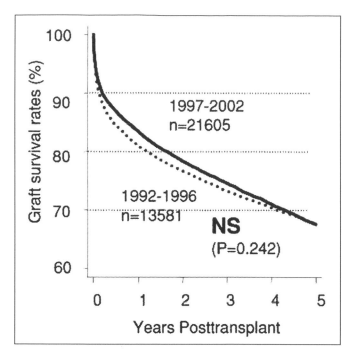

Figure 2. Liver transplant graft survival rates. Long-term graft survival rates for 1992-1996 and 1997-2002. Reprinted from: Futagawa Y et al. Am J Transplant 2006; 6(6):1398-406;[4] ©2006 with permission from Blackwell Publishing.

which has remained essentially unchanged.[20] Other studies have confirmed that CAV is the primary disease limiting long-term patient and allograft survival in heart transplantation, followed by malignancy, accounting for 30% of deaths, and infection other than cytomegalovirus (CMV), accounting for 10% of deaths.[21] The development of CAV is associated with both immunologic and non-immunologic factors. Risk of CAV increases with the number of HLA mismatches and the number and duration of acute rejection episodes. Classic cardiovascular risk factors such as smoking, diabetes mellitus, dyslipidemia and hypertension have also been linked to a higher risk of CAV, but the relationship is

less well characterized. Recipients with CMV infection tend to have a higher risk for CAV which is often more severe than in those without CMV infection.[21]

In 2005, the number of lung transplants performed in the USA reached 1405, the greatest number since the origin of the UNOS registry.[1] The number of single lung transplantations has been relatively constant internationally for almost a decade, but there has been a gradual increase in the number of double lung transplantations.[22] The volume of double lung transplants in 2004, over 1000, was twice that in 1994. The five- year graft survival rate for lung transplants in the U.S. is approximately 48%,[12] similar to the five-year survival rate of 49% reported by the ISHLT.[22] Similar to the survival trends observed in heart transplant recipients, the short-term survival rates in lung transplant recipients have improved over time but the gain has been concentrated in the first year after transplantation. Beyond the first year the survival slopes parallel those of earlier eras of lung transplantation (Fig. 4). In those who survive more than five years, bronchiolitis obliterans or bronchiolitis obliterans syndrome is prevalent. It affects up to 60% of lung transplant recipients who survive 10 years. Bronchiolitis obliterans accounts for 26.5% of all deaths in lung transplant recipients who survive beyond five years of transplantation. Other causes of death include infections in 18%, graft failure due to unknown causes in 17%, malignancy in 12%, cardiovascular diseases in 5.2%.[22] Bronchiolitis obliterans, which was first reported among recipients of heart-lung transplants at Stanford University who developed a progressive deterioration in forced expiratory volume in one second (FEV-1), is characterized by an inflammatory and fibrogenic process affecting the membranous and respiratory bronchioles and leading to cicatricial luminal narrowing and severe obstructive airways disease.[23] It is one of the most important factors limiting long-term survival among lung transplant recipients.[24] A prior history of acute rejection, presence of HLA-specific antibodies, exposure to environmental irritants and toxins, infections, airway ischemia, aspiration of gastrointestinal contents, preexisting connective tissue disorders, radiation injury and a variety of other donor and recipient factors have been implicated in the pathogenesis of this condition, suggesting a combined role for immunologic and non-immunologic injury as the underlying mechanism for chronic graft loss.[23-25]

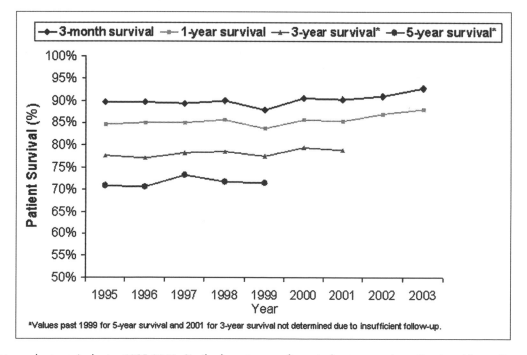

*Values past 1999 for 5-year survival and 2001 for 3-year survival not determined due to insufficient follow-up.

Figure 3. Heart transplant survival rates 1995-2003. Similar long-term graft survival rates over time. Reprinted from: Orens JB et al. Am J Transplant 2006; 6(5 Pt 2):1188-97;[2] ©2006 with permission from Blackwell Publishing.

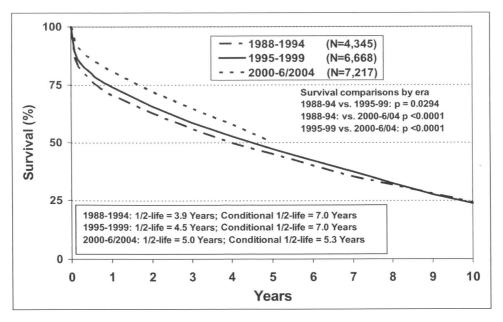

Figure 4. Lung transplant survival rates 1988-2004. Long-term survival in different transplant periods nearly identical. Reprinted from: Trulock EP et al. J Heart Lung Transplant 2006; 25(8):880-92;[22] ©2006 with permission from Elsevier.

It is evident from these data that long-term graft survival rates have reached a plateau in recent years, despite substantial improvements in short-term graft survival. The incidence of late graft failure in kidney, liver, heart and lung transplant recipients has remained at about 3% to 5% per year over the past decade. Since chronic allograft rejection is not the only underlying cause for graft loss in these cases, this incident rate defines the upper limit of rate of chronic allograft loss due to chronic rejection.[26]

## Improving Chronic Allograft Survival: Current Strategies and Future Prospects

### *Improving Patient Survival*

A clear understanding of the factors leading to chronic graft loss and a comprehensive approach to eradicate or minimize the impact of such factors could lead to better graft survival over time. Alternatively, improving the availability of transplantable organs or the introduction of other innovative approaches that may totally eliminate the need for organ transplantation could mitigate the current impact of graft loss. Cardiovascular disease, infection and malignancy are the main determinants of patient survival among transplant recipients. Currently, there is an almost complete lack of any prospective data examining the effect of risk factor modification, disease screening or other interventions which might modify mortality risk and improve survival among transplant recipients.[28,29]

### *Early Detection of Chronic Graft Injury: Surveillance Biopsies and Biologic Profiling*

Early detection of chronic allograft injury through surveillance biopsies could prove to be a viable approach to address chronic allograft loss. Protocol surveillance biopsies have shown to be useful in detecting sub-clinical pathologies such as acute rejection, but improvement in long-term graft function and survival following treatment of a sub-clinical disease entity has not been consistently observed in solid organ transplants. A role for surveillance biopsies in routine clinical practice for detection and management of chronic allograft injury with resultant improvement in long-term graft survival has not been validated.[30,31]

Functional genomic and proteomic approaches have begun to revolutionize transplant research. Advancements in powerful newer technologies, such as DNA microarrays, serial analysis of gene expression,

RNA interference and proteomics have greatly enhanced our ability to study the relationship between genetic profile and functional outcomes at an ever escalating pace. Integration of these technologies into clinical research and practice is expected to enhance basic understanding of the immunologic and non-immunologic processes leading to chronic graft failure. This knowledge may eventually lead to wider clinical applications, such as risk-profiling for susceptibility to chronic graft injury, establishment of surveillance parameters, or pharmacologic targeting of specific biologic processes, improving long-term graft life. The data emerging from such technologies are preliminary and have limited clinical applicability at present.

### *Advancements in Pharmacological Interventions*

The major breakthroughs in organ transplantation over the recent decades have largely been the consequence of the development and availability of new therapeutic agents to prevent and treat acute rejection. These agents have proven effective in reducing the incidence of acute rejection, resulting in remarkable improvement in short-term graft survival. Nonetheless, thus far a similar improvement in long-term graft survival has not been observed. There could be multiple factors to account for the disparity observed in the short- and long-term survival outcomes. The primary focus of drug development has been improvement in acute rejection rates and short-term graft survival. The recognition that these short-term advances have not resulted in improved long-term graft survival has been relatively recent. It could be anticipated that improvements in long-term survival would subsequently follow.[8]

Recent years have witnessed the introduction of new pharmacologic immunosuppressive agents and modifications in immunosuppressive protocols aimed at improving long-term outcome, through minimizing exposure to corticosteroid and the calcineurin inhibitors.

The newer agents primarily used to spare corticosteroid or the calcineurin inhibitors have been the depleting and non-depleting monoclonal antibodies, the anti-CD52 antibody alemtuzumab, the anti-CD20 antibody rituximab, the anti-CD25 antibodies daclizumab and basiliximab and newer agents such as CTLA-4-Ig (belatacept, also known as LEA29Y).[32] Early reports of the use of alemtuzumab for these purposes described higher rates of rejection and other adverse events in some steroid and calcineurin inhibitor minimization protocols but it has produced better results in several steroid-sparing protocols.[33] The consequences for long-term graft function and prevention of chronic allograft dysfunction remain to be seen. A recent report on recipients of de novo kidney allografts treated with belatacept (LEA29Y), a selective co-stimulation blocker, showed comparable efficacy in preventing acute rejection at six months posttransplant and better preservation of glomerular filtration rate and less chronic allograft nephropathy at twelve months, in comparison to a cyclosporine-based regimen.[34] Rituximab, a monoclonal antibody directed at the CD20 molecule on B-cells, has been used in organ transplant candidates to reduce preformed antibodies in sensitized patients and in ABO-incompatible transplantation of the kidney, liver and heart. Rituximab has also been successfully used in the treatment of acute rejection.[35] Early reports suggest that this drug may prove helpful in the treatment of de novo or recurrent glomerular disease in the transplanted kidney, an additional property which could enhance long-term outcome.[36] However, in our experience two adult kidney transplant recipients with recurrent focal segmental glomerulosclerosis (FSGS) in the transplanted kidney showed no response to treatment with rituximab.

Another approach directed at enhancing long-term graft outcome has employed use of the mammalian target of rapamycin (mTOR) antagonist, sirolimus. Use of this drug can allow avoidance or minimization of the calcineurin inhibitors in kidney transplant recipients and exploit its anti-proliferative property to prevent graft vasculopathy in heart transplant recipients. This drug has been available in the U.S. only since late 1999. Consequently, its impact on long-term graft survival has not been established.[37]

## Innovations in Organ Supply for Transplantation

The U.S. National Organ Transplant Act, passed in 1984, led to the institution of the Organ Procurement and Transplantation Network (OPTN) and the subsequent development of organ allocation policies. Since initiation of the OPTN about 400,000 patients have received organ transplants. The number of transplant recipients grew from around 10,000 per year to more than 28,000 per year in little over two decades. During this time, the number of patients on the wait list for transplantation grew from about 20,000 to over 93,000. While the number of organs recovered and transplanted has almost doubled to over 26,000 annually in the past fifteen years, the gap between organ supply and demand continues to widen despite recent remarkable success afforded by programs such as the Organ Donation Breakthrough Collaborative aimed to increase the deceased-donor organ consent rate from 50 to 75%.[38] The OPTN (UNOS) recently announced intent to increase the number of deceased-organ donors transplanted to 42,800 by year 2013. However, it has been argued that even if all potential brain-dead donors could become actual donors, the supply of organs from such donors would not meet the need of all waitlisted patients. Innovative approaches including use of non-heart beating donors, use of expanded-criteria donors and living donor exchange programs for kidney transplantation, are expected to diminish the widening gap. An increase in organs available for all transplant candidates, including those requiring repeat transplantation, will counter the problem of chronic graft failure to some extent.[38-40]

## Newer Technologies and Advancements

The gap between demand and supply in organ transplantation has encouraged other technologies which could address the shortfall. Given the success of organ transplantation, the case for embracing new approaches is even more compelling.[41] Xenotransplantation could offer a physiologic, expeditious and more economical way to replace dysfunctional human organs. However, no genetic or immunologic manipulation thus far has enabled a xenograft to survive in non-human primates. Stem cells, cloning and organogenesis could potentially be available for clinical use in the future, but the scientific and societal complexities of these innovations could hinder their eventual practical application. Mechanical devices such as hemodialysis machines, ventilators, ventricular assist devices, etc., are in clinical use throughout the world, but it is difficult to envision such devices fully substituting the

function of an organ for a prolonged period with the independence and reliability offered by an organ transplant. The clinical applications of such devices are limited at present and their ability to impact long-term survival remains speculative.[41]

In conclusion, chronic allograft dysfunction continues to be a significant problem among all solid organ transplant recipients and is one of the major causes of allograft failure. Improvements in short-term survival thus far have not consistently translated into enhanced long-term graft survival. Further studies looking into this complex problem would lead to better characterization of the factors that predispose to graft injury and decline. Collaborative efforts to develop better monitoring, diagnostic and interventional strategies are required to address this complex problem in a rational manner. Recently developed technologies show great promise in further characterizing the heterogeneity of the biologic responses. Newer pharmacologic approaches may provide the means to manipulate the immunologic and non-immunologic responses leading to graft failure. If the remarkable success in achieving excellent short-term graft survival rates over the past four decades is any reflection of future prospects of improving long-term graft survival, such success is assured.

## References

1. United Network for Organ Sharing (UNOS); http://www.unos.org
2. Starzl TE, Brettschneider L, Martin AJ Jr et al. Organ transplantation, past and present. Surg Clin North Am 1968; 48(4):817-38.
3. Meier-Kriesche HU, Schold JD, Srinivas TR et al. Lack of improvement in renal allograft survival despite a marked decrease in acute rejection rates over the most recent era. Am J Transplant 2004; 4(3):378-83.
4. Futagawa Y, Terasaki PI, Waki K et al. No improvement in long-term liver transplant graft survival in the last decade: an analysis of the UNOS data. Am J Transplant 2006; 6(6):1398-406.
5. Keck BM, White R, Breen TJ et al. Thoracic organ transplants in the United States: a report from the UNOS/ISHLT Scientific Registry for Organ Transplants. United Network for Organ Sharing. International Society for Heart and Lung Transplantation. Clin Transpl 1994; 7-46.
6. Kaplan B. Overcoming barriers to long-term graft survival. Am J Kidney Dis 2006; 47(4 Suppl 2):S52-64.
7. Cohen DJ, St Martin L, Christensen LL et al. Kidney and pancreas transplantation in the United States, 1995-2004. Am J Transplant 2006; 6(5 Pt 2):1153-69.
8. Hariharan S, Johnson CP, Bresnahan BA et al. Improved graft survival after renal transplantation in the United States, 1988 to 1996. N Engl J Med 2000; 342(9):605-12.
9. Meier-Kriesche HU, Schold JD, Kaplan B. Long-term renal allograft survival: have we made significant progress or is it time to rethink our analytic and therapeutic strategies? Am J Transplant 2004; 4(8):1289-95.
10. Port FK, Merion RM, Goodrich NP et al. Recent trends and results for organ donation and transplantation in the United States, 2005. Am J Transplant 2006; 6(5 Pt 2):1095-100.
11. Shiffman ML, Saab S, Feng S et al. Liver and intestine transplantation in the United States, 1995-2004. Am J Transplant 2006; 6(5 Pt 2):1170-87.
12. Orens JB, Shearon TH, Freudenberger RS et al. Thoracic organ transplantation in the United States, 1995-2004. Am J Transplant 2006; 6(5 Pt 2):1188-97.
13. Cecka JM. The UNOS Scientific Renal Transplant Registry. Clin Transpl 2000; 1-18.
14. Magee CC, Pascual M. Update in renal transplantation. Arch Intern Med 2004; 164(13):1373-88.
15. Pascual M, Theruvath T, Kawai T et al. Strategies to improve long-term outcomes after renal transplantation. N Engl J Med 2002; 346(8):580-90.
16. Chapman JR, O'Connell PJ, Nankivell BJ. Chronic renal allograft dysfunction. J Am Soc Nephrol 2005; 16(10):3015-26.
17. Hariharan S. BK virus nephritis after renal transplantation. Kidney Int 2006; 69(4):655-62.
18. Banff Working Group; Liver biopsy interpretation for causes of late liver allograft dysfunction. Hepatology 2006; 44(2):489-501.
19. Taylor DO, Edwards LB, Boucek MM et al. International Society for Heart and Lung Transplantation. Registry of the International Society for Heart and Lung Transplantation: twenty-third official adult heart transplantation report. J Heart Lung Transplant 2006; 25(8):869-79.
20. Taylor DO, Edwards LB, Mohacsi PJ et al. The registry of the International Society for Heart and Lung Transplantation: twentieth official adult heart transplant report. J Heart Lung Transplant 2003; 22(6):616-24.
21. Valantine H. Cardiac allograft vasculopathy after heart transplantation: risk factors and management. J Heart Lung Transplant 2004; 23(5 Suppl): S187-93.

22. Trulock EP, Edwards LB, Taylor DO et al. International Society for Heart and Lung Transplantation. Registry of the International Society for Heart and Lung Transplantation: twenty-third official adult lung and heart-lung transplantation report. J Heart Lung Transplant 2006; 25(8):880-92.

23. Burke CM, Theodore J, Dawkins KD et al. Post-transplant obliterative bronchiolitis and other late lung sequelae in human heart-lung transplantation. Chest 1984; 86(6):824-29.

24. Al-Githmi I, Batawil N, Shigemura N et al. Bronchiolitis obliterans following lung transplantation. Eur J Cardiothorac Surg 2006; 30(6):846-51.

25. Chan A, Allen R. Bronchiolitis obliterans: an update. Curr Opin Pulm Med 2004; 10(2):133-41.

26. Estenne M, Kotloff RM. Update in transplantation 2005. Am J Respir Crit Care Med 2006; 173(6):593-98.

27. Libby P, Pober JS. Chronic rejection. Immunity 2001; 14(4):387-97.

28. Schnitzler MA, Salvalaggio PR, Axelrod DA et al. Lack of interventional studies in renal transplant candidates with elevated cardiovascular risk. Am J Transplant 2007; 7(3):493-94.

29. Ojo AO. Cardiovascular complications after renal transplantation and their prevention. Transplantation 2006; 82(5):603-11.

30. Mengel M, Chapman JR, Cosio FG et al. Protocol biopsies in renal transplantation: insights into patient management and pathogenesis. Am J Transplant 2007; 7(3):512-17.

31. Stehlik J, Starling RC, Movsesian MA et al. Cardiac Transplant Research Database Group. Utility of long-term surveillance endomyocardial biopsy: a multi-institutional analysis. J Heart Lung Transplant 2006; 25(12):1402-9.

32. Halloran PF. Immunosuppressive drugs for kidney transplantation. N Engl J Med 2004; 351(26):2715-29.

33. Morris PJ, Russell NK. Alemtuzumab (Campath-1H): a systematic review in organ transplantation. Transplantation 2006; 81(10):1361-67.

34. Vincenti F, Larsen C, Durrbach A et al. Belatacept Study Group: Costimulation blockade with belatacept in renal transplantation. Engl J Med 2005; 353(8):770-81.

35. Pescovitz MD. Rituximab, an anti-cd20 monoclonal antibody: history and mechanism of action. Am J Transplant 2006; 6(5 Pt 1):859-66.

36. Pescovitz MD, Book BK, Sidner RA. Resolution of recurrent focal segmental glomerulosclerosis proteinuria after rituximab treatment. N Engl J Med 2006; 354(18):1961-63.

37. Augustine JJ, Bodziak KA, Hricik DE. Use of sirolimus in solid organ transplantation. Drugs 2007; 67(3):369-91.

38. Marks WH, Wagner D, Pearson TC et al. Organ donation and utilization, 1995-2004: entering the collaborative era. Am J Transplant 2006; 6 (5 Pt 2):1101-10.

39. Matas AJ, Sutherland DE. The importance of innovative efforts to increase organ donation. JAMA 2005; 294(13):1691-93.

40. Mitka M. Efforts under way to increase number of potential kidney transplant donors. JAMA 2006; 295(22):2588-89.

41. Cascalho M, Platt JL. The future of organ replacement: needs, potential applications and obstacles to application. Transplant Proc. 2006; 38(2):362-64.

# CHAPTER 3

# Solid Organ Transplantation—An Overview

Roy D. Bloom,* Lee R. Goldberg, Andrew Y. Wang, Thomas W. Faust and Robert M. Kotloff

## Introduction

Human solid organ transplantation became a reality in 1954 with the performance of the first successful kidney transplant by Dr. Joseph Murray and colleagues. The ensuing 15-20 years witnessed an expansion of the clinical science to encompass transplantation of heart, liver, pancreas and lung as well. A substantial advancement occurred during the 1980s with the advent of cyclosporine. More recently, the introduction into the therapeutic arena of tacrolimus, mycophenolic acid based therapies and proliferation signal inhibitors has brought about further declines in rates of acute rejection and the potential for enhanced graft and patient outcomes. Overall, solid organ transplantation has evolved into a major clinical discipline with the capacity to preserve, extend and enhance life. In 2005, over 27,000 new transplant procedures were performed in the United States alone;[1] currently nearly 95,000 are waitlisted to receive a transplanted organ.[2] In this chapter, we shall provide an overview of transplantation of the various solid organs, to include indications and contraindications, relevant surgical technical information, rationale for allocation of each organ, as well as appropriate clinical outcomes.

## Heart Transplantation

### Background/Current Status

Heart transplantation remains the treatment of choice for otherwise healthy, younger patients with intractable heart failure refractory to maximal medical and device therapy. In the United States for the 365 day period ending June 30, 2006, 2,125 cardiac transplants were performed. On that same date, 2,835 patients were on the waiting list for cardiac transplant.[3] The number of patients being listed for transplant since 1993 has steadily declined in the setting of improved medical and surgical management of advanced heart failure that has resulted in one year survival approaching that of transplantation.[4] In addition, a new status system has shifted the distribution of donor organs to sicker patients making early listing less imperative.[5] The annual mortality rate for patients on the heart transplant waiting list is approximately 18%, although this may be an underestimation as 2.6% of listed patients are de-listed due to "deterioration" prior to death. The median time from initial listing to cardiac transplant in adults is about 9.4 months.[3] Waiting times may vary significantly depending on the urgency status of the patient (1A, 1B, 2), blood group, body size and geographic location.[6] As an example, based on the US Organ Procurement and Transplantation Network (OPTN) data on January 19, 2007, the median waiting time for blood group O recipients was 290 days and 47 days for patients with blood group AB. Patients listed as a status 1A had a median time to transplant of 49 days, compared to those listed as a status 2 where the median waiting time of 392 days.

### Indications and Contraindications

The primary indications for cardiac transplant include refractory heart failure despite maximal medical support, refractory ventricular arrhythmias and refractory angina.[7] The one year survival of patients after transplant is about 86% and therefore patients listed for transplant should have an estimated one year mortality without transplantation of greater than 15%.[3] Several models have been proposed to help risk stratify patients with heart failure using both invasive and non-invasive methods.[7-10] The most potent predictor of outcome in ambulatory patients with heart failure is a symptom limited metabolic stress test to calculate peak oxygen consumption (VO2). Studies indicate that a peak VO2 of less than 10 ml/kg/min indicates a poor prognosis with a survival that is less than that of transplant.[11] Patients with peak VO2 of less than 12 ml/kg/min and refractory symptoms of heart failure have also been shown to have an improved quality of life after transplant.[7] Recently, the concept of a single VO2 "cut-off" for the determination for candidacy for cardiac transplant has been challenged, with the finding that gender differences as well as the use of beta-blockers impact on the peak VO2.[12,13] For this reason, most transplant centers now use peak VO2 in the context of other clinical markers in determining transplant candidacy and timing of listing. Non-ambulatory patients who require continuous intravenous inotropes that cannot be weaned or who require mechanical support to maintain an adequate cardiac index are also considered potential candidates for cardiac transplantation. On rare occasions, refractory ventricular arrhythmias or refractory angina despite maximal medical and surgical therapies are also indications for transplant.[14]

Contraindications to cardiac transplantation include any medical condition that would be expected to limit life expectancy following transplant, such as recent or active malignancies, active infections, or other chronic life threatening diseases. While evidence of other noncardiac end-organ damage precludes cardiac transplant alone, patients so afflicted may occasionally be considered for multiple organ transplantation. An example would be simultaneous heart-kidney transplantation for a patient with ESRD and indications for cardiac transplant as well. Pulmonary hypertension with pulmonary vascular resistance in excess of 4 wood units that cannot be reduced by either medical means or placement of a ventricular assist device is another absolute contraindication for isolated cardiac transplant; in this setting, heart-lung transplantation may be a consideration. Relative contraindications to heart transplantation include advancing age (greater than 65 years given worse outcomes at this age).[5,7] Because of the rigorous medical regimen posttransplant, psychosocial factors have to be carefully considered in the candidate evaluation.

*Corresponding Author: Roy D. Bloom—University of Pennsylvania Medical Center, 3400 Spruce Street, Philadelphia, PA 19104, USA.
Email: rdbloom@mail.med.upenn.edu

*Chronic Allograft Failure: Natural History, Pathogenesis, Diagnosis and Management*, edited by Nasimul Ahsan. ©2008 Landes Bioscience.

## *Allocation System*

In January 1999, the United Network for Organ Sharing (UNOS) implemented an acuity status system for patients awaiting cardiac transplantation in the United States. Under this system, donor hearts are prioritized to the sickest patients first in order to maximize waiting list survival. The current acuity system includes 3 levels, 1A, 1B and 2A. Patients at the greatest risk of death are listed as status 1A while status 2 patients are considered to have a lower risk. For each acuity status, a number of objective criteria must be met. The criteria for being listed as a status 1A include: mechanical circulatory support (intra-aortic balloon pump, total artificial heart or ECMO), implantation of a ventricular assist device (for 30 days once the center has determined the patient is stable), ventricular assist device complication including mechanical failure or infection, high dose or multiple inotropes (dobutamine, dopamine or milrinone) with an indwelling pulmonary artery catheter. 1A status can also be obtained via an exception review process in each region. This system is used when the patient has a high risk of death in the next 1 to 2 weeks without transplant but does not fit any of the established criteria. The 30 day period of 1A time following placement of a ventricular assist device can be applied at any time after implantation. This allows the patient to recover from the initial surgery as well as their heart failure state. Several studies have indicated that waiting several weeks after VAD surgery for end-organ function to normalize improves cardiac transplant outcomes. Therefore many centers will wait to activate the 30 days of 1A time at 2 to 6 weeks after implantation.[15]

1B status is for patients on a single inotrope that does not meet the criteria of "high dose". Patients can be ambulatory and in the community or hospitalized. 1B status can also be obtained either before or following the 30 day period after ventricular assist device placement. Status 2 patients are those who meet the indications for transplant and have an expected one year mortality of > 15% but are not on continuous inotropes or have a mechanical support device.

Within an ABO blood group and recipient size range donor organs are first offered to the highest priority patients and then to the lower risk groups until the organs are matched to a recipient. Hearts are offered geographically using the location of the donor. Hearts are first offered to local transplant centers and then to centers outside the region in a series of concentric circles of 500 miles in diameter until an organ is matched.[16] Heart transplants are limited to a cold ischemic time of approximately 4 hours, thereby restricting the distance from which a heart can be harvested.

## *Pretransplant Care of Heart Transplant Candidates*

Patients awaiting cardiac transplantation are managed with a variety of heart failure therapies including neurohormonal blocking agents (angiotensin converting enzyme inhibitors, beta blockers, aldosterone antagonists, angiotensin receptor blockers) and diuretics. In addition, eligible patients may receive a biventricular pacemaker and the vast majority of these patients will have an implantable defibrillator to protect against sudden cardiac death prior to transplant.[17] It is imperative that heart transplant candidates be regularly re-evaluated while on the waiting list as newer therapies may promote positive remodeling of the ventricle over time, precluding or delaying the need for transplant. Even patients on stable inotrope regimens should be aggressively managed with neurohormonal blocking agents as tolerated and periodically re-evaluated for improved clinical status.

Intravenous inotropic therapy is often initiated for patients with a low cardiac output state or refractory symptoms of congestion despite maximal medical support and biventricular pacing if indicated. The most commonly used chronic inotrope therapies include milrinone and dobutamine. Inotropes can significantly improve cardiac index, decrease symptoms and improve end-organ perfusion. However, since inotropes also increase the risk of arrhythmias (including fatal ventricular arrhythmias), patients on continuous inotropes have in the past, remained hospitalized until transplantation. Recent studies have shown that with the use of implantable defibrillators to treat dangerous ventricular arrhythmias, patients awaiting cardiac transplantation can be safely managed in the outpatient setting.[18]

Patients who have acute hemodynamic compromise or have a chronic low cardiac output state despite inotropic support may be candidates for ventricular assist device placement. Ventricular assist device technology has been used to bridge patients to transplant and there is evidence that in the appropriate population these devices can reverse end-organ dysfunction and allow for improved outcomes after transplant.[15]

The advent of an approved left ventricular assist device for permanent use has provided an option for some patients who are not candidates for cardiac transplant due to a recent malignancy or other chronic medical condition. Patients receiving permanent left ventricular assist devices also require a very strong care giver network in order to manage a challenging medical technology.[19]

## *Surgical Techniques*

There are three surgical techniques commonly used for cardiac transplantation. These include standard, bicaval or total techniques.[20] In the standard or biatrial technique, cuffs of recipient atria including the orifices of the pulmonary veins are left intact and then sewn to the donor atria. Advantages to this technique include a more rapid surgical time and no need to re-implant the pulmonary veins. Over the past several years the bicaval approach has gained favor as it has reduced atrial arrhythmias, sino-atrial nodal dysfunction and tricuspid regurgitation. In this technique, the recipient pulmonary veins are excised in a cuff of left atrium and then are attached to the donor left atrium. The entire recipient right atrium is removed. The superior and inferior vena cavae are attached as are the aorta and pulmonary arteries. Despite the longer ischemic time, this technique has been associated with improved short- and long-term outcomes. The total technique involves removal of the entire recipient heart with the exception of two small "buttons" of left atrial tissue containing the four pulmonary veins. The remainder of the anastomoses are identical to the bicaval technique with the exception that there are two anastomoses in the left atrium. In addition to longer operative times, this technique has not been demonstrated to improve outcomes.[7,20]

## *Cardiac Transplant Outcomes*

Cardiac transplant outcomes have continued to improve over time, with one year survival for patients transplanted from July 1, 2002 through June 30, 2003 being 86.8%. The three year survival from July 1, 2000 through June 30, 2003, is 79.2%.[4] At one year, 90% of surviving patients report no functional limitations; approximately 31% return to work while another 19% retire.[21] Dysfunction of the heart transplant is the most frequent cause of death in the first posttransplant year; beyond this time point, graft dysfunction and infectious complications more commonly lead to patient mortality, although malignancy and graft vasculopathy are important contributors as well.[21]

Graft vasculopathy is one of the major limitations to long term survival following cardiac transplantation. Several donor and recipient factors can influence the development of vasculopathy and lead to graft dysfunction and death. Graft vasculopathy differs from typical coronary disease in that it is diffuse in nature and can impact the small vessels first making it difficult to detect via coronary angiography. Intravascular ultrasound has become the gold standard for detecting and monitoring graft coronary disease but is not available routinely outside of research protocols. This form of vasculopathy appears to be immune mediated and regression may be achieved with newer or augmented immunosuppression. HMG-coA reductase inhibitors are routinely used in all cardiac transplant recipients regardless of lipid profile due to evidence suggesting that the anti-inflammatory property of statins may help prevent or delay vasculopathy and also may reduce the incidence of cellular rejection.[22]

The recent ISHLT registry report has indicated that heart recipients develop a significant burden of comorbidities over time, in concert with improved transplant outcomes. By 8 years, >97% of recipients have hypertension, >14% have significant chronic kidney disease (creatinine >2.5 in 10%, long-term dialysis in 4% and renal transplant in 1%), >90% have hyperlipidemia, >35% have diabetes and >40% have angiographic evidence of coronary artery vasculopathy.[21]

## Lung Transplantation

### Background/Current Status

Human lung transplantation was first attempted in 1963 but it was not until two decades later that extended survival was achieved. Following the initial technical successes of the 1980s, the field of lung transplantation experienced dramatic growth, such that by the late 1990s approximately 900 procedures were being performed annually in the United States. Subsequent growth has been constrained by a scarcity of suitable lung allografts. This scarcity reflects both the propensity of the lungs in the brain-dead donor to become compromised by aspiration, infection, edema, or trauma; and the relatively strict criteria that have been employed in identifying suitable lung donors. Consequently, only about 15 percent of potential deceased donors have historically been deemed to have lungs acceptable for transplantation. In 2003, the U.S. Department of Health and Human Services introduced the Organ Donation Breakthrough Collaborative, a national effort to increase the number of organ donors through increased consent for donation and through optimization and standardization of donor management practices.[3]

As a combined result of this initiative and liberalization of lung donor selection criteria, there has been a steady increase in the number of lung transplants performed in the past several years, reaching 1406 in 2005.[3]

### Candidate Selection

Lung transplantation is a therapeutic option for a broad spectrum of chronic, debilitating pulmonary disorders of the airways, parenchyma and vasculature.[23] Chronic obstructive pulmonary disease (COPD) represents the leading indication for lung transplantation, accounting for approximately half of all procedures performed worldwide.[24] Other leading indications include idiopathic pulmonary fibrosis (17% of cases) and cystic fibrosis (17% of cases). Once a common indication for transplantation, primary pulmonary hypertension now accounts for less than 5% of procedures, reflecting major advances in the medical management of these patients. Transplantation of patients with lung involvement due to collagen vascular disease (e.g., scleroderma) remains controversial due to concerns that extrapulmonary manifestations of the systemic disease could compromise the posttransplant course. Nonetheless, short-term functional outcomes and survival following transplantation are comparable to other patient populations and most centers are willing to offer transplantation to carefully selected patients without significant extrapulmonary organ dysfunction.[25] In contrast, lung transplantation for bronchoalveolar carcinoma, a subtype of lung cancer that tends to remain localized to the lung parenchyma, has largely been abandoned due to an unacceptably high rate of cancer recurrence.[26]

Listing for transplantation is considered at a time when the lung disease limits basic activities of daily living and is deemed to pose a high risk of death within several years. Disease-specific guidelines for timely referral and listing of patients, based on available predictive indices, have recently been updated.[27] The imprecise nature of these predictive indices often makes decisions about transplant listing problematic. This is particularly true for patients with COPD, where the disease tends to follow a protracted course even in advanced stages. The patient's perception of an unacceptably poor quality of life is an important additional factor to consider but should not serve as the sole justification for referral of a patient whose disease is not deemed to be at a life-threatening stage.

The scarcity of organs and the somewhat inferior outcomes achieved with increasing age have prompted the establishment of recommended age cutoffs: 55 years for heart-lung, 60 years for bilateral lung and 65 years for single lung transplantation. Candidates should be functionally disabled (New York Heart Association class III or IV) but still ambulatory. Many programs screen for and exclude profoundly debilitated patients by requiring a minimum distance on a standard six minute walk test, most frequently 600 feet.[28] The presence of significant renal, hepatic, or left ventricular dysfunction generally precludes isolated lung transplantation, but multi-organ transplantation has on rare occasion been considered in highly select patients. Other absolute contraindications include active infection with the human immunodeficiency virus, hepatitis B virus, or hepatitis C virus with histologic evidence of significant liver damage; active or recent cigarette smoking, drug abuse, or alcohol abuse; recent malignancy (other than non-melanotic skin cancers); and extremes of weight. The risk posed by other chronic medical conditions such as diabetes mellitus, osteoporosis, gastroesophageal reflux and limited coronary artery disease should be assessed individually based on severity of disease, presence of end-organ damage and ease of control with standard therapies. Among cystic fibrosis candidates, airways colonization with *Burkholderia cepacia* is considered a strong contraindication to lung transplantation by the majority of centers, due to the demonstrated propensity of this organism to cause lethal posttransplant infections.[28,29] Transplantation of patients on mechanical ventilation is associated with a increased risk of mortality within the first year and is therefore not commonly performed.[24,28]

### Allocation System

From 1990-2005, lung allocation in the United States was based upon a "seniority system" that prioritized candidates on the basis of the amount of time they had accrued on the waiting list, without regard to severity of illness. This system was ultimately called into question because it failed to accommodate those patients with a more rapidly progressive course who often could not tolerate the prolonged waiting times to transplantation and who were likely to die prior to receiving an organ. Indeed, excessive wait list mortality was documented among certain patient populations, such as those with idiopathic pulmonary fibrosis (IPF) and cystic fibrosis, as compared to patients with COPD.[30] In response to the perceived inequities of the time-based system and under mandate of the federal government, a new system was implemented in May 2005 where lungs are allocated on the basis of both medical urgency (i.e., risk of death without a transplant) and "net transplant benefit" (the difference between predicted posttransplant survival and survival with continued waiting).[16] By incorporating this latter concept, the system attempts to avoid the pitfall of preferentially allocating the limited donor organ pool to desperately ill patients with an unacceptably high posttransplantation mortality rate.

The new model, derived from a multivariate analysis of data from the comprehensive UNOS national database, identifies ten factors independently predictive of death on the wait list and seven predictive of death following transplantation.[16] For each patient, these factors are utilized to calculate predicted one year survival with and without transplantation. Patients who demonstrate a large net difference in survival (predicted post—pretransplant survival) in conjunction with a high degree of medical urgency (low predicted pretransplant survival) receive the highest allocation scores.

An analysis of data from the Penn Lung Transplant Program sheds light on the impact of this new system in the year since it was implemented. The system clearly prioritizes patients with IPF over those with COPD. Under the new system, the percentage of patients with IPF undergoing transplantation at our center has increased from

16% to 38% while the percentage of patients with COPD has fallen from 57% to 38%. Waiting times, which had previously ranged from 2-3 years, are now on the order of days to weeks for those with high allocation scores (e.g., patients with IPF). Additionally, since there is no longer an incentive to place patients on the active waiting list simply to accrue time (many of whom were ultimately deactivated rather than transplanted), the number of actively listed patients has fallen to approximately one-fourth of the previous number. This parallels national trends; the total number of actively listed patients on the UNOS lung transplant list fell from 2167 the year before the new system was implemented to only 800 currently.[1] Curiously, for the first time in almost two decades, the number of actively listed patients is less than the annual number of lung transplantation procedures performed.

Several concerns have been raised about the new system that will have to be addressed in an ongoing evaluation of its merits: (1) the predictive model employed has not been prospectively validated; (2) the calculation of net transplant benefit is based upon predicted one-year posttransplant survival, which is heavily influenced by differences in disease-specific perioperative mortality rates and does not truly reflect long-term outcomes; (3) net transplant benefit is defined exclusively in terms of survival and does not acknowledge dramatically improved functional status and quality of life as a net benefit.[31]

### Surgical Procedures

Heart-lung transplantation was the first procedure to be successfully performed but it has largely been supplanted by techniques to replace the lungs alone. Heart-lung transplantation is now principally restricted to Eisenmenger's syndrome with surgically uncorrectable cardiac lesions and advanced lung disease with concurrent left ventricular dysfunction or severe coronary artery disease. Previously, the presence of severe right ventricular dysfunction was deemed to be an indication for heart-lung transplantation. However, subsequent experience with isolated lung transplantation has demonstrated the remarkable ability of the right ventricle to recover once pulmonary artery pressures are normalized.

Single lung transplantation is the procedure of choice for pulmonary fibrosis and for older patients with chronic obstructive pulmonary disease but is contraindicated in patients with suppurative lung disorders such as cystic fibrosis. Most centers also consider severe pulmonary hypertension to be a contraindication to single lung transplant. In this setting, single lung transplantation results in diversion of nearly the entire cardiac output through the allograft, due to the high vascular resistance in the remaining native lung and this can contribute to exaggerated reperfusion pulmonary edema in the allograft. Major advantages of single lung transplantation are its technical ease and its efficient use of the limited donor pool, permitting two recipients to benefit from a single donor.

Bilateral sequential lung transplantation involves the performance of two single lung transplant procedures in succession during a single operative session. In the absence of severe pulmonary hypertension, cardiopulmonary bypass can often be avoided by sustaining the patient on the contralateral lung during implantation of each allograft. The primary indications for this procedure are cystic fibrosis, other forms of bronchiectasis and pulmonary vascular disorders. Additionally, an increasing number of programs are advocating its use for patients with emphysema, arguing that it offers functional and survival advantages over single lung transplantation.[32,33]

Living donor bilateral lobar transplantation is the newest procedure to be introduced and is still uncommonly performed. It involves the implantation of right and left lower lobes harvested from two living, blood group-compatible donors. The procedure has generally been reserved for candidates whose deteriorating status does not permit them to wait for a lung from a deceased donor. Given the inherently undersized nature of the grafts, it is preferable that the donors be considerably taller than the recipient. Patients with cystic fibrosis are particularly well suited as a target population since even as adults they tend to be of small stature. Concerns about excessive risk to the donor have thus far proven to be unfounded. In the largest experience reported to date, there were no deaths among 253 donors and only eight complications of sufficient magnitude to warrant surgical re-exploration.[34] Donation of a lobe results in an average decrement in vital capacity of 17 percent.

### Outcomes

Current one-, three- and five-year survival rates are 83%, 64% and 48%, respectively.[3] Survival rates have improved only modestly over the past decade.[3,24] While short-term survival approximates that achieved following heart and liver transplantation, five-year survival remains considerably below the 70% rate achieved following these other transplant procedures. Primary graft failure and infection represent the most common causes of early deaths while bronchiolitis obliterans is the leading cause of late deaths.

For patients with COPD, survival for recipients younger than 60 years of age is superior following bilateral compared to single lung transplantation,[35] but the opposite has been demonstrated for those with pulmonary fibrosis.[32] Survival following living donor transplantation is comparable to that achieved following deceased donor transplantation.[36]

Successful lung transplantation leads to dramatic improvement in lung function, gas exchange, exercise tolerance and quality of life.[33,37] By the end of the first posttransplant year, 85% of recipients report no limitations to activity, 14% require some assistance and only 1% require total assistance.[24] Among those patients with primary and secondary forms of pulmonary hypertension, elevated pulmonary pressures typically normalize within hours of transplantation while recovery of right ventricular function is more protracted, albeit ultimately complete.[38]

## Liver Transplantation

### Background

Since Dr. Thomas Starzl and colleagues performed the first successful human liver transplant in 1967,[39] the field has evolved at a remarkable pace. As of August 2006, 77,775 liver transplants had been performed in 145 institutions across the United States.[2] 6,443 liver transplants were performed in 2005 alone. The number of liver transplants in the United States has risen incrementally over the past decade. Part of this rise can be attributed to the increased use of expanded criteria donors, including donation after cardiac death. Another reason is the increased utilization of living liver donors.[3] Unfortunately, despite the increase in liver transplant volume, it has been insufficient to match the demand; currently over 17,000 patients are awaiting liver transplantation, with median waiting time ranging from 75 to 459 days according to blood type. Although pretransplant death rates have declined in recent years,[3] 1,818 patients died while awaiting liver transplantation in 2002, largely due to a shortage of suitable donor organs.[40]

### Indications for Liver Transplantation

Orthotopic liver transplantation (OLT) may be indicated for acute or chronic liver failure from nearly any cause.[41] In the United States, chronic hepatitis C infection is the most common indication for liver transplantation, accounting for 43% of all OLT in 2004,[2,42] followed by alcohol-induced liver disease and non-alcoholic steatohepatitis.[43]

Patients with chronic liver disease refractory to aggressive medical therapies should be referred for liver transplant evaluation at the first sign of hepatic decompensation. This may often present clinically as encephalopathy, ascites (with or without spontaneous bacterial peritonitis), or variceal hemorrhage. Hepatopulmonary syndrome is another late manifestation of chronic liver disease that is associated with significant morbidity and warrants referral for transplantation.

Underscoring the need for liver transplantation for patients with decompensated cirrhosis is the strong association of this condition with short-term morbidity and mortality. Two-year survival in patients with cirrhosis complicated by ascites is only 50%[44] while those with variceal hemorrhage have an inpatient mortality of 30 to 50%.[45,46] Hepatocellular carcinoma (HCC) can be found in patients with cirrhosis and in patients with chronic replicative hepatitis B infection, in the absence of significant fibrosis. Patients with a single, isolated HCC measuring between 2 and 5 cm in diameter or patients with fewer than three lesions, each less than 3 cm in diameter, should be referred for liver transplantation.[47]

Acute liver failure, typically in the setting of fulminant hepatic failure (FHF) is an indication for expedited transplant evaluation. Commonly observed causes of FHF include viral or autoimmune hepatitis, Wilson's disease, acute Budd-Chiari syndrome, or drug hepatotoxicity. In response to the severity of acute liver injury, patients with FHF are listed separately from those with chronic liver disease and are given Status 1 priority, which places them at the top of the waiting list. Refer to Table 1 for a summary of diseases that may necessitate liver transplantation.

Lastly, besides the classical indications listed above, liver transplantation may be indicated in patients with intractable pruritus, metabolic bone disease, recurrent bacterial cholangitis and progressive malnutrition. Although these patients may not have decompensated cirrhosis or FHF, transplantation should be considered to treat the extremely poor quality of life that is associated with these conditions.[47]

## Organ Allocation: Status 1 and MELD

At this time, there are no minimal listing criteria for liver transplantation.[48] Donor livers are allocated based on ABO-blood type. As described above, patients with FHF, who are listed as Status 1, are transplanted first. From 2003 to 2004, 47% of patients listed as Status 1 received a liver transplant within seven days.[3] After Status 1, livers are then allocated to adult patients via the modified Model for End-stage Liver Disease (MELD) score. The MELD score has been used in patients with chronic liver disease since early 2002. MELD was originally developed to predict outcomes in cirrhotic patients undergoing transjugular intrahepatic portosystemic shunt (TIPS).[49] Studies found the MELD score to be highly predictive of three-month mortality in hospitalized patients with cirrhosis. MELD scores from 20 to 35 are associated with 10 to 60% three-month mortality, whereas MELD scores greater than 35 are associated with 80% mortality at three months.[50,51]

The MELD score is currently used to identify and triage sicker patients for more urgent transplantation. The MELD score (range: 6 to 40) is derived from three widely available and easily repeatable lab values: total bilirubin, serum creatinine and international normalized ratio (INR). In situations where two or more patients of the same blood group have the same MELD score, patients who have the longest waiting-list time are given priority. At present, MELD exception points are conferred to patients who have T2-HCC (stage 2) or hepatopulmonary syndrome and meet all other transplant criteria.[2,48,52]

## Contraindications to Liver Transplantation

Active extrahepatic infections, poor social support, untreated psychiatric disorders, significant coronary artery disease, advanced chronic obstructive or restrictive pulmonary disease and active alcohol or drug abuse are problems that preclude liver transplantation.[43] Untreated HIV infection is also considered an absolute contraindication; however, patients with well-controlled HIV should be considered for ongoing clinical trials that will assess the role of transplantation in stable patients on highly active anti-retroviral therapy (HAART). Up to 21% of patients with primary sclerosing cholangitis will develop cholangiocarcinoma over time [53,54] and the discovery of this complication in the pretransplant setting has generally contraindicated OLT. However, recent evidence has suggested that neoadjuvant chemotherapy followed by liver transplantation can provide good long-term survival rates in this setting (82% at five years after transplantation).[55] Other extrahepatic malignancies as well as anatomic abnormalities are other factors that make transplantation unfeasible.[47]

Higher risk candidates for liver transplantation includes advancing age,[56] as well as pre-existing diabetes mellitus.[57,58] Although post-OLT outcomes are not as good in these patient populations, they can be safely transplanted following a thorough, multidisciplinary pretransplant assessment.

Renal failure, irrespective of its etiology, is another comorbidity that does not preclude liver transplantation. In many cases, successful liver transplantation leads to recovery of kidney function in association with resolution of Type 1 hepatorenal syndrome (HRS).[59,60] However, patients with HRS of prolonged duration (usually known as type 2 HRS) or with moderately advanced chronic kidney disease from other etiologies (such as hypertension, diabetes, or polycystic kidney disease) should be evaluated by a transplant nephrologist for combined liver-kidney transplantation.

## Table 1. Indications for liver transplantation

| Causes of Chronic Liver Failure | Causes of Fulminant Hepatic Failure |
|---|---|
| Alcohol (Laënnec cirrhosis) | Drug-induced hepatotoxicity (acetaminophen and others) |
| Hepatitis C | Acute viral hepatitis |
| Hepatitis B | Hepatitis A |
| Non-alcoholic steatohepatitis (NASH) | Hepatitis B (with or without Hepatitis D) |
| Autoimmune hepatitis (AIH) | Hepatitis C (rare) |
| Primary biliary cirrhosis (PBC) | Hepatitis E (outside of the US) |
| Primary sclerosing cholangitis (PSC) | Fulminant Wilson's disease |
| Chronic Budd-Chiari syndrome | Acute Budd-Chiari syndrome |
| Hemochromatosis | Autoimmune hepatitis (AIH) |
| $\alpha_1$-antitrypsin deficiency | Idiosyncratic drug reactions |
| Wilson's disease | |
| Crigler-Najjar syndrome, type 1 | |
| Polycystic liver disease | |
| Caroli's disease or syndrome | |
| Metabolic liver diseases (tyrosinemia) | |
| Sarcoidosis | |

### Deceased-Donor versus Living-Donor Liver Transplantation

The vast majority of transplanted livers come from deceased-donors. In 2005, 6,120 deceased-donor liver transplants (DDLT) were performed, whereas only 323 living-donor liver transplants (LDLT) were done during the same time period.[2] However, LDLT has been viewed as increasingly viable option to meet the demand for donor organs, especially given the relatively static deceased-donor pool. LDLT may also be particularly beneficial in patients who may have longer waiting times based on blood type, severity of disease, or etiology of liver failure.

The concept of LDLT originated from kidney transplantation and LDLT was found to be particularly applicable to pediatric patients as left lateral segment transplants could provide sufficient hepatic mass for small children. Unlike the pediatric patient, an adult recipient requires a larger liver volume, which may include a right hepatic lobe resection or a full left lobe graft in smaller adult recipients.[61-63]

Following live liver donation, up to 19 percent of donors may develop complications.[64] The most common complication, observed in approximately 10% of donors develop bile leaks and neuropraxia. Other complications include wound infection, small-bowel obstruction and incisional hernias. Although morbidity is a serious concern, most donors have demonstrated general acceptance of the procedure and have had favorable outcomes.[64] In order to minimize donor risk, careful medical and psychological screening of potential donors is essential.

### Surgical Procedure

The three major phases of surgery include native liver dissection (1 to 2 hours), the anhepatic phase (1.5 to 3 hours) and revascularization of the liver.[43] The most commonly used surgical technique was described by Starzl et al in 1963.[65] One of the major modifications in liver transplantation surgery has included preservation of retrohepatic vena cava to maintain venous return to the heart. This minimizes hemodynamic instability and eliminates the need for venovenous bypass.[66] This "piggyback" technique is now commonly used in most transplant centers and has been found to be safe and associated with few surgical complications.[43,67]

Although the more technical aspects of surgery involving DDLT and LDLT are beyond the scope of this review, certain peri-operative factors are of potential consequence in the posttransplantation period. Prolonged cold-ischemia time is associated with primary nonfunction (PNF) and hepatic artery thrombosis.[43] Cold-ischemia times greater than 15 hours are also associated with significantly higher rates of acute cellular rejection.[68]

### Complications after Liver Transplantation

Cirrhotic patients are prone to hyponatremia because of impaired free water excretion that occurs for many reasons, including splanchnic arterial vasodilation and decreased effective circulating arterial volume, reduced GFR and administration of diuretic therapy.[69] Hyponatremia is of particular importance in the perioperative period, as serum sodium rapidly normalizes hours after the donor liver is engrafted and patients are at risk for central pontine myelinolysis.[70] Therefore, careful management of water balance is required in the immediate pretransplant setting. Antagonists of the $V_2$-receptor of AVP have been shown to promote the urinary excretion of free water in rats and are being developed and studied in humans, but are not yet available for use in clinical practice.[69,71]

Many of the very early complications that occur posttransplantation are related to the surgical procedure. The most common post-operative surgical complication is intraabdominal bleeding, which may occur in 10 to 15% of patients.[72] Although a picture of ischemic liver injury is common immediately posttransplantation, the liver-associated enzymes usually normalize by two to three days after transplantation.

The presence of cholestasis more than a few days posttransplantation requires further investigation. An appropriate diagnostic strategy is to begin with an ultrasound examination to evaluate for biliary ductal dilation, which may signal a biliary stricture. Simultaneous doppler imaging should be obtained to evaluate for hepatic artery thrombosis, portal and hepatic vein thrombosis, or IVC obstruction. If the ultrasound is unrevealing, endoscopic retrograde cholangiopancreatography (ERCP) should be considered the gold standard to evaluate the biliary tree in transplanted patients who have had a duct-to-duct anastamosis. If a biliary stricture or leak is identified, ERCP also provides the opportunity for endoscopic treatment by the placement of plastic biliary stents. However, in those transplanted patients with Roux-en-Y choledochojejunostomy, percutaneous transhepatic cholangiography (PTC) with possible radiological intervention may be required, as ERCP may not be technically feasible. Liver biopsy is recommended for patients with persistently abnormal liver-associated enzymes to evaluate for acute cellular rejection or other causes of hepatic injury and dysfunction, such as drug-induced hepatotoxicity or infection.[47]

### Infections Associated with Liver Transplantation

In addition to the complications described above, infection is a common and dangerous complication that affects patients pre, peri and posttransplantation. Bacteremia has been shown to occur in approximately 70% of liver transplant recipients.[73] Torbenson et al found that infections were the cause of death in 64% of 321 patients who died after liver transplantation and underwent autopsy at the University of Pittsburgh.[74] Modern advances in surgical techniques, organ preservation, pretransplant evaluation by infectious disease experts, vaccination and judicious use of immunosuppressants have contributed to more favorable outcomes in liver transplant recipients. As a result, one-year mortality due to infection is now less than 10%.[75,76] Despite these improvements, infection remains a serious, life-threatening complication of liver transplantation.

Gram-negative bacteremia is a particularly serious complication in transplanted patients. In the early postoperative course, these patients are prone to bacterial infections (usually from enteric pathogens) that may persist for up to one month. Fever accompanied by cholestatic liver-associated enzymes (as described above) or positive blood cultures that yield gram-negative bacilli should prompt immediate evaluation of the biliary system to exclude biliary obstruction and cholangitis. Evidence of gram-negative bacteremia should also raise suspicion for intraabdominal and intrahepatic abscesses or urinary tract infections in the posttransplant patient.[47]

As with other solid organ transplants, cytomegalovirus (CMV) is the most important viral pathogen that affects liver transplant recipients. Without appropriate immunoprophylaxis, fifty percent of allograft recipients may develop symptomatic CMV infection two to six months posttransplantation.[76] CMV-naïve patients who receive CMV-positive livers are placed on prophylactic ganciclovir post-operatively, but CMV hepatitis, CMV syndrome and extrahepatic CMV infections can still occur.[43]

### Immunosuppression in Liver Transplantation

The field of liver transplant immunosuppression has evolved in concert with the introduction of new agents into the clinical arena. Despite the heterogeneity in management, immunosuppression in most transplantation centers is typically comprised of a calcineurin inhibitor and corticosteroids.[77] Mycophenolate mofetil, sirolimus and antibodies to T-cell receptors are also commonly used depending on the situation and institutional preference.[78] Steroids are usually weaned and eliminated anywhere from days to months following transplantation.[43,79] The goal of early steroid withdrawal is to limit or prevent the many well-described complications associated with this class of immunosuppressant.

## Outcomes

In general, the rates of graft and patient survival following liver transplantation are excellent in the United States. A recent OPTN/SRTR Annual Report 1995-2004 reported adjusted graft survival after DDLT of 82% at one year and 67% at five years, whereas adjusted patient survival for DDLT recipients was 87% at one year and 73% at five years posttransplantation. The same report found adjusted one-year patient survival of 88% and three-year survival of 80% for LDLT recipients.[3] These data support that LDLT is a reasonable option for patients who are unlikely to receive DDLT in a timely fashion.

## Kidney Transplantation

### Background

Chronic kidney disease (CKD) is a major risk factor for cardiovascular disease (CVD)[80] and the leading precursor to endstage renal disease (ESRD). Both of these consequences of CKD are associated with devastating outcomes that reduce life expectancy. Renal replacement therapy, in the form of either dialysis or transplantation, is required to prevent death from uremic complications. Kidney transplantation is the treatment of choice for afflicted patients and provides a substantially greater quality of life and longevity benefit compared to dialysis.[81] Kidney transplantation additionally offers an economic advantage, where the typical annualized per patient costs of around $17,000, pale in comparison to the $53,000 for dialysis.[82]

The first kidney transplant was carried out between a pair of identical twins under no immunosuppression.[83] Kidney transplantation has since evolved into a routine global practice, that has burgeoned in concert with the escalating rates of ESRD. In the United States alone, the prevalent population of ESRD patients now exceeds 470,000, more than double that in 1988.[84] Given changing population demographics, coupled with the escalation of diabetes and hypertension, this is projected to exceed 650,000 by the end of the decade.[85] Together with these mounting rates of ESRD, the success of transplantation has fueled an enormous demand-supply disparity and a surge in the number of patients waiting for a kidney. Despite the fact that well over 200,000 kidney transplants have been performed in the USA since 1988, almost 70,000 patients are now on the ever-expanding national kidney waiting list.[2]

### Indications for Kidney Transplantation

Any patient whose quality of life and/or lifespan will likely be improved after transplant should be considered a potential candidate. Continued advances have resulted in remarkably few absolute contraindications to kidney transplantation alone, generally limited to situations where the risks of sustaining harm with surgery and immunosuppression outweigh the benefits of transplantation. Such conditions include advanced liver or lung disease, intractable or advanced infection or unremitting malignancy. Since CVD is so rife in patients with advanced CKD,[80,86] assessment for coronary artery disease is a key component of the kidney transplant candidate evaluation.[87] As a rule, the presence of CVD does not preclude transplantation unless it is active and not amenable to any intervention, although there may be some variation in policy between different centers. Current guidelines recommend that waitlisted kidney candidates be periodically rescreened for cardiovascular disease until they are transplanted.[88]

### Types of Kidney Transplants

Kidneys for renal transplantation may be derived from both living, as well as deceased donors. Living donors may be either biologically related (parent, child, sibling) or unrelated (friend, spouse, altruistic donor) to the recipient. Kidney transplants from living donors are advantageous for several reasons: (i) living donation avoids ischemia-reperfusion injury associated with procurement, storage, transportation and implantation of kidneys from deceased donors;

**Table 2. Advantages of receiving a living donor kidney transplant**

| |
|---|
| Best quality kidney |
| • enhanced graft function |
| • superior graft survival |
| • superior patient survival |
| Minimize waiting time |
| • can avoid pretransplant dialysis |
| • better quality of life |
| • prolonged life expectancy |
| Elective surgery |
| • ability to plan for donor and recipient |
| • minimize disruption to lifestyle |

(ii) living donors are healthy at time of donation; (iii) the transplant is performed on an elective basis and can be timed to pre-empt the need for any dialysis in the recipient; (iv) largely, for the above reasons, kidneys from living donors are typically superior in terms of quality and function (see Table 2). Among living donor kidney transplants, the best results are observed with transplantation between HLA-identical siblings. Outcomes for non-HLA-identical living donor transplants are similar, irrespective of the biological relationship between donor and recipient and surpass results observed with even the best-matched deceased donor kidneys.[77]

With the growth of the waitlist and increasingly poor quality of available deceased donor kidneys, the emphasis on living donation has magnified over the past decade. As a result, living donation in the USA has expanded from about 2,500 in 1992 to over 6,000 in 2001.[89] Since the turn of the century, the number of living kidney donors has consistently exceeded the number of deceased kidney donors.[77] However, because two recipients can receive kidneys from each deceased donor, more deceased donor transplants are performed overall each year. From the perspective of living donors, living with one remaining kidney has no impact on life expectancy or lifestyle, child-bearing potential or access to medical care or insurability.

Efforts to increase the donor pool for kidney transplant candidates who don't have living donors have included the increasingly common use of organs from more "medically marginal" donors (expanded criteria donors [ECD]). Although results are worse in recipients of ECD kidneys than standard donor kidneys, they are nevertheless superior to outcomes in patients who remain on dialysis.[90,91] Kidneys procured from donors after cardiac death (DCD kidneys) are now also being used with growing success.[92]

### Placement on the Waiting List and Timing of Transplantation

According to current UNOS policy, patients can only be placed on the waitlist for a kidney transplant once their glomerular filtration rate (GFR) is less than 20 ml/minute. The median waiting times range between 3-4 years, with variation on a regional level as well as according to patient blood type.[89] Guidelines recommend referral of patients for transplantation once the GFR is between 20-25 ml/minute, in order to identify prospective living donors and to also start proactively accumulating time on the waitlist.[87] Since dialysis duration pretransplant is inversely related to posttransplant survival, transplantation prior to starting dialysis is desirable.[93,94]

### HLA Matching and Allocation

Incremental degrees of HLA mismatching are associated with higher rates of acute rejection and eventual graft loss.[77,95] Because zero-mismatched deceased donor kidneys have the lowest acute rejection rates, a national policy for sharing such organs has been in effect for two decades, though has been subject to revision during that time.[96]

At present, almost 17% of deceased donor kidneys are shared by this policy, although there is evidence that the benefits of optimal HLA matching may be mitigated by increased ischemia-reperfusion injury associated with shipment of such organs.[97]

One of the greatest challenges facing the transplant community is the mounting pool of patients who have acquired anti-HLA antibodies as a consequence of sensitization from prior blood transfusions or transplants, or through pregnancy. High anti-HLA antibody titers may make it nearly impossible to find compatible donors for affected patients. This has spawned enormous interest in the use of desensitization therapies administered pretransplant to reduce anti-HLA antibodies titers in potential recipients in order to facilitate subsequent transplantation, with promising short-term results.[82,98] Similar strategies have also been successfully implemented in the transplantation of kidneys across ABO incompatible barriers.[99,100]

Another novel and evolving option for recipients who have a positive cross-match against potential donors is the use of a paired-kidney exchange program.[101] In this setting, a potential, albeit incompatible, donor-recipient pairing, is paired with other incompatible donor-recipient pairing, such that the donor from each pairing is compatible with reciprocal recipient. This facilitates two living donor transplants without the need for desensitizing immunosuppression in either recipient.

## Kidney Transplant Surgery and the Perioperative Period

The kidney transplant is placed extraperitoneally in the right or left lower abdominal quadrant. The transplant renal artery and veins are generally anastamosed to the external iliac artery and vein respectively. The transplanted ureter is usually reimplanted to the recipient bladder utilizing a technique to prevent urine reflux. Immediate transplant kidney function is desirable, though does often not occur. Imaging studies including ultrasound and nuclear isotope flow scan are often utilized to assess kidney allograft blood flow and/or function in the immediate postoperative period. Vascular anastamotic catastrophes or hyperacute rejection may rarely result in immediate graft failure within hours of transplantation and the need for transplant nephrectomy.

More typically, most deceased donor and some live donor recipients experience early transient graft dysfunction. The term, delayed graft function (DGF), is defined by the requirement for dialysis within the first posttransplant week. This term is imperfect as it does not capture patients with early graft dysfunction who do not require dialysis and includes patients with immediate function dialyzed acutely for an indication such as hyperkalemia, yet it has persisted as a universal standard for clinical trial purposes. DGF is usually due to acute tubular necrosis and has a reported frequency from 2-50%.[102] Factors such as deceased donor source, procurement injury, ischemia-reperfusion injury, drug nephrotoxicity, volume depletion and acute rejection all predispose to DGF.[102] DGF typically recovers, although long-term graft survival may be compromised. Strategies to prevent DGF include: expediting the transplant to reduce ischemic injury, aggressive perioperative volume expansion, minimizing perioperative exposure to calcineurin inhibitors, sirolimus and other nephrotoxic agents. Where

DGF persists despite these above measures, kidney transplant biopsy is often performed within the first two to three posttransplant weeks to optimize therapy.

### Outcomes Following Kidney Transplantation

One-year patient survival rates exceed 95%, while one-year graft survival rates are now over 91% overall.[77] Current five-year patient and graft survival rates are 90% and 80%, 85% and 69%, 70% and 53% for recipients of living donor kidneys, non-ECD kidneys and ECD kidneys respectively.[77] However, graft half-life beyond the first posttransplant year has only marginally improved, with graft loss in this setting mainly due to either patient death with a functioning allograft or chronic allograft nephropathy, a process characterized by an inexorable decline in kidney function due to progressive fibrosis.[103] Both immunological (e.g., rejection, HLA mismatching) as well as non-immunological factors (e.g., quality of the donor organ, hypertension, calcineurin-inhibitor toxicity) predispose to chronic allograft nephropathy. CVD is by far the most common reason for patient mortality in long-term kidney transplant patients.[86,103]

## Pancreas Transplantation

### Background

Pancreas transplantation has evolved over the past forty years as a treatment for Type I diabetes. The first pancreas transplants were performed in 1966 and were associated with dismal results because of technical limitations and ineffective immunosuppression.[104] Increasing experience, procedural refinement and the emergence of superior immunosuppression has led to improving outcomes and ultimately, the wider acceptance of pancreas transplantation as an effective treatment option. Since the late 1980s, the field has enjoyed unabated growth.[105] In total, 23,043 pancreas transplants had been performed worldwide as of December 2004. The annual number of pancreas transplants has increased from approximately 200 in 1988 to over 1600 in 2003.[106]

### Transplant Options for Patients with Diabetes

Diabetes is the leading cause of kidney failure in the United States and is associated with a particularly ominous prognosis in patients on dialysis.[81] By comparison, kidney transplantation alone markedly improves outcomes for both Type I and Type II diabetics, especially if performed prior to dialysis initiation.[81,93] Pancreatic allografts are usually procured from deceased donors because reduced residual islet mass in living donors may increase the subsequent risk of glucose intolerance in such donors. Pancreas transplantation is almost always reserved for patients with Type I diabetes, though on occasion has been performed in Type II diabetics as well. As depicted in Table 3, pancreas transplantation takes place in one of three settings: (a) simultaneous pancreas-kidney transplantation (SPK), where the pancreas and a kidney are transplanted from the same donor into a recipient with advanced kidney disease, during one operation; (b) pancreas after kidney transplantation (PAK), in which a pancreas is transplanted into a patient who had previously received a kidney from a different living or deceased donor, or (c) pancreas transplant alone (PTA), usually reserved for patients with hyperlabile glycemic control and well preserved renal

---

*Table 3. Transplant options for patients with diabetes mellitus*

| Type of Transplant | Diabetes Type | Source of Organ | Advanced Kidney Disease |
|---|---|---|---|
| Kidney alone | I or II | Living or deceased donor | Yes |
| Simultaneous Pancreas-Kidney (SPK) | I (rarely II) | Same deceased donor for both organs; rarely living donor | Yes |
| Pancreas After Kidney (PAK) | I | Deceased donor pancreas after previous kidney from another living or deceased donor | Yes |
| Pancreas Alone (PTA) | I | Deceased donor | No |

function. There are currently 2400 patients waitlisted for SPK and over 1700 patients listed for isolated pancreas, either as PAK or PTA.[1]

### Goals and Rationale of Pancreas Transplantation

Studies have demonstrated that intensive insulin therapy in Type I diabetics is associated with fewer long-term complications and thereby, enhanced life expectancy.[107] Pancreas transplantation represents an alternative to chronic insulin therapy. A successful transplant outcome brings about chronic euglycemia whilst simultaneously avoiding the need for exogenous insulin administration. In this way, pancreas recipients with functioning allografts experience an improvement in quality of life and eliminate both the need for chronic glucose monitoring and the life-threatening risks of hypoglycemic unawareness. Additional benefits of pancreas transplantation include the prevention and reversal of diabetic nephropathy, as well as some amelioration of sensory, motor and autonomic neuropathy.[108]

### Indications for Pancreas Transplantation

As a rule, pancreas transplant candidates should be C-peptide deficient. Patients with advanced diabetic kidney disease are potentially eligible for SPK transplantation. Indications for isolated pancreas transplantation are less well established although the widely endorsed American Diabetes Association's position is that this procedure should be reserved for patients with life-threatening hypoglycemic unawareness, frequent acute metabolic complications associated with hyperlabile glycemic control, or failure of conventional insulin therapies to prevent acute complications.[109] Historically, candidate age above 45 years and the presence of cardiovascular disease were relative contraindications to pancreas transplantation, based on their association with worse outcomes.[110] Ongoing experience over the past decade and improving results has seen an increase in age, as well as comorbidity, of pancreas recipients.[106,111]

### Surgical Considerations in Pancreas Transplantation

Several surgical issues require consideration in pancreas transplantation. The transplant pancreas, together with a small segment of the adjacent duodenum (containing the sphincter of Oddi), is procured *en-bloc* from the donor. Although transplantation of the whole pancreas serves as the source of abundant functioning islets, the graft's exocrine drainage via the pancreatic duct has to be accommodated. The first widely adopted technique involved pancreatic duct drainage into the urinary bladder via a pancreaticoduodenocystostomy but this was associated with frequent complications.[112] Enteric drainage was developed as an alternative option to the bladder route.[113] With enteric drainage, the segment of transplant duodenum is attached to the small bowel either via a Roux-en-Y or directly end-to-side.[113,114]

Another surgical issue is the method of venous drainage for the transplant pancreas. Initially, most procedures were performed with venous drainage via the iliac vessels. With this technique, insulin released from the pancreas enters the systemic circulation directly, avoiding the physiological first pass effect in the liver. This results in systemic hyperinsulinemia, although there is no evidence that this form of chronic hyperinsulinemia has any adverse consequence.[115] More recently, the technique of draining the venous outflow directly into the portal vein has become increasingly common.[116] While this approach approximates the normal trafficking of insulin through the liver, registry data does not demonstrate any outcome differences between the two routes of venous drainage.[105]

The immediate posttransplant course is typically characterized by rapid improvements in blood glucose levels. Failure of this to occur should prompt a high index of suspicion for an early complication, such as graft thrombosis, rejection, pancreatitis, duct leaks and infection. Although technical failure has historically been an important outcome determinant, International Registry data now indicates that graft loss due to technical failure is below 10% for each of the different types of pancreas transplant performed.[106]

Monitoring for pancreas rejection remains imprecise. Hyperglycemia is a late marker only developing once most of the islet mass has been damaged. Elevations in amylase and lipase raise the suspicion of graft dysfunction and warrants imaging studies and consideration for biopsy.

### Immunosuppression in Pancreas Transplantation

Very few randomized controlled trials have examined immunosuppressive efficacy in pancreas recipients. Data from the Scientific Registry for Transplant Research indicates that the vast majority of patients receive induction therapy with either a T-cell depleting agent or an anti-CD-25 preparation.[77] Most maintenance regimens contain tacrolimus and mycophenolate mofetil, with or without low dose steroids.[77] Strategies involving withdrawal/avoidance of corticosteroids and/or tacrolimus continue to be explored. In pancreas transplantation in particular, both of these therapies have potentially diabetogenic toxicities that may negate the benefit of the pancreas transplant. Patients with suspected acute rejection are typically treated with antibody therapy.

### Outcomes Following Pancreas Transplantation

Historically, pancreas outcomes following SPK transplantation have been superior to those of either PAK or PTA. This has been attributed to increased immunological issues observed in isolated pancreas transplantation, as well as to the fact that creatinine is a reliable surrogate of graft function in patients receiving SPK.[117] Although superior immunosuppression and greater technical experience has helped to narrow the differences in one-year pancreas graft survival between the various types of pancreas transplant, SPK recipients continue to exhibit the best pancreas outcomes at five years at 71%.[77] Registry analyses further indicate that, compared to remaining on the waitlist, SPK transplantation is associated with lower mortality rates, though this has not been unequivocally established for PAK or PTA recipients.[118,119] SPK transplantation is also associated with better patient and kidney graft survival than transplantation of a deceased donor kidney alone, though offers no advantage over a kidney from a living donor.[120,121] Based on these above data, the following clinical approach to Type I diabetics with advanced chronic kidney disease is recommended. Pre-emptive transplantation should be pursued if at all possible to minimize or avoid the need for dialysis, a poor outcome determinant. For patients with acceptable comorbidity, a living donor kidney should be used if available, followed by subsequent PAK; patients without living donors should be waitlisted and transplanted with a deceased donor SPK. Patients in whom pancreas transplantation is contraindicated should receive a kidney alone, from either a living (preferable) or deceased donor.

Long-term pancreas graft failure is characterized by hyperglycemia and the need to restart conventional glucose-lowering therapies. In this setting, hyperglycemia may be a manifestation of either a) islet failure (low c-peptide and insulin levels) secondary to alloimmune (acute or chronic rejection) or recurrent autoimmune-mediated injury, or b) insulin resistance (high c-peptide and insulin levels), associated with posttransplant weight gain and some immunosuppressive therapies.

### Conclusion

Over the past half-century or so, organ transplantation has evolved from a medical curiosity into thriving clinical discipline with enormous potential to preserve and extend life expectancy, as well as to significantly improve quality of life. The astonishing accomplishments of organ transplantation have fueled demand in excess of organ supply to the point that many patients now die while awaiting a transplant. The result is that initiatives aimed at further optimizing organ outcomes are urgently needed. This will likely require a multidisciplinary approach incorporating the best possible allocation system, strategies to minimize procurement and ischemia-reperfusion injury, enhanced immunosuppression with less toxicity, continuing refinements in surgical technique, as well as ongoing efforts to provide the highest quality posttransplant medical care.

# References

1. The organ procurement and transplantation network database. (Accessed September 9, 2006, at http://www.optn.org/latestData/viewDataReports.asp.)
2. www.unos.org UNfOS. 2006.
3. 2005 Annual Report of the US Organ Procurement and Transplantation Network and the Scientific Registry of Transplant Recipients: Transplant Data 1994-2004. Department of Health and Human Services. 2006.
4. Orens JB, Shearon TH, Freudenberger RS et al. Thoracic organ transplantation in the United States, 1995-2004. Am J Transplant 2006; 6(5 Pt 2):1188-97.
5. Deng MC. Orthotopic heart transplantation: highlights and limitations. Surg Clin North Am 2004; 84(1):243-55.
6. Ellison MD, Edwards LB, Edwards EB et al. Geographic differences in access to transplantation in the United States. Transplantation 2003; 76(9):1389-94.
7. Kirklin JK, McGiffin DC, Pinderski LJ et al. Selection of patients and techniques of heart transplantation. Surg Clin North Am 2004; 84(1):257-87, xi-xii.
8. Frankel DS, Piette JD, Jessup M et al. Validation of prognostic models among patients with advanced heart failure. J Card Fail 2006; 12(6):430-38.
9. Lee DS, Austin PC, Rouleau JL et al. Predicting mortality among patients hospitalized for heart failure: derivation and validation of a clinical model. Jama 2003; 290(19):2581-87.
10. Levy WC, Mozaffarian D, Linker DT et al. The Seattle Heart Failure Model: prediction of survival in heart failure. Circulation 2006; 113(11):1424-33.
11. Mancini DM, Eisen H, Kussmaul W et al. Value of peak exercise oxygen consumption for optimal timing of cardiac transplantation in ambulatory patients with heart failure. Circulation 1991; 83(3):778-86.
12. Elmariah S, Goldberg LR, Allen MT et al. Effects of gender on peak oxygen consumption and the timing of cardiac transplantation. J Am Coll Cardiol 2006; 47(11):2237-42.
13. Peterson LR, Schechtman KB, Ewald GA et al. Timing of cardiac transplantation in patients with heart failure receiving beta-adrenergic blockers. J Heart Lung Transplant 2003; 22(10):1141-48.
14. Shanewise J. Cardiac transplantation. Anesthesiol Clin North America 2004; 22(4):753-65.
15. Stevenson LW, Rose EA. Left ventricular assist devices: bridges to transplantation, recovery and destination for whom? Circulation 2003; 108(25):3059-63.
16. Organ distribution: Allocation of thoracic organs (policy 3.7). (Accessed September 9, 2006, at http://www.unos.org/policiesandbylaws/policies.asp?resources=true.)
17. Hunt SA, Baker DW, Chin MH et al. ACC/AHA Guidelines for the Evaluation and Management of Chronic Heart Failure in the Adult: Executive Summary A Report of the American College of Cardiology/American Heart Association Task Force on Practice Guidelines (Committee to Revise the 1995 Guidelines for the Evaluation and Management of Heart Failure): Developed in Collaboration With the International Society for Heart and Lung Transplantation; Endorsed by the Heart Failure Society of America. Circulation 2001;104(24):2996-3007.
18. Brozena SC, Twomey C, Goldberg LR et al. A prospective study of continuous intravenous milrinone therapy for status IB patients awaiting heart transplant at home. J Heart Lung Transplant 2004; 23(9):1082-6.
19. Rose EA, Gelijns AC, Moskowitz AJ et al. Long-term mechanical left ventricular assistance for end-stage heart failure. N Engl J Med 2001; 345(20):1435-43.
20. Roselli EE, Smedira NG. Surgical advances in heart and lung transplantation. Anesthesiol Clin North America 2004; 22(4):789-807.
21. Taylor DO, Edwards LB, Boucek MM et al. Registry of the International Society for Heart and Lung Transplantation: twenty-third official adult heart transplantation report—2006. J Heart Lung Transplant 2006; 25(8):869-79.
22. Valantine H. Cardiac allograft vasculopathy after heart transplantation: risk factors and management. J Heart Lung Transplant 2004; 23(5 Suppl):S187-93.
23. Arcasoy SM, Kotloff RM. Lung transplantation. N Engl J Med 1999; 340(14):1081-91.
24. Trulock EP, Edwards LB, Taylor DO et al. Registry of the International Society for Heart and Lung Transplantation: twenty-third official adult lung and heart-lung transplantation report—2006. J Heart Lung Transplant 2006; 25(8):880-92.
25. Rosas V, Conte JV, Yang SC et al. Lung transplantation and systemic sclerosis. Ann Transplant 2000; 5(3):38-43.
26. Garver RI, Jr., Zorn GL, Wu X et al. Recurrence of bronchioloalveolar carcinoma in transplanted lungs. N Engl J Med 1999; 340(14):1071-4.
27. Orens JB, Estenne M, Arcasoy S et al. International guidelines for the selection of lung transplant candidates: 2006 update—a consensus report from the Pulmonary Scientific Council of the International Society for Heart and Lung Transplantation. J Heart Lung Transplant 2006; 25(7):745-55.
28. Levine SM. A survey of clinical practice of lung transplantation in North America. Chest 2004; 125(4):1224-38.
29. Chaparro C, Maurer J, Gutierrez C et al. Infection with Burkholderia cepacia in cystic fibrosis: outcome following lung transplantation. Am J Respir Crit Care Med 2001; 163(1):43-48.
30. Hosenpud JD, Bennett LE, Keck BM et al. Effect of diagnosis on survival benefit of lung transplantation for end-stage lung disease. Lancet 1998; 351(9095):24-27.
31. Egan TM, Kotloff RM. Pro/Con debate: lung allocation should be based on medical urgency and transplant survival and not on waiting time. Chest 2005; 128(1):407-15.
32. Meyer DM, Edwards LB, Torres F et al. Impact of recipient age and procedure type on survival after lung transplantation for pulmonary fibrosis. Ann Thorac Surg 2005; 79(3):950-57.
33. Pochettino A, Kotloff RM, Rosengard BR et al. Bilateral versus single lung transplantation for chronic obstructive pulmonary disease: intermediate-term results. Ann Thorac Surg 2000; 70(6):1813-18.
34. Bowdish ME, Barr ML, Schenkel FA et al. A decade of living lobar lung transplantation: perioperative complications after 253 donor lobectomies. Am J Transplant 2004; 4(8):1283-88.
35. Meyer DM, Bennett LE, Novick RJ et al. Single vs bilateral, sequential lung transplantation for end-stage emphysema: influence of recipient age on survival and secondary end-points. J Heart Lung Transplant 2001; 20(9):935-41.
36. Bowdish ME, Pessotto R, Barbers RG et al. Long-term pulmonary function after living-donor lobar lung transplantation in adults. Ann Thorac Surg 2004; 79(2):418-25.
37. Rodrigue JR, Baz MA, Kanasky Jr WF et al. Does lung transplantation improve health-related quality of life? The University of Florida experience. J Heart Lung Transplant 2005; 24(6):755-63.
38. Gammie JS, Keenan RJ, Pham SM et al. Single- versus double-lung transplantation for pulmonary hypertension. J Thorac Cardiovasc Surg 1998; 115(2):397-402.
39. Starzl TE, Groth CG, Brettschneider L et al. Extended survival in 3 cases of orthotopic homotransplantation of the human liver. Surgery 1968; 63(4):549-63.
40. 2003 Annual Report of the U.S. Organ Procurement and Transplantation Network and the Scientific Registry of Transplant Recipients: Transplant Data 1993-2002. Department of Health and Human Services. 2003.
41. Carithers RL, Jr. Liver transplantation. American Association for the Study of Liver Diseases. Liver Transpl 2000; 6(1):122-35.
42. Curry MP. Hepatitis B and hepatitis C viruses in liver transplantation. Transplantation 2004; 78(7):955-63.
43. Everson GT TJ. Transplantation of the Liver. In: Schiff ER SM, Maddrey WC, eds. Schiff's Diseases of the Liver. 9th ed. Philadelphia: Lippincott Williams & Wilkins; 2002:1585-614.
44. D'Amico G, Morabito A, Pagliaro L et al. Survival and prognostic indicators in compensated and decompensated cirrhosis. Dig Dis Sci 1986; 31(5):468-75.
45. D'Amico G, Pagliaro L, Bosch J. The treatment of portal hypertension: a meta-analytic review. Hepatology 1995; 22(1):332-54.
46. Smith JL, Graham DY. Variceal hemorrhage: a critical evaluation of survival analysis. Gastroenterology 1982; 82(5 Pt 1):968-73.
47. Bloom RD, Goldberg LR, Wang AY et al. An overview of solid organ transplantation. Clin Chest Med 2005; 26(4):529-43.
48. Olthoff KM, Brown RS, Jr., Delmonico FL et al. Summary report of a national conference: Evolving concepts in liver allocation in the MELD and PELD era. December 8, 2003, Washington, DC, USA. Liver Transpl 2004; 10(10 Suppl 2):A6-22.
49. Schepke M, Roth F, Fimmers R et al. Comparison of MELD, Child-Pugh and Emory model for the prediction of survival in patients undergoing transjugular intrahepatic portosystemic shunting. Am J Gastroenterol 2003; 98(5):1167-74.
50. Kamath PS, Wiesner RH, Malinchoc M et al. A model to predict survival in patients with end-stage liver disease. Hepatology 2001; 33(2):464-70.
51. Wiesner RH, McDiarmid SV, Kamath PS et al. MELD and PELD: application of survival models to liver allocation. Liver Transpl 2001; 7(7):567-80.

52. Yao FY. Selection criteria for liver transplantation in patients with hepatocellular carcinoma: beyond tumor size and number? Liver Transpl 2006; 12(8):1189-91.

53. Kaya M, de Groen PC, Angulo P et al. Treatment of cholangiocarcinoma complicating primary sclerosing cholangitis: the Mayo Clinic experience. Am J Gastroenterol 2001; 96(4):1164-69.

54. Solano E, Khakhar A, Bloch M et al. Liver transplantation for primary sclerosing cholangitis. Transplant Proc 2003; 35(7):2431-34.

55. Rea DJ, Heimbach JK, Rosen CB et al. Liver transplantation with neoadjuvant chemoradiation is more effective than resection for hilar cholangiocarcinoma. Ann Surg 2005; 242(3):451-8; discussion 8-61.

56. Keswani RN, Ahmed A, Keeffe EB. Older age and liver transplantation: a review. Liver Transpl 2004; 10(8):957-67.

57. John PR, Thuluvath PJ. Outcome of liver transplantation in patients with diabetes mellitus: a case-control study. Hepatology 2001; 34(5):889-95.

58. Yoo HY, Thuluvath PJ. The effect of insulin-dependent diabetes mellitus on outcome of liver transplantation. Transplantation 2002; 74(7):1007-12.

59. Arroyo V, Guevara M, Gines P. Hepatorenal syndrome in cirrhosis: pathogenesis and treatment. Gastroenterology 2002; 122(6):1658-76.

60. Iwatsuki S, Popovtzer MM, Corman JL et al. Recovery from "hepatorenal syndrome" after orthotopic liver transplantation. N Engl J Med 1973; 289(22):1155-59.

61. Bak T, Wachs M, Trotter J et al. Adult-to-adult living donor liver transplantation using right-lobe grafts: results and lessons learned from a single-center experience. Liver Transpl 2001; 7(8):680-86.

62. Marcos A, Ham JM, Fisher RA et al. Single-center analysis of the first 40 adult-to-adult living donor liver transplants using the right lobe. Liver Transpl 2000; 6(3):296-301.

63. Sorrell M. Transplantation. In: Schiff ER SM, Maddrey WC, eds. Schiff's Diseases of the Liver. 9th ed. Philadelphia: Lippincott Williams & Wilkins; 2002:1581-83.

64. Trotter JF, Talamantes M, McClure M et al. Right hepatic lobe donation for living donor liver transplantation: impact on donor quality of life. Liver Transpl 2001; 7(6):485-93.

65. Starzl TE, Marchioro TL, Vonkaulla KN et al. Homotransplantation of the Liver in Humans. Surg Gynecol Obstet 1963; 117:659-76.

66. Tzakis A, Todo S, Starzl TE. Orthotopic liver transplantation with preservation of the inferior vena cava. Ann Surg 1989; 210(5):649-52.

67. Parrilla P, Sanchez-Bueno F, Figueras J et al. Analysis of the complications of the piggy-back technique in 1,112 liver transplants. Transplantation 1999; 67(9):1214-17.

68. Wiesner RH, Demetris AJ, Belle SH et al. Acute hepatic allograft rejection: incidence, risk factors and impact on outcome. Hepatology 1998; 28(3):638-45.

69. Cardenas A GP, Rodes J. Renal complications. In: Schiff ER SM, Maddrey WC, eds. Schiff's Diseases of the Liver. 9th ed. Philadelphia: Lippincott Williams & Wilkins; 2002:498-509.

70. Abbasoglu O, Goldstein RM, Vodapally MS et al. Liver transplantation in hyponatremic patients with emphasis on central pontine myelinolysis. Clin Transplant 1998; 12(3):263-69.

71. Ros J, Fernandez-Varo G, Munoz-Luque J et al. Sustained aquaretic effect of the V2-AVP receptor antagonist, RWJ-351647, in cirrhotic rats with ascites and water retention. Br J Pharmacol 2005; 146(5):654-61.

72. Lebeau G, Yanaga K, Marsh JW et al. Analysis of surgical complications after 397 hepatic transplantations. Surg Gynecol Obstet 1990; 170(4):317-22.

73. Wagener MM, Yu VL. Bacteremia in transplant recipients: a prospective study of demographics, etiologic agents, risk factors and outcomes. Am J Infect Control 1992; 20(5):239-47.

74. Torbenson M, Wang J, Nichols L et al. Causes of death in autopsied liver transplantation patients. Mod Pathol 1998; 11(1):37-46.

75. George DL, Arnow PM, Fox AS et al. Bacterial infection as a complication of liver transplantation: epidemiology and risk factors. Rev Infect Dis 1991; 13(3):387-96.

76. Simon DM, Levin S. Infectious complications of solid organ transplantations. Infect Dis Clin North Am 2001; 15(2):521-49.

77. Cohen DJ, St Martin L, Christensen LL et al. Kidney and pancreas transplantation in the United States, 1995-2004. Am J Transplant 2006; 6(5 Pt 2):1153-69.

78. Cronin DC, 2nd, Faust TW, Brady L et al. Modern immunosuppression. Clin Liver Dis 2000; 4(3):619-55, ix.

79. Pageaux GP, Calmus Y, Boillot O et al. Steroid withdrawal at day 14 after liver transplantation: a double-blind, placebo-controlled study. Liver Transpl 2004; 10(12):1454-60.

80. Go AS, Chertow GM, Fan D et al. Chronic kidney disease and the risks of death, cardiovascular events and hospitalization. N Engl J Med 2004; 351(13):1296-305.

81. Wolfe RA, Ashby VB, Milford EL et al. Comparison of mortality in all patients on dialysis, patients on dialysis awaiting transplantation and recipients of a first cadaveric transplant. N Engl J Med 1999; 341(23):1725-30.

82. Jordan S, Cunningham-Rundles C, McEwan R. Utility of intravenous immune globulin in kidney transplantation: efficacy, safety and cost implications. Am J Transplant 2003; 3(6):653-64.

83. Merrill JP, Murray JE, Harrison JH et al. Successful homotransplantation of the human kidney between identical twins. J Am Med Assoc 1956; 160(4):277-82.

84. US Renal Data System, USRDS 2006 Annual Data Report, Atlas of End-Stage Renal Disease in the United States, National Institutes of Health, National Institute of Diabetes and Digestive and Kidney Diseases, Bethesda, MD. 2006.

85. Xue JL, Ma JZ, Louis TA et al. Forecast of the number of patients with end-stage renal disease in the United States to the year 2010. J Am Soc Nephrol 2001; 12(12):2753-58.

86. Ojo AO, Hanson JA, Wolfe RA et al. Long-term survival in renal transplant recipients with graft function. Kidney Int 2000; 57(1):307-13.

87. Kasiske BL, Cangro CB, Hariharan S et al. The evaluation of renal transplantation candidates: clinical practice guidelines. Am J Transplant 2001; 1 Suppl 2:3-95.

88. Gaston RS, Danovitch GM, Adams PL et al. The report of a national conference on the wait list for kidney transplantation. Am J Transplant 2003; 3(7).775-85.

89. Gaston RS, Alveranga DY, Becker BN et al. Kidney and pancreas transplantation. Am J Transplant 2003; 3 Suppl 4:64-77.

90. Ojo AO, Hanson JA, Meier-Kriesche H et al. Survival in recipients of marginal cadaveric donor kidneys compared with other recipients and wait-listed transplant candidates. J Am Soc Nephrol 2001; 12(3):589-97.

91. Merion RM, Ashby VB, Wolfe RA et al. Deceased-donor characteristics and the survival benefit of kidney transplantation. Jama 2005; 294(21):2726-33.

92. Weber M, Dindo D, Demartines N et al. Kidney transplantation from donors without a heartbeat. N Engl J Med 2002; 347(4):248-55.

93. Mange KC, Joffe MM, Feldman HI. Effect of the use or nonuse of long-term dialysis on the subsequent survival of renal transplants from living donors. N Engl J Med 2001; 344(10):726-31.

94. Meier-Kriesche HU, Port FK, Ojo AO et al. Effect of waiting time on renal transplant outcome. Kidney Int 2000; 58(3):1311-17.

95. Held PJ, Kahan BD, Hunsicker LG et al. The impact of HLA mismatches on the survival of first cadaveric kidney transplants. N Engl J Med 1994; 331(12):765-70.

96. Takemoto S, Terasaki PI, Cecka JM et al. Survival of nationally shared, HLA-matched kidney transplants from cadaveric donors. The UNOS Scientific Renal Transplant Registry. N Engl J Med 1992; 327(12):834-39.

97. Mange KC, Cherikh WS, Maghirang J et al. A comparison of the survival of shipped and locally transplanted cadaveric renal allografts. N Engl J Med 2001; 345(17):1237-42.

98. Montgomery RA, Hardy MA, Jordan SC et al. Consensus opinion from the antibody working group on the diagnosis, reporting and risk assessment for antibody-mediated rejection and desensitization protocols. Transplantation 2004; 78(2):181-85.

99. Stegall MD, Dean PG, Gloor JM. ABO-incompatible kidney transplantation. Transplantation 2004; 78(5):635-40.

100. Takahashi K, Saito K, Takahara S et al. Excellent long-term outcome of ABO-incompatible living donor kidney transplantation in Japan. Am J Transplant 2004; 4(7):1089-96.

101. Montgomery RA, Zachary AA, Ratner LE et al. Clinical results from transplanting incompatible live kidney donor/recipient pairs using kidney paired donation. Jama 2005; 294(13):1655-63.

102. Perico N, Cattaneo D, Sayegh MH et al. Delayed graft function in kidney transplantation. Lancet 2004; 364(9447):1814-27.

103. Pascual M, Theruvath T, Kawai T et al. Strategies to improve long-term outcomes after renal transplantation. N Engl J Med 2002; 346(8):580-90.

104. Kelly WD, Lillehei RC, Merkel FK et al. Allotransplantation of the pancreas and duodenum along with the kidney in diabetic nephropathy. Surgery 1967; 61(6):827-37.

105. Gruessner AC, Sutherland DE. Pancreas transplant outcomes for United States (US) and non-US cases as reported to the United Network for Organ Sharing (UNOS) and the International Pancreas Transplant Registry (IPTR) as of October 2002. Clin Transpl 2002:41-77.

106. International Pancreas Transplant Registry 2004 Annual Report. (Accessed at http://www.iptr.umn.edu/)

107. The Diabetes Control and Complications Trial Research Group, The effect of intensive treatment of diabetes on the development and progression of long-term complications in insulin-dependent diabetes mellitus. N Engl J Med 1993; 329(14):977-86.

108. Larsen JL. Pancreas transplantation: indications and consequences. Endocr Rev 2004; 25(6):919-46.
109. Robertson P, Davis C, Larsen J et al. Pancreas transplantation in type 1 diabetes. Diabetes Care 2004; 27 Suppl 1:S105.
110. Manske CL, Thomas W, Wang Y et al. Screening diabetic transplant candidates for coronary artery disease: identification of a low risk subgroup. Kidney Int 1993; 44(3):617-21.
111. Gruessner AC, Sutherland DE. Pancreas transplant outcomes for United States (US) and non-US cases as reported to the United Network for Organ Sharing (UNOS) and the International Pancreas Transplant Registry (IPTR) as of May 2003. Clin Transpl 2003:21-51.
112. Sollinger HW, Messing EM, Eckhoff DE et al. Urological complications in 210 consecutive simultaneous pancreas- kidney transplants with bladder drainage. Ann Surg 1993; 218(4):561-68.
113. Stephanian E, Gruessner RW, Brayman KL et al. Conversion of exocrine secretions from bladder to enteric drainage in recipients of whole pancreaticoduodenal transplants. Ann Surg 1992; 216(6):663-72.
114. Bloom RD, Olivares M, Rehman L et al. Long-term pancreas allograft outcome in simultaneous pancreas-kidney transplantation: a comparison of enteric and bladder drainage. Transplantation 1997; 64(12):1689-95.
115. Robertson RP, Abid M, Sutherland DE et al. Glucose homeostasis and insulin secretion in human recipients of pancreas transplantation. Diabetes 1989; 38 Suppl 1:97-98.
116. Kuo PC, Johnson LB, Schweitzer EJ et al. Simultaneous pancreas/kidney transplantation—a comparison of enteric and bladder drainage of exocrine pancreatic secretions. Transplantation 1997; 63(2):238-43.
117. Humar A, Ramcharan T, Kandaswamy R et al. Pancreas after kidney transplants. Am J Surg 2001; 182(2):155-61.
118. Gruessner RW, Sutherland DE, Gruessner AC. Mortality assessment for pancreas transplants. Am J Transplant 2004; 4(12):2018-26.
119. Venstrom JM, McBride MA, Rother KI et al. Survival after pancreas transplantation in patients with diabetes and preserved kidney function. Jama 2003; 290(21):2817-23.
120. Bunnapradist S, Cho YW, Cecka JM et al. Kidney allograft and patient survival in type I diabetic recipients of cadaveric kidney alone versus simultaneous pancreas kidney transplants: a multivariate analysis of the UNOS database. J Am Soc Nephrol 2003; 14(1):208-13.
121. Reddy KS, Stablein D, Taranto S et al. Long-term survival following simultaneous kidney-pancreas transplantation versus kidney transplantation alone in patients with type 1 diabetes mellitus and renal failure. Am J Kidney Dis 2003; 41(2):464-70.

# CHAPTER 4

# Analyzing Graft Failure in the Scientific Registry of Transplant Recipients:
## The Sources and Nature of the Data Available

David M. Dickinson,* Gregory N. Levine, Douglas E. Schaubel and Robert A. Wolfe

## Abstract

This chapter uses information from a series of articles published in previous editions of the SRTR Report on the State of Transplantation and the OPTN/SRTR Annual Report to assemble a practical background useful for a researcher becoming familiar with data available for analysis of transplant outcomes, collected and used by the Organ Procurement and Transplantation Network (OPTN) and the Scientific Registry of Transplant Recipients (SRTR). The chapter examines the timing and accuracy of primary data sources and the secondary data sources available to complement these. We finish by examining many of the methods available for analyzing outcomes on the waiting list and after transplant, with particular attention to caveats and biases pertinent to many of the analyses performed by the SRTR in support of OPTN policy-making committees.

## Introduction

Collecting, organizing and disseminating data for research on transplantation involves tremendous effort from professionals at transplant centers and organ procurement organizations (OPOs), as well as other data collection specialists. To put this wealth of data to its best use, researchers and readers of analyses, whether practitioners or patients, should understand the full process of collecting and organizing these data. Familiarity with the data's structures and sources—as well as their limitations—can help ensure that readers and researchers use data effectively and accurately.

The data described here are the foundation of the research activities performed by the Scientific Registry of Transplant Recipients (SRTR). These data are the basis for several types of research and reporting: analyses that support the policy committees of the Organ Procurement and Transplantation Network (OPTN); the source of figures, tables and analyses in the *OPTN/SRTR Annual Report*; program-specific reports for transplant centers and organ procurement organizations published regularly at www.ustransplant.org, and many peer-reviewed papers furthering knowledge about transplantation and organ allocation by researchers from the SRTR and throughout the community. These data are available to the public and various stakeholders in the field of transplantation: patients and their families, transplant professionals, policymakers, insurers and researchers wishing to contribute to the community's body of knowledge about transplantation.

Data collected by transplant centers and OPOs and submitted to the OPTN are primarily designed to facilitate the efficient allocation of organs to transplant candidates and to allow limited evaluation of the outcomes of this process. These data have become an increasingly rich source of information about the practice and outcomes of solid organ transplantation in the United States.

This chapter covers database and analytical issues. The first section of the chapter focuses on issues in transplant data sources and data collection. A summary of the scope of data available is presented, together with the value that the various extra ascertainment sources which the SRTR uses gives the data. The second half of this chapter centers on methods of analysis using these data for transplant research. Essential analytical approaches used by the SRTR are reviewed, with special attention placed on unadjusted and covariate-adjusted analyses.

Each year, the *SRTR Report on the State of Transplantation* is published in the *American Journal of Transplantation*, with similar materials mirrored in the OPTN/SRTR Annual Report. Material in this chapter reflects highlights taken from chapters in those reports focused on data sources and analytical methods. The SRTR is administered by the Arbor Research Collaborative for Health with the University of Michigan, under contract with the US Health Resources and Services Administration (HRSA).

## The Data Structure of the SRTR

A researcher seeking to fully understand the database design and the data structure of the SRTR may want to start with the "units of analysis". Figure 1 shows a useful method of organizing transplant data into such units. These units of analysis are designed to be of most use to researchers asking questions about the events or outcomes that may follow the placement of a candidate on the waiting list, organ donation, or a transplant itself. The data table in Figure 1 relates to specific subjects of interest for research: candidacies, donors, transplants and the components thereof. Also shown are some of the more specialized tables, ones from which researchers might analyze organ turndowns, use of immunosuppression medications, or changes in waiting list status prior to transplant.

Three tables in Figure 1 are the entry points for individual persons into the transplant process: the candidate registration table (which includes registrants who may become transplant recipients) and the living and deceased donor tables. Underlying these three individual level tables (and not shown in the figure) is a "Person Linking Table" (PLT) that is vital to the integration of multiple data sources discussed later. The PLT holds one record per person, establishes links on the basis of similarities in Social Security Numbers (SSNs), names

*Corresponding Author: David M Dickinson—Director of Information Systems, Arbor Research Collaborative for Health, 315 West Huron St, Suite 360, Ann Arbor, MI 48103, USA. Email: david.dickinson@arborresearch.org

*Chronic Allograft Failure: Natural History, Pathogenesis, Diagnosis and Management*, edited by Nasimul Ahsan. ©2008 Landes Bioscience.

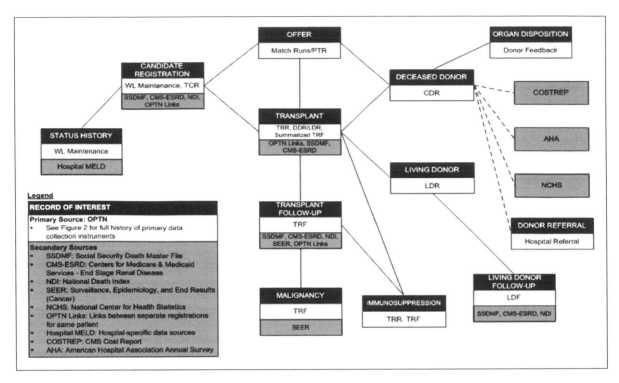

Figure 1. Transplantation research data organization, primary and secondary sources. Reprinted from reference 1.

and nicknames, dates of birth and other person-level information, while accounting for many of the mistakes that may occur in entering data in these fields. The maintenance of this identification roster, with aggregated identification information compiled from all data sources, facilitates a system of matching to both external data sources and other records within the OPTN data, such as for people who receive multiple transplants or even for donors who later become recipients.

In addition, this figure documents some of the primary and secondary data sources that may contribute to each table. Further detail regarding the specific data collection instruments, before the information is aggregated to records of interest, is shown in Figure 2.

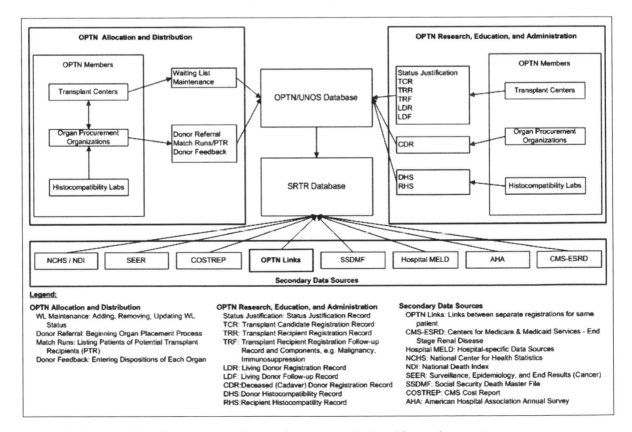

Figure 2. Data submission and data flow, primary and secondary sources. Reprinted from reference 1.

## Waiting List Data

In Figure 1, the "candidate registration" table holds records for potential transplant recipients: patients who are placed on the waiting list as well as patients who receive living donor transplants without having been wait-listed. Analytically, this table helps researchers describe the "demand" side of the transplant process, comparing characteristics of successful and unsuccessful transplant candidates and describing disease progression among prospective recipients while they are not transplanted, although the researcher must be cautious of the bias introduced by transplanting some of these patients, as discussed later. These candidates act as a useful comparison to those who do receive transplants: considering the consequences of not being transplanted can be helpful in evaluating the benefit of transplanting each type of patient.

The primary source of information about candidates for transplantation is the OPTN database, which stores information about all people on the national waiting lists. Transplant centers must continuously maintain their waiting lists by reporting on changes in severity of illness (for some organs) and other outcomes, such as transplant or death. Information in this table is taken from these waiting list maintenance records and the Transplant Candidate Registration (TCR) record completed soon after registration.

Because the maintenance of the waiting list is continuous, researchers should be able to report upon waiting list outcomes soon after they happen. In actuality, this depends on the outcome. Removal from the waiting list for transplant is linked to the generation of a transplant record, so reporting is nearly immediate. Reporting of death on the waiting list may display more lag in reporting, particularly among patients who are offered organs less frequently because of low severity of illness or accumulated waiting time, since turndown of offers often spurs waiting list maintenance.

## Transplant and Posttransplant Data

The transplants table shown in Figure 1 provides a collected source of information about each transplant event, including information about the donor, recipient, operation and follow-up information. This file contains the characteristics of transplant recipients, transplant and outcomes. The data for the transplant table are primarily taken from the Transplant Recipient Registration (TRR) form collected by the OPTN. Additional characteristics, from the donor and candidate files,

are added for ease of analysis, as are aspects of the interaction between donor and recipient characteristics (e.g., calculated HLA mismatch scores; ABO blood type compatibility; whether the organ was shared, based on the relationship between the OPO recovering the organ and the transplanting center).

The transplant follow-up data, collected primarily from the Transplant Recipient Follow-Up (TRF) record, may be summarized to the transplant level, creating indicators of death, graft failure and time to follow-up. The expected—and actual—timing of the follow-up forms are very important to cohort choice in analyses. After each transplant, follow-up forms are collected at the six-month (for nonthoracic organs) and yearly anniversaries (for thoracic and nonthoracic organs) of the transplant; these forms may also be submitted off-schedule to report such adverse events as graft failure or death. Transplant follow-ups may be useful on their own, or in conjunction with their own sub-tables for immunosuppression or malignancies. However, for analysis of specific events that occur during follow-up, they are most widely used in the summarized form for death and graft failure analyses discussed here. For such analyses, the timing is particularly important.

### *Timeliness of Follow-Up Forms*

Just as with events on the waiting list, it is important to consider the time lag until follow-up forms are filed when determining cohorts for analysis of posttransplant events. Implementation of new data collection mechanisms and stricter rules have shortened the time until validation. Table 1 shows that the time from the date of record generation until validation (when the form has been submitted and verified by the center) has grown shorter, but it is still nearly four months after each anniversary until four of five forms are submitted and six months before nine of ten are completed. However, the increase from 91% in 2003 to 97% in 2004 indicates that the timeliness of submission of routine follow-up forms continues to improve. If the trend continues, it is likely that more recent data could be used in analyses in the near future. However, a balance must be struck between the need for recent data and the need for complete data. Currently, the SRTR typically allows for between three and six months of lag time, depending on the need for analyzing data from the most recent cohort available.

---

**Table 1.  Timing for validation* of follow-up forms**

| | Cumulative Percent Validated* by Month | | | | | |
| | Routine Follow-Ups | | | Interim Follow-Ups | | |
| Year Added | 2002 | 2003 | 2004 | 2002 | 2003 | 2004 |
| --- | --- | --- | --- | --- | --- | --- |
| Month 1 | 26.0 | 30.6 | 32.7 | 43.9 | 52.8 | 56.0 |
| Month 2 | 51.7 | 60.3 | 67.3 | 60.2 | 70.6 | 76.1 |
| Month 3 | 68.3 | 72.0 | 80.7 | 72.2 | 78.4 | 84.3 |
| Month 4 | 77.1 | 79.3 | 87.7 | 79.4 | 83.7 | 89.5 |
| Month 5 | 82.2 | 86.4 | 93.3 | 83.6 | 88.3 | 93.5 |
| Month 6 | 85.9 | 90.8 | 97.0 | 86.5 | 91.6 | 96.5 |
| Month 7 | 89.0 | 93.8 | | 89.0 | 93.7 | |
| Month 8 | 91.6 | 95.8 | | 90.9 | 95.4 | |
| Month 9 | 93.5 | 97.1 | | 92.4 | 96.6 | |
| Month 10 | 94.9 | 98.0 | | 93.5 | 97.7 | |
| Month 11 | 95.8 | 98.7 | | 94.5 | 98.5 | |
| Month 12 | 96.5 | 99.3 | | 95.3 | 99.0 | |
| All unvalidated by 6 months | 14.1 | 9.2 | 3.0 | 13.5 | 8.5 | 3.5 |
| All unvalidated by 1 year | 3.5 | 0.7 | N/A | 4.7 | 1.0 | N/A |

Source: SRTR Analysis, July 2005. *The form has been submitted and verified as complete by the center. Reprinted from reference 1.

## Timing of Follow-Up Forms

In addition to the lag time until validation of follow-up forms after transplant, the pattern of form submission—often clustered soon after transplant anniversaries—has important implications for avoiding biases when analyzing recent data.

"Routine" follow-up forms are generated at each transplant anniversary, yet deaths occur on a continuous basis throughout the post-transplant period. When a patient dies during follow-up, the transplant center *may* file an "interim" follow-up form off the regular reporting schedule for that patient. This means that centers might report mortality more quickly and continuously than they report on surviving patients, for whom they must wait until the transplant anniversary.

For example, in an analysis of patients transplanted 18 months ago, patients currently alive will have a one-year follow-up form indicating their survival until the one-year point, with no information beyond that. Patients who have died, on the other hand, might have follow-up forms indicating death both during the first year and any interim follow-up forms filed between months 12 and 18. Therefore, all of the data reported during months 12 to 18 would be about patients who had died. If a researcher used the Kaplan-Meier method to take advantage of the most recent data available and censored at last follow-up, the portion of the survival curve calculated after the first year would be based inappropriately on over-reporting about patients who had died, thereby creating a bias in mortality reporting. This bias can be removed by waiting until the living patients are reported on at the two-year anniversary. Similarly, one-month survival rates cannot be reliably calculated until at least six months after transplant (one year for thoracic organs), after the anniversaries have prompted reporting on all patients.

The example given above is an extreme case. However, including these patients in a sample used for survival calculations, without appropriate censoring at transplant anniversaries, introduces the same bias into the average results. Further, these caveats are not limited to survival analyses: other analyses might over-represent outcomes associated with death in the final six-month period.

The above example describes the case when transplant centers may report deaths as they occur. If this were a reliable pattern of reporting, one analytical solution might be to assume that the patient is alive unless we know otherwise. This approach would be effective if the multiple sources of mortality reliably captured all deaths. However, all sources are not reliably complete during many periods, since many deaths are reported as they occur and many more are reported at the next reporting anniversary, as the following figure exhibits. Figure 3 depicts when transplant follow-up forms are filed, comparing those filed for patients who have died to those for patients who have not. The actual time of the follow-up event (death in the top panel or reported as alive in the lower panel) is shown on the y-axis and the time that the follow-up form was validated by the center is shown on the x-axis. If all events were reported as they occurred, points would fall only along the 45-degree diagonal dashed line. The horizontal distance, left to right, between this diagonal and each point represents the time lag between the event and notification to the OPTN.

The top panel shows this relationship for follow-up forms reporting deaths and the pattern of reporting along the diagonal shows deaths that were reported near the time of death itself. (In the earlier example of using a cohort of transplants from 18 months ago to calculate a survival curve, it is this pattern of reporting along the diagonal for dead patients that introduces a possible bias beyond the 12-month follow-up time.) There is a more obvious clustering to the right of each vertical line at 6, 12 and 24 months after transplant, showing deaths are most often reported with the timing of routine follow-up forms. The actual death dates are distributed vertically along the line, emphasizing the extent to which many centers wait until prompted by the reporting cycle to report mortality, no matter when the death actually occurred.

The lower panel of the figure shows a similar clustering after each reporting anniversary, but the vertical height of these clusters, close to the diagonal itself, indicates that the events being reported on—that the patient is alive—occurred more recently compared to the reporting date. This difference is also borne out in the median lag reporting times, shown by arrows of different sizes in the two panels, at 133 days for deceased patients and only 28 days for living patients.

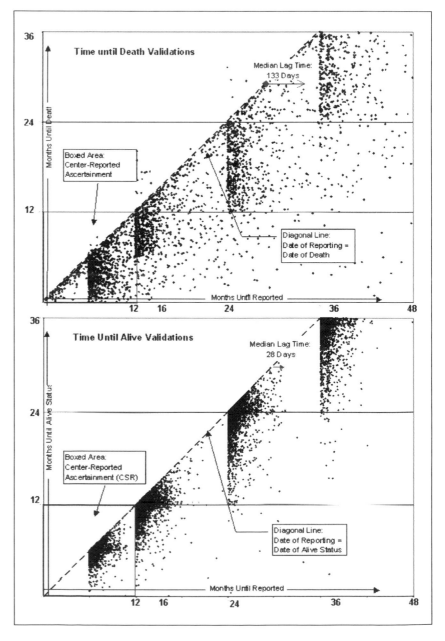

Figure 3. Time to validation of death and survivor records. Reprinted from reference 1.

## Extra Ascertainment Sources

A transplant center's reporting duties end upon each candidate's removal from the waiting list. However, events occurring in the months following removal—such as death or transplant at another center—are frequently interesting analytical endpoints to the researcher. Therefore, a candidate file may incorporate additional mortality sources or waiting list, transplant and follow-up information reported by other centers for the same person.

Many of the same additional sources of outcome ascertainment are used for both waiting list analyses and posttransplant analyses, particularly for mortality. Using the Person Linking Table (PLT, described above) to match patients, results may be incorporated from three other "secondary" sources:

**Patient linking between OPTN records** allows a researcher to tell that a transplant candidate at one center has had a death or transplant reported by a different center or that a graft has failed, on the basis of a retransplant at another center.

**The Social Security Death Master File (SSDMF)**, publicly available from the Social Security Administration (SSA), contains over 70 million records created from reports of death to the SSA, for beneficiaries and nonbeneficiaries alike.

**The CMS-ESRD Database** provides data primarily from Medicare records for ESRD patients and helps provide evidence of start of dialysis therapy, resumption of posttransplant maintenance dialysis indicating graft failure, or death.

In addition, the **National Death Index (NDI)** is available for validation of the completeness of these sources, though its use is not permitted for most analyses. The NDI, based on death certificate information submitted by state vital statistics agencies, misses only about 5% of all deaths in the US.

In 2002, the SRTR and OPTN jointly obtained data from the NDI for a sample of transplant candidates and patients to evaluate the completeness of mortality reporting in the other existing data sources. As the SRTR showed in 2002, the majority of deaths are reported by the main transplant center following the patient.[2] It continues to be important to use all of these available sources in mortality analyses: of patients receiving a transplant between July 1, 1999 and June 30, 2004, 78% of kidney and pancreas transplant recipient deaths were reported by the transplanting center. It is still the case that significant fractions of all the deaths are reported by other available sources, as 19% of these deaths were reported by the SSDMF and 3% of the deaths were reported first by another transplantation program. In cases where discrepancies arise among different death dates reported, the SRTR most often relies on what is reported by the center, first and foremost. The primary reason for this decision is that deaths are often reported to the SSDMF as occurring on either the first or last day of the month, or on the 15th of the month as an "average".

In 2003, the SRTR began using extra ascertainment from CMS-ESRD data for kidney graft failure for many types of analyses. A study was conducted to explore the possibility of supplementing existing SRTR data with CMS graft failure data for kidney recipients followed by the OPTN. The CMS data may provide additional information on recipients that are lost to follow-up, because CMS can be notified about a graft failure event through several possible mechanisms, not available to the OPTN. Further discussion of this work can be found in "Transplant Data: Sources, Collection and Caveats".[3]

## Extra Ascertainment Sources for Posttransplant

The importance of extra ascertainment of mortality is as great for posttransplant analyses as it is for the waiting list analyses described earlier. Patients are more prone to becoming lost to follow up (LTFU) after receiving transplant than they are while still on the waiting list. About 10% of recipients transplanted with kidneys, livers, hearts, or lungs were LTFU by the end of the third year after transplant; about

two-thirds of these had been coded as LTFU by the transplant center and the other third had no records completed for at least the last 1.5 years before the three-year anniversary. Above, we outlined arguments suggesting that even with LTFU, extra ascertainment of mortality makes it plausible to assume that all sources taken together provide reasonably complete ascertainment of death, such that less than 1% of deaths are missed.

## Multiple Sources of Data and Lag Time

For patient survival analyses, the SRTR often adopts a technique of assuming a patient is alive unless known otherwise, allowing us to follow patients after they become LTFU. It is important to continue to choose cohorts carefully, because the assumption of "alive unless we know otherwise" holds most true during periods when we expect all sources to be complete and unbiased. Because the lag time for extra sources such as the SSDMF is similar, if not a little shorter, than that for transplant follow-ups, the SRTR uses cohorts based on the timing of OPTN follow-up forms even when additional ascertainment is included. Researchers must also bear in mind other reporting patterns within the additional sources, such as the possible loss of Medicare eligibility three years after transplant for CMS data.

## Implications for Overall and PSR Death Analyses; Possibility of Relying on SSDMF

The SRTR Program-Specific Reports (PSRs) present an excellent example of the impact of extra ascertainment on survival. The assumption of "alive unless we know otherwise" allows us to add to survival calculations not only deaths, but also additional time at risk. Reported survival for many cohorts changes very little and may actually go up by using extra ascertainment. For smaller cohorts, such as at the center-specific level, differences introduced with extra ascertainment are more evident, though they may still go in either direction. Before the implementation of extra ascertainment in the PSRs, concern was expressed that low follow-up completion for some centers may bias results.[5,6] The use of extra ascertainment helps ensure more complete and reliable information for patients, families and administrators using the survival statistics in these reports, even if the overall average does not change.

Table 2 shows the one-year survival presented in the PSRs released at www.ustransplant.org in July 2004 and compares results to what would have been calculated without the use of extra ascertainment. While the total number of observed deaths, of course, increased with extra ascertainment, the mean center-specific survival rate for these organs remained virtually unchanged. Many individual programs and in fact the majority of heart programs, experienced no change in their center-specific one-year survival and many others had increased survival as a result of extra ascertainment. On the other hand, the majority of kidney programs (131) did experience a reduction in reported survival rate with extra ascertainment. Some of the center-specific changes in survival were sizeable. For heart programs, decreases in survival were as large as nearly 7 percentage points, with increases more than 9 percentage points; the ranges were smaller but still sizeable for the other organ programs, which likely have more stable center-specific outcomes due to larger sample sizes. These large changes at a center level point to the importance of extra ascertainment for these types of analysis. As we would expect, the largest changes occurred in centers that had a low percentage of their follow-up forms completed (not shown in this table).

## Graft Failure Analyses and Extra Ascertainment

In contrast to the options available for examining patient survival data, there is no "complete" source of graft failure data against which to test the completeness of OPTN/SRTR data. For many organs, retransplant is the only alternative therapy, so examining the transplant data file for retransplants for the same patient is sufficient for assuming

**Table 2.  PSR one-year survival difference due to extra ascertainment, July 2004**

|  | Heart | Kidney | Liver |
|---|---|---|---|
| **National Average of Center-Specific 1-year Patient Survival Rates:** | | | |
| Without Extra Ascertainment | 83.41 | 96.09 | 86.88 |
| With Extra Ascertainment | 83.51 | 95.98 | 86.25 |
| *Center-Specific Changes in Survival Pct Points with Extra Ascertainment:* | | | |
| Mean Pts Change | 0.11 | -0.10 | -0.63 |
| Largest Pts Decrease | -6.94 | -7.45 | -7.94 |
| Largest Pts Increase | 9.29 | 1.81 | 6.37 |
| *Direction of Center-Specific Survival Rate Changes with Extra Ascertainment:* | | | |
| Decrease | 32 | 131 | 34 |
| No Change | 87 | 60 | 22 |
| Increase | 7 | 45 | 51 |
| *Total Observed Deaths:* | | | |
| Without Extra Ascertainment | 624 | 1,104 | 1,206 |
| With Extra Ascertainment | 632 | 1,186 | 1,343 |

Source: SRTR analyses of Center-Specific Report data released at www.ustransplant.org in July 2004. Reprinted from reference 4.

complete follow-up. However, for kidney recipients the alternative of dialysis increases the possibility that graft failure may occur without the knowledge of the original transplanting center or any new (retransplanting) center. Some additional failure data may be available using CMS-ESRD data.

In 2003, the SRTR began using extra ascertainment for kidney graft failure for many types of analyses. Initially, a study was conducted to explore the possibility of supplementing existing SRTR data with CMS graft failure data for kidney recipients followed by the OPTN. The CMS data may provide additional information on recipients that are lost to follow-up, because CMS can be notified about a graft failure event through several possible mechanisms in addition to the OPTN, such as medical claims indicating return to (or initiation of) chronic dialysis treatment, the CMS 2728 medical evidence form and Standardized Information Management System (SIMS) ESRD Network reporting. One noteworthy limitation is that CMS data typically capture graft failure events only for those patients covered by Medicare, which accounts for about 65%-70% of all kidney transplant recipients. Because insurance status is not well documented for kidney recipients after transplant, determining which patients may have Medicare coverage during follow-up is difficult.

### Graft Failure Reporting Agreement Between Data Sources

For deceased donor kidney transplants that occurred between 1998 and 2001, graft failure events reported through October 31, 2003 were compared between OPTN and CMS. CMS failures reported after the end of the prescribed OPTN follow-up were not included in the calculations. Graft failures reported by the OPTN that resulted

in death or retransplant within 30 days were also excluded from the analysis, because they might not be eligible for detection by CMS via return-to-dialysis claims or the agency's other means of identifying graft failure. Furthermore, prior SRTR analyses of extra mortality ascertainment suggest that the OPTN data provide near-perfect ascertainment for graft failures that result in immediate retransplantation. The overall agreement between the two sources with respect to reporting of graft failure is shown in Table 3. This analysis assumed, for CMS follow-up, that if there was no evidence of a graft failure, then the graft was still functioning. Based on these results, CMS was found to have missed 22% of OPTN-reported failures and 20% of total failures reported by either source. The inclusion of CMS failure dates would increase the fraction found to have failed, compared to that based on OPTN reporting alone, from 12.9% to 14.5%.

### Unadjusted Analysis of Patient Survival and Graft Failure

Unadjusted (crude) methods, such as the "actuarial method", use death rates to compute the corresponding conditional survival probabilities for successive time intervals. These interval-specific conditional survival probabilities (i.e., the probability of surviving until the end of the interval, given that the patient was alive at the beginning of the interval) are multiplied to yield the cumulative survival probability for various time points (e.g., three-year survival). Depending on the question posed, these actuarial results are reported as either the fraction that died, the fraction still surviving, or the expected years of life through the end of the last interval.

Unadjusted posttransplant graft and patient survival outcomes are reported as cumulative "success" rates. These are calculated by Kaplan-Meier survival curves when the analyses are based on data

**Table 3. Graft failure status agreement for OPTN vs. CMS data, all patients (N = 31,265)**

|  |  | CMS | | |
|---|---|---|---|---|
|  |  | Functioning | Failed | Total |
| **OPTN** | Functioning | 26,712 | 513 | 27,225 |
|  | Failed | 893 | 3,147 | 4,040 |
|  | Total | 27,605 | 3,660 | **31,265** |

Source: SRTR analysis, August 2004. Reprinted from reference 4.

from a single cohort and they are shown at various time points after transplant. Results from different cohorts are sometimes shown at various time points after transplant, as in the Adjusted and Unadjusted Graft and Patient Survival tables in the *Annual Report*. However, since these results are from different groups of patients, the results computed across different time periods need not be consistent. For example, the five-year survival for the 10-year cohort is not reported and should not be assumed to be the same as the five-year survival that is reported for the five-year cohort.

## *Mortality*

Generally speaking, wait-listed registrants are not tracked by their former listing centers for mortality after removal from the waiting list. That is, mortality ascertainment stops when a recipient is lost to follow-up. Because of the incomplete follow-up available in the data, the actuarial methods described above must censor patients when they are lost to follow-up. If the death rates after LTFU are the same as the death rates among those still being followed, then the actuarial method estimates are appropriate, even though some observations were censored. However, if recipients at high risk for eventual death are disproportionately lost to follow-up before they die, then the estimated death rates will underestimate the overall death rates. When many subjects are LTFU, it is important to know if they were at high or low risk for subsequent unobserved events, compared with patients under observation.

OPTN death ascertainment, along with extra ascertainment from the SSDMF and the ESRD database, were used to compute death rates on the waiting list, as reported in each organ-specific section in the *Annual Report*. Such follow-up stops when a candidate is removed from the waiting list, because organ allocation is not affected by events after removal from the waiting list. The death rate per patient year at risk method includes events and time only while on the waiting list and is not affected by events after removal. However, the resulting death outcomes are difficult to interpret because registrants are often removed from the list if their health deteriorates to the extent that they are no longer suitable for a transplant. Thus, low death rates on a waiting list are likely to reflect an effective screening process which systematically removes (or transplants) patients when their health deteriorates. Rates based on patients not removed from the waiting list do not apply to registrants, in general, but to patients currently on (i.e., not removed from) the waiting list.

For the PSRs, mortality rates on the waiting list include extra ascertainment for death after removal from the waiting list or, in some cases, before removal. For these analyses, time at risk begins at the start of the observation period or the date of first wait-listing (latter thereof) and continues until the date of death, transplant, 60 days after removal for recovery, transfer to another center, or the end of the observation period (earliest thereof). To compute expected lifetimes on the waiting list, the SRTR uses information on deaths from other data sources, such as the SSDMF. This is especially important when comparing pretransplant mortality (which includes time after removal from the waiting list) to posttransplant mortality.

## *Graft Failure*

The analysis of graft failure is complicated by the potential for recipients to die. Death serves as a competing risk in the sense that the time of graft failure cannot be observed among patients who die with a functioning graft.[7] Death-censored graft failure estimates the "cause-specific" rate of graft failure; i.e., the rate of graft failure among patients who have not yet died. This is an interpretable measure that is frequently used. However, cause-specific rates, such as those estimated in an analysis of death-censored graft-failure, can only be combined to produce a meaningful survival curve if the competing risks are independent, an untenable assumption in the context of death and graft failure.

Frequently in analyses of graft failure, the end-point is defined as the minimum of the time until death and time until graft failure. This results in is a well-defined lifetime (i.e., survival, with a functioning graft). If only graft failure were specifically of interest, one could argue that the graft is, by definition, truly no longer functioning after the patient dies. In the regression setting, the trade-off for a cleanly-defined end-point is the interpretation of the covariate effects. For example, if a patient characteristic significantly increases the rate of graft failure, but not the rate of death, an analysis which combines graft failure and death may identify the covariate as being nonsignificant. In order to understand the mechanisms that lead to transplant failure, it is sometimes useful to count only failures of the transplanted organ itself, while not counting deaths from other causes. In addition to the issue of graft failure and death being competing risks, there is also the issue of determining exactly which events constitute graft failure. For example, when a graft failure is not explicitly recorded in the database, but a retransplant is recorded, the date of retransplantation can be used as the date of graft failure. In addition, for kidney transplant recipients, a reported return to dialysis can be counted as an organ failure.

## *Covariate-Adjusted Analyses*

Analyses with covariate adjustment, such as regression modeling, are intended to compare patient subgroups with "all other factors being equal". Many of the analyses performed by the SRTR involve comparisons of outcomes. For example, for tables comparing adjusted one-year survival over 10 years of transplantation, adjustment helps ensure that differences from year to year are not due to changes in case mix. Also, the PSRs use covariate adjustment to compare center-specific mortality rates with what would be expected for a given case mix, allowing the reader to separate which part of a good result, for example, is due to patient case mix.

The SRTR often uses an adjustment method based on regression models to compare the outcomes that would have resulted had the comparison groups been otherwise equivalent. Regression models can be used to compute expected outcomes given a patient's characteristics. The Cox proportional hazards regression model is commonly used for adjusted analyses of time-to-event data.[8] Similar to the Kaplan-Meier estimates described above, the Cox regression model can yield survival curve estimates for two or more groups of patients, adjusted to show the comparison that would result if the groups were equivalent with regard to particular factors, such as age and diagnosis.

Adjusted analyses are used extensively by the SRTR in the PSRs and in analyses based on data requests from OPTN policy-making committees. The choice of what to adjust for—or what to make equal in the comparison groups—is an important one that is under constant review by the SRTR and will differ according to the specific purpose of the analysis. For example, in a comparison involving patient characteristics (e.g., mortality rates by ethnicity), it would be prudent to adjust for variables reflecting therapeutic regimen, if available. However, in an analysis comparing center-specific transplant mortality rates, therapeutic regimen reflects a center's practices. To adjust for such factors amounts to adjusting away the difference that, if present, one wishes to discern. To make meaningful adjustments, relevant data must be available, complete and accurate. The choice of factors used when adjusting center-specific outcomes for the mix of characteristics at each center involves OPTN committees and SRTR analysts. The documentation of PSRs (available at www.ustransplant.org/csr/current/programs-report.cspx) includes detailed descriptions of the adjustment models they use.

Naturally, covariate adjustment is generally limited to patient characteristics for which data are collected. For these, data, limited information is available. The extent to which lack of comorbidity data biases the results of a regression analysis is an open question. For example, suppose that body mass index (BMI) is the covariate of interest in a kidney posttransplant model, with cardiovascular disease (CVD) being the potential confounder. The BMI regression coefficient, based on a model which does not contain a CVD covariate, would result in a biased estimate of the BMI effect only if CVD is both predictive of mortality and correlated with BMI after adjusting for all covariates which are included in the model. That CVD is a mortality risk factor alone would not mean that the BMI coefficient is biased if CVD were not included in the model. Although it is quite possible that CVD is correlated with BMI, the pair-wise correlation is of no relevance to the issue of bias; the pertinent correlation is that between BMI and CVD, adjusting for all other model covariates, which would be substantially less than the crude pair-wise correlation. In the assessment of potential residual confounding, it is often useful to compare the crude and covariate-adjusted analyses. For example, it would be encouraging if the unadjusted and covariate-adjusted hazard ratios for BMI were similar. That is, if there is little difference in the results which are unadjusted and the results which are adjusted for all available covariates, the hypothesis of residual confounding would be much less convincing. Nonetheless, the potential for residual confounding is frequently a consideration in SRTR analyses, mostly because it is impossible to verify its absence.

## Summary

This chapter focuses on data collection and organization schemes for transplant data and offers insights into implications of their timing and completeness. It also lays out caveats related to cohort choice, timing and timeliness of data submission and potential biases in follow-up data. The explanations are based on relevant discussions from chapters in various editions of the *OPTN/SRTR Annual Report* and *SRTR Report on the State of Transplantation*, as indicated in the references.

### Acknowledgement

The authors thank Valarie B. Ashby for assistance in preparing this chapter. The Scientific Registry of Transplant Recipients is funded by contract number 234-2005-37009C from the Health Resources and Services Administration, US Department of Health and Human Services. The views expressed herein are those of the authors and not necessarily those of the US Government.

### References

1. Levine GN, McCullough KP, Rodgers AM et al. Analytical methods and database design: implications for transplant researchers, 2005. Am J Transplant 2006; 6:1228-42.
2. Dickinson DM, Ellison MD, Webb RL. Data sources and structure. Am J Transplant 2003; 3 Suppl 4:13-28.
3. Dickinson DM, Bryant PC, Williams MC et al. Transplant data: sources, collection and caveats. Am J Transplant 2004; 4 Suppl 9:13-26.
4. Dickinson DM, Dykstra DM, Levine GN et al. Transplant data: sources, collection and research considerations, 2004. Am J Transplant 5, 2005; (Issue 4, Part 2):850-61.
5. Marchione M Transplant rate reports don't tell whole story. Milwaukee J Sentinel 2001; G1.
6. Cooper L. Survival Data: Do the numbers really mean anything? Transplant News and Issues 2001; 2(2):s9-s11, s13.
7. Kalbfleisch JD, Prentice RL. The Statistical Analysis of Failure Time Data. Second Edition. Hoboken: Wiley, 2002.
8. Cox DR. Regression models and life tables (with discussion). J Roy Stat Soc, Series B 1972 (34):197-220.

# CHAPTER 5

# The Immunology of Chronic Allograft Injury

Raphael Thuillier and Roslyn B. Mannon*

## Abstract

The causes of chronic graft injury are diverse and are dependant on the recipient, donor organ and immunosuppressive strategy. In this chapter, we explore the contribution of the immune system to this problem, recognizing that nonimmune factors have been shown to contribute to chronic injury as well. An immune response commences with the recognition of foreign antigen (allorecognition), which can occur by one of two pathways: direct and indirect. Although it is reported that the direct pathway has a role to play in the etiology of chronic graft injury, the indirect pathway is believed be a primary participant and a process that may not be specifically affected by immunosuppression. Once activated, requiring costimulatory signals, T-cells mediate a signaling cascade that will alter the milieu of the graft and activate other participants of the immune response. Not only are CD4+ and CD8+ T-cells participate, but monocytes, eosinophils and mast cells may also contribute to late graft injury. Blockade of this allorecognition process as well as blocking T-cell costimulation may ameliorate chronic graft injury. Moreover, activated CD4+ T-cells also have the capability to induce clonal differentiation of B-cells, resulting in alloantibody production—another critical mechanism of immune-mediated injury. In this chapter, we will review the mechanisms of immune cell activation and allorecognition, explore the role of both cellular and antibody mediated immune responses in chronic graft injury and explore the outcome of therapeutic intervention. The contribution of viral infection and its immune response within the graft will also be discussed. By understanding the immune mechanisms of chronic graft injury, we can further identify new markers and strategies to limit this injury in our recipients.

## Introduction

The etiology of chronic graft injury has been ascribed to two categories of instigating events: immunologic or antigen-dependent events and non-immunologic or antigen-independent events. Table 1 outlines this classification and the specific events that may impose injury to an allograft (reviewed in ref. 1). The relative importance of immunologic versus non-immunologic factors in contributing to injury has not been quantitated. Moreover, accumulating data suggest that it is the combination and interaction of these two types of events that can result in chronic and prolonged injury.[2] There is also a growing understanding that response to injury may be dysregulated after transplant and the institution of immunosuppression and thus the repair response may contribute to graft damage.[3] The focus of this chapter will be to highlight the immunological basis of chronic graft injury. While the term chronic implies a long term and progressive process, recent studies underscore that damage may be initiated and detected early after transplantation.[4] Moreover, while chronic injury occurs in all transplanted organs, much data have accumulated within the grafted kidney, which will be the primary emphasis in this chapter.

A number of insights noted years ago have supported a role for immunological events in chronic kidney graft failure. For example, HLA-identical kidney grafts enjoy a longer functional half-life.[5] Even in the context of newer immunosuppressive strategies, at 10 years, 52% of HLA-matched grafts survive compared to only 37% of HLA-mismatched grafts.[6] Secondly, acute rejection has been linked to chronic graft failure in kidney allografts, by a process called chronic allograft nephropathy (CAN).[7-9] The impact of acute rejection on CAN has been rising despite use of newer immunosuppressants.[10] Furthermore, there is improved graft survival for living related allografts compared to cadaveric grafts.[11,12] These clinical observations support the notion that immunological injury is a critical contributor to allograft failure.

To further understand the etiology of CAN, investigators have utilized a number of animal models. In general, these models of vascularized organ transplants have consisted of kidney[13] or heart transplants in rats[14] utilizing donor and recipient strain combinations with only minor antigen differences. The pathologic changes are seen over 4-6 months after transplantation making it a labor-intensive model to study. Using isolated carotid artery grafts in mice, investigators have demonstrated the development of arteriosclerosis in grafts over 30 days.[15] However, this model focuses solely on the vascular lesions of CAN, which are a small part of the changes seen in the kidney. Recently, we determined that mouse kidney allografts that are completely major histocompatibility complex (MHC) mismatched survive for prolonged periods of time in the absence of immunosuppression and these grafts develop reduced renal function and the typical histologic features of human CAN by 6 weeks post-transplant.[16] Transplanting kidneys that lack MHC class II antigens[17] or lack both class I and II antigens alters the immune response to the graft and alleviates the progression to CAN.[16] Finally, tracheal rings transplanted between MHC-disparate mice and develop the histological features of obliterative bronchiolitis, in the presence of cyclosporine A, over a 4 week period.[18] Thus, a number of animal models demonstrated reduced graft function and the typical histologic features of human disease. These animal models are useful tools and provide predictable and reproducible models of human chronic graft injury and are platforms that are amenable to studies of etiology and therapy. The contributions these studies have made to our understanding of the immunological features of the failing graft will be covered in more detail in the sections below.

*Corresponding Author: Roslyn B. Mannon—Transplantation Branch, NIDDK/NIH, 10 Center Drive, MSC 1450, CRC 5-5750, Bethesda, MD 20892, USA. Email: rozm@intra.niddk.nih.gov

*Chronic Allograft Failure: Natural History, Pathogenesis, Diagnosis and Management*, edited by Nasimul Ahsan. ©2008 Landes Bioscience.

*Table 1. Etiologies of chronic graft injury*

| Antigen-Dependent (Immunologic) | Antigen-Independent (Non-Immunologic) |
|---|---|
| Allorecognition | Drug Toxicity |
| Costimulatory signaling | Donor Senescence |
| Antibody-mediated rejection | Prolonged cold ischemic time; delayed graft function |
| Infection | Donor age |

## Overview of the Immune Response to Foreign Tissue

Rejection is triggered when the host immune system engages and activates humoral (antibody) and cellular effector responses, leading to graft injury and destruction. In the mouse, these foreign proteins or antigens are called major histocompatibility proteins or MHC and their genes are located on chromosome 17, while in man they are termed human leukocyte antigens or HLA and their genomic sequence is on chromosome 6. These consist of class I antigens expressed on all cells and class II antigens typically expressed only on antigen presenting cells (APCs) and in inflamed tissues.

Critical responders of this process are T-lymphocytes. While many cells participate in the cellular rejection process, T-cells are absolutely required [19,20] CD8+ T-cells or cytotoxic cells engage through class I MHC and function by directly lysing cells in the donor graft but they are not essential for the rejection process to occur.[21] CD4+ T-cells are important first responders, that engage in the context of class II MHC and are believed to be absolutely required for rejection to occur.[21] CD4+ T-cells produce a variety of cytokines that result not only in the amplification of the immune response in other cells but also act in an autocrine fashion as well.[22] Notably, recent studies in human recipients of kidney transplants using depletional induction antibody treatment have demonstrated that even in the absence or marked reductions in peripheral T-cell number, monocytes remain capable of instigating a functional rejection response.[23] By and large, most immunosuppressive strategies therefore focus their energies on abrogating T-cell responses through multiple signaling pathways (reviewed in ref. 24). Priming of T-cells may be of particular importance in chronic injury as demonstrated by Doege et al using the Lewis to Fisher rat model of chronic kidney injury.[25] In addition to their capacity for direct injury on the graft through cytotoxicity, T-cells may play another indirect role in graft injury by providing a microenvironment to stimulate proliferation of vascular smooth muscle within kidney grafts.[26]

Until recently, allorecognition was felt to occur primarily in the graft, with naïve T-cells activated while migrating through the blood supply, activated by donor APCs or donor endothelium.[27] The use of mice lacking secondary lymphoid organs such as splenectomized *aly/aly* alymphoplastic mice[28] and splenectomized lymphotoxin α and lymphotoxin β receptor deficient mice[29] have demonstrated a need for these secondary lymphoid sites to initiate and propagate a vigorous rejection response by presenting processed antigen to naïve T-cells. However, the relative contribution of each location isn't clear and it is presumed that immunosuppressive therapies should work effectively in either location.[30]

## T-Cell Activation

T-cell activation requires two signals. Signal one is engagement of the T-cell receptor (TCR) on host T-cells with contact of antigen presenting cells (APCs) and appropriate class of MHC molecule. Signaling occurs through the TCR via CD3 protein (Fig. 1). Following the binding of MHC to the TCR (Fig. 1), the appropriate coreceptor (CD8 for MHC class I, CD4 for MHC class II) binds the MHC

molecule as well (Fig. 1B). With engagement of costimulatory ligands (signal 2), the Lck kinase (bound to the coreceptor) and the Fyn kinase are activated by the change in conformation induced by the multiple bindings. These kinases then phosphorylate the immunoreceptor tyrosine-based activation motifs (ITAMs) located on the intracellular chains of the two CD3 molecules and the TCR, inducing the intracellular signaling and activation of the cell (Fig. 1C). While there are numerous costimulatory molecules that exist on T-cells (reviewed in ref. 31; Table 2), the focus of clinical transplantation has been on CD28, which binds CD80 (B7-1) and CD86 (B7-2) on activated APCs and CD40:CD154 (CD40 ligand), where CD154 is expressed on activated T-cells and CD40 is expressed on APCs and B-cells (Fig. 1B; reviewed in ref. 32). T-cells also possess CTLA4, a protein that is a negative regulator of the immune response[33] that can compete for binding with CD80 and CD86. It has been suggested that the absence of costimulatory signaling could lead to death of antigen specific T-cells or specific inactivation of such cells.

Numerous reports in rodent allograft transplant models have supported the notion that costimulatory blockade can interfere with acute rejection and thus prolong graft survival. Blockade of the B7/CD28 pathway has been associated with the prevention of acute rejection of mouse heart allografts[34] and rat kidney allografts.[35,36] Moreover, in some administrations, long term graft survival has been induced in mouse heart allografts with the absence of vascular lesions that are characteristic of allograft vasculopathy.[37,38] Blockade of this pathway has also been shown to attenuate chronic injury of the transplanted kidney[39] and transplanted heart[40] in the Fisher to Lewis rat models. Similar treatment induced either tolerance or attenuated chronic rejection in a rat model of LBNF1 hearts transplanted into Lewis rats sensitized with Brown Norway skin.[41]

Blockade of the CD40/CD154 pathway has also been investigated in numerous allograft models. Using anti-CD154 antibody alone, long term graft survival has been induced in primate renal[42] and islet allografts,[43] but the timing of therapy and its persistent application are critical. Alternatively, combined therapy against both pathways has induced long term acceptance of heart and skin allografts in mice[44] and kidney allografts in nonhuman primates.[45] In particular, dual pathway blockade prevented the development of cardiac vasculopathy in mice.[44]

*Table 2. Costimulatory pathways of T-cell activation*

| B7/CD28 Family | | |
|---|---|---|
| **Molecule (Location)** | **Ligand (Location)** | **Effect** |
| B7-1 (CD80) (APC, T-cell) | CD28 (T-cell) | Positive |
| B7-2 (CD86) (APC, T-cell) | CTLA-4 (T-cell) | Negative |
| B-7h (ICOS L) (APC, T-cell, fibroblast, endo and epithelium) | ICOS (T-cell, NK) | Positive |
| PD-L1 (B7-H1) (APC, T-cell, parenchyma) | PD-1 | Negative |
| PD-L2 (B7-H1) (APC) | PD-1 | Negative |
| B7-H3 (APC, T-cell, NK) | ? | ? |

| TNF/TNFR Family | | |
|---|---|---|
| **Molecule (Location)** | **Ligand (Location)** | **Effect** |
| CD40 (APC), endothelium | CD40-L (T-cell, NK) | Positive |
| LIGHT (APC) | HVEM (T-cell) | Positive |
| OX40-L (APC, endothelium) | OX40 (T-cell) | Positive |
| 4-1BB-L (APC) | 4-1BB (T-cell) | Positive |

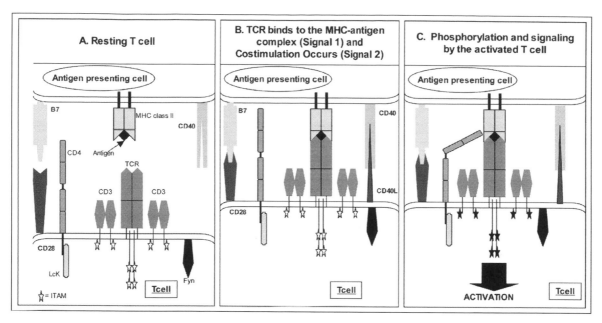

Figure 1. T-cell activation through the T-cell receptor (TCR) and CD3. Represented is the interaction of a CD4+ T-cell and its TCR with MHC class II molecule. Similar interactions are present with MHC class I molecules and CD8+ T-cells. A. In the resting state, TCR and the coreceptor CD4 are unbound and immunoreceptor tyrosine-based activation motifs (ITAMs) are not phosphorylated. B. In the presence of the antigen presenting cell, in which MHC is complexed with a foreign antigen, the TCR binds to the MHC, inducing the approach and subsequent binding of the CD4 coreceptor (signal 1). Signal 2 must also occur to facilitate activation. The interaction between CD40 and CD40L (CD154) is necessary to prevent apoptosis, facilitate the clonal differentiation of B-cells and induce CD80/CD86 (B7), while the interaction between CD28 and CD80/86 induces proliferation and prevents anergy. C. These extracellular events alter intracellular conditions and the kinases LcK and Fyn are now activated. They then phosphorylate the ITAMs and the signal for activation is sent.

Costimulatory blockade as a therapeutic avenue in human recipients is also under investigation. Clinical testing of anti-CD154 has been halted due to unexpected thrombotic complications associated with soluble CD154 on platelets. Additionally, a high rate of cellular rejection was seen in the first human patients treated with anti-CD154, while in the setting of low dose steroids and the absence of standard calcineurin inhibitor dosing. However, recent studies using belatacept, a second generation dimeric fusion protein consisting of the extracellular binding domain of CTLA4 linked to a modified Fc domain of human IgG$_1$, have shown significant clinical promise. Belatacept has shown superior binding of CD80 and CD86 in comparison with CTLA-4Ig. In a phase II multicenter trial in kidney transplant recipients,[46] low or high dose belatacept was compared to standard cyclosporine A in conjunction with basiliximab induction, prednisone and mycophenolate mofetil. At 6 months, there were no significant differences in biopsy proven acute rejection episodes between the groups. Moreover, at 12 months, renal function was significantly better in the belatacept groups than cyclosporine group and protocol biopsies at this timepoint showed a reduction in tubular atrophy and interstitial fibrosis, the hallmarks of chronic allograft injury, in the belatacept groups. Whether this represents an effect of calcineurin avoidance or supports the notion of a specific effect of costimulatory blockade is still under study. Phase III trials are underway with this agent investigating recipients of both standard criteria and extended criteria donors, with outcomes to focus not only on non-inferiority to cyclosporine based regimens but to indicate and identify the benefits of costimulatory blockade via the CD28/CD80-CD86 pathway.

## Allorecognition—Another Critical Variable in the Immune Response

Recognition of foreign MHC antigens occurs via one of two mechanisms (Fig. 2). In direct recognition, foreign antigens are presented in the context of donor APCs (Fig. 2B). In indirect allorecognition, recipient T-cells recognize foreign antigens processed and presented by recipient APCs, thus in the context of self-MHC (Fig. 2A). The direct pathway predominantly mediates early acute rejection, as the graft contains donor-derived APCs, expressing a high density of MHC molecules available for presentation (reviewed in ref. 22). The relative role of these processes in acute rejection and chronic injury is under debate. For example, several studies have suggested that indirect recognition may also be important in acute rejection. Animals immunized with donor MHC class II peptides rejected their grafts in an accelerated fashion compared to recipients that had not been immunized.[47] Further, skin allografts lacking MHC class II are rapidly rejected, via indirect recognition of donor MHC class I.[48] However, the absence of donor MHC molecules reduces the development of chronic injury in a mouse kidney transplant model.[49]

Indirect recognition may be more significant in chronic injury, where donor APCs are no longer present,[20,50] a process that may be resistant to standard immunosuppression. A number of studies have demonstrated the importance of indirect responses in the development of chronic graft injury. For example, transplant arteriopathy in a rat model of heart transplantation is associated with T-cells primed to either MHC class I[51] or class II peptides.[47] Moreover, CD4+ T-cells indirectly primed to MHC class I antigens are pathogenic in transplant arteriosclerosis and further provide help to CD8+ T-cells to promote cytotoxicity and alloantibody production to mediate chronic vascular injury.[52]

A number of investigators have focused on human transplant recipients and the implications of indirect recognition. Liu et al[50] found that T-cell alloreactivity to donor peptides was associated with acute rejection episodes in heart transplant recipients and that the frequency of allopeptide reactive T-cells was 10-50 fold more frequent in the graft, implicating indirect priming with acute rejection. Further studies by this group have demonstrated that the frequency of transplant arteriosclerosis following cardiac transplantation is higher in recipients that have persistent alloreactivity to donor peptides compared to those individuals who show no reactivity within the first 6 months of

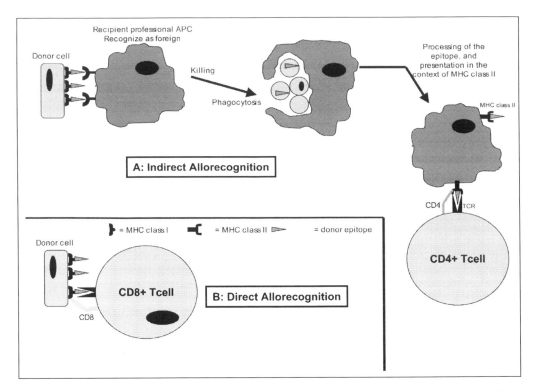

Figure 2. The process of allorecognition. A. In the indirect pathway, donor epitope, presented in the context of donor class I MHC is recognized as foreign by a recipient professional antigen presenting cell (APC). The donor cell is then killed and engulfed by the APC, which processes the epitope and presents it in the context of a recipient MHC class II. This APC will then activate a CD4⁺ T-cell. B. In the direct pathway, donor epitope presented in the context of a donor MHC molecule is recognized as foreign by a CD8⁺ T-cell.

transplantation.[53] This reactivity was associated with epitope spreading suggesting that a diversification of the immune response occurs and contributes to chronic graft injury. Similarly, T-cells from recipients of renal transplants with chronic graft injury demonstrate a higher frequency of response to donor allopeptides implicating that T-cell priming by the indirect pathway has a pathogenic role in man.[54] A direct comparison between direct and indirect pathway allorecogntion was performed by Baker et al.[55] Here, 22 longstanding renal transplant recipients, 9 of whom had chronic allograft nephropathy, were studied using standard mixed lymphocyte responses and limiting dilution analysis. Interestingly, all recipients showed donor specific hyporesponsiveness by the direct pathway, whereas indirect alloresponses were maintained and were significantly higher in recipients with CAN compared to those with "good function". In contrast, Coelho et al were not able to establish a relationship between indirect alloreactivity and recipient outcome, but their population was limited with only 2 of 12 studied recipients having chronic injury.[56] Finally, Poggio et al,[57] using a more sensitive IFNγ-ELISPOT assay has demonstrated that approximately 50% of kidney transplant recipients with chronic allograft nephropathy exhibited donor peptide specific responses compared to only 29% of recipients with stable function. These CAN recipients also had a higher rate of development of anti-HLA antibodies, although donor specificity was not tested. Thus, increased frequency of T-cells stimulated by allopeptides indirectly is associated with chronic graft injury in man. Further studies are underway to assess whether these recipients can be identified early in the post transplant period and to understand the role of immunosuppression in modifying this potentially harmful alloimmune response.

## The B-Cell Contribution to Graft Injury

As noted above, the indirect pathway provides T-cell help to B-cell function. Niiro et al[58] detailed that the initiation of this pathway requires the binding of antigen to the B-cell receptor, resulting in internalization of antigen and subsequent presentation in the context of MHC class II. This later complex interacts with the T-cell receptor (TCR) of a CD4⁺ helper T-cell. Costimulation in a positive activating context occurs, through CD40-CD154 and B7h-ICOS, activating the B-cell to produce antibodies.[22,59]

While blockade of CD40-CD154 improves graft survival in experimental models (reviewed in ref. 60), the ability to block alloantibody production has not been completely explored. For example, in a rat model of heart transplantation, CD40/CD154 blockade by gene transfer of CD40Ig in the graft decreased alloantibody production.[61] Moreover, simultaneous blockade of ICOS reinforces the inhibition of allospecific IgG production.[62] Combination therapy seems to be the most successful. Treatment of nonhuman primate kidney allograft recipients by abrogating CD40/CD154 and CD28/CD80:CD86 with CTLA-4Ig[63] or anti-CD86 antibody,[64,65] results not only in longer term kidney graft survival but also suppression of alloantibody production.

Alternatively, B7h/ICOS blockade has also been shown to prolong heart allograft survival both in MHC major and minor mismatched grafts and was associated with the suppression of alloantibody production as well as CD4⁺ T-cell expansion.[66] These responses were timing dependent and complex with a reduction in IL-10 producing T-cells and increase in expression of IL-4 and IFN-γ. Recent studies have also demonstrated the significance of the OX40/OX40L pathway in B-cell activation. Repeated transfusion of alloantigen by transfusion in OX40L deficient mice results in cytotoxic T-cell activity not unlike that in wild type controls, but with impaired antibody production.[67] Moreover, using Rag-2 deficient mice and combinations of wild type B and T-cells with OX40L deficient cells has demonstrated that antibody production is dependent on OX40L expression on B-cells and not T-cells. Thus, blockade of costimulatory pathways appears to ameliorate chronic graft injury in some models. These concepts need to be addressed fully in human transplant recipients.

## Alloantibody and the Role in Chronic Graft Injury

Recently, the importance of anti-donor antibody in the development of chronic injury and criteria to diagnose antibody-mediated rejection have been updated and added to the Banff pathology classification for kidney transplants.[68] The predictive value of anti-donor alloantibody production on poor graft outcome has long been emphasized by some groups (reviewed in ref. 69) and the mechanism examined in a number of animal models. Vascular lesions of transplant arteriosclerosis in murine heart allografts are dependent on the presence of anti-donor antibody[70] where adoptive transfer of immune serum in SCID mice (mice lacking T or B cells) results in the development of vascular obstructive lesions. In the absence of antibody production, heart allografts transplanted into immunoglobulin deficient mice with a targeted disruption of the μ chain are not rejected.[71] Moreover, studies using rat aortic allografts suggest that alloantibody binding to vascular smooth muscle are pathogenic and lead to neointimal injury and proliferation.[72]

In human recipients, early studies demonstrated that preexisting anti-HLA antibodies was associated with worse graft survival.[73] While the individual contribution of either class I or class II antibodies aren't certain, the combination of both antibodies appears to significantly worsen graft survival in kidney transplant recipients of HLA mismatched grafts.[74] The subsequent development of anti HLA antibodies after transplant also denotes significant negative impact on graft outcome. Prospective evaluation using flow cytometry of kidney transplant recipients demonstrates that the development of HLA antibodies is associated with a 6.6% graft loss at one year compared to 3.3% in recipients without antibody.[75] This alloantibody has been strongly associated not only with development of chronic rejection but also acute rejection,[76] such that the independent contribution to chronic graft injury is not entirely clear. Despite the increased risk of rejection and graft failure associated with alloantibody,[77] some studies have shown that only a minority of individuals (<20%) with chronic graft injury have detectable anti-donor HLA class II antibody detected by ELISA.[78] This has not been a consistent observation as some centers note a strong correlation with anti-HLA antibodies and chronic injury, although antibody may not be donor specific[79] and the development of alloantibody is also correlated with a strong level of anti-donor cellular immunity.[57] Similar observations have been made in recipients of other solid organs with late graft failure. Alloantibody is more common in recipients with bronchiolitis obliterans syndrome[80] and recipients with coronary arteriopathy have both cellular and humoral alloreactivity more commonly than in those without graft arteriosclerosis.[81] Understanding the role of alloantibody in the development of coronary arteriopathy is similarly still under investigation.[82]

Alloantibody can also induce platelet adhesion and subsequent thrombosis. In a rat model of CAN, glomerular deposition of non MHC specific IgM was associated with platelet containing thrombi which can fortify ongoing vascular injury.[83] Once antibody binds to its target antigen, it not only promotes the binding of cytotoxic cells, but also induces the binding of complement components, which leads to target cell death. This complement activation is represented by C4d production. To this end, Feutch et al were the first to demonstrate immunostaining for C4d in peritubular capillaries of kidney transplants indicating the presence of alloantibody within the graft.[84,85] This observation led to a flurry of studies and Cd4 deposition has become a diagnostic tool for acute antibody mediated rejection in which immunofluorescent staining in the basement membrane of the peritubular capillaries is seen in cases of antibody mediated rejection and was not detected in acute cellular rejection.[86] Indeed, Banff criteria have also been added to allow for the diagnosis of antibody mediated rejection and a number of potential pathologic variants not previously appreciated.[68] In the first 6 months post transplantation, C4d expression is associated with a 50% reduction in long term graft survival from an average of 8 years to 4 years[87] and the prevalence of staining may relate to whether the biopsy is performed for cause or in protocol settings where C4d positivity is seen relatively infrequently.[88] Not only has C4d been associated with acute antibody mediated rejection,[89] it is now appreciated in biopsies with chronic rejection. When all biopsies in one center (n = 38) with pathologic diagnosis of chronic rejection were analyzed for C4d deposition, 61% had C4d localized in peritubular capillaries compared to 2% of biopsies with cyclosporine nephrotoxicity or other nonspecific interstitial fibrosis.[90] 88% of recipients with C4d positive biopsies also had anti-donor antibodies compared to none of recipients with C4d negative chronic rejection on biopsy. These findings suggest that a subset of patients with chronic graft injury have alloantibody and this may lead to alternative therapeutic interventions.

In kidney grafts, another histological entity, chronic transplant glomerulopathy, has been recognized as a cause of late graft loss. This disorder is histologically characterized by widespread involvement of all glomeruli, with enlargement and duplication of the glomerular basement membrane and endothelial cell activation. By electron microscopy, there is subendothelial accumulation of electron lucent material, reduplication of the basement membrane and interposition of mesangial cells into the capillary wall.[91,92] Clinically, there is from 1 gram to greater than 3 grams of urinary protein excreted per day, progressive renal dysfunction and accelerated graft failure. Associated risk factors for development include the presence of anti-HLA antibodies at the time of transplant and late acute rejection.[93] Glomerular C4d staining may also be seen in this lesion, indicating antibody-mediated immune responses.[93] Endothelial C4d deposition may also be associated with chronic transplant glomerulopathy and its detection in peritubular capillaries in otherwise morphologically normal biopsies is associated with subsequent development of this lesion.[94] This antibody-associated immunity may also be a response to structural proteins in the kidney, rather than anti-HLA antibody. For example, Joosten et al demonstrated the presence of circulating anti-glomerular basement membrane (GBM) antibodies in the majority of recipients with transplant glomerulopathy, compared to recipients with typical chronic allograft nephropathy (11/16 compared to 3/16). These antibodies exhibit specificity to GBM heparan sulphate proteoglycan agrin in patients with transplant glomerulopathy.[95] The immune activation of this lesion is further supported by the observation that intraglomerular and periglomerular leukocytes express markers of T-cell activation CXCR3 and ICOS compared to their absence in biopsies with CAN lesions alone.[96] These studies support the need for further investigation into the prospective development of this lesion and the contributions of both cellular and antibody-mediated components.

## NonHLA Antibodies—What Are They and What Can They Do?

A number of groups have long speculated that nonHLA antibodies, antibodies against graft proteins, may mediate chronic graft injury. In a number of animal models, renal injury has been associated with the presence of nonHLA antibodies. For example, injection of monoclonal antibody against glomerular basement membrane heparan sulfate is associated with severe proteinuria in rats.[97] In the Fisher to Lewis rat model of chronic renal allograft rejection with transplant glomerulopathy, $IgG_1$ antibodies directed against GBM proteins perlecan, a heparan sulfate proteoglycan and alpha1(VI)/alpha5(IV) collagen can be detected.[98] Similarly, in rat recipients of cardiac allografts, antibodies to the donor vascular endothelial cell developed and were capable of binding donor origin endothelial cells and induce programmed cell death.[99]

In human kidney transplant recipients, antibodies directed against vascular endothelial cells have been associated with biopsy proven chronic rejection and worsened graft outcome.[100] Antibody against tubular basement membrane has also been found in 4 of 101 recipients from one transplant center, but their role in chronic tubulointerstitial injury isn't certain.[101] As noted above, transplant glomerulopathy may be associated with antibody against heparan sulfate proteoglycans. Recently, anti-angiotensin II type A receptor antibody has been detected in a subset of kidney transplant recipients with refractory vascular rejection episodes and malignant hypertension,[102] suggesting a protective role for angiotensin receptor blockade.

Antibodies against nonHLA graft proteins may also contribute to late graft failure in the lung and heart. Analysis of serial serum samples from recipients with bronchiolitis obliterans syndrome (BOS) has identified nonHLA antibodies in 5 of 16 recipients with BOS compared to none of the 11 without BOS.[103] The specificity of this anti-airway epithelial cell antibody is under investigation, but is associated with upregulation of TGF-β signaling in vitro. Complement activation is also seen in BOS biopsies, associated with anti-endothelial cell antibodies.[104] Immune responses against collagen V epitopes have also been implicated in the pathology of graft failure in BOS.[105] Similarly, in cardiac allograft vasculopathy, antibody responses towards endothelial cell antigens have long been appreciated to have a connection with chronic injury and rejection,[106,107] suggesting a link between endothelial injury and graft failure. Antibodies to vimentin, present also in leukocytes, fibroblasts and endothelial cells, have also been associated with the development of cardiac graft arteriopathy[108] and the detection of these antibodies may identify those at risk.[109] While anti-cardiac myosin[110] and phospholipid antibodies[111] have also been detected post heart transplantation, they have been predominantly associated with acute rejection episodes.

## Is There a Role for the Monocyte?

There is increasing appreciation for macrophage participation in acute graft rejection (reviewed in ref. 112). For example, dramatic T-lymphocyte depletion with anti-CD52 therapy following kidney transplantation is still associated with acute cellular rejection that is predominantly filled with macrophages.[23] However, the role of these cells in chronic graft injury is not entirely clear. Macrophages can be detected in chronically rejected kidney allografts. For example, semi-quantitation of macrophage infiltration using a modified Banff scoring has been suggested as a useful predictor of kidney function and survival.[113] The presence of both macrophages as well as myofibroblasts could be detected as early as 23 days post transplant and were associated with chronic rejection in kidney allografts.[114] The functional role of these invading cells is not clear but a number of animal studies suggest that macrophages residing in the graft are detrimental to graft function. For example, nitric oxide production via infiltrating macrophages may mediate endothelial injury and promote ongoing graft dysfunction.[115] Depletion of macrophages in recipients of kidney allografts, either by chemical means[116] or by adenoviral-mediated gene transfer of viral IL-10 or antagonists for TNFα and IL-12, ameliorated the development of CAN in the rat. Increased expression of matrix metalloproteinase II has been associated with macrophage infiltration and chronic allograft nephropathy,[117] suggesting that macrophages accentuate fibrosis through this mechanism. Magnetic resonance imaging utilizing supramagnetic iron oxide particles that are phagocytosed by macrophages can identify macrophage infiltration in allografts and an increasing magnetic resonance in the kidney cortex indicates progressive fibrosis and proteinuria in rat allografts with CAN.[118] Further studies are needed both in animal models and in man to understand the contribution of macrophages in chronic graft failure.

## Other Cellular Components of the Immune Response

While T and B-cells have considerable contributions to chronic graft failure, the role of other peripheral leukocyte populations are relatively unknown. NK cells and neutrophils may contribute to

acute rejection responses[119] but have not been characterized in so far as chronic injury responses. While eosinophils have been associated with acute rejection and are associated with poor prognosis for recovery,[120] they have been detected in biopsies with CAN.[121] Eosinophils may contribute to chronic injury by stimulating vascular smooth muscle cell activation[121] and by direct cytotoxicity following adherence to tubular epithelial cells.[122] Finally, activated mast cells produce histamine and serine proteases and heparan, which drive endothelial proliferation and stimulate fibroblast production of extracellular matrix (reviewed in ref. 123). Semi-quantitative assessment of mast cells within kidney allografts has correlated significantly with the extent of fibrosis and tubular atrophy.[124,125] While nearly absent in normal kidney biopsies, mast cell infiltration can also be seen in biopsies with membranous, IgA and diabetic nephropathy as well as in CAN and chronic cyclosporine toxicity[126] but were not detected in allograft biopsies with acute cellular rejection or cyclosporine toxicity. Mast cell infiltration is associated with expression of the protease chymase and the extent of chymase expression correlates with the extent of fibrosis.[127] Mast cell invasion in renal allograft biopsies at 100 days post transplant has also been associated with long term loss of function and fibrosis.[128] Similarly, mast cells have also been associated with progressive fibrosis of bronchiolitis obliterans.[129] In heart allografts, intramyocardial mast cell invasion and basic fibroblast growth factor expression, a pro-fibrotic growth factor, has been associated with arteriosclerosis and intimal thickening which may be blocked in part by cyclosporine.[130] Thus, further studies are needed to determine if blocking the recruitment or function of the eosinophil or mast cell may ameliorate chronic graft injury.

## Infection and Graft Injury: A Primary or Secondary Effect?

Infection of an allograft by virus, bacteria, or fungus, can result in a substantial inflammatory response, which in turn, may trigger acute rejection episodes. The antimicrobial responses may inadvertently injure the graft in the process of controlling infection. Additionally, microbial infection may have direct effects on the parenchyma, either promoting alterations in normal cell function or by direct cell lysis and death. Infection may also result in alterations in baseline immunosuppression by the clinician, which in turn may alter the balance between rejection and immune surveillance. Thus, reductions in basal therapy may result in a rejection episode that requires additional therapy and can worsen infection.[131]

Cytomegalovirus infection has been well-studied in solid organ transplant recipients and effective anti-viral therapy and screening methods for infection are available at most medical centers. Analysis of ~17,000 recipients of kidney allografts recorded in the United States Renal Data Systems from 1995-1997 demonstrated that CMV seropositive donors are associated with higher rates of CMV disease and graft loss,[132] but is unclear if CMV infection in the kidney graft can actually by itself cause allograft dysfunction. Viral glomerulopathy in kidney allografts has been reported[133] but is infrequently seen. Prevalence of disease may be related to the sensitivity of detection using immunohistochemistry versus molecular techniques.[134] Persistent CMV infection in the kidney allograft, detected by immunohistochemical detection of CMV antigens and/or DNA, has been associated with worse renal function and shorter graft survival.[135] CMV infection has similarly been linked to vascular disease in cardiac allografts (reviewed in ref. 136) and prophylaxis reduces CMV disease and the incidence of acute rejection and vascular disease in the heart[137] and bronchiolitis obliterans syndrome in the lung allograft.[138]

The mechanisms for CMV-mediated graft loss have been studied in rodent models. In kidney allografts, CMV accelerates chronic rejection and is associated with upregulation of TGFβ and platelet derived growth factor (PDGF) in endothelial cells and connective tissue growth factor within fibroblasts.[139] CMV infection has also been associated with accelerated vascular disease in kidney allografts, with upregulation of ICAM-1 in peritubular capillary endothelium and PDGF in arteriolar endothelium.[140] In rat cardiac allografts, CMV infection reduces allograft survival from an average of 90 days to 45 days[141] and is associated with accelerated atherosclerosis, marked inflammation and upregulation of CC chemokines RANTES, MCP-1 and MIP-1α.[142] Alternatively, CMV infection may mediate an autoimmune response that results in vascular damage and/or engages the innate immune system to upregulate chemokines and cytokines that accelerate atherogenesis.[143]

Over the past decade, BK polyomavirus infection of kidney allografts has been recognized as a significant cause of late graft loss,[144,145] occurring in 40-70% of infected kidneys.[146,147] Strongly implicated in the pathogenesis of this disease is over-immunosuppression with concomitant tubular injury.[148] Whether tubular damage is solely due to viral effects, or is also augmented by the virally-directed and perhaps bystander immune response remains uncertain. Infection is manifested by a strong inflammatory response, consisting of T-cells, B-cells and macrophages,[146,149] and in particular, a strong cytotoxic T-cell response that is significantly upregulated when compared to acute rejection.[149] Also associated with graft loss is progressive fibrosis and atrophy of the kidney graft.[150] Molecular evaluation of BK nephropathy biopsies demonstrates a marked induction of transcripts associate with fibrogenesis and epithelial-mesenchymal transformation implicating the virus as a direct mediator of mesenchymal transformation.[149] Endothelial cell infection in vitro alters gene transcript expression dramatically compared to uninfected cells, particularly genes associated with cell division, DNA replication and transcription and immune responses.[151] Using a murine model of infection and kidney transplantation, infection with murine polyomavirus at the time of transplant causes rapid decline in renal function and ensuing death due to graft failure.[152] This is accompanied by a marked anti-donor cytotoxic T-cell response. Thus, this model may provide a useful platform to understand the mechanisms for graft failure following infection and may provide new insights and therapeutic strategies to limit graft loss.

## Conclusion

Following transplantation, the immune response to an allograft can mediate acute and early responses that are injurious to the graft. If left unabated, these responses can result in progressive destruction and graft loss. Both direct and indirect recognition may occur and lead to graft loss; disruption of these responses by immunosuppression or other novel strategies have been and are under further testing. While we classically focus on T-cells in alloimmune responses, both macrophages and B-cells may play a role. The contribution of humoral immunity cannot be understated but difficulties lie in monitoring for these responses and invocating effective therapy which is still under study. Viral infection in the allograft is also an important contributor to graft injury, not only by the ensuing anti-viral immune response that may cause bystander damage to the graft, but also through direct injury to the graft parenchyma. Identifying strategies that can detect chronic and persistent immune responses following solid organ transplantation remains an important task for the coming decade.

### *Acknowledgements*

RT and RBM are supported by the Division of Intramural Research, National Institute of Diabetes and Digestive and Kidney Diseases, of the National Institutes of Health.

## References

1. Cornell LD, Colvin RB. Chronic allograft nephropathy. Curr Opin Nephrol Hypertens 2005; 14(3):229-234.
2. Tullius SG, Tilney NL. Both alloantigen-dependent and -independent factors influence chronic allograft rejection. Transplantation 1995; 59(3):313-318.
3. Mannon RB. Therapeutic targets in the treatment of allograft fibrosis. Am J Transplant 2006; 6(5 Pt 1):867-875.
4. Nankivell BJ, Borrows RJ, Fung CL et al. The natural history of chronic allograft nephropathy. N Engl J Med 2003; 349(24):2326-2333.
5. Cecka JM. The UNOS Scientific Renal Transplant Registry. Clin Transpl 1996; 1-14.
6. Takemoto SK, Terasaki PI, Gjertson DW et al. Twelve years' experience with national sharing of HLA-matched cadaveric kidneys for transplantation. N Engl J Med 2000; 343(15):1078-1084.
7. Almond PS, Matas A, Gillingham K et al. Risk factors for chronic rejection in renal allograft recipients. Transplantation 1993; 55(4):752-756.
8. Basadonna GP, Matas AJ, Gillingham KJ et al. Early versus late acute renal allograft rejection: impact on chronic rejection. Transplantation 1993; 55(5):993-995.
9. Humar A, Hassoun A, Kandaswamy R et al. Immunologic factors: the major risk for decreased long-term renal allograft survival. Transplantation 1999; 68(12):1842-1846.
10. Meier-Kriesche HU, Ojo AO, Hanson JA et al. Increased impact of acute rejection on chronic allograft failure in recent era. Transplantation 2000; 70(7):1098-1100.
11. Cecka JM, Terasaki PI. The UNOS scientific renal transplant registry. In: Cecka J, Terasaki P, eds. Clinical Transplants 2001. Los Angeles: UCLA Tissue Typing Laboratory, 2001:1-14.
12. Meier-Kriesche HU, Ojo AO, Hanson JA et al. Increased impact of acute rejection on chronic allograft failure in recent era. Transplantation 2000; 70(7):1098-1100.
13. White E, Hildemann WH, Mullen Y. Chronic kidney allograft reactions in rats. Transplantation 1969; 8(5):602-17.
14. Adams DH, Tilney NL, Collins JJ, Jr. et al. Experimental graft arteriosclerosis. I. The Lewis-to-F-344 allograft model. Transplantation 1992; 53(5):1115-19.
15. Shi C, Russell ME, Bianchi C et al. Murine model of accelerated transplant arteriosclerosis. Circ Res 1994; 75(2):199-207.
16. Mannon RB, Kopp JB, Ruiz P et al. Chronic rejection of mouse kidney allografts. Kidney Int 1999; 55(5):1935-1944.
17. Mannon RB, Doyle C, Griffiths R et al. Altered intragraft immune responses and improved renal function in MHC class II-deficient mouse kidney allografts. Transplantation 2000; 69(10):2137-43.
18. Neuringer IP, Mannon RB, Coffman TM et al. Immune cells in a mouse airway model of obliterative bronchiolitis. Am J Respir Cell Mol Biol 1998; 19(3):379-86.
19. Hricik DE, Rodriguez V, Riley J et al. Enzyme linked immunosorbent spot (ELISPOT) assay for interferon-gamma independently predicts renal function in kidney transplant recipients. Am J Transplant 2003; 3(7):878-84.
20. Sayegh MH. Why do we reject a graft? Role of indirect allorecognition in graft rejection. Kidney Int 1999; 56(5):1967-79.
21. Krieger NR, Yin DP, Fathman CG. CD4+ but not CD8+ cells are essential for allorejection. J Exp Med 1996; 184(5):2013-18.
22. Sayegh MH, Turka LA. The role of T-cell costimulatory activation pathways in transplant rejection. N Engl J Med 1998; 338(25):1813-21.
23. Kirk AD, Hale DA, Mannon RB et al. Results from a human renal allograft tolerance trial evaluating the humanized CD52-specific monoclonal antibody alemtuzumab (CAMPATH-1H). Transplantation 2003; 76(1):120-29.
24. Halloran PF. Immunosuppressive drugs for kidney transplantation. N Engl J Med 2004; 351(26):2715-29.
25. Doege C, Koch M, Heratizadeh A et al. Chronic allograft nephropathy in athymic nude rats after adoptive transfer of primed T-lymphocytes. Transpl Int 2005; 18(8):981-91.
26. Yamada K, Hatakeyama E, Sakamaki T et al. Involvement of platelet-derived growth factor and histocompatibility of DRB 1 in chronic renal allograft nephropathy. Transplantation 2001; 71(7):936-41.
27. Krensky AM, Weiss A, Crabtree G et al. T-lymphocyte-antigen interactions in transplant rejection. N Engl J Med 1990; 322(8):510-17.
28. Lakkis FG, Arakelov A, Konieczny BT et al. Immunologic 'ignorance' of vascularized organ transplants in the absence of secondary lymphoid tissue. Nat Med 2000; 6(6):686-88.
29. Zhou P, Hwang KW, Palucki D et al. Secondary lymphoid organs are important but not absolutely required for allograft responses. Am J Transplant 2003; 3(3):259-66.

30. Lakkis FG. Where is the alloimmune response initiated? Am J Transplant 2003; 3(3):241-42.

31. Snanoudj R, de Preneuf H, Creput C et al. Costimulation blockade and its possible future use in clinical transplantation. Transpl Int 2006; 19(9):693-704.

32. Larsen CP, Knechtle SJ, Adams A et al. A new look at blockade of T-cell costimulation: a therapeutic strategy for long-term maintenance immunosuppression. Am J Transplant 2006; 6(5 Pt 1):876-83.

33. Walunas TL, Bakker CY, Bluestone JA. CTLA-4 ligation blocks CD28-dependent T-cell activation. J Exp Med 1996; 183(6):2541-2550.

34. Turka LA, Linsley PS, Lin H et al. T-cell activation by the CD28 ligand B7 is required for cardiac allograft rejection in vivo. Proc Natl Acad Sci USA 1992; 89(22):11102-05.

35. Laskowski IA, Pratschke J, Wilhelm MJ et al. Anti-CD28 monoclonal antibody therapy prevents chronic rejection of renal allografts in rats. J Am Soc Nephrol 2002; 13(2):519-27.

36. Sayegh MH, Akalin E, Hancock WW et al. CD28-B7 blockade after alloantigenic challenge in vivo inhibits Th1 cytokines but spares Th2. J Exp Med 1995; 181(5):1869-74.

37. Glysing-Jensen T, Raisanen-Sokolowski A, Sayegh MH et al. Chronic blockade of CD28-B7-mediated T-cell costimulation by CTLA4Ig reduces intimal thickening in MHC class I and II incompatible mouse heart allografts. Transplantation 1997; 64(12):1641-45.

38. Sayegh MH, Zheng XG, Magee C et al. Donor antigen is necessary for the prevention of chronic rejection in CTLA4Ig-treated murine cardiac allograft recipients. Transplantation 1997; 64(12):1646-50.

39. Azuma H, Chandraker A, Nadeau K et al. Blockade of T-cell costimulation prevents development of experimental chronic renal allograft rejection. Proc Natl Acad Sci USA 1996; 93(22):12439-44.

40. Russell ME, Hancock WW, Akalin E et al. Chronic cardiac rejection in the LEW to F344 rat model. Blockade of CD28-B7 costimulation by CTLA4Ig modulates T-cell and macrophage activation and attenuates arteriosclerosis. J Clin Invest 1996; 97(3):833-38.

41. Onodera K, Chandraker A, Volk HD et al. Distinct tolerance pathways in sensitized allograft recipients after selective blockade of activation signal 1 or signal 2. Transplantation 1999; 68(2):288-93.

42. Kirk AD, Burkly LC, Batty DS et al. Treatment with humanized monoclonal antibody against CD154 prevents acute renal allograft rejection in nonhuman primates. Nat Med 1999; 5(6):686-93.

43. Kenyon NS, Chatzipetrou M, Masetti M et al. Long-term survival and function of intrahepatic islet allografts in rhesus monkeys treated with humanized anti-CD154. Proc Natl Acad Sci USA 1999; 96(14):8132-37.

44. Larsen CP, Elwood ET, Alexander DZ et al. Long-term acceptance of skin and cardiac allografts after blocking CD40 and CD28 pathways. Nature 1996; 381(6581):434-38.

45. Kirk AD, Harlan DM, Armstrong NN et al. CTLA4-Ig and anti-CD40 ligand prevent renal allograft rejection in primates. Proc Natl Acad Sci USA 1997; 94(16):8789-94.

46. Vincenti F, Larsen C, Durrbach A et al. Costimulation blockade with belatacept in renal transplantation. N Engl J Med 2005; 353(8):770-81.

47. Vella JP, Magee C, Vos L et al. Cellular and humoral mechanisms of vascularized allograft rejection induced by indirect recognition of donor MHC allopeptides. Transplantation 1999; 67(12):1523-32.

48. Auchincloss H, Jr., Lee R, Shea S et al. The role of "indirect" recognition in initiating rejection of skin grafts from major histocompatibility complex class II-deficient mice. Proc Natl Acad Sci USA 1993; 90(8):3373-77.

49. Mannon RB, Griffiths R, Ruiz P et al. Absence of donor MHC antigen expression ameliorates chronic kidney allograft rejection. Kidney Int 2002; 62(1):290-300.

50. Liu Z, Colovai AI, Tugulea S et al. Indirect recognition of donor HLA-DR peptides in organ allograft rejection. J Clin Invest 1996; 98(5):1150-57.

51. Womer KL, Stone JR, Murphy B et al. Indirect allorecognition of donor class I and II major histocompatibility complex peptides promotes the development of transplant vasculopathy. J Am Soc Nephrol 2001; 12(11):2500-2506.

52. Ensminger SM, Spriewald BM, Witzke O et al. Indirect allorecognition can play an important role in the development of transplant arteriosclerosis. Transplantation 2002; 73(2):279-86.

53. Ciubotariu R, Liu Z, Colovai AI et al. Persistent allopeptide reactivity and epitope spreading in chronic rejection of organ allografts. J Clin Invest 1998; 101(2):398-405.

54. Vella JP, Spadafora-Ferreira M, Murphy B et al. Indirect allorecognition of major histocompatibility complex allopeptides in human renal transplant recipients with chronic graft dysfunction. Transplantation 1997; 64(6):795-800.

55. Baker RJ, Hernandez-Fuentes MP, Brookes PA et al. Loss of direct and maintenance of indirect alloresponses in renal allograft recipients: implications for the pathogenesis of chronic allograft nephropathy. J Immunol 2001; 167(12):7199-7206.

56. Coelho V, Spadafora-Ferreira M, Marrero I et al. Evidence of indirect allorecognition in long-term human renal transplantation. Clin Immunol 1999; 90(2):220-29.

57. Poggio ED, Clemente M, Riley J et al. Alloreactivity in renal transplant recipients with and without chronic allograft nephropathy. J Am Soc Nephrol 2004; 15(7):1952-60.

58. Niiro H, Clark EA. Regulation of B-cell fate by antigen-receptor signals. Nat Rev Immunol 2002; 2(12):945-56.

59. Mak TW, Shahinian A, Yoshinaga SK et al. Costimulation through the inducible costimulator ligand is essential for both T helper and B-cell functions in T-cell-dependent B-cell responses. Nat Immunol 2003; 4(8):765-72.

60. Yamada AA, Sayegh MH. The CD154-CD40 costimulatory pathway in transplantation. Transplantation 2002; 73(1 Suppl):S36-39.

61. Guillot C, Guillonneau C, Mathieu P et al. Prolonged blockade of CD40-CD40 ligand interactions by gene transfer of CD40Ig results in long-term heart allograft survival and donor-specific hyporesponsiveness, but does not prevent chronic rejection. J Immunol 2002; 168(4):1600-09.

62. Guillonneau C, Aubry V, Renaudin K et al. Inhibition of chronic rejection and development of tolerogenic T-cells after ICOS-ICOSL and CD40-CD40L costimulation blockade. Transplantation 2005; 80(4):546-54.

63. Pearson TC, Trambley J, Odom K et al. Anti-CD40 therapy extends renal allograft survival in rhesus macaques. Transplantation 2002; 74(7):933-40.

64. Haanstra KG, Ringers J, Sick EA et al. Prevention of kidney allograft rejection using anti-CD40 and anti-CD86 in primates. Transplantation 2003; 75(5):637-43.

65. Haanstra KG, Sick EA, Ringers J et al. Costimulation blockade followed by a 12-week period of cyclosporine A facilitates prolonged drug-free survival of rhesus monkey kidney allografts. Transplantation 2005; 79(11):1623-26.

66. Harada H, Salama AD, Sho M et al. The role of the ICOS-B7h T-cell costimulatory pathway in transplantation immunity. J Clin Invest 2003; 112(2):234-43.

67. Kato H, Kojima H, Ishii N et al. Essential role of OX40L on B-cells in persistent alloantibody production following repeated alloimmunizations. J Clin Immunol 2004; 24(3):237-48.

68. Racusen LC, Colvin RB, Solez K et al. Antibody-mediated rejection criteria—an addition to the Banff 97 classification of renal allograft rejection. Am J Transplant 2003; 3(6):708-14.

69. McKenna RM, Takemoto SK, Terasaki PI. Anti-HLA antibodies after solid organ transplantation. Transplantation 2000; 69(3):319-26.

70. Joosten SA, van Kooten C, Paul LC. Pathogenesis of chronic allograft rejection. Transpl Int 2003; 16(3):137-45.

71. Wasowska BA, Qian Z, Cangello DL et al. Passive transfer of alloantibodies restores acute cardiac rejection in IgKO mice. Transplantation 2001; 71(6):727-36.

72. Thaunat O, Louedec L, Dai J et al. Direct and indirect effects of alloantibodies link neointimal and medial remodeling in graft arteriosclerosis. Arterioscler Thromb Vasc Biol 2006; 26(10):2359-65.

73. Terasaki PI, Kreisler M, Mickey RM. Presensitization and kidney transplant failures. Postgrad Med J 1971; 47(544):89-100.

74. Susal C, Opelz G. Kidney graft failure and presensitization against HLA class I and class II antigens. Transplantation 2002; 73(8):1269-73.

75. Terasaki PI, Ozawa M. Predicting kidney graft failure by HLA antibodies: a prospective trial. Am J Transplant 2004; 4(3):438-43.

76. Piazza A, Poggi E, Borrelli L et al. Impact of donor-specific antibodies on chronic rejection occurrence and graft loss in renal transplantation: posttransplant analysis using flow cytometric techniques. Transplantation 2001; 71(8):1106-12.

77. Worthington JE, Martin S, Al Husseini DM et al. Posttransplantation production of donor HLA-specific antibodies as a predictor of renal transplant outcome. Transplantation 2003; 75(7):1034-40.

78. Supon P, Constantino D, Hao P et al. Prevalence of donor-specific anti-HLA antibodies during episodes of renal allograft rejection. Transplantation 2001; 71(4):577-80.

79. Lee PC, Terasaki PI, Takemoto SK et al. All chronic rejection failures of kidney transplants were preceded by the development of HLA antibodies. Transplantation 2002; 74(8):1192-94.

80. Jaramillo A, Smith MA, Phelan D et al. Development of ELISA-detected anti-HLA antibodies precedes the development of bronchiolitis obliterans

syndrome and correlates with progressive decline in pulmonary function after lung transplantation. Transplantation 1999; 67(8):1155-1161.

81. Poggio ED, Roddy M, Riley J et al. Analysis of immune markers in human cardiac allograft recipients and association with coronary artery vasculopathy. J Heart Lung Transplant 2005; 24(10):1606-1613.

82. Soleimani B, Lechler RI, Hornick PI et al. Role of alloantibodies in the pathogenesis of graft arteriosclerosis in cardiac transplantation. Am J Transplant 2006; 6(8):1781-1785.

83. Duijvestijn AM, Breda Vriesman PJ. Chronic renal allograft rejection. Selective involvement of the glomerular endothelium in humoral immune reactivity and intravascular coagulation. Transplantation 1991; 52(2):195-202.

84. Feucht HE, Felber E, Gokel MJ et al. Vascular deposition of complement-split products in kidney allografts with cell-mediated rejection. Clin Exp Immunol 1991; 86(3):464-470.

85. Feucht HE, Schneeberger H, Hillebrand G et al. Capillary deposition of C4d complement fragment and early renal graft loss. Kidney Int 1993; 43(6):1333-1338.

86. Collins AB, Schneeberger EE, Pascual MA et al. Complement activation in acute humoral renal allograft rejection: diagnostic significance of C4d deposits in peritubular capillaries. J Am Soc Nephrol 1999; 10(10):2208-2214.

87. Lederer SR, Kluth-Pepper B, Schneeberger H et al. Impact of humoral alloreactivity early after transplantation on the long-term survival of renal allografts. Kidney Int 2001; 59(1):334-341.

88. Mengel M, Bogers J, Bosmans JL et al. Incidence of C4d stain in protocol biopsies from renal allografts: results from a multicenter trial. Am J Transplant 2005; 5(5):1050-1056.

89. Mauiyyedi S, Crespo M, Collins AB et al. Acute humoral rejection in kidney transplantation: II. Morphology, immunopathology and pathologic classification. J Am Soc Nephrol 2002; 13(3):779-787.

90. Mauiyyedi S, Pelle PD, Saidman S et al. Chronic humoral rejection: identification of antibody-mediated chronic renal allograft rejection by C4d deposits in peritubular capillaries. J Am Soc Nephrol 2001; 12(3):574-582.

91. Habib R, Zurowska A, Hinglais N et al. A specific glomerular lesion of the graft: allograft glomerulopathy. Kidney Int Suppl 1993; 42:S104-S111.

92. Maryniak RK, First MR, Weiss MA. Transplant glomerulopathy: evolution of morphologically distinct changes. Kidney Int 1985; 27(5):799-806.

93. Sijpkens YW, Joosten SA, Wong MC et al. Immunologic risk factors and glomerular C4d deposits in chronic transplant glomerulopathy. Kidney Int 2004; 65(6):2409-2418.

94. Regele H, Bohmig GA, Habicht A et al. Capillary deposition of complement split product C4d in renal allografts is associated with basement membrane injury in peritubular and glomerular capillaries: a contribution of humoral immunity to chronic allograft rejection. J Am Soc Nephrol 2002; 13(9):2371-2380.

95. Joosten SA, Sijpkens YW, van HV et al. Antibody response against the glomerular basement membrane protein agrin in patients with transplant glomerulopathy. Am J Transplant 2005; 5(2):383-393.

96. Akalin E, Dikman S, Murphy B et al. Glomerular infiltration by CXCR3+ ICOS+ activated T-cells in chronic allograft nephropathy with transplant glomerulopathy. Am J Transplant 2003; 3(9):1116-1120.

97. van den BJ, van den Heuvel LP, Bakker MA et al. A monoclonal antibody against GBM heparan sulfate induces an acute selective proteinuria in rats. Kidney Int 1992; 41(1):115-123.

98. Joosten SA, van Dixhoorn MG, Borrias MC et al. Antibody response against perlecan and collagen types IV and VI in chronic renal allograft rejection in the rat. Am J Pathol 2002; 160(4):1301-1310.

99. Wu GD, Jin YS, Salazar R et al. Vascular endothelial cell apoptosis induced by anti-donor nonMHC antibodies: a possible injury pathway contributing to chronic allograft rejection. J Heart Lung Transplant 2002; 21(11):1174-1187.

100. Ball B, Mousson C, Ratignier C et al. Antibodies to vascular endothelial cells in chronic rejection of renal allografts. Transplant Proc 2000; 32(2):353-354.

101. Paul LC, van Es LA, Stuffers-Heiman M et al. Antibodies directed against tubular basement membranes in human renal allograft recipients. Clin Immunol Immunopathol 1979; 14(2):231-237.

102. Dragun D, Muller DN, Brasen JH et al. Angiotensin II type 1-receptor activating antibodies in renal-allograft rejection. N Engl J Med 2005; 352(6):558-569.

103. Jaramillo A, Naziruddin B, Zhang L et al. Activation of human airway epithelial cells by nonHLA antibodies developed after lung transplantation: a potential etiological factor for bronchiolitis obliterans syndrome. Transplantation 2001; 71(7):966-976.

104. Magro CM, Ross P, Jr., Kelsey M et al. Association of humoral immunity and bronchiolitis obliterans syndrome. Am J Transplant 2003; 3(9):1155-1166.

105. Yoshida S, Haque A, Mizobuchi T et al. Anti-type V collagen lymphocytes that express IL-17 and IL-23 induce rejection pathology in fresh and well-healed lung transplants. Am J Transplant 2006; 6(4):724-735.

106. Dunn MJ, Crisp SJ, Rose ML et al. Anti-endothelial antibodies and coronary artery disease after cardiac transplantation. Lancet 1992; 339(8809):1566-1570.

107. Ferry BL, Welsh KI, Dunn MJ et al. Anti-cell surface endothelial antibodies in sera from cardiac and kidney transplant recipients: association with chronic rejection. Transpl Immunol 1997; 5(1):17-24.

108. Wheeler CH, Collins A, Dunn MJ et al. Characterization of endothelial antigens associated with transplant-associated coronary artery disease. J Heart Lung Transplant 1995; 14(6 Pt 2):S188-S197.

109. Jurcevic S, Ainsworth ME, Pomerance A et al. Antivimentin antibodies are an independent predictor of transplant-associated coronary artery disease after cardiac transplantation. Transplantation 2001; 71(7):886-892.

110. Warraich RS, Pomerance A, Stanley A et al. Cardiac myosin autoantibodies and acute rejection after heart transplantation in patients with dilated cardiomyopathy. Transplantation 2000; 69(8):1609-1617.

111. Laguens RP, Vigliano CA, Argel MI et al. Anti-skeletal muscle glycolipid antibodies in human heart transplantation as predictors of acute rejection: comparison with other risk factors. Transplantation 1998; 65(10):1345-1351.

112. Wyburn KR, Jose MD, Wu H et al. The role of macrophages in allograft rejection. Transplantation 2005; 80(12):1641-1647.

113. Srinivas TR, Kubilis PS, Croker BP. Macrophage index predicts short-term renal allograft function and graft survival. Transpl Int 2004; 17(4):195-201.

114. Pilmore HL, Painter DM, Bishop GA et al. Early up-regulation of macrophages and myofibroblasts: a new marker for development of chronic renal allograft rejection. Transplantation 2000; 69(12):2658-2662.

115. Elahi MM, Matata BM, Hakim NS. Quiescent interplay between inducible nitric oxide synthase and tumor necrosis factor-alpha: influence on transplant graft vasculopathy in renal allograft dysfunction. Exp Clin Transplant 2006; 4(1):445-450.

116. Azuma H, Nadeau KC, Ishibashi M et al. Prevention of functional, structural and molecular changes of chronic rejection of rat renal allografts by a specific macrophage inhibitor. Transplantation 1995; 60(12):1577-1582.

117. Palomar R, Ruiz JC, Mayorga M et al. The macrophage infiltration index and matrix metalloproteinase-II expression as a predictor of chronic allograft rejection. Transplant Proc 2004; 36(9):2662-2663.

118. Beckmann N, Cannet C, Zurbruegg S et al. Macrophage infiltration detected at MR imaging in rat kidney allografts: early marker of chronic rejection? Radiology 2006; 240(3):717-724.

119. Meehan SM, McCluskey RT, Pascual M et al. Cytotoxicity and apoptosis in human renal allografts: identification, distribution and quantitation of cells with a cytotoxic granule protein GMP-17 (TIA-1) and cells with fragmented nuclear DNA. Lab Invest 1997; 76(5):639-649.

120. Weir MR, Hall-Craggs M, Shen SY et al. The prognostic value of the eosinophil in acute renal allograft rejection. Transplantation 1986; 41(6):709-712.

121. Nolan CR, Saenz KP, Thomas CA,III et al. Role of the eosinophil in chronic vascular rejection of renal allografts. Am J Kidney Dis 1995; 26(4):634-642.

122. Makhlouf HR, Drachenberg CB, Trifillis A et al. Cytotoxic effects of eosinophils on renal proximal tubular epithelial cells: implications for renal allograft rejection. J Submicrosc Cytol Pathol 1999; 31(4):533-541.

123. Galli SJ, Wershil BK. The two faces of the mast cell. Nature 1996; 381(6577):21-22.

124. Goto E, Honjo S, Yamashita H et al. Mast cells in human allografted kidney: correlation with interstitial fibrosis. Clin Transplant 2002; 16 Suppl 8:7-11.

125. Pardo J, Diaz L, Errasti P et al. Mast cells in chronic rejection of human renal allografts. Virchows Arch 2000; 437(2):167-172.

126. Roberts IS, Brenchley PE. Mast cells: the forgotten cells of renal fibrosis. J Clin Pathol 2000; 53(11):858-862.

127. Yamada M, Ueda M, Naruko T et al. Mast cell chymase expression and mast cell phenotypes in human rejected kidneys. Kidney Int 2001; 59(4):1374-1381.

128. Ishida T, Hyodo Y, Ishimura T et al. Mast cell numbers and protease expression patterns in biopsy specimens following renal transplantation from living-related donors predict long-term graft function. Clin Transplant 2005; 19(6):817-824.

129. Yousem SA. The potential role of mast cells in lung allograft rejection. Hum Pathol 1997; 28(2):179-182.

130. Koskinen PK, Kovanen PT, Lindstedt KA et al. Mast cells in acute and chronic rejection of rat cardiac allografts—a major source of basic fibroblast growth factor. Transplantation 2001; 71(12):1741-47.

131. Pouteil-Noble C, Ecochard R, Landrivon G et al. Cytomegalovirus infection—an etiological factor for rejection? A prospective study in 242 renal transplant patients. Transplantation 1993; 55(4):851-57.

132. Schnitzler MA, Lowell JA, Hardinger KL et al. The association of cytomegalovirus sero-pairing with outcomes and costs following cadaveric renal transplantation prior to the introduction of oral ganciclovir CMV prophylaxis. Am J Transplant 2003; 3(4):445-51.

133. Richardson WP, Colvin RB, Cheeseman SH et al. Glomerulopathy associated with cytomegalovirus viremia in renal allografts. N Engl J Med 1981; 305(2):57-63.

134. Liapis H, Storch GA, Hill DA et al. CMV infection of the renal allograft is much more common than the pathology indicates: a retrospective analysis of qualitative and quantitative buffy coat CMV-PCR, renal biopsy pathology and tissue CMV-PCR. Nephrol Dial Transplant 2003; 18(2):397-402.

135. Helantera I, Koskinen P, Finne P et al. Persistent cytomegalovirus infection in kidney allografts is associated with inferior graft function and survival. Transpl Int 2006; 19(11):893-900.

136. Valantine HA. The role of viruses in cardiac allograft vasculopathy. Am J Transplant 2004; 4(2):169-77.

137. Potena L, Holweg CT, Chin C et al. Acute rejection and cardiac allograft vascular disease is reduced by suppression of subclinical cytomegalovirus infection. Transplantation 2006; 82(3):398-405.

138. Ruttmann E, Geltner C, Bucher B et al. Combined CMV prophylaxis improves outcome and reduces the risk for bronchiolitis obliterans syndrome (BOS) after lung transplantation. Transplantation 2006; 81(10):1415-20.

139. Inkinen K, Soots A, Krogerus L et al. Cytomegalovirus enhance expression of growth factors during the development of chronic allograft nephropathy in rats. Transpl Int 2005; 18(6):743-49.

140. Helantera I, Loginov R, Koskinen P et al. Persistent cytomegalovirus infection is associated with increased expression of TGF-beta 1, PDGF-AA and ICAM-1 and arterial intimal thickening in kidney allografts. Nephrol Dial Transplant 2005; 20(4):790-96.

141. Orloff SL, Streblow DN, Soderberg-Naucler C et al. Elimination of donor-specific alloreactivity prevents cytomegalovirus-accelerated chronic rejection in rat small bowel and heart transplants. Transplantation 2002; 73(5):679-88.

142. Streblow DN, Kreklywich C, Yin Q et al. Cytomegalovirus-mediated upregulation of chemokine expression correlates with the acceleration of chronic rejection in rat heart transplants. J Virol 2003; 77(3):2182-94.

143. Stassen FR, Vega-Cordova X, Vliegen I et al. Immune activation following cytomegalovirus infection: more important than direct viral effects in cardiovascular disease? J Clin Virol 2006; 35(3):349-53.

144. Hirsch HH. Polyomavirus BK nephropathy: a (re-)emerging complication in renal transplantation. Am J Transplant 2002; 2(1):25-30.

145. Nickeleit V, Mihatsch MJ. Polyomavirus nephropathy in native kidneys and renal allografts: an update on an escalating threat. Transpl Int 2006; 19(12):960-73.

146. Ahuja M, Cohen EP, Dayer AM et al. Polyoma virus infection after renal transplantation. Use of immunostaining as a guide to diagnosis. Transplantation 2001; 71(7):896-99.

147. Nickeleit V, Hirsch HH, Binet IF et al. Polyomavirus infection of renal allograft recipients: from latent infection to manifest disease. J Am Soc Nephrol 1999; 10(5):1080-89.

148. Fishman JA. BK virus nephropathy—polyomavirus adding insult to injury. N Engl J Med 2002; 347(7):527-30.

149. Mannon RB, Hoffmann SC, Kampen RL et al. Molecular evaluation of BK polyomavirus nephropathy. Am J Transplant 2005; 5(12):2883-93.

150. Liptak P, Kemeny E, Ivanyi B. Primer: histopathology of polyomavirus-associated nephropathy in renal allografts. Nat Clin Pract Nephrol 2006; 2(11):631-36.

151. Grinde B, Gayorfar M, Rinaldo CH. Impact of a polyomavirus (BKV) infection on mRNA expression in human endothelial cells. Virus Res 2006.

152. Han Lee ED, Kemball CC, Wang J et al. A mouse model for polyomavirus-associated nephropathy of kidney transplants. Am J Transplant 2006; 6(5 Pt 1):913-22.

# Ischemia-Reperfusion Injury:
## Pathophysiology and Clinical Approach

Maria Teresa Gandolfo and Hamid Rabb*

## Abstract

Significant ischemia-reperfusion injury (IRI) occurs in every deceased donor organ transplant and in some live donor ones. In renal transplants, it remains the leading contributor to delayed graft function (DGF), which in turn predisposes to increased acute rejection and worse long-term allograft function. Donor organ quality, modalities and time of organ preservation and recipient risk factors can all contribute to the generation of IRI in the graft. IRI pathogenesis involves a complex interplay between microvascular, cellular and systemic factors. Activation of inflammatory and immune responses is considered a major factor in the generation and amplification of damage. Activation of T-cells, not only by antigen binding but also by antigen-independent mechanisms, can play an important role in this process. As a result of better understanding of IRI pathogenesis, improved diagnostics and therapeutics have been developed to reduce the extent of the ischemic insult in the graft. This has enabled a safer expansion of the donor pool in order to better cope with the always increasing demand for organs for transplantation.

## Indroduction

Ischemia-reperfusion injury (IRI) is a major pathophysiologic process during organ transplantation from deceased donors. IRI is also implicated in a variety of nontransplant conditions, including myocardial ischemia, shock and stroke. Tissue ischemia and reperfusion represent a complex interplay between biochemical, cellular, vascular endothelial and tissue-specific factors.[1] A common feature of this process is activation of inflammatory and immune responses. In the kidney, IRI remains the leading cause of intrinsic acute renal failure (ARF), is the main contributor to early renal dysfunction in deceased donor allograft recipients and decreases long-term allograft survival. In this chapter we will provide an overview of both the clinical and pathophysiologic aspects of IRI, focusing primarily on the kidney.

## Delayed Graft Function

Delayed graft function (DGF) is a clinical diagnosis. The term describes the acute transplant kidney dysfunction which occurs in the immediate postoperative phase after the transplantation procedure. The most common definition is the requirement for dialysis in the first week after kidney transplantation. However, there are many limitations to this definition, including that it does not properly apply to some patients with residual native kidney function or non-oliguric acute renal failure. The term "slow graft function" (SGF) is sometimes used to describe those milder clinical situations that don't meet the DGF criteria; SGF indicates a delayed fall in serum creatinine after the transplantation procedure. Many other definitions have been used

in the literature to define DGF. Rodrigo et al[2] described the creatinine reduction rate at day 2 (CCR2) as an earlier parameter of renal allograft function; in their study, the patients with DGF as defined by CCR2 had a worse outcome, even if they weren't dialyzed. This demonstrates the presence of a continuum of early graft dysfunction, which also decreases long-term the graft survival. The importance of this is often missed when DGF is only thought of as the early need for dialysis.

The rate of DGF after kidney transplantation is usually between 15% and 25%, but it can vary from 2 to 50%, depending on the registry and center practice. DGF is an independent risk factor for both acute rejection and impaired renal function at one year after transplantation and is also associated with decreased one-year and long-term graft survival. DGF is one of the most important independent risk factors for the development of chronic allograft nephropathy (CAN).[3]

The etiology of DGF can be roughly divided in prerenal, intrarenal and postrenal (Table 1). Frequently more than one cause occurs in a single patient. Ischemic acute tubular necrosis (ATN) is by far the most common cause of DGF; sometimes the two terms are used interchangeably, but it is important to note that they are not equivalent. The term acute kidney injury (AKI) is increasingly used for IRI in native kidneys and should be used to describe allograft IRI as well. Risk factors for DGF (Table 2) promote both ischemia-reperfusion injury

---

**Table 1. Causes of delayed graft function**

**Prerenal**
- Intravascular volume contraction
- Severe hypotension/ reduced cardiac contractility
- Renal arterial occlusion/venous thrombosis

**Intrarenal**
- Acute tubular necrosis (ATN)
- Hyperacute rejection
- Accelerated or acute rejection superimposed on ATN
- Nephrotoxicity
- Thrombotic microangiopathy

**Postrenal**
- Ureteric obstruction/urine leak
- Catheter obstruction

Modified from: Magee CC, Milford E. Clinical aspects of renal transplantation. In: Brenner BM, Rector FC, eds. The kidney 7th ed., Philadelphia: WB Saunders, 2004:2811.

---

*Corresponding Author: Hamid Rabb—Division of Nephrology, Johns Hopkins University, Ross 965, 720 Rutland Avenue, Baltimore, MD, 21205 USA. E-mail: hrabb1@jhmi.edu

*Chronic Allograft Failure: Natural History, Pathogenesis, Diagnosis and Management*, edited by Nasimul Ahsan. ©2008 Landes Bioscience.

***Table 2. Factors affecting delayed graft function***

**1. Donor Organ Quality**
- Expanded criteria donors
- Gender mismatch: female to male
- Nonheart-beating donors

**2. Brain Death and Related Stress**
- Cause of death:
  - cerebrovascular accident
  - penetrating cerebral trauma and disseminated intravascular coagulation
- Severe donor hypotension/ shock syndrome
- Catecholamines (endogenous and exogenous)
- Nephrotoxic drugs/ donor acute renal dysfunction

**3. Organ Procurement Surgery**
- Hypotension
- Trauma to renal vessels
- Inadequate flushing and cooling
- Prolonged first warm ischemia time (nonheart-beating donors)

**4. Organ Transport and Storage**
- Type of flushing solution
- Prolonged storage (cold ischemia time)
- Pulsatile perfusion/ simple cold storage

**5. Transplantation of Recipient**
- Prolonged second warm ischemia time
- Trauma to renal vessels
- Hypovolemia/ hypotension
- Hemodialysis with ultrafiltration close to transplantation

**6. Post-Operative Period**
- Calcineurin inhibitor toxicity/ rapamycin immunosuppression
- Acute heart failure, myocardial infarction, impaired effective circulating volume

**7. Immune Factors**
- HLA mismatch
- Preformed anti-donor antibodies: - anti-ABO- anti- HLA
- Panel reactive antibodies
- Previous transfusions
- Previous transplants
- Acute rejection

Modified from Peeters P, Terryn W, Vanholder R et al. Delayed graft function in renal transplantation. Curr Opin Crit Care 2004; 10(6):492.

and immunological mechanisms. We will discuss factors that affect the ischemic injury, provide an overview of the pathophysiology and also propose mechanisms by which DGF leads to immunological damage to the kidney, both alloimmune and autoimmune.

## Donor Organ Quality

In recent years, the shortage of organs for transplantation has led to expansion of the donor pool. The United Network for Organ Sharing (UNOS) has defined expanded-criteria donors (ECD) as donors at 60 years of age or older or donors between 50 and 59 years and with at least two of the following characteristics: donor history of arterial hypertension, elevated serum creatinine levels (>1.5 mg/dl) or donor history of cerebrovascular accident.[4]

Donor age is an important risk factor for DGF and an independent predictor of CAN[5] and long-term graft survival. The effect of increased donor age even holds in rejection-free patients.[6] Risk of DGF is double in patients who receive a kidney from a donor older than 55 years.[7] With advancing age, kidneys generally develop progressive glomerulosclerosis, tubular atrophy and interstitial fibrosis, with consequent decrease in renal blood flow and glomerular filtration rate and deterioration of renal function. The frail kidney seems to be more susceptible to IRI, thus making long cold ischemia time more deleterious in transplantations from old donors.[8]

Despite lower graft survival, many centers have utilized older donors' kidneys with acceptable results. New strategies have been adopted recently to optimize the utilization of older kidneys, that, even if thought to have reduced survival, are less prone to rejection and the recipient could be more likely to die with a functioning allograft.[9] The "Eurotransplant Senior Program" (ESP), which has allocated cadaveric donors above the age of 65 years to kidney transplant recipients of the same age group, has obtained successful results. There was no significant difference in the outcome between young and old kidney transplants in old recipients after a 5 year-observation period.[10] One important approach is to select older donors' kidneys according to their histological characteristics before implantation.[11]

Donor diabetes mellitus, even lasting more than 10 years, isn't necessarily an overwhelming risk factor for graft and patient survival;[12] on the other hand, hypertension is a significant independent risk factor for graft survival, especially if it lasts for more than 10 years.[13] In both conditions, a biopsy could be an useful tool to define the quality of the organ before transplant.

Nonheart-beating donation is being increasingly used to expand the door pool. These kidneys are exposed to an obligate period of warm ischemia (the time between cardiopulmonary arrest and initiation of resuscitation), in which there is no perfusion in the kidney. Once transplanted, this results in an increased rate of primary nonfunction and delayed graft function; however long-term results of transplantation might not differ from other cadaveric kidneys.[14] Serum creatinines in recipients from nonheart-beating donors (NHBDs) are relatively higher in the initial period after transplant, but later on renal function achieves levels comparable to the kidneys from heart-beating donors.[15,16] Conflicting results about the risk of acute rejection are obtained from different studies.[17,18] The increased risk of DGF is probably due directly to ATN secondary to long warm ischemia time, even if other factors are involved, such as long duration of pretransplant recipient dialysis or increased body weight.[19]

## Brain Death and Related Stress

Donor brain death is an independent factor for graft failure[20] and is one of the potential causes for the inferior outcome after cadaveric versus living donor transplantation. It has been also associated with an increased risk of acute vascular rejection.[21] Different studies have been conducted to evaluate the effects of brain death on donor organ quality. Brain death seems to act by at least two different and complementary pathways. First, the increase in intracranial pressure associated with cerebrovascular accidents leads to activation of the parasympathetic system, which is soon opposed by release of catecholamines (autonomic storm). This results in initial hypertension, excessive vasoconstriction, relative hypoperfusion of organs and onset of ischemic damage. The parasympathetic activity soon prevails, with appearance of hemodynamic instability and hypotension, which aggravate the already present ischemic injury in kidney and other organs.[22] Thus, brain death per se causes IRI prior to organ retrieval. A high incidence of post-transplant ATN has been observed in kidneys obtained from brain-dead donors that had experienced unstable hemodynamic status.[23]

Another pathway of damage involves a progressive inflammatory and stress-related response in organs which results in histological lesions and decreased function (cytokine storm). These observations derive from several studies in animals and seem independent from hemodynamic instability. In fact as Van Der Hoeven et al[24] have demonstrated, a progressive inflammatory activation, exemplified by the expression of ICAM-1 and VCAM-1, and influx of leucocytes are observed in the kidneys of brain-dead rats, regardless of hemodynamic status. Similarly, in human kidney biopsy specimens from brain-dead donors, Nijboer et al[25] showed an increased expression of E-selectin and interstitial leukocyte invasion, as well as upregulation of protective proteins such as Hsp-70 and heme-oxygenase-1 (HO-1). After 1 and 3 years of transplantation, high ICAM-1 and VCAM-1 expression were associated with a deleterious effect on kidney function.[26]

## Organ Transport and Storage

### Cold Ischemia Time

During donor kidney retrieval, the organ is flushed with a cold preservation solution and kept in hypothermic storage at 0-4 °C. Cold ischemia time (CIT) is the time between initial cold perfusion and the start of the surgical procedure for blood vessel anastomosis to the recipient.

Several studies have shown that CIT is an important risk factor for DGF and long-term renal graft survival; it also impacts on renal survival independently from DGF.[3] A still unanswered question is what a reasonable short CIT is in kidney transplantation. CIT negatively impacts long-term graft failure if a 12-hour time increase in organ-preservation time is compared to the range 0-12 hours[27] or if a 10-hour increase is considered.[28] However, DGF is strongly influenced by CIT with an effect that seems continuous.[29] As Ojo et al[3] pointed out, CIT is strongly associated with DGF with an increase in risk for every 6 hours of CIT; nevertheless the risk of DGF increases by more than twofold only when CIT is greater than 24 hours and becomes even higher after 36 hours. Similarly, Boom et al[30] have found that CIT longer than 28 hours is independently associated with DGF. All efforts should be made to reduce CIT, at least under 24 hours, after which its effects are more pronounced.

### Preservation Solution

Kidneys have traditionally been flushed with Collins solution or its modifications. Collins solution is rich in potassium, hyperosmolar and presents an intracellular-like composition that seems to stabilize cell membrane. This solution has some disadvantages, among which is the presence of glucose to obtain hyperosmolar condition; in fact, even under hypothermia, glucose is broken down to lactate, thus increasing the intracellular substrate molecules and eventually leading to the cellular swelling Collins solution is supposed to prevent.[31] University of Wisconsin (UW) solution was introduced later on; it has been shown to be superior to Collins for liver, pancreas and kidney transplantation and has allowed extended period of preservation without compromising allograft function.[32-35] In UW solution, osmotic concentration is obtained by the use of lactobinat and raffinose, which are inert substrates. It also contains the colloid carrier hydroxyethylstarch (HES) and oxygen radical scavengers, like adenosine, glutathione and allopurinol.[31]

HKT solution is a very low viscosity solution, that contains histidine, tryptophan and ketoglutarate, thus creating a potent buffer system. It is comparable to UW solution in liver and pancreas.[36,37] Several kidney transplantation studies suggest similar results. In 1999 Eurotransplant conducted a multicenter randomized trial comparing HKT and UW; the two solutions showed the same results when DGF was used as endpoint.[38] Agarwal et al[39] have recently compared UW to HTK solution in kidney transplants with CIT superior to 16 and 24 hours, analyzing DGF, graft and patient survival. In CIT greater than 16 hours, graft and patient survival were comparable; HTK cohort had lower DGF. Considering only CIT superior to 24 hours, there were no difference in patient survival, a trend towards improved graft survival in HTK and decreased rate of DGF in HTK.

### Machine Perfusion

This technique has been introduced to allow longer preservation times. Kidneys are flushed and separated and then placed on a specifically designed perfusion machine that pumps cold colloid solution continuously through the renal artery until the transplantation. In 2003, Wight et al[40] conducted a meta-analysis about the effectiveness of machine perfusion and cold storage techniques in reducing DGF and improving graft survival in patients receiving kidneys from beating and nonheart-beating donors. Machine perfusion caused a reduced relative risk of DGF when compared with cold storage alone (20% protection). No statistically significant difference between the two techniques was observed on graft survival at 1 year, even if predictions based upon quantifying the link between DGF and graft survival suggested improvements of between 0 and 6% at 10 years. In another analysis, Schold et al[41] examined rates of pulsatile perfusion (PP) in deceased donor kidney transplantation in the United States (US) from 1994 to 2003 and evaluated long-term effects of PP. They realized that there was a significantly higher utilization rate with PP when ECD donors were considered. In all patients, DGF rates were significantly lower with PP (19.6% vs. 27.6%) and this technique was also associated with a modest benefit on death censored graft survival.

## Recipient Risk Factors and Postoperative Period

Persistent hypoperfusion to the kidney is one the most common causes of ischemic injury in the transplanted kidney. The reduced blood flow to the kidney enhances the already present damage, secondary to warm and cold ischemia time. The protective role of ample hydration derives from several clinical observations. Pretreatment of recipients with nocturnal hemodialysis is associated with higher rate of DGF than routine diurnal one;[42] patients in peritoneal dialysis experience less DGF compared to the ones in hemodialysis regimen before transplantation.[43]

The use of calcineurin inhibitors (CNIs) during the first week posttransplantation may result in abnormalities in early graft function. The recovery from ischemic injury can be delayed and, even in patients who show initial good renal function, these drugs can cause a sudden deterioration that has to be distinguished from early rejection or graft thrombosis. CNIs in fact induce dose-related renal vasoconstriction, particularly affecting the afferent arteriole and consequently a picture similar to prerenal dysfunction. Renal vasoconstriction seems to be related to alteration of the arachidonic acid metabolism in favour of the vasoconstrictor thromboxane, increase in endothelin and alterations of NO pathway.[44]

Rapamycin was initially considered an alternative drug to the calcineurin inhibitors in the early posttransplant period, but several evidences suggest that it may retard the improvement from DGF. Simon et al[45] have conducted a retrospective cohort study of US deceased donor kidney transplantation recipients in the United States Renal Data System (USRDS) from January 1, 2000 to May 31, 2001 and have analysed the relationship between induction therapy with rapamycin and DGF. Their results have showed that induction rapamycin was independently associated with DGF in US deceased donor kidney transplantation recipients, adjusted for all other factors previously shown to be associated with DGF. In another study, McTaggart et al[46] have compared transplant outcomes of rapamycin-based induction immunosuppression with other calcineurin-inhibitor sparing regimens (depleting antibodies or other therapies) in the DGF setting. In their analysis of a large cohort of adult cadaveric transplant recipients with DGF (n=132), induction immunosuppression with a depleting antibody preparation reduced rejection, whereas rapamycin prolonged DGF duration. The negative impact on recovery from DGF could be due to the anti-proliferative effects of this drug on human renal epithelial cells (HRECs). In studies in vitro, rapamycin has been shown to inhibit the proliferative response of HRECs to mitogenic stimuli and to cause cell cycle arrest in the early G(1) phase, not only by a nonspecific process due to inhibition of the p70(S6k) pathway, but also by a direct effect on cyclin D3 mRNA stability.[47]

## Pathophysiology

The pathogenesis of IRI is attributed to several inter-linked processes, triggered by an initial ischemic insult and worsened by reperfusion products. Our knowledge about these processes derives primarily from animal studies and has been in part confirmed by observations in human transplantations. Ischemic organs are characterized by deprivation of oxygen and nutrients along with accumulation of metabolic waste products. The absence of oxygen induces depletion of ATP, inactivation of Na/K ATPase and increase of anaerobic glycolysis. As a consequence, the ischemic organ accumulates lactic acid, a final product of anaerobic glycolysis.[48] The reduced intracellular pH induced by lactic acid leads to activation of lysosomal enzymes and increase of free metals, such as iron, that are inhibited in their linking to binding proteins and catalyze production of oxygen radicals. From this, a cascade of events is activated that leads to the formation of other radicals, such as nitric oxide (NO).

NO is a labile gaseous free radical, which plays an important role in ischemic conditions. For some time, NO has been considered protective against renal ischemia due to its vasodilatatory properties. Several more recent animals studies have suggested that increased inducible nitric oxide synthase (iNOS) derived NO production during renal ischemia is deleterious to the kidney and that inhibition of iNOS activity (by antisense deoxynucleotides targeting iNOS or α-melanocyte-stimulating hormone) can ameliorate IRI.[49,50] iNOS knockout mice are protected against ischemia-reperfusion injury.[51] Under ischemic conditions, NO is supposed to react with superoxide anion to form peroxynitrite, which induces renal injury via direct oxidant injury and protein tyrosine nitration.[52]

A destructive cascade of events leads to endothelial dysfunction, enhanced vasomotor tone, reduction in renal blood flow, hypoxic injury to renal tubule epithelial cells, tubular obstruction by casts and backleak of glomerular filtrate through an ischemic epithelium. Renal injury is further complicated by the fact that reperfusion, although essential for tissue survival, causes additional damage (reperfusion injury). Reperfusion leads to reoxygenation, return to aerobic metabolism and new production of ATP, but, on the other hand, reperfused tissue also has an increased production of free oxygen radicals (superoxide anion and hydrogen peroxide), that induce membrane lipo-peroxidation, with subsequent cell damage and death. Reactive oxygen species generation is also accompanied by production of cytokines and chemokines and by the expression of adhesion molecules on the endothelial cells of the damaged kidney. Leukocytes are attracted and enter the organ, leading to amplification of ongoing inflammation and cytotoxic injury. Chemokines are upregulated by ROS and inflammatory cytokines, such as IL-1 and TNF-α.[53] Transgenic mice that overproduce the antioxidants, intracellular and extracellular human glutathione peroxidases, have less induction of the chemokines KC and macrophage inflammatory protein-2 (MIP-2), less neutrophil infiltration and less functional injury after ischemia.[54] Exposure of leukocytes to cytokines can reduce their deformability and enhance the tendency for them to be sequestered.[55] Sequestered leukocytes can then enhance the injury generating further ROS and eicosanoids. Functional and morphologic protection of the kidney with agents designed to prevent leukocyte-endothelial adhesive interactions (monoclonal antibodies against CD11a and CD11b or ICAM-1) or in animals that do not produce specific adhesion molecules (ICAM-1 deficient mice) is associated with reduction in leukocyte sequestration.[56-58] Increased ICAM-1 staining in human liver biopsies at the end of cold storage has also been reported.[59]

An excess of thromboxane and a reduction of prostacyclin characterize the damaged endothelium, leading to vasoconstriction. In the ischemic kidneys both endothelin mRNA and protein are increased, mediating vasospasm and contributing to posttransplant renal damage.[60] In rats, endothelin-receptor antagonists reduce IRI;[61] patients with DGF present high serum levels of endothelin-1.[62]

Complement can also play a role in renal IRI. Complement can induce endothelial upregulation of adhesion molecules and resultant neutrophil accumulation in the microvasculature[63] and to damage epithelial cells as a direct effect of the membrane attack complex.[64]

Renal tubular cells are not only innocent bystanders of leukocyte-induced damage but can also have an active role in IRI. They can produce a number of proinflammatory cytokines, including TNF-α, IL-6, TGF-β and chemokines, such as RANTES, monocyte chemotactic protein-1 (MCP-1) and IL-8.[65,66] Human proximal tubule cells have been demonstrated to express CD40 that links to CD154 (CD40L) present on T-lymphocytes. CD40-CD40L interactions induce the translocation of the complex to the cytoplasm together with TNF-α receptor activating factor (TRAF-6). This factor is able to activate SAPK/JNK and p38 MAPK phosphorylation and, by them, to promote the production of IL-8 and MCP-1.[67] CD40 activation leads to upregulation of ICAM-1 expression with concomitant enhanced adhesion of mononuclear cells, which is mediated again via the p38 MAPK signal transduction pathway.[68]

Different studies have focused on the relative importance of the different types of leukocytes in IRI. The role of neutrophil accumulation has been considered of primary importance for many years, while, more recently, macrophages and T-lymphocytes have been demonstrated to have a more prominent role.[69-72] In rat ischemic renal allografts, cold ischemia resulted in increased expression of P-selectin, ICAM-1 and MHC class II molecules on renal endothelial and proximal tubular cells, followed by infiltration by CD4+ T-cells and macrophages.[73] Increased expression of MHC class II molecules on tubular epithelial cells was described in a model of unilateral ischemic ATN in the mouse.[74] Increased expression of MHC I has been also observed in other experimental models.[75]

In several studies, T-cells have been identified in the kidney following IRI.[76-79] So, in addition to activation of the classic innate immune response, components of the adaptive immune response, including T and B-cells, are involved in the pathogenesis of IRI. Originally, activation of T-cells was thought to be only dependent on antigen binding; however, recent evidence suggests that T-cells can also be activated by antigen-independent mechanisms involving RANTES and combinations of cytokines. T-lymphocytes are hypothesized to contribute to a "no-flow" state within the small renal blood vessels by engagement with adhesion molecules to macrophages, platelets, neutrophils and endothelium. Within the interstitium, T-cells interact with antigen-presenting cells or they can directly damage epithelial cells and regulate epithelial cell function. T-cells may also act from a distant site (thymus, bone marrow, lymph nodes, or spleen) by releasing cytokines and chemokines that modulate inflammatory mediators within the kidney.[78] To define the role of T-cell subsets in renal IRI, the effects of combined CD4 and CD8 deficiency in a murine model were examined;[80] these animals were relatively protected both functionally and histologically from injury. The athymic nu/nu mouse, a T-cell knockout animal, was significantly protected from renal IRI, while adoptive transfer of T-cells from wild-type mice restored the full expression of ATN. Adoptive transfer of wild-type CD4+ cells into these knockout mice also led to a worsening injury phenotype, defining the importance of CD4+ cells in renal IRI. Intriguingly, neutrophil and macrophage infiltration into post-ischemic kidney was not affected by either the nu/nu or CD4 knockout or the adoptive transfer. The CD4+ T-cell effect was found to require IFN-γ and the B7-CD28 pathway.[81] Alternative approaches using CTLA4Ig (that binds B7 with high affinity) to block the B7-CD28 interaction significantly attenuated renal dysfunction in a rat renal IRI model.[82] It appears that B7-1 and not the B7-2 pathway is the important T-cell costimulatory pathway in IRI.[83] A specific and time-course analysis of kidney lymphocytes after ischemia has been effectuated by immunohistochemistry and flow cytometry. In ischemic kidneys, there is an increased infiltration of CD3+ T-cells and CD19+ B-cells at 3 hours after ischemia and of natural killer (NK1.1+) and natural killer T-cells (CD4+ NK1.1+) cells at 3 and 24 h respectively.[84]

The role for T-cells in renal IRI is becoming increasingly complex, with the surprising finding that specific T-cells can serve a protective function, which could be dependent whether one examines early injury, extension or repair. It appears that the Th1 phenotype of T-cells is deleterious and the Th2 phenotype is protective. These data have been obtained studying STAT6 and STAT4 deficient mice which have impaired Th2 and Th1 responses respectively.[85] In addition, RAG1 mice, deficient in both T and B-cells, are not protected from renal IRI.[86] This may be in part due to enhanced innate immunity in these mice, with up-regulation of natural killer cells. Finally, isolation and transfer of kidney-infiltrating lymphocytes 24 h after renal IRI into T-cell-deficient mice reduce their functional and histological injury after renal IRI, suggesting once again that kidney-infiltrating lymphocytes could also have a protective function.[84] The newly demonstrated role of immune cells along with the increased expression of histocompatibility antigens observed in ischemic injury explains the tight connection between DGF and occurrence of rejection that is observed in the clinical setting.[87]

## Management of Delayed Graft Function

The presence of oliguria or kidney dysfunction in a newly transplanted patient must be investigated carefully and quickly. The causes can be several (Table 1) and involve both renal and extrarenal etiologies. The diagnostic approach requires a combination of clinical, radiological and sometimes histological techniques. Hemodynamic assessment is necessary along with a careful review of donor kidney procurement and preservation and transplant surgery procedure. Urine output is a crucial point, but it is influenced by the amount (or absence) of urine output before transplant from native kidneys. The algorithm in Figure 1 describes an approach to use for a patient with oliguria after transplantation. A fluid and/or diuretic challenge, according to the clinical examination, is performed; if this fails to improve function, other investigations are necessary. Doppler ultrasonography is commonly used and allows investigation of blood flow in the renal artery and vein; the absence of flow to the graft requires surgical exploration to evaluate for the presence of vascular obstruction or compression and to exclude hyperacute rejection. Ultrasound is also useful to reveal obstruction at the level of the ureter or bladder outlet. Though ultrasound might demonstrate also the presence of an urinary leak, scintigraphy with DTPA is more sensitive to reveal this problem. Ureter obstruction and urinary leak often involve the ureterovesical junction; surgical intervention may be necessary and the use of an ureteral stent reduces their incidence.

DGF can be attributed to an intrarenal cause once the extrarenal ones are excluded. Ischemic ATN is the most common cause of DGF and its treatment is usually conservative, with careful hemodynamic evaluation (fluid balance, blood pressure) and support and possible hemodialysis. Often biocompatible membranes are used in this setting, though there is limited evidence about their superiority.[88] ATN usually improves in several days in most patients, but can persist for weeks with still functional recovery of the graft. The other causes of intrarenal DGF are more threatening and have to be ruled out and distinguished from simple ATN. Periodical ultrasound examinations should be performed and a core biopsy is necessary at 7-10 days, if improvement has not occurred (3-5 days if the patient is at high risk for hyperacute or accelerated rejection).

Figure 1. Management of kidney allograft early dysfunction.

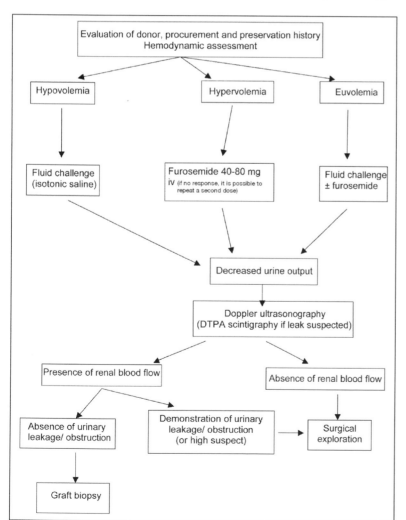

Ischemic AKI/ATN has immediate sequelae in the management of immunosoppressive therapy; many centers change their immunosuppression in the presence of ATN or if it is expected with high probability. Calcineurin inhibitors are frequently avoided or used in lower doses in these situations because they are known to delay recovery from ATN.[89] However, more recent studies show no difference in renal function at 3 and 12 months in renal transplant patients subjected to early or delayed introduction of cyclosporine microemulsion, regardless of DGF risk level.[90] Depleting antibodies (thymoglobulin) are frequently used and they are generally started prior to graft reperfusion in the surgical setting to reduce the risk for DGF and acute rejection.[91] In recent studies, nondepleting antibodies (basiliximab and daclizumab) appear to have some effect in reducing DGF in high-risk patients.[92,93] In a recent study, Brennan et al[94] have compared the use of induction therapy with a 5-day course of antithymocyte globulin to basiliximab, in recipients from deceased donors at high risk for acute rejection or delayed graft function. Depleting antibodies have been demonstrated to reduce the incidence and severity of acute rejection but not the incidence of delayed graft function. Patient and graft survival were similar in the two groups. As cited above, rapamycin is no longer recommended in the early posttransplant phase because of its negative impact on the recovery from DGF.[46,95]

## Prevention of Delayed Graft Function and New Perspectives

DGF remains a key problem in the early management of transplanted patients. More attention has to be paid to prevent its occurrence, rather than try to treat it.

First of all, an aggressive management of brain-dead donors is necessary; relative hypovolemia is often present in these patients, because of previous attempts to dehydrate them in order to reduce cerebral edema. Diabetes insipidus can also occur as consequence of head trauma. Hypovolemia has to be counteracted in order to maintain arterial blood pressure above 100 mmHg or central venous pressure above 10 mmHg, in order to ensure correct renal blood flow. If fluid management is not sufficient, inotropic agents (dopamine, dobutamine, norepinephine) can be used to sustain circulation. Analyzing a large database of solid organ transplants (the registry of the Eurotransplant International Foundation), the use of catecholamines in brain-dead donors was associated with increased 4-year graft survival in kidney transplants.[96]

Aggressive donor management with hormonal resuscitation (methylprednisolone, vasopressin, triiodothyronine or L-thyroxine) resulted in significant increases in solid organs transplanted per donor in a retrospective analysis of all brain-dead donors recovered in a period of 20 months between 2000 and 2001.[97] Brain-death induced kidney injury is not only due to hemodynamic instability, but also to the cytokine storm and inflammation. Therefore it is believed that the induction of cytoprotective and anti-oxidant agents can in part reduce IRI in the brain-dead donor kidney. HO-1 heat-shock protein has anti-oxidant, anti-inflammatory and anti-apoptotic functions. HO-1 induction exerts beneficial effects in several living-donor models of solid organ transplantion.[98,99] In an experimental model of rat kidney transplant with brain-dead donor, the administration of cobalt protoporphyrin (CoPP), a selective inducer of HO-1, led to enhanced graft survival, which was reversed by the blockade of HO-1 by zinc- PP.[100] Induction of HO-1 in our patients could be a future option to reduce IRI in kidneys. Dopamine has been shown to increase the presence of inducible HO-1. The effects are not limited to increased cytoprotection, but dopamine is also able to reduce renal monocyte infiltration and major histocompatibility class II and P-selectin expression. It also prevents further up-regulation of the inflammatory markers TNF-α and MCP-1 in brain-dead animals.[101]

Steroid therapy in brain-dead human donors can also reduce serum and tissue expression of proinflammatory cytokines to levels similar to the ones of living donors.[102]

Modification of preservation solutions have been conducted in order to reduce the deleterious effects of increased cold ischemia times. The addition of compounds, such as nitric oxide donors, inhibitors of reactive oxygen species and 21-aminosteroids, has been shown to provide some protection against IRI in animal models.[103-105] Cold storage induced injury of renal tubular cells is substantially ameliorated by adding selected flavonoids directly to UW and Collins preservation solutions in an in vitro model.[106] In human studies, a new preservation solution has been used only recently; it is called SCOT (the Solution de Conservation des Organes et des Tissus) and combines an extracellular-like composition with 20 kDa polyethylene-glycol (PEG), known for its cell-protection capacity and immunocamouflage properties in animal models. In autotransplanted pig kidneys, preserved with this PEG-solution, there was less tubular cell damage, attenuated expression of major MHC class II on tubule epithelial cells, significantly reduced number of CD4+ T-lymphocytes and infiltration of macrophages and monocytes; besides the progression of interstitial fibrosis in the 8- to 12-week post-transplanted kidneys was inferior.[107] In a preliminary study involving 29 transplants, SCOT has shown good results in reducing DGF, when compared to the classic preservation solution.[108]

Vasodilatory agents have been used during the transplant procedure or in the immediate postoperative period in experimental models and clinical studies. Atrial natriuretic peptide (ANP) seems to prevent ATN in human kidney transplants and to reduce the need for dialysis in nontransplant ATN.[109,110] Dopamine is of limited efficacy in the prevention of early renal dysfunction in the setting of liver transplant procedure, however, new promising results have been obtained with fenoldopam, a selective DA1 selective agonist.[111] In kidney transplantation, the positive role of dopamine remains uncertain; in a preliminary study, the use of fenoldopam has not shown statistically significant differences with the administration of low-dose dopamine.[112]

New drugs have been introduced to target the inflammatory and immune responses involved in the pathogenesis of IRI. CTLA4Ig blocks the costimulatory pathway in T-cell activation and has been approved for human treatment of rheumatoid arthritis. In renal transplantation, belatacept (an improved form of CTLA4Ig that binds with high affinity to the CD28 ligands) was as effective as cyclosporine in preventing acute rejection at 6 months when administered in combination with basiliximab, mycophenolate mofetil and corticosteroids; besides at 12 months GFR was higher in belatacept groups.[113] Inhibitors of the Janus protein tyrosine kinase (JAK)-3 show some selectivity for cells of the lymphoid lineage and have been effective in late preclinical transplant models. The most frequent adverse effects are due to nonspecific binding to JAK2 kinases. CP-690550, a JAK3 inhibitor is currently in phase II clinical trials.[114] Phase III trials of enlimomab, the anti-ICAM-1 molecule, have not demonstrated a reduction in acute rejection and DGF when compared to traditional regimens.[115] There are ongoing human studies targeting selectin ligands as well as integrins to decrease IRI and incidence of DGF.

## Conclusion

IRI remains the leading contributor to DGF in kidney transplantation. Early renal dysfunction is an important predictor of short- and long-term allograft survival. An improved understanding of IRI pathogenesis has been achieved in the last few years. These new findings have led to improved diagnostics and therapeutics to reduce the extent of the ischemic insult in the graft. This has already enabled a safer expansion of the donor pool in order to better cope with the increasing demand for organs for transplantation.

# References

1. Brady HR, Clarkson MR, Lieberthal W. Acute renal failure. In: Brenner BM, Rector FC, eds. The Kidney. 7th ed. Philadelphia: WB Saunders, 2004:1212-70.

2. Rodrigo E, Ruiz JC, Pinera C et al. Creatinine reduction rate on post-transplant day two as criterion in defining delayed graft function. Am J Transplant 2004; 4(7):1163-69.

3. Ojo AO, Wolfe RA, Held PJ et al. Delayed graft function: risk factors and implications for renal allograft survival. Transplantation 1997; 63(7):968-74.

4. Metzger RA, Delmonico FL, Feng S et al. Expanded-criteria donors for kidney transplantation. Am J Transplant 2003; 3 suppl 4:114-25.

5. Oppenheimer F, Aljama P, Asensio Peinado C et al. The impact of donor age on the results of renal transplantation. Nephrol Dial Transplant 2004; 19 suppl 3:iii11-15.

6. Moreso F, Seron D, Gil-Vernet S et al. Donor age and delayed graft function as predictors of renal allograft survival in rejection-free patients. Nephrol Dial Transplant 1999; 14(4):930-35.

7. Halloran PF, Hunsicker LG. Delayed graft function: state of the art, November 10-11, 2000. Summit meeting, Scottsdale, Arizona, USA. Am J Transplant 2001; 1(2):115-20.

8. Perico N, Ruggenenti P, Scalamogna M et al. Tackling the shortage of donor kidneys: how to use the best that we have. Am J Nephrol 2003; 23(4):245-59.

9. Tesi RJ, Elkhammas EA, Davies EA et al. Renal transplantation in older people. Lancet 1994; 343(8895): 461-64.

10. Bodingbauer M, Pakrah B, Steininger R et al. The advantage of allocating kidneys from old cadaveric donors to old recipients: a single-center experience. Clin Transplant 2006; 20(4):471-75.

11. Remuzzi G, Cravedi P, Perna A et al. Long-term outcome of renal transplantation from older donors. N Engl J Med 2006; 354(4):343-52.

12. Ojo AO, Hanson JA, Meier-Kriesche H et al. Survival in recipients of marginal cadaveric donor kidneys compared with other recipients and wait-listed transplant candidates. J Am Soc Nephrol 2001; 12(3):589-97.

13. Carter JT, Lee CM, Weinstein RJ et al. Evaluation of the older cadaveric kidney donor: the impact of donor hypertension and creatinine clearance on graft performance and survival. Transplantation 2000; 70(5):765-71.

14. Brook NR, Waller JR, Nicholson ML. Nonheart-beating kidney donation: current practice and future developments. Kidney Int 2003; 63(4):1516-29.

15. Castelao AM, Grino JM, Gonzalez C et al. Update of our experience in long-term renal function of kidneys transplanted from nonheart-beating cadaver donors. Transplant Proc 1993; 25(1):1513-15.

16. Wijnen RM, Booster MH, Stubenitsky BM et al. Outcome of transplantation of nonheart-beating donor kidneys. Lancet 1995; 345(8957):1067-70.

17. Gulanikar AC, MacDonald AS, Sungurtekin U et al. The incidence and impact of early rejection episodes on graft outcome in recipients of first cadaver kidney transplants. Transplantation 1992; 53(2):323-28.

18. Cho YW, Terasaki PI, Cecka JM et al. Transplantation of kidneys from donors whose hearts have stopped beating. N Engl J Med 1998; 338(4):221-25.

19. Yokoyama I, Uchida K, Hayashi S et al. Factors affecting graft function in cadaveric renal transplantation from nonheart-beating donors using a double balloon catheter. Transplant Proc 1996; 28(1):116-17.

20. Roodnat JI, van Riemsdijk IC, Mulder PG et al. The superior results of living-donor renal transplantation are not completely caused by selection or short cold ischemia time: a single-center, multivariate analysis. Transplantation 2003; 75(12):2014-18.

21. Sanchez-Fructuoso AI, Prats D, Marques M et al. Does donor brain death influence acute vascular rejection in the kidney transplant? Transplantation 2004; 78(1):142-46.

22. Pratschke J, Wilhelm MJ, Kusaka M et al. Brain death and its influence on donor organ quality and outcome after transplantation. Transplantation 1999; 67(3):343-48.

23. Lagiewska B, Pacholczyk M, Szostek M et al. Hemodynamic and metabolic disturbances observed in brain-dead organ donors. Transplant Proc 1996; 28(1):165-66.

24. Van Der Hoeven JA, Molema G, Ter Horst GJ et al. Relationship between duration of brain death and hemodynamic (in)stability on progressive dysfunction and increased immunologic activation of donor kidneys. Kidney Int 2003; 64(5):1874-82.

25. Nijboer WN, Schuurs TA, van Der Hoeven JA et al. Effect of brain death on gene expression and tissue activation in human donor kidneys. Transplantation 2004; 78(7):978-86

26. Nijboer WN, Schuurs TA, van Der Hoeven JA et al. Effects of brain death on stress and inflammatory response in the human donor kidney. Transplant Proc 2005; 37(1):367-69.

27. Held PJ, Kahan BD, Hunsicker LG et al. The impact of HLA mismatches on the survival of first cadaveric kidney transplants. N Engl J Med 1994; 331(12):765-70.

28. Salahudeen AK, Haider N, May W. Cold ischemia and the reduced long-term survival of cadaveric renal allografts. Kidney Int 2004; 65(2):713-18.

29. Quiroga I, McShane P, Koo DD et al. Major effects of delayed graft function and cold ischaemia time on renal allograft survival. Nephrol Dial Transplant 2006; 21(6):1689-96.

30. Boom H, Mallat MJK, De Fijter JW et al. Delayed graft function influences renal function, but not survival. Kidney Int 2000; 58(2):859-66.

31. Mühlbacher F, Langer F, Mittermayer C. Preservation solutions for transplantation. Transplant Proc 1999; 31(5):2069-70.

32. Jamieson NV, Sundberg R, Lindell S et al. Preservation of the canine liver for 24-48 hours using simple cold storage with UW solution. Transplantation 1988; 46(4):517-22.

33. Olthoff KM, Millis JM, Imagawa DK et al. Comparison of UW solution and Euro-Collins solutions for cold preservation of human liver grafts. Transplantation 1990; 49(2):284-90.

34. Belzer FO, D'Alessandro AM, Hoffmann RM et al. The use of UW solution in clinical transplantation. A 4-year experience. Ann Surg 1992; 215(6):579-83.

35. Ploeg RJ, van Bockel JH, Langendijk PT et al. Effect of preservation solution on results of cadaveric kidney transplantation. The European Multicentre Study Group. Lancet 1992; 340(8812):129-37.

36. Potdar S, Malek S, Eghtesad B et al. Initial experience using histidine-tryptophan-ketoglutarate solution in clinical pancreas transplantation. Clin Transplant 2004; 18(6):661-65.

37. Mangus RS, Tector AJ, Agarwal A et al. Comparison of histidine-tryptophan-ketoglutarate solution (HTK) and University of Wisconsin solution (UW) in adult liver transplantation. Liver Transpl 2006; 12(2):226-30.

38. de Boer J, De Meester J, Smits JM et al. Eurotransplant randomized multicenter kidney graft preservation study comparing HTK with UW and Euro-Collins. Transpl Int 1999; 12(6):447-53.

39. Agarwal A, Murdock P, Fridell JA. Comparison of histidine-tryptophan ketoglutarate solution and University of Wisconsin solution in prolonged cold preservation of kidney allografts. Transplantation 2006; 81(3):480-82.

40. Wight JP, Chilcott JB, Holmes MW et al. Pulsatile machine perfusion vs. cold storage of kidneys for transplantation: a rapid and systematic review. Clin Transplant 2003; 17(4):293-307.

41. Schold JD, Kaplan B, Howard RJ et al. Are we frozen in time? Analysis of the utilization and efficacy of pulsatile perfusion in renal transplantation. Am J Transplant 2005; 5(7):1681-88.

42. McCormick BB, Pierratos A, Fenton S et al. Review of clinical outcomes in nocturnal haemodialysis patients after renal transplantation. Nephrol Dial Transplant 2004; 19(3):714-19.

43. Snyder JJ, Kasiske BL, Gilbertson DT et al. A comparison of transplant outcomes in peritoneal and hemodialysis patients. Kidney Int 2002; 62(4):1423-30.

44. Olyaei AJ, de Mattos AM, Bennett WM. Nephrotoxicity of immunosuppressive drugs: new insight and preventive strategies. Curr Opin Crit Care 2001; 7(6):384-89.

45. Simon JF, Swanson SJ, Agodoa LY et al. Induction sirolimus and delayed graft function after deceased donor kidney transplantation in the United States. Am J Nephrol 2004; 24(4):393-401.

46. McTaggart RA, Tomlanovich S, Bostrom A et al. Comparison of outcomes after delayed graft function: sirolimus-based versus other calcineurin-inhibitor sparing induction immunosuppression regimens. Transplantation 2004; 78(3):475-80.

47. Pallet N, Thervet E, Le Corre D et al. Rapamycin inhibits human renal epithelial cell proliferation: effect on cyclin D3 mRNA expression and stability. Kidney Int 2005; 67(6):2422-33.

48. Perico N, Cattaneo D, Sayegh MH et al. Delayed graft function in kidney transplantation. Lancet 2004; 364(9447):1814-27.

49. Noiri E, Peresleni T, Miller T et al. In vivo targeting of inducible NO synthase with oligodeoxynucleotides protects rat kidney against ischemia. J Clin Invest 1996; 97(10):2377-83.

50. Chiao H, Kohda Y, McLeroy P et al. Alpha-melanocyte-stimulating hormone protects against renal injury after ischemia in mice and rats. J Clin Invest 1997; 99(6):1165-72.

51. Ling H, Edelstein C, Gengaro P et al. Attenuation of renal ischemia-reperfusion injury in inducible nitric oxide synthase knockout mice. Am J Physiol 1999; 277(3 Pt 2):F383-90.

52. Lieberthal W. Biology of ischemic and toxic renal tubular cell injury: role of nitric oxide and the inflammatory response. Curr Opin Nephrol Hypertens 1998; 7(3):289-95.

53. Bonventre JV, Weinberg JM. Recent advances in the pathophysiology of ischemic acute renal failure. J Am Soc Nephrol 2003; 14(8):2199–2210.

54. Ishibashi N, Weisbrot-Lefkowitz M, Reuhl K et al. Modulation of chemokine expression during ischemia/reperfusion in transgenic mice overproducing human glutathione peroxidases. J Immunol 1999; 163(10):5666–5677.

55. Suwa T, Hogg JC, Hards J et al. Interleukin-6 changes deformability of neutrophils and induces their sequestration in the lung. Am J Resp Crit Care Med 2001; 163(4):970–976.

56. Kelly KJ, Williams WW Jr, Colvin RB et al. Intercellular adhesion molecule-1-deficient mice are protected against ischemic renal injury. J Clin Invest 1996; 97(4):1056–1063.

57. Rabb H, Mendiola CC, Dietz J et al. Role of CD11a and CD11b in ischemic acute renal failure in rats. Am J Physiol 1994; 267(6 Pt 2): F1052-1058.

58. Rabb H, Mendiola CC, Saba SR et al. Antibodies to ICAM-1 protect kidneys in severe ischemic reperfusion injury. Biochem Biophys Res Commun 1995; 211(1):67-73.

59. Hoang V, El Wahsh M, Butler P et al. Donor liver preservation and stimulation of adhesion molecules: cold preservation initiates the cascade. Cryo Lett 2003; 24(6):359–364

60. Wilhelm SM, Simonson MS, Robinson AV et al. Endothelin up-regulation and localization following renal ischemia and reperfusion. Kidney Int 1999; 55(3):1011-1018.

61. Huang C, Huang C, Hestin D et al. The effect of endothelin antagonists on renal ischaemia-reperfusion injury and the development of acute renal failure in the rat. Nephrol Dial Transplant 2002; 17(9):1578-1585.

62. Schilling M, Holzinger F, Friess H et al. Pathogenesis of delayed kidney graft function: role of endothelin-1, thromboxane B2 and leukotriene B4. Transplant Proc 1996; 28(1):304-5.

63. Homeister JW, Lucchesi BR. Complement activation and inhibition in myocardial ischemia and reperfusion injury. Annu Rev Pharmacol Toxicol 1994; 34:17-40.

64. Zhou W, Farrar CA, Abe K et al. Predominant role for C5b-9 in renal ischemia/reperfsuion injury. J Clin Invest 2000; 105(10):1363-1371.

65. Segerer S, Nelson PJ, Schlondorff D. Chemokines, chemokine receptors and renal disease: from basic science to pathophysiologic and therapeutic studies. J Am Soc Nephrol 2000; 11(1):152-176.

66. Kapper S, Beck G, Riedel S et al. Modulation of chemokine production and expression of adhesion molecules in renal tubular epithelial and endothelial cells by catecholamines. Transplantation 2002; 74(2):253-260.

67. Li H, Nord EP. CD40 ligation stimulates MCP-1 and IL-8 production, TRAF6 recruitment and MAPK activation in proximal tubule cells. Am J Physiol 2002; 282(6):F1020-1033.

68. Li H, Nord EP. CD40/CD154 ligation induces mononuclear cell adhesion to human renal proximal tubule cells via increased ICAM-1 expression. Am J Physiol 2005; 289(1):F145-53

69. Paller MS. Effect of neutrophil depletion on ischemic renal injury in the rat. J Lab Clin Med 1989; 113(3):379-386.

70. Hellberg PO, Kallskog TO. Neutrophil-mediated post-ischemic tubular leakage in the rat kidney. Kidney Int 1989; 36(4):555-561.

71. De Greef KE, Ysebaert DK, Ghielli M et al. Neutrophils and acute ischemia-reperfusion injury. J Nephrol 1998; 11(3):110-122.

72. Burne-Taney MJ, Rabb H. The role of adhesion molecules and T-cells in ischemic renal injury. Curr Opin Nephrol Hypertens 2003; 12(1):85-90.

73. Kouwenhoven EA, de Bruin RW, Bajema IM et al. Cold ischemia augments allogeneic-mediated injury in rat kidney allografts. Kidney Int 2001; 59(3):1142-1148.

74. Shoskes DA, Parfrey NA, Halloran PF. Increased major histocompatibility complex antigen expression in unilateral ischemic acute tubular necrosis in the mouse. Transplantation 1990; 49(1):201-207

75. Suranyi MG, Bishop GA, Clayberger C et al. Lymphocyte adhesion molecules in T-cell-mediated lysis of human kidney cells. Kidney Int 1991; 39(2):312-319.

76. Azuma H, Nadeau K, Takada M et al. Cellular and molecular predictors of chronic renal dysfunction after initial ischemia/reperfusion injury of a single kidney. Transplantation 1997; 64(2):190-197.

77. De Broe ME. Apoptosis in acute renal failure. Nephrol Dial Transplant 2001; 16(Suppl 6):23-26.

78. Rabb H. The T-cell as a bridge between innate and adaptive immune systems: implications for the kidney. Kidney Int 2002; 61(6):1935-1946.

79. Takada M, Nadeau KC, Shaw GD et al. The cytokine-adhesion molecule cascade in ischemia/reperfusion injury of the rat kidney. Inhibition by a soluble P-selectin ligand. J Clin Invest 1997; 99(11):2682-2690.

80. Rabb H, Daniels F, O'Donnell M et al. Pathophysiological role of T-lymphocytes in renal ischemia-reperfusion injury in mice. Am J Physiol Renal Physiol 2000; 279(3):F525-F531.

81. Burne MJ, Daniels F, El Ghandour A et al. Identification of the CD4(+) T-cell as a major pathogenic factor in ischemic acute renal failure. J Clin Invest 2001; 108(9):1283-1290.

82. Takada M, Chandraker A, Nadeau KC et al. The role of the B7 costimulatory pathway in experimental cold ischemia/reperfusion injury. J clin invest 1997; 100(5):1199-1203.

83. De Greef KE, Ysebaert DK, Dauwe S et al. Anti-B7-1 blocks mononuclear cell adherence in vasa recta after ischemia. Kidney Int 2001; 60(4):1415-1427.

84. Ascon DB, Lopez-Briones S, Liu M et al. Phenotypic and functional characterization of kidney-infiltrating lymphocytes in renal ischemia reperfusion injury. J Immunol 2006; 117(5):3380-3387.

85. Yokota N, Burne-Taney M, Racusen L et al. Contrasting roles for STAT4 and STAT6 signal transduction pathways in murine renal ischemia-reperfusion injury. Am J Physiol Renal Physiol 2003; 285(2): F319-F325.

86. Burne-Taney MJ, Yokota-Ikeda N, Rabb H. Effects of combined T- and B-cell deficiency on murine ischemia reperfusion injury. Am J Transplant 2005; 5(6):1186-1193.

87. Howard RJ, Pfaff WW, Brunson ME et al. Delayed graft function is associated with an increased incidence of occult rejection and results in poorer graft survival. Transplant Proc 1993; 25(1 Pt 2):884.

88. Woo YM, Craig AM, King BB et al. Biocompatible membranes do not promote graft recovery following cadaveric renal transplantation. Clin Nephrol 2002; 57(1):38-44.

89. Novick AC, Hwei HH, Steinmuller D et al. Detrimental effect of cyclosporine on initial function of cadaver renal allografts following extended preservation. Results of a randomized prospective study. Transplantation 1986; 42(2):154-158.

90. Kamar N, Garrigue V, Karras A et al. Impact of early or delayed cyclosporine on delayed graft function in renal transplant recipients: a randomized, multicenter study. Am J Transplant 2006; 6(5 Pt 1):1042-1048.

91. Goggins WC, Pascual MA, Powelson JA et al. A prospective, randomized, clinical trial of intraoperative versus postoperative Thymoglobulin in adult cadaveric renal transplant recipients. Transplantation 2003; 76(5):798-802.

92. Fernandez Rivera C, Alonso Hernandez A, Villaverde Verdejo P et al. Basiliximab (Simulect) in renal transplantation with high risk for delayed graft function. Transplant Proc 2005; 37(3):1435-1437.

93. Sandrini S. Use of IL-2 receptor antagonists to reduce delayed graft function following renal transplantation: a review. Clin Transplant 2005; 19(6):705-710.

94. Brennan DC, Daller JA, Lake KD et al. Rabbit antithymocyte globulin versus basiliximab in renal transplantation. N Engl J Med 2006; 355(19):1967-1977.

95. Boratynska M, Banasik M, Patrzalek D et al. Sirolimus delays recovery from posttransplant renal failure in kidney graft recipients. Transplant Proc 2005; 37(2):839-842.

96. Schnuelle P, Berger S, de Boer J et al. Effects of catecholamine application to brain-dead donors on graft survival in solid organ transplantation. Transplantation 2001; 72(3):455-463.

97. Rosendale JD, Kauffman HM, McBride MA et al. Aggressive pharmacologic donor management results in more transplanted organs. Transplantation 2003; 75(4):482-487.

98. Amersi F, Buelow R, Kato H et al. Up-regulation of heme-oxigenase-1 protects genetically fat Zucker rat livers from ischemia/reperfusion injury. J Clin Invest 1999; 104(11):1631-1639.

99. Soares MP, Lin Y, Anrather J et al. Expression of heme oxygenase-1 can determine cardiac xenograft survival. Nat Med 1998; 4(9):1392-1398.

100. Kotsch K, Francuski M, Pascher A et al. Improved long-term graft survival after HO-1 induction in brain-dead donors. Am J Transplant 2006; 6(3):477-486.

101. Schaub M, Ploetz CJ, Gerbaulet D et al. Effect of dopamine on inflammatory status in kidneys of brain-dead rats. Transplantation 2004; 77(9):1333-1340.

102. Kuecuek O, Mantouvalou L, Klemz R et al. Significant reduction of proinflammatory cytokines by treatment of the brain-dead donor. Transplant Proc 2005; 37(1):387-388.

103. Shoskes DA, Xie Y, Gonzalez-Cadavid NF. Nitric oxide synthase activity in renal ischemia-reperfusion injury in the rat: implications for renal transplantation. Transplantation 1997; 63(4):495-500.

104. Shoskes DA. Effect of bioflavonoids quercetin and curcumin on ischemic renal injury: a new class of renoprotective agents. Transplantation 1998; 66(2):147-152.

105. Shoskes DA, Jones E, Garras N et al. Effect of the lazaroid U-74389G on chemokine gene expression and apoptosis in renal ischemia-reperfusion injury. Transplant Proc 1998; 30(4):974-975.

106. Ahlenstiel T, Burkhardt G, Kohler H et al. Improved cold preservation of kidney tubular cells by means of adding bioflavonoids to organ preservation solutions. Transplantation 2006; 81(2):231-39.

107. Hauet T, Goujon JM, Baumert H et al. Polyethylene glycol reduces the inflammatory injury due to cold ischemia/reperfusion in autotransplanted pig kidneys. Kidney Int 2002; 62(2):654-67.

108. Billault C, Vaessen C, Van Glabeke E et al. Use of the SCOT solution in kidney transplantation: preliminary report. Transplant Proc 2006; 38(7):2281-82.

109. Gianello P, Carlier M, Jamart J et al. Effect of 1-28 alpha-h atrial natriuretic peptide on acute renal failure in cadaveric renal transplantation. Clin Transplant 1995; 9(6):481-89.

110. Sward K, Valsson F, Odencrants P et al. Recombinant human atrial natriuretic peptide in ischemic acute renal failure: a randomized placebo-controlled trial. Crit Care Med 2004; 32(6):1310-15.

111. Biancofiore G, Della Rocca G, Bindi L et al. Use of fenoldopam to control renal dysfunction early after liver transplantation. Liver Transpl 2004; 10(8):986-92.

112. Fontana I, Germi MR, Beatini M et al. Dopamine "renal dose" versus fenoldopam mesylate to prevent ischemia-reperfusion injury in renal transplantation. Transplant Proc 2005; 37(6):2474-75.

113. Vincenti F, Larsen C, Durrbach A et al. Costimulation blockade with belatacept in renal transplantation. N Engl J Med 2005; 353(8):770-81.

114. Tedesco Silva H, Pinheiro Machado P, Rosso Felipe C et al. Immunotherapy for De Novo Renal Transplantation: What's in the Pipeline? Drugs 2006; 66(13):1665-84.

115. Salmela K, Wramner L, Ekberg H et al. A randomized multicenter trial of the anti-ICAM-1 monoclonal antibody (enlimomab) for the prevention of acute rejection and delayed onset of graft function in cadaveric renal transplantation: a report of the European Anti-ICAM-1 Renal Transplant Study Group. Transplantation 1999; 67(5):729-36.

# CHAPTER 7

# Heat Shock Protein 47 in Chronic Allograft Nephropathy

Takashi Taguchi and Mohammed Shawkat Razzaque*

## Abstract

Chronic allograft nephropathy (CAN), associated with late allograft dysfunction is caused by alloantigen-dependent and -independent mechanisms that eventually progresses to irreversible interstitial fibrosis. Heat shock protein 47 (HSP47) is a collagen-specific molecular chaperone and is closely associated with progression of fibroproliferative lesions in various organs, by facilitating increased production of collagens; a correlation between the increased expression of HSP47 and extent of interstitial fibrosis is found in the biopsy tissues obtained from patients with CAN. The activation of complement cascade is a poor prognostic indicator of graft survival; a correlation between the expression of C4d (a marker of complement activation) and HSP47 has also been noted in renal biopsy tissues of patients with CAN. A detailed and comprehensive review of all aspects of interstitial fibrosis in patients with CAN is beyond the scope of this chapter; rather we will briefly summarize the lessons learned from recent studies on roles of HSP47 during the progression of fibrotic diseases and its possible implication in CAN.

## Chronic Allograft Nephropathy (CAN)

Our understanding of immunological mechanisms and availability of new effective immunosuppressants has significantly diminished the incidence of renal allograft dysfunction during the early postrenal transplantation period. However, no such dramatic improvement has been made with regard to the incidence of late allograft dysfunction. CAN is usually associated with progressive decline in glomerular filtration rate, in conjunction with proteinuria and arterial hypertension and is believed to be the single most common cause of long-term allograft failure. In this chapter, the term CAN is mostly used to describe the non-specific morphologic changes in chronically dysfunctioning allografts. In the majority of the patients with histological evidence of CAN, a linear relationship usually exists between the time of diagnosis of CAN and the deterioration of renal allograft function.[1-3] The pathological changes in CAN, which are classified according to the Banff classification,[4] could be alloantigen-dependent and/or alloantigen-independent phenomena and usually progress to irreversible interstitial fibrosis. A better understanding of cellular and molecular events of CAN will provide the specific therapeutic targets and/or options to intervene the progression of interstitial fibrosis.

Peritubular capillary (PTC) deposition of C4d reflects complement activation via the classical pathway[5] and represents a trace of the remaining alloantibodies;[6] presence of C4d is considered to be a reliable marker of antibody-mediated allograft rejection. C4d is a suitable marker of acute humoral rejection;[7,8] studies have also documented C4d in renal allograft biopsies with morphologic features of chronic rejection.[9] Presence of C4d in a substantial number cases reemphasize the fact that immunologically mediated chronic renal allograft rejection is indeed the major cause of CAN. The capillary C4d staining has also found to be associated with poor graft survival.[10-12] Interestingly, a correlation between the renal expression of C4d and HSP47 (a molecule that helps in intracellular maturation of collagens) has also been documented in CAN.[13]

## Heat Shock Protein 47 (HSP47)

Heat shock proteins (HSPs) provide cellular defense against a wide range of injuries; a number of HSPs are constitutively expressed and actively involved in maintaining cellular homeostasis, by acting as molecular chaperones.[14-16] HSPs in mammalian cells are transcriptionally regulated by the heat shock transcription factor (HSF), which can selectively bind to the heat shock promoter element (HSE).[17,18] In normal, unstressed cells, HSF is present in the cytoplasm, but under stressful microenvironment, HSF converts from an inactive monomeric form to an active trimeric DNA-binding form, which then translocates to the nucleus and interacts with HSE to induce transcription of HSP genes[19,20] (Fig. 1).

Recently, HSP47, a collagen-binding molecular chaperone, has found to be involved in the molecular maturation of procollagen molecules. Collagen-binding molecular chaperone was first characterized from murine parietal endoderm cells and was termed as colligin.[21] Subsequent studies from different laboratories have identified species-specific collagen-binding proteins in human and rat as gp46,[22] in the mouse as J6[23] and in the chick and rabbit as HSP47;[24,25] all these proteins were later found to be the same group of molecules with a common collagen binding abilities and now generally refers as HSP47.

Collagen biosynthesis is a complex multi-step process that has both intracellular and extracellular events; the synthesis of the alpha polypeptide chains, their hydroxylation and formation of stable triple-helical procollagen molecules are intracellular events of collagen synthesis. HSP47, a 47-kDa glycoprotein protein, resides in the endoplasmic reticulum (ER) of collagen-producing cells and help in the assembly and correct folding of triple helical procollagen molecules,[26] which is eventually transported to the extracelluar space across the Golgi complex, where N and C propeptides are cleaved by procollagen N- and C- proteinases to assemble into collagen fibrils.[27,28] An in vivo essential role of HSP47 in collagen synthesis and subsequent organogenesis has

*Corresponding Author: M. Shawkat Razzaque—Department of Developmental Biology, Harvard School of Dental Medicine, Research and Educational Building, Room: 304, 190 Longwood Avenue, Boston, MA 02115, USA. Email: mrazzaque@hms.harvard.edu

*Chronic Allograft Failure: Natural History, Pathogenesis, Diagnosis and Management*, edited by Nasimul Ahsan. ©2008 Landes Bioscience.

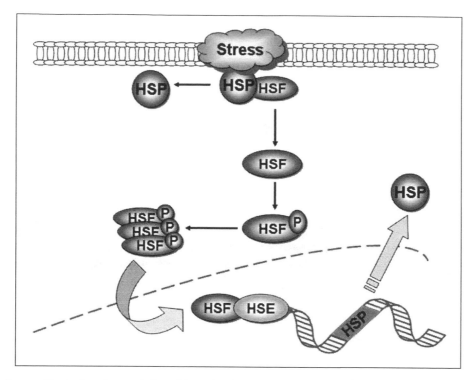

Figure 1. Simplified diagram illustrating the regulation of heat shock proteins (HSPs). Heat shock transcription factors (HSF) are normally bound to HSPs as inactive molecules in the cytosol. Upon exposure to stressors, HSFs are phosphorylated (P) by protein kinases, rapidly form trimmers and translocate to the nucleus where HSFs interact with heat shock promoter element (HSE) to induce the transcription of HSPs, which are then transcribed and relocate to the cytosol. We have only included the essential steps of regulation of HSP to keep the diagram simple.

been demonstrated in *hsp47* knockout mice; ablation of *hsp47* from mouse genome has resulted in abnormal collagen formation and impaired organogenesis. Complete ablation of *hsp47* from the mouse was embryonically lethal and knockout mice died prenatally at embryonic 11.5 day.[29] These studies suggest that, in absence of HSP47, there is abnormal molecular maturation of its substrate protein, collagen. In fibrotic diseases, it is expected that by targeting HSP47, it might be possible to generate unstable collagen which is more likely to degrade and thereby reducing accumulation of collagen in the fibrotic mass.

## HSP47 and Fibrotic Diseases

One of the common features of pathologic tissue scarring is excessive accumulation of matrix proteins due to uncontrolled synthesis and/or degradation.[30-32] Most of the fibrotic diseases are progressive in nature and gradual expansion of fibrotic mass eventually leads to the destruction of normal structure of the affected tissues; the size and extent of fibrotic mass usually influence the functionality of the affected organs and uncontrolled fibrosis progresses to end-stage organ failure. A close association between increased expression of HSP47 and deposition of collagens has been documented in fibrotic diseases affecting various human and experimental animals.[33-40] For instance, antithymocyte serum (ATS)-induced nephritis, is a widely used experimental model of mesangial cell proliferation and glomerulosclerosis;[33] an increased glomerular expression of HSP47 has found to be closely associated with excessive accumulation of collagens in the scleroproliferative glomeruli;[33] phenotypically altered collagen producing glomerular myofibroblasts (α-smooth muscle actin positive) and glomerular epithelial cells (desmin-positive) are the main HSP47-producing cells in the sclerotic glomeruli.[33,34,41] Blocking, in vivo, the bioactivities of HSP47 by treating the nephritic animals with antisense oligodeoxynucleotides against HSP47 could delay the progression of glomerulosclerosis.[42] These studies provide an experimental basis of why we believe that increased expression of HSP47 in fibrotic diseases is pathogenic rather than an epiphenomenon. Since HSP47 is intimately involved in the

molecular maturation of procollagens, it is likely that high levels of glomerular HSP47 might help in increased production of collagens, and thus contribute to the glomerular sclerotic process. A similar induction in the expression of HSP47 and excessive accumulation of collagens in the glomeruli was also noted in other experimental models of glomerulosclerosis, including in hypertensive nephrosclerosis (Fig. 2) and diabetic nephropathy.[39,43]

The exact molecular mechanism of renal tubulointerstitial fibrosis is not yet clear and is believed to be characterized by interstitial accumulation of collagens, produced by phenotypically altered interstitial cells and tubular epithelial cells. Expression of α-smooth muscle actin (α-SMA) in renal interstitial cells is indicative of acquiring myofibroblastic phenotype, while expression of intermediate filament vimentin in tubular epithelial cells is suggestive of phenotypical alteration of renal tubular epithelial cells. Studies have demonstrated that increased synthesis of interstitial collagens by phenotypically altered interstitial myofibroblasts and tubular epithelial cells play an important role in the initiation and progression of tubulointerstitial fibrotic process in various experimental models of tubulointerstitial fibrosis, including in cisplatin nephropathy, aged-associated nephropathy and hypertensive nephrosclerosis.[35,41,44-47] In these above-mentioned experimental models, elevated expression of HSP47 was often associated with excessive accumulation of collagens in areas around the interstitial fibrosis.

A very few studies have examined the role of HSP47 in human fibrotic diseases. The first such human study was conducted using renal biopsy tissues of IgA nephropathy and diabetic nephropathy.[48] In adult human kidneys, expression of HSP47 was very weak. In contrast, enhanced expression of HSP47 was detected in the early sclerotic glomeruli of IgA nephropathy and diabetic nephropathy. HSP47 was also expressed in tubulointerstitial cells in areas around interstitial fibrosis in both IgA nephropathy and diabetic nephropathy.[48] The glomerular and tubulointerstitial expression of HSP47 in renal biopsy tissues was closely associated with glomerular accumulation of type IV collagen and interstitial accumulation of types I and III collagens, respectively.

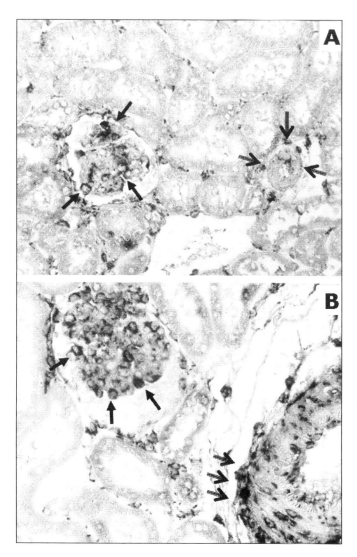

Figure 2. Immunohistochemical expression of HSP47 in normotensive (A) and hypertensive (B) Dahl rat kidneys. Compared to normotensive rats, there is increased expression of HSP47 in the glomerulus of hypertensive rats (arrows). Note increased expression of HSP47 also in the thickened blood vessels in the hypertensive kidney (arrows). It is presumed that increased expression of HSP47 in the hypertensive kidney is associated with eventual nephrovascular sclerosis, by facilitating increased production of collagens.

One of the unique features of HSP47 is that its overexpression is consistently observed in all the studied fibrotic diseases involving the lung, liver, heart, eye and skin.[24,36-40,49-60] It appears likely that irrespective of primary diseases, upregulation of HSP47 is a common phenomenon during collagenization of the involved tissues/organs. It is therefore reasonable to speculate that monitoring the expression of HSP47 might help in defining those patients at risk for developing fibrotic complications and in assessing the response to the conventional and selective therapies in various fibrotic diseases, including in patients with CAN.

## Role of C4d and HSP47 in CAN

The possible roles of HSP47 in developing the tubulointerstitial fibrotic lesions in patients with CAN need comprehensive studies. Preliminary studies, however, suggest a potential role of HSP47 in CAN; an association between the expression of C4d and HSP47 in the renal biopsy tissues, obtained from the patients with CAN has been reported. Recently, in a study that comprised 48 renal allografts (30 male, 18 women; 45 living related donors and 3 cadaveric origin), were

retrospectively analyzed for the renal expression of C4d and HSP47.[13] C4d was very weakly detected in the PTC cells of the control kidney, in contrast, in 16 out of 48 (33.3%) patients, C4d deposition was detected along endothelial cells of PTC; C4d expression significantly correlated with the presence of proteinuria (p = 0.002) in these group of patients. Moreover, the level of serum creatinine in C4d-positive patients was significantly higher than in C4d-negative patients (p = 0.004). More importantly, as expected, C4d staining correlated significantly with subsequent graft loss (p = 0.004). When the expression of C4d was related to Banff classification,[4] the C4d-positive cases were found to have advanced histological features of interstitial fibrosis and tubular atrophy than C4d-negative cases.[13] In the same group of patients, the correlation analysis showed that the median proportion of cell area positive for HSP47 in C4d-positive group of patients (average 2.70%) was significantly higher (p = 0.038) than in C4d-negative group (average 1.64%). Moreover, an association between the enhanced expressions of C4d and HSP47 has been detected in renal biopsy sections of patients with CAN; the interstitial expression of HSP47 was mostly located in and around C4d-positive PTCs in the interstitium. These observations, though preliminary, suggest a possible role of HSP47 in the pathogenesis of CAN. In addition, compared to control, increased interstitial expression of TGF-β1, a well-know fibrogenic factor was also detected in the renal biopsy sections positive for C4d stained patients with CAN.[13] This study has clearly demonstrated an association between the expression of C4d and HSP47 in patients with CAN. Further studies with larger population of patients with various stages of disease process will be needed to define the role of HSP47 in CAN; at this stage HSP47 appears to have a fibrogenic role in patients with CAN.

C4d is a complement split product generated through complement degradation, activated by antigen-antibody complexes; it is considered as an indicator of humoral activity in allografts and is a useful marker of humoral immunoreaction.[61,62] In recent reports, capillary deposition of C4d was identified in 30-34% of allograft biopsies[6] and chronic rejection seems to be distinguished from alloantigen-independent CAN by C4d staining. Furthermore, Mróz et al[62] reported that specific histological changes of chronic rejection, such as chronic transplant glomerulopathy and/or arterial intimal thickening with lymphocyte infiltration, were present in about 83% of C4d-positive cases and that such deposits were noted in PTC, again suggesting that staining of C4d in the biopsy tissue is a specific marker for the rejection process.[62] In recent reports, capillary deposition of C4d was recommended as a marker of CAN caused by immunological reaction (chronic rejection), although the exact role of C4d remains unclear.[6,7,62]

HSP47, a collagen-binding protein that binds with newly synthesized procollagen and its overexpression correlates with tubulointerstitial fibrosis, a morphological feature of CAN that result in renal dysfunction. Recent studies have shown that HSP47 is produced by activated fibroblasts and myofibroblasts, the main cell types that are also responsible for increased synthesis of collagens during fibrosis in various organs.[38,39,47,50] The transformation of fibroblasts to myofibroblasts and infiltration of macrophages are common features of CAN; these infiltrating and transformed cells are source of fibrogenic factors including TGF-β1 and platelet-derived growth factor (PDGF) to regulate interstitial fibrosis in CAN.[63,64]

TGF-β1 mediates the attainment of myofibroblast features including α-SMA expression by cultured skin fibroblasts[65] and PDGF-induced tubulointerstitial myofibroblast formation in experimental animals.[66] In a recent study, an increased numbers of CD68-positive macrophages, TGF-β1-positive cells and α-SMA-positive interstitial cells were detected in renal biopsy sections of patients with CAN that was positive for C4d staining in PTC; the results of this study and those of previous studies suggest that the immunopathological changes of CAN is mostly association with C4d deposition in PTC; moreover, infiltration of macrophages and production of TGF-β1 and PDGF might help in transforming myofibroblast in the tubulointerstitium of patients with CAN. The activated interstitial cells, including myofibroblasts could produce an increased level of HSP47 and TGF-β1; both these fibrogenic molecules could induce excessive synthesis and interstitial accumulation of collagen in CAN, by regulating transcriptional and posttranslational processing of collagens. Furthermore, increased expression of TGF-β1 might in turn could induce the expression of HSP47 in CAN; recent in vitro studies have shown such induction of HSP47 by TGF-β1.[67-69] Since inhibition of HSP47 expression in fibrotic models resulted in less fibrogenic changes in the affected organs,[70,71] similar approach might help in preventing or delaying the progression of tubulointerstitial fibrosis in patients with CAN.

## Clinical Potential of HSP47 as an Antifibrotic Target

Despite a number of important fibrogenic molecules that have been identified in recent years, most of these molecules are not for suitable therapeutic targets because of their widespread vital systemic effects. In addition, regardless of the in vitro efficacy, some of the circulating fibrogenic molecules were not effective in the complex in vivo microenvironment. Since HSP47 is involved in the molecular maturation of collagens, a selective blockade of its activities in fibrotic diseases might be clinically useful. In vivo suppression of HSP47 expression by the administration of antisense oligodeoxynucleotides against HSP47 delayed the progression of glomerular sclerotic process by reducing glomerular accumulation of collagens in rats with experimental nephritis.[42] Similar antifibrotic response using antisense HSP47 therapy has shown to be effective in reducing wound-related scarring.[70,71] These preliminary observations suggest a profibrotic role of HSP47 and more importantly, provide the in vivo evidence of HSP47 as a potential antifibrotic therapeutic target;[42,70] a phenomenon that has enormous clinical importance and applicability in a wide range of fibrotic diseases, including in CAN.

## Conclusion

How feasible HSP therapy is for clinical use? It is particularly useful in various neurodegenerative disorders, where aberrant protein aggregation and neuron degeneration are the common pathologic features; induction of HSPs, particularly HSP70 by gene transfer can reduce the aberrant protein misfolding and inhibit the apoptotic deletion of cells to attenuate dopaminergic neuron degeneration in Parkinson's disease.[72] Recent understanding of structural basis of the HSP47-collagen interaction can also form the conceptual templates for possible drug designing by taking the advantage of pharmacophore-based strategies.[73-75] In contrast to most, if not all of the, molecular chaperones that recognize several target proteins, HSP47 has a single substrate protein, collagen. HSP47, therefore, provides a very selective target to manipulate collagen production, a phenomenon that might have enormous clinical application in controlling fibrotic diseases, including CAN. Preliminary observations suggest a strong rationale for blocking the bioactivities of collagen-binding HSP47, as one of the options to control the progression of fibrotic diseases;[76] further controlled in vivo studies to determine both favorable and adverse effects of blocking the bioactivities of HSP47.

### *Acknowledgements*

We are grateful to Miss Kanako Egashira for her assistance in preparing of histological sections. Part of this chapter is based on a recently published review article, entitled, "The collagen-specific molecular chaperone HSP47: is there a role in fibrosis?" in the journal "Trends in Molecular Medicine". Our apology goes to the authors whose original work might be inadvertently oversighted from the reference list.

## References

1. Shin GT et al. Effect of nifedipine on renal allograft function and survival beyond one year. Clin Nephrol 1997; 47:33-36.
2. Cheigh JS et al. Hypertension in kidney transplant recipients. Effect on long-term renal allograft survival. Am J Hypertens 1989; 2:341-48.
3. Cheigh JS et al. Kidney transplant nephrotic syndrome: relationship between allograft histopathology and natural course. Kidney Int 1980; 18:358-65.
4. Racusen LC et al. The Banff 97 working classification of renal allograft pathology. Kidney Int 1999; 55:713-23.
5. Zwirner J, Felber E, Herzog V et al. Classical pathway of complement activation in normal and diseased human glomeruli. Kidney Int 1989; 36:1069-77.
6. Regele H et al. Capillary deposition of complement split product C4d in renal allografts is associated with basement membrane injury in peritubular and glomerular capillaries: a contribution of humoral immunity to chronic allograft rejection. J Am Soc Nephrol 2002; 13:2371-80.
7. Bohmig GA et al. Capillary C4d deposition in kidney allografts: a specific marker of alloantibody-dependent graft injury. J Am Soc Nephrol 2002; 13:1091-99.
8. Collins AB et al. Complement activation in acute humoral renal allograft rejection: diagnostic significance of C4d deposits in peritubular capillaries. J Am Soc Nephrol 1999; 10:2208-14.
9. Mauiyyedi S et al. Chronic humoral rejection: identification of antibody-mediated chronic renal allograft rejection by C4d deposits in peritubular capillaries. J Am Soc Nephrol 2001; 12:574-82.
10. Feucht HE. Complement C4d in graft capillaries—the missing link in the recognition of humoral alloreactivity. Am J Transplant 2003; 3:646-52.
11. Feucht HE et al. Capillary deposition of C4d complement fragment and early renal graft loss. Kidney Int 1993; 43:1333-38.
12. Feucht HE, Mihatsch MJ. Diagnostic value of C4d in renal biopsies. Curr Opin Nephrol Hypertens 2005; 14:592-98.
13. Ohba K et al. Interstitial expression of heat-shock protein 47 correlates with capillary deposition of complement split product C4d in chronic allograft nephropathy. Clin Transplant 2005; 19:810-16.
14. Becker J, Craig EA. Heat-shock proteins as molecular chaperones. Eur J Biochem 1994; 219:11-23.
15. Gething MJ, Sambrook J. Protein folding in the cell. Nature 1992; 355:33-45.
16. Hendrick JP, Hartl FU. Molecular chaperone functions of heat-shock proteins. Annu Rev Biochem 1993; 62:349-84.
17. Ohtsuka K, Hata M. Molecular chaperone function of mammalian Hsp70 and Hsp40--a review. Int J Hyperthermia 2000; 16:231-45.
18. Perisic O, Xiao H, Lis JT. Stable binding of Drosophila heat shock factor to head-to-head and tail-to-tail repeats of a conserved 5 bp recognition unit. Cell 1989; 59:797-806.
19. Bonner JJ, Ballou C, Fackenthal DL. Interactions between DNA-bound trimers of the yeast heat shock factor. Mol Cell Biol 1994; 14:501-8.
20. Kroeger PE, Morimoto RI. Selection of new HSF1 and HSF2 DNA-binding sites reveals difference in trimer cooperativity. Mol Cell Biol 1994; 14:7592-603.
21. Kurkinen M, Taylor A, Garrels JI et al. Cell surface-associated proteins which bind native type IV collagen or gelatin. J Biol Chem 1984; 259:5915-22.
22. Clarke EP, Sanwal BD. Cloning of a human collagen-binding protein and its homology with rat gp,46 chick hsp47 and mouse J6 proteins. Biochim Biophys Acta 1992; 1129:246-48.
23. Takechi H et al. Molecular cloning of a mouse 47-kDa heat-shock protein (HSP47), a collagen-binding stress protein and its expression during the differentiation of F9 teratocarcinoma cells. Eur J Biochem 1992; 206:323-29.
24. Hart DA, Reno C, Hellio Le Graverand MP et al. Expression of heat shock protein 47(Hsp47) mRNA levels in rabbit connective tissues during the response to injury and in pregnancy. Biochem Cell Biol 2000; 78:511-18.
25. Hirayoshi K et al. HSP47: a tissue-specific, transformation-sensitive, collagen-binding heat shock protein of chicken embryo fibroblasts. Mol Cell Biol 1991; 11:4036-44.
26. Nagata K. Hsp:47 a collagen-specific molecular chaperone. Trends Biochem Sci 1996; 21:22-26.
27. Kagan HM, Li W. Lysyl oxidase: properties, specificity and biological roles inside and outside of the cell. J Cell Biochem 2003; 88:660-72.
28. Hulmes DJ. Building collagen molecules, fibrils and suprafibrillar structures. J Struct Biol 2002; 137:2-10.
29. Nagai N et al. Embryonic lethality of molecular chaperone hsp47 knockout mice is associated with defects in collagen biosynthesis. J Cell Biol 2000; 150:1499-506.

30. Razzaque MS, Taguchi T. Pulmonary fibrosis: Cellular and molecular events. Pathol Int 2003; 53:133-45.
31. Razzaque MS, Taguchi T. Factors that influence and contribute to the regulation of fibrosis. Contrib Nephrol 2003; 139:1-11.
32. Razzaque MS, Taguchi T. Cellular and molecular events leading to renal tubulointerstitial fibrosis. Med Electron Microsc 2002; 35:68-80.
33. Razzaque MS, Taguchi T. Collagen-binding heat shock protein (HSP) 47 expression in antithymocyte serum (ATS)-induced glomerulonephritis. J Pathol 1997; 183:24-29.
34. Razzaque MS, Foster CS, Ahmed AR. Tissue and molecular events in human conjunctival scarring in ocular cicatricial pemphigoid. Histol Histopathol. 2001; 16:1203-12.
35. Razzaque MS, Le VT, Taguchi, T. Heat shock protein 47 and renal fibrogenesis. Contrib Nephrol 2005; 148:57-69.
36. Razzaque MS, Kumari S, Foster CS et al. Expression profiles of collagens, HSP47, TGF-beta,1 MMPs and TIMPs in epidermolysis bullosa acquisita. Cytokine 2003; 21:207-13.
37. Masuda H, Fukumoto M, Hirayoshi K et al. Coexpression of the collagen-binding stress protein HSP47 gene and the alpha 1(I) and alpha 1(III) collagen genes in carbon tetrachloride-induced rat liver fibrosis. J Clin Invest 1994; 94:2481-88.
38. Liu D, Razzaque MS, Nazneen A et al. Role of heat shock protein 47 on tubulointerstitium in experimental radiation nephropathy. Pathol Int 2002; 52:340-47.
39. Liu D, Razzaque MS, Cheng M et al. The renal expression of heat shock protein 47 and collagens in acute and chronic experimental diabetes in rats. Histochem J 2001; 33:621-28.
40. Kuroda K, Tsukifuji R, Shinkai H. Increased expression of heat-shock protein 47 is associated with overproduction of type I procollagen in systemic sclerosis skin fibroblasts. J Invest Dermatol 1998; 111:1023-28.
41. Razzaque MS, Shimokawa I, Nazneen A et al. Age-related nephropathy in the Fischer 344 rat is associated with overexpression of collagens and collagen-binding heat shock protein 47. Cell Tissue Res 1998; 293:471-78.
42. Sunamoto, M et al. Antisense oligonucleotides against collagen-binding stress protein HSP47 suppress collagen accumulation in experimental glomerulonephritis. Lab Invest 1998; 78:967-72.
43. Razzaque MS, Azouz A, Shinagawa T et al. Factors regulating the progression of hypertensive nephrosclerosis. Contrib Nephrol 2003; 139:173-86.
44. Taguchi T, Nazneen A, Abid MR et al. Cisplatin-associated nephrotoxicity and pathological events. Contrib Nephrol 2005; 148:107-21.
45. Razzaque MS et al. Life-long dietary restriction modulates the expression of collagens and collagen-binding heat shock protein 47 in aged Fischer 344 rat kidney. Histochem J 1999; 31:123-32.
46. Razzaque MS, Ahsan N, Taguchi T. Heat shock protein 47 in renal scarring. Nephron 2000; 86:339-41.
47. Razzaque MS, Taguchi T. The possible role of colligin/HSP47, a collagen-binding protein, in the pathogenesis of human and experimental fibrotic diseases. Histol Histopathol 1999; 14:1199-212.
48. Razzaque MS, Kumatori A, Harada T et al. Coexpression of collagens and collagen-binding heat shock protein 47 in human diabetic nephropathy and IgA nephropathy. Nephron 1998; 80:34-43.
49. Brown KE, Broadhurst KA, Mathahs MM et al. Expression of HSP47, a collagen-specific chaperone, in normal and diseased human liver. Lab Invest 2005; 85:789-97.
50. Cheng M, Razzaque MS, Nazneen A et al. Expression of the heat shock protein 47 in gentamicin-treated rat kidneys. Int J Exp Pathol 1998; 79:125-32.
51. Coletta RD, Almeida OP, Ferreira LR et al. Increase in expression of Hsp47 and collagen in hereditary gingival fibromatosis is modulated by stress and terminal procollagen N-propeptides. Connect Tissue Res 1999; 40:237-49.
52. Dafforn TR, Della M, Miller AD. The molecular interactions of heat shock protein 47 (Hsp47) and their implications for collagen biosynthesis. J Biol Chem 2001; 276:49310-19.
53. Hirai K et al. Immunohistochemical distribution of heat shock protein 47 (HSP47) in scirrhous carcinoma of the stomach. Anticancer Res 2006; 26:71-78.
54. Kaur J et al. Co-expression of colligin and collagen in oral submucous fibrosis: plausible role in pathogenesis. Oral Oncol 2001; 37:282-87.
55. Kawada N et al. Expression of heat-shock protein 47 in mouse liver. Cell Tissue Res 1996; 284:341-46.
56. Naitoh M et al. Upregulation of HSP47 and collagen type III in the dermal fibrotic disease, keloid. Biochem Biophys Res Commun 2001; 280:1316-22.
57. Razzaque MS, Foster CS, Ahmed AR. Role of collagen-binding heat shock protein 47 and transforming growth factor-beta1 in conjunctival scarring in ocular cicatricial pemphigoid. Invest Ophthalmol Vis Sci 2003; 44:1616-21.
58. Razzaque MS, Hossain MA, Kohno S et al. Bleomycin-induced pulmonary fibrosis in rat is associated with increased expression of collagen-binding heat shock protein (HSP) 47. Virchows Arch 1998; 432:455-60.
59. Razzaque MS, Nazneen A, Taguchi T. Immunolocalization of collagen and collagen-binding heat shock protein 47 in fibrotic lung diseases. Mod Pathol 1998; 11:1183-88.
60. Razzaque MS, Taguchi T. Role of glomerular epithelial cell-derived heat shock protein 47 in experimental lipid nephropathy. Kidney Int Suppl 1999; 71:S256-59.
61. Chowdhury P, Zhou W, Sacks SH. Complement in renal transplantation. Nephron Clin Pract 2003; 95:c3-8.
62. Mroz A et al. C4d complement split product expression in chronic rejection of renal allograft. Transplant Proc 2003; 35:2190-92.
63. Diamond JR, Kees-Folts D, Ding G et al. Macrophages, monocyte chemoattractant peptide-1 and TGF-beta 1 in experimental hydronephrosis. Am J Physiol 1994; 266:F926-33.
64. Plemons JM et al. PDGF-B producing cells and PDGF-B gene expression in normal gingival and cyclosporine A-induced gingival overgrowth. J Periodontol 1996; 67:264-70.
65. Desmouliere A, Geinoz A, Gabbiani F et al. Transforming growth factor-beta 1 induces alpha-smooth muscle actin expression in granulation tissue myofibroblasts and in quiescent and growing cultured fibroblasts. J Cell Biol 1993; 122:103-11.
66. Tang WW et al. Platelet-derived growth factor-BB induces renal tubulointerstitial myofibroblast formation and tubulointerstitial fibrosis. Am J Pathol 1996; 148:1169-80.
67. Razzaque MS, Ahmed BS, Foster CS et al. Effects of IL-4 on conjunctival fibroblasts: possible role in ocular cicatricial pemphigoid. Invest Ophthalmol Vis Sci 2003; 44:3417-23.
68. Razzaque MS, Ahmed AR. Collagens, collagen-binding heat shock protein 47 and transforming growth factor-beta 1 are induced in cicatricial pemphigoid: possible role(s) in dermal fibrosis. Cytokine 2002; 17:311-16.
69. Yamamura I, Hirata H, Hosokawa N et al. Transcriptional activation of the mouse HSP47 gene in mouse osteoblast MC3T3-E1 cells by TGF-beta 1. Biochem Biophys Res Commun 1998; 244:68-74.
70. Wang Z, Inokuchi T, Nemoto TK et al. Antisense oligonucleotide against collagen-specific molecular chaperone 47-kDa heat shock protein suppresses scar formation in rat wounds. Plast Reconstr Surg 2003; 111:1980-87.
71. Ohba S, Wang ZL, Baba TT et al. Antisense oligonucleotide against 47-kDa heat shock protein (Hsp47) inhibits wound-induced enhancement of collagen production. Arch Oral Biol 2003; 48:627-33.
72. Benn SC, Woolf CJ. Adult neuron survival strategies--slamming on the brakes. Nat Rev Neurosci 2004; 5:686-700.
73. Koide T et al. Specific recognition of the collagen triple helix by chaperone HSP47: minimal structural requirement and spatial molecular orientation. J Biol Chem 2006; 281:3432-38.
74. Balakin KV, Kozintsev AV, Kiselyov AS et al. Rational design approaches to chemical libraries for hit identification. Curr Drug Discov Technol 2006; 3:49-65.
75. Guner O, Clement O, Kurogi Y. Pharmacophore modeling and three dimensional database searching for drug design using catalyst: recent advances. Curr Med Chem 2004; 11:2991-3005.
76. Taguchi T, Razzaque MS. The collagen-specific molecular chaperone HSP47: is there a role in fibrosis? Trends Mol Med 2007:13:45-53..

# CHAPTER 8

# Dendritic Cell-Based Approaches to Organ Transplantation

Andrea Meinhardt and Giuseppe Vassalli*

## Abstract

Dendritic cells (DCs) take up antigens at peripheral sites and migrate to T-cell areas of lymph nodes and spleen, where they present antigenic peptides to T-cells. As such, DCs initiate innate and adaptive immune responses to microorganisms and other antigens, including alloantigens in organ transplantation. Accumulating evidence suggests DCs play dichotomous roles including both immune stimulation and regulation, largely depending on the subsets of DCs involved and their maturation status. Maturation of myeloid DCs is associated with a shift away from a regulatory to a stimulatory phenotype. Plasmacytoid DCs migrating preferentially to draining lymph nodes, rather than spleen, have been shown to induce allograft tolerance in rodents through generation of regulatory T-cells. Multiple different approaches have been evaluated for enhancing the regulatory effects of DCs in organ transplantation. These approaches include ex vivo cell culture in the presence of T-helper type-2 (Th2) cytokines, as well as inhibitors of chemokines that drive DC migration, pharmacological inhibitors of DC maturation and genetically modified DCs. An alternate strategy includes donor-derived apoptotic cells for delivery in situ of donor alloantigen to quiescent DCs of graft recipients. Several agents commonly used in human transplant recipients such as corticosteroids, cyclosporine A and mycophenolate mofetil, as well as aspirin and vitamin $D_3$, inhibit DC maturation. In proof-of-principle studies, DCs genetically engineered to express immunoregulatory factors (e.g., TGF-β, IL-10, CTLA4.Ig, soluble TNF receptor, galectin-1), death receptors (e.g., Fas ligand) and mediators of lymphoid homing (e.g., chemokine receptor CCR7, chimeric E/L-selectin) elicited antigen-specific T-cell hyporesponsiveness and prolonged allograft survival. However, overall results in vivo have been less consistent than those in vitro possibly due, in part, to the eventual maturation of DCs in the graft recipient. Recent data suggest that regulatory DCs can be generated from rhesus monkey monocytes conditioned with vitamin $D_3$ and IL-10, and that these DCs are resistant to maturation induced by a potent combination of inflammatory cytokines. Collectively, these findings suggest a significant potential for regulatory DC therapy in organ transplantation.

## Introduction

Induction of donor-specific operational immune tolerance—clinically defined as the lack of destructive immune reactions to the graft in the absence of chronic immunosuppression and with preserved general immunocompetence—remains the ultimate goal in organ transplantation.[1] Indeed, achieving this goal would obviate the need for chronic immunosuppression and, presumably, prevent chronic rejection, at least in part. The paradigm of T-cell costimulatory activation provides a conceptual framework for the interactions between antigen-presenting cells (APCs) including dendritic cells (DCs) and T-cells.[2,3] This paradigm dictates that three distinct signals are required for effective T-cell stimulation (Fig. 1). The first signal is provided by presentation of antigenic peptides associated with major histocompatibility complex (MHC) class II molecules to T-cells through the T-cell receptor (TCR). The second signal arises from interactions between costimulatory receptors expressed on the surface of the APC and the respective ligands on the T-cell surface. The third signal consists of stimulatory cytokines, such as interleukin (IL)-12p70, released from APCs interacting with T-cells. Antigen presented (signal 1) in the absence of appropriate costimulation (signals 2 and 3) fails to activate the T-cell but induces a state of antigen-specific T-cell hyporesponsiveness,

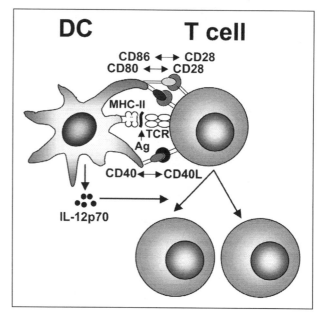

Figure 1. Schematic view of the DC–T-cell interaction in transplantation. Three signals are required for strong T-cell activation. Signal 1: Presentation of donor antigen (either directly by donor DCs or indirectly as donor antigen peptide bound to MHC molecules on recipient DCs) to the T-cell via the T-cell receptor (TCR). Signal 2: Interactions between pairs of receptors and ligands expressed on the surface of the DC and on the T-cell surface, respectively (e.g., CD80 [B7-1] and CD86 [B7-2] interacting with CD28, CD40 interacting with CD154 [CD40L]). Signal 3: Secretion of activatory cytokines such as IL-12p70 leading to full amplification of T-cell proliferation.

*Corresponding Author: Giuseppe Vassalli—Dept. of Cardiology, University Hospital Center and University of Lausanne, 1011 Lausanne, Switzerland. Email: giuseppe.vassalli@chuv.ch

*Chronic Allograft Failure: Natural History, Pathogenesis, Diagnosis and Management*, edited by Nasimul Ahsan. ©2008 Landes Bioscience.

or anergy.[4] Under these conditions, DCs may also induce apoptotic cell death of activated T-cell clones via expression of death ligands such as Fas ligand (CD95L), which binds the Fas receptor (CD95) on activated T-cells.[5,6] In addition, interactions between antigen-specific T-cells and DCs expressing low levels of costimulatory molecules may induce the formation of regulatory T-cells ($T_{reg}$).[7]

T-cells can be physiologically silenced by various mechanisms including anergy, central deletion in the thymus, deletion in the peripheral immune system and immune suppression.[8] An improved understanding of these mechanisms may lead to novel strategies to induce transplant tolerance by inhibiting alloreactive T-cells.[1] Solid organ transplants contain rare nonparenchymal cells, i.e., interstitial "passenger" leukocytes that can migrate into the recipient and establish long-lasting microchimerism, defined as the persistence of a small number of donor bone marrow (BM)-derived cells in the recipient.[9] Although such microchimerism is often associated with graft acceptance and tolerance, it has been difficult to demonstrate a true causal link. In a skin transplantation model using mutant mice deficient for leukocyte subsets, donor T-cell chimerism resulted in very different outcomes

depending on the host's immunological maturity and the antigenic disparities involved.[10] In immunologically mature hosts, chimerism enhanced immunity and graft rejection, whereas in immature hosts, it induced tolerance to the chimeric T-cells, but not to graft antigens not expressed by the chimeric cells.

## DC Maturation

DCs are present in the interstitium of many nonlymphoid organs, including donor organs used in transplantation. Under normal conditions, organ-resident DCs are immature, as reflected by low expression of surface MHC molecules and other molecules involved in cell adhesion (e.g., intercellular adhesion molecule [ICAM]-1; CD54) and T-cell costimulation (e.g., CD40, CD80 [B7-1] and CD86 [B7-2]). These immature DCs are poor stimulators of naive T-cells. Acting as sentinels of the immune system, they take up and process antigen at peripheral sites, load antigen fragments onto MHC class II molecules for export to the cell surface and migrate to T-cell areas of draining lymph nodes and spleen where they present antigen to T-cells (Fig. 2). These processes generally are associated with a certain degree of DC maturation.

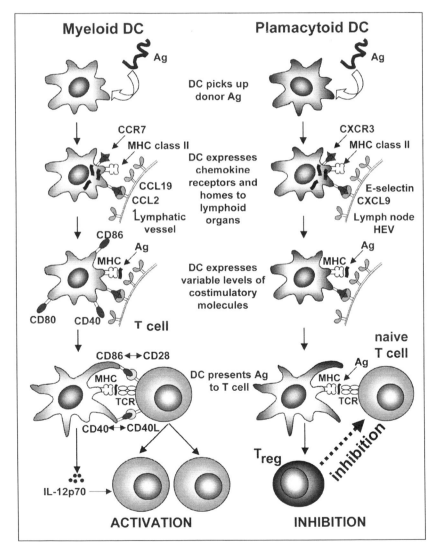

Figure 2. Schematic view of stimulatory and regulatory roles of DCs. It has generally been assumed that mature myeloid DCs induce allograft rejection, whereas immature and, possibly, relatively mature plasmacytoid DCs promote graft survival. However, the relationships between DC subtypes and functions are complex and only incompletely understood. The immature DC picks up antigen at peripheral sites and up-regulates chemokine receptors that allow for migration into peripheral lymphoid tissues. Among those receptors, CCR7 (which binds CCL19 and CCL21 on lymphatic vessel endothelium) is particularly important for myeloid cells, while CXCR3 is important for lymphoid homing of plasmacytoid cells. In mice, the latter cell subset can enter lymphoid tissues directly from blood by transmigrating across high endothelial venules (HEVs) of inflamed lymph nodes in a CXCL9 and E-selectin–dependent manner. Mature myeloid cells express high level of costimulatory molecules. Plasmacytoid cells express low level of costimulatory molecules leading to T-cell anergy and generation of $T_{reg}$ cells.

DC maturation can be induced by a number of stimuli derived from ischemia/reperfusion injury and graft inflammation at the time of transplantation.[11] Nuclear translocation of the transcription factor, nuclear factor-κB (NF-κB), promotes the activation of a number of pro-inflammatory genes leading to DC maturation. Mature DCs express high levels of MHC class II and costimulatory molecules (e.g., CD40, CD80, CD86) and secrete high amounts of IL-12p70, a cytokine that mediates clonal amplification of T-cell proliferation. Upon maturation, DCs up-regulate the C-C chemokine receptor (CCR) 7 that drives their migration through the lymphatics (see below).[12] In T-cell areas of draining lymph nodes, mature DCs acting as professional APCs initiate antigen-specific responses by naive CD4+ and CD8+ T-cells.[13]

## DCs and Pathways of Allorecognition

Three general pathways of alloantigen recognition by the host immune system have been described: the "direct" pathway, the "indirect" pathway and the "semi-indirect" pathway (or "cross-presentation"). At the time of transplantation, "passenger" DCs and monocytes/macrophages present in the donor organ are transplanted simultaneously with the allograft. After transplantation, donor-derived DCs migrate from the graft to recipient lymph nodes and spleen where they directly present donor antigens to recipient T-cells.[14,15] In contrast, indirect alloantigen presentation refers to recipient DCs that migrate into the graft, take up and process donor antigen and then present donor peptides to recipient T-cells. Finally, mature live DCs can transfer cell membranes to immature DCs by cell-cell contacts.[16] In such "semi-indirect" pathway, recipient DCs can "cross-present" peptides derived from transferred membranes of donor DCs to T-cells and generate cytotoxic T-lymphocytes (CTLs).

It has generally been assumed that the direct pathway plays a predominant role in acute rejection, while indirect antigen presentation may be more important in chronic rejection. In long-surviving rat renal allografts that had been re-transplanted, immunogenicity could be restored by infusion of donor-strain DCs.[17] This observation suggested a key role of intra-renal DCs in graft rejection. In a mouse model of skin allograft rejection, ≈90% of responding T-cells were activated via the direct pathway and < 10% via the indirect pathway.[18] Nevertheless, it has been shown using MHC class II-deficient mice that the indirect pathway by itself is able to elicit acute rejection.[19] In MHC-mismatched rat cardiac allografts in a strain combination characterized by donor DC-dependent tolerance induction, exhaustive depletion of graft resident DCs by parking the graft in an intermediate recipient delayed acute rejection only marginally, although it markedly reduced graft infiltration by inflammatory cells early after transplantation.[20]

Important insights into the role of DCs in transplantation have been gained from experiments using the fms-like tyrosine kinase 3 ligand (Flt3L), a potent endogenous hematopoietic growth factor that dramatically increases various DC subsets in lymphoid and non-lymphoid organs. Treatment of donors with Flt3L caused loss of liver allograft acceptance and of donor-specific tolerance.[21] These effects were accompanied by a marked increase in donor-derived DCs that matured rapidly ex vivo into potent inducers of T-helper (CD4+) type-1 (Th1) responses. Conversely, mice lacking Flt3L (Flt3L−/−) exhibited severe reductions in both myeloid (CD8α−) and lymphoid-related (CD8α+) DC subsets (see below) in the thymus and secondary lymphoid tissues, as well as in interstitial cardiac DCs and dermal (but not epidermal) DCs.[22] Flt3L−/− recipients rejected wild-type donor heart and skin allografts with kinetics similar to normal recipients, presumably reflecting a preserved function of the direct pathway. Conversely, Flt3L−/− heart, but not skin, allografts placed in wild-type recipients induced a decreased trafficking of donor DCs into host spleens and draining lymph nodes, which was accompanied with attenuated rejection.[23] However, rejection of Flt3L−/− hearts could be precipitated by infusion of wild-type donor DCs at the time of transplantation. Thus, severe depletion of

donor heart DCs secondary to Flt3L gene deficiency was associated with improved cardiac allograft survival.

The contribution that direct allorecognition makes to the progression of chronic cardiac transplant rejection in humans was evaluated by assessing donor-specific tolerance in T-cells displaying direct anti-donor allospecificity.[24] Donor-specific hyporesponsiveness was demonstrated in ≈50% of recipients in both the helper and CTL cell compartments of the direct alloresponse. This observation suggests that direct allorecognition is not primarily responsible for the progression of chronic rejection, which implies a predominant role of the indirect pathway in this condition.

## DC Subsets

Five distinct subsets of human blood DCs that express unique phenotypes have been identified based on their expression of the CD123, CD1c/b, BDC3 (CD3), CD34 and CD16 markers.[25,26] In humans, the two subsets of particular interest to transplantation are the myeloid and the plasmacytoid subsets (also referred to as DC1 and DC2 because they promote Th1 and Th2 cytokine responses, respectively). The two subsets display differential chemokine receptor expression and chemokine receptor activities, as well as distinct chemokine and cytokine expression profiles.[27,28] In mice, expression of the surface marker CD8α discriminates the CD8α+ subset, which is related to human DC2, and CD8α− DCs, which has been linked to human DC1. In addition, the CD11c+b220+gr-1+ subset represents a murine plasmacytoid DC subset present in lymph nodes and spleen.[29] However, the relationships between murine and human DC subsets are incompletely understood.

## Regulatory DCs in Organ Transplantation

While the central role of mature DCs in initiating immune responses against foreign antigen (including alloantigen) has been known for a long time, only recently have immunoregulatory roles of immature DCs been recognized. These dichotomous roles of DCs are largely dependent on the subsets of DCs involved and their maturation status. Mature myeloid DCs are generally assumed to polarize T-helper cytokine responses towards Th1 and to accelerate allograft rejection, while plasmacytoid DCs are thought to promote Th2 responses, allogeneic T-cell hyporesponsiveness, and graft survival.[30] In the absence of CD40 costimulation, immature DCs induced transient, antigen-specific T-cell activation followed by T-cell deletion and anergy in vivo.[31] Immature, but not mature, donor BM-derived CD8α− DCs delivered intravenously (i.v.) before transplantation extended cardiac allograft survival in non-immunosuppressed recipient mice.[32,33] Conversely, both immature and comparatively mature CD8α+ (putative "lymphoid-related") DCs delivered 7 days before transplantation improved cardiac allograft survival in fully MHC disparate mice.[34]

A recent study showed a key role of plasmacytoid DCs in establishing and maintaining immune tolerance to vascularized mouse cardiac allografts.[35] A tolerogenic treatment consisting of donor-specific transfusion plus anti-CD40L antibody resulted in the "preferential" migration of alloantigen-specific plasmacytoid DCs into T-cell areas of lymph nodes, rather than spleen. In the lymph nodes, plasmacytoid DCs stimulated antigen-specific CD4+CD25+Foxp3+ Treg cells, which was associated with prolonged graft survival. It is worth noting that ≈90% of alloantigen-presenting cells were plasmacytoid DCs both in tolerized and in rejecting mice. In tolerized mice, however, these cells were distributed systemically throughout lymph nodes yet were almost completely absent from the spleen. Conversely, in rejecting mice alloantigen-presenting cells were localized mainly in the spleens but were almost completely absent from lymph nodes. Passive transfer of blood plasmacytoid DCs from tolerized hosts conferred donor-specific tolerance on naive allograft recipients, whereas depletion of plasmacytoid DCs or prevention of their homing to lymph nodes suppressed

T$_{reg}$ cell induction and tolerance. These findings indicate a key role of plasmacytoid DCs homing to lymph nodes in allograft tolerance through stimulation of T$_{reg}$ cells.[36]

## Circulating DCs after Organ Transplantation in Humans

Organ transplantation in humans is associated with changes in circulating precursors of DCs. It has been shown that renal transplantation reverses functional deficiencies in circulating precursor DC subsets observed in chronic renal failure patients.[37] On the other hand, surgery directly caused a sharp decline in the number of precursors of myeloid and plasmacytoid DCs both in renal and in heart transplant recipients.[38,39] These cell subsets subsequently recovered slowly over several months but failed to reach normal levels. Surgery or rejection episodes did not appear to be associated with activation of DC precursors in the circulation. Long-term immunosuppression, especially involving prednisolone, was associated with a significant decrease in precursors of myeloid and plasmacytoid DCs in human kidney transplant recipients compared with healthy controls.[40] In heart transplant recipients, precursors of myeloid DCs positive for the maturation marker CD83 and the lymphoid homing receptor CCR7 were decreased during acute rejection episodes, whereas plasmacytoid DCs were unchanged.[41] It has been speculated that these data may reflect peripheral blood depletion of myeloid DCs due to their selective homing to lymphoid organs during rejection. In human heart transplant recipients, a low circulating myeloid/plasmacytoid DC subset ratio correlated with increased risk of acute rejection at 3 months posttransplant, independent of anti-rejection therapy.[42] In human liver transplant recipients, the myeloid/plasmacytoid DC ratio in those patients on minimal calcineurin inhibitor monotherapy undergoing successful weaning, as well as in those off all anti-rejection therapy was lower compared to patients on maintenance immunosuppression.[43] In human kidney transplant recipients, numbers of circulating IL-10– and IL-12–producing DCs correlated with time after transplantation, dosage of immunosuppression and plasma cytokines.[44] IL-10–producing DCs predominated late after transplantation in the presence of Th1 cytokines and low plasma IL-10, possibly reflecting immunoregulatory processes. However, the precise relevance of changes in circulating subsets of DC progenitors in transplanted patients remains unclear.

## General DC-Based Approaches to Organ Transplantation

DCs have been evaluated as potential tools for the therapeutic modulation of immune responses in autoimmune diseases, cancer and transplantation. The peculiar ability of DCs to migrate into T-cell–dependent areas of lymphoid tissues, the very sites where T-cell responses are initiated, can be exploited to target T-cell activation in an antigen-specific manner.[45-48]

Two general strategies can be used to generate regulatory DCs in humans. The first approach consists of the generation of immature DCs from monocyte precursors in vitro.[49] An alternate modality includes the mobilization of donor-derived immature precursors of DCs in the live donor using human growth factors (e.g., Flt3L, granulocyte-colony stimulating factor; G-CSF),[21,50] followed by their collection by leukapheresis, purification and administration to the recipient before transplantation. In humans, mobilization of BM cells with G-CSF selectively increased the number of circulating precursors of plasmacytoid DCs that mediated Th2 cytokine responses.[50] In two human volunteers, a single subcutaneous injection with autologous, immature-monocyte-derived DCs pulsed with influenza matrix peptide or keyhole limpet haemocyanin (KLH) stimulated the generation of IL-10–producing, peptide-specific CD8$^+$ T$_{reg}$ cells, while suppressing the peptide-specific killing activity of CD8$^+$ T-cells.[51] This important study demonstrated that injection of human immature DCs in the absence of inflammation induces T$_{reg}$ cells that inhibit effector T-cell functions in an antigen-specific manner in humans.

Various regulatory DC therapies including modification of chemokine signaling pathways, pharmacological DC inhibitors and genetically modified DCs have been evaluated in proof-of-principle studies in transplantation models. These modalities are briefly discussed in the following sections.

## Targeting Chemokines Mediating DC Lymphoid Homing

Chemokines recruit leukocytes including alloreactive T-cells and DCs to the graft, which act as cellular mediators of acute rejection.[52] The primary source of chemokines released at the time of transplantation is damaged graft tissue itself. A few hours later, leukocytes infiltrating the graft in the context of innate immune reactions secrete high amounts of chemokines.[53] In human heart transplants, a number of chemokines including CXCL8/IL-8 (designations refer to the new and old classification, respectively), CCL5/RANTES, CCL2/MCP-1 (monocyte chemoattractant protein-1), CCL7/MCP-3 and CXCL10/IP-10 (interferon-γ–inducible protein of 10 kD) are up-regulated in the graft early after transplantation.[54,55] Distinct patterns and kinetics of chemokine production regulate DC functions.[56] As mentioned above, human myeloid and plasmacytoid DC subsets display differential chemokine receptor expression and chemokine receptor activities, as well as distinct chemokine and cytokine expression profiles.[27,28] While most chemokine receptors expressed on blood myeloid DCs are functional, as assessed in cell migration assays in vitro, most chemokine receptors expressed on freshly isolated plasmacytoid DCs are nonfunctional.[27] In a mouse model, bacterial infection induced a significant number of precursors of plasmacytoid DCs and myeloid DCs into the circulation.[57] Both subsets expressed a common set of chemokine receptors except CXCR3, displayed parallel mobilization into the blood, but showed distinct trafficking pathways to the lymph nodes. In a short-term homing assay, myeloid DC precursors migrated to peripheral tissues and subsequently to draining lymph nodes, while plasmacytoid DC precursors directly entered the lymph nodes in a CXCL9- and E-selectin–dependent manner by transmigration across high endothelial venules (HEVs) of inflamed lymph nodes.[57]

Lack of CCR7 or its ligands leads to dramatic changes in DC and lymphocyte traffic, along with an abnormal lymph node architecture.[58,59] DCs from CCR7$^{-/-}$ mice failed to migrate to draining lymph nodes and to induce the rapid increase in lymph node cellularity that normally precedes T-cell proliferation in CCR7$^{+/+}$ mice.[12] While many inducible chemokines of the CC subfamily attract immature DCs,[27,60] mature DCs migrate to fewer, if any, inducible chemokines and respond instead to constitutively expressed chemokines such as CCL19/MIP-3β (macrophage inflammatory protein-3β) and CCL21/SLC (secondary lymphoid tissue chemokine).[60,61] CCL19/MIP-3β is constitutively expressed in T-cell areas of secondary lymphoid organs, while CCL21/SLC is constitutively expressed in lymphatic vessels[57] and secondary lymphoid organs.

Observations in mice lacking single chemokines or chemokine receptors illustrate the important roles of chemokine signaling pathways in graft rejection. Mice lacking either CCR1, CCR5, or CXCR-3 accepted cardiac allografts from wild-type donors for extended periods of time.[62-64] Moreover, heart transplants from CXCL10/IP-10–deficient donors survived long-term in wild-type recipients.[65] Treatment with a blocking antibody against CXCL9/Mig (monokine induced by interferon-γ)[66] or a CCR1-specific nonpeptide antagonist,[67] as well as gene transfer of the virally encoded chemokine antagonists, vMIP-II and MC148,[68] or a mutated CCL5/RANTES peptide[69] prolonged allograft survival moderately in skin, kidney and heart transplantation models, respectively.

Mice with the paucity of lymph node T-cell (plt) mutation lack secondary lymphoid expression of the CCR7 ligands, CCL19/MIP-3β and CCL21/SLC. plt recipients displayed normal allogeneic T-cell responses but deficient migration of donor DCs to draining lymph nodes, leading to permanent acceptance of fully allogeneic islets engrafted under the kidney capsule.[70] Peri-transplant i.v. injection of donor splenocytes caused plt recipients to reject their allografts. Islet allografts transplanted intrahepatically in plt mice were rejected normally, as were primarily revascularized cardiac allografts. These results suggest a central role of chemokine-directed lymphoid homing of donor DCs in host sensitization and islet allograft rejection. A separate study[71] showed a marked reduction in DC trafficking from heart, but not skin, transplants to secondary lymphoid tissues in plt recipients, which was associated with a marked prolongation of heart, but not skin, allograft survival. Anti-CXCL9/Mig antibody administered for 2 weeks post-transplant to inhibit migration of activated T-cells extended cardiac allograft survival further in plt, but not in wild-type, mice. These results suggest that targeting both DCs and the migration of activated T-cells simultaneously may have additive effects on allograft survival.

## Generation of Regulatory DCs by Differential Cell Culture Conditions

The first report on in vitro generation of regulatory DCs described immature DCs derived from murine BM cells cultured in the presence of GM-CSF.[72] These DCs induced antigen-specific hyporesponsiveness in allogeneic T-cells in vitro. A subsequent study[33] showed that immature DCs generated with low doses of GM-CSF in the absence of IL-4 were maturation-resistant and induced prolonged cardiac allograft survival. Monocyte or BM-derived, immature donor DCs cultured in the presence of GM-CSF have been combined with agents that target the expression of costimulatory receptors on DC surfaces, or expression of their ligands on T-cells, to prevent DC maturation in a pro-inflammatory environment. Immature donor DCs in conjunction with anti-CD40L (CD154) mAb,[73] anti-lymphocyte serum,[74] or anti-CD54 mAb plus CTLA4.Ig[75] achieved long-term (>100 days) cardiac allograft survival in rodents. Moreover, immature donor DCs combined with anti-CD40L mAb prevented the development of intimal hyperplasia and arterial lesions in mouse aortic allografts.[76]

Immature DCs derived from the recipient have also been utilized to facilitate allograft acceptance.[77] In several studies, recipient DCs were pulsed with donor peptide or donor cell lysate to influence the regulation of the indirect pathway of allorecognition.[78,79] Immature recipient DCs generated ex vivo, given in conjunction with a short course of a deoxyspergualin derivative to block NF-κB activity, induced indefinite cardiac allograft survival in rats.[80]

## Pharmacological Modulation of DC Maturation

As mentioned, immature DCs may induce antigen-specific T-cell anergy in the absence of adequate costimulation, while also stimulating the generation of $T_{reg}$ cells in vivo. In contrast, maturation of myeloid DCs is associated with upregulation of surface costimulatory molecules leading to T-cell activation. Several pharmacological agents can be used to block DC maturation in organ transplantation.[81] Corticosteroids have been shown to inhibit DC differentiation and maturation both in vitro and in vivo.[82-87] Dexamethasone (DEXA) preferentially suppressed the differentiation of plasmacytoid DCs and mediated their apoptotic death.[87] In another study, DEXA induced DCs characterized by decreased CD83 and CD86 expression, de novo expression of plasmacytoid CD123 and myeloid CD14 cell markers and sustained expression of Toll-like receptor 2 (TLR2).[88] Myeloid DCs generated in the presence of the calcineurin inhibitors cyclosporin A (CsA) or tacrolimus (FK-506) exhibited up-regulation of the costimulatory molecule B7-H2 (ICOS-ligand) and decreased stimulation of T-cell proliferation and IFN-γ secretion in vitro. These results suggested that myeloid DC subsets induced in the presence of DEXA, CsA or FK-506 in cell cultures have reduced allostimulatory properties but are equipped with different molecular repertoires to exert these functions.

Both CsA and FK-506, but not rapamycin (sirolimus), inhibited MHC-restricted antigen presentation pathways in DCs.[89] In mouse BM-derived DCs, CsA and FK-506 each blocked efficiently the bidirectional (DC→T-cell and T-cell→DC) activatory and maturation changes resulting from intercellular signaling.[90] In addition, CsA impaired DC migration by regulating chemokine receptor expression and by inhibiting cyclooxygenase-2 expression.[91] A recent study analyzed the effects of immunosuppressive drugs on the modulation of chemokine receptors in maturing human DCs.[92] DEXA and IL-10, but not CsA and FK506, inhibited human DC migration in response to the CCL19 chemokine in vitro and mouse DC migration to lymph nodes in vivo. This effect was associated with decreased CCR7 expression. In contrast, rapamycin increased DC migration to CCL19 in vitro and to lymph nodes in vivo by enhancing CCR7 expression.

Mycophenolate mofetyl (MMF), an immunosuppressive agent routinely used in transplant recipients, is a reversible inhibitor of inosine monophosphate dehydrogenase, which is essential to the de novo synthesis of purines required for the generation of RNA and DNA. The proliferation of T- and B-cells is particularly dependent on this mechanism. MMF has been shown to inhibit the maturation of DCs and their ability to stimulate cellular immune responses in vivo.[93]

Additional potent DC inhibitors include aspirin[94-96] and deoxyspergualin[97,98] (two NF-κB inhibitors[97]), as well as vitamin $D_3$.[99,100] Importantly, low-dose aspirin therapy was associated with improved allograft function and survival in a clinical study is renal transplant recipients.[96] Peri-transplant treatment with deoxyspergualin in conjunction with anti-CD3 immunotoxin to induce T-cell depletion induced tolerance to kidney allografts in a rhesus macaque model.[98]

These data illustrate potent modulation of DC maturation and function by a number of agents, including immunosuppressive drugs currently used in the clinical setting. Although much attention has been focused on their T-cell inhibitory activities, these drugs show important DC-modulating properties that contribute to their immunosuppressive effects.

## Genetically Engineered DCs

Immature DCs genetically engineered to express immunoregulatory cell surface molecules, Th2 cytokines, death ligands, chemokine receptors or adhesion molecules mediating lymphoid homing, or other therapeutic molecules permit to selectively target antigen-specific T-cells. A practical advantage of this approach is that DCs are genetically manipulated in vitro, which minimizes immune responses to the gene transfer vector in vivo.

Initial studies focused on genetically engineered DCs expressing Th2-polarized cytokines, such as IL-10, IL-4 and TGF-β. Retroviral and adenoviral vectors were used to transduce genes into replicating immature DCs. Adenoviral vectors (at a multiplicity of infection of 50) induced a moderate increase in the expression of the costimulatory molecules CD40 and CD86 and enhanced the poor allostimulatory activity of DC progenitors. However, these effects were not observed using DCs genetically engineered to express TGF-β.[101] Immature mouse myeloid DCs retrovirally transduced with the TGF-β gene were weak stimulators of alloreactive T-cells in vitro and prolonged cardiac allograft survival moderately.[102] TGF-β1–transduced mature monkey DCs suppressed alloimmune responses in vitro in an antigen-specific manner.[103] In another study, TGF-β1–transduced, immature murine DCs displayed enhanced tolerogenicity but caused cardiac allograft fibrosis in vivo.[104]

Human DCs adenovirally transduced with the IL-10 gene inhibited Th1 responses, while selectively activating IL-10–producing CD4+ T-cells.[105] Human myeloid DCs transduced with an adenoviral IL-10 gene construct inhibited human skin graft rejection in humanized NOD-scid chimeric mice.[106] Portal venous infusion with immature DCs adenovirally transduced with the IL-10 gene prolonged cardiac allograft survival in mice.[107] To the contrary, administration of the same DCs via the tail vein accelerated graft rejection. The investigators speculated that the diverse effects of IL-10–transduced DCs infused through different routes might be due to the Th2-polarized responses and poor T-cell stimulating activity of liver DCs.

Viral IL-10 is highly homologous to mammalian IL-10 but lacks certain of its T-cell stimulatory activities. Myeloid DCs retrovirally transduced with either the viral or mammalian IL-10 gene exhibited differential effects regarding CTL and natural killer (NK) cell functions.[108] DCs expressing viral IL-10 displayed decreased CTL induction and enhanced tumor growth in a tumor transplant model, reflecting immunosuppression. Conversely, DCs overexpressing mammalian IL-10 showed enhanced CTL induction and NK cell activity and inhibited tumor growth, indicating immunostimulation.

Other studies have questioned the in vivo efficacy of DC-based approaches. BM-derived immature DCs primed in vivo alloreactive T-cells for IL-4–dependent rejection of MHC class II antigen-disparate cardiac allografts.[109] BM-derived DCs retrovirally engineered to express viral IL-10 showed inhibitory activities in vitro but failed to clearly down-regulate allogeneic responses after single or multiple cell injections in vivo.[110] Myeloid DCs adenovirally engineered to express mammalian IL-10 or IL-4 showed enhanced IL-12p70 production in response to CD40 ligation and worsened allograft rejection.[111,112] Adenovirus-induced DC maturation via NF-κB–dependent pathways contributed to accelerated graft rejection.[113] Indeed, systemic administration of immature recipient DCs treated with antisense oligodeoxynucleotide (ODN) with specific affinity for NF-κB resulted in prolonged cardiac allograft survival.[114] Moreover, DCs transfected with an anti–NF-κB ODN decoy and adenovirally transduced with the immunosuppressive CTLA4.Ig gene extended cardiac allograft survival further.[115] Finally, allogeneic DCs rendered immature by adenoviral transduction of a kinase-defective dominant negative form of IKK2 (dnIKK2), which blocks NF-κB activation, induced in vitro differentiation of naive T-cells into potent $T_{reg}$ cells.[116] These observations illustrate the central role of NF-κB in DC maturation.

Myeloid DCs adenovirally or retrovirally transduced with the CTLA4.Ig gene displayed decreased allostimulatory potency in vitro and prolonged heart and islet allograft survival in vivo.[115,117-120] Concomitant blockade of the CD40 pathway enhanced immune tolerance by CTLA4.Ig-transduced immature DCs.[120] However, it is worth mentioning that these results were not substantially superior to those achieved using recombinant CTLA4.Ig protein treatment or direct CTLA4.Ig gene transfer into the graft in other studies.[121-124]

Engagement of Fas (CD95) on the surface of an activated T-cell by FasL on the surface of an APC triggers apoptosis of the activated T-cell. Fas-FasL interactions play a central role in the maintenance of peripheral T-cell tolerance, including immune privilege in certain organs such as the eye and testis. FasL-expressing killer-DCs have been shown to eliminate activated, but not resting, primary human CD4+ and CD8+ T-cells.[125] FasL-transduced DCs induced antigen-specific T-cell hyporesponsiveness and extended survival of fully MHC-mismatched vascularized cardiac allografts.[126,127] Alternate APCs such as macrophages engineered to express FasL also induced profound and specific T-cell unresponsiveness to alloantigen, along with profound clonal deletion of antigen-specific, peripheral T-cells in mice.[128] On the other hand, distinct studies have suggested that FasL-transduced DCs can elicit vigorous allospecific T-cell responses[129] and also induce pulmonary vasculitis in mice.[130] In one of those studies, BM-derived DCs retrovirally transduced with the FasL gene died quickly by a Fas-dependent mechanism, as indicated by the fact that only DCs from Fas-deficient lpr mice survived after FasL gene transduction.[129] After subcutaneous

injection in MHC class II-disparate mice, FasL-transduced lpr DCs hyperactivated the allospecific proliferation of T-cells in draining lymph nodes.

Most recent approaches included DCs genetically modified with antisense ODN targeting the costimulatory molecules CD80 or CD86,[131] or with a soluble TNF-α receptor type I (sTNFRI) gene construct,[132] which achieved prolonged cardiac allograft survival in rodents. Murine DCs adenovirally transduced with the galectin (gal)-1 gene, an endo-genous lectin that binds to glycoproteins and exerts potent regulatory effects on T-cells, induced rapid apoptosis of activated T-cells.[133] In vivo, gal-1-DCs delayed the onset of autoimmune diabetes in mice.[134]

A single infusion of DCs coexpressing CCR7, which mediates lymphoid homing, and viral IL-10 prolonged cardiac allograft survival for >100 days, whereas either gene alone was ineffective.[135] DC targeting to T-cell areas of peripheral lymph nodes was also achieved by endowing DCs with a novel E/L-selectin chimeric receptor for peripheral node addressin, an adhesion molecule present on the lymph node venular endothelium.[136] In retrovirally transduced DCs, the transgenic receptor was expressed at the tip of microvilli and mediated rolling of DCs on peripheral node addressin both in vitro and in vivo. Such genetically engineered DCs could extravasate directly from blood through the lymph node endothelium, whereas untransduced DCs could not.

## Donor-Derived Apoptotic Cells and Exosomes

Alternate DC-based strategies to induce transplantation tolerance include the delivery in situ of donor alloantigen to quiescent DCs of graft recipients by means of donor-derived apoptotic cells or exosomes.[137] Exosomes are nanovesicles (<100 nm) produced by DCs, which are rich in MHC molecules that can be employed to target DCs in situ. Donor leukocytes in early apoptosis are rich in alloantigen, are internalized efficiently by recipient DCs in vivo and deliver immunosuppressive signals to DCs. Once intravenously injected, exosomes carrying donor MHC molecules were captured by recipient's DCs and exerted a profound downregulatory effect on anti-donor T-cell responses, resulting in extended cardiac allograft survival in mice.[138]

## Concluding Remarks

Accumulating evidence suggests dichotomous roles for DCs including both immune stimulation and regulation. The outcome may vary depending on the subsets of DCs involved and their maturation status. In a transplantation model, "preferential" migration of alloantigen-specific plasmacytoid DCs to T-cell areas of lymph nodes, rather than spleen, induced antigen-specific $T_{reg}$ cells leading to transplant tolerance.[35] DCs conditioned with pharmacological agents that inhibit their maturation and function, or genetically engineered to express immunomodulatory molecules, have been shown to inhibit allogeneic T-cell responses in vitro. However, results in vivo have been less consistent so far. This divergence might be accounted for by several factors including poor migration of the immature DCs through lymphatics, their eventual maturation in the graft recipient and ineffective generation of $T_{reg}$ cells in vivo. Therefore, combined approaches that impact multiple different components of regulatory DC therapies may be advantageous, as suggested by several studies mentioned above. As an example, dual gene transduction of DCs with CCR7 and viral IL-10, but not with either gene alone, mediated long-term cardiac allograft survival.[135] Similar effects were achieved by dual CTLA4.Ig and antisense NF-κB gene transfer into DCs, but not with adenoviral CTLA4.Ig gene transfer alone.[115] Likewise, blockade of the CD40 pathway enhanced immune modulation by immature DCs expressing CTLA4.Ig.[120] Moreover, immature donor DCs in conjunction with anti-CD40L mAb,[73] anti-lymphocyte serum,[74] or anti-CD54 mAb plus CTLA4.Ig[75] achieved long-term cardiac allograft survival, while immature donor DCs combined with anti-CD40L mAb prevented graft arteriopathy in mouse aortic allografts.[76] A promising alternate approach includes in situ delivery of donor alloantigen to quiescent recipient DCs by means of donor-derived apoptotic cells and exosomes.[137]

Critical steps towards the development of regulatory DC therapies include a more precise identification of the DC subset(s) involved (most likely plasmacytoid cells[30,35]) and an improved understanding of the molecular signals that drive their lymphoid homing. Other aspects of DC-based therapies that need to be evaluated include cell viability, dose, frequency of cell delivery, its timing relative to transplantation and route of administration (i.v. injection seemingly being a suitable route). Because tolerance induction is more easily achievable in small animal models compared to monkeys and humans, preclinical studies in nonhuman primate models are needed. Recent studies have shown that rhesus monkey DCs transduced with the TGF-β1 gene suppress immune responses in vitro in an antigen-specific manner,[103] and that regulatory DCs can be generated from rhesus monkey monocytes conditioned with vitamin D and IL-10.[48] Such monocyte-derived regulatory DCs cells expressed low levels of surface MHC and co-stimulatory molecules and were poor stimulators of allogeneic T-cells in vitro. Particularly encouraging was the observation that these DCs were resistant to maturation induced by a potent combination of inflammatory cytokines (IL-1β, TNF-α, IFN-κ, IL-6, PGE2), suggesting than the problem of DC maturation in vivo could be circumvented. This was also achieved by "alternatively activated" DCs generated in the cell cultures containing IL-10 and TGF-β, then exposed to LPS, which retained their ability to regulate T cell functions in vitro and in vivo, even under inflammatory conditions.[139] These cells expressed comparatively low levels of surface MHC class II, CD80 and CD86, secreted high levels of IL-10, induced alloantigen-specific hyporesponsive T-cell proliferation, and expanded CD4+, CD25+, FOXp3, functional T reg cells. "Altenatively activated" DC promoted long-term allograft survival in combination with CTLAY-Ig in vivo.

Because generation of regulatory DCs derived from circulating monocytes in cell cultures takes several days and because regulatory DCs seem to be most effectively delivered a few days before transplantation, regulatory DC therapies could be more easily applied in live-donor human transplant recipients, primarily in kidney transplant recipients. The fact that several immunosuppressive drugs routinely used in human transplant recipients inhibit DC maturation and function offers interesting opportunities. An improved understanding of the roles of distinct DC subsets in alloimmunity and their differential modulation by pharmacological agents may lead to a more targeted utilization of these drugs. Clearly, the fact that future (hypothetical) clinical applications in DC therapy for transplantation would be initiated in combination with routinely used immunosuppressive regimens that also inhibit DC maturation represents a practical advantage. In addition, it should be emphasized that DC vaccine trials for cancer have been started more than 10 years ago and have included more than 1000 patients so far.[140] These approaches have generally proven to be safe and, in some cases, beneficial. The overall safety of DC vaccine trials for cancer is reassuring regarding the clinical translation of regulatory DC therapy in transplantation.

## Acknowledgements

The support of the Swiss Science Foundation, the Swiss Cardiology Foundation, the Lausanne Transplant Foundation, the Teo Rossi di Montelera Foundation and the Novartis Research Foundation is gratefully acknowledged.

## References

1. Salama AD, Remuzzi G, Harmon WE et al. Challenges to achieving clinical transplantation tolerance. J Clin Invest 2001; 108:943-48.
2. Billingham RE, Brent L, Medawar PB. "Actively acquired tolerance" of foreign cells. 1953. Transplantation 2003; 76:1409-12.
3. Sayegh MH, Turka LA. The role of T-cell costimulatory activation pathways in transplant rejection. N Engl J Med 1998; 338:1813-21.
4. Jenkins MK, Chen CA, Jung G et al. Inhibition of antigen-specific proliferation of type 1 murine T-cell clones after stimulation with immobilized anti-CD3 monoclonal antibody. J Immunol 1990; 144:16-22.
5. Suss G, Shortman K. A subclass of dendritic cells kills CD4 T-cells via Fas/Fas-ligand- induced apoptosis. J Exp Med 1996; 183:1789-96.
6. Lu L, Qian S, Hershberger PA et al. Fas ligand (CD95L) and B7 expression on dendritic cells provide counter-regulatory signals for T-cell survival and proliferation. J Immunol 1997; 158:5676-84.
7. Jonuleit H, Schmitt E, Schuler G et al. Induction of interleukin 10-producing, nonproliferating CD4(+) T-cells with regulatory properties by repetitive stimulation with allogeneic immature human dendritic cells. J Exp Med 2000; 192:1213-22.
8. Lechler RI, Garden OA, Turka LA. The complementary roles of deletion and regulation in transplantation tolerance. Nat Rev Immunol 2003; 3:147-58.
9. Starzl TE, Demetris AJ, Murase N et al. Cell migration, chimerism and graft acceptance. Lancet 1992; 339:1579-82.
10. Anderson CC, Matzinger P. Immunity or tolerance: opposite outcomes of microchimerism from skin grafts. Nat Med 2001; 7:80-87.
11. Pulendran B. Immune activation: death, danger and dendritic cells. Curr Biol 2004; 14:R30-32.
12. Sallusto F, Schaerli P, Loetscher P et al. Rapid and coordinated switch in chemokine receptor expression during dendritic cell maturation. Eur J Immunol 1998; 28:2760-69.
13. Martin-Fontecha A, Sebastiani S, Hopken UE et al. Regulation of dendritic cell migration to the draining lymph node: impact on T-lymphocyte traffic and priming. J Exp Med 2003; 198:615-21.
14. Larsen, CP, Morris PJ, Austyn JM. Migration of dendritic leukocytes from cardiac allografts into host spleens. A novel pathway for initiation of rejection. J Exp Med 1990; 171:307-14.
15. Austyn JM, Larsen CP. Migration patterns of dendritic leukocytes. Implications for transplantation. Transplantation 1990; 49:1-7.
16. Harshyne LA, Watkins SC, Gambotto A et al. Dendritic cells acquire antigens from live cells for cross-presentation to CTL. J Immunol 2001; 166:3717-23.
17. Lechler RI, Batchelor JR. Restoration of immunogenicity to passenger cell-depleted kidney allografts by the addition of donor strain dendritic cells. J Exp Med 1982; 155:31-41.
18. Benichou G, Valujskikh A, Heeger PS. Contributions of direct and indirect T-cell alloreactivity during allograft rejection in mice. J Immunol 1999; 162:352-58.
19. Auchincloss H Jr, Lee R, Shea S et al. The role of "indirect" recognition in initiating rejection of skin grafts from major histocompatibility complex class II-deficient mice. Proc Natl Acad Sci USA 1993; 90:3373-77.
20. Roussey-Kesler G, Brouard S, Ballet C et al. Exhaustive depletion of graft resident dendritic cells: marginally delayed rejection but strong alteration of graft infiltration. Transplantation 2005; 80:506-13.
21. Morelli AE, O'Connell PJ, Khanna A et al. Preferential induction of Th1 responses by functionally mature hepatic (CD8a− and CD8a+) dendritic cells: association with conversion from liver transplant tolerance to acute rejection. Transplantation 2000; 69:2647-57.
22. McKenna HJ, Stocking KL, Miller RE et al. Mice lacking flt3 ligand have deficient hematopoiesis affecting hematopoietic progenitor cells, dendritic cells and natural killer cells. Blood 2000; 95:3489-97.
23. Wang Z, Castellaneta A, De Creus A et al. Heart, but not skin, allografts from donors lacking Flt3 ligand exhibit markedly prolonged survival time. J Immunol 2004; 172:5924-30.
24. Hornick PI, Mason PD, Yacoub MH et al. Assessment of the contribution that direct allorecognition makes to the progression of chronic cardiac transplant rejection in humans. Circulation 1998; 97:1257-63.
25. MacDonald KP, Munster DJ, Clark GJ et al. Characterization of human blood dendritic cell subsets. Blood 2002; 100:4512-20.
26. Lindstedt M, Lundberg K, Borrebaeck CA. Gene family clustering identifies functionally associated subsets of human in vivo blood and tonsillar dendritic cells. J Immunol 2005; 175:4839-46.
27. Penna G, Vulcano M, Sozzani S et al. Differential migration behavior and chemokine production by myeloid and plasmacytoid dendritic cells. Hum Immunol 2002; 63:1164-71.
28. Zabel BA, Silverio AM, Butcher EC. Chemokine-like receptor 1 expression and chemerin-directed chemotaxis distinguish plasmacytoid from myeloid dendritic cells in human blood. J Immunol 2005; 174:244-51.
29. Nakano H, Yanagita M, Gunn MD. Cd11c(+)b220(+)gr-1(+) cells in mouse lymph nodes and spleen display characteristics of plasmacytoid dendritic cells. J Exp Med 2001; 194:1171-78.
30. Abe M, Wang Z, de Creus A et al. Plasmacytoid dendritic cell precursors induce allogeneic T-cell hyporesponsiveness and prolong heart graft survival. Am J Transplant 2005; 5:1808-19.

31. Hawiger D, Inaba K, Dorsett Y et al. Dendritic cells induce peripheral T-cell unresponsiveness under steady state conditions in vivo. J Exp Med 2001; 194:769-79.

32. Fu F, Li Y, Qian S et al. Costimulatory molecule-deficient dendritic cell progenitors (MHC class II+, CD80dim, CD86-) prolong cardiac allograft survival in non-immunosuppressed recipients. Transplantation 1996; 62:659-65.

33. Lutz MB, Suri RM, Niimi M et al. Immature dendritic cells generated with low doses of GM-CSF in the absence of IL-4 are maturation resistant and prolong allograft survival in vivo. Eur J Immunol 2000; 30:1813-22.

34. O'Connell PJ, Li W, Wang Z et al. Immature and mature CD8alpha+ dendritic cells prolong the survival of vascularized heart allografts. J Immunol 2002; 168:143-54.

35. Ochando JC, Homma C, Yang Y et al. Alloantigen-presenting plasmacytoid dendritic cells mediate tolerance to vascularized grafts. Nat Immunol 2006; 7:652-62.

36. Tang Q, Bluestone JA. Plasmacytoid DCs and T(reg) cells: casual acquaintance or monogamous relationship? Nat Immunol 2006; 7:551-53.

37. Lim WH, Kireta S, Thomson AW et al. Renal transplantation reverses functional deficiencies in circulating dendritic cell subsets in chronic renal failure patients. Transplantation 2006; 81:160-68.

38. Hesselink DA, Vaessen LM, Hop WC et al. The effects of renal transplantation on circulating dendritic cells. Clin Exp Immunol 2005; 140:384-93.

39. Athanassopoulos P, Vaessen LM, Maat AP et al. Peripheral blood dendritic cells in human end-stage heart failure and the early posttransplant period: evidence for systemic Th1 immune responses. Eur J Cardiothorac Surg 2004; 25:619-26.

40. Hackstein H, Renner FC, Bohnert A et al. Dendritic cell deficiency in the blood of kidney transplant patients on long-term immunosuppression: results of a prospective matched-cohort study. Am J Transplant 2005; 5:2945-53.

41. Athanassopoulos P, Vaessen LM, Maat AP et al. Preferential depletion of blood myeloid dendritic cells during acute cardiac allograft rejection under controlled immunosuppression. Am J Transplant. 2005; 5:810-20.

42. Athanassopoulos P, Vaessen LM, Balk AH et al. Impaired circulating dendritic cell reconstitution identifies rejecting recipients after clinical heart transplantation independent of rejection therapy. Eur J Cardiothorac Surg 2005; 27:783-89.

43. Mazariegos GV, Zahorchak AF, Reyes J et al. Dendritic cell subset ratio in tolerant, weaning and nontolerant liver recipients is not affected by extent of immunosuppression. Am J Transplant 2005; 5:314-22.

44. Daniel V, Naujokat C, Sadeghi M et al. Association of circulating interleukin (IL)-12- and IL-10-producing dendritic cells with time posttransplant, dose of immunosuppression and plasma cytokines in renal-transplant recipients. Transplantation 2005; 79:1498-506.

45. Coates PTH, Colvin BL, Hackstein H et al. Manipulation of dendritic cells as an approach to improved outcomes in transplantation. Expert Rev Mol Med 2002; 2002:1-21.

46. Coates PT, Colvin BL, Kaneko K et al. Pharmacologic, biologic and genetic engineering approaches to potentiation of donor-derived dendritic cell tolerogenicity. Transplantation 2003; 75(9 Suppl):32S-36S.

47. Enk AH. Dendritic cells in tolerance induction. Immunol Lett 2005; 99:8-11.

48. McCurry KR, Colvin BL, Zahorchak AF et al. Regulatory dendritic cell therapy in organ transplantation. Transpl Int 2006; 19:525-38.

49. Caux C, Dezutter-Dambuyant C, Schmitt D et al. GM-CSF and TNF-alpha cooperate in the generation of dendritic Langerhans cells. Nature 1992; 360:258-61.

50. Arpinati M, Green CL, Heimfeld S et al. Granulocyte-colony stimulating factor mobilizes T helper 2-inducing dendritic cells. Blood 2000; 95:2484-90.

51. Dhodapkar MV, Steinman RM, Krasovsky J et al. Antigen-specific inhibition of effector T-cell function in humans after injection of immature dendritic cells. J Exp Med 2001; 193:233-38.

52. el-Sawy T, Fahmy NM, Fairchild RL. Chemokines: directing leukocyte infiltration into allografts. Curr Opin Immunol 2002; 14:562-68.

53. el-Sawy T, Miura M, Fairchild R. Early T-cell response to allografts occurring prior to alloantigen priming up-regulates innate-mediated inflammation and graft necrosis. Am J Pathol 2004; 165:147-57.

54. de Groot-Kruseman HA, Baan CC, Loonen EH et al. Failure to down-regulate intragraft cytokine mRNA expression shortly after clinical heart transplantation is associated with high incidence of acute rejection. J Heart Lung Transplant 2001; 20:503-10.

55. Azzawi M, Hasleton PS, Geraghty PJ et al. RANTES chemokine expression is related to acute cardiac cellular rejection and infiltration by CD45RO T-lymphocytes and macrophages. J Heart Lung Transplant 1998; 17:881-87.

56. Sallusto F, Palermo B, Lenig D et al. Distinct patterns and kinetics of chemokine production regulate dendritic cell function. Eur J Immunol 1999; 29:1617-25.

57. Yoneyama H, Matsuno K, Zhang Y et al. Evidence for recruitment of plasmacytoid dendritic cell precursors to inflamed lymph nodes through high endothelial venules. Int Immunol 2004; 16:915-28.

58. Forster R, Schubel A, Breitfeld D et al. CCR7 coordinates the primary immune response by establishing functional microenvironments in secondary lymphoid organs. Cell 1999; 99:23-33.

59. Gunn MD, Kyuwa S, Tam C et al. Mice lacking expression of secondary lymphoid organ chemokine have defects in lymphocyte homing and dendritic cell localization. J Exp Med 1999; 189:451-60.

60. Allavena P, Sica A, Vecchi A et al. The chemokine receptor switch paradigm and dendritic cell migration: its significance in tumor tissues. Immunol Rev 2000; 177:141-49.

61. Saeki H, Moore AM, Brown MJ et al. Cutting edge: secondary lymphoid-tissue chemokine (SLC) and CC chemokine receptor 7 (CCR7) participate in the emigration pathway of mature dendritic cells from the skin to regional lymph nodes. J Immunol 1999; 162:2472-75.

62. Gao W, Topham PS, King JA et al. Targeting of the chemokine receptor CCR1 suppresses development of the acute and chronic cardiac allograft rejection. J Clin Invest 2000; 105:35-44.

63. Gao W, Faia KL, Csizmadia V et al. Beneficial effects of targeting CCR5 in allograft recipients. Transplantation 2001; 72:1199-205.

64. Hancock WW, Lu B, Gao W et al. Requirement of the chemokine receptor CXCR3 for acute allograft rejection. J Exp Med 2000; 192:1515-20.

65. Hancock WW, Gao W, Csizmadia V et al. Donor-derived IP-10 initiates development of acute allograft rejection. J Exp Med 2001; 193:975-80.

66. Koga S, Kobayashi H, Novick AC et al. Prolonged class II MHC disparate skin allograft survival by treatment with antibodies to the chemokine Mig. Transplant Proc 2001; 33:549-50.

67. Horuk R, Shurey S, Ng HP et al. CCR1-specific nonpeptide antagonist: efficacy in a rabbit allograft rejection model. Immunol Lett 2001; 76:193-201.

68. DeBruyne LA, Li K, Bishop DK et al. Gene transfer of virally encoded chemokine antagonists vMIP-II and MC148 prolongs cardiac allograft survival and inhibits donor-specific immunity. Gene Ther 2000; 7:575-82.

69. Vassalli G, Simeoni E, Li J et al. Lentiviral gene transfer of the chemokine antagonist RANTES 9-68 prolongs heart graft survival. Transplantation 2006; 81:240-46.

70. Wang L, Han R, Lee I et al. Permanent survival of fully MHC-mismatched islet allografts by targeting a single chemokine receptor pathway. J Immunol 2005; 175:6311-18.

71. Colvin BL, Wang Z, Nakano H et al. CXCL9 antagonism further extends prolonged cardiac allograft survival in CCL19/CCL21-deficient mice. Am J Transplant 2005; 5:2104-13.

72. Lu L, McCaslin D, Starzl TE et al. Bone marrow-derived dendritic cell progenitors (NLDC 145+, MHC class II+, B7-1dim, B7-2-) induce alloantigen-specific hyporesponsiveness in murine T-lymphocytes. Transplantation 1995; 60:1539-45.

73. Lu L, Li W, Zhong C et al. Increased apoptosis of immunoreactive host cells and augmented donor leukocyte chimerism, not sustained inhibition of B7 molecule expression are associated with prolonged cardiac allograft survival in mice preconditioned with immature donor dendritic cells plus anti-CD40L mAb. Transplantation 1999; 68:747-57.

74. DePaz HA, Oluwole OO, Adeyeri AO et al. Immature rat myeloid dendritic cells generated in low-dose granulocyte macrophage-colony stimulating factor prolong donor-specific rat cardiac allograft survival. Transplantation 2003; 75:521-28.

75. Wang Q, Zhang M, Ding G et al. Anti-ICAM-1 antibody and CTLA-4Ig synergistically enhance immature dendritic cells to induce donor-specific immune tolerance in vivo. Immunol Lett 2003; 90:33-42.

76. Wang Z, Morelli AE, Hackstein H et al. Marked inhibition of transplant vascular sclerosis by in vivo-mobilized donor dendritic cells and anti-CD154 mAb. Transplantation 2003; 76:562-71.

77. Peche H, Trinite B, Martinet B et al. Prolongation of heart allograft survival by immature dendritic cells generated from recipient type bone marrow progenitors. Am J Transplant 2005; 5:255-67.

78. Garrovillo M, Ali A, Depaz HA et al. Induction of transplant tolerance with immunodominant allopeptide-pulsed host lymphoid and myeloid dendritic cells. Am J Transplant 2001; 1:129-37.

79. Taner T, Hackstein H, Wang Z et al. Rapamycin-treated, alloantigen-pulsed host dendritic cells induce ag-specific T-cell regulation and prolong graft survival. Am J Transplant 2005; 5:228-36.

80. Tiao MM, Lu L, Tao R et al. Prolongation of cardiac allograft survival by systemic administration of immature recipient dendritic cells deficient in NF-kappaB activity. Ann Surg 2005; 241:497-505.

81. Hackstein H, Morelli AE, Thomson AW. Designer dendritic cells for tolerance induction: guided not misguided missiles. Trends Immunol 2001; 22:437-42.

82. Moser M, De Smedt T, Sornasse T et al. Glucocorticoids down-regulate dendritic cell function in vitro and in vivo. Eur J Immunol 1995; 25:2818-24.

83. Piemonti L, Monti P, Allavena P et al. Glucocorticoids affect human dendritic cell differentiation and maturation. J Immunol 1999; 162:6473-81.

84. Vanderheyde N, Verhasselt V, Goldman M et al. Inhibition of human dendritic cell functions by methylprednisolone. Transplantation 1999; 67:1342-47.

85. Matyszak MK, Citterio S, Rescigno M. Differential effects of corticosteroids during different stages of dendritic cell maturation. Eur J Immunol 2000; 30:1233-42.

86. Woltman AM, de Fijter JW, Kamerling SW et al. The effect of calcineurin inhibitors and corticosteroids on the differentiation of human dendritic cells. Eur J Immunol 2000; l 30:1807-12.

87. Abe M, Thomson AW. Dexamethasone preferentially suppresses plasmacytoid dendritic cell differentiation and enhances their apoptotic death. Clin Immunol 2006; 118:300-6.

88. Duperrier K, Velten FW, Bohlender J et al. Immunosuppressive agents mediate reduced allostimulatory properties of myeloid-derived dendritic cells despite induction of divergent molecular phenotypes. Mol Immunol 2005; 42:1531-40.

89. Lee YR, Yang IH, Lee YH et al. Cyclosporin A and tacrolimus, but not rapamycin, inhibit MHC-restricted antigen presentation pathways in dendritic cells. Blood 2005; 105:3951-55.

90. Matsue H, Yang C, Matsue K et al. Contrasting impacts of immunosuppressive agents (rapamycin, FK506, cyclosporin A and dexamethasone) on bidirectional dendritic cell-T-cell interaction during antigen presentation. J Immunol 2002; 169:3555-64.

91. Chen T, Guo J, Yang M et al. Cyclosporin A impairs dendritic cell migration by regulating chemokine receptor expression and inhibiting cyclooxygenase-2 expression. Blood 2004; 103:413-21.

92. Sordi V, Bianchi G, Buracchi C et al. Differential effects of immunosuppressive drugs on chemokine receptor CCR7 in human monocyte-derived dendritic cells: selective upregulation by rapamycin. Transplantation 2006; 82:826-34.

93. Mehling A, Grabbe S, Voskort M et al. Mycophenolate mofetil impairs the maturation and function of murine dendritic cells. J Immunol 2000; 165:2374-81.

94. Matasic R, Dietz AB, Vuk-Pavlovic S. Cyclooxygenase-independent inhibition of dendritic cell maturation by aspirin. Immunology 2000; 101:53-60.

95. Hackstein H, Morelli AE, Larregina AT et al. Aspirin inhibits in vitro maturation and in vivo immunostimulatory function of murine myeloid dendritic cells. J Immunol 2001; 166:7053-62.

96. Grotz W, Siebig S, Olschewski M et al. Low-dose aspirin therapy is associated with improved allograft function and prolonged allograft survival after kidney transplantation. Transplantation. 2004; 77:1848-53.

97. Tepper MA, Nadler SG, Esselstyn JM et al. Deoxyspergualin inhibits kappa light chain expression in 70Z/3 pre-B cells by blocking lipopolysaccharide-induced NF-kappa B activation. J Immunol 1995; 155:2427-36.

98. Thomas JM, Contreras JL, Jiang XL et al. Peritransplant tolerance induction in macaques: early events reflecting the unique synergy between immunotoxin and deoxyspergualin. Transplantation 1999; 68:1660-73.

99. Piemonti L, Monti P, Sironi M et al. Vitamin D3 affects differentiation, maturation and function of human monocyte-derived dendritic cells. J Immunol 2000; 164:4443-51.

100. Griffin MD, Lutz WH, Phan VA et al. Potent inhibition of dendritic cell differentiation and maturation by vitamin D analogs. Biochem Biophys Res Commun 2000; 270:701-8.

101. Lee WC, Zhong C, Qian S et al. Phenotype, function and in vivo migration and survival of allogeneic dendritic cell progenitors genetically engineered to express TGF-beta. Transplantation 1998; 66:1810-17.

102. Takayama T, Kaneko K, Morelli AE et al. Retroviral delivery of transforming growth factor-beta1 to myeloid dendritic cells: inhibition of T-cell priming ability and influence on allograft survival. Transplantation 2002; 74:112-19.

103. Asiedu C, Dong SS, Pereboev A et al. Rhesus monocyte-derived dendritic cells modified to over-express TGF-beta1 exhibit potent veto activity. Transplantation 2002; 74:629-37.

104. Sun W, Wang Q, Zhang L et al. TGF-beta(1) gene modified immature dendritic cells exhibit enhanced tolerogenicity but induce allograft fibrosis in vivo. J Mol Med 2002; 80:514-23.

105. Rea D, Laface D, Hutchins B et al. Recombinant adenovirus-transduced human dendritic cells engineered to secrete interleukin-10 (IL-10) suppress Th1-type responses while selectively activating IL-10-producing CD4+ T-cells. Hum Immunol 2004; 65:1344-55.

106. Coates PT, Krishnan R, Kireta S et al. Human myeloid dendritic cells transduced with an adenoviral interleukin-10 gene construct inhibit human skin graft rejection in humanized NOD-scid chimeric mice. Gene Ther 2001; 8:1224-33.

107. Zhang M, Wang Q, Liu Y et al. Effective induction of immune tolerance by portal venous infusion with IL-10 gene-modified immature dendritic cells leading to prolongation of allograft survival. J Mol Med 2004; 82:240-49.

108. Takayama T, Tahara H, Thomson AW. Differential effects of myeloid dendritic cells retrovirally transduced to express mammalian or viral interleukin-10 on cytotoxic T-lymphocyte and natural killer cell functions and resistance to tumor growth. Transplantation 2001; 71:1334-40.

109. Buonocore S, Flamand V, Goldman M et al. Bone marrow-derived immature dendritic cells prime in vivo alloreactive T-cells for interleukin-4-dependent rejection of major histocompatibility complex class II antigen-disparate cardiac allograft. Transplantation 2003; 75:407-13.

110. Buonocore S, Van Meirvenne S, Demoor FX et al. Dendritic cells transduced with viral interleukin 10 or Fas ligand: no evidence for induction of allotolerance in vivo. Transplantation 2002; 73(1 Suppl):S27-30.

111. Lee WC, Qiani S, Wan Y et al. Contrasting effects of myeloid dendritic cells transduced with an adenoviral vector encoding interleukin-10 on organ allograft and tumor rejection. Immunology 2000; 101:233-41.

112. Kaneko K, Wang Z, Kim SH et al. Dendritic cells genetically engineered to express IL-4 exhibit enhanced IL-12p70 production in response to CD40 ligation and accelerate organ allograft rejection. Gene Ther 2003; 10:143-52.

113. Morelli AE, Larregina AT, Ganster RW et al. Recombinant adenovirus induces maturation of dendritic cells via an NF-kappaB-dependent pathway. J Virol 2000; 74:9617-28.

114. Tiao MM, Lu L, Tao R et al. Prolongation of cardiac allograft survival by systemic administration of immature recipient dendritic cells deficient in NF-kappaB activity. Ann Surg 2005; 241:497-505.

115. Bonham CA, Peng L, Liang X et al. Marked prolongation of cardiac allograft survival by dendritic cells genetically engineered with NF-kappa B oligodeoxyribonucleotide decoys and adenoviral vectors encoding CTLA4Ig. J Immunol 2002; 169:3382-91.

116. Tomasoni S, Aiello S, Cassis L et al. Dendritic cells genetically engineered with adenoviral vector encoding dnIKK2 induce the formation of potent CD4+ T-regulatory cells. Transplantation 2005; 79:1056-61.

117. Lu L, Gambotto A, Lee WC et al. Adenoviral delivery of CTLA4Ig into myeloid dendritic cells promotes their in vitro tolerogenicity and survival in allogeneic recipients. Gene Ther 1999; 6:554-63.

118. Takayama T, Morelli AE, Robbins PD et al. Feasibility of CTLA4Ig gene delivery and expression in vivo using retrovirally transduced myeloid dendritic cells that induce alloantigen-specific T-cell anergy in vitro. Gene Ther 2000; 7:1265-73.

119. O'Rourke RW, Kang SM, Lower JA et al. A dendritic cell line genetically modified to express CTLA4-IG as a means to prolong islet allograft survival. Transplantation 2000; 69:1440-46.

120. Sun W, Wang Q, Zhang L et al. Blockade of CD40 pathway enhances the induction of immune tolerance by immature dendritic cells genetically modified to express cytotoxic T-lymphocyte antigen 4 immunoglobulin. Transplantation 2003; 76:1351-59.

121. Lenschow DJ, Zeng Y, Thistlethwaite JR et al. Long-term survival of xenogeneic pancreatic islet grafts induced by CTLA4Ig. Science 1992; 257:789-92.

122. Guillot C, Mathieu P, Coathalem H et al. Tolerance to cardiac allografts via local and systemic mechanisms after adenovirus-mediated CTLA4Ig expression. J Immunol 2000; 164:5258-68.

123. Olthoff KM, Olthoff KM, Judge TA et al. Adenovirus-mediated gene transfer into cold-preserved liver allografts: survival pattern and unresponsiveness following transduction with CTLA4Ig. Nat Med 1998; 4:194-200.

124. Yamashita K, Masunaga T, Yanagida N et al. Long-term acceptance of rat cardiac allografts on the basis of adenovirus mediated CD40Ig plus CTLA4Ig gene therapies. Transplantation 2003; 76:1089-96.

125. Hoves S, Krause SW, Herfarth H et al. Elimination of activated but not resting primary human CD4+ and CD8+ T-cells by Fas ligand (FasL/CD95L)-expressing Killer-dendritic cells. Immunobiology 2004; 208:463-75.
126. Matsue H, Matsue K, Walters M et al. Induction of antigen-specific immunosuppression by CD95L cDNA-transfected 'killer' dendritic cells. Nat Med 1999; 5:930-37.
127. Min WP, Gorczynski R, Huang XY et al. Dendritic cells genetically engineered to express Fas ligand induce donor-specific hyporesponsiveness and prolong allograft survival. J Immunol 2000; 164:161-67.
128. Zhang HG, Su X, Liu D et al. Induction of specific T-cell tolerance by Fas ligand-expressing antigen-presenting cells. J Immunol 1999; 162:1423-30.
129. Buonocore S, Paulart F, Le Moine A et al. Dendritic cells overexpressing CD95 (Fas) ligand elicit vigorous allospecific T-cell responses in vivo. Blood 2003; 101:1469-76.
130. Buonocore S, Flamand V, Claessen N et al. Dendritic cells overexpressing Fas-ligand induce pulmonary vasculitis in mice. Clin Exp Immunol 2004; 137:74-80.
131. Liang X, Lu L, Chen Z et al. Administration of dendritic cells transduced with antisense oligodeoxyribonucleotides targeting CD80 or CD86 prolongs allograft survival. Transplantation 2003; 76:721-29.
132. Wang Q, Liu Y, Wang J et al. Induction of allospecific tolerance by immature dendritic cells genetically modified to express soluble TNF receptor. J Immunol 2006; 177:2175-85.
133. Perone MJ, Larregina AT, Shufesky WJ et al. Transgenic galectin-1 induces maturation of dendritic cells that elicit contrasting responses in naive and activated T-cells. J Immunol 2006; 176:7207-20.
134. Perone MJ, Bertera S, Tawadrous ZS et al. Dendritic cells expressing transgenic galectin-1 delay onset of autoimmune diabetes in mice. J Immunol 2006; 177:5278-89.
135. Garrod KR, Chang CK, Liu FC et al. Targeted lymphoid homing of dendritic cells is required for prolongation of allograft survival. J Immunol 2006; 177:863-64.
136. Robert C, Klein C, Cheng G et al. Gene therapy to target dendritic cells from blood to lymph nodes. Gene Ther 2003; 10:1479-86.
137. Morelli AE. The immune regulatory effect of apoptotic cells and exosomes on dendritic cells: its impact on transplantation. Am J Transplant 2006; 6:254-61.
138. Wang Z, Larregina AT, Shufesky WJ et al. Use of the inhibitory effect of apoptotic cells on dendritic cells for graft survival via T-cell deletion and regulatory T-cells. Am J Transplant 2006; 6:1297-311.
139. Lan YY, Wang Z, Raimondi G et al. "Alternatively activated" dendritic cells preferentially secrete IL-10, expand FOXp3=CD4= T cells, and induce long-term organ allograft survival in combination with CTLAY-Ig. J Immunol 2006; 177:5868-77.
140. Figdor CG, de Vries IJ, Lesterhuis WJ et al. Dendritic cell immunotherapy: mapping the way. Nat Med 2004; 10:475-80.

# CHAPTER 9

# Cardiac Allograft Vasculopathy

Jignesh K. Patel and Jon A. Kobashigawa*

## Cardiac Allograft Vasculopathy

Over the last four decades, cardiac transplantation has been the preferred therapy for select patients with end-stage heart disease. Improvements in immunosuppression, donor procurement, surgical techniques and posttransplant care over this period have resulted in a substantial decrease in acute allograft rejection which had previously significantly limited survival of transplant recipients (Fig. 1). In contrast, long-term allograft survival has been limited by the development of cardiac allograft vasculopathy (CAV) and there has been only a modest corresponding decrease in its prevalence over the years. According to the registry of the International Society of Heart and Lung transplantation (ISHLT), after 5 years, CAV and late graft failure (likely due to allograft vasculopathy) together account for 30% of deaths.[1]

CAV is a process akin to native atherosclerosis whereby coronary arteries develop progressive intimal thickening which ultimately compromises myocardial perfusion and function. CAV was noted clinically in the 1970s, when histologic studies of coronary artery lesions in transplant hearts revealed extensive disease in vessels that had appeared to be normal angiographically at the time of transplantation.

## Incidence and Prognosis

Despite its limited sensitivity, coronary angiography remains the preferred method of the clinical detection of CAV[2] and routine annual angiography is performed at most centers due to the insidious nature of the disease. The incidence of CAV in reported studies has varied widely due to differences in definition of disease and patient populations. In one of the largest cohorts studied of over 6000 angiograms performed in over 2600 patients from 39 institutions, angiographically significant CAV was noted in 42% of the patients at 5 years. CAV-related events (death or retransplantation) had an actuarial incidence of 7% at 5 years.[3] In a more recent study, while only 10% of heart transplant recipients developed CAV at 5 years, there was a substantial increase in incidence thereafter, with 50% having developed disease by 10 years.[4] CAV may occur as early as one year after transplantation and disease appearing early following transplantation is more aggressive and associated with a worse prognosis.[5] In one study, those with angiographic disease had a relative risk of any cardiac event of 3.4 and of death of 4.6 compared with those without disease over a mean follow up of 3.5 years.[6] In patients without apparent angiographic epicardial disease, microvascular abnormalities may be present and even this extent of disease has been associated with adverse outcomes.[7]

## Pathology

An understanding of the pathology of CAV requires comparison with processes which lead to nontransplant atherosclerosis. Conventional atherosclerosis is a manifestation of a response to injury of the endothelium as first hypothesized by Ross.[8] This leads to chronic inflammation affecting the vessel wall with accumulation of inflammatory cells including macrophages, T-cells and smooth muscle cells. The initial injury to the endothelium may be caused by a number of factors including cholesterol, hypertension and oxidative stress. Similarly, CAV is thought to result from an initial injury to the allograft endothelium. However, here there are wider variety of determinants which may be involved. These include preservation injury, alloimmune response and possibly chronic CMV infection, in addition to the conventional risk factors for atherosclerosis. Lipid accumulation in allograft arteries may be prominent, with lipoprotein entrapment in the subendothelial tissue, through interactions with proteoglycans. Smooth muscle cell (SMC) activation leads to migration from the media into the intima, proliferation and release of cytokines and extracellular matrix proteins, resulting in luminal narrowing and impaired vascular function.[9] A number of features distinguish CAV from conventional atherosclerosis (Table 1); CAV lesions are more diffuse nature with frequent involvement of large and medium sized vessels as well as the microvasculature. Many transplant recipients with CAV however also have lesions typical of more conventional atherosclerosis but generally, lesions tend to be lipid poor and calcification is less prominent and occurs late. A wide spectrum of lesion-type is therefore apparent. The disease not only affects the intima but the media and adventitia frequently undergo fibrous infiltration. As a consequence compensatory remodeling of the artery is inhibited (the Glagov phenomenon) and the artery may even undergo constriction,[10]

**Table 1. Distinguishing features of CAV from nontransplant atherosclerosis**

| Nontransplant Atherosclerosis | CAV |
| --- | --- |
| Mostly epicardial disease | Panvascular disease (including microvasculature) |
| Slow progression | Rapid progression |
| Eccentric lesions | Concentric lesions (generally) |
| Lipid rich | Generally lipid poor |
| Early calcification | Calcification occurs late |
| Compensatory remodeling with early dilation (Glagov phenomenon) | Arterial constriction |

*Corresponding Author: Jon A. Kobashigawa—Medical Director, UCLA Heart Transplant Program David Geffen School of Medicine at UCLA, 100 UCLA Medical Plaza #630 Los Angeles, CA 90095, USA. Email: jonk@mednet.ucla.edu

*Chronic Allograft Failure: Natural History, Pathogenesis, Diagnosis and Management*, edited by Nasimul Ahsan. ©2008 Landes Bioscience.

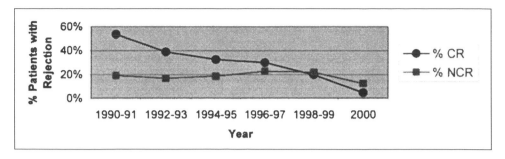

Figure 1. Incidence of cellular and noncellular (humoral) rejection following heart transplantation over a 10 year period at a single institution.[133] Reprinted from Subherwal S et al. 2004; 36(10):3171-3172; [133] ©2004 with permission from Elsevier.

a phenomenon which may contribute to the earlier manifestation of clinically apparent disease. In some cases, subepicardial inflammatory infiltrates are noted even in the absence of myocardial interstitial inflammatory infiltrates and intravascular thrombus may be found at autopsy or[11] following explant for retransplantation.

Angiographic lesions have been classified to describe the spectrum of lesions seen in CAV. Type A lesions have features of conventional coronary atherosclerotic disease. Type B and C lesions represent more typical transplant-related disease (Fig. 2). However, as CAV lesions are frequently concentric due to subintimal cellular proliferation, conventional coronary angiography generally underestimates the extent of disease (Fig. 3). The development of intravascular ultrasound (IVUS) has allowed direct imaging of the vessel wall and determine extent of intimal and medial disease. The technique has been useful in determining characteristics of donor-related lesions (conventional atherosclerosis) by examination early following transplant and comparing transplant-acquired lesions later on in the same regions.[12,13] IVUS therefore is able to provide more detailed assessment with increased sensitivity.[14]

## Clinical Presentation

The typical clinical presentation of cardiac ischemia or infarction with chest pain seen in nontransplanted patients is frequently not seen in heart transplant recipients due to the denervated state of the allograft.[15] Many patients in due course however do develop evidence for cardiac reinnervation and chest pain due to ischemia and infarction in transplant patients has been documented.[16-18]

Electrocardiographic changes with myocardial infarction may be atypical due to baseline abnormalities or heterogenous disease resulting from diffuse vasculopathy.[19] In general, the atypical presentation often leads to lower utilization for revascularization therapies and consequently worse outcomes,[6,19] including heart failure, arrhythmia or sudden death.

## Immune Mechanisms in the Development of CAV

Despite a cohort of immunosuppressive strategies employed following cardiac transplantation, the allograft remains a major stimulus to the recipient immune system. A substantial portion of circulating donor T-lymphocytes are able to recognize allo-antigens (Major Histocompatibility Complex) and hence mount a robust immune response which leads to production of a variety of cytokines (interleukins, interferion γ and tumor necrosis factor α). These cytokines allow development of effector mechanisms including cytotoxic T-cells, infiltrating macrophages and antibody production. Whilst much of this immune response is responsible for the allograft dysfunction seen in acute rejection, there is evidence to suggest that it also plays an important contributory role in the development of CAV. This phenomenon is therefore also frequently termed chronic rejection, a process not only seen in the transplanted heart but also in renal, lung and liver allografts. Retrospective analysis suggests a correlation between the extent of HLA-DR mismatch and the subsequent development of CAV and long-term survival.[20] The development of posttransplant HLA Class II antibodies was also associated with CAV by IVUS in another study.[21]

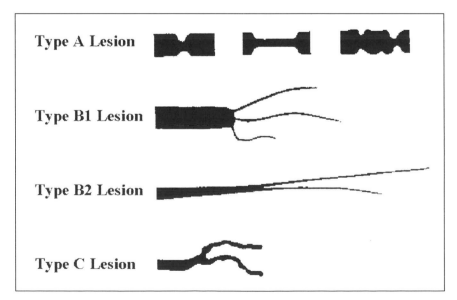

Figure 2. Types of angiographic lesions seen in CAV. Type A lesion: discrete tubular or multiple stenoses. Type B lesions: Distal disease with either normal epicardial morphology (B1) or gradual tapering of epicardial vessels (B2). Type C: diseased irregular vessel with abrupt termination. Reprinted from: Patel J, Kobashigawa J. Advanced Therapy of Cardiac Surgery, 2nd ed. 2003:581; with permission from Elsevier.

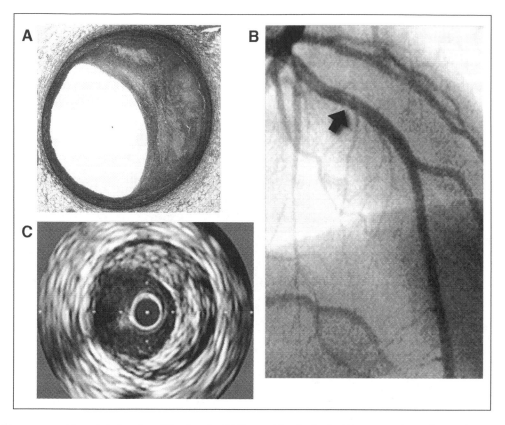

Figure 3. Concentric or eccentric sub-intimal proliferation in CAV seen histologically (A) may under-estimate the severity of the lesion angiographically (B) but is better appreciated by IVUS (C).

Endothelial cells likely play a pivotal role in mediating both acute and chronic rejection.[22,23] They form the initial interface between donor and the host circulating lymphocytes and maintain a barrier to the egress of inflammatory cells into the interstitium. They are also highly responsive to cytokines,[24] express Class II antigens particularly in the microvasculature[25,26] and coronary arteries.[26] They may even act as antigen presenting cells (APCs).[27,28] Although T-cell injury to the myocardium in acute cellular rejection is limited by immunosuppressive agents, it appears that these agents have a more limited effect on the development of CAV. One explanation may be that these agents are more effective at suppressing IL-2 production and less effective at suppressing the production of cytokines which lead to antibody production.[29] The majority of patients following solid organ transplantation continue to make antibodies to the allograft and anti-HLA antibody production in recipients has been associated with a higher mortality.[30] More specifically, production of antiendothelial antibodies has been shown to correlate with the development of CAV.[31] The development of CAV has been particularly associated with the formation of antibodies against the intermediate filament vimentin,[32] a protein characteristic but not restricted to endothelial cells. Vimentin is diffusely expressed in the intima and media of normal and diseased coronary arteries. Early endothelial injury following transplantation, for example, by ischemia and reperfusion, may lead to release of vimentin into the circulation. The protein may be taken up by APCs and presented as an autoantigen as it is not normally exposed to the immune system. Interestingly, these antibodies do not seem to mediate complement mediated cytotoxity to endothelial cells in vitro, a process associated with hyperacute rejection. Anti-vimentin antibodies may therefore exert a more subtle form of low grade damage in a process in keeping with the chronic progressive course of CAV development. One possible mechanism may be the up-regulation of endothelial cell adhesion molecules[33] which over a period of time would allow adherence and trans-migration of inflammatory cells into the intima.

## Endothelial Activation

Endothelial activation is a phenomenon seen in allografts whereby the cells express HLA Class II antigens and adhesion molecules such as ICAM-1 and VCAM and likely represents a response to injury.[34,35] These molecules are not usually expressed on endothelium of normal hearts. Although expression of adhesion molecules on endothelium is a feature of nontransplant atherosclerosis, the expression of HLA antigens seems to be unique to the transplant endothelium. A consequence of this allo-antigen expression is that it provides a sustained stimulus for lymphocyte proliferation and production of interleukin-2 and thereby maintains a chronic immunologic response.[35] Allografts with expression of ICAM-1 and HLA-DR on arterial and arteriolar endothelium have been shown to develop transplant vasculopathy.[36]

A number of factors may lead to endothelial activation (Fig. 4). Ischemia-reperfusion injury at the time of transplantation likely plays an important role in endothelial activation. Human endothelial cells subjected to hypoxia experimentally do show upregulation of ICAM-1.[37] In animal models, the induction of adhesion molecules on allograft endothelium prior to transplantation has also been shown to be associated with the subsequent development of CAV in the recipient.[38] In humans, the assessment of circulating soluble ICAM-1 following transplantation has been suggested to be a marker for the subsequent development of CAV.[39] Cytokines released following an episode of acute rejection may lead to this process. An increased incidence of CAV has been correlated with the frequency of acute allograft rejection episodes.[40] Other stimuli for endothelial activation include injury by endothelial-specific T-lymphocytes or the generation of recipient antibodies to donor endothelium such as anti-vimentin antibodies.

One of the distinguishing features of CAV is the diffuse nature of the disease affecting not only the arterial network but also the capillary network and the venous system.[41] Panvascular endothelial injury likely accounts for this phenomenon. The normal cardiac microvasculature is highly resistant to thrombosis and an intact endothelium plays an

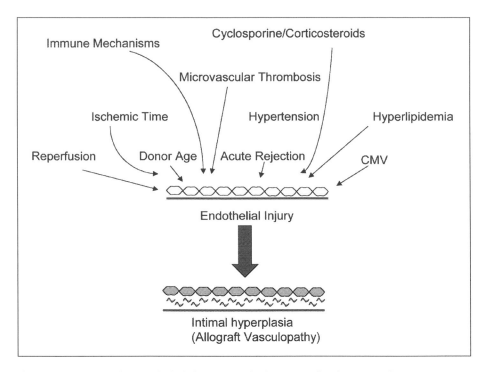

Figure 4. Factors contributing to posttransplant endothelial injury and subsequent development of CAV.

important role in maintaining the balance between procoagulant and anti-thrombotic forces. As early as one month following cardiac transplantation, however, endomyocardial biopsies begin to show evidence of microvascular fibrin deposition and this subsequently significantly correlates with the development of CAV.[42] In a rat model of heart transplantation, a hypercoagulable microvasculature was associated with the development of CAV.[43] Microvascular thrombosis also results in acute myocardial cell damage from ischemia and cardiac troponins, sensitive markers for myocardial damage, are indeed elevated in transplant patients with evidence of microvascular fibrin deposition on endomyocardial biopsies.[44] Furthermore, patients with persistently elevated troponins have a significantly higher likelihood of developing graft failure or severe CAV. Elevated troponins in the first month following cardiac transplantation are very common, presumably related to the ischemia-reperfusion injury which occurs at the time of transplantation. The persistent elevation in the following months which occurs in some patients and which also correlates with ongoing microvascular thrombosis presumably relates to a more persistent thrombogenic insult resulting from the failure of a normally functioning endothelium. One possibility is that this relates to cellular rejection. Certainly, severe acute allograft rejection is frequently associated with microvascular thrombosis. However, patients with myocardial fibrin deposits and elevated troponins do not show more cellular rejection episodes than those without.[42] Other possibilities include the development of anti-endothelial cell antibodies including those against vimentin. Multiple time-dependent factors may be involved with ischemia-reperfusion injury being important early following transplantation and development of antibodies or cytokine-dependent endothelial cell activation being more important later on. The early events may serve to release endothelial cell antigens such as vimentin to which a humoral response is subsequently mounted.

The propensity for vascular thrombosis depends upon the balance between procoagulant status and endogenous fibrinolytic activity. Decreased fibrinolytic activity would prevent removal of fibrin and in a prothrombotic microvasculature, would facilitate further production of thrombin and fibrin. Both increased plasma fibrinogen concentration and decreased plasma fibrinolytic activity have been associated with increased arterial intimal thickening in cardiac transplant patients.[34,45] Labarrere noted a significantly increased incidence of angiographic

coronary artery disease in transplant patients with lower levels of tissue plasminogen activator (t-PA) on endomyocardial biopsies.[34] These patients developed earlier and more aggressive disease than those patients with normal t-PA levels within the arteriolar microvasculature. Circulating levels of t-PA and plasminogen activator inhibitor-1 (PAI-1) are also associated with the development of CAV.[46]

Certain membrane proteins (integrins) and cytokines which induce directed chemotaxis (chemokines) may also contribute to transplant vasculopathy.[47] Studies imply a correlation between sustained expression of the vitronectin receptor (integrin α-V,β-3) and progression of CAV by IVUS.[48] Histologic studies of CAV lesions in humans show chemokine and chemokine receptor expression suggestive of a memory T Helper cell (TH1) response which is often associated with a strong inflammatory response.[49] Expression of the chemokine RANTES (regulated on activation, normal T-cell expressed and secreted), a chemokine that selectively chemoattracts T-lymphocytes, NK cells, monocytes and eosinophils has been shown to be expressed in human CAV.[50] In an animal model, sustained RANTES production was required for both monocyte recruitment and the development of intimal thickening and this required the presence of CD4+ cells.[51]

## Non-Immune Factors in the Development of CAV

Given that living unrelated donors in renal transplantation have survival comparable to grafts from related donors and consistently superior survival to matched grafts from cadaveric donors,[52] both immune and non-immune factors clearly play a significant role in the development of chronic graft dysfunction. One possibility is that immune factors may be important in the development of early graft vasculopathy, whilst non-immune factors may determine the propensity to develop CAV late following transplantation.[53]

Cadaveric organs are procured often under profound physiologic derangements. Donors frequently have sustained massive central nervous system injuries with often periods of labile blood pressure requiring prolonged inotropic support. Organ harvesting may take place under less than ideal conditions before being perfused with sub-physiologic electrolyte preservation solutions. Organs are then stored in the cold for several hours prior to final engraftment. Further changes during reperfusion then take place.

## Table 2. Risk factors for the development of CAV[54,58]

Donor age
Male donor
Ischemic etiology for transplant
Recipient history of smoking
Recipient history of gout
Black recipient
Positive donor CMV status
Posttransplant hypertension
Posttransplant hyperlipidema
Posttransplant diabetes

To determine specific risk factors for fatal donor CAV, a multi-institutional analysis of over 7000 cardiac transplants over a period of 10 years was performed by Costanzo.[54] By multivariable analysis, risk factors for fatal allograft vasculopathy (Table 2) included older donor age, male donor, younger recipient age, earlier date of transplant, ischemic etiology, history of recipient cigarette use, history of gouty arthritis, black recipient and positive donor CMV serology. Allografts from donors older than 50 years had a 50% chance of developing moderate to severe CAV at 5 years. This presumably relates to the extent of pre-existing atherosclerosis in the older donor. In contrast however, programs which actively pursue the use of older grafts have not seen a significant difference in overall survival or the development of CAV when compared to recipients with younger allografts.[55] Furthermore the presence of pre-existing coronary artery disease does not seem to accelerate the progression of CAV.[56] In a recent study,[57] patients underwent IVUS one month following transplant and again annually for upto 3 years to determine the influence of pre-existing donor disease in the subsequent development of CAV. Pre-existing donor lesions did not seem to act as a nidus for accelerating the progression of intimal hyperplasia. However, patients with donor lesions did have a higher incidence of angiographic CAV but this did not affect the long-term survival of patients up to three years.

In a more recent multi-center retrospective analysis of cardiovascular risk factors posttransplant revealed no significant covariates for the development of CAV, although the presence of posttransplant diabetes and hypercholesterolemia were associated with nonfatal adverse cardiac events.[58]

A number of important changes occur following brain death which may impact upon the performance of the donor heart. Circulating neuroendocrine hormonal levels may be increased whilst levels of other hormones may decrease.[59] Inflammatory markers including cytokines and adhesion molecules may be upregulated leading to apoptosis and cardiac dysfunction.[60] Interestingly however, the use of adrenergic agents in optimizing donors prior to transplantation has been associated with improved renal allograft survival but impaired cardiac allograft function following transplantation.[61]

Ischemia occurring during organ procurement and storage and subsequent reperfusion following engraftment may produce endothelial changes which may be central to the subsequent development of CAV over the ensuing years.[62] Acutely, changes include desquamation and retraction of endothelial cells, increasing their permeability. A number of endothelial genes can be induced by hypoxia including VEGF, b-FGF and PDGF.[63-65] Both b-FGF and PDGF may induce intimal smooth muscle migration and proliferation whilst VEGF stimulates endothelial proliferation and enhances vascular permeability.[66] Angiotensin II is also a potent mitogen for smooth muscle cell proliferation and experimentally, its inhibition has been shown to be effective in inhibiting the development of CAV.[67,68] Similarly, inhibition of angiotensin converting enzyme (ACE), whose expression is increased in injured vessels, is also effective experimentally in ameliorating transplant vasculopathy and ACE inhibitors may be of some value as preventive agents clinically.[67,69,70]

Ischemia may also lead to increased expression of CD11/CD18 adhesion molecules on circulating leukocytes and corresponding increase in expression of ICAM-1 and MHC molecules on endothelial cells. As a result, avidity of CD11/CD18 for ICAM-1 is promoted, leukocytes become activated and this leads to release of proinflammatory mediators. Experimentally, anti-ICAM-1 antibodies have been shown to limit myocardial damage following ischemia/reperfusion injury.[71]

## Hyperlipidemia

Approximately half of all cardiac transplants are performed on patients with ischemic cardiomyopathy. A substantial portion of these patients have a history of hyperlipidemia as a major contributor to their disease. Lipid levels have also been shown to be widely elevated following cardiac transplantation.[72] A number of factors are thought to contribute to posttransplant hyperlipidemia. Steroids contribute to increased apolipoprotein B production and also contribute to posttransplant obesity. Cyclosporine may enhance this effect and also independently increase hepatic lipase activity and decrease lipoprotein lipase activity.[73] The effect would be impaired very low-density lipoprotein (VLDL) and low-density lipoprotein (LDL) clearance.

Hyperlipidemia is a well established risk factor for nontransplant atherosclerosis. Both clinical and experimental observations suggest that it may be important in the development of transplant vasculopathy.[74-76] In a small retrospective study, elevated lipid values six months following transplantation had a strong predictive value for the development of CAV at three years.[75] In another study, posttransplant elevation of LDL at one year was the only predictor for the development or progression CAV by IVUS.[77] Experimentally, rabbits fed a high cholesterol diet following cardiac transplantation develop accelerated graft atherosclerosis.[78] The allogeneic state of the transplanted heart which also leads to endothelial activation may further augment the vascular response to hyperlipidemia. In this respect, greater intimal thickening, more intimal angiogenesis and a greater accumulation of T-cells is seen in transplanted vasculature compared to native vessels in animals exposed to the same level of hyperlipidemia.[79,80] Interestingly, treatment initiated within two weeks of transplantation with an inhibitor of the rate-limiting enzyme in the cholesterol biosynthetic pathway, hydroxymethylglutryl coenzyme A (HMG Co-A) reductase, is associated not only with decreased development of coronary intimal thickening, but also a lower frequency of hemodynamically compromising rejection episodes and improved survival.[81] These agents likely have an immunosuppressive effect in addition to their lipid-lowering activity. As hyperlipidemia is so common following transplantation, these findings suggest that all cardiac transplant recipients should receive HMG Co-A reductase inhibitors where tolerated.

## Hypertension

There is a clear link between hypertension and conventional coronary atherosclerosis. Hypertension is a common problem following cardiac transplantation, related in part to the use of corticosteroids, frequent associated weight gain and use of calcineurin inhibitors. Although no studies have specifically addressed the issue of whether hypertension is a risk factor for the development of CAV, treatment seems to have a beneficial effect. In one study, angiographic prevalence of CAV was significantly reduced in patients randomized to diltiazem.[82] In this study, at five years there was also a significant difference in freedom from both death and CAV. Similarly, in 32 cardiac transplant recipients, intimal thickness at 1 year measured by IVUS was significantly greater in the untreated control group than in those who received calcium channel blockers, angiotensin-converting enzyme inhibitors or both.[83] In a more recent study, the combined use of these agents was more effective than either drug alone at reducing IVUS indices of CAV.[70]

In a rat heterotopic transplant model, amlodipine was shown to significantly decrease the development of allograft vasculopathy.[84] In vitro, calcium channel blockers have been shown to stabilize endothelial cell function, inhibit platelet aggregation and decrease the release of platelet-derived growth factors.[85]

## Diabetes Mellitus

Given that diabetes mellitus is a major cardiovascular risk factor leading to the development of end-stage heart disease, a large number of patients with this condition undergo evaluation for heart transplantation. The use of certain agents posttransplant including steroids and tacrolimus may lead to the development or exacerbate diabetes posttransplant. Posttransplant diabetes in heart transplant recipients is common and may reach 32% at 5 years.[86] However, there are inadequate data correlating post—heart transplant diabetes with poor outcome. The literature is mixed with some studies showing increased risk for infections, decreased survival and more CAV in diabetics and other studies demonstrating no differences in outcome between diabetics and nondiabetic heart transplant recipients.[87-89] In an analysis from the United Network of Organ Sharing database of over 20 000 first-time heart transplant recipients, posttransplantation survival among patients with uncomplicated diabetes was not significantly different than that among nondiabetics. However, when stratified by disease severity and pretransplant diabetes-related complications, recipients with more severe diabetes had significantly worse survival than nondiabetics.[90] The presence of diabetes was associated with an increased risk of CAV in this study. In a recent multi-center study,[58] the presence of posttransplant diabetes was associated with the risk of nonfatal major adverse cardiac events.

## Immunosuppression and CAV

Immunosuppressive agents may affect the development of CAV in a number of ways. The indirect effects of long-term steroid therapy on lipid metabolism has already been described. This effect is enhanced with the use of cyclosporine, although there is some evidence that substitution of cyclosporine for the newer calcineurin inhibitor tacrolimus may abrogate the rise in serum lipids.[91] The number and severity of episodes of acute rejection likely contributes to the subsequent development of CAV.[40,53] More effective immunosuppressive regimens may therefore have a beneficial impact on the development of allograft vasculopathy.

Tacrolimus, a calcineurin inhibitor with actions similar to cyclosporine, has been studied extensively in liver transplantation and shown to be superior to cyclosporine in preventing chronic rejection.[92] In cardiac transplantation, whilst it has been demonstrated to cause less hypertension and hyperlipidemia than cyclosporine,[91] data on preventing CAV has been less conclusive. In one study treatment with tacrolimus has been associated with the lower incidence of anti-endothelial antibodies.[93] In another study, there was a trend towards more CAV by IVUS in patients treated with tacrolimus compared to cyclosporine.[94] In a long-term follow-up of patients randomized to tacrolimus or cyclosporine under a standardized non-induction protocol, there was no difference in the development of angiographgic CAV between these two agents.[95]

Mycophenolate mofetil (MMF), an inhibitor of the de novo pathway for purine biosynthesis, has been shown to be more effective at reducing cardiac allograft rejection and posttransplant mortality.[96] More recent studies suggest that it may also be more effective at reducing allograft vasculopathy.[97-99] This effect may be related at least in part to the ability of MMF to reduce B cell responses as patients treated with this agent developed lower anti-vimentin titers and this was correlated with the lower incidence of CAV by IVUS. MMF may also inhibit the production of reactive oxidant species which contributes to endothelial cell dysfunction.[100]

Another class of agents, the proliferation signal inhibitors, or target of rapamycin (TOR) inhibitors have significantly expanded therapeutic options. These agents are effective at not only reducing allograft rejection but also have a significant beneficial effect on the development of CAV. Sirolimus and the related agent everolimus block activation of T-cells following autocrine stimulation by interleukin-2. Their action is therefore complementary to the calcineurin inhibitors. Sirolimus has been shown to effectively prevent acute graft rejection and inhibit refractory acute graft rejection in heart transplant recipients.[101] In the Australian randomized study of an open-label trial of sirolimus compared to azathioprine, sirolimus was shown to reduce allograft rejection.[102] However, concerns over delayed wound healing and renal dysfunction prompted a change in therapeutic targets midway into the study. No differences in 1 year mortality were noted but sirolimus significantly worsened renal function and its use was associated with a higher incidence of hypertension. Other adverse outcomes with sirolimus included hyperlipidemia, skin malignancies, leucopenia, thrombocytopenia and anemia. Importantly, sirolimus, which is also known to have anti-proliferative effects on smooth muscle cells, was shown to decrease the development of CAV as assessed by IVUS at 6 months. At 2 years, this benefit was sustained and sirolimus was associated with markedly less intimal and medial proliferation and reduced lumen encroachment compared to azathioprine. In another angiographic study of cardiac transplant patients with established transplant vasculopathy, sirolimus was also shown to slow down disease progression.[103]

In a randomized, double-blind, clinical trial comparing everolimus with azathioprine in recipients of a first heart transplant,[104] a total of 634 patients were randomly assigned to receive one of two doses of everolimus, or azathioprine, in combination with cyclosporine, corticosteroids and statins. The primary efficacy end point was a composite of death, graft loss or retransplantation, loss to follow-up, biopsy-proven acute rejection of ISHLT grade 3A or greater, or rejection with hemodynamic compromise. At six months, the percentage of patients who had reached the primary efficacy end point was significantly smaller in the groups receiving everolimus compared with the group receiving azathioprine. IVUS showed that the average increase in maximal intimal thickness 12 months after transplantation was significantly smaller in the two everolimus groups than in the azathioprine group. The rates of CMV infection were significantly lower in the everolimus groups than in the azathioprine group. Rates of bacterial infection were, however, significantly higher in the high dose everolimus group than in the azathioprine group. Serum creatinine levels were also significantly higher in the two everolimus groups than in the azathioprine group. The authors concluded that everolimus was more efficacious than azathioprine in reducing the severity and incidence of cardiac-allograft vasculopathy.

Leflunomide is another immunosuppressive agent shown to be effective in both allo and xenotransplantation in animal models. In vitro, it is able to inhibit smooth muscle cell proliferation. In animal models it has been shown to prevent allograft vasculopathy.[105] In one study the agent was even able to reverse transplant vasculopathy when the agent was started late and this correlated with a decrease in antibody formation.[105]

## Cytomegalovirus

Among the many possible etiological factors implicated in the development of CAV, CMV infection has held a prominent position over the years. Human CMV (CMV) has a wide global distribution and exposure to it increases with age such that by the age of 60, up to 90% have evidence of prior exposure. Both clinical and experimental studies suggest that CMV infection may play an important role in the development of CAV. In one of the earlier studies of 102 transplant recipients, 16% developed transplant vasculopathy of which 62% had evidence of CMV infection compared to only 25% of patients without CAV.[106] Similar findings were confirmed by Loebe et al.[107] In animal studies, rat aortic allografts have been shown to develop neointimal proliferation when the recipients have been infected with CMV at the time of transplantation.[108] In humans, active CMV infection in heart transplant recipients has been associated with the development of endothelial dysfunction.[109] The routine use of prophylactic ganciclovir following cardiac transplantation may however significantly abrogate the influence of CMV in the development of CAV.[110]

The mechanisms by which CMV infection may contribute to transplant vasculopathy are unclear. Endothelial cell infection leads to upregulation of adhesion molecules on both infected and non-infected cells. Production of interleukin-1 beta has been implicated in this paracrine effect.[111] Coculture experiments suggest that CMV infection of T-cells results in production of interferon-( and TNF-(, which are then able induce expression of Class I and II antigens, VCAM-1 and ICAM-1. CMV infection of smooth muscle cells may also contribute to neointimal proliferation by inhibition of apoptosis.[112]

## Detecting Cardiac Allograft Vasculopathy

CAV is usually very advanced and beyond therapeutic intervention by the time symptoms develop. Surveillance is therefore the major approach to monitoring the development of CAV. Both non-invasive and invasive approaches have been utilized for diagnosis. In general, non-invasive methods are unable to detect early disease as these techniques depend upon the presence of hemodynamically significant lesions. Ultrafast CT has been used to detect coronary calcification and may be useful in the detection of CAV[113] but still does not provide detailed information about the vessel lumen or wall. 64-detector CT scanning has recently made non-invasive evaluation of coronaries possible. Limited studies have suggested the feasibility of this technology in the evaluation of CAV despite the limitation of a high resting heart rate and elevated body mass index in heart transplant recipients.[114,115] Other non-invasive attempts have included echocardiographic techniques and stress testing. In a recent report, measurement of deceleration time of diastolic flow velocity and assessment of coronary flow reserve in the left anterior descending artery with adenosine using contrast-enhanced transthoracic echocardiography were reliable markers for CAV-related events, including death, the need for stenting and heart failure.[116] In a 4 year prospective evaluation of treadmill stress testing, dobutamine stress echocardiography, thallium scintigraphy and angiography, positive dobutamine stress echocardiography and angiography were associated with adverse cardiac events or death by univariate analysis.[117] Abnormal dobutamine stress echocardiography has been shown to correlate with both angiographic disease and adverse clinical outcomes.[118] Studies have also shown dobutamine myocardial scintigraphy to be reliable for the detection of clinically relevant CAV.[119,120] Invasive methods however continue to remain the mainstay of CAV detection. Coronary angiography relies upon the ability to compare normal segments of the vessel with diseased segments. The diffuse nature of CAV often results in underestimation of disease because there is no reference segment in which the normal diameter of the vessel can be assessed. Minimal luminal irregularities may suggest the presence of early disease and is therefore often commented upon. Comparison with prior studies may help determine development of disease but requires the use of the same angiographic protocol at each study for precise evaluation to avoid confounding by technical factors such as angiographic projections and magnification. Moreover this allows use of computer-assisted quantitative coronary angiography (QCA) which improves sensitivity of detection of CAV. However, QCA also has its limitations as it does not allow evaluation of the vessel wall and may miss early disease where compensatory dilatation of the vessel may occur to preserve luminal area.

IVUS is currently the only technique which allows evaluation of the vessel wall. Cross-sectional images of the coronary vessel wall comparable to histological sections are obtained (Fig. 3). Intimal area can be quantitatively planimetered to accurately assess even early plaque burden. Sequential images are usually obtained as the catheter is pulled back to determine the extent of disease along a vessel wall. Several studies have shown IVUS is more sensitive than angiography in detecting CAV.[12,121,122] The technique however has several limitations—it is highly invasive, requires anticoagulation, use of expensive single-use catheters and evaluation is mainly limited to the major epicardial vessels. IVUS has however be shown to have prognostic value. In a multicenter study, progression of intimal thickening $\geq 0.5$ mm in the first year after transplantation appeared to be a reliable surrogate marker for subsequent mortality, nonfatal major adverse cardiac events and development of angiographic CAV through five years after heart transplantation.[5]

Determination of coronary flow reserve with intracoronary Doppler flow measurement may be useful in assessing CAV, although the clinical importance of this information has yet to be determined. Intracoronary flow velocities are determined using a Doppler transducer mounted on a guide wire. Pharmacologic interventions such as intracoronary adenosine can be used to measure maximal coronary flow and calculate flow reserve. Coronary flow reserve has been shown to be reduced in patients with CAV and deteriorates with increasing time after transplantation.[123] Measurement of coronary flow reserve reflects changes in the microvasculature as well as the epicardial vessels.

Production of vasodilatory mediators by the endothelium plays a major role in the regulation of vascular tone. Endothelial dysfunction is not only an early feature in the development of atherosclerosis, but likely also plays a fundamental role in the development of CAV. Nitroglycerine is a direct vasodilator acting on vascular smooth muscle. Acetylcholine works indirectly through its action on the endothelium. Its vasodilatory property is therefore dependent upon an intact functional endothelium. The loss or attenuation of endothelium-dependent vasodilatation can be used to determine endothelial cell dysfunction. Both QCA and IVUS can be used to assess changes in vessel diameter in response to pharmacologic challenge. Abnormal responses to acetylcholine have been reported after transplantation.[124]

## Treatment of CAV

Clinically apparent disease is associated with a poor prognosis and therefore prevention is an important strategy in addressing transplant vasculopathy (Table 3). Agents used in the treatment and prevention of conventional atherosclerosis are also utilized for CAV. Aspirin is widely used given its widely established use in nontransplant coronary disease. Control of hypertension and hyperlipidemia has already been discussed. The use of HMG Co-A reductase inhibitors is particularly important as it also helps prevent allograft rejection. Newer immunosuppressive agents, such as MMF and the TOR inhibitors, show significant promise in impacting the natural course of CAV and have been discussed above.

Once clinically significant CAV is apparent, a number of approaches are available to relieve ischemia. For focal disease, percutaneous coronary intervention (PCI) with balloon angioplasty has been successful, although restenosis is particularly common in the transplant setting.[125] The availability of drug-eluting stents has helped to address this problem to some extent but restenosis rates continue to be higher than for similar interventions in the nontransplant population.[126,127] Atherectomy techniques and laser therapy have been reported in a small number of patients.[128-130] No studies are available to show whether percutaneous coronary intervention alters the prognosis of CAV and since many patients with significant disease are asymptomatic, intervention often presents a dilemma.

Patients with multi-vessel focal disease with adequate distal target vessels may be candidates for coronary artery bypass surgery (CABG). Efficacy is difficult to determine as relatively small numbers have been reported, reflecting the many patients who do not have adequate targets and the preferential use of PCI.

**Table 3. Treatment options for CAV**

| Prevention |
| --- |
| Aspirin |
| Control of hypertension |
| (calcium channel blockers, ACE inhibitors) |
| HMG CoA reductase inhibitors |
| Control of diabetes |
| MMF |
| Proliferation of signal inhibitors |

| Treatment |
| --- |
| Drug eluting stents |
| Proliferation of signal inhibitors |
| CABG |
| Retransplantation |

Retransplantation may be a consideration for many patients with advanced CAV not amenable to PCI or CABG. Survival rates reported after retransplantation have been consistently lower than those after primary transplantation. Retransplantation within the first six months after original transplant have the worst outcomes with a one year survival of only 38%.[131] These were however most commonly performed for treatment of refractory rejection. In a Stanford study, the 1-year actuarial survival of those patients who underwent retransplantation for specifically CAV was 69 ± 10%, which approached the 1-year survival following primary transplants.[132] This study also showed that patients having a second heart transplant do not have an increased risk for development of CAV in the second donor heart. The actuarial freedom from CAV in the entire heart retransplantation population at 5 years was 89 ± 7% and, in patients who underwent retransplantation for CAV, the actuarial freedom from this disease process in the retransplanted heart at 5 years was 91 ± 9%. Such data suggest that heart transplant patients who develop severe CAV may be suitable candidates for retransplantation. The scarcity of donor hearts, however, creates an ethical dilemma. Some have argued that it is better to use organs to give more patients the opportunity of a first transplant rather than allocate two organs to the same individual. Others hold the argument that patients needing a second transplant should be considered on the same basis as those being evaluated for first transplants.

## Conclusion

CAV is a significant complication which limits the log-term survival of the cardiac allograft. Although its pathology shares many similarities with nontransplant atheroscelorsis, factors related to organ procurement, the host immune response and posttransplant medications affords the disease unique characteristics. Clinical presentation is atypical and subtle due to surgical denervation of the allograft. Diagnosis therefore predominantly relies on careful monitoring which generally requires invasive testing. The development of CAV is associated with a poor prognosis and treatment options are limited. However, recent studies with mycophenolate mofetil and particularly the proliferation signal inhibitors suggest that these agents may have a significant impact on the course of this disease. Aggressive treatment of posttransplant risk factors including diabetes mellitus, hypertension and hyperlipidemia likely has a beneficial effect. Conventional revascularization techniques for palliation remain an option for those with established disease but results are generally less satisfactory than with nontransplant atherosclerosis due to higher restenosis rates. Ultimately retransplantation may be considered for select individuals.

## References

1. Taylor DO, Edwards LB, Boucek MM et al. Registry of the International Society for Heart and Lung Transplantation: twenty-third official adult heart transplantation report. J Heart Lung Transplant 2006; 25(8):869-79.
2. Gao SZ, Hunt SA, Schroeder JS. Accelerated transplant coronary artery disease. Semin Thorac Cardiovasc Surg 1990; 2(3):241-49.
3. Costanzo MR, Naftel DC, Pritzker MR et al. Heart transplant coronary artery disease detected by coronary angiography: a multiinstitutional study of preoperative donor and recipient risk factors. Cardiac Transplant Research Database. J Heart Lung Transplant 1998; 17(8):744-53.
4. Syeda B, Roedler S, Schukro C et al. Transplant coronary artery disease: Incidence, progression and interventional revascularization. International journal of cardiology 2005; 104(3):269-74.
5. Kobashigawa JA, Tobis JM, Starling RC et al. Multicenter intravascular ultrasound validation study among heart transplant recipients: outcomes after five years. Journal of the American College of Cardiology 2005; 45(9):1532-37.
6. Uretsky BF, Kormos RL, Zerbe TR et al. Cardiac events after heart transplantation: incidence and predictive value of coronary arteriography. J Heart Lung Transplant 1992; 11(3 Pt 2):S45-51.
7. Potluri SP, Mehra MR, Uber PA et al. Relationship among epicardial coronary disease, tissue myocardial perfusion and survival in heart transplantation. J Heart Lung Transplant 2005; 24(8):1019-25.
8. Ross R, Glomset J, Harker L. Response to injury and atherogenesis. The American journal of pathology 1977; 86(3):675-84.
9. Rahmani M, Cruz RP, Granville DJ et al. Allograft vasculopathy versus atherosclerosis. Circulation research 2006; 99(8):801-15.
10. Kobashigawa J, Wener L, Johnson J et al. Longitudinal study of vascular remodeling in coronary arteries after heart transplantation. J Heart Lung Transplant 2000; 19(6):546-50.
11. Arbustini E, Roberts WC. Morphological observations in the epicardial coronary arteries and their surroundings late after cardiac transplantation (allograft vascular disease). The American journal of cardiology 1996; 78(7):814-20.
12. Kapadia SR, Nissen SE, Ziada KM et al. Development of transplantation vasculopathy and progression of donor- transmitted atherosclerosis: comparison by serial intravascular ultrasound imaging. Circulation 1998; 98(24):2672-78.
13. Kapadia SR, Nissen SE, Tuzcu EM. Impact of intravascular ultrasound in understanding transplant coronary artery disease. Current opinion in cardiology 1999; 14(2):140-50.
14. St Goar FG, Pinto FJ, Alderman EL et al. Intracoronary ultrasound in cardiac transplant recipients. In vivo evidence of "angiographically silent" intimal thickening. Circulation 1992; 85(3):979-87.
15. Aranda Jr JM, Hill J. Cardiac transplant vasculopathy. Chest 2000; 118(6):1792-800.
16. Stark RP, McGinn AL, Wilson RF. Chest pain in cardiac-transplant recipients. Evidence of sensory reinnervation after cardiac transplantation. N Engl J Med 1991; 324(25):1791-94.
17. Ramsdale DB, Bellamy CM. Angina and threatened acute myocardial infarction after cardiac transplantation. American heart journal 1990; 119(5):1195-97.
18. Schroeder JS, Hunt SA. Chest pain in heart-transplant recipients. N Engl J Med 1991; 324(25):1805-7.
19. Gao SZ, Schroeder JS, Hunt SA et al. Acute myocardial infarction in cardiac transplant recipients. Am J Cardiol 1989; 64(18):1093-97.
20. Kaczmarek I, Deutsch MA, Rohrer ME et al. HLA-DR matching improves survival after heart transplantation: is it time to change allocation policies? J Heart Lung Transplant 2006; 25(9):1057-62.
21. Tambur AR, Pamboukian SV, Costanzo MR et al. The presence of HLA-directed antibodies after heart transplantation is associated with poor allograft outcome. Transplantation 2005; 80(8):1019-25.
22. Rose ML. Role of endothelial cells in allograft rejection. Vascular medicine (London, England) 1997; 2(2):105-14.
23. Bishop DK, Shelby J, Eichwald EJ. Mobilization of T-lymphocytes following cardiac transplantation. Evidence that CD4-positive cells are required for cytotoxic T-lymphocyte activation, inflammatory endothelial development, graft infiltration and acute allograft rejection. Transplantation 1992; 53(4):849-57.
24. Shirwan H. Chronic allograft rejection. Do the Th2 cells preferentially induced by indirect alloantigen recognition play a dominant role? Transplantation 1999; 68(6):715-26.
25. Daar AS, Fuggle SV, Fabre JW et al. The detailed distribution of MHC Class II antigens in normal human organs. Transplantation 1984; 38(3):293-98.
26. Page C, Rose M, Yacoub M et al. Antigenic heterogeneity of vascular endothelium. Am J Pathol 1992; 141(3):673-83.

27. Rose ML, Page C, Hengstenberg C et al. Identification of antigen presenting cells in normal and transplanted human heart: importance of endothelial cells. Hum Immunol 1990; 28(2):179-85.
28. Rose ML. Endothelial cells as antigen-presenting cells: role in human transplant rejection. Cell Mol Life Sci 1998; 54(9):965-78.
29. Han CW, Imamura M, Hashino S et al. Differential effects of the immunosuppressants cyclosporin A, FK506 and KM2210 on cytokine gene expression. Bone marrow transplantation 1995; 15(5):733-39.
30. Suciu-Foca N, Reed E, Marboe C et al. The role of anti-HLA antibodies in heart transplantation. Transplantation 1991; 51(3):716-24.
31. Dunn MJ, Crisp SJ, Rose ML et al. Anti-endothelial antibodies and coronary artery disease after cardiac transplantation. Lancet 1992; 339(8809):1566-70.
32. Jurcevic S, Ainsworth ME, Pomerance A et al. Antivimentin antibodies are an independent predictor of transplant- associated coronary artery disease after cardiac transplantation. Transplantation 2001; 71(7):886-92.
33. Pidwell DJ, Heller MJ, Gabler D et al. In vitro stimulation of human endothelial cells by sera from a subpopulation of high-percentage panel-reactive antibody patients. Transplantation 1995; 60(6):563-69.
34. Labarrere CA, Pitts D, Nelson DR et al. Vascular tissue plasminogen activator and the development of coronary artery disease in heart-transplant recipients. The New England journal of medicine 1995; 333(17):1111-16.
35. Salomon RN, Hughes CC, Schoen FJ et al. Human coronary transplantation-associated arteriosclerosis. Evidence for a chronic immune reaction to activated graft endothelial cells. Am J Pathol 1991; 138(4):791-98.
36. Labarrere CA, Nelson DR, Faulk WP. Endothelial activation and development of coronary artery disease in transplanted human hearts. Jama 1997; 278(14):1169-75.
37. Zund G, Uezono S, Stahl GL et al. Hypoxia enhances induction of endothelial ICAM-1: role for metabolic acidosis and proteasomes. The American journal of physiology 1997; 273(5 Pt 1):C1571-80.
38. Poston Jr RS, Billingham ME, Pollard J et al. Effects of increased ICAM-1 on reperfusion injury and chronic graft vascular disease. The Annals of thoracic surgery 1997; 64(4):1004-12.
39. Labarrere CA, Nelson DR, Miller SJ et al. Value of serum-soluble intercellular adhesion molecule-1 for the noninvasive risk assessment of transplant coronary artery disease, posttransplant ischemic events and cardiac graft failure. Circulation 2000; 102(13):1549-55.
40. Kobashigawa JA, Miller L, Yeung A et al. Does acute rejection correlate with the development of transplant coronary artery disease? A multicenter study using intravascular ultrasound. Sandoz/CVIS Investigators. J Heart Lung Transplant 1995; 14(6 Pt 2):S221-26.
41. Oni AA, Ray J, Hosenpud JD. Coronary venous intimal thickening in explanted cardiac allografts. Evidence demonstrating that transplant coronary artery disease is a manifestation of a diffuse allograft vasculopathy. Transplantation 1992; 53(6):1247-51.
42. Labarrere CA, Nelson DR, Faulk WP. Myocardial fibrin deposits in the first month after transplantation predict subsequent coronary artery disease and graft failure in cardiac allograft recipients. The American journal of medicine 1998; 105(3):207-13.
43. Labarrere CA, Ortiz MA, Ruzmetov N et al. Microvascular thrombosis and cardiac allograft vasculopathy in rat heart transplantation. J Heart Lung Transplant 2006; 25(10):1213-22.
44. Labarrere CA, Nelson DR, Cox CJ et al. Cardiac-specific troponin I levels and risk of coronary artery disease and graft failure following heart transplantation. Jama 2000; 284(4):457-64.
45. Meckel CR, Anderson TJ, Mudge GH et al. Hemostatic/fibrinolytic predictors of allograft coronary artery disease after cardiac transplantation. Vascular medicine (London, England) 1997; 2(4):306-12.
46. Warshofsky MK, Wasserman HS, Wang W et al. Plasma levels of tissue plasminogen activator and plasminogen activator inhibitor-1 are correlated with the presence of transplant coronary artery disease in cardiac transplant recipients. The American journal of cardiology 1997; 80(2):145-49.
47. Kao J, Kobashigawa J, Fishbein MC et al. Elevated serum levels of the CXCR3 chemokine ITAC are associated with the development of transplant coronary artery disease. Circulation 2003; 107(15):1958-61.
48. Yamani MH, Masri S, Ratliff NB et al. The role of vitronectin receptor and tissue factor in the pathogenesis of transplant coronary vasculopathy. J Heart Lung Transplant 2001; 20(2):185.
49. van Loosdregt J, van Oosterhout MF, Bruggink AH et al. The chemokine and chemokine receptor profile of infiltrating cells in the wall of arteries with cardiac allograft vasculopathy is indicative of a memory T-helper 1 response. Circulation 2006; 114(15):1599-607.
50. Pattison JM, Nelson PJ, Huie P et al. RANTES chemokine expression in transplant-associated accelerated atherosclerosis. J Heart Lung Transplant 1996; 15(12):1194-99.
51. Yun JJ, Fischbein MP, Laks H et al. Rantes production during development of cardiac allograft vasculopathy. Transplantation 2001; 71(11):1649-56.
52. Gjertson DW, Cecka JM. Living unrelated donor kidney transplantation. Kidney international 2000; 58(2):491-99.
53. Hornick P, Smith J, Pomerance A et al. Influence of acute rejection episodes, HLA matching and donor/recipient phenotype on the development of 'early' transplant- associated coronary artery disease. Circulation 1997; 96(9 Suppl):II-148-53.
54. Costanzo MR, Eisen HJ, Brown RN et al. Are there specific risk factors for fatal allograft vasculopathy? An analysis of over 7,000 cardiac transplant patients. J Heart Lung Transplant 2001; 20(2):152.
55. Drinkwater DC, Laks H, Blitz A et al. Outcomes of patients undergoing transplantation with older donor hearts. J Heart Lung Transplant 1996; 15(7):684-91.
56. Botas J, Pinto FJ, Chenzbraun A et al. Influence of preexistent donor coronary artery disease on the progression of transplant vasculopathy. An intravascular ultrasound study. Circulation 1995; 92(5):1126-32.
57. Li H, Tanaka K, Anzai H et al. Influence of pre-existing donor atherosclerosis on the development of cardiac allograft vasculopathy and outcomes in heart transplant recipients. Journal of the American College of Cardiology 2006; 47(12):2470-76.
58. Kobashigawa JA, Starling RC, Mehra MR et al. Multicenter retrospective analysis of cardiovascular risk factors affecting long-term outcome of de novo cardiac transplant recipients. J Heart Lung Transplant 2006; 25(9):1063-69.
59. Arita K, Uozumi T, Oki S et al. The function of the hypothalamo-pituitary axis in brain dead patients. Acta Neurochir (Wien) 1993; 123(1-2):64-75.
60. Birks EJ, Yacoub MH, Burton PS et al. Activation of apoptotic and inflammatory pathways in dysfunctional donor hearts. Transplantation 2000; 70(10):1498-506.
61. Schnuelle P, Berger S, de Boer J et al. Effects of catecholamine application to brain-dead donors on graft survival in solid organ transplantation. Transplantation 2001; 72(3):455-63.
62. Day JD, Rayburn BK, Gaudin PB et al. Cardiac allograft vasculopathy: the central pathogenetic role of ischemia-induced endothelial cell injury. J Heart Lung Transplant 1995; 14(6 Pt 2):S142-49.
63. Kourembanas S, Hannan RL, Faller DV. Oxygen tension regulates the expression of the platelet-derived growth factor-B chain gene in human endothelial cells. J Clin Invest 1990; 86(2):670-74.
64. Levy AP, Levy NS, Wegner S et al. Transcriptional regulation of the rat vascular endothelial growth factor gene by hypoxia. The Journal of biological chemistry 1995; 270(22):13333-40.
65. Lindner V, Reidy MA. Proliferation of smooth muscle cells after vascular injury is inhibited by an antibody against basic fibroblast growth factor. Proceedings of the National Academy of Sciences of the United States of America 1991; 88(9):3739-43.
66. Dvorak HF, Brown LF, Detmar M et al. Vascular permeability factor/vascular endothelial growth factor, microvascular hyperpermeability and angiogenesis. The American journal of pathology 1995; 146(5):1029-39.
67. Furukawa Y, Matsumori A, Hirozane T et al Angiotensin II receptor antagonist TCV-116 reduces graft coronary artery disease and preserves graft status in a murine model. A comparative study with captopril. Circulation 1996; 93(2):333-39.
68. Richter M, Skupin M, Grabs R et al. New approach in the therapy of chronic rejection? ACE- and AT1-blocker reduce the development of chronic rejection after cardiac transplantation in a rat model. J Heart Lung Transplant 2000; 19(11):1047-55.
69. Mehra MR, Ventura HO, Smart FW et al. Impact of converting enzyme inhibitors and calcium entry blockers on cardiac allograft vasculopathy: from bench to bedside. J Heart Lung Transplant 1995; 14(6 Pt 2): S246-49.
70. Erinc K, Yamani MH, Starling RC et al. The effect of combined Angiotensin-converting enzyme inhibition and calcium antagonism on allograft coronary vasculopathy validated by intravascular ultrasound. J Heart Lung Transplant 2005; 24(8):1033-38.
71. Yamazaki T, Seko Y, Tamatani T et al. Expression of intercellular adhesion molecule-1 in rat heart with ischemia/reperfusion and limitation of infarct size by treatment with antibodies against cell adhesion molecules. Am J Pathol 1993; 143(2):410-18.
72. Stamler JS, Vaughan DE, Rudd MA et al. Frequency of hypercholesterolemia after cardiac transplantation. Am J Cardiol 1988; 62(17):1268-72.
73. Superko HR, Haskell WL, Di Ricco CD. Lipoprotein and hepatic lipase activity and high-density lipoprotein subclasses after cardiac transplantation. Am J Cardiol 1990; 66(15):1131-34.
74. Kobashigawa JA, Kasiske BL. Hyperlipidemia in solid organ transplantation. Transplantation 1997; 63(3):331-38.

75. Eich D, Thompson JA, Ko DJ et al. Hypercholesterolemia in long-term survivors of heart transplantation: an early marker of accelerated coronary artery disease. J Heart Lung Transplant 1991; 10(1 Pt 1):45-49.

76. Esper E, Glagov S, Karp RB et al. Role of hypercholesterolemia in accelerated transplant coronary vasculopathy: results of surgical therapy with partial ileal bypass in rabbits undergoing heterotopic heart transplantation. J Heart Lung Transplant 1997; 16(4):420-35.

77. Kapadia SR, Nissen SE, Ziada KM et al. Impact of lipid abnormalities in development and progression of transplant coronary disease: a serial intravascular ultrasound study. Journal of the American College of Cardiology 2001; 38(1):206-13.

78. Alonso DR, Starek PK, Minick CR. Studies on the pathogenesis of atheroarteriosclerosis induced in rabbit cardiac allografts by the synergy of graft rejection and hypercholesterolemia. The American journal of pathology 1977; 87(2):415-42.

79. Tanaka H, Sukhova GK, Libby P. Interaction of the allogeneic state and hypercholesterolemia in arterial lesion formation in experimental cardiac allografts. Arterioscler Thromb 1994; 14(5):734-45.

80. Raisanen-Sokolowski A, Tilly-Kiesi M, Ustinov J et al. Hyperlipidemia accelerates allograft arteriosclerosis (chronic rejection) in the rat. Arterioscler Thromb 1994; 14(12):2032-42.

81. Kobashigawa JA, Katznelson S, Laks H et al. Effect of pravastatin on outcomes after cardiac transplantation. The New England journal of medicine 1995; 333(10):621-7.

82. Schroeder JS, Gao SZ, Alderman EL et al. A preliminary study of diltiazem in the prevention of coronary artery disease in heart-transplant recipients. N Engl J Med 1993; 328(3):164-70.

83. Mehra MR, Ventura HO, Smart FW et al. An intravascular ultrasound study of the influence of angiotensin- converting enzyme inhibitors and calcium entry blockers on the development of cardiac allograft vasculopathy. The American journal of cardiology 1995; 75(12):853-54.

84. Atkinson JB, Wudel JH, Hoff SJ et al. Amlodipine reduces graft coronary artery disease in rat heterotopic cardiac allografts. J Heart Lung Transplant 1993; 12(6 Pt 1):1036-43.

85. Betz E, Weiss HD, Heinle H et al. Calcium antagonists and atherosclerosis. J Cardiovasc Pharmacol 1991; 18(Suppl 10):S71-75.

86. Hertz MI, Taylor DO, Trulock EP et al. The registry of the international society for heart and lung transplantation: nineteenth official report-2002. J Heart Lung Transplant 2002; 21(9):950-70.

87. Marelli D, Laks H, Patel B et al. Heart transplantation in patients with diabetes mellitus in the current era. J Heart Lung Transplant 2003; 22(10):1091-97.

88. Lang CC, Beniaminovitz A, Edwards N et al. Morbidity and mortality in diabetic patients following cardiac transplantation. J Heart Lung Transplant 2003; 22(3):244-49.

89. Czerny M, Sahin V, Fasching P et al. The impact of diabetes mellitus at the time of heart transplantation on long-term survival. Diabetologia 2002; 45(11):1498-508.

90. Russo MJ, Chen JM, Hong KN et al. Survival after heart transplantation is not diminished among recipients with uncomplicated diabetes mellitus: an analysis of the United Network of Organ Sharing database. Circulation 2006; 114(21):2280-87.

91. Taylor DO, Barr ML, Radovancevic B et al. A randomized, multicenter comparison of tacrolimus and cyclosporine immunosuppressive regimens in cardiac transplantation: decreased hyperlipidemia and hypertension with tacrolimus. J Heart Lung Transplant 1999; 18(4):336-45.

92. Williams R, Neuhaus P, Bismuth H et al. Two-year data from the European multicentre tacrolimus (FK506) liver study. Transpl Int 1996; 9 Suppl 1:S144-50.

93. Jurcevic S, Dunn MJ, Crisp S et al. A new enzyme-linked immunosorbent assay to measure anti-endothelial antibodies after cardiac transplantation demonstrates greater inhibition of antibody formation by tacrolimus compared with cyclosporine. Transplantation 1998; 65(9):1197-202.

94. Klauss V, Konig A, Spes C et al. Cyclosporine versus tacrolimus (FK 506) for prevention of cardiac allograft vasculopathy. The American journal of cardiology 2000; 85(2):266-69.

95. Kobashigawa JA, Patel J, Furukawa H et al. Five-year results of a randomized, single-center study of tacrolimus vs microemulsion cyclosporine in heart transplant patients. J Heart Lung Transplant 2006; 25(4):434-39.

96. Kobashigawa J, Miller L, Renlund D et al. A randomized active-controlled trial of mycophenolate mofetil in heart transplant recipients. Mycophenolate Mofetil Investigators. Transplantation 1998; 66(4):507-15.

97. Kobashigawa JA, Tobis JM, Mentzer RM et al. Mycophenolate mofetil reduces intimal thickness by intravascular ultrasound after heart transplant: reanalysis of the multicenter trial. Am J Transplant 2006; 6(5 Pt 1):993-97.

98. Kaczmarek I, Ertl B, Schmauss D et al. Preventing cardiac allograft vasculopathy: long-term beneficial effects of mycophenolate mofetil. J Heart Lung Transplant 2006; 25(5):550-6.

99. Rose ML, Danskine A, Smith JD et al. Mycophenolate mofetil (MMF) depresses antibody production after cardiac transplantation. Circulation 2000; 102(18S):II.490.

100. Krotz F, Keller M, Derflinger S et al. Mycophenolic acid inhibits endothelial NAD(P)H oxidase activity and superoxide formation by a Rac1-dependent mechanism. Hypertension 2007; 49(1):201-8.

101. Radovancevic B, Vrtovec B. Sirolimus therapy in cardiac transplantation. Transplantation proceedings 2003; 35(3 Suppl):171S-76S.

102. Keogh A, Richardson M, Ruygrok P et al. Sirolimus in de novo heart transplant recipients reduces acute rejection and prevents coronary artery disease at 2 years: a randomized clinical trial. Circulation 2004; 110(17):2694-700.

103. Mancini D, Pinney S, Burkhoff D et al. Use of rapamycin slows progression of cardiac transplantation vasculopathy. Circulation 2003; 108(1):48-53.

104. Eisen HJ, Tuzcu EM, Dorent R et al. Everolimus for the prevention of allograft rejection and vasculopathy in cardiac-transplant recipients. The New England journal of medicine 2003; 349(9):847-58.

105. Hwang MW, Matsumori A, Furukawa Y et al. FTY720, a new immunosuppressant, promotes long-term graft survival and inhibits the progression of graft coronary artery disease in a murine model of cardiac transplantation. Circulation 1999; 100(12):1322-29.

106. MacDonald AS, Sabr K, MacAuley MA et al. Effects of leflunomide and cyclosporine on aortic allograft chronic rejection in the rat. Transplant Proc 1994; 26(6):3244-45.

107. Loebe M, Schuler S, Zais O et al. Role of cytomegalovirus infection in the development of coronary artery disease in the transplanted heart. J Heart Transplant 1990; 9(6):707-11.

108. Lemstrom KB, Bruning JH, Bruggeman CA et al. Cytomegalovirus infection enhances smooth muscle cell proliferation and intimal thickening of rat aortic allografts. J Clin Invest 1993; 92(2):549-58.

109. Petrakopoulou P, Kubrich M, Pehlivanli S et al. Cytomegalovirus infection in heart transplant recipients is associated with impaired endothelial function. Circulation 2004; 110(11 Suppl 1):II207-12.

110. Luckraz H, Charman SC, Wreghitt T et al. Does cytomegalovirus status influence acute and chronic rejection in heart transplantation during the ganciclovir prophylaxis era? J Heart Lung Transplant 2003; 22(9):1023-27.

111. Dengler TJ, Raftery MJ, Werle M et al. Cytomegalovirus infection of vascular cells induces expression of pro-inflammatory adhesion molecules by paracrine action of secreted interleukin-1beta. Transplantation 2000; 69(6):1160-68.

112. Zhu H, Shen Y, Shenk T. Human cytomegalovirus IE1 and IE2 proteins block apoptosis. J Virol 1995; 69(12):7960-70.

113. Farzaneh-Far A. Electron-beam computed tomography in the assessment of coronary artery disease after heart transplantation. Circulation 2001; 103(10):E60.

114. Iyengar S, Feldman DS, Cooke GE et al. Detection of coronary artery disease in orthotopic heart transplant recipients with 64-detector row computed tomography angiography. J Heart Lung Transplant 2006; 25(11):1363-66.

115. Sigurdsson G, Carrascosa P, Yamani MH et al. Detection of transplant coronary artery disease using multidetector computed tomography with adaptive multisegment reconstruction. Journal of the American College of Cardiology 2006; 48(4):772-78.

116. Tona F, Caforio AL, Montisci R et al. Coronary flow velocity pattern and coronary flow reserve by contrast-enhanced transthoracic echocardiography predict long-term outcome in heart transplantation. Circulation 2006; 114(1 Suppl):I49-55.

117. Bacal F, Moreira L, Souza G et al. Dobutamine stress echocardiography predicts cardiac events or death in asymptomatic patients long-term after heart transplantation: 4-year prospective evaluation. J Heart Lung Transplant 2004; 23(11):1238-44.

118. Spes CH, Angermann CE. Stress echocardiography for assessment of cardiac allograft vasculopathy. Zeitschrift fur Kardiologie 2000; 89(Suppl 9):IX/50-53.

119. Hacker M, Tausig A, Romuller B et al. Dobutamine myocardial scintigraphy for the prediction of cardiac events after heart transplantation. Nuclear medicine communications 2005; 26(7):607-12.

120. Wu YW, Yen RF, Lee CM et al. Diagnostic and prognostic value of dobutamine thallium-201 single-photon emission computed tomography after heart transplantation. J Heart Lung Transplant 2005; 24(5):544-50.

121. Rickenbacher P. Role of intravascular ultrasound versus angiography for diagnosis of graft vascular disease. Transplantation proceedings 1998; 30(3):891-92.

122. Konig A, Theisen K, Klauss V. Intravascular ultrasound for assessment of coronary allograft vasculopathy. Zeitschrift fur Kardiologie 2000; 89(Suppl 9):IX/45-49.
123. Mullins PA, Chauhan A, Sharples L et al. Impairment of coronary flow reserve in orthotopic cardiac transplant recipients with minor coronary occlusive disease. British heart journal 1992; 68(3):266-71.
124. Hartmann A, Mazzilli N, Weis M et al. Time course of endothelial function in epicardial conduit coronary arteries and in the microcirculation in the long-term follow-up after cardiac transplantation. Int J Cardiol 1996; 53(2):127-36.
125. Sharifi M, Siraj Y, O'Donnell J et al. Coronary angioplasty and stenting in orthotopic heart transplants: a fruitful act or a futile attempt? Angiology 2000; 51(10):809-15.
126. Tanaka K, Li H, Curran PJ et al. Usefulness and safety of percutaneous coronary interventions for cardiac transplant vasculopathy. The American journal of cardiology 2006; 97(8):1192-97.
127. Simpson L, Lee EK, Hott BJ et al. Long-term results of angioplasty vs stenting in cardiac transplant recipients with allograft vasculopathy. J Heart Lung Transplant 2005; 24(9):1211-17.
128. Strikwerda S, Umans V, van der Linden MM et al. Percutaneous directional atherectomy for discrete coronary lesions in cardiac transplant patients. American heart journal 1992; 123(6):1686-90.
129. Patel VS, Radovancevic B, Springer W et al. Revascularization procedures in patients with transplant coronary artery disease. Eur J Cardiothorac Surg 1997; 11(5):895-901.
130. Topaz O, Bailey NT, Mohanty PK. Application of solid-state pulsed-wave, mid-infrared laser for percutaneous revascularization in heart transplant recipients. J Heart Lung Transplant 1998; 17(5):505-10.
131. Hosenpud JD, Novick RJ, Breen TJ et al. The Registry of the International Society for Heart and Lung Transplantation: twelfth official report. J Heart Lung Transplant 1995; 14(5):805-15.
132. Smith JA, Ribakove GH, Hunt SA et al. Heart retransplantation: the 25-year experience at a single institution. J Heart Lung Transplant 1995; 14(5):832-39.
133. Subherwal S, Kobashigawa JA, Cogert G et al. Incidence of acute cellular rejection and noncellular rejection in cardiac transplantation. Transplantation proceedings 2004; 36(10):3171-72.

# The Pathology of Heart Allograft Rejection

Jon Carthy, Heather Heine, Alice Mui and Bruce McManus*

## Historical Context

In 1905, the innovative French surgeon Alexis Carrel performed the first heterotopic canine heart transplant with Charles Guthrie.[1] Twenty years later, the concept of cardiac allograft rejection was proposed by Frank Mann at the Mayo Clinic to explain the eventual failure of heterotopic canine allografts. He described the rejection process as the biologic incompatibility between donor and recipient manifested by an impressive leukocytic infiltration of the rejecting myocardium.[2] The first human heart transplant was a chimpanzee xenograft performed in 1964 by James Hardy at the University of Mississippi. Although the surgical procedure was technically satisfactory, the heart was too small to maintain independent circulation and functioned for only 90 minutes before failing.[3] On December 3, 1967, South African Christiaan Barnard surprised the world when he successfully performed the first human-to-human heart transplant. Unfortunately the recipient, Mr. Louis Washkansky, died 18 days later of pneumonia.[4] By the end of 1968, over 100 patients had received heart transplants at 50 different institutions in 17 countries around the world. The results of these early transplant programs were generally miserable and were perceived as premature experiments by the medical community and the public. During the 1970's, only a few institutions continued clinical cardiac transplantation, with Dr. Norman Shumway leading the way at Stanford. During that decade, the 1-year survival after transplantation increased from 22% to 65%.[5,6] In 1973, Philip Caves redeveloped the transvenous endomyocardial bioptome which finally provided a reliable means for monitoring allograft function and rejection.[7] In 1981, the advent of the immunosuppressive agent cyclosporine dramatically increased patient survival and marked the beginning of the modern era of successful cardiac transplantation. Transplantation is now a widely accepted therapeutic option for patients with end-stage cardiac failure.

## Immunopathology of Cardiac Rejection

The specifics of the immune response in allogeneic transplantation are discussed in Chapter 5 of this book. Briefly, in the setting of heart transplantation, the host's T-lymphocytes interaction with graft endothelial and antigen-presenting cells starts and sustains the immune and subsequent inflammatory response against the graft tissue.[8] The host's T-cells are triggered by two signals: (a) foreign major histocompatibility complex (MHC) plus peptide (antigen) and (b) simultaneous stimulatory signals from antigen-presenting cells. The antigen-presenting cells, which present the donor antigen shed from the graft as MHC molecules and other alloantigens to T-cells, can either be classified as donor (direct antigen presentation) or as recipient (indirect antigen presentation).[9] The principal signal from the antigen-presenting cell arises from a molecule called B7, a signal which is received by a receptor on the T-cell called CD28 or cytotoxic T-lymphocyte antigen 4.[10]

The appropriate signals from the antigen-presenting cell, together with antigen (MHC plus peptidic antigen), activate T-cells. On activation, CD4 + lymphocytes produce interleukin (IL)-2, IL-4 and IL-5 and other cytokines. IL-2 further activates CD4 + lymphocytes and induces CD8 + differentiation (direct cytotoxicity). IL-4 and IL-5 activate B-cells (humoral immunity) and other cytokines increase vascular permeability and regional accumulation of mononuclear cells. IL-2-activated cytotoxic T-lymphocytes also lyse allograft cells.[11] Cardiac allograft rejection is a culmination of this immunologic response within the cardiac allograft and its blood vessels. Without the use of immunosuppressive drugs, these rejection events result in the release of powerful biologic effectors of myocyte injury and death. While the exact pathogenesis and the multitude of underlying contributors to cardiac rejection are still not fully appreciated, evidence of cellular injury in the presence of immunologic effector cells can be detected and graded by histological examination of myocardial tissue. Thus, the histological analysis of endomyocardial biopsies remains central to the identification of rejection and the overall monitoring of heart transplant recipients.

## Classification of Allograft Rejection in Heart Biopsies by Histopathologic Criteria

The original classification of heart allograft rejection as proposed by Dr. Margaret Billingham and widely applied by institutions throughout the world has fundamentally stood the test of time.[12-15] Refinements and modified expression of the original classification have been found useful by some observers, although each of these systems relies on the same constellation of features for general classification as the original conception.[16,17] In 1990, the International Society of Heart and Lung Transplantation (ISHLT) commissioned the development of a common grading scale in an attempt to establish uniform descriptions and grading criteria of various transplant histologies to refine communication and comparison of treatment regimens and outcomes between transplant centers.[18] In 2004, under direction of the ISHLT, a multidisciplinary review of the cardiac biopsy grading system was undertaken to address challenges and inconsistencies in its use and to address recent advances in the knowledge of antibody-mediated rejection.[19] This recently revised grading scale is now becoming widely accepted and applied in transplant programs around the world.

## Technical Considerations

Proper handling of endomyocardial biopsy specimens obtained by standard bioptome catheters from the right ventricular septum is imperative in efforts to issue critical diagnoses. Multiple myocardial

*Corresponding Author: Bruce McManus—The James Hogg iCAPTURE Centre for Cardiovascular and Pulmonary Research, St. Paul's Hospital, Vancouver, British Columbia. Email: bmcmanus@mrl.ubc.ca

*Chronic Allograft Failure: Natural History, Pathogenesis, Diagnosis and Management*, edited by Nasimul Ahsan. ©2008 Landes Bioscience.

biopsy samples should be obtained from different right ventricle sites with a minimum of three and a preference for four or more evaluable pieces of myocardium recommended for grading acute rejection. An evaluable piece of myocardium contains at least 50% myocardium. At least 3 levels through the tissue samples are stained with hematoxylin and eosin and additional slides may be saved unstained for further study if needed. This approach represents the minimum recommendation for adequate assessment of acute rejection. Special stains are not routinely required although a trichrome stain may be helpful in selected instances for assessing myocyte damage and fibrosis, such as may occur in the early postoperative period. The nature of the specimens available for diagnostic pathologic examination will change considerably over time in a given patient eventually including many chronic changes and it will behoove the pathologist to be in constant contact with transplant physicians regarding each patient's clinical progress.

## Acute Cellular Rejection

In essence, rejection severity is histopathologically defined by increasing numbers of infiltrating inflammatory cells and increasing numbers of inflammatory foci, associated with increasing parenchymal and vascular damage and, in most severe forms, is often accompanied by extracellular edema and evidence of hemorrhage. As such, the histopathologic severity of inflammation in acute rejection follows a logical geographical algorithm, progressing from perivascular clusters to perivascular and interstitial aggregates and, ultimately, to widespread mixed cellular infiltrates. Myocyte damage is an important but sometimes difficult feature to identify. Myocyte damage is frequently accompanied by encroachment of inflammatory cells at the plasmalemma of myocytes and is often characterized by myocytolysis and not contraction band or coagulation necrosis in milder forms of rejection; cell death may be an obvious feature of the most severe forms of rejection. The rate of resorption of necrotic myocytes may be slowed in the setting of immunosuppression.[20] Generally, contraction bands are not a meaningful indicator of rejection-related injury because of their ubiquity in biopsy specimens. Focal regions of coagulative necrosis with polymorphonuclear leukocytes sparing a subendocardial cardiac myocytes and associated with a polymorphic infiltrate should be considered ischemic or "pressor" in origin. The ISHLT grades of rejection are as follows:

### Grade 0 R (No Acute Cellular Rejection)
No Evidence of Mononuclear Inflammation or Myocyte Damage is Present in Grade 0 R.

### Grade 1 R (Mild, Low-Grade, Acute Cellular Rejection)
Mild or low-grade rejection is diagnosed if either of the following is found: (1) presence of perivascular and/or interstitial mononuclear cells that respect myocyte borders, do not encroach on adjacent myocytes and do not distort the normal architecture; (2) presence of one focus of mononuclear cells with associated myocyte damage.

### Grade 2 R (Moderate, Intermediate-Grade, Acute Cellular Rejection)
Grade 2 R is characterized by the presence of two or more foci of mononuclear cells with associated myocyte damage found in one or more biopsy fragments. Eosinophils may be present. Areas of intervening of uninvolved myocardium are present between the foci of rejection.

### Grade 3 R (Severe, High-Grade, Acute Cellular Rejection)
In grade 3 R, multiple biopsy fragments will include a diffuse inflammatory process, either predominantly lymphocytes and macrophages or a polymorphous infiltrate, with multiple areas of associated myocyte damage. In the most severe forms of rejection, edema, interstitial hemorrhage and vasculitis may be present.

## Acute Antibody-Mediated (Humoral) Rejection

Since the introduction of the endomyocardial biopsy in the mid 1970s, the treatment of cell mediated immunity has led to excellent improvement in heart transplant patient outcomes.[21,22] This progress led to the impression that heart transplant rejection was primarily cell mediated.[23] In 1970 Ellis et al[24] described a connection between anti-myocyte antibodies and rejection, but this finding was not investigated further. When the number of heart transplant recipients increased during the 1980s, some patients emerged with "biopsy-negative" heart transplant rejection. In 1989, Hammond et al[25] described "vascular (humoral) rejection" and methods for its diagnosis based on observations of endothelial cell swelling and/or vasculitis and the deposition of immunoglobulin and complement as seen by fluorescent microscopy. Several reports have now shown that humoral rejection (HR) can be more severe than T cell mediated rejection.[26-29] It is infrequent, but a potentially fatal type of acute allograft rejection. Some studies have shown that HR occurs soon after heart transplant or many months afterwards.[25,30-32] The method of choice for humoral rejection management among centres remains unclear.[33]

Acute antibody-mediated rejection (AMR) appears to have an incidence rate of up to 15% in the first year posttransplantation but it remains controversial due to a highly varied incidence rate between different centers.[25,32,34,35] Patients who have previously been allosensitized, including those with previous transplantation, transfusion or pregnancy, are more susceptible to AMR. The 2004 ISHLT consensus recommends that every endomyocardial biopsy should undergo critical histologic evaluation for features suggestive of AMR. Biopsy features of AMR include myocardial capillary injury with endothelial-cell swelling and intravascular macrophage accumulation. Interstitial edema and hemorrhage, neutrophils in and around capillaries, intravascular thrombi and myocyte necrosis without cellular infiltration can also be present. If these features are not seen, the biopsy should be designated negative for antibody-mediated rejection, or AMR 0. However, if biopsies do show evidence of these features and the patient suffers from unexplained cardiac dysfunction, immunostaining for CD68, CD31 and C4d should be performed and the patient should be tested for donor-specific antibody to confirm a diagnosis of antibody-mediated rejection (AMR 1).

Recently Leech et al[36] have reported that the combination of plasmapheresis and intravenous immunoglobulin were useful in lowering circulating alloantibody levels to improve heart transplantation, even across a positive lymphocyte mismatch. These observations provide more evidence that the humoral immune system requires more attention in the documentation and treatment of transplant rejection.

## Other Biopsy Findings, Artifacts, Mimickers and Caveats

The standard range of bioptome and biopsy-related artifacts apply to the study of specimens from heart allografts. Numerous of the most common have been described in detail previously.[37-39] This discussion will focus on findings that are specifically relevant to the assessment of heart allografts. Recollection of the normal constituents of the endomyocardium is, of course, essential to the process of evaluating biopsy specimens.

Ischemic injury of the heart allograft may take two forms pertinent to biopsy evaluation. Global ischemia at the time of organ procurement may result from prolonged time from harvest to implant and from undue warming of the heart in transport. This peri-operative ischemia may have a severe impact on early postoperative function of the graft, including on contractility and electrical activity.[40] Ischemic injury is a common biopsy finding in the early posttransplant period (up to 6 weeks) and may initially manifest as contraction band necrosis or coagulative myocyte vacuolization and fat necrosis, frequently extending to the endocardial surface. As healing progresses, biopsies may show a mixed inflammatory cell infiltrate that is difficult to distinguish from acute rejection. Recognition of posttransplant ischemia is very important as it will avert the unnecessary augmentation of immunosuppression for wrongly presumed severe rejection.

Ischemia may also occur months to years after transplantation due to lesions in the coronary arteries as a result of transplant arteriopathy. Such ischemia produces classical, multi focal lesions with coagulative necrosis and polymorphonuclear leukocyte infiltration (early) and granulation and fibrosis (late). In contrast to procurement injury which occurs early after transplant, coronary-based ischemia will appear months to years after transplantation as a sequela of chronic rejection. This type of ischemia represents a significant challenge, as it is infrequently identified in biopsies of failing allograft hearts because it is rarely associated with chest pain.[41]

The prominent, at times nodular, endocardial infiltrates originally coined the Quilty effect by Margaret Billingham occur in approximately 10%-20% of posttransplant endocardial biopsies. These infiltrates may be confined to the endocardium or may extend into the underlying myocardium where associated myocyte damage may be present. They are typically well vascularized, with cellularity that is primarily mononuclear and may vary from very small to very large in size. Immunohistochemical studies of these infiltrates have revealed the prominence of B cells as compared to the underlying myocardium, but numerous T cells and macrophages are also present.[42,43] They can evolve to be a typical lymphoid follicle and are rich in dendritic cells. If these lesions are restricted to the endocardial surface, differentiation from acute rejection is not a problem. However, if the Quilty lesion projects deeply into the underlying myocardium, it may cause confusion in analyzing the biopsy.[44] Currently, the relationship of these endocardial infiltrates to acute rejection remains unresolved.

Opportunistic infections may masquerade as alloreactive inflammatory processes and remain important causes of posttransplant morbidity and mortality, but for the most part are relatively rare in posttransplant biopsies. The broad range of bacterial, fungal, viral and protozoal infections may involve more than 50% of the transplant population, resulting in death in approximately 5%.[45] Systemic infections derived from respiratory and urinary tracts or skin are most common. While bacterial and fungal infection typically elicit a mixed inflammatory infiltrate predominated by polymorphonuclear leukocytes, the infiltrate associated with viral or protozoal infections may be quite reminiscent of rejection. Most notable among these are cytomegalovirus and toxoplasmosis.

In addition to these other biopsy findings, particular awareness of recent, healing and healed biopsy sites must be emphasized in allograft specimens. The average patient will undergo a dozen biopsies in the first year after transplantation alone. Therefore it is not surprising that previous biopsy sites are frequently encountered in biopsy specimens for rejection surveillance. The histopathology of the biopsy site reflects the acute injury of the bioptome followed by healing and repair. As time increases between the first and second biopsy, changes indicative of progressive healing will be seen. If the repeat biopsy is performed within a few days of the first biopsy, fresh microthrombus may be found overlying areas of acute myocyte necrosis, focal hemorrhage and leukocyte and leukocyte infiltration. As time progresses, hemosiderin-laden phagocytes will be present along with variable numbers of fibroblasts, lymphocytes and plasma cells within a loose connective tissue stoma. As compared with rejection-related injury, biopsy sites have pathognomic marginal disorientation of atrophic and dysmorphic, "frayed" cardiac myocytes which often lack a normal parallel relationship to the endocardial surface.

## Major Limitations of Endomyocardial Biopsy

One of the major problems facing clinical caregivers in the management of transplant recipients is determining whether a transplanted organ is undergoing rejection. While the current grading system for cardiac allograft biopsies has allowed for better consistency in assessment of rejection severity, there are still inherent limitations in its usage. In particular, variability in the assessment of rejection severity, particularly regarding what were previously known as grade 2 lesions, the presence of Quilty lesions that tend to cause overestimation of rejection severity and unstandardized methods for detecting humoral rejection are just beginning to be addressed. Due to the invasive nature of the biopsy, this procedure causes both emotional and physical discomfort to the patient and thereby minimizes the frequency at which samples can be obtained to monitor graft function. Therefore, while the biopsy is currently considered the gold standard for detecting acute rejection episodes, a more practical, accurate and non-invasive method for detecting rejection is necessary to improve our ability to monitor and treat heart transplant recipients.

## Non-Invasive Methods of Detecting Rejection

The endomyocardial biopsy (EMB) has long been considered the "gold standard" in the determination of cardiac transplant rejection. Currently it is the most informative and well-established technique available to diagnose acute rejection episodes in transplant patients. While the technique is an essential diagnostic tool, the clinical reality is that a number of patients with hemodynamic compromise exhibit low-grade or no rejection on biopsy.[46] This could be due to sampling error, since an EMB taken from the right ventricle could reflect poorly other myocardial areas undergoing more obvious rejection. It is also possible that many of the hemodynamically compromised patients have acute AMR. The possibility of AMR is now being histologically investigated with light microscopy for the presence of capillary endothelial swelling and by immmunofluorescence for the deposition of complement and immunoglobulin. ISHLT grading now also includes assessment of AMR.[19]

A variety of methods are employed in attempt to detect both acute and chronic rejection posttransplant. Magnetic resonance imaging (MRI) allows noninvasive functional assessment and visualization with very high spatial resolution. MRI has unique advantages over other imaging modalities, in its relatively high soft tissue contrast and in that it avoids the need to utilize radioisotopes. Cellular MRI studies rely on superparamagnetic iron oxide (SPIO) or ultra-small superparamagnetic iron oxide (USPIO) nanoparticles for imaging contrast. SPIOs are 50-100 nm in size and have also been used as a diagnostic tool for cancers of the liver and spleen.[47,48] USPIOs have a long half-life in vessels and are taken up by macrophages in a variety of organs. They have been used to identify lymph node metastasis in head and neck, retroperitoneal and pelvic cancers.[49,50]

In a rat heterotopic heart transplant model, micrometer-sized paramagnetic iron oxide (MPIO) particles were used to monitor individual immune cells during acute rejection using MRI.[51] Immune cells were labeled in vivo by directly injecting the iron oxide particles intravenously and relying on the cells to endocytose the particles and migrate to sites of rejection. The degree of spatial distribution of MPIO-labeled cells, confirmed histologically, was dependent on the degree of rejection. While the effects of MPIO endocytosis on immune cells has yet to be thoroughly investigated, animals injected have not shown adverse effects after two months. The physiochemical pathways and toxicity of very similar USPIO and SPIO molecules are well characterized.[52-55] Other groups have found that other blood pooling agents make it possible to differentiate syngeneic and allogeneic heart grafts in rats by MRI relying on the permeability of unformulated agents NC100150 or AMI-277.[56] In this study, the rejection process is thought to cause injury to the microvasculature leading to more leakage of USPIOs into the extravasular space. The degree of edema seen in the allogeneic grafts may have increased the volume of contrast agent entering the grafts as compared to the syngenic grafts.

Cardiac allograft vasculopathy (CAV) remains the most important predictor of recipient mortality beyond five years following heart transplant. CAV develops silently in heart transplant patients and careful surveillance is needed with coronary-angiography or intravascular ultrasound.[57] Non-invasive investigation of CAV involves echocardiography and radionucleide imaging techniques[58] and most recently multi-slice computed tomography.[59] To diagnose CAV in heart transplant patients some have attempted to use the ratio of the levels of phosphocreatine and adenosine tri-phosphate (PCr/ATP) by observing $31^P$ chemical shift. A 93% negative predictive value in patients with PCr/ATP ratios over 1.59 suggests that one might justify a reduction in the number of coronary angiograms for such patients.[60] No single predictor or combination of predictors was significant enough to rule out the need for biopsy for monitoring acute rejection, however.

Changes in heart tissue constituency upon rejection has led to the creation of implant devices included in the transplant that allow intramyocardial electocardiograms to be performed. Several studies have indicated that use of these devices may result in negative predictive values of 98 and 97%[61-63] and may reduce the need for biopsies significantly. Others have seen lower negative predictive values and, as such, the clinical value of the technique is still in question.[64]

One's breath may be found as a useful indicator of rejection. Tissue rejection involves oxidative stress that causes degradation of membrane polyunsaturated fatty acids into alkanes and methylalkanes. These metabolites are released in the breath as organic compounds that can be detected by gas chromatography and mass spectrometry. One group has shown that the breath methylated alkane contour in 539 heart transplant patients had 78.6% sensitivity and 62.4% specificity for detecting ISHLT grade 3 rejection.[65] Unfortunately, the negative predictive values were similar.

Inflammation results in cell necrosis and release of a variety of proteins. Elevated blood troponin has been detected in severe rejection in specific populations.[66] B-type natriuretic peptide (BNP), a 32 amino acid neurohormone, is released from the ventricular myocardium upon increased wall stress. Blood BNP has been found to be elevated in acute and chronic rejection in heart transplant recipients[67-69] and levels have been shown to drop upon treatment.[67] Additional work has shown that in microarray analysis of PBMCs in heart transplant recipients with otherwise clinically quiescent conditions, BNP levels correlate with molecular patterns of gene expression associated with structural remodeling, vascular injury, inflammation and alloimmune reactivity.[70] This finding suggests that BNP is not only a marker of hemodynamic processes, but also perhaps of other processes going on in the context of rejection.

Inhibition of cell mediated immunity has improved heart transplant patient outcomes,[21,22] indicating that much of heart transplant rejection may be cell mediated.[23] Thus efforts have been directed towards non-invasive methods for monitoring the anti-graft immune response of cells of both the innate and adaptive immune systems. For example, activation and graft infiltration of neutrophils contributes to graft rejection and upregulation of adhesion protein CD11b on circulating neutrophils was found to be significantly correlated to rejection grade at the first postoperative biopsy.[71] This finding suggests that monitoring of peripheral blood neutrophil activation status could provide a means of monitoring their activity in the graft if activation due to other causes (i.e., infection) can be ruled out. Natural Killer (NK) cells are similarly implicated in allograft rejection and down-regulation of NK cell function by the presence of a soluble form of major histocompatability complex class I chain-related molecule A (sMICA) during the first year postheart transplant may contribute to allograft acceptance.[72] Thus monitoring sMICA levels in serum may provide another marker of anti-graft immune status. However, most attention has focused on developing methods of quantifying non-invasively the number and activation status of T cells that react with the allograft (alloreactive T cells). Although most of the anti-graft, alloreactive T cells are found infiltrating the graft itself, increased frequencies of circulating alloreac-

tive T cells have been detected after human cardiac transplantation.[73] The number of these cells increases during rejection[74] and conversely decreases with allograft acceptance.[75] Since activation of alloreactive T cells precedes and/or enhances all the other immune responses, including humoral, methods for quantifying both the frequency and functional state of the alloreactive T cell population would be valuable for monitoring the host response against the graft. Currently, methods exist only to determine the frequency,[76-79] but not the activation state of alloreactive T cells in the patient.

Hammond et al[25] first described "vascular (humoral) rejection" and methods for its diagnosis based on observations of endothelial cell swelling and/or vasculitis and the deposition of immunoglobulin and complement as seen by fluorescent microscopy. Others have reported on the incidence, pathological features and clinical outcomes[28,30,32,34,80] and have attempted to improve upon the histopathological criteria described by Hammond. Some studies have shown rejection correlates with high levels of antibodies to human vascular endothelial cells,[81] cleavage products in the complement system such as C4, C4d,[31] immunoglobulins (IgG, IgM, IgA), fibrinogen and/or HLA-DR.[25,31,34,82-86] Crespo-Leiro et al (2005) found that although testing for IgM, IgG, C1q, C3 and fibrin were of no diagnostic value for heart rejection, levels of C4d correlated well with rejection in 6 patients, suggesting a potential utility for this marker.[30] Plasmapheresis has been used to battle humoral aspects of posttransplant rejection.[30,87-91] This approach has been found to be effective in treating severe humoral rejection episodes.[80,92]

Complications due to infection from organisms such as cytomegalovirus, bacteria and fungi still are common causes of pneumonia and septicemia and a significant source of morbidity and mortality in heart transplant patients. Sarmiento et al[93,94] recently reported that pretransplant IgG and posttransplant IgG levels at day 7 were associated with an increased risk for developing infections. Monitoring immunoglobulin levels by a rapid and standardized nephelometric assay may improve infection outcomes in heart transplant recipients.

## The Road Ahead

In the four decades since the first human-to-human heart transplant was successfully performed, we have moved dramatically from what was considered an "experimental" treatment to one that has widespread acceptance as a major therapy for patients with end-stage heart failure. Advances in immunosuppressive drugs have resulted in a significant drop in the rate of acute rejection episodes and early graft loss, however, despite these advances, the long-term success of heart transplantation is limited by the development of allograft coronary disease. As such, we are in need of an effective, cost-efficient and minimally invasive method to monitor heart transplant patients over the long-term. While the endomyocardial biopsy remains the "gold standard" for monitoring heart transplant patients for acute rejection and while the development and refinement of a universal grading scheme to consistently employ this strategy at different institutions has greatly improved the ability of transplant pathologists to accurately monitor acute rejection, there are still major limitations associated with usage of recently revised ISHLT grading system. Perhaps the biggest limitation is the invasiveness of this procedure, which causes both emotional and physical discomfort to the patient and thereby minimizes the frequency at which samples can be obtained to monitor graft integrity. The future of efforts to optimize the monitoring of patients with heart allografts will focus on the discovery of an effective, non-invasive method to better monitor for rejection. A biomarker set to monitor patients for signs of immune accommodation or preclinical rejection would revolutionize the way we manage transplant patients and would minimize the cost and side-effects associated with the over-prescription of immunosuppressive drugs. There have been encouraging early results in several largescale initiatives aimed at identifying and validating genomic, proteomic, or metabolomic biomarkers of rejection in peripheral blood or urine.

## Acknowledgement

The authors appreciate the support of other members of the Biomarkers in Transplantation Team. We also thank the Canadian Institutes of Health Research and Genome Canada for competitive awards.

## References

1. Carrel A: The surgery of blood vessels. Johns Hopkins Hosp Bull 1907; 18:18.
2. Mann F, Priestly, JT, Markowitz, J et al. Transplantation of the intact mammalian heart. Arch Surg 1933; 26:219-24.
3. Hardy JD, Kurrus FD, Chavez CM et al. Heart transplantation in man. Developmental studies and report of a case. JAMA 1964; 188:1132-40.
4. Barnard CN: The operation. A human cardiac transplant: an interim report of a successful operation performed at Groote Schuur Hospital, Cape Town. S Afr Med J 1967; 41(48):1271-4.
5. Stinson EB, Dong E, Jr., Schroeder JS et al. Initial clinical experience with heart transplantation. Am J Cardiol 1968; 22(6):791-803.
6. Griepp RB: A decade of human heart transplantation. Transplant Proc 1979; 11(1):285-92.
7. Caves PK, Schulz WP, Dong E et al. New instrument for transvenous cardiac biopsy. Am J Cardiol 1974; 33(2):264-67.
8. Libby P, Swanson SJ, Tanaka H et al. Immunopathology of coronary arteriosclerosis in transplanted hearts. J Heart Lung Transplant 1992; 11(3 Pt 2):S5-6.
9. Wahlers T, Mugge A, Oppelt P et al. Coronary vasculopathy following cardiac transplantation and cyclosporine immunosuppression: preventive treatment with angiopeptin, a somatostatin analog. Transplant Proc 1994; 26(5):2741-42.
10. Linsley PS, Greene JL, Brady W et al. Human B7-1 (CD80) and B7-2 (CD86) bind with similar avidities but distinct kinetics to CD28 and CTLA-4 receptors. Immunity 1994; 1(9):793-801.
11. Potter B: Structure-activity relationships of adenophostin A and related molecules at the 1-D-myo-inositol 1,4,5-trisphosphate receptor. Chicago, USA: OUP, 1999. (Bruzik K, ed. Phosphoinositides—Chemistry, Biochemistry and Biomedical Applications).
12. Billingham ME, Caves PK, Dong Jr E et al. The diagnosis of canine orthotopic cardiac allograft rejection by transvenous endomyocardial biopsy. Transplant Proc 1973; 5(1):741-43.
13. Caves PK, Stinson EB, Billingham ME et al. Serial transvenous biopsy of the transplanted human heart. Improved management of acute rejection episodes. Lancet 1974; 1(7862):821-26.
14. Billingham M: Diagnosis of cardiac rejection by endomyocardial biopsy. Heart Transplant 1982; 1:25-30.
15. Billingham ME: Some recent advances in cardiac pathology. Hum Pathol 1979; 10(4):367-86.
16. Kemnitz J, Cohnert T, Schafers HJ et al. A classification of cardiac allograft rejection. A modification of the classification by Billingham. Am J Surg Pathol 1987; 11(7):503-15.
17. McAllister HA, Jr., Schnee MJ, Radovancevic B et al. A system for grading cardiac allograft rejection. Tex Heart Inst J 1986; 13(1):1-3.
18. Billingham ME, Cary NR, Hammond ME et al. A working formulation for the standardization of nomenclature in the diagnosis of heart and lung rejection: Heart Rejection Study Group. The International Society for Heart Transplantation. J Heart Transplant 1990; 9(6):587-93.
19. Stewart S, Winters GL, Fishbein MC et al. Revision of the 1990 working formulation for the standardization of nomenclature in the diagnosis of heart rejection. J Heart Lung Transplant 2005; 24(11):1710-20.
20. Carricr M, Paplanus SH, Graham AR et al. Histopathology of acute myocardial necrosis: effects of immunosuppression therapy. J Heart Transplant 1987; 6(4):218-21.
21. Caves PK, Stinson EB, Billingham M et al. Percutaneous transvenous endomyocardial biopsy in human heart recipients. Experience with a new technique. Ann Thorac Surg 1973; 16(4):325-36.
22. Ellis RJ, Lillehei CW, Fischetti VA et al. Heart-reactive antibody: an index of cardiac rejection in human heart transplantation. Circulation 1970; 41(5 Suppl): II91-97.
23. Billingham ME: The diagnosis of acute cardiac rejection by endomyocardial biopsy. Bibl Cardiol 1988(43):83-102.
24. Schuurman HJ, Jambroes G, Borleffs JC et al. Acute humoral rejection after heart transplantation. Transplantation 1988; 46(4):603-5.
25. Hammond EH, Yowell RL, Nunoda S et al. Vascular (humoral) rejection in heart transplantation: pathologic observations and clinical implications. J Heart Transplant 1989; 8(6):430-43.
26. Cherry R, Nielsen H, Reed E et al. Vascular (humoral) rejection in human cardiac allograft biopsies: relation to circulating anti-HLA antibodies. J Heart Lung Transplant 1992; 11(1 Pt 1):24-29; discussion 30.
27. Ensley RD, Hammond EH, Renlund DG et al. Clinical manifestations of vascular rejection in cardiac transplantation. Transplant Proc 1991; 23(1 Pt 2):1130-32.
28. Miller LW, Wesp A, Jennison SH et al. Vascular rejection in heart transplant recipients. J Heart Lung Transplant 1993; 12(2):S147-52.
29. Olsen SL, Wagoner LE, Hammond EH et al. Vascular rejection in heart transplantation: clinical correlation, treatment options and future considerations. J Heart Lung Transplant 1993; 12(2): S135-42.
30. Crespo Leiro MG, Portela Torron F, Vazquez Gonzalez N et al. [Corticoid-resistant rejection and humoral rejection. The diagnostic criteria and therapeutic management]. Rev Esp Cardiol 1995; 48 Suppl 7:77-85.
31. Hammond EH, Hansen JK, Spencer LS et al. Immunofluorescence of endomyocardial biopsy specimens: methods and interpretation. J Heart Lung Transplant 1993; 12(2):S113-24.
32. Michaels PJ, Espejo ML, Kobashigawa J et al. Humoral rejection in cardiac transplantation: risk factors, hemodynamic consequences and relationship to transplant coronary artery disease. J Heart Lung Transplant 2003; 22(1):58-69.
33. Trento A, Hardesty RL, Griffith BP et al. Role of the antibody to vascular endothelial cells in hyperacute rejection in patients undergoing cardiac transplantation. J Thorac Cardiovasc Surg 1988; 95(1):37-41.
34. Lones MA, Czer LS, Trento A et al. Clinical-pathologic features of humoral rejection in cardiac allografts: a study in 81 consecutive patients. J Heart Lung Transplant 1995; 14(1 Pt 1):151-62.
35. Loh E, Bergin JD, Couper GS et al. Role of panel-reactive antibody cross-reactivity in predicting survival after orthotopic heart transplantation. J Heart Lung Transplant 1994; 13(2):194-201.
36. Leech SH, Mather PJ, Eisen HJ et al. Donor-specific HLA antibodies after transplantation are associated with deterioration in cardiac function. Clin Transplant 1996; 10(6 Pt 2):639-45.
37. Nippoldt TB, Edwards WD, Holmes Jr DR et al. Right Ventricular endomyocardial biopsy: clinicopathologic correlates in 100 consecutive patients. Mayo Clin Proc 1982; 57(7):407-18.
38. Adomian GE, Laks MM, Billingham ME. The incidence and significance of contraction bands in endomyocardial biopsies from normal human hearts. Am Heart J 1978; 95(3):348-51.
39. Foerster A, Simonsen S, Froysaker T. Heart transplantation in Norway. Morphological monitoring of cardiac allograft rejection. A 3-year follow-up. APMIS 1988; 96(1):14-24.
40. Fyfe B, Loh E, Winters GL et al. Heart transplantation-associated perioperative ischemic myocardial injury. Morphological features and clinical significance. Circulation 1996; 93(6):1133-40.
41. Gao SZ, Schroeder JS, Hunt SA et al. Acute myocardial infarction in cardiac transplant recipients. Am J Cardiol 1989; 64(18):1093-97.
42. Radio SJ, McManus BM, Winters GL et al. Preferential endocardial residence of B-cells in the "Quilty effect" of human heart allografts: immunohistochemical distinction from rejection. Mod Pathol 1991; 4(5):654-60.
43. Kottke-Marchant K, Ratliff NB. Endomyocardial lymphocytic infiltrates in cardiac transplant recipients. Incidence and characterization. Arch Pathol Lab Med 1989; 113(6):690-98.
44. Fishbein MC, Bell G, Lones MA et al. Grade 2 cellular heart rejection: does it exist? J Heart Lung Transplant 1994; 13(6):1051-57.
45. Linder J. Infection as a complication of heart transplantation. J Heart Transplant 1988; 7(5):390-94.
46. Fishbein MC, Kobashigawa J. Biopsy-negative cardiac transplant rejection: etiology, diagnosis and therapy. Curr Opin Cardiol 2004; 19(2): 166-69.
47. Stark DD, Weissleder R, Elizondo G et al. Superparamagnetic iron oxide: clinical application as a contrast agent for MR imaging of the liver. Radiology 1988; 168(2):297-301.
48. Weissleder R, Hahn PF, Stark DD et al. Superparamagnetic iron oxide: enhanced detection of focal splenic tumors with MR imaging. Radiology 1988; 169(2):399-403.
49. Bellin MF, Lebleu L, Meric JB. Evaluation of retroperitoneal and pelvic lymph node metastases with MRI and MR lymphangiography. Abdom Imaging 2003; 28(2):155-63.
50. Sigal R, Vogl T, Casselman J et al. Lymph node metastases from head and neck squamous cell carcinoma: MR imaging with ultrasmall superparamagnetic iron oxide particles (Sinerem MR)—results of a phase-III multicenter clinical trial. Eur Radiol 2002; 12(5):1104-13.
51. Wu YL, Ye Q, Foley LM et al. In situ labeling of immune cells with iron oxide particles: an approach to detect organ rejection by cellular MRI. Proc Natl Acad Sci USA 2006; 103(6):1852-57.
52. Arbab AS, Bashaw LA, Miller BR et al. Intracytoplasmic tagging of cells with ferumoxides and transfection agent for cellular magnetic resonance imaging after cell transplantation: methods and techniques. Transplantation 2003; 76(7):1123-30.

53. Arbab AS, Yocum GT, Kalish H et al. Efficient magnetic cell labeling with protamine sulfate complexed to ferumoxides for cellular MRI. Blood 2004; 104(4):1217-23.

54. Frank JA, Miller BR, Arbab AS et al. Clinically applicable labeling of mammalian and stem cells by combining superparamagnetic iron oxides and transfection agents. Radiology 2003;228(2):480-87.

55. Lewin M, Carlesso N, Tung CH et al. Tat peptide-derivatized magnetic nanoparticles allow in vivo tracking and recovery of progenitor cells. Nat Biotechnol 2000; 18(4):410-14.

56. Penno E, Johnsson C, Johansson L et al. Comparison of ultrasmall superparamagnetic iron oxide particles and low molecular weight contrast agents to detect rejecting transplanted hearts with magnetic resonance imaging. Invest Radiol 2005; 40(10):648-54.

57. Young JB. Allograft vasculopathy: diagnosing the nemesis of heart transplantation. Circulation 1999; 100(5):458-60.

58. Fang JC, Rocco T, Jarcho J et al. Noninvasive assessment of transplant-associated arteriosclerosis. Am Heart J 1998; 135(6 Pt 1):980-87.

59. Romeo G, Houyel L, Angel CY et al. Coronary stenosis detection by 16-slice computed tomography in heart transplant patients: comparison with conventional angiography and impact on clinical management. J Am Coll Cardiol 2005; 45(11):1826-31.

60. Caus T, Kober F, Marin P et al. Non-invasive diagnostic of cardiac allograft vasculopathy by 31P magnetic resonance chemical shift imaging. Eur J Cardiothorac Surg 2006; 29(1):45-49.

61. Auer T, Schreier G, Hutten H et al. Intramyocardial electrograms for the monitoring of allograft rejection after heart transplantation using spontaneous and paced beats. Transplant Proc 1995; 27(5):2621-24.

62. Bourge R, Eisen H, Hershberger R et al. Non-invasive rejection monitoring of cardiac transplants using high resolution intramyocardial electrograms: initial US multicenter experience. Pacing Clin Electrophysiol 1998; 21(11 Pt 2):2338-44.

63. Grasser B, Iberer F, Schaffellner S et al. Non-invasive graft monitoring after heart transplantation: rationale to reduce the number of endomyocardial biopsies. Transpl Int 2000; 13 Suppl 1:S225-27.

64. Bainbridge AD, Cave M, Newell S et al. The utility of pacemaker evoked T wave amplitude for the non-invasive diagnosis of cardiac allograft rejection. Pacing Clin Electrophysiol 1999; 22(6 Pt 1):942-46.

65. Phillips M, Boehmer JP, Cataneo RN et al. Heart allograft rejection: detection with breath alkanes in low levels (the HARDBALL study). J Heart Lung Transplant 2004; 23(6):701-8.

66. Gleissner CA, Klingenberg R, Nottmeyer W et al. Diagnostic efficiency of rejection monitoring after heart transplantation with cardiac troponin T is improved in specific patient subgroups. Clin Transplant 2003; 17(3):284-91.

67. Hammerer-Lercher A, Mair J, Antretter H et al. B-type natriuretic peptide as a marker of allograft rejection after heart transplantation. J Heart Lung Transplant 2005; 24(9):1444.

68. Mehra MR, Uber PA, Potluri S et al. Usefulness of an elevated B-type natriuretic peptide to predict allograft failure, cardiac allograft vasculopathy and survival after heart transplantation. Am J Cardiol 2004; 94(4):454-8.

69. Park MH, Scott RL, Uber PA et al. Usefulness of B-type natriuretic peptide levels in predicting hemodynamic perturbations after heart transplantation despite preserved left ventricular systolic function. Am J Cardiol 2002; 90(12):1326-29.

70. Mehra MR, Uber PA, Walther D et al. Gene expression profiles and B-type natriuretic peptide elevation in heart transplantation: more than a hemodynamic marker. Circulation 2006; 114(1 Suppl):I21-26.

71. Healy DG, Watson RW, O'Keane C et al. Neutrophil transendothelial migration potential predicts rejection severity in human cardiac transplantation. Eur J Cardiothorac Surg 2006; 29(5):760-66.

72. Suarez-Alvarez B, Lopez-Vazquez A, Diaz-Molina B et al. The predictive value of soluble major histocompatibility complex class I chain-related molecule A (MICA) levels on heart allograft rejection. Transplantation 2006; 82(3):354-61.

73. Welters MJ, Oei FB, Vaessen LM et al. Increased numbers of circulating donor-specific T-helper lymphocytes after human heart valve transplantation. Clin Exp Immunol 2001; 124(3):353-58.

74. DeBruyne LA, Ensley RD, Olsen SL et al. Increased frequency of alloantigen-reactive helper T-lymphocytes is associated with human cardiac allograft rejection. Transplantation 1993; 56(3):722-27.

75. DeBruyne LA, Renlund DG, Bishop DK. Evidence that human cardiac allograft acceptance is associated with a decrease in donor-reactive helper T-lymphocytes. Transplantation 1995; 59(5):778-83.

76. Chung SW, Yoshida EM, Cattral MS et al. Donor-specific stimulation of peripheral blood mononuclear cells from recipients of orthotopic liver transplants is associated, in the absence of rejection, with type-2 cytokine production. Immunol Lett 1998; 63(2):91-96.

77. Chen Y, McKenna GJ, Yoshida EM et al. Assessment of immunologic status of liver transplant recipients by peripheral blood mononuclear cells in response to stimulation by donor alloantigen. Ann Surg 1999; 230(2):242-50.

78. Molajoni ER, Cinti P, Orlandini A et al. Mechanism of liver allograft rejection: the indirect recognition pathway. Hum Immunol 1997; 53(1):57-63.

79. Molajoni ER, Cinti P, Ho E et al. Allospecific T-suppressor cells in liver transplantation. Transplant Proc 2001; 33(1-2):1381-83.

80. Grauhan O, Muller J, v Baeyer H et al. Treatment of humoral rejection after heart transplantation. J Heart Lung Transplant 1998; 17(12):1184-94.

81. Fredrich R, Toyoda M, Czer LS et al. The clinical significance of antibodies to human vascular endothelial cells after cardiac transplantation. Transplantation 1999; 67(3):385-91.

82. Tambur AR, Winkel E, Heroux A et al. Flow panel reactive antibody monitoring following heart transplantation. Transplant Proc 2001; 33(7-8):3295-97.

83. Cook DJ, Bishay ES, Yamani M. The use and misuse of immunologic monitoring after transplantation: approaches that have proved useful. Curr Opin Cardiol 2000; 15(2):104-7.

84. McCarthy JF, Cook DJ, Massad MG et al. Vascular rejection post heart transplantation is associated with positive flow cytometric cross-matching. Eur J Cardiothorac Surg 1998; 14(2):197-200.

85. Kimball PM, Radovancevic B, Isom T et al. The paradox of cytokine monitoring-predictor of immunologic activity as well as immunologic silence following cardiac transplantation. Transplantation 1996; 61(6):909-15.

86. Reed EF, Hong B, Ho E et al. Monitoring of soluble HLA alloantigens and anti-HLA antibodies identifies heart allograft recipients at risk of transplant-associated coronary artery disease. Transplantation 1996; 61(4):566-72.

87. Castro P, Arriagada G, Moreno M et al. [Humoral rejection in heart transplantation. Report of 2 cases]. Rev Med Chil 2000; 128(11):1245-49.

88. Dzemeshkevich S, Ragimov A, Mikhaylov Y et al. Plasmapheresis in the treatment of posttransplant cardiomyopathy. Artif Organs 1998; 22(3):197-202.

89. Malafa M, Mancini MC, Myles JL et al. Successful treatment of acute humoral rejection in a heart transplant patient. J Heart Lung Transplant 1992; 11(3 Pt 1):486-91.

90. Partanen J, Nieminen MS, Krogerus L et al. Heart transplant rejection treated with plasmapheresis. J Heart Lung Transplant 1992; 11(2 Pt 1):301-5.

91. Ratkovec RM, Hammond EH, O'Connell JB et al. Outcome of cardiac transplant recipients with a positive donor-specific cross-match--preliminary results with plasmapheresis. Transplantation 1992; 54(4):651-55.

92. Grauhan O, Knosalla C, Ewert R et al. Plasmapheresis and cyclophosphamide in the treatment of humoral rejection after heart transplantation. J Heart Lung Transplant 2001; 20(3):316-21.

93. Sarmiento E, Rodriguez-Molina J, Munoz P et al. Decreased levels of serum immunoglobulins as a risk factor for infection after heart transplantation. Transplant Proc 2005; 37(9):4046-49.

94. Sarmiento E, Rodriguez-Molina JJ, Fernandez-Yanez J et al. IgG monitoring to identify the risk for development of infection in heart transplant recipients. Transpl Infect Dis 2006; 8(1):49-53.

# CHAPTER 11

# Experimental Gene Therapy of Heart Transplantation

Giuseppe Vassalli,* Charles Seydoux, Pierre Vogt, Manuel Pascual and Ludwig K. von Segesser

## Abstract

Maintenance of a functional graft requires life-long immuno-suppression to prevent rejection by the immune system. Unfortunately, current immunosuppressive agents do not effectively prevent chronic rejection and are associated with significant comorbidity. Therefore, improved strategies that target alloimmune responses more specifically are needed. Over the last decade, gene therapy has attracted considerable interest as a novel approach to transplantation. In fact, organ transplantation offers the unique opportunity of delivering a therapeutic gene to the donor organ during the preservation time before engraftment. Gene transfer into the donor organ is performed ex vivo under controlled conditions and the therapeutic factor is then produced by the graft itself in vivo during extended periods of time. Proof-of-principle studies in animal models have suggested that localized overexpression of cytoprotective and anti-inflammatory factors in the graft may attenuate ischemia-reperfusion injury and acute rejection, resulting in prolonged graft survival. Interestingly, exogenous overexpression of immunoregulatory genes such as inhibitors of T-cell costimulatory activation in the graft may result in local, rather than systemic, immunosuppression. Genetically engineered host bone marrow-derived cells expressing donor major histocompatibility complex (MHC) antigens have a potential for inducing donor-specific immune tolerance. Despite encouraging results obtained in animal models, however, gene therapy of organ transplantation has not entered the clinical arena yet. Several important issues including the development of flawless and regulatable vectors, as well as of delivery protocols adapted to clinical transplantation remain to be solved. An improved understanding of the mechanisms underlying the activation of alloreactive cells, as well as the generation of regulatory T-cells, will be key to identifying critical molecular targets. A major hurdle to the translation of gene therapy for transplantation from the bench to the bedside is that most studies have been carried out in rodents; however, transplant models in rodents are of limited usefulness in humans. Future preclinical studies will need to compare gene therapy with established immunosuppressive treatments in nonhuman primates.

## Introduction

Life-long immunosuppression is required to prevent rejection of allogeneic transplants by the immune system and maintain a functional graft. Immunosuppressive agents currently used in transplanted patients mainly act by indiscriminately blocking T-cell activation, the central mechanism of immune rejection. Although these agents have been quite effective in preventing acute rejection, they have failed to successfully prevent accelerated graft vascular disease and chronic allograft rejection.[1,2] In addition, current immunosuppressive agents are associated with significant comorbidity including dyslipidemia, diabetes, renal toxicity, opportunistic infections and an increased risk of developing cancer. For these reasons, considerable efforts have been made to develop improved therapeutic strategies that target alloreactive cells more directly.

Over the last decade, gene therapy has attracted considerable interest as a potential approach to organ transplantation. In fact, transplantation offers the unique opportunity of delivering a therapeutic gene to the donor organ after procurement before engraftment. Consequently, gene delivery to the donor organ is carried out ex vivo during the preservation time under well-controlled conditions.[3] This opportunity obviates the need for technically difficult in vivo gene delivery protocols, which often achieve only sub-optimal gene transfer efficiencies. As an example, the catheter-based infusion of a gene containing solution into a coronary artery in vivo is limited by a short dwelling time of the delivered gene in the coronary circulation, as blood flow can only be interrupted for a short time ($\approx 1$ minute). In a pig model of adenovirus-mediated in vivo gene transfer into the heart, the cardiac uptake of the delivered gene after intracoronary administration was $\approx 7$-fold lower than after direct intramyocardial injection.[4] Low cardiac uptake correlates with increased amounts of the vector that pass through the coronary circulation and are disseminated systemically. Conversely, ex vivo gene delivery to the donor organ during the preservation time results in the production of high amounts of the therapeutic factor at the very site where the protective effect is needed. The case for local immunosuppression of organ transplants, independent of gene therapy, was discussed some 20 years ago.[5] However, gene transfer technologies offer the additional opportunity of producing the therapeutic factor for extended periods of time, potentially even for a lifetime. Clearly, this is an appealing aspect in transplantation as a lifelong condition. In the case when the protective factor is a secreted peptide, gene therapy may be viewed as an alternate approach to the systemic administration of the recombinant peptide. However, many peptides have short half-lives in vivo and require daily injections over long periods of time. In contrast, a single dose of the encoding gene can achieve sustained production and release of the peptide in vivo. In the case when the therapeutic factor is an intracellular protein, administration of the recombinant protein is not feasible and gene transfer-based approaches are mandatory.

Proof-of-principle studies in animal models have established gene transfer technologies as a powerful tool for investigating the biological effects of genes of interest both in vitro and in vivo. Moreover, they have demonstrated that gene therapy approaches to organ transplantation are feasible, even though several issues will need to be addressed

---

*Corresponding Author: Giuseppe Vassalli—Dept. of Cardiology, BH-10, CHUV, 1011 Lausanne, Switzerland. Email: giuseppe.vassalli@chuv.ch

*Chronic Allograft Failure: Natural History, Pathogenesis, Diagnosis and Management*, edited by Nasimul Ahsan. ©2008 Landes Bioscience.

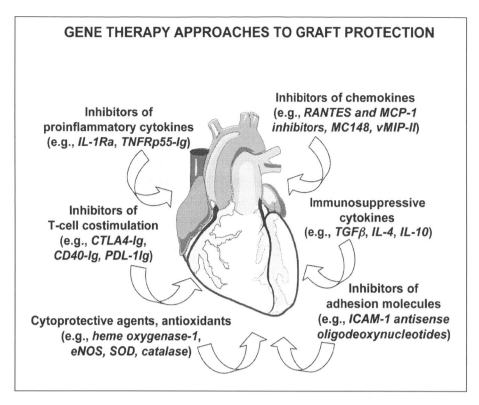

Figure 1. Multiple different mechanisms of action of gene therapy strategies in heart transplantation (non-exhaustive list). A sample of candidate genes that have been tested in experimental studies are indicated for each mechanism of action (see text for abbreviations).

before clinical applications can be envisioned. This chapter is devoted to a concise review of experimental gene therapy approaches to heart transplantation. A sample of studies illustrating the multiplicity of gene transfer-based strategies (Fig. 1) will be discussed briefly. Major hurdles that need to be overcome before gene therapy of heart transplantation can be translated from the bench to the bedside will be addressed in the last part of this review.

## Gene Transfer Vectors

Central to any gene therapy strategy is a gene vehicle capable of delivering and expressing the therapeutic gene with high efficiency and, ideally, specificity in a target tissue. Gene vehicles can be subdivided into viral and nonviral vectors. Recombinant viral vectors exploit the ability of their parent viruses to infect susceptible cells. However, they also stimulate immune reactions similar to those induced by the corresponding wild-type viruses (see below). Recombinant viral vectors are deleted in gene sequences that are required for virus replication. Therefore, they infect and deliver genes to target cells but do not replicate in these cells. Cell lysis and uncontrolled virus spreading as a result of productive infection are avoided.

Several types of vectors exist, each one with distinct advantages and disadvantages. Recombinant adenovirus vectors have been employed in a majority of gene therapy studies in organ transplantation because they efficiently infect a wide range of tissues and also because high-titer virus preparations can be easily produced. However, standard adenovirus vectors suffer from several limitations including immune responses to viral proteins and transient gene expression (typically lasting for 1 to 4 weeks in the absence of immunosuppression).[6] In a rat model of heart transplantation, adenovirus-mediated expression of a *lacZ* reporter gene in the donor heart was short-lived but could be prolonged up to 12 weeks after transplantation by treatment with cyclosporine A.[7] In the case when the delivered gene encodes an immunosuppressive factor, immune responses to the adenovirus vector and allograft rejection are inhibited at once and the immunosuppressive gene is expressed for

extended periods of time.[8] Newly developed adenovirus vectors deleted in all viral genes (so-called "gutless" or helper-dependent vectors) are less immunogenic than first-generation adenovirus vectors. They are capable of expressing genes long-term in several tissues including skeletal muscle and liver.[9] However, transgene expression after direct intra-myocardial injection of gutless adenovirus vectors in rat hearts in vivo was short-lived ($\approx$2 weeks) in one study.[10]

Unlike adenovirus, both adeno-associated virus (AAV) and retrovirus vectors integrate their genomes into the chromosomes of host cells, thereby providing a potential for long-term expression of the foreign gene. As an example, AAV-mediated expression of the enhanced green fluorescent protein (*EGFP*) reporter gene lasted for more than 1 year in mouse hearts, with negligible tissue inflammation.[11] AAV vectors also achieved long-term gene expression in rat cardiac isografts.[3] Vectors derived from different serotypes of AAV have disparate tissue tropisms. Traditional AAV vectors based on serotype 2 show a broad tissue tropism resulting in the predominant transduction of the liver after vector infusion into a peripheral vein. AAV-2 vector genomes crosspackaged into capsids from AAV serotypes 1,4,5 and 6 achieved a $\approx$10-fold increase in cardiac gene transfer efficiency.[12]

Among retroviruses, lentiviruses including human immunodeficiency virus type-1 (HIV-1) have the unique ability to integrate their genomes into the chromosomes of nondividing cells. As a result, lentivirus vectors—unlike traditional retrovirus vectors derived from the Moloney murine leukemia virus—efficiently deliver and express genes in nondividing cells including adult cardiomyocytes[13] and quiescent endothelial cells.[14] Multiply attenuated, self-inactivating lentivirus vectors have been employed to deliver genes of interest to donor hearts in transplant models and functional effects have been reported (see below).[15]

In general, nonviral vectors including cationic lipids, cationic polymers (e.g., polylysine) and liposomes combined with fusogenic viral proteins (e.g., from the hemagglutinating virus of Japan; HVJ) are less efficient than viral vectors. However, they offer potential advantages including the capacity to accommodate for large gene sequences, negligible tissue inflammation and a favorable safety profile. Liposomes have been employed to deliver protective genes to cardiac grafts and beneficial effects have been observed.[16-20]

## Gene Transfer of Cytoprotective and Anti-Inflammatory Factors

Oxygen deprivation of the donor organ during the preservation time results in increased oxidative stress, which is exacerbated by reperfusion.[21] Vascular endothelium is highly susceptible to ischemia-reperfusion injury. Inflammatory activation of endothelial cells results in the upregulation of cell adhesion molecules leading to leukocyte adhesion. The complement system is activated and neutrophils are recruited to the graft, followed by natural killer (NK) cells and macrophages. Early antigen-independent inflammatory reactions mediated by the innate immune system are followed by allospecific responses that culminate in massive graft infiltration by T- and B-cells, macrophages and dendritic cells.

Local overexpression of cytoprotective genes and inhibitors of inflammatory cytokines in the graft has been evaluated as a means of mitigating ischemia-reperfusion injury and inflammatory events that take place in the graft early after transplantation. A practical advantage of this approach is that it does not require long-term expression of the delivered gene. Endothelial nitric oxide (NO) synthase (eNOS), the key enzyme in enzymatic NO production, has been tested as a cytoprotective gene in transplant models. Ex vivo eNOS gene transfection into donor hearts using liposomes reduced graft ischemia-reperfusion injury and improved allograft survival in a rabbit model.[22] The protective effect of eNOS overexpression was associated with inhibition of NFκB, a central molecule in molecular cascades of inflammatory activation. Direct targeting of NFκB with specific double-stranded oligodeoxynucleotides also attenuated ischemia-reperfusion injury in rat hearts.[23] These results suggest that local eNOS overexpression and NFκB inhibition may attenuate graft ischemia-reperfusion injury, resulting in prolonged graft survival.

Endogenous scavengers neutralize toxic reactive oxygen species (ROS) generated during ischemia and reperfusion. Their neutralizing capacity is limited, however, suggesting that local overexpression of ROS scavengers in the graft may be beneficial. Consistent with this assumption, transgenic mice that overexpress human copper-zinc superoxide dismutase (SOD) in cardiac myocytes and cardiac endothelial cells neutralized the superoxide burst after 30 minutes of myocardial ischemia almost completely, showing improved functional recovery and decreased infarct size as compared with normal mice.[24] Using a gene transfer-based approach, adenovirus-mediated gene transfer of Mn-SOD into rabbit donor hearts mitigated ischemia-reperfusion injury after organ preservation and transplantation.[25] Adenovirus-mediated gene transfer of human Cu-Zn-SOD into rat liver allografts resulted in improved survival of graft recipients.[26] These observations suggest that local overexpression of free radical scavengers may neutralize the elevated oxidative stress, protecting the graft against ischemia-reperfusion injury.

Heme oxygenase-1, the enzyme that degrades heme into carbon monoxide, biliverdine and free iron, is a cytoprotective, anti-oxidant and anti-apoptotic factor. In a mouse model of heart transplantation, systemic adenovirus-mediated gene transfer of heme oxygenase-1 achieved long-term ( >100 days) allograft survival in 80% of mice.[27] Immune responses to cognate antigens were essentially unaffected, indicating lack of general immunosuppression. In line with these results, transgenic mice overexpressing heme oxygenase-1 accepted normal donor hearts for prolonged periods of time.[28] Interestingly, the heme oxygenase-1 gene was more effective after systemic as compared to local administration to the graft. It was postulated that immunoregulatory effects of heme oxygenase-1 in the spleen are responsible for this observation.[27]

Interleukin (IL)-1β and tumor necrosis factor (TNF)-α are among the most potent proinflammatory cytokines and mediators of ischemia-reperfusion injury and early graft inflammation. The activity of IL-1β is regulated by the naturally occurring IL-1 receptor antagonist (IL-1Ra). Adenovirus-mediated IL-1Ra gene transfer reduced inflammatory cell infiltrates in rat hearts after ischemia and reperfusion.[29] Adenovirus-mediated gene transfer of a TNF-α receptor antagonist (TNFRp55-Ig) attenuated allograft rejection in a rat heart transplant model.[30] These data suggest that gene transfer of IL-1 and TNF-α inhibitors into donor hearts may be beneficial.

Upregulation of adhesion molecules on endothelial cell surfaces during ischemia and reperfusion promotes binding of leukocytes, followed by their transendothelial migration and accumulation in the graft. Therefore, adhesion molecules are important therapeutic targets in transplantation. Hyperbaric transfection of antisense oligodeoxynucleotides with specific affinity for intercellular adhesion molecule (ICAM)-1 mitigated ischemia-reperfusion injury and prolonged allograft survival in a rat model of heart transplantation.[31]

## Gene Transfer of Immunoregulatory Cytokines

Increased production of interferon (IFN)-γ and other T helper type 1 (Th1) cytokines has been correlated with acute graft rejection. On the other hand, T helper type 2 (Th2) cytokines such as IL-4, IL-10, IL-13, and transforming growth factor (TGF)-β have been associated with enhanced graft survival. Therefore, skewing the Th1/Th2 balance toward Th2 cytokines ("immune deviation") may have beneficial effects. Gene transfer of IL-4, IL-10, IL-13, or TGF-β into donor hearts improved cardiac allograft survival in rodent models.[17,32-37] The viral homologue of IL-10 encoded by the Ebstein-Barr virus (EBV) lacks some of the immunostimulatory activities of cellular IL-10. Local expression of viral IL-10 in donor hearts using feline immunodeficiency virus (FIV), retrovirus or plasmid vectors resulted in improved cardiac allograft survival in rodent models.[38-40] Hematopoietic stem cells genetically engineered to express viral IL-10 and delivered to syngeneic mice 6 weeks before heart transplantation induced prolonged acceptance of cardiac allografts.[41] These results suggest that local overexpression of immunoregulatory cytokines may delay graft rejection, even though it usually fails to prevent it. This is presumably due, at least in part, to the functional redundancy of proinflammatory cytokine cascades. Consistent with this assumption, dual IL-4 and IL-10 gene transfection synergistically achieved sustained expression of the two transgenes and long-term cardiac allograft survival in a rabbit model.[37]

## Gene Transfer of Inhibitors of Chemokines

Leukocytes are recruited to foci of inflammation along gradients of chemoattractant cytokines (i.e., chemokines).[42] A number of chemokines including IL-8, the regulated upon activation and T-cell-secreted (RANTES) chemokine, monocyte chemoattractant proteins (MCP)-1 and -3 and IFN-γ–inducible protein of 10 kD (IP-10) are upregulated in the graft early after heart transplantation in humans.[43,44] Adenovirus- or lentivirus-mediated gene transfer of an NH₂-terminally deleted RANTES mutant gene encoding a peptide inhibitor of multiple different C-C chemokines delayed cardiac allograft rejection modestly in a rat model.[15,45] Interestingly, some viruses produce chemokine homologues that act as functional antagonists of endogenous chemokines. Chemokine homologues are thought to help viruses escape immunosurveillance. Adenovirus-mediated gene transfer of two viral chemokine homologues, MC148 and vMIP-II, prolonged cardiac allograft survival in mice.[46] Similarly to inhibition of proinflammatory cytokines, however, gene transfer of chemokine inhibitors improved graft survival only modestly.[15,45,46] In principle, this might be accounted for by several factors including low gene transfer efficiency, weak inhibitory activity of the chemokine antagonist and the "molecular promiscuity" of chemokines for their receptors. An individual chemokine can bind to multiple different receptors, while each leukocyte subpopulation expresses multiple chemokine receptors.[42] Given the complex interactions of chemokines with their cellular receptors, blocking an individual chemokine receptor may not be sufficient to prevent the recruitment of leukocytes to the graft. It is interesting to note, however, that gene-targeted mice lacking either the C-C chemokine receptor (CCR)-1, CCR-5, or the CXC chemokine receptor (CXCR)-3 accept normal cardiac allografts for extended periods of time.[47-49] In addition, cardiac grafts from IP-10-deficient mice survive long-term in normal hosts.[50] These findings highlight the important role of chemokines in graft rejection, suggesting that effective inhibition of one or more chemokine receptors may profoundly impact graft survival.

## T-Cell Costimulatory Blockade

Graft rejection requires full activation of host CD4[+] and CD8[+] T-cells, which acquire effector functions including the ability to produce stimulatory cytokines. Multiple different molecular signals regulate T-cell activation. The first signal is provided by the T-cell receptor (TCR) upon recognition of the MHC-peptide complex on antigen-presenting cells. Additional recognition signals arise from interactions of costimulatory receptors and ligands expressed on the surface of T-cells and antigen-presenting cells (APCs).[51] Pairs of costimulatory molecules include CD28 and B7-1/B7-2 (CD80/86), CD40 ligand (CD154) and CD40, inducible costimulator (ICOS) and B7RP-1, CD134 ligand and CD134, as well as CD27 and CD70. Antigenic peptides that occupy the TCR in the absence of appropriate costimulatory signals may induce antigen-specific T-cell anergy or prime T-cells for apoptosis. Upon ligation by B7-1/B7-2, T-cells also upregulate the CTLA-4 surface antigen that modulates T-cell activation.

Several attempts have been made to suppress T-cell costimulatory signals using blocking antibodies, recombinant proteins, pharmacological agents[52] and gene transfer technology. In initial studies, treatment with a soluble recombinant CTLA-4Ig fusion protein induced donor-specific unresponsiveness and long-term acceptance of pancreatic islet xenografts.[53] Subsequent studies focused on local CTLA-4Ig expression in the graft to avoid systemic immunosuppression. After adenovirus-mediated CTLA-4Ig ex vivo gene transfer into rat donor hearts, CTLA-4Ig was detectable in the serum 120 days after transplantation and allografts survived long-term (>100 days). Recipients of long-term surviving grafts expressing CTLA-4Ig accepted a second cardiac graft from the original donor strain in the absence of any immunosuppressive treatment, while rejecting third-party grafts. These results suggested that intragraft CTLA-4Ig gene transfer induced a donor-specific operational tolerance in the rat model used.[8] However, this approach was associated with a moderate inhibition of T-cell responses to unrelated third-party antigens, indicating some degree of general immunosuppression. A direct comparison with recombinant CTLA-4Ig protein administration in the same model showed that systemic CTLA-4Ig treatment induced a higher degree of systemic immunosuppression, while improving graft survival only moderately. This study illustrates the potential advantages of expressing immunosuppressive genes locally in transplant organs.[8] However, this approach may not be well-suited to all candidate genes. As mentioned above, systemic gene transfer of heme oxygenase-1 was more effective than the direct delivery of the same adenovirus vector to the graft in a mouse model of heart transplantation.[27]

In another study, adenovirus-mediated CTLA-4Ig gene transfer into cold-preserved rat donor livers attenuated reperfusion injury and induced donor-specific tolerance with indefinite liver graft survival.[54] However, it has been suggested that rodent cardiac and liver allografts exert direct tolerogenic effects[55] and that the graft itself may have contributed to the sustained unresponsiveness to donor antigens initially

induced by CTLA-4Ig in these models.[8,54] Therefore, these results in rodents should be interpreted cautiously as evidence for donor-specific T-cell unresponsiveness, rather than true immune tolerance.

CD40-CD154 interactions provide important costimulatory stimuli, leading to the production of TNF-α, IL-1β, IL-6, IL-8, IL-12 and other cytokines by APCs. CD40 is expressed on professional APCs such as dendritic cells and macrophages and on nonprofessional APCs such as B-cells, endothelial cells and fibroblasts. CD154 is upregulated on T-cells upon their initial activation. Blockade of CD40-CD154 interactions by adenovirus-mediated intragraft expression of a soluble CD40Ig fusion protein prolonged cardiac allograft survival for limited periods of time in a rat model.[56] However, the protective effect was accompanied by nonspecific inhibition of immune responses to third-party antigens. Although general immune responses were restored at later time points, donor-specific T-cell responses reappeared at the same time causing graft rejection. Importantly, dual CTLA-4Ig and CD40Ig gene therapy was more effective than either gene alone.[57]

Programmed death (PD)-1, a CD28 homologue, is expressed on activated T-cells, B-cells and myeloid cells.[58] Engagement of PD-1 by its ligands, PD-L1 and PD-L2, inhibits proliferation and cytokine production by activated T-cells. Systemic treatment with a soluble PD-L1Ig fusion protein improved cardiac allograft survival in CD28-deficient mice.[59] In conjunction with an immunosuppressive drug, PD-L1Ig protein treatment also improved graft survival in fully MHC-disparate, normal mice, in some cases inducing permanent engraftment.[59] Local adenovirus-mediated PD-L1Ig expression improved cardiac allograft survival moderately in a rat model.[60]

Collectively, these data suggest that gene transfer-based blockade of T-cell costimulatory pathways attenuates acute cardiac allograft rejection.

## Gene Transfer of Donor MHC Antigens

MHC mismatch between the donor and recipient is a major determinant of graft rejection. Attempts have been made to genetically modify host-derived cells to express donor MHC genes, which are reintroduced in the host before organ transplantation, to induce immune tolerance to donor MHC antigens. In initial studies, injection of host-derived fibroblasts genetically modified to express donor MHC class I or II genes achieved prolonged cardiac allograft survival in mice.[61] More recently, reconstitution of the host bone marrow with autologous hematopoietic stem cells genetically modified to express donor MHC genes induced donor-specific tolerance to cardiac allografts in several studies.[62-65] Remarkably, adenovirus-mediated expression of a single donor MHC class I gene in syngeneic bone marrow cells, in conjunction with a nondepleting anti-T-cell antibody, induced long-term acceptance of fully allogeneic cardiac grafts.[63]

A central question with these tolerogenic strategies is how stable is the T-cell unresponsiveness induced by "molecular chimerism", since this state can be abolished by provision of T-cell help.[66] Encouraging results have been obtained using bone marrow-derived cells retrovirally transduced with an allogeneic MHC class I gene.[67] This approach induced indefinite acceptance of MHC class I-mismatched skin grafts, but normal rejection of third-party grafts, in mice. Cytotoxic T-cells capable of lysing donor cells in vitro were undetectable even after challenge with the antigen.

An important study in rhesus macaques evaluated hematopoietic stem cells retrovirally transduced with a xenogeneic protein.[68] Progeny lymphocytes expressing the foreign gene were still detectable two years after stem cell transplantation. In contrast, infused mature lymphocytes expressing the foreign gene were no longer detectable within several weeks of infusion in untreated monkeys. These observations suggested that genetically engineered hematopoietic stem cells promoted immune tolerance to a xenogeneic protein in nonhuman primates.

## Gene Therapy of Graft Vasculopathy

Accelerated coronary arteriosclerosis and chronic graft dysfunction are among the most discouraging aspects in heart transplantation, as they are not effectively prevented by current immunosuppressive regimens.[1] However, some agents including statins, mycophenolate mofetil, sirolimus and everolimus have been associated with potential beneficial effects.[69-72] Both alloantigen-dependent and independent factors contribute to the pathogenesis of graft vasculopathy.[73,74]

Adenovirus-mediated gene transfer of heme oxygenase-1 delayed the development of antibody-induced arterial lesions in transplanted rat aortas.[75] Conversely, adenovirus-mediated gene delivery of another anti-apoptotic factor, the mitochondrial protein Bcl-2, worsened the coronary artery disease in rat cardiac allografts despite inhibition of apoptosis.[76] Interestingly, the same vector delivered to the donor heart 4 days before allotransplantation reduced oxidative stress in graft tissue and coronary artery lesions 90 days after transplantation.[77] The divergent results in these two studies suggest that Bcl-2 gene transfer a few days before allotransplantation may be critical to achieve therapeutic Bcl-2 levels at the time of ischemia and reperfusion. However, gene delivery to the donor heart a few days before transplantation would be hardly compatible with clinical applications. Local adenovirus-mediated CD40Ig gene transfer reduced alloreactive antibody levels, leukocyte infiltration of the vessel wall and intimal thickening in rat aortic allografts.[78] However, the same vector did not prevent accelerated coronary vasculopathy in cardiac allografts.[78]

Gene transfer of a soluble Fas molecule reduced lesion formation in rat aortic allografts, possibly by interacting with Fas ligand (FasL) expressed on the surfaces of macrophages within arterial lesions.[79] Somewhat paradoxically, FasL overexpression on graft vascular endothelium also reduced inflammatory cell infiltrates and transplant-associated intimal hyperplasia in another study.[80] The reason for the similar outcomes of FasL blockade and FasL overexpression in the two studies is unclear, but it is presumably related to the use of soluble Fas or cell-membrane bound FasL molecules. While cell-membrane bound FasL induces apoptosis, soluble FasL variants mediate more complex biological effects.

Gene transfer of tissue-type plasminogen activator, a vasoprotective factor that is downregulated in graft vasculopathy, delayed the development of coronary lesions in cardiac allografts.[81] Ribozymes against metalloproteinase-2, a matrix-degrading enzyme that is persistently expressed in the vessel wall throughout the progression of the disease, similarly inhibited graft coronary vasculopathy in a separate study.[20] Systemic gene delivery of an MCP-1 inhibitory peptide blocking monocyte chemoattraction reduced the number of mononuclear cells accumulating in the lumen of graft coronary arteries and decreased arterial lesion formation 8 weeks after heart transplantation in mice.[82] Finally, double-stranded DNA with specific affinity for the E2F transcription factor, a key activator of cell-cycle regulatory genes, prevented the development of coronary lesions up to 8 weeks after heart transplantation in nonhuman primates.[83] Clearly, these results in a preclinical model directly relevant to humans are particularly important. However, longer follow-up studies will be required to evaluate the potential usefulness of gene therapy of graft vasculopathy.

## Future Perspectives

Whether or not clinical gene therapy applications in heart transplantation can be envisioned in the near future depends on a number of factors, but chiefly on the development of a sound gene transfer vector and the identification of a highly effective therapeutic gene. Currently, AAV vectors are among the most promising candidate vectors. They achieve efficient gene transfer and sustained gene expression in many tissues and lack significant proinflammatory effects. Moreover, AAV vectors are relatively safe, as the parent virus is not a known human pathogen. However, a problem shared by all viral vectors that integrate their genomes randomly in the chromosomes is the risk of insertional mutagenesis. This problem is illustrated by a recent clinical trial in children with X-linked severe combined immunodeficiency (SCID-X1).[84] Almost 3 years after retroviral gene therapy, uncontrolled exponential clonal proliferation of transduced T-cells occurred in three young patients. In these patients, T-cell clones showed retrovirus vector integration in proximity to the LMO2 proto-oncogene promoter, leading to aberrant transcription and expression of LMO2 and deregulated premalignant cell proliferation. The development of viral vectors that integrate in a nonrandom manner is an important area of ongoing research.

Another key requirement for future vectors is the ability to adjust gene expression ("gene dosage") to match the patient's needs. This also would enable physicians to suppress transgene expression should toxic effects be observed. Although regulatable vectors controlled by oral intake of doxycycline or other agents have been successfully tested in vivo,[85,86] they need to be further improved for clinical applications.

The most effective gene(s) to be delivered remains to be identified. Encouraging results have been achieved by gene transfer of T-cell costimulatory inhibitors or donor MHC molecules. Dual blockade of different pairs of T-cell costimulatory molecules appears to have synergistic potential.[56,87,88] Given the complexity of alloimmune responses, combined strategies targeting multiple different components of the immune system are intuitively appealing. An improved understanding of the mechanisms underlying the activation of alloreactive T-cells and the generation of regulatory T-cells will be important to identify critical molecular targets. Clearly, induction of donor-specific immune tolerance remains the ultimate goal, as it would obviate the need for life-long immunosuppression.[89] Beneficial effects of costimulation blockade in nonhuman primates are documented by inhibition of kidney allograft rejection after short-term treatment with humanized anti-CD154 antibody.[90] Recently, a randomized clinical trial of belatacept, a selective costimulation blocker, suggested that this agent is not inferior to cyclosporine with respect to the prevention of acute graft rejection in kidney transplant recipients.[52]

The scarcity of preclinical studies in nonhuman primates has been a major impediment to the translation of gene therapy for transplantation from the bench to the bedside.[91] Well-designed, controlled studies directly comparing gene therapy with established immunosuppressive treatments in primate models will be required. Because current treatments are fairly effective at preventing acute graft rejection, the new approaches will need to demonstrate superior protection especially from graft vasculopathy and chronic rejection.

### Acknowledgements

The support of the Swiss Science Foundation, the Swiss Cardiology Foundation, the Lausanne Transplant Foundation, the Teo Rossi di Montelera Foundation and the Novartis Research Foundation is gratefully acknowledged. The Lausanne Transplant Center is supported by the "Plan stratégique 2004-2007 des Hôspices-CHUV", Lausanne, Switzerland.

### References

1. Pascual M, Theruvath T, Kawai T et al. Strategies to improve long-term outcomes after renal transplantation. N Engl J Med 2002; 346:580-90.
2. Kirklin JK, Pambukian SV, McGiffin DC et al. Current outcomes following heart transplantation. Semin Thorac Cardiovasc Surg 2004; 16:395-403.
3. Asfour B, Baba HA, Scheld HH et al. Uniform long-term gene expression using adeno-associated virus (AAV) by ex vivo recirculation in rat cardiac isografts. Thorac Cardiovasc Surg 2002; 50:347-50.
4. Lee LY, Patel SR, Hackett NR et al. Focal angiogen therapy using intramyocardial delivery of an adenovirus vector coding for vascular endothelial growth factor 121. Ann Thorac Surg 2000; 69:14-23.
5. Gruber SA. The case for local immunosuppression. Transplantation 1992; 54:1-11.
6. Schulick AH, Vassalli G, Dunn PF et al. Established immunity precludes adenovirus-mediated gene transfer in rat carotid arteries. Potential for immunosuppression and vector engineering to overcome barriers of immunity. J Clin Invest 1997; 99:209-19.

7. Yap J, O'Brien T, Tazelaar HD et al. Immunosuppression prolongs adenoviral mediated transgene expression in cardiac allograft transplantation. Cardiovasc Res 1997; 35:529-35.

8. Guillot C, Mathieu P, Coathalem H et al. Tolerance to cardiac allografts via local and systemic mechanisms after adenovirus-mediated CTLA4Ig expression. J Immunol 2000; 164:5258-68.

9. Schiedner G, Morral N, Parks RJ et al. Genomic DNA transfer with a high-capacity adenovirus vector results in improved in vivo gene expression and decreased toxicity. Nat Genet 1998; 18:180-83.

10. Fleury S, Driscoll RA, Simeoni E et al. Helper-dependent adenovirus vectors devoid of all viral genes cause less myocardial inflammation compared with first-generation adenovirus vectors. Basic Res Cardiol 2004; 99:247-56.

11. Vassalli G, Büeler H, Dudler J et al. Adeno-associated virus (AAV) vectors achieve prolonged transgene expression in mouse myocardium and arteries in vivo: A comparative study with adenovirus vectors. Int J Cardiol 2003; 90:229-38.

12. Muller OJ, Leuchs B, Pleger ST et al. Improved cardiac gene transfer by transcriptional and transductional targeting of adeno-associated viral vectors. Cardiovasc Res 2006; 70:70-78.

13. Fleury S, Simeoni E, Zuppinger C et al. Multiply attenuated, self-inactivating lentiviral vectors efficiently deliver and express genes for extended periods of time in adult rat cardiomyocytes in vivo. Circulation 2003; 107:2375-82.

14. Cefaï D, Simeoni E, Ludunge KM et al. Multiply attenuated, self-inactivating lentiviral vectors efficiently transduce human coronary artery cells in vitro and rat arteries in vivo. J Mol Cell Cardiol 2005; 38:333-44.

15. Vassalli G, Simeoni E, Li J et al. Lentiviral gene transfer of the chemokine antagonist RANTES 9-68 prolongs heart graft survival. Transplantation 2006; 81:240-46.

16. Sen L, Hong YS, Luo H et al. Efficiency, efficacy and adverse effects of adenovirus- vs. liposome-mediated gene therapy in cardiac allografts. Am J Physiol Heart Circ Physiol 2001; 281:H1433-41.

17. Chan SY, Goodman RE, Szmuszkovicz JR et al. DNA-liposome versus adenoviral mediated gene transfer of transforming growth factor beta1 in vascularized cardiac allografts: differential sensitivity of CD4+ and CD8+ T-cells to transforming growth factor beta1. Transplantation 2000; 70:1292-304.

18. Hong YS, Laks H, Cui G et al. Localized immunosuppression in the cardiac allograft induced by a new liposome-mediated IL-10 gene therapy. J Heart Lung Transplant 2002; 21:1188-200.

19. Sawa Y, Suzuki K, Bai HZ et al. Efficiency of in vivo gene transfection into transplanted rat heart by coronary infusion of HVJ liposome. Circulation 1995; 92(9 Suppl):II479-82.

20. Tsukioka K, Suzuki J, Fujimori M et al. Expression of matrix metalloproteinases in cardiac allograft vasculopathy and its attenuation by anti MMP-2 ribozyme gene transfection. Cardiovasc Res 2002; 56:472-78.

21. Grinyo JM. Reperfusion injury. Transplant Proc 1997; 29:59-62.

22. Iwata A, Sai S, Nitta Y et al. Liposome-mediated gene transfection of endothelial nitric oxide synthase reduces endothelial activation and leukocyte infiltration in transplanted hearts. Circulation 2001; 103:2753-59.

23. Sakaguchi T, Sawa Y, Fukushima N et al. A novel strategy of decoy transfection against nuclear factor-kappaB in myocardial preservation. Ann Thorac Surg 2001; 71:629-30.

24. Wang P, Chen H, Qin H et al. Overexpression of human copper, zinc-superoxide dismutase (SOD1) prevents postischemic injury. Proc Natl Acad Sci USA 1998; 95:4556-60.

25. Abunasra HJ, Smolenski RT, Yap J et al. Multigene adenoviral therapy for the attenuation of ischemia-reperfusion injury after preservation for cardiac transplantation. J Thorac Cardiovasc Surg 2003; 125:998-1006.

26. Lehmann TG, Wheeler MD, Schoonhoven R et al. Delivery of Cu/Zn-superoxide dismutase genes with a viral vector minimizes liver injury and improves survival after liver transplantation in the rat. Transplantation 2000; 69:1051-57.

27. Braudeau C, Bouchet D, Tesson L et al. Induction of long-term cardiac allograft survival by heme oxygenase-1 gene transfer. Gene Ther 2004; 11:701-19.

28. Araujo JA, Meng L, Tward AD et al. Systemic rather than local heme oxygenase-1 overexpression improves cardiac allograft outcomes in a new transgenic mouse. J Immunol 2003; 171:1572-80.

29. Suzuki K, Murtuza B, Smolenski RT et al. Overexpression of interleukin-1 receptor antagonist provides cardioprotection against ischemia-reperfusion injury associated with reduction in apoptosis. Circulation 2001; 104 (Suppl 1):I308-13.

30. Ritter T, Schroder G, Risch K et al. Ischemia/reperfusion injury-mediated down-regulation of adenovirus-mediated gene expression in a rat heart transplantation model is inhibited by co-application of a TNFRp55-Ig chimeric construct. Gene Ther 2000; 7:1238-43.

31. Poston RS, Mann MJ, Hoyt EG et al. Antisense oligodeoxynucleotides prevent acute cardiac allograft rejection via a novel, nontoxic, highly efficient transfection method. Transplantation 1999; 68:825-32.

32. Brauner R, Nonoyama M, Laks H et al. Intracoronary adenovirus-mediated transfer of immunosuppressive cytokine genes prolongs allograft survival. J Thorac Cardiovasc Surg 1997; 114:923-33.

33. David A, Chetritt J, Guillot C et al. Interleukin-10 produced by recombinant adenovirus prolongs survival of cardiac allografts in rats. Gene Ther 2000; 7:505-10.

34. Oshima K, Sen L, Cui G et al. Localized interleukin-10 gene transfer induces apoptosis of alloreactive T-cells via FAS/FASL pathway, improves function and prolongs survival of cardiac allografts. Transplantation 2002; 73:1019-26.

35. Ke B, Ritter T, Kato H et al. Regulatory cells potentiate the efficacy of IL-4 gene transfer by up-regulating Th2-dependent expression of protective molecules in the infectious tolerance pathway in transplant recipients. J Immunol 2000; 164:5739-45.

36. Ke B, Shen XD, Zhai Y et al. Heme oxygenase 1 mediates the immunomodulatory and antiapoptotic effects of interleukin 13 gene therapy in vivo and in vitro. Hum Gene Ther 2002; 13:1845-57.

37. Furukawa H, Oshima K, Tung T et al. Liposome-mediated combinatorial cytokine gene therapy induces localized synergistic immunosuppression and promotes long-term survival of cardiac allografts. J Immunol 2005; 174:6983-92.

38. Fu S, Chen D, Mao X et al. Feline immunodeficiency virus-mediated viral interleukin-10 gene transfer prolongs nonvascularized cardiac allograft survival. Am J Transplant 2003; 3:552-61.

39. Zuo Z, Wang C, Carpenter D et al. Prolongation of allograft survival with viral IL-10 transfection in a highly histoincompatible model of rat heart allograft rejection. Transplantation 2001; 71:686-91.

40. Qin L, Ding Y, Tahara H et al. Viral IL-10-induced immunosuppression requires Th2 cytokines and impairs APC function within the allograft. J Immunol 2001; 166:2385-93.

41. Salgar SK, Yang D, Ruiz P et al. Viral interleukin-10-engineered autologous hematopoietic stem cell therapy: a novel gene therapy approach to prevent graft rejection. Hum Gene Ther 2004; 15:131-44.

42. Baggiolini M, Loetscher P. Chemokines in inflammation and immunity. Immunol Today 2000; 21:418-20.

43. de Groot-Kruseman HA, Baan CC, Loonen EH et al. Failure to down-regulate intragraft cytokine mRNA expression shortly after clinical heart transplantation is associated with high incidence of acute rejection. J Heart Lung Transplant 2001; 20:503-10.

44. Azzawi M, Hasleton PS, Geraghty PJ et al. RANTES chemokine expression is related to acute cardiac cellular rejection and infiltration by CD45RO T-lymphocytes and macrophages. J Heart Lung Transplant 1998; 17:881-87.

45. Fleury S, Li J, Simeoni E et al. Gene transfer of RANTES and MCP-1 chemokine receptor antagonists prolongs cardiac allograft survival. Gene Ther 2006; 13:1104-9.

46. DeBruyne LA, Li K, Bishop DK et al. Gene transfer of virally encoded chemokine antagonists vMIP-II and MC148 prolongs cardiac allograft survival and inhibits donor-specific immunity. Gene Ther 2000; 7:575-82.

47. Gao W, Topham PS, King JA et al. Targeting of the chemokine receptor CCR1 suppresses development of the acute and chronic cardiac allograft rejection. J Clin Invest 2000; 105:35-44.

48. Gao W, Faia KL, Csizmadia V et al. Beneficial effects of targeting CCR5 in allograft recipients. Transplantation 2001; 72:1199-205.

49. Hancock WW, Lu B, Gao W et al. Requirement of the chemokine receptor CXCR3 for acute allograft rejection. J Exp Med 2000; 192:1515-20.

50. Hancock WW, Gao W, Csizmadia V et al. Donor-derived IP-10 initiates development of acute allograft rejection. J Exp Med 2001; 193:975-80.

51. Sayegh MH, Turka LA. The role of T-cell costimulatory activation pathways in transplant rejection. N Engl J Med 1998; 338:1813-21.

52. Vincenti F, Larsen C, Durrbach A et al. Costimulation blockade with belatacept in renal transplantation. N Engl J Med 2005 Aug 25; 353(8):770-81.

53. Lenschow DJ, Zeng Y, Thistlethwaite JR et al. Long-term survival of xenogeneic pancreatic islet grafts induced by CTLA4Ig. Science 1992; 257:789-92.

54. Olthoff KM, Olthoff KM, Judge TA et al Adenovirus-mediated gene transfer into cold-preserved liver allografts: survival pattern and unresponsiveness following transduction with CTLA4Ig. Nat Med 1998; 4:194-200.

55. Bagley J, Jacomini J. Gene therapy in organ transplantation. Gene Ther 2003; 10:605-11.

56. Guillot C, Guillonneau C, Mathieu P et al. Prolonged blockade of CD40-CD40 ligand interactions by gene transfer of CD40Ig results in long-term heart allograft survival and donor-specific hyporesponsiveness, but does not prevent chronic rejection. J Immunol 2002; 168:1600-9.

57. Yamashita K, Masunaga T, Yanagida N et al. Long-term acceptance of rat cardiac allografts on the basis of adenovirus mediated CD40Ig plus CTLA4Ig gene therapies. Transplantation 2003; 76:1089-96.

58. Freeman GJ, Long AJ, Iwai Y et al. Engagement of the PD-1 immunoinhibitory receptor by a novel B7 family member leads to negative regulation of lymphocyte activation. J Exp Med 2000; 192:1027-34.

59. Özkaynak E, Wang L, Goodearl A et al. Programmed death-1 targeting can promote allograft survival. J Immunol 2002; 169:6546-53.

60. Dudler J, Li J, Pagnotta M et al. Gene transfer of programmed death ligand-1.Ig prolongs cardiac allograft survival. Transplantation 2006; 82:1733-37.

61. Madsen JC, Superina RA, Wood KJ et al. Immunological unresponsiveness induced by recipient cells transfected with donor MHC genes. Nature 1988; 331:161-64.

62. Wong W, Stranford SA, Morris PJ et al. Retroviral gene transfer of a class I MHC gene to recipient bone marrow induces tolerance to alloantigens in vivo. Transplant Proc 1997; 29:1130.

63. Fry JW, Morris PJ, Wood KJ. Adenoviral transfer of a single donor-specific MHC class I gene to recipient bone marrow cells can induce specific immunological unresponsiveness in vivo. Gene Ther 2002; 9:220-26.

64. Sonntag KC, Emery DW, Yasumoto A et al. Tolerance to solid organ transplants through transfer of MHC class II genes. J Clin Invest 2001; 107:65-71.

65. Bracy J, Iacomini J. Engraftment of genetically modified bone marrow cells in sensitized hosts. Mol Ther 2002; 6:252-57.

66. Bagley J, Bracy JL, Tian C et al. Establishing immunological tolerance through the induction of molecular chimerism. Front Biosci 2002; 7:1331-37.

67. Bagley J, Tian C, Sachs DH et al. Induction of T-cell tolerance to an MHC class I alloantigen by gene therapy. Blood 2002; 99:4394-99.

68. Heim DA, Hanazono Y, Giri N et al. Introduction of a xenogeneic gene via hematopoietic stem cells leads to specific tolerance in a rhesus monkey model. Mol Ther 2000; 1:533-44.

69. Mahle WT, Vincent RN, Berg AM et al. Pravastatin therapy is associated with reduction in coronary allograft vasculopathy in pediatric heart transplantation. J Heart Lung Transplant 2005; 24:63-66.

70. Eisen HJ, Kobashigawa J, Keogh A et al. Three-year results of a randomized, double-blind, controlled trial of mycophenolate mofetil versus azathioprine in cardiac transplant recipients. J Heart Lung Transplant 2005; 24:517-25.

71. Keogh A, Richardson M, Ruygrok P et al. Sirolimus in de novo heart transplant recipients reduces acute rejection and prevents coronary artery disease at 2 years: a randomized clinical trial. Circulation 2004; 110:2694-700.

72. Valentine H. Prevention of cardiac allograft vasculopathy with Certican (everolimus): the Stanford University experience within the Certican Phase III clinical trial. J Heart Lung Transplant 2005; 24(a Suppl): S191-95.

73. Libby P, Tanaka H. The pathogenesis of coronary arteriosclerosis ("chronic rejection") in transplanted hearts. Clin Transplant 1994; 8:313-18.

74. Vassalli G, Gallino A, Weis M et al. Alloimmunity and non-immunologic risk factors in cardiac allograft vasculopathy. Eur Heart J 2003; 24:1180-88.

75. Bouche D, Chauveau C, Roussel JC et al. Inhibition of graft arteriosclerosis development in rat aortas following heme oxygenase-1 gene transfer. Transplant Immunol 2002; 9:235-38.

76. Kown MH, Miniati DN, Jahncke CL et al. Bcl-2-mediated inhibition of apoptosis in rat cardiac allografts worsens development of graft coronary artery disease. J Heart Lung Transplant 2003; 22:986-92.

77. Miniati DN, Lijkwan MA, Murata S et al. Effects of adenoviral up-regulation of bcl-2 on oxidative stress and graft coronary artery disease in rat heart transplants. Transplantation 2003; 76:382-86.

78. Mathieu P, Guillot C, Gerdes C et al. Adenovirus-mediated CD40Ig expression attenuates chronic vascular rejection lesions in an aorta allotransplantation model. Transplant Proc 2002; 34:743-44.

79. Wang T, Dong C, Stevenson SC et al. Overexpression of soluble fas attenuates transplant arteriosclerosis in rat aortic allografts. Circulation 2002; 106:1536-42.

80. Sata M, Luo Z, Walsh K. Fas ligand overexpression on allograft endothelium inhibits inflammatory cell infiltration and transplant-associated intimal hyperplasia. J Immunol 2001; 166:6964-71.

81. Scholl FG, Sen L, Drinkwater DC et al. Effects of human tissue plasminogen gene transfer on allograft coronary atherosclerosis. J Heart Lung Transplant 2001; 20:322-29.

82. Saiura A, Sata M, Hiasa K et al. Antimonocyte chemoattractant protein-1 gene therapy attenuates graft vasculopathy. Arterioscler Thromb Vasc Biol 2004; 24:1886-90.

83. Kawauchi M, Suzuki J, Morishita R et al. Gene therapy for attenuating cardiac allograft vasculopathy using ex vivo E2F decoy transfection by HVJ-AVE-liposome method in mice and nonhuman primates. Circ Res 2000; 87:1063-68.

84. Hacein-Bey-Abina S, Von Kalle C, Schmidt M et al. LMO2-associated clonal T-cell proliferation in two patients after gene therapy for SCID-X1. Science 2003; 302:415-9. Erratum in: Science 2003; 302:568.

85. Ye X, Rivera VM, Zoltick P et al. Regulated delivery of therapeutic proteins after in vivo somatic cell gene transfer. Science 1999; 283:88-91.

86. Blau HM, Rossi FMV. Tet B or not Tet B: Advances in tetracycline-inducible gene expression. Proc Natl Acad Sci USA 1999; 96:797-99.

87. Larsen CP, Elwood ET, Alexander DZ et al. Long-term acceptance of skin and cardiac allografts after blocking CD40 and CD28 pathways. Nature 1996; 381:434-38.

88. Elwood ET, Larsen CP, Cho HR et al. Prolonged acceptance of concordant and discordant xenografts with combined CD40 and CD28 pathway blockade. Transplantation 1998; 65:1422-28.

89. Salama AD, Remuzzi G, Harmon WE et al. Challenges to achieving clinical transplantation tolerance. J Clin Invest 2001; 108:943-48.

90. Kirk AD, Burkly LC, Batty DS et al. Treatment with humanized monoclonal antibody against CD154 prevents acute renal allograft rejection in nonhuman primates. Nat Med 1999; 5:686-93.

91. Hamawy MM, Knechtle SJ. Strategies for tolerance induction in nonhuman primates. Curr Opin Immunol 1998; 10:513-17.

# The Basic Science of Lung Allograft Failure

Trudie Goers, Ryan Fields and Thalachallour Mohanakumar*

## Abstract

Although great strides have been made in the field of lung transplantation with respect to surgical technique, treatment of acute rejection and post-operative management, the long term success of lung transplantation is limited by the development of chronic rejection as manifest by the bronchiolitis obliterans syndrome (BOS). In this chapter, we discuss the current basic scientific understanding of lung transplant rejection as it relates to the development of acute and chronic rejection. An overview of the basic immunological responses that lead to the development of BOS is presented, including the respective roles of HLA, non-HLA and lung specific antigens, humoral and cellular immune responses and regulatory T-cells. Taken together, these mechanisms represent our current knowledge for chronic lung transplant rejection and, hopefully, will result in novel therapeutics to combat this rate-limiting step in the long-term success of lung transplantation.

## Introduction

Lung transplantation is recognized as a viable treatment option in a variety of end-stage pulmonary disease. Improvements in immuno-suppression, surgical technique, lung preservation and management of ischemia-reperfusion injury and infection have greatly increased the one year survival rate after lung transplantation to 70-80%.[1] However, the long-term survival after lung transplantation is limited by the development of obliterative bronchiolitis, which is considered to represent chronic lung allograft rejection. Obliterative bronchiolitis and its clinical correlate bronchiolitis obliterans syndrome (BOS), affects up to 50-60% of patients at 4 years after transplantation.[2,3] In addition, the prevalence of BOS increases over time to 90-100% by nine years after transplantation.[4] In the majority of the patients, BOS is a progressive process that responds poorly to immunosuppression. It accounts for more than 30% of the mortality occurring after the third year following transplantation.[1] Furthermore, survival at 5 years after the onset of BOS is only 30-40% and survival at 5 years after transplantation is 20-40% lower in patients with BOS than in patients without BOS.[5]

Human organ transplantation provides life-saving organs to patients with end-stage organ failure that have exhausted all other therapeutic options. The complexity of the transplant process arises from the rejection immune response to the transplanted organ, when this organ is derived from another genetically non-identical individual of the same species (allograft). Allograft rejection is a unique immunologic disease because the onset and the target antigens can be easily identified. In spite of these opportunities, the most basic aspects of the disease are still not well-understood: the nature of the immunogenic peptides (epitopes), the identity of the effector immune mechanisms and the factors responsible for recipient differences in the immune response to allografts, etc. Therefore, although there have been important advances in the prevention, diagnosis and treatment protocols, allograft rejection still occurs frequently, is incompletely controlled by current treatment protocols and represents a major treatment cost.

This chapter will describe the nature of alloantigens in lung transplantation, the laboratory methods implemented to identify these alloantigens, the types of rejection and their pathogenesis. We will conclude by discussing the future directions and frontiers of basic investigation of lung transplantation.

## Nature of Alloantigens

### Major Histocompatibility Antigens

The most important of the antigens expressed by lung allografts are the human leukocyte antigens (HLA) encoded within the major histocompatibility complex (MHC) on chromosome 6. In normal immune responses, T-cells recognize foreign antigens when presented as peptides in association with self MHC molecules. In the transplant response, the precise molecular nature of the recognition of the foreign MHC alloantigen is less clear. Most likely, nonself MHC alloantigens can be directly recognized, with or without peptide.[6] Despite these fundamental differences, the cascade of events that occur during the process of allograft rejection appears to be essentially the same as that which occurs against foreign antigens such as bacteria or viruses. These processes lead to the recognition and destruction of the invader, in this case, the allograft.

### Allorecognition

In 1955, Billingham et al[7,8] reported that isografts (organs transplanted between members of the same inbred mouse strain) were not rejected. However, allografts were rapidly rejected in 8-10 days. Furthermore, a recipient who received a second skin allograft from the same donor, the second allograft is rejected more rapidly. However, if the second skin allograft were from a different donor strain, this rejection is not accelerated. Hence, these results were suggestive of specific immunologic recognition and memory in the rejection process. This was confirmed by transferring the ability to reject allografts by adoptively transferred lymphocytes. When an organ from one individual is transplanted into another non-identical individual of the same species, several cellular and molecular immunological processes are activated, which if left untreated result in rejection of the allograft. This allogenic immune response consists of a variety of mechanisms directed against the non-self foreign antigens expressed by the graft.

*Corresponding Author: Thalachallour Mohanakumar—Department of Surgery, Pathology and Immunology, Washington University School of Medicine, St. Louis, Missouri, USA. Email: kumart@msnotes.wustl.edu

*Chronic Allograft Failure: Natural History, Pathogenesis, Diagnosis and Management*, edited by Nasimul Ahsan. ©2008 Landes Bioscience.

Two models of alloantigen recognition have been proposed.[9-12] In direct alloantigen recognition, the recipient's T-cells recognize foreign MHC molecules directly, without processing by self antigen-presenting cells (APC).[11,12] Recognition is due to the inherent natural affinity of the T-cell receptor for MHC molecules. This process is explained in the context of negative and positive selection in the thymus where T-cells are educated to distinguish self from non-self. T-cells with too high an affinity for self MHC molecules plus self peptides are deleted (negative selection), whereas T-cells with low affinity for self MHC molecules plus self peptides are selected for maturation and exit the thymus for the peripheral immune organs (positive selection). This process ensures the deletion of high affinity auto-reactive T-cells leaving behind low affinity auto-reactive but non-pathogenic T-cells with the potential to recognize self MHC molecules plus foreign peptides. Since developing T-cells with high affinity for non-self MHC molecules would not have been eliminated in the thymus, they should be present in the periphery. It is these T-cells that recognize foreign MHC molecules expressed by parenchymal cells of the allograft (Fig. 1).

In indirect alloantigen recognition, soluble foreign MHC molecules, most likely shed by apoptotic or necrotic parenchymal cells in the allograft, are phagocytosed and processed by self APCs infiltrating the allograft during the rejection process. Subsequently, peptides derived from these foreign MHC molecules are recognized by CD4+ T-cells as epitopes presented by self MHC class II molecules expressed on self APCs.[9,10] This is the same mechanism of presentation involved in the recognition of natural foreign peptides. Several studies have shown that indirect allorecognition occurs and is relevant to the process of allograft rejection.[13-15] In addition, a growing body of evidence indicates that during rejection, this pathway of alloantigen recognition by CD4+ T-cells has a level of importance comparable to the direct pathway by CD8+ cytotoxic T-cells. Further, CD4+ T-cells may be more important in chronic rejection.

### Minor Histocompatibility Antigens

As mentioned above, an allograft transplanted between a donor and recipient matched at the MHC genes have a higher acceptance rate. However, allograft rejection can still occur. Such rejection was shown to be caused by the disparities in polymorphic gene products called minor histocompatibility antigens (MiHA).[9,16-18] It is estimated that the murine genome may contain up to 720 MiHA genes. In humans, the number is estimated to be even larger. The majority of these loci are located on autosomal chromosomes, but a few of them also map to mitochondrial DNA and the Y-chromosome.[19] Due to the high number of MiHA genes, only identical twins can share all of their transplantation antigens, both MHCs and MiHAs. Siblings, even if they are MHC identical, are generally matched for only about half the total number of MiHAs. The importance of MiHAs in transplantation is manifested by several clinical findings that in MHC-matched solid organ transplantations, the disparities in MiHAs often lead to graft rejection.[10,20]

Both CD4+ and CD8+ T-cell cooperation is required for MiHA-mediated graft rejection. Both cells recognize their specific MiHAs in a MHC-restricted manner. This cooperation was implicit from early studies on responses to H-Y antigens in which permissive and nonpermissive MHC class I and class II alleles could be identified that affected CD8+ cytotoxic T-cell generation and skin graft rejection.[19,21] The mouse strains expressing permissive alleles allow the generation of both CD4+ T-cells and CD8+ cytotoxic T effector cells; thus the H-Y antigen can elicit a functional immune response. While the mouse strains expressing non-permissive alleles have the precursors for the CD8+ cytotoxic T effector cells, however, no in vivo responses to H-Y are generated due to the lack of appropriate CD4+ T helper function.

The major features that distinguish MiHAs from other transplantation antigens are that MiHAs are recognized by T-cells but not by antibodies and that MiHAs are recognized in a MHC-restricted manner. It is now clear that MiHAs are peptides derived from polymorphic proteins that are presented by MHC molecules on the allograft.[22,23] The fact that the MiHAs are peptides recognized in association with MHC molecules has explained many of the features of these antigens that were known, but poorly understood for a long time. First, it is difficult, if not impossible, to detect humoral responses to MiHAs. This is probably because most MiHAs come from intracellular proteins, thus, even if an antibody response were to occur, it could not be detected at the cell surface. Second, MiHAs do not stimulate a primary in vitro cell-mediated response, where as MHC antigens evoke a powerful primary response in both mixed lymphocyte reaction and cell-mediated lympholysis assays. This is in agreement with the general difficulty in detecting in vitro T-cell responses to peptides of nominal protein antigens unless they have been primed in vivo. Third, the recognition of MiHAs is MHC-restricted.

### Lung-Specific Antigens

A growing body of evidence suggests that the immune response developed against antigenic determinants expressed by epithelial cells of the human lung allograft can play a crucial role in the development of BOS.[11,24,25] The identities of the tissue-specific antigens that are targets of the antibodies capable of triggering intracellular cascades resulting in the fibrosis and dysfunction leading to graft failure remain largely unknown. However, work from our laboratory has produced some insight into this question. Our studies have shown that sera from approximately 32% of lung transplant recipients recognized a ~60-kDa antigen expressed by airway epithelial cell lines.[26] Binding of antibodies in this subpopulation sera to this antigen was shown to induce intracellular calcium influx, tyrosine phosphorylation, cellular proliferation and up-regulation of transforming growth factor-β (TGF-β) and heparin-binding epidermal growth factor (HB-EGF) mRNA transcription in airway epithelial cells.[27] Similarly, multiple studies have also shown that upon activation, epithelial cells from different tissues have the ability to produce high quantities of several growth factors including epidermal growth factor (EGF),[28] HB-EGF,[29] endothelin (ET)-1,[30] basic fibroblast growth factor (bFGF), granulocyte-monocyte colony-stimulating factor (GM-CSF),[31] insulin-like growth factor (IGF)-1,[32] platelet-derived growth factor (PDGF),[29,33] and transforming growth factor (TGF)-β.[29,33] All of these growth factors are involved in the activation and proliferation of fibroblasts and smooth muscle cells as seen in BOS.

Our laboratory has also identified another subset of HLA-antibody-negative patients *status post* lung transplantation that developed immunoreactivity to a ~48 kDa airway epithelial cell surface antigen. The antibodies were not present in the patients' sera pre-transplant. Binding of the antibodies to airway epithelial cells in vitro result in cellular upregulation of heat-shock proteins 27 and 90, protein kinase C pathways, transcription factor modulator c-Myc and transcription factor family protein TCF-5. Also, similar to previous studies, the antibody binding caused production of vascular endothelial growth factor (VEGF), HB-EGF and TGF-β. The identity of this membrane-bound protein was found to be Kα1-tubulin. It is noteworthy that this tubulin isotype has been reportedly associated with both autoimmune disorders (relapsing polychondritis) as well as pulmonary malignancies.[34] The significance of this identification of an intracellular autoantigen is BOS following lung transplantation and its clinical impact remains under investigation.

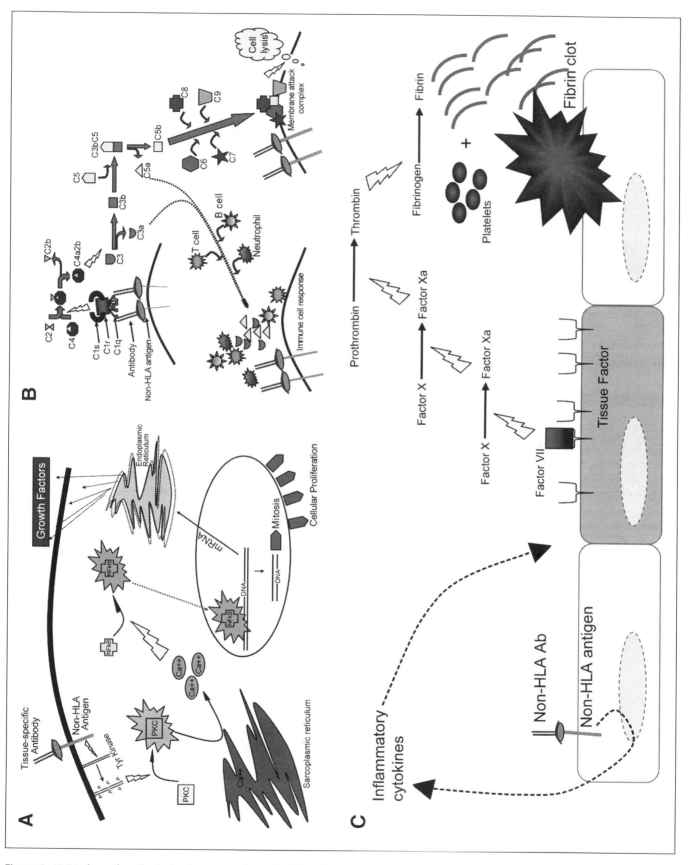

Figure 1. A) Binding of antibody leads to activation of calcium homeostasis pathway resulting in production of fibrotic growth factors and cellular proliferation. B) Binding of antibody to the cell-surface antigen causes complement deposition and resultant assembly of the membrane attach complex which leads to cellular destruction. C) Binding of antibody to the cell-surface antigen results in release of inflammatory cytokines which causes activation of the hemostatic cascade.

# Mechanisms and Pathogenesis of Lung Allograft Rejection

## Types of Allograft Rejection

There are three main types of allograft rejection: Hyperacute, acute and chronic.

Hyperacute rejection occurs within minutes to 48 hours. Complement-fixing antibodies bind to the vascular endothelium and activate complement. Such complement activation may induce thrombosis by attracting polymorphonuclear leukocytes and platelets and activate the coagulation cascade.[35-37] Initially, this type of rejection was the first obstacle to successful transplantation. This condition is not treatable and demands emergency removal of the allograft. Histopathologic examination can differentiate the immunologic etiology of the condition, in contrast to thrombosis from a surgical technical error.[38] Antibodies reactive to incompatible blood ABO antigens or donor HLA antigens can cause hyperacute rejection. Therefore, all allografts are typically matched for the blood group antigens. Furthermore, as mentioned above, a crossmatch is performed prior to transplantation to rule out prior sensitization to donor HLA antigens. As a result of blood group matching and the crossmatch test, hyperacute rejection is rarely seen in clinical allograft transplantation.[38] The liver is an interesting exception to the above discussion. At least in the short term, livers have been successfully transplanted despite a blood group mismatch.[39,40] This could be due to better immunosuppression, the decreased relative immunogenicity of the liver and/or the capacity of the liver to regenerate.

The next barrier to successful transplantation is acute rejection, which occurs between 6 and 90 days post-transplantation, but it may also occur later during the post-transplant period, as result of treatment non-compliance.[39,40] This condition is characterized histologically by cellular infiltrates with CD8+ cytotoxic T-cells and an inflammatory response produced in situ cytokine production by CD4+ Th1 cells activated through the indirect pathway of antigen presentation by self APCs.[6,41-43] Many animal models, later correlated with human histology, point to the importance of the CD8+ cytotoxic T-cell. Although a perivascular lymphocyte infiltrate characteristic of DTH reactions is frequently present, the characteristic acute rejection lesion is thought to be mediated by a direct interaction of lymphocytes with parenchymal cells. Once activated, these cells damage both the vascular endothelium and the graft parenchyma. Allograft dysfunction and failure would inevitably ensue without more immunosuppression.

Antibodies do not play a very important role in the first episode of acute rejection. However, subsequent severe episodes of acute rejection frequently include an antibody-mediated immune response that induces endovasculitis with binding of lymphoid cells on swollen, ischemic endothelial cells.[38] For the development of these antibodies, the CD4+ T-cell response must necessarily develop first. Acute cellular rejection may therefore have both cellular and humoral components.

Chronic rejection usually appears after 60 to 90 days post-transplantation and induces the typical obliterative disease.[39,40] Clinically, patients have a slow but progressive course of allograft deterioration over a period of several months to years. This condition is not responsive to available immunosuppressive strategies. Risk factors for kidney allograft rejection include several acute rejection episodes within the first 90 days post-transplantation and the presence of anti-HLA antibodies.[39,40] Despite the common characteristic of fibrosis, the pathology of chronic rejection differs slightly between the different organ allografts. In the kidney and heart, pathologists note increasing arteriosclerosis in glomeruli and coronary arteries, respectively. In the lung, the fibrosis centers around the airways.[40] This suggests an important role of the lung epithelium as an immunologic target during the chronic rejection of the lung allograft.

While the etiology and pathogenesis of acute rejection are well understood, the immunopathology of chronic rejection is still being elucidated. Several theories have been proposed and will be discussed in the next section. The fibrosis in chronic rejection may be due to healing after acute rejection. After all, a strong predictor of chronic rejection is a prior episode of acute rejection. However, chronic rejection can occur in the absence of acute rejection. Other theories include ischemic injury and a chronic delayed type hypersensitivity reaction.[44] In lung transplantation, the development of anti-donor HLA antibodies precede the onset of chronic lung allograft rejection.[4] Furthermore, such antibodies have the ability to activate the proliferation of both vascular endothelial cells as well as lung epithelial cells.[45,46] Presumably, such anti-donor HLA antibodies may induce the diffuse arteriosclerosis noted in chronically rejected glomeruli and coronary arteries. However, the mechanism of antibody induction of graft fibrosis is still unclear.

While incompletely understood, the details of the pathogenesis of chronic rejection are emerging. In adult lung transplant recipients, antibodies to unmatched donor HLA molecules precede the onset of bronchiolitis obliterans by about twenty months.[4] It has been shown that anti-donor HLA antibodies may induce calcium influx and tyrosine phosphorylation upon binding to lung epithelial cells.[45,46] Such changes activate the epithelial cells and leads to cytokine secretion. Such cytokines could activate underlying smooth muscle cells and fibroblasts to create the characteristic fibrosis of chronic rejection. A significantly increased precursor frequency of T-cells reactive to peptides derived from mismatched donor class I or II HLA and presented on recipient MHC has also been seen in lung transplant recipients with bronchiolitis obliterans syndrome (BOS).[14,47] This suggests a role for CD4+ T-cells and indirect antigen presentation of donor HLA peptides in the pathogenesis of BOS. Further studies to elucidate the pathogenesis of bronchiolitis obliterans may improve the treatment and prognosis of lung transplants.

## Airway Epithelial Cells in the Pathogenesis of BOS

Several studies have demonstrated that airway epithelial cells (AEC) can serve as important immunological targets during the process of lung allograft rejection.[25,29,48] AECs are able to express MHC class II molecules and this expression is up-regulated during chronic allograft rejection.[49-51] These results are corroborated by several animal studies showing that the expression of MHC class I and class II molecules is up-regulated by AECs in rejecting lung allografts but not in normal lung epithelium.[24,25] Despite their ability to express both MHC class I and class II molecules, very little data is available on the antigen-presenting ability of AECs. Related studies have shown that AECs constitutively express adhesion molecules such as CD54 and CD58, costimulatory molecules such as CD80 and CD86 and MHC class II molecules and the expression of these molecules as well as MHC class I molecules is up-regulated by interferon (IFN)-γ.[29,48,52] Thus, treatment with IFN-γ render AECs capable of inducing a strong CD4+ T-cell proliferative response.[29] Overall, these results suggest that during the development of BOS, AECs are capable of antigen presentation and activation of CD4+ T-cells.

Several studies from our laboratory have strongly indicated that allograft rejection is targeted at the bronchial epithelium.[53,54] Lysis of AECs by bronchoalveolar lymphocytes from lung transplant patients was greater during acute rejection episodes as compared to non-rejection periods. Conversely, lysis of donor MHC class I-matched B-lymphoblastoid cell lines was not detected during acute rejection episodes. These results indicate that the airway epithelium is an important target in the process of lung allograft rejection.[53] A subsequent study has further confirmed that AECs are immunologic targets during the process of BOS development.[54] In this study, it was demonstrated that a subgroup of lung transplant recipients with BOS developed de novo AEC-specific antibodies. A significant correlation between development of AEC-specific antibodies and development of BOS after lung transplantation was observed. In addition, the onset of AEC-specific antibody development preceded the onset of BOS with a significant difference of 22.5 months. Sera from these patients recognized a 60 kDa membrane antigen on AECs. These antibodies failed to react with lymphocytes, monocytes, and granulocytes indicating that the reactivity was not directed to MHC molecules. Further, these antibodies showed no reactivity to a panel of cell lines from different origin including endothelial cells, fibroblasts, and smooth muscle cells indicating that the reactivity was specific for the airway epithelium. Although it is possible that the de novo production of AECs-specific antibodies occurs as an epi-phenomenon as a result of the activation of cellular immune responses during the development of BOS, we have obtained compelling evidence that AEC-specific antibodies may play an active role in the immunopathogenesis of BOS. We have demonstrated that AEC-specific antibodies induce intracellular signal transduction in AECs resulting in the production of transforming growth factor (TGF)-β and heparin-binding epidermal growth factor (HB-EGF) as well as cellular proliferation.

## Humoral Immune Mechanisms in the Pathogenesis of BOS

Numerous studies have shown that development of anti-MHC antibodies are associated with the development of transplant atherosclerosis and graft loss after heart and kidney allograft transplantation.[55,56] Related studies have also suggested that pre-existing anti-MHC antibodies increase the rate of early kidney and heart allograft failure.[57] Furthermore, passive transfer of serum containing anti-MHC antibodies has been shown to accelerate the development of cardiac allograft atherosclerosis in an experimental murine transplantation model.[58] Other studies have also shown that cardiac allografts transplanted into B-cell-deficient mice failed to develop atherosclerotic lesions observed in cardiac allografts transplanted into normal mice.[59]

Previous studies have demonstrated that lung allograft recipients with pre-existing anti-MHC antibodies have a significantly increased risk for early allograft failure and death.[60] Subsequent studies from our laboratory have also demonstrated that development of de novo anti-MHC class I antibodies is a predisposing factor for the development of BOS and lower survival after lung transplantation.[61-63] A significant correlation between anti-MHC class I antibodies and early development of BOS after lung transplantation was observed.[62] In addition, these studies showed that the onset of anti-MHC class I antibodies preceded the onset of BOS by a significant difference of 20.1 months.[63] These studies also showed that patients who developed anti-MHC class I antibodies in the first 24 months after transplantation have a significantly lower survival rate than patients that did not develop anti-MHC class I antibodies within the same time period.[61] Subsequent studies by Palmer et al[64] showed that the development of de novo anti-MHC class II antibodies are also a predisposing factor for the development of BOS and lower survival after lung transplantation. Taken together, these findings indicate that anti-MHC antibodies may play in an important role in the pathogenesis of BOS after lung transplantation.

## Cellular Immune Mechanisms in the Pathogenesis of BOS

During allograft rejection, intact donor MHC class I and class II molecules are recognized by recipient CD8+ and CD4+ T-cells, respectively.[65,66] This recognition results in an immune response in which parenchymal cells in the allograft are specifically killed leading to the organ dysfunction. Recognition of MHC molecules by CD4+ T-cells occurs by two mechanisms: Direct recognition of donor MHC molecules expressed by parenchymal cells of the allograft and indirect recognition of processed donor MHC molecules presented as peptides bound to MHC class II molecules expressed by the recipient antigen-presenting

cells. Although it is well established that both pathways are active early after transplantation, the relative contribution of each individual pathway to the process of allograft rejection remains unclear. It has been suggested that acute allograft rejection is predominantly mediated through the direct pathway of antigen recognition because the graft contains a significant number of "passenger" donor antigen-presenting cells that express high levels of MHC class I and class II molecules and can provide co-stimulatory signals to T-cells. However, the succeeding decline in the number of donor antigen-presenting cells present within the graft suggests that the direct allorecognition pathway may not play a major role in the process of chronic allograft rejection. Hence, this process may be predominantly mediated by CD4+ T-cells activated through the indirect pathway of antigen presentation. These alloreactive CD4+ T-cells may provide the cytokines required for the generation and expansion of alloreactive B cells and CD8+ cytotoxic T-cells as well as the development of a delayed-type hypersensitivity response.

## Regulatory T-Cells in the Pathogenesis of BOS

Our laboratory has also investigated the role of regulatory T-cells (T$_{regs}$) in the development of BOS. Our previous studies have demonstrated significant involvement of donor specific MHC class-I and -II alloreactivity in the development of BOS.[26,63] CD4+ CD25+ T-cells from BOS patients are less efficient suppressors of both mitogenic or allogeneic conventional T-cells proliferation. Although, we did not find a statistically significant difference in the percentage of CD4+CD25+ T-cells in the peripheral blood of BOS positive and negative patients, ongoing studies have demonstrated that BOS is associated with a decrease in CD4+CD25+CD45RO+ T$_{regs}$. Further, BOS positive patients demonstrated a higher frequency of alloreactive IFN-γ producing CD4+ T-cells and high proliferation to mismatched donor peptides (Th1 response) while BOS negative patients revealed predominantly IL-5 production and very low IFN-γ (Th2).

Following transplantation, lung allografts sustain multiple injuries due to ischemia-reperfusion, alloimmunity, external pathogens and gastro-esophageal reflux. Such an inflammatory milieu would be conducive for the development of autoimmunity. Using animal models, it has been recently demonstrated that collagen type V (col-V) may represent a "sequestered" self-protein localized in the lung tissue which can induce autoimmune responses and contribute to the development of lung allo- and isograft rejection.[67] We investigated the role of T$_{regs}$ in the development of col-V autoimmunity in patients with chronic lung allograft rejection. Following transplantation, there was an increase in the frequency of IL-10 producing T-cells reactive to col-V. The expansion of IL-10 producing col-V specific T-cells was dependent on direct cellular contact and signaling through CTLA-4 on T$_{regs}$. These IL-10 producing T-cells were found to significantly suppress the proliferation of IFN-γ producing col-V reactive CD4+ T-cells. Furthermore, BOS was associated with a decline in T$_{regs}$ and the col-V specific IL-10 producing T-cells with a concomitant increase in IFN-γ producing T-cells.[68,69] A hypothetical modeling for this regulatory pathway is illustrated (Fig. 2).

Although there exists a plethora of studies that support the relationship of anti-HLA antibodies and the rejection of solid organ transplants, an increasing number of studies have emphasized the clinical importance of antibodies against non-HLA antigen.[55-57,62,63,70] With the knowledge that some lung transplant patients develop BOS while testing negative for anti-HLA antibodies, recent studies from our laboratory indicate that a subset of BOS+, anti-HLA antibody negative patients *status post* lung transplantation (33%) develop antibodies against a cell surface antigen located on AECs. Other cell lines tested by FACS were unreactive; therefore, the specificity to AECs was supported.

Despite previous studies supporting a role for humoral immune response in the pathogenesis of chronic rejection of allografts,[54,71-73] the exact character of the antibody and its respective antigenic target has

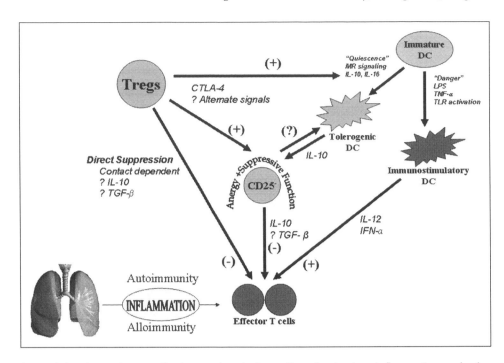

Figure 2. Alloimmunity and Autoimmunity contributing to chronic lung allograft rejection. Inflammation and release of immunoactive compounds at the time of lung transplant lead the recruitment of effector T-cells that cause end organ damage to the transplanted lung. Regulatory T-cells (Tregs) and tolerogenic dendritic cells (DC) have the capability to inhibit these effector T-cells and prevent damage. This can be accomplished via direct cell-cell interactions via ligands such as CTLA-4, or via the immunoinhibitory cytokines IL-10 and TGF-β.

been ill-defined. Our work determined that the BOS+, FACS-positive sera showed immunoreactivity with a protein ~48 kDa in size. BOS- sera did not show this reactivity. Similar to previous post-transplant antibody studies,[54-57,62,63,70] in our study the appearance of anti-AEC antibodies appeared well in advance to the onset of clinical BOS. This would support the assertion that these non-HLA antibodies have a role in the pathologic processes that lead to lung allograft failure and BOS development. The alternate hypothesis is that the indolent inflammatory environment that is present in the post-transplant patient causes an injury-recovery pattern at the level of the airway epithelium. This chronic injury phenomenon could lead to constant destruction and turnover of the AECs, thereby exposing AEC proteins to the surveillance immune system with resultant antibody production.

BOS is associated histologically with epithelial injury, bronchocentric mononuclear inflammation, fibrosis and proliferation of the lamina propria and obliteration of small airways.[74] Various growth factors have been found to be elevated in lung transplant patients. TGF-β, PDGF and ET-1 have been found significantly elevated in human patients and VEGF in rat allograft models.[75-78] The results from our gene array and luminex assays indicated a significant upregulation of heat shock proteins (HSPs) 27 and 90 and constituents of the calcium homeostasis pathway and upregulation in transcription factors TCF5 and cMyc. In turn, there was an increase in HB-EGF, TGF-β and VEGF production by AECs after exposure to the BOS+, non-HLA-antibody positive serum. The increases in these proteins and growth factors would indicate that the binding of the non-HLA antibody activates a PKC-driven calcium maintenance pathway that is regulated by HSPs 27 and 90. These pathways culminate in increased cellular mitosis, proliferation and growth factor production (Fig. 1A).

Within other fields of transplantation, the identities of tissue-specific non-HLA antigens have been elucidated. In cardiac transplant, autoantibodies against the vimentin molecule has been shown to cause accelerated onset of allograft vasculopathy.[79] In some patients with kidney allografts, Dragun and associates identified that the non-HLA angiotensin II Type 1 receptor was the target of antibodies that caused refractory vascular rejection.[72] The sequence of isolated ~48 kDa

protein using the BOS positive serum from lung transplant recipients matched that of Kα1-tubulin with 100% homology. Kα1-tubulin is a ~50 kDa protein in its usual polyglycosylated, post-translational form. It is found in cells as one of six different isotypes. Its usual cellular functions include GTP binding, GTPase activity, maintaining cellular structure in the form of microtubules and microtubule-based intracellular movement. Even though most of these known functions would indicate that Kα1-tubulin is not prone to act at the cell surface. However, other investigators in the fields of oncology, autoimmunity and transplantation have identified autoantibodies against Kα1-tubulin in small cell lung cancer and breast cancer,[80] relapsing polychondritis,[34] and post-cardiac transplant fatal cardiomyopathy.[81] In the post-transplant pathogenesis of BOS, there is a high rate of epithelial cell apoptosis and turnover. With this process, intracellular proteins are exposed to the extracellular milleau inducing antibody formation. With the continued exposure of Kα1-tubulin to the environment, the AECs are activated with production of resultant growth factors, cellular proliferation and potential airway occlusion.

## Conluding Remarks and Future Directions

Our current understanding of chronic lung allograft rejection is clearly limited. The complex relationship between inflammation, immune responses to both alloantigens and autoantigens, injury, fibrosis and infection dictates the course of graft survival. However, the relative contributions from each of these entities (and likely others) are not fully understood. Further, it is likely that different patients with different underlying pathologies will have various contributions from each of these etiological factors. New research is adding even more complexity to this field as the role of regulatory T-cells in transplantation immunology is being elucidated. As the pathways, cellular and molecular "players" in chronic lung allograft rejection are delineated, treating physicians will hopefully be armed with new strategies to combat graft loss.

Although direct and indirect allorecognition of MHC antigens clearly is responsible for a large part of lung allograft rejection, recent studies also demonstrate that the early inflammatory environment surrounding lung and resulting autoimmunity, affects the fate of the

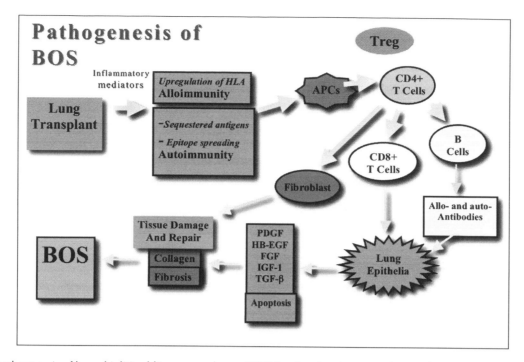

Figure 3. The pathogenesis of bronchiolitis obliterans syndrome (BOS) is related to the upregulation of an autoimmune response against the transplanted organ. This leads to the activation of professional antigen presenting cells (APC), such as dendritic cells (DC), that go on to activate cytotoxic CD8+ cells and B-cells that cause end-organ damage. Further, the activation of CD4+ helper T-cells augments the CD8+ and B-cell responses and via cytokine production, activates fibroblasts. These fibroblasts then proliferate and produce large amounts of extracellular matrix proteins that are characteristic of the fibroproliferative BOS lesion. The treatment of BOS must address all of these pathways in order to reduce end organ damage.

allograft. In addition, we are only now realizing the important role that regulatory T-cells perform in this process. The contributions from each of these factors and likely other undiscovered variables, affects the fate of any transplanted allograft as shown schematically in Figure 3.

One of the critical shortcomings facing physicians who treat post-lung transplant patients is the inability to diagnose chronic lung allograft rejection before clinical signs of irreversible graft damage are present, which occurs when the $FEV_1$ begins to decline. Due to the clinical nature of the diagnosis of BOS, this is inherent to current definitions of chronic lung allograft rejection. Thus, another intense area of research is centered around predicting and diagnosing BOS early so that treatments may be instituted earlier with the goal of preventing substantial end organ damage. Studies analyzing antibodies and other markers in the serum, BAL and biopsy in patients suspected of chronic rejection all show promise in both animal and pre-clinical models.

### Acknowledgements
This work was supported by NIH HL56543 and HL66452 (TM) and NIH Training Grant AI07163 (TG).

## References
1. Trulock EP, Edwards LB, Taylor DO et al. The Registry of the International Society for Heart and Lung Transplantation: Twentieth Official adult lung and heart-lung transplant report-2003. J Heart Lung Transpl 2003; 22(6):625-35.
2. Estenne M, Maurer JR, Boehler A et al. Bronchiolitis obliterans syndrome 2001: an update of the diagnostic criteria. J Heart Lung Transplant 2002; 21297-310.
3. Boehler A, Kesten S, Weder W et al. Bronchiolitis obliterans after lung transplantation: a review Chest 1998; 114(5):1411-26.
4. Smith MA, Sundaresan S, Mohanakumar T et al. Effect of development of antibodies to HLA and cytomegalovirus mismatch on lung transplantation survival and development of bronchiolitis obliterans syndrome. J Thorac Cardiovasc Surg 1998; 116(5):812-20.
5. Valentine VG, Robbins RC, Berry GJ et al. Actuarial survival of heart-lung and bilateral sequential lung transplant recipients with obliterative bronchiolitis. J Heart Lung Transplant 1996; 15(4):371-83.
6. Shoskes D, Wood K. Indirect presentation of MHC antigens in transplantation. Immunol Today 1994; 15:32.
7. Billingham R, Brent L, Medawar P et al. Quantitative Studies on Tissue Transplantation Immunity. I. The Survival Times of Skin Homografts Exchanged between Members of Different Inbred Strains of Mice. Proceedings of the Royal Society of London. Series B, Biological Sciences (1934-1990). 1954; 143(910):43-58.
8. Billingham R, Brent L, Medawar P. Quantitative Studies on Tissue Transplantation Immunity. II. The Origin, Strength and Duration of Actively and Adoptively Acquired Immunity. Proceedings of the Royal Society of London. Series B, Biological Sciences (1934-1990). 1954; 143(910):58-80.
9. Goulmy E. Human minor histocompatibility antigens. Curr Opin Immunol 1996; 875.
10. Beck Y, Sekimata M, Nakayama S et al. Expression of human minor histocompatibility antigen on cultured kidney cells. Eur J Immunol 1993; 23:467-72.
11. Spurzem JR, Sacco O, Rossi GA et al. Regulation of major histocompatibility complex class II gene expression on bovine bronchial epithelial cells. J Lab Clin Med 1992; 120:94-102.
12. Krensky AM WA, Crabtree G, Davis MM et al. Mechanisms of disease: T-lymphocyte-antigen interactions in transplant rejection. New England Journal of Medicine 1990; 322:510-17.
13. Liu Z, Harris PE, Colovai AI et al. Indirect recognition of donor MHC class II antigens in human transplantation. Clin Immunol Immunopath 1996; 78(3):228-35.
14. SivaSai KSR, Smith MA, Poindexter N et al. Indirect recognition of donor HLA class I peptides in lung transplantation recipients with bronchiolitis obliterans syndrome. Transplantation 1999; 67:1094-98.
15. Vella JP, Spadafora-Ferreira M, Murphy B et al. Indirect allorecognition of major histocompatibility complex allopeptides in human renal transplant recipients with chronic graft dysfunction. Transplantation 1997; 64(6):795-800.
16. Abe R HR. T-cell recognition of minor lymphocyte stimulating (Mls) gene products. Ann Rev Immunol 1989; 7:683-708.
17. Simpson E, Roopenian D. Minor histocompatibility antigens. Curr Opin Immunol 1997; 9:655.
18. Perreault C RD, Fortin C. Immunodominant minor histocompatibility antigens: the major ones. Immunol Today 1998; 19(2):69-74.
19. Rotzschke O, Falk K, Wallney HJ et al. Characterization of naturally occuring minor histocompatibility peptides including H-Y and H-4. Science 1990; 249:283-87.

20. Poindexter NJ, Shenoy S, Howard T et al. Allograft infiltrating cytotoxic T-lymphocytes recognize kidney specific human minor histocompatibility antigens. Clinical Transplantation 1996; 11:174-77.

21. Rotzschke O, Falk K, Faath S et al. On the nature of peptides involved in T-cell alloreactivity. J Exp Med 1991; 174:1059.

22. Mendoza LM, Paz P, Zuberi A et al. Minors held by Majors: The H13 minor histocompatibility locus defined as a peptide/MHC class I complex. Immunity 1997; 7:461-72.

23. Wallny HJ, Ramensee HG. Identification of classical minor histocompatibility antigens as cell derived peptides. Nature 1990; 343:275-78.

24. Romaniuk A, Prop J, Petersen AH et al. Expression of class II major histocompatibility complex antigens by bronchial epithelium in rat lung allografts. Transplantation 1987; 44(3):209-14.

25. Kubota H, Yagyu K, Takeshita OM et al. Importance of bronchus-associated lymphoid tissue and major histocompatibility complex class I and II antigen expression on bronchial epithelium in acute lung allograft rejection and lung infection in rats. Transplant Proc 1994; 26:1856-58.

26. Reznik SI, Jaramillo A, SivaSai KSR et al. Indirect allorecognition of mismatched donor HLA class II peptides in lung transplant recipients with bronchiolitis obliterans syndrome. Am J Transpl 2001; 1:228-35.

27. Jaramillo A, Smith CR, Zhang L et al. Anti-HLA class I antibody binding to airway epithelial cells induces production of fibrogenic growth factors and apoptotic cell death: A possible mechanism for bronchiolitis obliterans syndrome. Human Immunol 2003; 64:521-29.

28. Zhao J, Sime PJ, Bringas P et al. Epithelium-specific adenoviral transfer of a dominant-negative mutant TGF-á type II receptor stimulates embryonic lung branching morphogenesis in culture and potentiates EGF and PDGF-AA. Mech Dev 1998; 72(1-2):89-100.

29. Mauck KA, Hosenpud JD. The bronchial epithelium: A potential allogeneic target for chronic rejection after lung transplantation. J Heart Lung Transplant 1996; 15:709-14.

30. Markewitz BA, Kohan DE, Michael JR. Endothelin-1 synthesis, receptors and signal transduction in alveolar epithelium: evidence for an autocrine role. Am J Physiol 1995; 268:L192-L200.

31. O'Brien AD, Standiford TJ, Christensen PJ et al. Chemotaxis of alveolar macrophages in response to signals derived from alveolar epithelial cells. J Lab Clin Med 1998; 131:417-24.

32. Cambrey AD, Kwon OJ, Gray AJ et al. Insulin-like growth factor-II is a major fibroblast mitogen produced by primary cultures of human airway eipthelial cells. Clin Sci (Colch) 1995; 89(6):611-17.

33. Wagner CR, Morris TE, Shipley GD et al. Regulation of human aortic endothelial cell-derived mesenchymal growth factors by allogeneic lymphocytes in vitro. A potential mechanism for cardiac allograft vasculopathy. J Clin Invest 1993; 92:1269-77.

34. Tanaka Y NM, Matsui T, Iizuka N et al. Proteomic surveillance of autoantigens in relapsing polychondritis. Microbiology and Immunology 2006; 50(2):117-26.

35. Auchincloss H. Xenogeneic transplantation: A review. Transplantation Surg 1988; 46:1-20.

36. Kissmeyer-Nielsen F OS, Petersen VP, Fjeldborg O. Hyperacute rejection of kidney allografts, associated with pre-existing humoral antibodies against donor cells. Lancet 1966; 2:662-65.

37. Williams GM HD, Hudson RP, Morris PJ et al. "hyperacute" renal-homograft rejection in men. New England Journal of Medicine 1968; 279:611-18.

38. Sibley RK PW. Morphologic findings inthe renal allograft biopsy. Seminars in Nephrology 1985; 5(4):294-306.

39. The Washington Manual of Surgery. New York: Lippincott Williams & Wilkins; 1999.

40. Norman DJ. Primer on Transplantation. New Jersey: American Society of Transplant Physicians; 1998.

41. Abbas AK, Lichtman AK, Pober JS et al. Cellular and Moleculer Immunology. Philadelphia: WB Saunders; 1991.

42. Janeway CA TP, Walport M. et al. Immunobiology The Immune System in Health and Disease. 4th edition ed. New York: Garland Publishing; 1999.

43. Paul WE. Fundamental Immunology. 4th ed. New York: Lippincott-Raven; 1999.

44. Abbas AK LA, Pober JS, eds. Cellular and Molecular Immunology. Vol 3rd edition. Philadelphia: WB Saunders Company; 1997.

45. Bian H, Harris PE, Mulder A et al. Anti-HLA antibody ligation to HLA class I molecules expressed by endothelial cells stimulates tyrosine phosphorylation, inositol phosphate generation and proliferation. Human Immunol 1997; 53(1):90-97.

46. Reznik SI, Jaramillo A, Zhang L et al. Anti-HLA antibody binding to HLA class I molecules induces proliferation of airway epithelial cells: A potential mechanism for bronchiolitis obliterans syndrome. J Thorac Cardiovasc Surg 2000; 119:39-45.

47. Lu KC, Jaramillo A, Mendeloff EN et al. Concomitant allorecognition of mismatched donor HLA class I- and II-derived peptides in pediatric lung transplant recipients with bronchiolitis obliterans syndrome. J Heart Lung Transplant 2003; 22:35-43.

48. Smith CR, Jaramillo A, Duffy BF et al. Airway epithelial cell damage mediated by antigen-specific T-cells: Implications in lung allograft rejection. Human Immunol 2000; 61:985-92.

49. Taylor PM, Rose ML, Yacomb MH. Expression of MHC antigens in normal human lungs and transplanted lungs with obliterative bronchiolitis. Transplantation 1989; 48:506-10.

50. Milne DS, Gascoigne A, Wilkes J et al. The immunohistopathology of obliterative bronshiolitis following lung transplantation. Transplantation 1992; 54(748):750.

51. Milne DS, Gascoigne AD, Wilkes J et al. MHC class II and ICAM-I expression and lymphocyte subsets in transbronchial biopsies from lung transplant recipients. Transplantation 1994; 57:1762-66.

52. Nakajima J, Ono M, Takeda M et al. Role of costimulatory molecules on airway epithelial cells acting as alloantigen-presenting cells. Transplant Proc 1997; 29:2297-2300.

53. Nakajima J, Poindexter NJ, Hillemeyer PB et al. Cytotoxic T-lymphocytes directed against donor HLA class I antigens on airway epithelial cells are present in bronchoalveolar lavage fluid from lung transplant recipients during acute rejection. J Thorac Cardiovasc Surg 1999; 117(3):565-71.

54. Jaramillo A, Naziruddin B, Zhang L et al. Activation of human airway epithelial cells by non-HLA antibodies developed after lung transplantation: a potential etiological factor for bronchiolitis obliterans syndrome. Transplantation 2001; 71(7):966-76.

55. Davenport A, Younie ME, Parson JEM et al. Development of cytotoxic antibodies following renal allograft transplantation is associated with reduced graft survival due to chronic vascular rejection. Nephrol Dial Transplant 1994; 9:1315-19.

56. Reed EF, Hong B, Ho E et al. Monitoring of soluble HLA alloantigens and anti-HLA antibodies identifies heart allograft recipients at risk of transplant-associated coronary artery disease. Transplantation 1996; 61:566-72.

57. Cherry R, Nielsen H, Reed E et al. Vascular (humoral) rejection in human cardiac allograft biopsies: Relation to circulating anti-HLA antibodies. J Heart Lung Transplant 1992; 11:24-29.

58. Russell PS, Chase CM, Winn HJ et al. Coronary atherosclerosis in transplanted mouse hearts. II. Importance of humoral immunity. J Immunol 1994; 152:5135.

59. Russell PS, Chase CM, Colvin RB. Alloantibody and T-cell-mediated immunity in the pathogenesis of transplant arteriosclerosis: lack of progression to sclerotic lesions in B-cell-deficient mice. Transplantation 1997; 64:1531-36.

60. Lau CL, Palmer SM, Posther KE et al. Influence of panel-reactive antibodies on posttransplant outcomes in lung transplant recipients. Ann Thorac Surg 2000; 69(5):1520-24.

61. Smith MA, Sundaresan RS, Mohanakumar T et al. Effect of development of antibodies to HLA and CMV mismatch on lung transplant survival and development of bronchiolitis obliterans syndrome. American J Thoracic Surgery 1998; 116:812-20.

62. Sundaresan S, Mohanakumar T, Smith MA et al. HLA-A locus mismatches and development of antibodies to HLA after lung transplantation correlate with the development of bronchiolitis obliterans syndrome. Transplantation 1998; 65(5):648-53.

63. Jaramillo A, Smith MA, Phelan D et al. Development of ELISA-detected anti-HLA antibodies precedes the development of bronchiolitis obliterans syndrome and correlates with progressive decline in pulmonary function after lung transplantation. Transplantation 1999; 67(8):1155-61.

64. Palmer SM, Davis RD, Hadjiliadis D et al. Development of an antibody specific to major histocompatibility antigens detectable by flow cytometry after lung transplant is associated with bronchiolitis obliterans syndrome. Transplantation 2002; 74(6):799-804.

65. Rogers NJ, Lechler RI. Allorecognition. Am J Transplant 2001; 1(2): 97-102.

66. Heeger PS. T-cell allorecognition and transplant rejection: a summary and update. Am J Transplant 2003; 3(5):525-33.

67. Haque MA MT, Yasufuku K et al. Evidence for immune response to a self-antigen in lung transplantation: role of type V collagen-specific T-cells in the pathogenesis of lung allograft rejection. J Immunol 2002; 169(3):1542-49.

68. Bharat A FR, Trulock EP, Patterson GA et al. Induction of IL-10 suppressors in lung transplant patients by CD4+ 25+regulatory T-cells through CTLA-4 signaling. J Immunol 2006; 177:5631-38.

69. Bharat A SN, Trulock EP, Patterson GA et al. CD4+ 25+regulatory T-cells limit Th1-autoimmunity by inducing IL-10 producing T-cells following human lung transplantation. Am J Transpl 2006; 6:1799-1808.
70. Duijvestijn AM, Van Breda Vriesman PJC. Chronic renal allograft rejection: Selective involvement of the glomerular endothelium in humoral immune reactivity and intravascular coagulation. Transplantation 1991; 52:195-202.
71. Dubel L, Farges O, Johanet C et al. High incidence of antitissue antibodies in patients experiencing chronic liver allograft rejection. Transplantation 1998; 65(8):1072-75.
72. Dragun D MD, Brasen JH, Fritsche L et al. Angiotensin II type 1-receptor activating antibodies in renal-allograft rejection. N Engl J Med 2005; 352(6):558-69.
73. Behr TM SC, Pongratz DE, Weiss M et al. Adult human cardiomyocytes coexpress vimentin and Ki67 in heart transplant rejection and in dilated cardiomyopathy. J Heart Lung Transplant 1998; 17(8):795-800.
74. Martinu T HD, Davis RD, Steele MP et al. Pathologic correlates of bronchiolitis obliterans syndrome in pulmonary retransplant recipients. Chest 2006; 129(4):1016-23.
75. Hertz MI, Henke CA, Nakhleh RE et al. Obliterative bronchiolitis after lung transplantation: A fibroproliferative disorder associated with platelet-derived growth factor. Proc Natl Acad Sci USA 1992; 89:10385-89.
76. Hirabayashi T, Demertzis S, Schafers J et al. Chronic rejection in lung allografts: immunohistological analysis of fibrogenesis. Transpl Int 1996; 9(Suppl 1):S293-95.
77. Meyer KC CA, Ziang Z, Cornwell RD et al. Vascular endothelial growth factor in human lung transplantation. Chest 2001; 119(1):137-43.
78. Krebs R TJ, Nykanen AI, Wood J et al. Dual role of vascular endothelial growth factor in experimental obliterative bronchiolitis. Am J Respir Crit Care Med 2005; 171(12):1421-29.
79. Ationu A. Identification of endothelial antigens relevant to transplant coronary artery disease from a human endothelial cell cDNA expression library. Int J Mol Med 1998; 1(6):1007-10.
80. Rao S AF, Nieves E, Band Horwitz S et al. Identification by mass spectrophotometry of a new alpha-tubulin isotype expressed in human breast and lung carcinoma cell lines. Biochemistry 2001; 40(7):2096-2103.
81. Hein S ST, Schaper J. Ischemia induces early changes to cytoskeletal and contractile proteins in diseased human myocardium. J Thorac Cardiovasc Surg 98 1995;110(1):89.

# CHAPTER 13

# The Role for Cytokine Responses in the Pathogenesis of Lung Allograft Dysfunction

John A. Belperio,* Brigette Gomperts, Samuel Weigt and Michael P. Keane

## Introduction

Lung transplantation is now considered to be a therapeutic option for patients with end-stage pulmonary disorders.[1,2] However, due to problems of allograft dysfunction, 5 year survival rates are only 42%, as compared to greater than 70% for other solid organ transplantations.[1,2] There are three distinct histopathologic features of lung allograft dysfunction. The first and earliest feature is postlung transplantation ischemia reperfusion injury (IRI). This is followed by acute lung allograft rejection and bronchiolitis obliterans syndrome (BOS), a form of chronic lung allograft rejection.

Early lung allograft dysfunction continues to be the most common cause of early mortality postlung transplantation.[3-5] Studies suggest that up to 97% of lung transplantation recipients develop at least mild IRI while moderate to severe injury can occur in up to 30% of recipients. The mortality rate for more severe episodes of IRI can be as high as 41%.[4,6-8] Early lung allograft dysfunction results from a "multi-hit" mechanism that includes the complications of brain death, mechanical ventilation, hypotension, cold storage and reperfusion injury. Reperfusion of blood results in reoxygenation of ischemic tissue, which generates reactive oxygen intermediates (ROI).[9-11] These ROIs cause cell damage and the release of inflammatory mediators followed by the recruitment of leukocytes resulting in acute lung injury.[12-19] Furthermore, the occurrence and magnitude of IRI is not without consequence, as it may be a significant risk factor for both acute and chronic lung allograft rejection.[20-21]

Rejection is a recipient (host) response to a foreign antigen (i.e., the newly transplanted allograft). Allograft major histocompatibility complex (MHC)/human leukocyte antigen (HLA) is the surface antigen that is recognized by the recipients immune system as foreign, stimulating an intense inflammatory/immune response. This is further escalated early on by a reverse response, in which the allograft immune system reacts against the recipient, intensifying the inflammatory/immune cascade. Despite immunosuppressive therapy, rejection in the form of inflammation, immunity and fibroproliferation continues to be a problem.

Acute rejection occurs to some degree in almost all lung transplantation recipients.[1,22] Histologically, it is characterized by a peri-vascular/bronchiolar mononuclear cell infiltration that extends into the interstium and alveolar space causing parenchymal injury.[1,22] BOS represents chronic lung allograft rejection and is the major limitation to survival posttransplantation with a prevalence as high as 80% by ten years postlung transplantation.[1,2,23-26] Its major risk factor is the number/severity of acute rejections.[21,27-31] Histopathologic features begin with a peri-bronchiolar leukocyte infiltration that invades/disrupts the basement membrane, submucosa and lumenal epithelium.[23,24] This is followed by increased numbers of mesenchymal cells, extracellular matrix (ECM) deposition and granulation tissue formation within/around the lumen of the allograft airway.[23,24] Ultimately, myofibroblasts and mature collagen obliterate the airway.[23,24] Unfortunately, treatment with augmented immunosuppressive regimens has been disappointing.

## The Role of Cytokines in the Pathogenesis of Lung Allograft Dysfunction

Critical to lung allograft wound repair is a delicate balance between pro- and anti-inflammatory cytokines. Changes in this balance can influence lung allograft remodeling. The specific mechanisms that lead to lung allograft dysfunction may involve the interactions between Type 1 and 2 cells/immune responses. Naive cells (i.e., CD4+ T-cells (Th), CD8+ T-cells (Tc), mononuclear phagocytes (M) and natural killer cells (NK)) can all differentiate into at least two distinct cell subsets, [Type 1 immune cells (i.e., $Th_1$, $Tc_1$, $M_1$ and $NK_1$) or Type 2 immune cells (i.e., $Th_2$, $Tc_2$, $M_2$ and $NK_2$)]; which have distinct functions and cytokine expression profiles.[32-35] The Type 1 immune response is mainly associated with the cell-mediated immune response and is identified by the production of IL-2, IL-12, IFN-γ and lymphotoxin which drives a CTL and DTH response. The Type 2 immune response is identified by the production of IL-4, IL-5, IL-6, IL-9 and IL-13 and promotes mucosal, allergic and humoral immunity.[32-35]

The nature of the antigen and the pattern of cytokines released into the microenvironment are considered to be the most important factors dictating whether the immune response is directed toward a Type 1 or 2 response. In addition, during specific disease states Type 1 and 2 immune cells can cross regulate each other through their respective cytokine profiles[36,37] (Fig. 1). However, recent data suggest this cross regulation does not hold true for allograft dysfunction.[38-49] For instance, while Type 1 cytokines are considered the predominate regulators of rejection by promoting CTL and DTH responses, simply driving an allogeneic response toward a Type 2 cytokine cascade may not be as beneficial as initially thought and may accelerate rejection.[38-49] In fact, total inhibition of lung allograft dysfunction may depend on the down regulation of both Type 1 and 2 immune responses.[50]

*Corresponding Author: John A. Belperio—Department of Medicine, Division of Pulmonary and Critical Care Medicine, David Geffen-School of Medicine at UCLA, Los Angeles, CA 90024-1922, USA. Email: jbelperio@mednet.ucla.edu

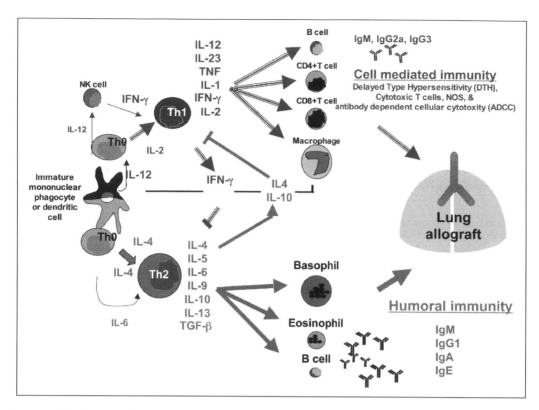

Figure 1. Development of a Type 1 and Type 2 immune response. Immature mononuclear phagocytes and dendritic cells present alloantigen to the T-helper cells (Th0 cells) allowing for polarization (Type 1 versus 2) of T-cells and mononuclear phagocytes depending on the cytokine microenvironment produced by the alloantigen and donor/recipient genetic factors. Type 1 differentiation is developed by an IL-12 rich microenvironment predominately produced by mononuclear phagocytes and dendritic cells. The Type 1 cells produce IL-2, IL-23, IFN-γ, lymphotoxin, IL-1, TNF-α, promoting the development of cytotoxic T-cells, delayed type hypersensitivity reaction and facilitate antibody dependent cellular cytotoxicity (i.e., the effector mechanisms against the lung allograft). Type 2 differentiation is produced by an IL-4 rich microenvironment polarizing T-cells and mononuclear phagocytes toward a Type 2 cell which produces IL-3, IL-4, IL-5, IL-6, IL-9, IL-10 and IL-13 effecting immunoglobulin isotypes and stimulating IgE and IgG1 production by B-cells and eosinophils and ultimately creating a pro-fibrotic allograft airway microenvironment. Type 1 cells secreting IFN-γ can inhibit differentiation of Type 2 cells and Type 2 cells secreting IL-4 and IL-10 can inhibit differentiation of Type 1 cells. Both the Type 1 and 2 responses are likely to be involved in the pathogenesis of lung allograft dysfunction.

## The Role of the IL-1 Cytokine Family in the Pathogenesis of Lung Allograft Dysfunction

The IL-1 family consists of two agonists (IL-1α and IL-1β), two receptors (biologically active IL-1RI and inert IL-1RII) and a naturally occurring receptor antagonist (IL-1Ra).[51] IL-1Ra and IL-1 are known to have near equal affinity for the active IL-1 receptor (IL-1RI).[52,53] IL-1 is a proximal pro-inflammatory cytokine that can upregulate many pro-inflammatory molecules including chemokines and adhesion molecules and has been implicated in many inflammatory/fibroproliferative diseases such as rheumatoid arthritis. However, excess IL-1Ra can enhance a local pro-fibrotic environment through the inhibition of the normal 'fibrolytic activity' mediated by IL-1. For instance, the inhibition of IL-1 by IL-1Ra has been shown to cause a reduction in the production of $PGE_2$, nitric oxide and metalloproteinase (MMP) resulting in the promotion of excess deposition of ECM.[54,55]

Elevated levels of IL-1 were associated with "warm" IRI (i.e., simply cross-clamping the pulmonary artery of a rat for a period of time followed by reperfusion).[56] Neutralizing antibodies to IL-1 resulted in a reduction in lung vascular permeability, neutrophil sequestration, BALF leukocytes and overall warm IRI.[56]

With regard to human acute lung allograft rejection, cultured alveolar macrophages (AM) were found to have increased expression of IL-1α, IL-1β and IL-1Ra.[57] In contrast, only elevated levels of IL-1Ra, not IL-1β in human bronchoalveolar lavage fluid (BALF) was found to be associated with BOS and future BOS (FBOS) (i.e., last BALF just prior to the diagnosis of BOS).[58] Furthermore, cardiac transplantation

recipients with an intronic VNTR polymorphism (IL1RN) of the gene encoding IL-1Ra, which is suspected to be associated with increased expression of IL-1Ra with or without decreased IL-1 levels has been shown to be an important risk factor for the development of chronic cardiac rejection.[59,60] Thus, the elevated ratio of (IL-1Ra to IL-1β) is related to chronic rejection and is consistent with an ongoing alteration in IL-1 activity in the allograft airways producing a pro-fibrotic allograft airway micro-environment.

Collectively, the above findings support a contention that inhibition of IL-1 biology early on may help decrease early allograft dysfunction. However, persistent (i.e., during the development of BOS) inhibition of IL-1 biology may impair the ability of the allograft airway to repair/remodel appropriately and lead to the development of BOS. Future studies involving the role of Receptor/IL-1 biology may be fruitful in identify a novel mechanism that can inhibit lung allograft dysfunction.

## The Role of TNF-α in the Pathogenesis of Lung Allograft Dysfunction

TNF-α is a proximal pro-inflammatory cytokine with numerous effects on inflammatory/immunologic responses including enhanced NK cell cytolytic activity, augmented T-cell proliferation and the up-regulation of MHC class II antigen and IL-2 receptors on lymphocytes.[61,62] In addition, TNF-α has been shown to play a key role in inflammatory/fibroproliferative disorders such as pulmonary and hepatic fibrosis.[63,64]

Investigators found an association between elevated levels of TNF-α and warm IRI using a rat model.[56] Moreover, neutralizing antibodies to TNF-α resulted a reduction in lung vascular permeability, neutrophil sequestration, BALF leukocytes and overall warm IRI.[56] Similarly, TNF-α was found to have a bi-modal expression pattern when examined in a rat orthotopic single lung transplantation model. The first peak in TNF-α paralleled "cold" IRI, while the second peak paralleled maximal acute lung allograft rejection.[65] Importantly, in vivo neutralization of TNF-α attenuated acute lung allograft rejection.[65]

Using animal models of BOS investigators have found increased expression of TNF-α associated with tracheal fibroplasias.[66] Importantly, when TNFR:Fc; an inhibitor of TNF-α, or neutralizing antibodies to TNF-α were administered to recipients of allografts there was significant attenuation of animal BOS.[66,67]

Surprisingly, translational studies in humans have not found elevated level of TNF-α in BALF from patients with BOS or any association between TNF-α and chronic liver or renal rejection.[58,68,69] In contrast, a large human study involving first and repeat cadaver kidney transplant recipients demonstrated that the TNF-α single nucleotide polymorphism (SNP) homozygous for the high TNF-α producer (genotype-308) was associated with decreased kidney survival.[70] However, this effect was only found in retransplants and not in primary grafts. Collectively, this suggests that recipients with TNF-α high responsiveness may require pre-immunization/immune (i.e., priming by rejection of a previous graft) in order to exert a true detrimental effect.[70] Future human and animal studies will be required to determine the exact role of TNF-α during the continuum of IRI to acute to chronic lung allograft rejection.

## The Role of IL-12 in the Pathogenesis of Lung Allograft Dysfunction

IL-12–related cytokines are encoded by five independently regulated genes: p40, p35, Epstein-Barr virus-induced gene-3 (EBI3), p19 and p28. Because of alternative heterodimeric partnering and monomer secretion, this family is composed of seven secreted proteins: IL-12 (a p40 and p35 heterodimer), IL-12 p80 (a p40 homodimer), p40 (a p40 monomer), IL-23 (a p40 and p19 heterodimer), EBI3 (an EBI3 monomer), IL-27 (a p28 and EBI3 heterodimer) and EBI3-p35 (an EBI3 and p35 heterodimer).[71,72] The classic IL-12 protein exists as a p70 heterodimer consisting of p35 and p40 subunits.[73,74] These subunits individually interact with distinct components of the IL-12R.[75,76] IL-12 is a pivotal cytokine for the promotion of cell mediated (Type 1) immune response and has been shown to induced IFN-γ during an allogeneic response.[77-89]

The exact role of the IL-12 family during allograft rejection remains controversial. For instance, in a Type 1 immune mediated model of graft verses host disease (GVHD), neutralizing IL-12 or using 40-/- mice (cells from BALB/c p40-/- donors transferred into C57BL/6 p40-/- recipients) was associated with recipient cells having enhanced IL-5 and IL-10 and reduced IFN-γ production (driving a Type 1 → 2 response) and attenuated acute GVHD.[90-92] Conversely, administration of IL-12 in this model exacerbated acute GVHD and was able to convert a chronic (Type 2) GVHD to an acute (Type 1) GVHD.[82,93] Alternatively, neutralization of IL-12 (either with polyclonal antibodies to IL-12 or by IL-12 receptor blockade using a p40 homdimer) in a murine model of cardiac transplantation led to an increased intragraft expression of IL-4 and IL-10 (Type 2 response) and accelerated rejection.[94] Moreover, the administration of IL-12 in this same model system markedly increased IFN-γ levels without augmentation of the rejection response.[77,78,88,89] These studies suggest that high levels of IL-12 can induce superphysiologic levels of IFN-γ leading an inhibition of the immune response as a result of IFN-γ anti-proliferative effects on emerging Th2 cells (i.e., IFN-γ receptors are expressed only on Th2 cells).[95,96]

With regard to lung allograft dysfunction, investigators using the murine model of BOS compared uninfected syngeneic, allogeneic, to infected (i.e., murine parainfluenza virus-1 (Sendai virus)). They found that infected allografts had enhanced epithelial cell injury with incomplete re-epithelialization of the basement membrane. In addition they detected p.40 immunolabeling (predominantly on epithelial cells) in infected allografts, but not uninfected allograft or syngeneic controls. Furthermore, allogeneic transplantation induced p.80 expression and concurrent viral infection resulted in a synergistic increase in p.80, but not IL-12 or IL-23. Importantly, blockade of p.80 function attenuated allograft dysfunction as well as virus-driven allograft dysfunction while overexpression of p.80 accelerated allograft dysfunction. Translational studies demonstrated enhanced expression of epithelial p.80 in human transplant recipients correlated with transplant bronchitis.[97] Moreover, in a translation/longitudinal study of 44 human lung recipients with elevated levels of IL-12 in BALF was predictive of developing BOS.[98] These findings support the contention that IL-12 plays a role in the modulation of the local pro-/anti-fibrotic balance of the lung allograft. Further mechanisms involving receptor/IL-12 cytokine family may lead to novel ways to attenuated lung allograft dysfunction.

## The Role of IFN-γ in the Pathogenesis of Lung Allograft Dysfunction

IFN-γ is a Type 1 pleiotropic cytokine that can be induced by IL-12 and inhibited by IL-10.[99-101] With regards to transplantation, IFN-γ can induce class I and II expression, yet inhibit T (Th2) cell proliferation.[102-105] While IRI is considered to be a predominately neutrophil mediated disease recent studies suggest that recruited mononuclear cells are just as important.[106] While there are no specific studies of the role of IFN-γ and postlung transplantation IRI, Day and associates evaluated the contribution of IFN-γ during renal IRI. They first demonstrated that Rag-1-/- mice (lacking T and B cells) were protected in comparison to wild-type mice when subjected to IRI.[107] Because IFN-γ is produced by kidney cells and T-cells they then performed their renal model of IRI in BM chimeras (IFN-γ-/- → WT) and found a reduction in IRI supporting a role for IFN-γ expressing leukocytes during IRI. In addition, the protection from renal IRI that was observed in Rag-1-/- mice was reversed in BM chimeras (IFN-γ+/+ CD4+ → Rag-1-/-), as compared to chimeras (IFN-γ-/- CD4+ → Rag-1-/-). These studies demonstrate that CD4+ IFN-γ expressing cells are critical to IRI.[107]

Using a rat heterotopic tracheal transplantation model of BOS, IFN-γ was found to be persistently elevated during fibro-obliterative events.[108] However, other animal models of acute and chronic rejection have demonstrated a few unanticipated results. Using an immunosuppressed, murine cardiac transplantation model, IFN-γ-/- recipients demonstrated decreased graft integrin and MHC class II expression, yet accelerated acute parenchymal rejection.[109-111] Interestingly, these allografts demonstrated decreased chronic transplantation coronary artery rejection (i.e., graft arterial/vascular disease which involves intimal lesions containing smooth muscle cells, leukocytes and ECM deposition).[109-111] Collectively, these studies suggest a role for IFN-γ, in part, limiting acute allograft rejection, due to its antiproliferative effects on T-cells and its ability to down regulate the CTL lytic activity against allogeneic targets. In contrast, IFN-γ secretion by activated infiltrating cells may, in part, worsen chronic rejection by upregulating integrin and MHC class II on the surface of allograft endothelial cells, culminating in the accumulation of activated macrophages secreting profibrotic mediators.

Pertaining to human lung transplantation, elevated expression of IFN-γ from BALF was associated with acute and refractory lung allograft rejection, both risk factors for the development of BOS.[112,113] In addition, a significant correlation was detected between the presence of a high expressing human polymorphism at position +874 of the IFN-γ gene and BOS.[114] Similarly, Tambur and associates found, in a small study of approximately 60 heart transplantation recipients, that low producers of IFN-γ have a low risk of TCAD, whereas intermediate and high producer of IFN-γ have a high risk of TCAD.[115] In contrast, two other studies did not show this association.[116,117] The disparity in outcomes likely has to due with differences in study design, different definition of TCAD angiographic abnormalities and differences in studies endpoints. More animal and human studies will be required to determine the exact influence of Receptor/INF-γ biology during the continuum of early to late lung allograft dysfunction

## The Role of IL-10 in the Pathogenesis of Lung Allograft Dysfunction

IL-10 is a pleiotropic Type 2 cytokine with immunomodulatory bioactivity including the inhibition of cytotoxicity and the down-regulation of MHC class II antigens and proinflammatory cytokine production.[101,118-131] The role of IL-10 in modulating the response to allograft rejection has been controversial. Elevated expression of IL-10 in allografts undergoing both rejection and tolerance suggests IL-10 can promote or inhibit alloimmune destruction.[132-134] Animal experiments have demonstrated that IL-10 administration or overexpression either accelerated or had no effect on islet cell allograft failure.[135,136] Yet, augmentation of IL-10 was found to be protective in multiple models of lung IRI including a pig model of cold IRI.[137-139]

With regard to rejection, systemic administration of superphysiologic doses of IL-10 was found to exacerbate murine cardiac allograft rejection.[110,140] Similarly, immunosupressed IL-10-/- mice rejected cardiac allografts twice as rapidly as controls.[110,140] In contrast, the administration of IL-10 at physiologic doses or by liposome-mediated ex-vivo intracoronary IL-10 gene transfer in animal models of cardiac rejection prolonged survival by attenuating intragraft mononuclear cell recruitment and increasing cytotoxic T cell apoptosis secondary to the Fas/FasL pathway.[141,142] In addition, either retro- or adeno-viral vector delivery of viral IL-10, not cellular (cIL10) caused prolonged cardiac allograft survival through similar mechanisms.[143,144] Similarly, over-expression of IL-10 attenuated acute lung allograft rejection when used in the rat orthotopic single lung transplantation model.[145] More profound results were seen when the over-expression was performed in conjunction with cyclosporin.[145] Moreover, neutralization of IL-10 in a rat heterotopic tracheal transplantation model accelerated airway obliteration, whereas the administration of physiologic doses of IL-10 or local Sedai virus (SeV)-mediated IL-10 gene transfer targeted to the airway or autologous hematopoetic stem cell-enriched mouse bone marrow transduced with retrovirus encoding vIL-10 attenuated airway fibroplasias.[146-149]

The discrepancy in the effect of IL-10 during allograft dysfunction may be related to dose (physiologic verses super-physiologic), type (viral verses cellular IL-10), timing of the dose and compartment (systemic or local administration) at which IL-10 is delivered. The effects of systemic IL-10 appear to depend on dosing and timing (i.e., superphysiologic doses augment rejections and physiologic doses given at the right time attenuate rejection) and suggest superphysiologic doses of IL-10 may reflect desensitization of IL-10 biology by unabated pro-inflammatory effects. With regard to vIL-10 verses cIL-10, vIL-10 does not possess the T-cell costimulation activities of cIL-10, but vIL-10 has a 100-fold lower affinity than cIL-10 for binding to the IL-10 receptor.[140,143,144,150] Therefore, when vIL-10 or cIL-10 are delivered in equivalent concentrations they will not have the same biological effects.[140,143] Lastly, direct transfer of IL-10 into the transplanted organ may precede local intragraft levels of IL-10 from systemic administration making the timing of the administration of IL-10 another variable to be considered.[151] Hence the efficacy of IL-10 manipulation coupled with differences in type of IL-10 and transplantation microenvironment may explain the inconsistent effects seen in

graft survival to date.[151] Further investigations of timing, dose and type of IL-10 in animal models may clarify these controversies.

Pertaining to human allograft studies, the polymorphism in the IL-10 gene promoter (position −1082) (high IL-10 producers) was associated with acute renal allograft rejection.[152] In contrast, another high IL-10 producing polymorphism was shown to be protective in heart and kidney transplantation and was associated with the tolerant state in pediatric liver transplantation recipients not receiving immunosuppression therapy.[153-157] With regard to human lung transplantation recipients, the increased IL-10 production genotype (GCC/GCC) protected recipients against acute rejection when compared with the intermediate or decreased IL-10 production genotypes.[158] Surprisingly, IL-10 levels in BALF from patients with BOS did not demonstrate differences when compared nonBOS controls.[58]

## The Role of IL-15 in the Pathogenesis of Lung Allograft Dysfunction

IL-15 is another pro-inflammatory cytokine that is predominately produced by activated mononuclear phagocytes.[159-162] Under physiological conditions it is involved in the differentiation and proliferation of NK and T-cells.[159-162] In addition, IL-15 stimulates the activation, proliferation, survival and effector functions of immune cells.[163-167] While IL-15 has not been specifically associated with lung allograft dysfunction it has been associated with both autoimmune disorders and other solid organ allograft dysfunction.[168,169] For example, increased intragraft expression of IL-15 was associated with human acute rejection in both kidney and cardiac allografts.[168,170-172]

Zheng and associates evaluated the contribution of IL-15 interactions with IL-15 Receptor (IL-15R) positive cells during cardiac allograft rejection.[173] They constructed a recombinant fusion protein consisting of a mutated IL-15 at the constant region of murine IgG2a (mIL-15/Fc fusion protein). This mIL-15/Fc fusion protein binds the IL-15R alpha chain, which is expressed on activated, but not resting mononuclear cells and deletes them by antibody mediated immune effector mechanisms.[174-176] This mIL-15/Fc fusion protein prolonged the survival of fully mismatched heterotopic cardiac allografts by reducing the number of infiltrating CD4, CD8 T-cells and mononuclear phagocytes. In addition, this mIL-15/Fc fusion protein caused a reduction in intragraft expression of CTL markers Fas Ligand, perforin and granzyme B as well as pro-inflammatory cytokines IL-1β, TNF-α and IFN-γ. Surprisingly, this group did not find similar effects on attenuating cardiac allograft rejection when they used a nonlytic variant of mIL-15/Fc (IL-15 pure antagonist) suggesting that deletion of IL-15R expressing cells is important for the protective effect of mIL-15/Fc. This may explain, in part, why Smith and associates found that recombinant soluble IL-15R alpha subunit (sIL-15Rα) was only effective in attenuating cardiac rejection on in minor histocompatibility mismatch but not a fully mismatched model of cardiac rejection.[177] Overall these studies suggest a potential role for IL-15R/IL-15 biological axis during lung allograft dysfunction.

## The Role of IL-17 in the Pathogenesis of Lung Allograft Dysfunction

Interleukin-17 (IL-17) is a proinflammatory cytokine that is predominately produced by TCRα/β+CD4−CD8− T-cells as well as activated CD8+, CD8+CD45RO+ memory cells and CD4+ cells.[178-181] IL-17 stimulates multiple cell types to express IL-6, IL-1β, CXCL8, TNF-α, G-CSF, CCL2 and PGE2.[181,182] In addition, IL-17 stimulates proliferation of alloreactive T-cells and the maturation of DC.[183] The high affinity receptor for IL-17 is IL-17R 5. IL-17 deficient mice have been shown to exhibit abnormal contact, delayed-type and airway hypersensitivity responses, along with impaired T-dependent antibody production[184] suggesting it has a critical role in specific T-cell mediated immune responses.

While there are no specific studies involving IL-17 and lung transplantation recipients, elevated levels of IL-17 were found to be associated with human renal allograft rejection.[185,186] In addition, in a non fully mismatched murine model of cardiac allograft rejection intraperitoneal administration of an IL-17 inhibitor (soluble IL-17R-Ig fusion protein) resulted in prolonged graft survival.[183] More recently, Li and associates using a gene transfer based approach (i.e., donor graft production of the IL-17 inhibitor (soluble IL-17R-Ig fusion protein))[187] and demonstrated a reduction in allograft expression of INF-γ and TGF-β as well as infiltrating mononuclear phagocytes and CD4 cells, but not CD8 cells. Importantly this led to a delay in allograft acute rejection. These results suggest that further investigation on IL-17 inhibition as a potential adjuvant therapy in transplantation is warranted.

Interestingly, lung rejection has been associated with immunity to a native protein, minor type V collagen (col (V)) Anti-type V collagen lymphocytes that express IL-17 and IL-23 induce rejection pathology in fresh and well-healed lung transplants. Col (V) is intercalated within type I collagen, a major collagen in the lung.[188-190] In fact, col (V) is considered to be a sequestered antigen in the normal lung and is located in the perivascular and peribronchiolar connective tissue (i.e., the same sites of lung allograft dysfunction activity). Importantly, IRI postlung transplantation allows for antigenic fragments of col (V) to be released locally.[191] Yoshida and associates demonstrated that col (V)-reactive T-cells that develop within lung allografts when adoptively transferred into isografts cause a "rejection-like" histopathology only in isografts (i.e., with a previous IRI phase) and not naïve controls and this was associated with the upregulation of IL-17 transcripts.[192] Future studies of Receptor/IL-17 biological axis during allograft dysfunction may lead to immunosuppression sparing novel therapies.

## The Role of IL-6 in the Pathogenesis of Lung Allograft Dysfunction

IL-6 is a cytokine with both proinflammatory and anti-inflammatory properties and has been associated with a Type 2 profile and fibrogenesis.[193-197] With respect to an allogeneic response, in vitro studies have demonstrated that allogeneic CD2+ lymphocytes are capable of activating airway-derived epithelial cells to produce high levels of IL-6 possibly promoting airway fibrosis.[198]

The orchestration of IL-6 signal transduction predominately involves two parallel pathways that may regulate one another. The first pathway is the is the Janus-associated kinase signal transducer and activators of transcription (STAT) family and the second is through the mitogen activated protein kinases and nuclear factor kappa B (NK-kB).[199] Importantly, target genes for the STAT families are similar to NK-kB (i.e., inducing cytokines and adhesion molecules).

Interestingly, giving IL-6 to rat donors in an intestinal IRI model that also results in acute lung injury prolonged graft viability and reduced neutrophil extravasation within the lung.[199] Similar results were seen in models of liver and retinal ischemia injury.[199] With regard to lung allograft dysfunction using a rat lung transplantation model IL-6 was found to have a bimodal distribution with early expression during IRI and then later at maximal rejection.[200] Favivar and associates evaluated the effects intra-tracheally administered IL-6 during "warm" IRI and found a protective effect (decreased lung permeability, neutrophil sequestration, BALF total cell counts and hypoxemia).[199] In addition, they found that IL-6 inhibited NF-κB nuclear translocation and un-expectantly, they also found that IL-6 inhibited STAT-3 nuclear translocation. Furthermore, they found that IL-6 also caused a reduction in TNF-α and CCL2 expression. Overall this studies shows that during early allograft dysfunction IL-6 may be protective.

Elevated levels of human IL-6 in BALF have been associated with refractory acute lung allograft rejection.[199] Similarly, genetic polymorphisms of the IL-6 gene has been shown to be associate with

an earlier onset of BOS.[199] In addition, a prospective cohort study of human BALF, analyzed within 2 months postlung transplantation, demonstrated that increased levels of IL-6 were predictive for the development of BOS.[201] Furthermore, ex vivo data demonstrated that alveolar macrophages (AM) from acutely rejecting patients secrete high levels of IL-6 while TGF-β1 secretion remained normal. Subsequently, at the time of BOS onset, AM production of IL-6 was normal and TGF-β1 production increased. This suggests IL-6 activity is involved in the development of BOS, perhaps by priming resident AM to upregulate their production of the profibrotic cytokine TGF-β. Lastly, the presence of high expression polymorphism IL-6 at position −174 significantly increased the risk for the development of BOS. These studies underscore the possible role for IL-6 during the pathogenesis of BOS.[114]

More recently Kaneda and associates, collected biopsies from 169 donor lungs before lung transplantation into recipients.[202] Unfortunately 17 lung transplantation recipients died within 30 days. No donor factor was associated with 30-day mortality. Univariate analysis of the 84 cases initially used for the development of the prediction model showed that donor lung cytokines expression of IL-6, IL-8, TNF-α and IL-1β were risk factors for mortality and IL-10 and IFN-γ were protective factors. Furthermore, they analyzed the cytokine expression ratios of risk to protective cytokines. A stepwise logistic regression for 30-day mortality demonstrated that a model containing the ratio of IL-6/IL-10 was the most predictive. When applied to the remaining 85 cases for validation, the test of model fit was significant. Overall they show that IL-6/IL-10 ratio is important in predicting who will have trouble postlung transplantation. Ultimately, more work will have to be performed to determine the specific role of IL-6 and lung allograft dysfunction.

## The Role of TGF-β in the Pathogenesis of Lung Allograft Dysfunction

While TGF-β has a strong history of immunosuppressive activity, it is the most potent inducer of collagen synthesis, fibroblast proliferation and fibroblast chemotaxis.[203,204] Studies involving solid organ transplantation have demonstrated that TGF-β has beneficial effects during acute allograft rejection.[205,206] Overexpression of the TGF-β transgene led to a reduction in acute rejection and prolonged graft survival in experimental heart and lung transplantation.[205,206] In addition, elevated serum levels of TGF-β has been associated with tolerance induction in an experimental model of lung transplantation. Importantly, these animals allo-immune responses were restored with the administration of neutralizing antibodies to TGF-β.[207]

More recent animal studies have found TGF-β to be localized to infiltrating mononuclear cells and fibrotic tissue in a rodent model of BOS.[208] The administration of adenoviral mediated soluble TGF-βIIIR; a functional TGF-β antagonist, topically on day 5 posttransplantation attenuated rodent BOS.[208] However, if soluble TGF-βIIIR was given on day 0 or 10, or intramuscularly posttransplantation no significant effect on BOS was distinguished.[208] These studies demonstrate that compartmentalization and timing of TGF-β augmentation may be important factors with regard to its fibro-obliterative effects during BOS.

Smad3 is a member of the highly conserved Smad family of intracellular signaling proteins, which mediate many of the effects of TGF-β1. Smad3 is directly phosphorylated by the ligand activated TGF-β type I receptor (TGF-βRI).[209] After joining with a common mediator SMAD4, the heteromeric complex translocates into the nucleus and regulates gene transcription of such molecules as platelet derived growth factor (PDGF), connective tissue growth factor (CTGF), MMP's, fibronectin and collagen.[210-213] Ramirez and associates using a murine model of BOS demonstrated early expression of TGF-β and CTGF localized to inflammatory cells then later during fibro-obliteration both were predominately localized to fibroblast.[214] In addition, the TGF-β intracellular signal transducer, Smad3 was detected in these fibroblasts throughout the fibrous tissue. Importantly, Smad activation was confirmed by demonstrating phosphorylated Smad2/3 in these fibroblasts. Using Smad3-/- recipient mice these investigators demonstrated a reduction in fibronectin, collagen and BOS thus establishing Smad3 and possible other downstream mediators of TGF-β1 may provide possible therapeutic options in the prevention and treatment of BOS.

Human data has demonstrated increased expression of TGF-β during the pathogenesis of chronic rejection involving liver, kidney and heart transplantation.[215-217] Homozygous TT/GG in the condons 10 and 25 (single gene polymorphism) have been identified as TGF-β1 high producers,[5,6] yet studies have failed to find a correlation between TGF-β1 single gene polymorphism TGF-β1-codon 10 or TGF-β1-codon 25 and cardiac rejection.[8] However, multiple studies have demonstrated that TGF-β1-codon10-25 (double gene polymorphisms) (high-TGF-β1 producers) were associated with acute and chronic cardiac rejection.[217,218] Similarly, studies involving lung transplantation have shown augmented expression of TGF-β to be an early marker of BOS and could be correlated with the severity of luminal fibrosis.[219-221] In contrast, other studies have demonstrated no difference in BALF protein levels or cellular expression of TGF-β between BOS and nonBOS groups.[58,219] More human studies involving lung transplantation recipients will be required to determine if there is an association between elevated levels of TGF-β and lung allograft dysfunction.

## The Role of Growth Factors in the Pathogenesis of Lung Allograft Dysfunction

Insulin like growth factor-1 (IGF-1) is a potent profibrogenic mediator acting as a mitogen/stimulator of collagen synthesis by fibroblasts.[222-225] The local bioactivity of IGF-1 in the lung is regulated by a system of multiple high affinity IGF binding proteins (IGFBP).[226] Some of which act as inhibitors while others can potentate IGF-1 cellular response.[226] In a human study of sequential BALF from lung transplantation recipients, IGF-1 expression was found to be increased prior to the development of BOS, yet was not affected by acute rejection episodes or CMV infection.[227] Similarly, IGFBP-3 was markedly increased from patients that later developed BOS, as compared to those that did not. This suggests IGF-1 and IGFBP-3 interaction could have a critical role in the pathogenesis of BOS through potentiation of IGF-1 profibrotic activities.[227] Moreover, IGF-1 has been found to be associated with murine BOS and future studies on the inhibition of IGF-1 and the manipulation of its multiple IFGBP may add insight to the role of IGF-1/IFGBP during BOS.

PDGF is another mitogen for mesenchymal cells including fibroblasts and smooth muscle cells. PDGF ligands consist of two polypeptides, the PDGF-A and PDGF-B chains, which can be expressed as homodimers (PDGF-AA and PDGF-BB) or as a heterodimer PDGF-AB.[228] The isoforms of PDGF have different affinity to their related receptors, PDGF-Rα and PDGF-Rβ.[229,230] PDGF-Rβ binds the PDGF-B chain and PDGF-Rα binds both PDGF-A and -B chains.[229,230] In humans elevated levels of biologically active PDGF from BALF was associated with BOS.[231] Furthermore, the expression PDGF-AA and PDGF-Rα >> PDGF-Rβ were found to elevated in an animal model of BOS.[232] Treatment with a protein tyrosine kinase inhibitor specific for PDGF receptor significantly reduced the myofibroproliferation associated with BOS without effecting inflammation or immune activation.[232] These studies suggest a regulatory role for PDGF on fibroplasia during the development of BOS.

Hepatocyte growth factor (HGF) has been found to be highly expressed in the lung and serum during injurious conditions and has been found to help stimulate epithelial cell proliferation and prevent aberrant wound repair.[233,234] Aharinejah and associates demonstrated in human lung transplantation recipients that elevated serum HGF concentration was a significant predictor of rejection.[235] In contrast, serum HGF concentration was not a predictor for infection allowing the authors to discriminate between patients with rejection from infection. Thus elevated serum levels of HGF may be a marker for acute lung allograft rejection. Further studies will be required to find its role in predicting BOS.

## The Role of Receptors/Chemokines in the Pathogenesis of Lung Allograft Dysfunction

### Chemokines and Chemokine Receptors

The hallmark of allograft dysfunction is the infiltration of leukocytes to the lung allograft. The ability to maintain leukocyte recruitment throughout the continuum early (IRI and acute lung allograft rejection) to late (BOS) lung allograft dysfunction despite aggressive immunosuppression is pivotal in the transition from inflammation/immune response to fibroplasia of the allograft. This persistent elicitation of mononuclear cells requires intercellular communication between infiltrating leukocytes, endothelium, parenchymal cells and components of the ECM. These events are mediated via the generation of adhesion molecules, cytokines and chemokines. The chemokines, by virtue of their specific cell surface receptor expression, can selectively mediate the local recruitment/activation of distinct leukocytes/cells allowing for migration across the endothelium and beyond the vascular compartment along established chemotactic gradients.

The chemokine superfamily is divided into four subfamilies (C, CC, CXC and CX₃C) based on the presence of a conserved cysteine residue at the NH₂-terminus.[236-238] CXC chemokines have been further subdivided on the basis of the presence or absence of the sequence glutamic acid-leucine-arginine (ELR) near the NH₂-terminal. ELR+ chemokines are neutrophil chemoattractants with angiogenic properties. ELR-CXC chemokines are chemoattractants of lymphocytes with angiostatic properties.[239-243] CC chemokines predominantly recruit mononuclear cells.[236,244] The C subfamily consists of XCL1 and XCL2, which attract lymphocytes, while CX₃CL1 is the only member of the CX₃C subfamily and its domain sits on a mucin stalk allowing for cellular adhesion.[245-248]

All chemokine action is mediated through seven transmembrane spanning G protein coupled receptors.[249] These heterotrimeric G proteins are composed of α (defines the identity of the protein), β and γ subunit. The chemokine receptors generally undergo internalization and phosphorylation following ligand binding. Interaction of a ligand with its receptor leads to exchange of GTP for GDP and the dissociation of the α subunit from the βγ subunit. The dissociated $G_α$ and $G_{βγ}$ can activate downstream signal transduction events.[250,251]

## The Role of Receptors/CC Chemokines in the Pathogenesis of Lung Allograft Dysfunction

The inflammatory cascade involved in lung allograft dysfunction results in leukocyte recruitment and proliferation. This makes the CC chemokines desirable proteins to study during lung allograft dysfunction. CCL3 and CCL5 recruit leukocytes, activates lymphocytes and causes cellular proliferation.[252-256] Their major mononuclear cell receptors are CCR1 and CCR5.[257-265] With regard to warm lung IRI in rats both CCL3 and CCL5 were found to be upregulated and in vivo neutralization of either chemokine reduced lung injury.[266]

Pertaining to acute lung allograft rejection, CCL5 was the first chemokine found to be elevated.[267,268] Animal studies have also demonstrated a correlation between elevated levels of CCL5 from rejecting lungs and the recruitment of intragraft mononuclear cells expressing CCR1 and CCR5.[269-271] Interestingly, dynamic changes of these receptors related to internalization and cell surface expression were noted

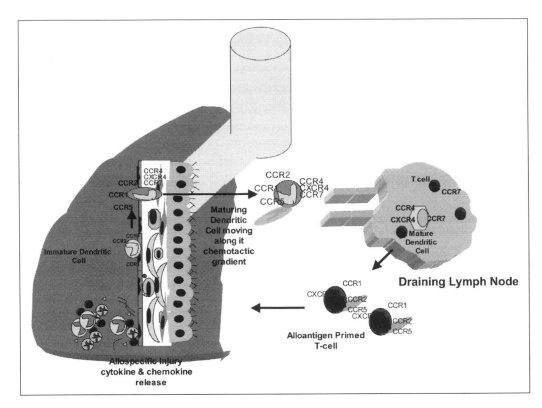

Figure 2. Involvement of CC chemokines during lung allograft dysfunction. Allogeneic injury stimulates the release of cytokines/chemokines from endothelial cells, leukocytes and stromal cells. Immature dendritic cells expressing CCR1, CCR2, CCR5 are recruited by CC chemokines to the injured area where they sample antigens. As these dendritic cells mature they lose CCR1, CCR2 and CCR5 expression and begin to express CCR4, CCR7 and CXCR4, allowing for homing to regional lymph nodes. Here they present and activate naïve T-cells. These primed T-cells have upregulated expression of CCR1, CCR2, CCR5 and CXCR3 facilitating the migration to the sites of immune injury.

throughout the rejection process.[271] Two other CC chemokines, which share receptors with CCL5, were also analyzed (CCL3 and CCL4). Although CCL3 levels were elevated, they were 170-fold less than CCL5, while CCL4 levels were surprisingly lower as compared to syngeneic controls. In vivo depletion of CCL5 attenuated lung rejection by delaying the recruitment of infiltrating mononuclear cells expressing CCR1 and CCR5.[268] CCL5 depletion did not effect T-cell proliferation, but did reduce allospecific responsiveness of recipient cells, presumably by decreasing the recruitment of mononuclear cells and immature DC expressing CCR1 and CCR5.[271-274] A reduction in immature DC sampling of allospecific antigens leads to a decrease in activation/priming of effector and memory cells in regional lymph nodes (Fig. 2).[273,275]

CCL5 has also been associated with human and animal renal allograft rejection.[263,276-278] Inhibition of CCL5 interaction with its receptors using a CCL5 analog/antagonist (i.e., Met-RANTES), led to a reduction in renal graft injury.[278] Furthermore, CCL5 was found to be important in upregulating leukocyte integrins allowing for firm adhesion of leukocytes to activated endothelium (Fig. 3). Similar, results and mechanisms were demonstrated when a CCR1 antagonist was used to attenuate renal and cardiac rejection.[279,280]

Gao and associates also found upregulated expression of CCL5 and CCL3 associated with intra-graft leukocytes expressing CCR1 and CCR5 during murine cardiac rejection.[281,282] They found a moderate increase in allograft survival when using either CCR1-/- or CCR5-/- recipient mice. Moreover, when given a short course of cyclosporin A (CsA), both CCR1-/- and CCR5-/- recipients remarkably accepted their allografts. Mechanistically, allografts from either CCR1-/- or CCR5-/- mice demonstrated a reduction in infiltrating leukocytes and cytokine/chemokine/receptor expression. Both knockouts had minimal effects on cell proliferation suggesting CCR1

and CCR5 knockouts have defects in cellular recruitment/activation with impaired cytokine/chemokine amplification, as compared to minimal augmentation in proliferation responses. Furthermore, using a heterotopic murine model of pancreatic islet cell allograft rejection, CCR5-/- recipient were able to prolong allograft survival from 10 (control recipients) to 38 days (recipient CCR5-/- mice).[283] Collectively, these studies demonstrate an important role for CCR1/ligand and CCR5/ligand biology during the pathogenesis of allograft dysfunction.

In both human and animal models of BOS, inflammation has been shown to be a requirement for fibro-obliteration.[284-294] CCL5 was found to be elevated during human and animal BOS.[108,271,295] Neutralization of endogenous CCL5 reduced the numbers of infiltrating graft CD4 T-cells, preserved lumen patency and attenuated early epithelial injury.[295,296] Similarly in a murine model of chronic cardiac rejection CCR1-/- recipients mice completely ablated chronic rejection.[281] These studies suggest CC chemokines are important during chronic rejection.

Unfortunately, to date there are no studies that demonstrate whether specific Receptors/CC chemokines polymorphisms are associated with lung allograft dysfunction. However, there are studies in humans of chemokine receptor polymorphisms that impact on renal allograft survival.[297,298] CCR5Δ32 is a nonfunctional mutant allele of CCR5[299-301] and in a multicenter study was found to be associated with prolonged renal allograft survival.[297] These studies suggest CC chemokines may be a potential target for inflammatory/fibroproliferative disorders such as transplantation allograft dysfunction.

CCL2 is known to be produced de novo by human islets and its expression is increased in the presence of TNF-α, IL-1β.[302-304] CCR2 is the receptor for CCL2, CCL7, CCL13 and CCL12, which are potent chemoattractants for mononuclear cells and maybe important for the

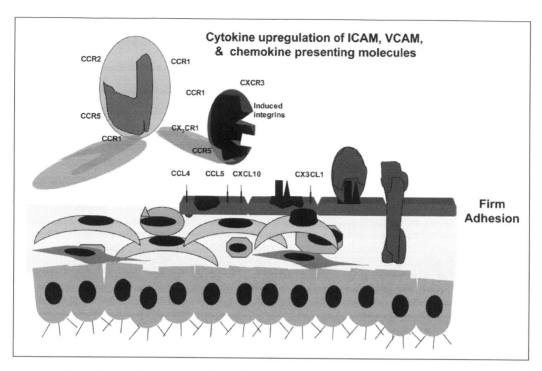

Figure 3. Involvement of chemokine activation during lung allograft dysfunction. Chemokines are presented to rolling leukocytes by stimulated endothelial cells. These chemokines activate leukocytes, upregulating integrins allowing adherence to endothelium expressing adhesion molecules. $CX_3CL1$ by itself can cause leukocyte firm adhesion.

promotion of a Th2 response.[302,304-306] Using a rat model of warm IRI, neutralization of CCL2 led to a reduction in lung injury.[266]

Abdi and colleges using a fully mismatched heterotopic islet cell allograft rejection model (pancreatic islet allografts placed under the kidney capsules of recipient mice rendered diabetic by streptozotocin) found increased expression of CCR2 and its ligands CCL2 and CCL12 during the rejection process.[307] When islet cells were transplanted into CCR2-/- recipients there was a prolonged allograft median survival from 12 days (wild type recipients) to 24 days (CCR2-/- recipients). Furthermore, 25% of the CCR2-/- recipients exhibited long-term graft survival. CCR2-/- recipients had decreased IFN-γ and increased IL-4 producing splenocytes when exposed to donor cells in vitro. In addition, recovered allografts from CCR2-/- recipients expressing increase IL-4, IL-5, IL-10 and decreased levels of IFN-γ and IL-12). Together, these results demonstrate that the CCR2-/- recipients were driving a Th2 alloresponse while attenuating allograft rejection. Similarly, Lee and associates using the same model system demonstrated decreased islet cell allograft rejection in recipient CCR2-/- mice. Mechanistically, they found that the interaction of CCL2/CCR2 contributed to alloreactive T-cell clonal expansion/proliferation and differentiation. In addition, prolonged islet allograft survival was achieved by blockade of the CCL2/CCR2 pathway in conjunction with low dose rapamycin therapy via a sown-regulation of the programmed death-1 (PD-1: PD-1/PD-L1) pathway.[302] Lastly, using a murine model of cardiac allograft rejection, CCR2-/- recipient led to a survival benefit from 8 days (wild type) recipients to 12 days (CCR2-/-) recipients.[307] Collectively, these studies demonstrate that the abrogation of CCR2/ligand biology has different effects on allograft dysfunction depending on the type of solid organ transplanted.

With regard to lung allograft dysfunction, CCL2 levels in BALF from patients with acute rejection and BOS were markedly elevated, biologically active and localized to airway epithelium and mononuclear cells.[271] This suggests CCL2 is important in the continuum from human acute to chronic allograft rejection by causing persistent accumulation of peribronchiolar leukocytes. Translational studies using a murine

model of BOS was consistent with human data demonstrating CCL2 localizing to airway columnar epithelium and infiltrating mononuclear cells. Furthermore, CCL2 levels paralleled the recruitment of mononuclear cells and cellular expression of its receptor, CCR2. A genetic approach was used to determine the effects of inhibiting CCR2 biology on BOS. Allografts from CCR2-/- mice demonstrated significant reductions in mononuclear phagocytes that were not accompanied by significant reductions in lymphocytes. Histopathological assessment of these allografts demonstrated significantly less matrix deposition, airway obliteration and epithelial injury. This suggests that a phenotypically distinct mononuclear phagocyte expressing CCR2 (i.e., producing more IL-13, TGF-β1 and PDGF) is pivotal during the pathogenesis of BOS (Fig. 4). These results were corroborated in a rat heterotopic subcutaneous model of BOS.[296]

CCL17 and CCL22 bind to CCR4, a highly expressed receptor on Type 2 cells and used for homing memory T-cells.[308-310] This receptor has also been identified on monocytes, Langerhans DC cells, monocytes, NK cells and platelets. While no studies have been performed with regard to lung allograft dysfunction, Alferink and colleagues demonstrated a significant prolongation of cardiac allograft survival when fully mismatch hearts were transplanted into immunosuppressed CCL17-/- recipients.[311] Follow up studies by Huser and associates using CCR4-/- recipient mice demonstrated prolonged cardiac survival and more profound effects were found with the addition of gallium nitrate.[312] Mechanistically, they demonstrated a reduction in allograft infiltrating CD4 cells and NK cells. Collectively these studies suggest that inhibition of CCR4/ligand biology increases cardiac allograft survival presumably by attenuating the interaction of DC-T-cells in secondary lymph nodes and the inhibition of recruited CCR4 expressing monocytes and NK cells.

One of the best inducers of rodent tolerance is the inhibition of CD40-CD154 interactions in conjunction with donor specific transfusion.[313-316] In fact, treatment with (anti-CD154 with donor specific transfusion) leads to well functioning allografts despite intragraft infiltrating mononuclear cells. This paucity of allograft injury suggests an active process of regulating infiltrating mononuclear

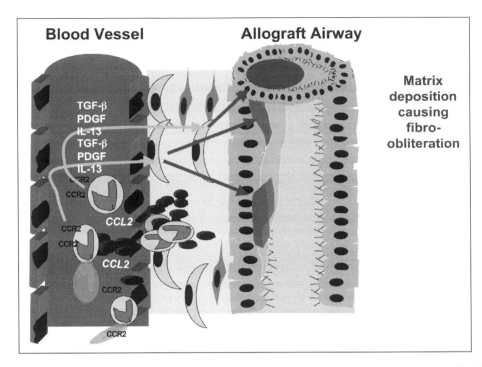

Figure 4. CCR2/CCL2 during lung allograft dysfunction. CCL2 is critical in maintaining persistent phenotypically distinct mononuclear phagocytes expressing CCR2, which eventually secrete proinflammatory/immunoregulatory cytokines/growth factors causing fibroplasia during lung allograft dysfunction.

cells within the allograft (i.e., Treg cells). This is consistent with data demonstrating the presence of Foxp3+ Treg cells in tolerized skin allografts after CD4 Ab blockade. Foxp3 is a gene that encodes a forkhead/winged helix transcription factor Scurfin, which is required for Treg cell development and thus is a marker of Treg cells.[317,318] Based on this data Lee and associates evaluated the presence of Treg cells (allograft Foxp3 expression) in tolerized (anti-CD154 with donor specific transfusion) allografts, as compared to untreated allografts, as compared to isografts and naïve hearts. They found a hierarchy of Foxp3 and CCR4 expression (anti-CD154 with donor specific transfusion >>> untreated allografts >> isografts > naïve hearts). Importantly, cardiac allografts from CCR4-/- recipients treated with (anti-CD154 with donor specific transfusion therapy) had significant reductions in Foxp3 expression and rejected their cardiac grafts at a normal rate. This suggests, CCR4, in this model system is critical for Treg cell recruitment. Overall the above studies suggest CCR4/ligand biology has multiple functions during allograft rejection (important in DC-T cell attraction for T-cell priming, recruits injurious CD4 and NK cells, yet can recruit protective Treg cells). It will be important to determine the specific detrimental and protective mechanism of CCR4/ligand biology during the pathogenesis of lung allograft dysfunction.

CCR7 and its ligands CCL19 and CCL21 play a role, in part, in the localization of antigen loaded DC and antigen specific T-cells within T-cell rich zones, which is critical for the initiation of an adaptive immune response.[311,319-321] CCR7 has been demonstrated on naïve and memory T-lymphocytes, on a subset of activated T-cells within secondary lymphoid organs, on mature DC and on mature B-cells.[322] Hopken and associates dissected the role of CCR7 in T-cell activation and differentiation in the process of allograft rejection, using two different model systems.[323] They adapted a class I mismatch animal model of acute allogeneic tumor rejection to address whether CCR7 expression was required for the initiation of an acute CD8+ T-cell mediated alloantigen response and confirmed their finding using a cardiac vascularized model of organ transplantation rejection. CCR7-/- recipients completely failed to reject tumor

allografts and demonstrated minimal leukocyte infiltration and CTL response. In addition, using the CCR7-/- they demonstrated the ligand/CCR7 interactions mediate, in part, the migration of DC from inflamed/tumor sites into draining lymphoid organs.[323] Using adoptive transfer experiments of alloantigen stimulated T-lymphocytes they also demonstrated that CCR7-/- mice fail to reject the allograft because naïve CTL precursor cells do not develop into functional circulating effector cells in vivo. These data suggest in vivo priming of naïve T-lymphocytes is completely abolished in CCR7-/- mice indicating the importance of CCR7 expression for efficient T-cell homing and T-cell priming to and within secondary lymph organs. They then investigated these finding in a fully MCH class I and II mismatched vascularized heterotopic cardiac allograft model. Specifically, CCR7-/- BALB/c donor to CCR7-/- C57BL/6 recipients prolonged survival to 14 days and the addition of low dose CsA prolonged survival to 16 days. Overall, they demonstrated that CCR7 is a key regulator of DC/T-cell interaction within secondary lymphoid tissue/organs for both initiation of an antigen specific T-cell response and for the initiation of an alloantigen induced cytotoxic T-cell response.

Studies using immature DC subsets have shown promise in vitro with regard to inducing tolerance.[324-327] Thus the concept that T-cell-DC interactions within secondary lymphoid organs seems to be critical for the induction of both immunity (T-cell-mature DC interactions) and tolerance (T-cell-immature DC interactions).[328,329] However, while mature DC express CCL7, immature DC do not and thus immature DC traffic poorly to secondary (CCL19 expressing) lymph organs.[330,331] Garrod and associates used a novel retroviral transduction of bone marrow precursors DC prior to their differentiation into immature DC. This technique allowed them to engineer the expression of CCR7 and vIL-10 from immature DCs. Using a murine cardiac allograft model they found that the infusion of these immature DC (expressing CCR7 and vIL-10) markedly prolonged cardiac allograft survival to > 100 days. This study suggests that the genetic manipulation of immature DC (with CCR7 and vIL-10) increase homing (via CCR7) to lymphoid organs and increases interaction of immature DC expressing vIL-10 and T-cells ultimately decreasing allospecific reactivity and possibly, the induction of tolerogenic Treg cells. More studies will be required to determine the role of CCR7 and lung allograft dysfunction.

## The Role of CX₃CL1/CX₃CR1 in the Pathogenesis of Lung Allograft Dysfunction

CX₃CL1 is involved in direct leukocyte activation, chemotaxis and adhesion through its interaction with CX₃CR1.[245-248,332-336] Although there are currently no studies of CX₃CL1 and lung rejection others have found it to be important during murine cardiac rejection. In these studies CX₃CL1 localized to endothelium, epicardium, endocardium, myocardium and infiltrating mononuclear cells in allografts.[245,337] In vivo antibody depletion of CX₃CL1 increased allograft survival without gross disruption in graft leukocyte infiltration.[245] Interestingly, when CX₃CR1-/- recipients were used, there was no effect on allograft survival, even though there was a reduction in graft infiltrating NK cells.[337] The addition of CsA to the CX₃CR1-/- mice, prolonged graft survival by 18 days. Together these studies suggest that NK cells of the innate immune system and T-cells of the adaptive immune system act synergistically to cause rejection and CX₃CL1 (both soluble and tethered from recipient/donor cells)/CX₃CR1 interaction is important for the recruitment of NK cells to the allograft. Future studies involving CX₃CL1/CX₃CR1 interaction may demonstrate import mechanisms involving in lung allograft dysfunction.

## The Role of Receptor/CXC Chemokines in the Pathogenesis of Lung Allograft Dysfunction

Interferon inducible ELR- CXC chemokines are important in the pathogenesis of rejection due to their chemoattraction to activated T-cells.[338-343] These chemokines (CXCL9, CXCL10 and CXCL11) bind and activate through their shared receptor CXCR3.[344-346]

The recently highlighted role of T-cells in IRI raises a question as to how T-lymphocytes are recruited to the newly transplanted organ. Zhai and associates dissected the mechanism of innate immune-induced T-cell recruitment and activation in a rat syngeneic orthotopic liver transplantation (OLT) cold IRI model.[347] They fund increased expression of CXCL9, CXCL10 and CXCL11 during IRI. Importantly, in vivo neutralization of CXCR3 led to a reduction in recruited CXCR3 + CD4 T-cells and ameliorated hepatocellular damage and improved liver allograft survival.

An association between ligand/CXCR3 has been demonstrated in both human and animal cardiac rejection.[348,349,350-352] In fact, using CXCR3-/- mice as recipients of heterotopic hearts prolonged graft survival by 2-months and the addition of CsA allowed for permanent engraftment.[351,352] Furthermore, in vivo neutralization of CXCL10 doubled cardiac allograft survival. However, when CXCL10-/- mice were the recipients of allografts there was no prolongation of graft survival. When the converse experiments were performed (CXCL10-/- hearts used as donors) there was prolongation of graft survival to greater than 40-days.[352] Therefore, the anti-CXCL10 antibody prolonged allograft survival by neutralizing CXCL10 produced by the allograft endothelial cells, leukocytes and other stromal cells. In another cardiac rejection model increased expression of CXCL9 was 5-fold greater than CXCL10.[353] Depletion of CXCL10 prolonged allograft survival by 3 days, however neutralization of CXCL9 increased survival by 11 days. Increased expression of CXCL9 was also seen in a murine class II mismatch skin allograft model.[354] Neutralization of CXCL9 prolonged graft survival by reducing intragraft CD4 T-cells and macrophages. They extended these studies and demonstrated that allografts from IFN-γ-/- recipient mice survived indefinitely, had limited graft production of CXCL9 and minimal graft leukocyte infiltration. Intradermal injections of recombinant CXCL9 directly into the skin allografts restored the ability of the IFN-γ-/- recipients to recruit

leukocytes and reject grafts. This suggests IFN-γ is not only required for the up-regulation of class II MHC expression, but is also required to induce the production of intragraft CXCL9, which is critical for the recruitment of primed/alloreactive lymphocytes to the allografts which leads to rejection. Similar results were found in a heterotopic murine model of pancreatic islet allograft rejection, were murine A/J islet cells were transplanted into H-2 disparate C57BL/6 recipients with streptozotocin induced diabetic. There was prolonged allograft survival when CXCL10 was neutralized or CXCR3-/- recipients were used. Mechanistically they found a reduction in gene expression for mononuclear cells.[355]

Elevated levels of CXCL10, CXCL9, CXCL11 and IFN-γ were found in BALF from rejecting human lung allografts.[112,356,357] Lymphocytes from these BALF expressed high levels of CXCR3 and IFN-γ, while supernatants from cultured alveolar macrophages were chemotactic for CXCR3 positive cells. Using biopsies from rejecting lungs CXCL10 localized to infiltrating macrophages and epithelial cells.[356] This suggests these interferon inducible ELR- CXC chemokines are produced by lung macrophages/epithelial cells which recruits CXCR3 and IFN-γ expressing T-cells, having paracrine effects on recruited macrophages to express more interferon inducible ELR- CXC chemokines augmenting a feedback loop and perpetuating rejection (Fig. 5).

Importantly, human studies of lung allograft dysfunction have demonstrated that the levels of CXCL9 and CXCL10 were a magnitude greater than CXCL11 and were better predictors of rejection.[357] Translational experiments using a fully mismatched rat orthotopic single left lung transplantation model were performed to further evaluate the role of these chemokines during acute lung allograft rejection. Both CXCL9 and CXCL10 expression paralleled the inflammatory process and CXCR3 expression during acute lung allograft rejection. However, whole lung allograft homogenates demonstrated CXCL9 levels were more than 15-fold greater than CXCL10 at the time of maximal allograft rejection. Anti-CXCL9 depletion reduced intragraft mononuclear cells (i.e., CD3, CD4, CD8, NK cells, B-cells and mononuclear phagocytes) and intragraft CXCR3 cellular expression. This was accompanied by a decrease in lung allograft rejection out to 10 days and the addition of low dose CsA attenuated the rejection process out to 20 days. Taken together, these studies demonstrate the importance of ligand/CXCR3 interactions during acute allograft rejection.

Elevated levels of human plasma CXCL11 were associated with severe transplantation coronary artery disease (TCAD) as compared to long-term survivors of cardiac transplantation without TCAD.[350] Immunohistochemical localization confirmed the presence of infiltrating CXCR3 expressing mononuclear cells within TCAD lesion and the presence of CXCL11, which was found on the surface of endothelial cells within rejected areas of the allograft.[350] Similarly, elevated levels of interferon inducible ELR- CXC chemokines CXCL9, CXCL10 and CXCL11 in human BALF were found to be associated with the continuum from acute to chronic rejection.[356,357] Translational studies using a murine model of BOS demonstrated increased expression of interferon-inducible ELR- CXC chemokines paralleling the recruitment of CXCR3 expressing mononuclear cells. In vivo neutralization of CXCR3 or its ligands CXCL9 and CXCL10 decreased intragraft recruitment of CXCR3 expressing mononuclear cells and attenuated BOS.[357] Similar results were seen using murine models of cardiac allograft rejection.[351-354,358-360] However, Medoff and associates found that the deletion of either CXCL9 or CXCL10 alone did not effect T-cell recruitment into the airway allograft during murine BOS.[361] However, the deletion of CXCR3 did reduce mononuclear cell recruitment and fibro-obliteration. The difference between the above studies with regard to inhibition of either CXCL9 or CXL10 interaction with CXCR3 individually and mononuclear cell recruitment likely has to do with limitation in each study. In the former study, there is the limitation

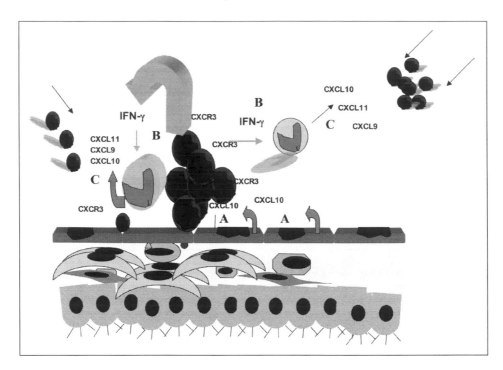

Figure 5. Involvement of ELR- CXC chemokines during lung allograft dysfunction. A) Interferon-inducible ELR- CXC chemokines expressed from endothelial and mononuclear cells recruit alloantigen primed T-cells. B) These primed T-cells express IFN-γ and CXCR3. C) IFN-γ stimulates macrophages to secrete more interferon-inducible ELR-CXC chemokines creating a destructive cycle of alloreactive T-cells into the allograft.

of using passive immunization with polyclonal antibodies and in the later study, there is the limitation of using nonconditional knockouts of CXCL9 and CXCL10 (i.e., allowing these mice to develop, over time, compensatory mechanisms and pathways to overcome their gene deletion). Only a study using conditional knockouts of CXCL9, CXCL10 and CXCL11 in both recipients and donors will truly answer the important question of whether inhibition of one CXCR3 ligand is enough to attenuate lung allograft dysfunction.

There is now evidence to suggest that Treg cells are important in the maintenance of immune tolerance by limiting the responses of effector CD4 and CD8 T-cells. However there are multiple subsets of Treg cells that have been identified in both animal models and humans including CD4+ CD25+ Treg cells, IL-10 producing type I Treg cells, CD4+ CD103+ T-cells, CD4+ CD25− Treg cells, CD8+ CD28− T-cells, γδ T-cells, NK T-cells, αβTCR+ CD3+ NK1.1− CD4−CD8− double negative (DN) T-cells.[7-18] One particular sub-population of Treg cells, specifically the DN Tregs have been shown to down regulate immune responses.[362-365] In addition, pretransplantation donor lymphocyte infusion (DLI) have been shown to activate recipient DN Treg cells and induce long-term survival in single class I or class II mismatch skin allografts.[363,365-367] Furthermore, DN Treg cell clones from DLI treated tolerant mice were shown to prevent the rejection of donor specific skin and cardiac allografts in the absence of exogenous immunosuppression.[367-369] However, from these DN Treg clones arose spontaneous mutants that lost suppressive capacity.[368,369] Interestingly, the chemokine receptor CXCR5 was highly expressed in the DN Treg clones compared with their natural mutants.[368] CXCR5 was originally identified from Burkitt's lymphoma as a G protein coupled receptor and found in B-cells and subsets of CD4+ cells.[370-376] CXCL13, thus far has been determined to be CXCR5 exclusive ligand.[374] Lee and associates demonstrated that both DN Treg clones and primary DN Treg cells express a high level of CXCR5 protein on their surface and the DN Treg cells, (*but not CD4 or CD8 T-cells*) and can use this receptor to migrate to CXCL13 in vitro.[377] Both in vivo and in vivo neutralizing studies of CXCL13 demonstrated that cell migration could be inhibited. Specifically, in vivo neutralization

of CXCL13 led to a 9 fold reduction of graft infiltrating DN Treg cells while homing into the spleen and lymph nodes appeared to be unaffected. Moreover, CXCR5+ Treg clones, but not their mutants were able to migrate to and accumulate in cardiac allografts and prevented rejection. Overall, this study demonstrates the CXCR5/CXCL13 biological axis plays an important role in DN Treg cell homing into cardiac allografts, but not secondary lymphoid organs suggesting other chemokine receptors are important.

ELR+ CXC chemokines and their ability to recruit neutrophils during murine cardiac rejection has been examined.[378] Increased expression of CXCL1/2/3 was found early in both allo- and iso-grafts, suggesting its involvement during IRI. Neutralization of CXCL1/2/3 prolonged allograft survival by 15 days, decreased endothelial degeneration and reduced leukocyte infiltration. Similarly, neutrophil depletion using anti-Ly6G antibodies extended graft survival by 13 days.

With regard to lung transplantation proof of concept studies using a rat orthotropic lung transplantation model of "cold" IRI demonstrated an increase in lung graft neutrophil sequestration and injury. In addition, lung expression of CXCL1, CXCL2/3 and their shared receptor CXCR2 paralleled lung neutrophil infiltration and injury. Importantly, inhibition of CXCR2/CXCR2 ligand interactions in vivo led to a marked reduction in lung neutrophil sequestration and graft injury. Taken together these experiments support the notion that increased expression of ELR+ CXC chemokines and their interaction with CXCR2 plays an important role in the pathogenesis of postlung transplantation cold IRI. Importantly, translational studies in human lung transplantation has demonstrated increased levels of CXCL8 associated with human IRI, while high levels of IL-8 in human donor lungs were associated with poor graft function and early mortality posttransplantation.[379,380] These studies were corroborated by a recent study demonstrating that elevated levels of multiple ELR+CXC chemokines in human bronchoalveolar lavage fluid are associated with IRI. Together these studies suggest an important role for the neutrophil, through their recruitment (via CXCR2/ligand interaction) into the allograft and their interactions with local cells to initiate downstream events involved in the immune responses during allograft dysfunction.

Concentrating on airway inflammation multiple studies have demonstrated increased CXCL8 levels in BALF from patients with BOS that correlated with airway wall neutrophilia.[289-294] CXCL8 localized to actin positive smooth muscle cells of BOS lesions.[294] However, the ELR+ CXC chemokines not only mediate neutrophil recruitment, but can also promote angiogenesis. Their shared endothelial cell receptor is CXCR2. Elevated levels of multiple ELR+ CXC chemokines correlated with the presence of BOS. Proof-of-concept studies using a murine model of BOS not only demonstrated an early neutrophil infiltration but also marked vascular remodeling in the tracheal allografts.[381] In addition, tracheal allograft ELR+ CXC chemokines were persistently expressed even in the absence of significant neutrophil infiltration and were temporally associated with vascular remodeling during fibro-obliteration of the tracheal allograft. Furthermore, treatment with anti-CXCR2 Abs inhibited early neutrophil infiltration and later vascular remodeling, which resulted in the attenuation of murine BOS. A more profound attenuation of fibro-obliteration was seen when CXCR2-/- mice received cyclosporin A. This supports the notion that the CXCR2/CXCR2 ligand biological axis has a bimodal function during the course of BOS: early, it is important for neutrophil recruitment and later, during fibro-obliteration, it is important for vascular remodeling independent of neutrophil recruitment.

## Conclusion

In summary, the above human studies have demonstrated the importance of receptors/cytokines in transplantation allograft dysfunction. Furthermore, translational studies in animal models of allograft dysfunction have demonstrated proof of principle that receptors/cytokines interactions play a pivotal role in mediating the leukocyte infiltration that leads to the continuum of early to late allograft dysfunction. The future studies of other receptors/cytokines will lead to the development of new paradigms to understand the pathogenesis of lung allograft dysfunction. Furthermore, they should pave the way for the development of pharmaceutical agents that will target cytokine biology and provide new treatments that will ultimately enhance long-term allograft survival.

## References

1. Trulock EP. Lung transplantation. Am J Respir Crit Care Med 1997; 155(3):789-818.
2. Arcasoy SM, Kotloff RM. Lung transplantation. N Engl J Med 1999; 340(14):1081-91.
3. Kirk AJ, Colquhoun IW, Dark JH. Lung preservation: a review of current practice and future directions. Ann Thorac Surg 1993; 56(4):990-1100.
4. King RC, Binns OA, Rodriguez F et al. Reperfusion injury significantly impacts clinical outcome after pulmonary transplantation. Ann Thorac Surg 2000; 69(6):1681-85.
5. Novick RJ, Gehman KE, Ali IS et al. Lung preservation: the importance of endothelial and alveolar type II cell integrity. Ann Thorac Surg 1996; 62(1):302-14.
6. Anderson DC, Glazer HS, Semenkovich JW et al. Lung transplant edema: chest radiography after lung transplantation—the first 10 days. Radiology 1995; 195(1):275-81.
7. Fiser SM, Kron IL, McLendon Long S et al. Early intervention after severe oxygenation index elevation improves survival following lung transplantation. J Heart Lung Transplant 2001; 20(6):631-36.
8. Meyers BF, Sundt TM, 3rd, Henry S et al. Selective use of extracorporeal membrane oxygenation is warranted after lung transplantation. J Thorac Cardiovasc Surg 2000; 120(1):20-26.
9. McCord JM. Oxygen-derived radicals: a link between reperfusion injury and inflammation. Fed Proc 1987; 46(7):2402-6.
10. Friedl HP, Smith DJ, Till GO et al. Ischemia-reperfusion in humans. Appearance of xanthine oxidase activity. Am J Pathol 1990; 136(3):491-95.
11. Welbourn CR, Goldman G, Paterson IS et al. Pathophysiology of ischaemia reperfusion injury: central role of the neutrophil. Br J Surg 1991; 78(6):651-55.
12. Engler RL, Dahlgren MD, Morris DD et al. Role of leukocytes in response to acute myocardial ischemia and reflow in dogs. Am J Physiol 1986; 251(2 Pt 2):H314-23.
13. Eppinger MJ, Jones ML, Deeb GM et al. Pattern of injury and the role of neutrophils in reperfusion injury of rat lung. J Surg Res 1995; 58(6):713-18.
14. Eppinger MJ, Deeb GM, Bolling SF et al. Mediators of ischemia-reperfusion injury of rat lung. Am J Pathol 1997; 150(5):1773-84.
15. Zwacka RM, Zhang Y, Halldorson J et al. CD4(+) T-lymphocytes mediate ischemia/reperfusion-induced inflammatory responses in mouse liver. J Clin Invest 1997; 100(2):279-89.
16. Sommers KE, Griffith BP, Hardesty RL et al. Early lung allograft function in twin recipients from the same donor: risk factor analysis. Ann Thorac Surg 1996; 62(3):784-90.
17. Steimle CN, Guynn TP, Morganroth ML et al. Neutrophils are not necessary for ischemia-reperfusion lung injury. Ann Thorac Surg 1992; 53(1):64-72; discussion -3.
18. Lu YT, Hellewell PG, Evans TW. Ischemia-reperfusion lung injury: contribution of ischemia, neutrophils and hydrostatic pressure. Am J Physiol 1997; 273(1 Pt 1):L46-54.
19. Fiser SM, Tribble CG, Long SM et al. Lung transplant reperfusion injury involves pulmonary macrophages and circulating leukocytes in a biphasic response. J Thorac Cardiovasc Surg 2001; 121(6):1069-75.
20. Girgis RE, Tu I, Berry GJ et al. Risk factors for the development of obliterative bronchiolitis after lung transplantation. J Heart Lung Transplant 1996; 15(12):1200-8.
21. Bando K, Paradis IL, Similo S et al. Obliterative bronchiolitis after lung and heart-lung transplantation. An analysis of risk factors and management. J Thorac Cardiovasc Surg 1995; 110(1):4-13; discussion-4.
22. Trulock EP. Management of lung transplant rejection. Chest 1993; 103(5):1566-76.
23. Kelly K, Hertz MI. Obliterative bronchiolitis. Clin Chest Med 1997; 18(2):319-38.
24. Paradis I, Yousem S, Griffith B. Airway obstruction and bronchiolitis obliterans after lung transplantation. Clinics in Chest Medicine 1993; 14(4):751-63.
25. Heng D, Sharples LD, McNeil K et al. Bronchiolitis obliterans syndrome: incidence, natural history, prognosis and risk factors. J Heart Lung Transplant 1998; 17(12):1255-63.
26. Valentine VG, Robbins RC, Berry GJ et al. Actuarial survival of heart-lung and bilateral sequential lung transplant recipients with obliterative bronchiolitis. J Heart Lung Transplant 1996; 15(4):371-83.
27. Milne DS, Gascoigne AD, Ashcroft T et al. Organizing pneumonia following pulmonary transplantation and the development of obliterative bronchiolitis. Transplantation 1994; 57(12):1757-62.
28. Reichenspurner H, Girgis RE, Robbins RC et al. Obliterative bronchiolitis after lung and heart-lung transplantation. Ann Thorac Surg 1995; 60(6):1845-53.
29. Yousem SA, Martin T, Paradis IL et al. Can immunohistological analysis of transbronchial biopsy specimens predict responder status in early acute rejection of lung allografts? Hum Pathol 1994; 25(5):525-29.
30. Estenne M, Maurer JR, Boehler A et al. Bronchiolitis obliterans syndrome 2001: an update of the diagnostic criteria. J Heart Lung Transplant 2002; 21(3):297-310.
31. Sharples LD, McNeil K, Stewart S et al. Risk factors for bronchiolitis obliterans: a systematic review of recent publications. J Heart Lung Transplant 2002;21(2):271-81.
32. Gordon S. Alternative activation of macrophages. Nat Rev Immunol 2003; 3(1):23-35.
33. Zingoni A, Sornasse T, Cocks BG et al. NK cell regulation of T-cell-mediated responses. Mol Immunol 2005; 42(4):451-54.
34. Mosmann TR, Sad S. The expanding universe of T-cell subsets: Th1, Th2 and more. Immunol Today 1996; 17(3):138-46.
35. Mantovani A, Allavena P, Sica A. Tumour-associated macrophages as a prototypic type II polarised phagocyte population: role in tumour progression. Eur J Cancer 2004; 40(11):1660-7.
36. Zhai Y, Kupiec-Weglinski JW. What is the role of regulatory T-cells in transplantation tolerance? Curr Opin Immunol 1999; 11(5):497-503.
37. Mosmann TR, Cherwinski H, Bond MW et al. Two types of murine helper T-cell clone. I. Definition according to profiles of lymphokine activities and secreted proteins. J Immunol 1986; 136(7):2348-57.
38. VanBuskirk AM, Wakely ME, Orosz CG. Transfusion of polarized TH2-like cell populations into SCID mouse cardiac allograft recipients results in acute allograft rejection. Transplantation 1996; 62(2):229-38.
39. Piccotti JR, Li K, Chan SY et al. Cytokine regulation of chronic cardiac allograft rejection: evidence against a role for Th1 in the disease process. Transplantation 1999; 67(12):1548-55.

40. Piccotti JR, Li K, Chan SY et al. Alloantigen-reactive Th1 development in IL-12-deficient mice. J Immunol 1998; 160(3):1132-38.

41. Nocera A, Tagliamacco A, De Palma R et al. Cytokine mRNA expression in chronically rejected human renal allografts. Clin Transplant 2004; 18(5):564-70.

42. Beauregard C, Stevens C, Mayhew E et al. Cutting edge: atopy promotes Th2 responses to alloantigens and increases the incidence and tempo of corneal allograft rejection. J Immunol 2005; 174(11):6577-81.

43. Piccotti JR, Chan SY, VanBuskirk AM et al. Are Th2 helper T-lymphocytes beneficial, deleterious, or irrelevant in promoting allograft survival? Transplantation 1997; 63(5):619-24.

44. Steiger JU, Nickerson PW, Hermle M et al. Interferon-gamma receptor signaling is not required in the effector phase of the alloimmune response. Transplantation 1998; 65(12):1649-52.

45. Alexander J, Bryson K. T helper (h)1/Th2 and Leishmania: paradox rather than paradigm. Immunol Lett 2005; 99(1):17-23.

46. Koshiba T, Giulietti A, Van Damme B et al. Paradoxical early upregulation of intragraft Th1 cytokines is associated with graft acceptance following donor-specific blood transfusion. Transpl Int 2003; 16(3):179-85.

47. Mhoyan A, Wu GD, Kakoulidis TP et al. Predominant expression of the Th2 response in chronic cardiac allograft rejection. Transpl Int 2003; 16(8):464-73.

48. Zhai Y, Ghobrial RM, Busuttil RW et al. Th1 and Th2 cytokines in organ transplantation: paradigm lost? Crit Rev Immunol 1999;19(2):155-72.

49. Shirwan H. Chronic allograft rejection. Do the Th2 cells preferentially induced by indirect alloantigen recognition play a dominant role? Transplantation 1999; 68(6):715-26.

50. Kunzendorf U, Tran TH, Bulfone-Paus S. The Th1-Th2 paradigm in 1998: law of nature or rule with exceptions. Nephrol Dial Transplant 1998; 13(10):2445-48.

51. Ruth JH, Bienkowski M, Warmington KS et al. IL-1 receptor antagonist (IL-1ra) expression, function and cytokine- mediated regulation during mycobacterial and schistosomal antigen- elicited granuloma formation. J Immunol 1996; 156(7):2503-9.

52. Dinarello CA. Biologic basis for interleukin-1 in disease. Blood 1996; 87(6):2095-147.

53. Arend WP, Welgus HG, Thompson RC et al. Biological properties of recombinant human monocyte-derived interleukin 1 receptor antagonist. J Clin Invest 1990; 85(5):1694-97.

54. Wilborn J, Crofford LJ, Burdick MD et al. Cultured lung fibroblasts isolated from patients with idiopathic pulmonary fibrosis have a diminished capacity to synthesize prostaglandin E2 and to express cyclooxygenase-2. J Clin Invest 1995; 95(4):1861-68.

55. Naruse K, Shimizu K, Muramatsu M et al. Long-term inhibition of NO synthesis promotes atherosclerosis in the hypercholesterolemic rabbit thoracic aorta. PGH2 does not contribute to impaired endothelium-dependent relaxation [see comments]. Arterioscler Thromb 1994; 14(5):746-52.

56. Krishnadasan B, Naidu BV, Byrne K et al. The role of proinflammatory cytokines in lung ischemia-reperfusion injury. J Thorac Cardiovasc Surg 2003; 125(2):261-72.

57. Rizzo M, SivaSai KS, Smith MA et al. Increased expression of inflammatory cytokines and adhesion molecules by alveolar macrophages of human lung allograft recipients with acute rejection: decline with resolution of rejection. J Heart Lung Transplant 2000; 19(9):858-65.

58. Belperio JA, DiGiovine B, Keane MP et al. Interleukin-1 receptor antagonist as a biomarker for bronchiolitis obliterans syndrome in lung transplant recipients. Transplantation 2002; 73(4):591-99.

59. Vamvakopoulos J, Green C, Metcalfe S. Genetic control of IL-1beta bioactivity through differential regulation of the IL-1 receptor antagonist. Eur J Immunol 2002; 32(10):2988-96.

60. Vamvakopoulos JE, Taylor CJ, Green C et al. Interleukin 1 and chronic rejection: possible genetic links in human heart allografts. Am J Transplant 2002; 2(1):76-83.

61. Beutler BA. The role of tumor necrosis factor in health and disease. J Rheumatol 1999; 26 Suppl 57:16-21.

62. Ostensen ME, Thiele DL, Lipsky PE. Tumor necrosis factor-alpha enhances cytolytic activity of human natural killer cells. J Immunol 1987; 138(12):4185-91.

63. Piguet PF, Collart MA, Grau GE et al. Tumor necrosis factor/cachectin plays a key role in bleomycin-induced pneumopathy and fibrosis. J Exp Med 1989; 170(3):655-63.

64. He Y, Liu W. The preliminary research on the relationship between TNF- and egg-induced granuloma and hepatic fibrosis of schistosomiasis japonica. J Tongji Med Univ 1996; 16(4):205-8.

65. DeMeester SR, Rolfe MW, Kunkel SL et al. The bimodal expression of tumor necrosis factor-alpha in association with rat lung reimplantation and allograft rejection. J Immunol 1993; 150(6):2494-505.

66. Aris RM, Walsh S, Chalermskulrat W et al. Growth factor upregulation during obliterative bronchiolitis in the mouse model. Am J Respir Crit Care Med 2002; 166(3):417-22.

67. Smith CR, Jaramillo A, Lu KC et al. Prevention of obliterative airway disease in HLA-A2-transgenic tracheal allografts by neutralization of tumor necrosis factor. Transplantation 2001; 72(9):1512-18.

68. Hayashi M, Martinez OM, Garcia-Kennedy R et al. Expression of cytokines and immune mediators during chronic liver allograft rejection. Transplantation 1995; 60(12):1533-38.

69. Noronha IL, Eberlein-Gonska M, Hartley B et al. In situ expression of tumor necrosis factor-alpha, interferon-gamma and interleukin-2 receptors in renal allograft biopsies. Transplantation 1992; 54(6):1017-24.

70. Mytilineos J, Laux G, Opelz G. Relevance of IL10, TGFbeta1, TNFalpha and IL4Ralpha gene polymorphisms in kidney transplantation: a collaborative transplant study report. Am J Transplant 2004; 4(10):1684-90.

71. Langrish CL, McKenzie BS, Wilson NJ et al. IL-12 and IL-23: master regulators of innate and adaptive immunity. Immunol Rev 2004; 202:96-105.

72. Trinchieri G, Pflanz S, Kastelein RA. The IL-12 family of heterodimeric cytokines: new players in the regulation of T-cell responses. Immunity 2003; 19(5):641-44.

73. Kobayashi M, Fitz L, Ryan M et al. Identification and purification of natural killer cell stimulatory factor (NKSF), a cytokine with multiple biologic effects on human lymphocytes. J Exp Med 1989; 170(3):827-45.

74. Stern AS, Podlaski FJ, Hulmes JD et al. Purification to homogeneity and partial characterization of cytotoxic lymphocyte maturation factor from human B-lymphoblastoid cells. Proc Natl Acad Sci USA 1990; 87(17):6808-12.

75. Presky DH, Yang H, Minetti LJ et al. A functional interleukin 12 receptor complex is composed of two beta-type cytokine receptor subunits. Proc Natl Acad Sci USA 1996; 93(24):14002-7.

76. Presky DH, Minetti LJ, Gillessen S et al. Evidence for multiple sites of interaction between IL-12 and its receptor. Ann NY Acad Sci 1996; 795:390-93.

77. Gately MK, Renzetti LM, Magram J et al. The interleukin-12/interleukin-12-receptor system: role in normal and pathologic immune responses. Annu Rev Immunol 1998; 16:495-521.

78. Trinchieri G. Interleukin-12: a proinflammatory cytokine with immunoregulatory functions that bridge innate resistance and antigen-specific adaptive immunity. Annu Rev Immunol 1995; 13:251-76.

79. Seder RA, Gazzinelli R, Sher A et al. Interleukin 12 acts directly on CD4+ T-cells to enhance priming for interferon gamma production and diminishes interleukin 4 inhibition of such priming. Proc Natl Acad Sci USA 1993; 90(21):10188-92.

80. McKnight AJ, Zimmer GJ, Fogelman I et al. Effects of IL-12 on helper T-cell-dependent immune responses in vivo. J Immunol 1994; 152(5):2172-79.

81. Marshall JD, Secrist H, DeKruyff RH et al. IL-12 inhibits the production of IL-4 and IL-10 in allergen-specific human CD4+ T-lymphocytes. J Immunol 1995; 155(1):111-17.

82. Via CS, Rus V, Gately MK, Finkelman FD. IL-12 stimulates the development of acute graft-versus-host disease in mice that normally would develop chronic, autoimmune graft-versus-host disease. J Immunol 1994; 153(9):4040-47.

83. Hsieh CS, Macatonia SE, Tripp CS et al. Development of TH1 CD4+ T-cells through IL-12 produced by Listeria-induced macrophages. Science 1993; 260(5107):547-49.

84. Oswald IP, Caspar P, Jankovic D et al. IL-12 inhibits Th2 cytokine responses induced by eggs of Schistosoma mansoni. J Immunol 1994; 153(4):1707-13.

85. Pearlman E, Heinzel FP, Hazlett FE et al. IL-12 modulation of T helper responses to the filarial helminth, Brugia malayi. J Immunol 1995; 154(9):4658-64.

86. Gazzinelli RT, Giese NA, Morse HC, 3rd. In vivo treatment with interleukin 12 protects mice from immune abnormalities observed during murine acquired immunodeficiency syndrome (MAIDS). J Exp Med 1994; 180(6):2199-208.

87. Zhou P, Sieve MC, Bennett J et al. IL-12 prevents mortality in mice infected with Histoplasma capsulatum through induction of IFN-gamma. J Immunol 1995; 155(2):785-95.

88. Rosenberg AS, Singer A. Cellular basis of skin allograft rejection: an in vivo model of immune-mediated tissue destruction. Annu Rev Immunol 1992; 10:333-58.

89. Piccotti JR, Li K, Chan SY et al. Interleukin-12 (IL-12)-driven alloimmune responses in vitro and in vivo: requirement for beta1 subunit of the IL-12 receptor. Transplantation 1999; 67(11):1453-60.

90. Williamson E, Garside P, Bradley JA et al. Neutralizing IL-12 during induction of murine acute graft-versus-host disease polarizes the cytokine profile toward a Th2-type alloimmune response and confers long term protection from disease. J Immunol 1997; 159(3):1208-15.

91. Orr DJ, Bolton EM, Bradley JA. Neutralising IL-12 activity as a strategy for prolonging allograft survival and preventing graft-versus-host disease. Scott Med J 1998; 43(4):109-11.

92. Welniak LA, Blazar BR, Wiltrout RH et al. Role of interleukin-12 in acute graft-versus-host disease(1). Transplant Proc 2001; 33(1-2):1752-53.

93. Williamson E, Garside P, Bradley JA et al. IL-12 is a central mediator of acute graft-versus-host disease in mice. J Immunol 1996; 157(2):689-99.

94. Piccotti JR, Chan SY, Goodman RE et al. IL-12 antagonism induces T helper 2 responses, yet exacerbates cardiac allograft rejection. Evidence against a dominant protective role for T helper 2 cytokines in alloimmunity. J Immunol 1996; 157(5):1951-57.

95. Konieczny BT, Dai Z, Elwood ET et al. IFN-gamma is critical for long-term allograft survival induced by blocking the CD28 and CD40 ligand T-cell costimulation pathways. J Immunol 1998; 160(5):2059-64.

96. Tau GZ, von der Weid T, Lu B et al. Interferon gamma signaling alters the function of T helper type 1 cells. J Exp Med 2000; 192(7):977-86.

97. Mikols CL, Yan L, Norris JY et al. IL-12 p80 is an innate epithelial cell effector that mediates chronic allograft dysfunction. Am J Respir Crit Care Med 2006; 174(4):461-70.

98. Meloni F, Vitulo P, Cascina A et al. Bronchoalveolar lavage cytokine profile in a cohort of lung transplant recipients: a predictive role of interleukin-12 with respect to onset of bronchiolitis obliterans syndrome. J Heart Lung Transplant 2004; 23(9):1053-60.

99. Farrar MA, Schreiber RD. The molecular cell biology of interferon-gamma and its receptor. Annu Rev Immunol 1993; 11:571-611.

100. Thierfelder WE, van Deursen JM, Yamamoto K et al. Requirement for Stat4 in interleukin-12-mediated responses of natural killer and T-cells. Nature 1996; 382(6587):171-74.

101. Fiorentino DF, Zlotnik A, Mosmann TR et al. IL-10 inhibits cytokine production by activated macrophages. J Immunol 1991; 147(11):3815-22.

102. Hobart M, Ramassar V, Goes N et al. The induction of class I and II major histocompatibility complex by allogeneic stimulation is dependent on the transcription factor interferon regulatory factor 1 (IRF-1): observations in IRF-1 knockout mice. Transplantation 1996; 62(12):1895-901.

103. Hobart M, Ramassar V, Goes N et al. IFN regulatory factor-1 plays a central role in the regulation of the expression of class I and II MHC genes in vivo. J Immunol 1997; 158(9):4260-69.

104. Hassan AT, Dai Z, Konieczny BT et al. Regulation of alloantigen-mediated T-cell proliferation by endogenous interferon-gamma: implications for long-term allograft acceptance. Transplantation 1999; 68(1):124-29.

105. Hidalgo LG, Halloran PF. Role of IFN-gamma in allograft rejection. Crit Rev Immunol 2002; 22(4):317-49.

106. de Perrot M, Young K, Imai Y et al. Recipient T-cells mediate reperfusion injury after lung transplantation in the rat. J Immunol 2003; 171(10):4995-5002.

107. Day YJ, Huang L, Ye H et al. Renal ischemia-reperfusion injury and adenosine 2A receptor-mediated tissue protection: the role of CD4+ T-cells and IFN-gamma. J Immunol 2006; 176(5):3108-14.

108. Boehler A, Bai XH, Liu M et al. Upregulation of T-helper 1 cytokines and chemokine expression in posttransplant airway obliteration. Am J Respir Crit Care Med 1999; 159(6):1910-17.

109. Nagano H, Mitchell RN, Taylor MK et al. Interferon-gamma deficiency prevents coronary arteriosclerosis but not myocardial rejection in transplanted mouse hearts. J Clin Invest 1997; 100(3):550-57.

110. Raisanen-Sokolowski A, Mottram PL, Glysing-Jensen T et al. Heart transplants in interferon-gamma, interleukin 4 and interleukin 10 knockout mice. Recipient environment alters graft rejection. J Clin Invest 1997; 100(10):2449-56.

111. Raisanen-Sokolowski A, Glysing-Jensen T, Russell ME. Leukocyte-suppressing influences of interleukin (IL)-10 in cardiac allografts: insights from IL-10 knockout mice. Am J Pathol 1998; 153(5):1491-500.

112. Moudgil A, Bagga A, Toyoda M et al. Expression of gamma-IFN mRNA in bronchoalveolar lavage fluid correlates with early acute allograft rejection in lung transplant recipients. Clin Transplant 1999; 13(2):201-7.

113. Iacono A, Dauber J, Keenan R et al. Interleukin 6 and interferon-gamma gene expression in lung transplant recipients with refractory acute cellular rejection: implications for monitoring and inhibition by treatment with aerosolized cyclosporine. Transplantation 1997; 64(2):263-69.

114. Lu KC, Jaramillo A, Lecha RL et al. Interleukin-6 and interferon-gamma gene polymorphisms in the development of bronchiolitis obliterans syndrome after lung transplantation. Transplantation 2002; 74(9):1297-302.

115. Tambur AR, Pamboukian S, Costanzo MR et al. Genetic polymorphism in platelet-derived growth factor and vascular endothelial growth factor are significantly associated with cardiac allograft vasculopathy. J Heart Lung Transplant 2006; 25(6):690-98.

116. Holweg CT, Baan CC, Balk AH et al. The transforming growth factor-beta1 codon 10 gene polymorphism and accelerated graft vascular disease after clinical heart transplantation. Transplantation 2001; 71(10):1463-67.

117. Densem CG, Hutchinson IV, Yonan N et al. Influence of IFN-gamma polymorphism on the development of coronary vasculopathy after cardiac transplantation. Ann Thorac Surg 2004; 77(3):875-80.

118. de Waal Malefyt R, Abrams J, Bennett B et al. Interleukin 10(IL-10) inhibits cytokine synthesis by human monocytes: an autoregulatory role of IL-10 produced by monocytes. J Exp Med 1991; 174(5):1209-20.

119. de Waal Malefyt R, Haanen J, Spits H et al. Interleukin 10 (IL-10) and viral IL-10 strongly reduce antigen-specific human T-cell proliferation by diminishing the antigen-presenting capacity of monocytes via downregulation of class II major histocompatibility complex expression. J Exp Med 1991;174(4):915-24.

120. de Waal Malefyt R, Yssel H, Roncarolo MG et al. Interleukin-10. Curr Opin Immunol 1992; 4(3):314-20.

121. de Waal Malefyt R, Yssel H, de Vries JE. Direct effects of IL-10 on subsets of human CD4+ T-cell clones and resting T-cells. Specific inhibition of IL-2 production and proliferation. J Immunol 1993; 150(11):4754-65.

122. te Velde AA, de Waal Malefijt R, Huijbens RJ et al. IL-10 stimulates monocyte Fc gamma R surface expression and cytotoxic activity. Distinct regulation of antibody-dependent cellular cytotoxicity by IFN-gamma, IL-4 and IL-10. J Immunol 1992; 149(12):4048-52.

123. Fiorentino DF, Zlotnik A, Vieira P et al. IL-10 acts on the antigen-presenting cell to inhibit cytokine production by Th1 cells. J Immunol 1991; 146(10):3444-51.

124. Bogdan C, Vodovotz Y, Nathan C. Macrophage deactivation by interleukin 10. J Exp Med 1991; 174(6):1549-55.

125. Ralph P, Nakoinz I, Sampson-Johannes A et al. IL-10, T-lymphocyte inhibitor of human blood cell production of IL-1 and tumor necrosis factor. J Immunol 1992; 148(3):808-14.

126. Hsu DH, de Waal Malefyt R, Fiorentino DF et al. Expression of interleukin-10 activity by Epstein-Barr virus protein BCRF1. Science 1990; 250(4982):830-32.

127. Hsu DH, Moore KW, Spits H. Differential effects of IL-4 and IL-10 on IL-2-induced IFN-gamma synthesis and lymphokine-activated killer activity. Int Immunol 1992;4(5):563-69.

128. Chen WF, Zlotnik A. IL-10: a novel cytotoxic T-cell differentiation factor. J Immunol 1991; 147(2):528-34.

129. MacNeil IA, Suda T, Moore KW et al. IL-10, a novel growth cofactor for mature and immature T-cells. J Immunol 1990; 145(12):4167-73.

130. Hu S, Chao CC, Ehrlich LC et al. Inhibition of microglial cell RANTES production by IL-10 and TGF-beta. J Leukoc Biol 1999; 65(6):815-21.

131. Bejarano MT, de Waal Malefyt R, Abrams JS et al. Interleukin 10 inhibits allogeneic proliferative and cytotoxic T-cell responses generated in primary mixed lymphocyte cultures. Int Immunol 1992; 4(12):1389-97.

132. Bromberg JS. IL-10 immunosuppression in transplantation. Curr Opin Immunol 1995; 7(5):639-43.

133. Zuo XJ, Matsumura Y, Prehn J et al. Cytokine gene expression in rejecting and tolerant rat lung allograft models: analysis by RT-PCR. Transpl Immunol 1995; 3(2):151-61.

134. Maeda H, Takata M, Takahashi S et al. Adoptive transfer of a Th2-like cell line prolongs MHC class II antigen disparate skin allograft survival in the mouse. Int Immunol 1994; 6(6):855-62.

135. Zheng XX, Steele AW, Nickerson PW et al. Administration of noncytolytic IL-10/Fc in murine models of lipopolysaccharide-induced septic shock and allogeneic islet transplantation. J Immunol 1995; 154(10):5590-600.

136. Lee MS, Wogensen L, Shizuru J et al. Pancreatic islet production of murine interleukin-10 does not inhibit immune-mediated tissue destruction. J Clin Invest 1994; 93(3):1332-38.

137. de Perrot M, Fischer S, Liu M et al. Impact of human interleukin-10 on vector-induced inflammation and early graft function in rat lung transplantation. Am J Respir Cell Mol Biol 2003; 28(5):616-25.

138. Kozower BD, Kanaan SA, Tagawa T et al. Intramuscular gene transfer of interleukin-10 reduces neutrophil recruitment and ameliorates lung graft ischemia-reperfusion injury. Am J Transplant 2002; 2(9):837-42.

139. Martins S, de Perrot M, Imai Y et al. Transbronchial administration of adenoviral-mediated interleukin-10 gene to the donor improves function in a pig lung transplant model. Gene Ther 2004; 11(24):1786-96.

140. Qian S, Li W, Li Y et al. Systemic administration of cellular interleukin-10 can exacerbate cardiac allograft rejection in mice. Transplantation 1996; 62(12):1709-14.

141. Mulligan MS, Warner RL, McDuffie JE et al. Regulatory role of Th-2 cytokines, IL-10 and IL-4, in cardiac allograft rejection. Exp Mol Pathol 2000; 69(1):1-9.

142. Oshima K, Sen L, Cui G et al. Localized interleukin-10 gene transfer induces apoptosis of alloreactive T-cells via FAS/FASL pathway, improves function and prolongs survival of cardiac allograft. Transplantation 2002; 73(7):1019-26.

143. Qin L, Chavin KD, Ding Y et al. Retrovirus-mediated transfer of viral IL-10 gene prolongs murine cardiac allograft survival. J Immunol 1996; 156(6):2316-23.

144. Qin L, Ding Y, Pahud DR, Robson ND et al. Adenovirus-mediated gene transfer of viral interleukin-10 inhibits the immune response to both alloantigen and adenoviral antigen. Hum Gene Ther 1997; 8(11):1365-74.

145. Pierog J, Gazdhar A, Stammberger U et al. Synergistic effect of low dose cyclosporine A and human interleukin 10 overexpression on acute rejection in rat lung allotransplantation. Eur J Cardiothorac Surg 2005; 27(6):1030-35.

146. Naidu B, Krishnadasan B, Whyte RI et al. Regulatory role of IL-10 in experimental obliterative bronchiolitis in rats. Exp Mol Pathol 2002; 73(3):164-70.

147. Yonemitsu Y, Kitson C, Ferrari S et al. Efficient gene transfer to airway epithelium using recombinant Sendai virus. Nat Biotechnol 2000; 18(9):970-73.

148. Shoji F, Yonemitsu Y, Okano S et al. Airway-directed gene transfer of interleukin-10 using recombinant Sendai virus effectively prevents posttransplant fibrous airway obliteration in mice. Gene Ther 2003; 10(3):213-18.

149. Salgar SK, Yang D, Ruiz P et al. Viral interleukin-10-engineered autologous hematopoietic stem cell therapy: a novel gene therapy approach to prevent graft rejection. Hum Gene Ther 2004; 15(2):131-44.

150. Berman RM, Suzuki T, Tahara H et al. Systemic administration of cellular IL-10 induces an effective, specific and long-lived immune response against established tumors in mice. J Immunol 1996; 157(1):231-38.

151. Lowry RP, Konieczny B, Alexander D et al. Interleukin-10 eliminates anti-CD3 monoclonal antibody-induced mortality and prolongs heart allograft survival in inbred mice. Transplant Proc 1995; 27(1):392-94.

152. Sankaran D, Asderakis A, Ashraf S et al. Cytokine gene polymorphisms predict acute graft rejection following renal transplantation. Kidney Int 1999; 56(1):281-88.

153. Awad MR, Webber S, Boyle G et al. The effect of cytokine gene polymorphisms on pediatric heart allograft outcome. J Heart Lung Transplant 2001;20(6):625-30.

154. Hutchinson IV, Turner D, Sankaran D et al. Cytokine genotypes in allograft rejection: guidelines for immunosuppression. Transplant Proc 1998; 30(8):3991-92.

155. Mazariegos GV, Reyes J, Webber SA et al. Cytokine gene polymorphisms in children successfully withdrawn from immunosuppression after liver transplantation. Transplantation 2002; 73(8):1342-45.

156. Turner D, Grant SC, Yonan N et al. Cytokine gene polymorphism and heart transplant rejection. Transplantation 1997; 64(5):776-9.

157. Uboldi de Capei M, Dametto E, Fasano ME et al. Cytokines and chronic rejection: a study in kidney transplant long-term survivors. Transplantation 2004; 77(4):548-52.

158. Zheng HX, Burckart GJ, McCurry K et al. Interleukin-10 production genotype protects against acute persistent rejection after lung transplantation. J Heart Lung Transplant 2004; 23(5):541-46.

159. Grabstein KH, Eisenman J, Shanebeck K et al. Cloning of a T-cell growth factor that interacts with the beta chain of the interleukin-2 receptor. Science 1994; 264(5161):965-68.

160. Lodolce JP, Boone DL, Chai S et al. IL-15 receptor maintains lymphoid homeostasis by supporting lymphocyte homing and proliferation. Immunity 1998; 9(5):669-76.

161. Kennedy MK, Glaccum M, Brown SN et al. Reversible defects in natural killer and memory CD8 T cell lineages in interleukin 15-deficient mice. J Exp Med 2000; 191(5):771-80.

162. Doherty TM, Seder RA, Sher A. Induction and regulation of IL-15 expression in murine macrophages. J Immunol 1996; 156(2):735-41.

163. Alleva DG, Kaser SB, Monroy MA et al. IL-15 functions as a potent autocrine regulator of macrophage proinflammatory cytokine production: evidence for differential receptor subunit utilization associated with stimulation or inhibition. J Immunol 1997; 159(6):2941-51.

164. Fehniger TA, Suzuki K, Ponnappan A et al. Fatal leukemia in interleukin 15 transgenic mice follows early expansions in natural killer and memory phenotype CD8+ T-cells. J Exp Med 2001; 193(2):219-31.

165. Carson WE, Giri JG, Lindemann MJ et al. Interleukin (IL) 15 is a novel cytokine that activates human natural killer cells via components of the IL-2 receptor. J Exp Med 1994; 180(4):1395-403.

166. Kanegane H, Tosato G. Activation of naive and memory T-cells by interleukin-15. Blood 1996; 88(1):230-35.

167. Bulfone-Paus S, Ungureanu D, Pohl T et al. Interleukin-15 protects from lethal apoptosis in vivo. Nat Med 1997; 3(10):1124-28.

168. Pavlakis M, Strehlau J, Lipman M et al. Intragraft IL-15 transcripts are increased in human renal allograft rejection. Transplantation 1996; 62(4):543-45.

169. Fehniger TA, Caligiuri MA. Interleukin 15: biology and relevance to human disease. Blood 2001; 97(1):14-32.

170. Strehlau J, Pavlakis M, Lipman M et al. Quantitative detection of immune activation transcripts as a diagnostic tool in kidney transplantation. Proc Natl Acad Sci USA 1997; 94(2):695-700.

171. van Gelder T, Baan CC, Balk AH et al. Blockade of the interleukin (IL)-2/IL-2 receptor pathway with a monoclonal anti-IL-2 receptor antibody (BT563) does not prevent the development of acute heart allograft rejection in humans. Transplantation 1998; 65(3):405-10.

172. Baan CC, Knoop CJ, van Gelder T et al. Anti-CD25 therapy reveals the redundancy of the intragraft cytokine network after clinical heart transplantation. Transplantation 1999; 67(6):870-76.

173. Zheng XX, Gao W, Donskoy E et al. An antagonist mutant IL-15/Fc promotes transplant tolerance. Transplantation 2006; 81(1):109-16.

174. Kim YS, Maslinski W, Zheng XX et al. Targeting the IL-15 receptor with an antagonist IL-15 mutant/Fc gamma2a protein blocks delayed-type hypersensitivity. J Immunol 1998; 160(12):5742-48.

175. Zheng XX, Steele AW, Hancock WW et al. IL-2 receptor-targeted cytolytic IL-2/Fc fusion protein treatment blocks diabetogenic autoimmunity in nonobese diabetic mice. J Immunol 1999; 163(7):4041-48.

176. Ferrari-Lacraz S, Zheng XX, Kim YS et al. An antagonist IL-15/Fc protein prevents costimulation blockade-resistant rejection. J Immunol 2001; 167(6):3478-85.

177. Smith XG, Bolton EM, Ruchatz H et al. Selective blockade of IL-15 by soluble IL-15 receptor alpha-chain enhances cardiac allograft survival. J Immunol 2000; 165(6):3444-50.

178. Shin HC, Benbernou N, Esnault S et al. Expression of IL-17 in human memory CD45RO+ T-lymphocytes and its regulation by protein kinase A pathway. Cytokine 1999; 11(4):257-66.

179. Rouvier E, Luciani MF, Mattei MG et al. CTLA-8, cloned from an activated T-cell, bearing AU-rich messenger RNA instability sequences and homologous to a herpesvirus saimiri gene. J Immunol 1993; 150(12):5445-56.

180. Infante-Duarte C, Horton HF, Byrne MC et al. Microbial lipopeptides induce the production of IL-17 in Th cells. J Immunol 2000; 165(11):6107-15.

181. Kennedy J, Rossi DL, Zurawski SM et al. Mouse IL-17: a cytokine preferentially expressed by alpha beta TCR + CD4-CD8-T-cells. J Interferon Cytokine Res 1996; 16(8):611-17.

182. Jovanovic DV, Di Battista JA, Martel-Pelletier J et al. IL-17 stimulates the production and expression of proinflammatory cytokines, IL-beta and TNF-alpha, by human macrophages. J Immunol 1998; 160(7):3513-21.

183. Antonysamy MA, Fanslow WC, Fu F et al. Evidence for a role of IL-17 in organ allograft rejection: IL-17 promotes the functional differentiation of dendritic cell progenitors. J Immunol 1999; 162(1):577-84.

184. Nakae S, Komiyama Y, Nambu A et al. Antigen-specific T-cell sensitization is impaired in IL-17-deficient mice, causing suppression of allergic cellular and humoral responses. Immunity 2002; 17(3):375-87.

185. Loong CC, Hsieh HG, Lui WY et al. Evidence for the early involvement of interleukin 17 in human and experimental renal allograft rejection. J Pathol 2002; 197(3):322-32.

186. Hsieh HG, Loong CC, Lui WY et al. IL-17 expression as a possible predictive parameter for subclinical renal allograft rejection. Transpl Int 2001; 14(5):287-98.

187. Li J, Simeoni E, Fleury S et al. Gene transfer of soluble interleukin-17 receptor prolongs cardiac allograft survival in a rat model. Eur J Cardiothorac Surg 2006; 29(5):779-83.

188. Madri JA, Furthmayr H. Collagen polymorphism in the lung. An immunochemical study of pulmonary fibrosis. Hum Pathol 1980; 11(4):353-66.

189. Yousem SA, Suncan SR, Ohori NP et al. Architectural remodeling of lung allografts in acute and chronic rejection. Arch Pathol Lab Med 1992; 116(11):1175-80.

190. Zheng L, Ward C, Snell GI et al. Scar collagen deposition in the airways of allografts of lung transplant recipients. Am J Respir Crit Care Med 1997; 155(6):2072-77.

191. Haque MA, Mizobuchi T, Yasufuku K et al. Evidence for immune responses to a self-antigen in lung transplantation: role of type V collagen-specific T-cells in the pathogenesis of lung allograft rejection. J Immunol 2002; 169(3):1542-49.

192. Yoshida S, Haque A, Mizobuchi T et al. Anti-type V collagen lymphocytes that express IL-17 and IL-23 induce rejection pathology in fresh and well-healed lung transplants. Am J Transplant 2006; 6(4):724-35.

193. Kishimoto T. Interleukin-6 and its receptor in autoimmunity. J Autoimmun 1992; 5 Suppl A:123-32.

194. Hirano T, Matsuda T, Turner M et al. Excessive production of interleukin 6/B-cell stimulatory factor-2 in rheumatoid arthritis. Eur J Immunol 1988; 18(11):1797-801.

195. Taga T, Kishimoto T. Gp130 and the interleukin-6 family of cytokines. Annu Rev Immunol 1997; 15:797-819.

196. Shahar I, Fireman E, Topilsky M et al. Effect of IL-6 on alveolar fibroblast proliferation in interstitial lung diseases. Clin Immunol Immunopathol 1996; 79(3):244-51.

197. Horii Y, Muraguchi A, Iwano M et al. Involvement of IL-6 in mesangial proliferative glomerulonephritis. J Immunol 1989; 143(12):3949-55.

198. Borger P, Kauffman HF, Scholma J et al. Human allogeneic CD2+ lymphocytes activate airway-derived epithelial cells to produce interleukin-6 and interleukin-8. Possible role for the epithelium in chronic allograft rejection. J Heart Lung Transplant 2002; 21(5):567-75.

199. Farivar AS, Merry HE, Fica-Delgado MJ et al. Interleukin-6 regulation of direct lung ischemia reperfusion injury. Ann Thorac Surg 2006; 82(2):472-78.

200. Rolfe MW, Kunkel S, Lincoln P et al. Lung allograft rejection: role of tumor necrosis factor-alpha and interleukin-6. Chest 1993; 103(2 Suppl):133S.

201. Scholma J, Slebos DJ, Boezen HM et al. Eosinophilic granulocytes and interleukin-6 level in bronchoalveolar lavage fluid are associated with the development of obliterative bronchiolitis after lung transplantation. Am J Respir Crit Care Med 2000; 162(6):2221-25.

202. Kaneda H, Waddell TK, de Perrot M et al. Pre-implantation multiple cytokine mRNA expression analysis of donor lung grafts predicts survival after lung transplantation in humans. Am J Transplant 2006; 6(3):544-51.

203. Penttinen RP, Kobayashi S, Bornstein P. Transforming growth factor beta increases mRNA for matrix proteins both in the presence and in the absence of changes in mRNA stability. Proc Natl Acad Sci USA 1988; 85(4):1105-8.

204. Roberts AB, Sporn MB, Assoian RK et al. Transforming growth factor type beta: rapid induction of fibrosis and angiogenesis in vivo and stimulation of collagen formation in vitro. Proc Natl Acad Sci USA 1986; 83(12):4167-71.

205. Mora BN, Boasquevisque CH, Boglione M et al. Transforming growth factor-beta1 gene transfer ameliorates acute lung allograft rejection. J Thorac Cardiovasc Surg 2000; 119(5):913-20.

206. Qin L, Chavin KD, Ding Y et al. Gene transfer for transplantation. Prolongation of allograft survival with transforming growth factor-beta 1. Ann Surg 1994; 220(4):508-18; discussion 18-19.

207. Yasufuku K, Heidler KM, O'Donnell PW et al. Oral tolerance induction by type V collagen downregulates lung allograft rejection. Am J Respir Cell Mol Biol 2001; 25(1):26-34.

208. Liu M, Suga M, Maclean AA et al. Soluble transforming growth factor-beta type III receptor gene transfection inhibits fibrous airway obliteration in a rat model of Bronchiolitis obliterans. Am J Respir Crit Care Med 2002; 165(3):419-23.

209. Massague J, Wotton D. Transcriptional control by the TGF-beta/Smad signaling system. EMBO J 2000; 19(8):1745-54.

210. Taylor LM, Khachigian LM. Induction of platelet-derived growth factor B-chain expression by transforming growth factor-beta involves transactivation by Smads. J Biol Chem 2000; 275(22):16709-16.

211. Holmes A, Abraham DJ, Sa S et al. CTGF and SMADs, maintenance of scleroderma phenotype is independent of SMAD signaling. J Biol Chem 2001; 276(14):10594-601.

212. Isono M, Chen S, Hong SW et al. Smad pathway is activated in the diabetic mouse kidney and Smad3 mediates TGF-beta-induced fibronectin in mesangial cells. Biochem Biophys Res Commun 2002; 296(5):1356-65.

213. Yuan W, Varga J. Transforming growth factor-beta repression of matrix metalloproteinase-1 in dermal fibroblasts involves Smad3. J Biol Chem 2001; 276(42):38502-10.

214. Ramirez AM, Takagawa S, Sekosan M et al. Smad3 deficiency ameliorates experimental obliterative bronchiolitis in a heterotopic tracheal transplantation model. Am J Pathol 2004; 165(4):1223-32.

215. Demirci G, Nashan B, Pichlmayr R. Fibrosis in chronic rejection of human liver allografts: expression patterns of transforming growth factor-TGFbeta1 and TGF-beta3. Transplantation 1996; 62(12):1776-83.

216. Shihab FS, Yamamoto T, Nast CC et al. Transforming growth factor-beta and matrix protein expression in acute and chronic rejection of human renal allografts. J Am Soc Nephrol 1995; 6(2):286-94.

217. Aziz T, Hasleton P, Hann AW et al. Transforming growth factor beta in relation to cardiac allograft vasculopathy after heart transplantation. J Thorac Cardiovasc Surg 2000; 119(4 Pt 1):700-8.

218. Di Filippo S, Zeevi A, McDade KK et al. Impact of TGFbeta1 gene polymorphisms on acute and chronic rejection in pediatric heart transplant allografts. Transplantation 2006; 81(6):934-39.

219. Elssner A, Jaumann F, Dobmann S et al. Elevated levels of interleukin-8 and transforming growth factor-beta in bronchoalveolar lavage fluid from patients with bronchiolitis obliterans syndrome: proinflammatory role of bronchial epithelial cells. Munich Lung Transplant Group. Transplantation 2000; 70(2):362-67.

220. El-Gamel A, Sim E, Hasleton P et al. Transforming growth factor beta (TGF-beta) and obliterative bronchiolitis following pulmonary transplantation. J Heart Lung Transplant 1999; 18(9):828-37.

221. Charpin JM, Valcke J, Kettaneh L et al. Peaks of transforming growth factor-beta mRNA in alveolar cells of lung transplant recipients as an early marker of chronic rejection. Transplantation 1998; 65(5):752-55.

222. Cambrey AD, Kwon OJ, Gray AJ et al. Insulin-like growth factor I is a major fibroblast mitogen produced by primary cultures of human airway epithelial cells. Clin Sci (Lond) 1995; 89(6):611-17.

223. Goldstein RH, Poliks CF, Pilch PF et al. Stimulation of collagen formation by insulin and insulin-like growth factor I in cultures of human lung fibroblasts. Endocrinology 1989; 124(2):964-70.

224. Rom WN, Basset P, Fells GA et al. Alveolar macrophages release an insulin-like growth factor I-type molecule. J Clin Invest 1988; 82(5):1685-93.

225. Homma S, Nagaoka I, Abe H et al. Localization of platelet-derived growth factor and insulin-like growth factor I in the fibrotic lung. Am J Respir Crit Care Med 1995; 152(6 Pt 1):2084-89.

226. Jones JI, Clemmons DR. Insulin-like growth factors and their binding proteins: biological actions. Endocr Rev 1995; 16(1):3-34.

227. Charpin JM, Stern M, Grenet D et al. Insulinlike growth factor-1 in lung transplants with obliterative bronchiolitis. Am J Respir Crit Care Med 2000;161(6):1991-98.

228. Ross R, Raines EW, Bowen-Pope DF. The biology of platelet-derived growth factor. Cell 1986; 46(2):155-69.

229. Williams LT. Signal transduction by the platelet-derived growth factor receptor. Science 1989; 243(4898):1564-70.

230. Heldin CH, Westermark B. Platelet-derived growth factor: three isoforms and two receptor types. Trends Genet 1989; 5(4):108-11.

231. Hertz MI, Henke CA, Nakhleh RE et al. Obliterative bronchiolitis after lung transplantation: a fibroproliferative disorder associated with platelet-derived growth factor. Proc Natl Acad Sci USA 1992; 89(21):10385-89.

232. Kallio EA, Koskinen PK, Aavik E et al. Role of platelet-derived growth factor in obliterative bronchiolitis (chronic rejection) in the rat. Am J Respir Crit Care Med 1999; 160(4):1324-32.

233. Defrances MC, Wolf HK, Michalopoulos GK et al. The presence of hepatocyte growth factor in the developing rat. Development 1992; 116(2):387-95.

234. Matsumoto K, Tajima H, Hamanoue M et al. Identification and characterization of "injurin," an inducer of expression of the gene for hepatocyte growth factor. Proc Natl Acad Sci USA 1992; 89(9):3800-4.

235. Aharinejad S, Taghavi S, Klepetko W et al. Prediction of lung-transplant rejection by hepatocyte growth factor. Lancet 2004; 363(9420):1503-8.

236. Rollins BJ. Chemokines. Blood 1997; 90(3):909-28.

237. Luster AD. Review Articles: Mechanisms of Disease: Chemokines—Chemotactic Cytokines That Mediate Inflammation. N Engl J Med 1998; 338(7):436-45.

238. Strieter RM, Kunkel SL. Chemokines and the lung. In: Crystal R, West J, Weibel E, Barnes P, eds. Lung: Scientific Foundations, 2nd edition. New York: Raven Press, 1997:155-86.

239. Bleul CC, Fuhlbrigge RC, Casasnovas JM et al. A highly efficacious lymphocyte chemoattractant, stromal cell-derived factor 1 (SDF-1) [see comments]. J Exp Med 1996; 184(3):1101-9.

240. Loetscher M, Gerber B, Loetscher P et al. Chemokine receptor specific for IP10 and mig: structure, function and expression in activated T-lymphocytes [see comments]. J Exp Med 1996; 184(3):963-69.

241. Keane MP, Belperio JA, Arenberg DA et al. IFN-gamma-inducible protein-10 attenuates bleomycin-induced pulmonary fibrosis via inhibition of angiogenesis [In Process Citation]. J Immunol 1999; 163(10):5686-92.

242. Keane MP, Belperio JA, Moore TA et al. Neutralization of the CXC chemokine, macrophage inflammatory protein-2, attenuates bleomycin-induced pulmonary fibrosis. J Immunol 1999; 162(9):5511-18.

243. Moore BB, Arenberg DA, Addison CL et al. Tumor angiogenesis is regulated by CXC chemokines. J Lab Clin Med 1998; 132(2):97-103.

244. Baggiolini M, Dewald B, Moser B. Human chemokines: an update. Annu Rev Immunol 1997; 15:675-705.

245. Robinson LA, Nataraj C, Thomas DW et al. A role for fractalkine and its receptor (CX3CR1) in cardiac allograft rejection. J Immunol 2000; 165(11):6067-72.

246. Fong AM, Robinson LA, Steeber DA et al. Fractalkine and CX3CR1 mediate a novel mechanism of leukocyte capture, firm adhesion and activation under physiologic flow. J Exp Med 1998; 188(8):1413-19.

247. Pan Y, Lloyd C, Zhou H et al. Neurotactin, a membrane-anchored chemokine upregulated in brain inflammation [published erratum appears in Nature 1997; 389(6646):100]. Nature 1997; 387(6633):611-17.

248. Bazan JF, Bacon KB, Hardiman G et al. A new class of membrane-bound chemokine with a CX3C motif. Nature 1997; 385(6617):640-44.

249. Zlotnik A, Yoshie O. Chemokines: a new classification system and their role in immunity. Immunity 2000; 12(2):121-27.

250. Murphy PM, Baggiolini M, Charo IF et al. International union of pharmacology. XXII. Nomenclature for chemokine receptors. Pharmacol Rev 2000; 52(1):145-76.

251. Segerer S, Nelson PJ, Schlondorff D. Chemokines, chemokine receptors and renal disease: from basic science to pathophysiologic and therapeutic studies. J Am Soc Nephrol 2000; 11(1):152-76.

252. Taub DD, Conlon K, Lloyd AR et al. Preferential migration of activated CD4+ and CD8+ T-cells in response to MIP-1 alpha and MIP-1 beta. Science 1993; 260(5106):355-58.

253. Bacon KB, Premack BA, Gardner P et al. Activation of dual T-cell signaling pathways by the chemokine RANTES. Science 1995; 269(5231):1727-30.

254. Taub D. C-C chemokines: An overview. In: Koch A, Strieter R, eds. Chemokines in Disease. Austin: R.G. Landes, Co., 1996:27-54.

255. Vaddi K, Newton RC. Regulation of monocyte integrin expression by beta-family chemokines. J Immunol 1994; 153(10):4721-32.

256. Tanaka Y, Adams DH, Hubscher S et al. T-cell adhesion induced by proteoglycan-immobilized cytokine MIP-1 beta. Nature 1993; 361(6407):79-82.

257. Witt DP, Lander AD. Differential binding of chemokines to glycosaminoglycan subpopulations. Curr Biol 1994; 4(5):394-400.

258. Heath H, Qin S, Rao P et al. Chemokine receptor usage by human eosinophils. The importance of CCR3 demonstrated using an antagonistic monoclonal antibody. J Clin Invest 1997; 99(2):178-84.

259. Yamada H, Hirai K, Miyamasu M et al. Eotaxin is a potent chemotaxin for human basophils. Biochem Biophys Res Commun 1997; 231(2):365-68.

260. Hadida F, Vieillard V, Autran B et al. HIV-specific T-cell cytotoxicity mediated by RANTES via the chemokine receptor CCR3. J Exp Med 1998; 188(3):609-14.

261. Broxmeyer HE, Kim CH. Regulation of hematopoiesis in a sea of chemokine family members with a plethora of redundant activities. Exp Hematol 1999; 27(7):1113-23.

262. Schall TJ. Biology of the RANTES/SIS cytokine family. Cytokine 1991; 3:165.

263. Pattison J, Nelson PJ, Huie P et al. RANTES chemokine expression in cell-mediated transplant rejection of the kidney [see comments]. Lancet 1994; 343(8891):209-11.

264. von Luettichau I, Nelson PJ, Pattison JM et al. RANTES chemokine expression in diseased and normal human tissues. Cytokine 1996; 8(1):89-98.

265. Snowden N, Hajeer A, Thomson W et al. RANTES role in rheumatoid arthritis. Lancet 1994; 343(8896):547-48.

266. Krishnadasan B, Farivar AS, Naidu BV et al. Beta-chemokine function in experimental lung ischemia-reperfusion injury. Ann Thorac Surg 2004; 77(3):1056-62.

267. Monti G, Magnan A, Fattal M et al. Intrapulmonary production of RANTES during rejection and CMV pneumonitis after lung transplantation. Transplantation 1996; 61(12):1757-62.

268. Belperio JA, Burdick MD, Keane MP et al. The role of the CC chemokine, RANTES, in acute lung allograft rejection. J Immunol 2000; 165(1):461-72.

269. Panoskaltsis-Mortari A, Strieter RM, Hermanson JR et al. Induction of monocyte- and T-cell-attracting chemokines in the lung during the generation of idiopathic pneumonia syndrome following allogeneic murine bone marrow transplantation. Blood 2000; 96(3):834-39.

270. Sekine Y, Yasufuku K, Heidler KM et al. Monocyte chemoattractant protein-1 and RANTES are chemotactic for graft infiltrating lymphocytes during acute lung allograft rejection. Am J Respir Cell Mol Biol 2000; 23(6):719-26.

271. Belperio JA, Keane MP, Burdick MD et al. Critical role for the chemokine MCP-1/CCR2 in the pathogenesis of bronchiolitis obliterans syndrome. J Clin Invest 2001; 108(4):547-56.

272. Randolph GJ, Beaulieu S, Lebecque S et al. Differentiation of monocytes into dendritic cells in a model of transendothelial trafficking [see comments]. Science 1998; 282(5388):480-83.

273. Sallusto F, Lanzavecchia A. Mobilizing dendritic cells for tolerance, priming and chronic inflammation [comment]. J Exp Med 1999; 189(4):611-14.

274. Sallusto F, Schaerli P, Loetscher P et al. Rapid and coordinated switch in chemokine receptor expression during dendritic cell maturation. Eur J Immunol 1998; 28(9):2760-69.

275. Kurts C, Kosaka H, Carbone FR et al. Class I-restricted cross-presentation of exogenous self-antigens leads to deletion of autoreactive CD8(+) T-cells. J Exp Med 1997; 186(2):239-45.

276. Nadeau KC, Azuma H, Tilney NL. Sequential cytokine dynamics in chronic rejection of rat renal allografts: roles for cytokines RANTES and MCP-1. Proc Natl Acad Sci USA 1995; 92(19):8729-33.

277. Nagano H, Nadeau KC, Takada M et al. Sequential cellular and molecular kinetics in acutely rejecting renal allografts in rats. Transplantation 1997; 63(8):1101-8.

278. Grone HJ, Weber C, Weber KS et al. Met-RANTES reduces vascular and tubular damage during acute renal transplant rejection: blocking monocyte arrest and recruitment. FASEB J 1999; 13(11):1371-83.

279. Horuk R, Clayberger C, Krensky AM et al. A nonpeptide functional antagonist of the CCR1 chemokine receptor is effective in rat heart transplant rejection. J Biol Chem 2001; 276(6):4199-204.

280. Horuk R, Shurey S, Ng HP et al. CCR1-specific nonpeptide antagonist: efficacy in a rabbit allograft rejection model. Immunol Lett 2001; 76(3):193-201.

281. Gao W, Topham PS, King JA et al. Targeting of the chemokine receptor CCR1 suppresses development of acute and chronic cardiac allograft rejection. J Clin Invest 2000; 105(1):35-44.

282. Gao W, Faia KL, Csizmadia V et al. Beneficial effects of targeting CCR5 in allograft recipients. Transplantation 2001; 72(7):1199-205.

283. Abdi R, Smith RN, Makhlouf L et al. The role of CC chemokine receptor 5 (CCR5) in islet allograft rejection. Diabetes 2002; 51(8):2489-95.

284. Uyama T, Winter JB, Groen G et al. Late airway changes caused by chronic rejection in rat lung allografts. Transplantation 1992; 54:809-12.

285. Uyama T, Sakiyama S, Fukumoto T et al. Graft-infiltrating cells in rat lung allograft with late airway damage. Transplant Proc 1995; 27(3):2118-19.

286. Matsumura Y, Marchevsky A, Zuo XJ et al. Assessment of pathological changes associated with chronic allograft rejection and tolerance in two experimental models of rat lung transplantation. Transplantation 1995; 59(11):1509-17.

287. Tazelaar HD, Prop J, Nieuwenhuis P, Billingham ME et al. Airway pathology in the transplanted rat lung. Transplantation 1988; 45(5):864-69.

288. Hertz MI, Jessurun J, King MB et al. Reproduction of the obliterative bronchiolitis lesion after heterotopic transplantation of mouse airways. American Journal of Pathology 1993; 142(6):1945-51.

289. Riise GC, Williams A, Kjellstrom C et al. Bronchiolitis obliterans syndrome in lung transplant recipients is associated with increased neutrophil activity and decreased antioxidant status in the lung. Eur Respir J 1998; 12(1):82-88.

290. Riise GC Andersson BA, Kjellstrom C et al. Persistent high BAL fluid granulocyte activation marker levels as early indicators of bronchiolitis obliterans after lung transplant. Eur Respir J 1999; 14(5):1123-30.

291. Zheng L, Walters EH, Ward C et al. Airway neutrophilia in stable and bronchiolitis obliterans syndrome patients following lung transplantation. Thorax 2000; 55(1):53-59.

292. Elssner A, Jaumann F, Dobmann S et al. Elevated levels of interleukin-8 and transforming growth factor-beta in bronchoalveolar lavage fluid from patients with bronchiolitis obliterans syndrome: proinflammatory role of bronchial epithelial cells. Munich Lung Transplant Group. Transplantation 2000; 70(2):362-67.

293. Elssner A, Vogelmeier C. The role of neutrophils in the pathogenesis of obliterative bronchiolitis after lung transplantation. Transpl Infect Dis 2001; 3(3):168-76.

294. DiGiovine B, Lynch JP, 3rd, Martinez FJ et al. Bronchoalveolar lavage neutrophilia is associated with obliterative bronchiolitis after lung transplantation: role of IL-8. J Immunol 1996; 157(9):4194-202.

295. Suga M, Maclean AA, Keshavjee S et al. RANTES plays an important role in the evolution of allograft transplant-induced fibrous airway obliteration. Am J Respir Crit Care Med 2000; 162(5):1940-48.

296. Farivar AS, Krishnadasan B, Naidu BV et al. The role of the beta chemokines in experimental obliterative bronchiolitis. Exp Mol Pathol 2003; 75(3):210-16.

297. Fischereder M, Luckow B, Hocher B et al. CC chemokine receptor 5 and renal-transplant survival. Lancet 2001; 357(9270):1758-61.

298. Abdi R, Tran TB, Sahagun-Ruiz A et al. Chemokine receptor polymorphism and risk of acute rejection in human renal transplantation. J Am Soc Nephrol 2002; 13(3):754-58.

299. Martinson JJ, Chapman NH, Rees DC et al. Global distribution of the CCR5 gene 32-basepair deletion. Nat Genet 1997; 16(1):100-3.

300. Libert F, Cochaux P, Beckman G et al. The deltaccr5 mutation conferring protection against HIV-1 in Caucasian populations has a single and recent origin in Northeastern Europe. Hum Mol Genet 1998; 7(3):399-406.

301. Stephens JC, Reich DE, Goldstein DB et al. Dating the origin of the CCR5-Delta32 AIDS-resistance allele by the coalescence of haplotypes. Am J Hum Genet 1998; 62(6):1507-15.

302. Lee I, Wang L, Wells AD et al. Blocking the monocyte chemoattractant protein-1/CCR2 chemokine pathway induces permanent survival of islet allografts through a programmed death-1 ligand-1-dependent mechanism. J Immunol 2003; 171(12):6929-35.

303. Chen MC, Proost P, Gysemans C et al. Monocyte chemoattractant protein-1 is expressed in pancreatic islets from prediabetic NOD mice and in interleukin-1 beta-exposed human and rat islet cells. Diabetologia 2001; 44(3):325-32.

304. Piemonti L, Leone BE, Nano R et al. Human pancreatic islets produce and secrete MCP-1/CCL2: relevance in human islet transplantation. Diabetes 2002; 51(1):55-65.

305. Gu L, Tseng S, Horner RM et al. Control of TH2 polarization by the chemokine monocyte chemoattractant protein-1. Nature 2000; 404(6776):407-11.

306. Luster AD. The role of chemokines in linking innate and adaptive immunity. Curr Opin Immunol 2002; 14(1):129-35.

307. Abdi R, Means TK, Ito T et al. Differential role of CCR2 in islet and heart allograft rejection: tissue specificity of chemokine/chemokine receptor function in vivo. J Immunol 2004; 172(2):767-75.

308. Imai T, Baba M, Nishimura M et al. The T-cell-directed CC chemokine TARC is a highly specific biological ligand for CC chemokine receptor 4. J Biol Chem 1997; 272(23):15036-42.

309. Campbell JJ, Haraldsen G, Pan J et al. The chemokine receptor CCR4 in vascular recognition by cutaneous but not intestinal memory T-cells. Nature 1999; 400(6746):776-80.

310. Luther SA, Cyster JG. Chemokines as regulators of T-cell differentiation. Nat Immunol 2001; 2(2):102-7.

311. Alferink J, Lieberam I, Reindl W et al. Compartmentalized production of CCL17 in vivo: strong inducibility in peripheral dendritic cells contrasts selective absence from the spleen. J Exp Med 2003; 197(5):585-99.

312. Huser N, Tertilt C, Gerauer K et al. CCR4-deficient mice show prolonged graft survival in a chronic cardiac transplant rejection model. Eur J Immunol 2005; 35(1):128-38.

313. Lee I, Wang L, Wells AD et al. Recruitment of Foxp3+ T regulatory cells mediating allograft tolerance depends on the CCR4 chemokine receptor. J Exp Med 2005; 201(7):1037-44.

314. Hancock WW, Sayegh MH, Zheng XG et al. Costimulatory function and expression of CD40 ligand, CD80 and CD86 in vascularized murine cardiac allograft rejection. Proc Natl Acad Sci USA 1996; 93(24):13967-72.

315. Graca L, Honey K, Adams E et al. Cutting edge: anti-CD154 therapeutic antibodies induce infectious transplantation tolerance. J Immunol 2000; 165(9):4783-86.

316. Jarvinen LZ, Blazar BR, Adeyi OA et al. CD154 on the surface of CD4+CD25+ regulatory T-cells contributes to skin transplant tolerance. Transplantation 2003; 76(9):1375-79.

317. Hori S, Nomura T, Sakaguchi S. Control of regulatory T-cell development by the transcription factor Foxp3. Science 2003; 299(5609):1057-61.

318. Fontenot JD, Gavin MA, Rudensky AY. Foxp3 programs the development and function of CD4+CD25+ regulatory T-cells. Nat Immunol 2003; 4(4):330-36.

319. Forster R, Schubel A, Breitfeld D et al. CCR7 coordinates the primary immune response by establishing functional microenvironments in secondary lymphoid organs. Cell 1999; 99(1):23-33.

320. Sallusto F, Mackay CR, Lanzavecchia A. The role of chemokine receptors in primary, effector and memory immune responses. Annu Rev Immunol 2000; 18:593-620.

321. Cyster JG. Chemokines and cell migration in secondary lymphoid organs. Science 1999; 286(5447):2098-102.

322. Muller G, Lipp M. Concerted action of the chemokine and lymphotoxin system in secondary lymphoid-organ development. Curr Opin Immunol 2003; 15(2):217-24.

323. Hopken UE, Droese J, Li JP et al. The chemokine receptor CCR7 controls lymph node-dependent cytotoxic T-cell priming in alloimmune responses. Eur J Immunol 2004; 34(2):461-70.

324. Garrod KR, Chang CK, Liu FC et al. Targeted lymphoid homing of dendritic cells is required for prolongation of allograft survival. J Immunol 2006; 177(2):863-68.

325. Coates PT, Thomson AW. Dendritic cells, tolerance induction and transplant outcome. Am J Transplant 2002; 2(4):299-307.

326. Fu F, Li Y, Qian S et al. Costimulatory molecule-deficient dendritic cell progenitors induce T-cell hyporesponsiveness in vitro and prolong the survival of vascularized cardiac allografts. Transplant Proc 1997; 29(1-2):1310.

327. Rastellini C, Lu L, Ricordi C et al. Granulocyte/macrophage colony-stimulating factor-stimulated hepatic dendritic cell progenitors prolong pancreatic islet allograft survival. Transplantation 1995; 60(11):1366-70.

328. Bai Y, Liu J, Wang Y et al. L-selectin-dependent lymphoid occupancy is required to induce alloantigen-specific tolerance. J Immunol 2002; 168(4):1579-89.

329. Lakkis FG, Arakelov A, Konieczny BT et al. Immunologic 'ignorance' of vascularized organ transplants in the absence of secondary lymphoid tissue. Nat Med 2000; 6(6):686-88.

330. Lappin MB, Weiss JM, Delattre V et al. Analysis of mouse dendritic cell migration in vivo upon subcutaneous and intravenous injection. Immunology 1999; 98(2):181-88.

331. Takayama T, Morelli AE, Onai N et al. Mammalian and viral IL-10 enhance C-C chemokine receptor 5 but down-regulate C-C chemokine receptor 7 expression by myeloid dendritic cells: impact on chemotactic responses and in vivo homing ability. J Immunol 2001; 166(12):7136-43.

332. Imai T, Hieshima K, Haskell C et al. Identification and molecular characterization of fractalkine receptor CX3CR1, which mediates both leukocyte migration and adhesion. Cell 1997; 91(4):521-30.

333. Combadiere C, Gao J, Tiffany HL et al. Gene cloning, RNA distribution and functional expression of mCX3CR1, a mouse chemotactic receptor for the CX3C chemokine fractalkine. Biochem Biophys Res Commun 1998; 253(3):728-32.

334. Maciejewski-Lenoir D, Chen S, Feng L et al. Characterization of fractalkine in rat brain cells: migratory and activation signals for CX3CR-1-expressing microglia. J Immunol 1999; 163(3):1628-35.

335. Harrison JK, Jiang Y, Chen S et al. Role for neuronally derived fractalkine in mediating interactions between neurons and CX3CR1-expressing microglia. Proc Natl Acad Sci USA 1998; 95(18):10896-901.

336. Feng L, Chen S, Garcia GE et al. Prevention of crescentic glomerulonephritis by immunoneutralization of the fractalkine receptor CX3CR1 rapid communication. Kidney Int 1999; 56(2):612-20.

337. Haskell CA, Hancock WW, Salant DJ et al. Targeted deletion of CX(3)CR1 reveals a role for fractalkine in cardiac allograft rejection. J Clin Invest 2001; 108(5):679-88.

338. Taub DD, Lloyd AR, Conlon K et al. Recombinant human interferon-inducible protein 10 is a chemoattractant for human monocytes and T-lymphocytes and promotes T-cell adhesion to endothelial cells. J Exp Med 1993; 177(6):1809-14.

339. Sallusto F, Lenig D, Mackay CR et al. Flexible programs of chemokine receptor expression on human polarized T helper 1 and 2 lymphocytes. J Exp Med 1998; 187(6):875-83.

340. Bonecchi R, Bianchi G, Bordignon PP et al. Differential expression of chemokine receptors and chemotactic responsiveness of type 1 T helper cells (Th1s) and Th2s. J Exp Med 1998; 187(1):129-34.

341. Maghazachi AA, Skalhegg BS, Rolstad B et al. Interferon-inducible protein-10 and lymphotactin induce the chemotaxis and mobilization of intracellular calcium in natural killer cells through pertussis toxin-sensitive and -insensitive heterotrimeric G-proteins. FASEB J 1997; 11(10):765-74.

342. Liao F, Rabin RL, Yannelli JR et al. Human Mig chemokine: biochemical and functional characterization. J Exp Med 1995; 182(5):1301-14.

343. Rabin RL, Park MK, Liao F et al. Chemokine receptor responses on T-cells are achieved through regulation of both receptor expression and signaling. J Immunol 1999; 162(7):3840-50.

344. Qin S, Rottman JB, Myers P et al. The chemokine receptors CXCR3 and CCR5 mark subsets of T-cells associated with certain inflammatory reactions. J Clin Invest 1998; 101(4):746-54.

345. Loetscher M, Gerber B, Loetscher P et al. Chemokine receptor specific for IP10 and mig: structure, function and expression in activated T-lymphocytes. J Exp Med 1996; 184(3):963-69.

346. Piali L, Weber C, LaRosa G et al. The chemokine receptor CXCR3 mediates rapid and shear-resistant adhesion-induction of effector T-lymphocytes by the chemokines IP10 and Mig. Eur J Immunol 1998; 28(3):961-72.

347. Zhai Y, Shen XD, Hancock WW et al. CXCR3+CD4+ T-cells mediate innate immune function in the pathophysiology of liver ischemia/reperfusion injury. J Immunol 2006; 176(10):6313-22.

348. Zhao DX, Hu Y, Miller GG et al. Differential expression of the IFN-gamma-inducible CXCR3-binding chemokines, IFN-inducible protein 10, monokine induced by IFN and IFN-inducible T-cell alpha chemoattractant in human cardiac allografts: association with cardiac allograft vasculopathy and acute rejection. J Immunol 2002; 169(3):1556-60.

349. Melter M, Exeni A, Reinders ME et al. Expression of the chemokine receptor CXCR3 and its ligand IP-10 during human cardiac allograft rejection. Circulation 2001; 104(21):2558-64.

350. Kao J, Kobashigawa J, Fishbein MC et al. Elevated serum levels of the CXCR3 chemokine ITAC are associated with the development of transplant coronary artery disease. Circulation 2003; 107(15):1958-61.

351. Hancock WW, Lu B, Gao W et al. Requirement of the chemokine receptor CXCR3 for acute allograft rejection. J Exp Med 2000; 192(10):1515-20.

352. Hancock WW, Gao W, Csizmadia V et al. Donor-derived IP-10 initiates development of acute allograft rejection. J Exp Med 2001; 193(8):975-80.

353. Miura M, Morita K, Kobayashi H et al. Monokine Induced by IFN-gamma Is a Dominant Factor Directing T-cells into Murine Cardiac Allografts During Acute Rejection. J Immunol 2001; 167(6):3494-504.

354. Koga S, Auerbach MB, Engeman TM et al. T-cell infiltration into class II MHC-disparate allografts and acute rejection is dependent on the IFN-gamma-induced chemokine Mig. J Immunol 1999; 163(9):4878-85.

355. Baker MS, Chen X, Rotramel AR et al. Genetic deletion of chemokine receptor CXCR3 or antibody blockade of its ligand IP-10 modulates posttransplantation graft-site lymphocytic infiltrates and prolongs functional graft survival in pancreatic islet allograft recipients. Surgery 2003; 134(2):126-33.

356. Agostini C, Calabrese F, Rea F et al. Cxcr3 and its ligand CXCL10 are expressed by inflammatory cells infiltrating lung allografts and mediate chemotaxis of T-cells at sites of rejection. Am J Pathol 2001; 158(5):1703-11.

357. Belperio JA, Keane MP, Burdick MD et al. Critical role for CXCR3 chemokine biology in the pathogenesis of Bronchiolitis obliterans syndrome. J Immunol 2002; 169(2):1037-49.

358. Yun JJ, Fischbein MP, Whiting D et al. The role of MIG/CXCL9 in cardiac allograft vasculopathy. Am J Pathol 2002; 161(4):1307-13.

359. Whiting D, Hsieh G, Yun JJ et al. Chemokine monokine induced by IFN-gamma/CXC chemokine ligand 9 stimulates T-lymphocyte proliferation and effector cytokine production. J Immunol 2004; 172(12):7417-24.

360. Kapoor A, Morita K, Engeman TM et al. Early expression of interferon-gamma inducible protein 10 and monokine induced by interferon-gamma in cardiac allografts is mediated by CD8+ T-cells. Transplantation 2000; 69(6):1147-55.

361. Medoff BD, Wain JC, Seung E et al. CXCR3 and its ligands in a murine model of obliterative bronchiolitis: regulation and function. J Immunol 2006; 176(11):7087-95.

362. Chen W, Ford MS, Young KJ et al. Role of double-negative regulatory T-cells in long-term cardiac xenograft survival. J Immunol 2003; 170(4):1846-53.

363. Ford MS, Young KJ, Zhang Z et al. The immune regulatory function of lymphoproliferative double negative T-cells in vitro and in vivo. J Exp Med 2002; 196(2):261-67.

364. Fischer K, Voelkl S, Heymann J et al. Isolation and characterization of human antigen-specific TCR alpha beta+ CD4(−)CD8- double-negative regulatory T-cells. Blood 2005; 105(7):2828-35.

365. Young KJ, Yang L, Phillips MJ et al. Donor-lymphocyte infusion induces transplantation tolerance by activating systemic and graft-infiltrating double-negative regulatory T-cells. Blood 2002; 100(9):3408-14.

366. Zhang ZX, Stanford WL, Zhang L. Ly-6A is critical for the function of double negative regulatory T-cells. Eur J Immunol 2002; 32(6):1584-92.

367. Zhang ZX, Yang L, Young KJ et al. Identification of a previously unknown antigen-specific regulatory T-cell and its mechanism of suppression. Nat Med 2000; 6(7):782-89.

368. Lee BP, Mansfield E, Hsieh SC et al. Expression profiling of murine double-negative regulatory T-cells suggest mechanisms for prolonged cardiac allograft survival. J Immunol 2005; 174(8):4535-44.

369. Chen W, Ford MS, Young KJ et al. Infusion of in vitro-generated DN T regulatory cells induces permanent cardiac allograft survival in mice. Transplant Proc 2003; 35(7):2479-80.

370. Dobner T, Wolf I, Emrich T et al. Differentiation-specific expression of a novel G protein-coupled receptor from Burkitt's lymphoma. Eur J Immunol 1992; 22(11):2795-99.

371. Lim HW, Hillsamer P, Kim CH. Regulatory T-cells can migrate to follicles upon T-cell activation and suppress GC-Th cells and GC-Th cell-driven B-cell responses. J Clin Invest 2004; 114(11):1640-49.

372. Ansel KM, McHeyzer-Williams LJ, Ngo VN et al. In vivo-activated CD4 T-cells upregulate CXC chemokine receptor 5 and reprogram their response to lymphoid chemokines. J Exp Med 1999; 190(8):1123-34.

373. Forster R, Emrich T, Kremmer E et al. Expression of the G-protein—coupled receptor BLR1 defines mature, recirculating B-cells and a subset of T-helper memory cells. Blood 1994; 84(3):830-40.

374. Ebert LM, Schaerli P, Moser B. Chemokine-mediated control of T-cell traffic in lymphoid and peripheral tissues. Mol Immunol 2005; 42(7):799-809.

375. Breitfeld D, Ohl L, Kremmer E et al. Follicular B helper T-cells express CXC chemokine receptor 5, localize to B-cell follicles and support immunoglobulin production. J Exp Med 2000; 192(11):1545-52.

376. Chtanova T, Tangye SG, Newton R et al. T follicular helper cells express a distinctive transcriptional profile, reflecting their role as nonTh1/Th2 effector cells that provide help for B-cells. J Immunol 2004; 173(1):68-78.

377. Lee BP, Chen W, Shi H et al. CXCR5/CXCL13 interaction is important for double-negative regulatory T-cell homing to cardiac allografts. J Immunol 2006; 176(9):5276-83.

378. Morita K, Miura M, Paolone DR et al. Early chemokine cascades in murine cardiac grafts regulate T-cell recruitment and progression of acute allograft rejection. J Immunol 2001; 167(5):2979-84.

379. Fisher AJ, Donnelly SC, Hirani N et al. Elevated levels of interleukin-8 in donor lungs is associated with early graft failure after lung transplantation. Am J Respir Crit Care Med 2001; 163(1):259-65.

380. De Perrot M, Sekine Y, Fischer S et al. Interleukin-8 release during early reperfusion predicts graft function in human lung transplantation. Am J Respir Crit Care Med 2002; 165(2):211-15.

381. Belperio JA, Keane MP, Burdick MD et al. Role of CXCR2/CXCR2 ligands in vascular remodeling during bronchiolitis obliterans syndrome. J Clin Invest 2005; 115(5):1150-62.

# CHAPTER 14

# Diagnosis of Chronic Graft Failure after Lung Transplantation

David B. Erasmus, Andras Khoor and Cesar A. Keller*

## Abstract

Since 1984, when bronchiolitis obliterans (BO) was recognized as the main factor influencing long-term survival after lung and heart-lung transplantation, this condition has remained the main cause of morbidity and mortality one year after transplant. It is characterized by submucosal lymphocytic infiltration of the airways, epithelial disruption, fibromyxoid granulation and ultimately partial or complete occlusion of bronchioli. Despite attempts to refine the diagnostic process, decline in spirometry and the so-called bronchiolitis obliterans syndrome (BOS) remains the most clinically relevant surrogate marker for BO. CT remains the most useful radiographic adjunct in the diagnosis of BO but there are no pathognomonic findings and results should be interpreted within clinical context. The value of a bronchoscopic study including BAL and trans-bronchial biopsies in the diagnosis of BO is to exclude other potential causes of functional decline, rather than establishing histological diagnosis. Recurrent episodes of acute rejection are the most widely accepted risk factor for BO but even mild rejection or a single episode of rejection may be significant. Acid or non-acid reflux disease, infection especially with viruses, HLA mismatching, lymphocytic bronchitis and possibly primary graft dysfunction may also predispose to BO. Effective measures to prevent or treat BO will undoubtedly have the greatest impact on long-term survival in lung transplant recipients.

## Introduction

The first human lung transplantation performed in 1963 by Dr James Hardy at the University of Mississippi survived 18 days.[1] His immunosuppression regimen consisted of steroids, azathioprine and radiation therapy. The recipient died from renal failure 18 days later and a postmortem study did not reveal evidence of rejection or vascular injury. Over the following 18 years, all lung transplant recipients died acutely within a year, falling victim to an array of acute complications, largely related to the use of massive doses of corticosteroids.[2] The addition of cyclosporine as the main immunosuppressive agent, sparing large doses of steroids and accompanied by improved operative techniques resulted in measurements of long term survival for the first time after heart lung transplant in 1981 and single-lung transplantation in 1983.[3,4]

By 1984, it was recognized that progressive and relentless airflow obstruction was a complication among heart-lung transplant recipients which became increasingly recognized as the main complication among patients surviving over one year.[5]

Biopsies of these lesions revealed histological presence of obliterative bronchiolitis or bronchiolitis obliterans (BO), characterized by epithelial injury, bronchocentric mononuclear inflammation and fibrosis of distal airways.

Chronic rejection in solid organ transplantation is characterized by progressive fibrous obliteration of luminal structures within the graft. Coronary graft vasculopathy characterizes chronic graft failure in heart transplantation. Endothelial and myofibroblast proliferation in kidney transplants produces glomerulosclerosis and nephrosclerosis and a similar process will produce the vanishing bile duct syndrome in liver transplant recipients.[6]

It was initially believed BO was exclusively observed in heart-lung transplant recipients, but by 1989 cases of obliterative bronchiolitis were recognized also among long term survivors of single lung transplants. By the mid 1990s, it became clear that 40-50% of long term survivors of either heart-lung, double-lung or single-lung transplantation will later develop chronic airflow obstruction.[7] BO was recognized as the main cause of morbidity and mortality among lung transplant recipients surviving more than one year.

Soon it was realized that the presence of acute rejection early after transplant was the most common factor associated with development of BO. Bronchoscopically obtained transbronchial biopsies became the instrument of choice to screen and diagnose acute rejection and in 1996 a histological grading system was refined to grade acute rejection ranging from normal to minimal, mild, moderate and severe (A0 to A4).[8] It was recognized however, that transbronchial biopsy specimens were not sufficiently sensitive to establish diagnosis in cases with BO. In 1993 a committee sponsored by the International Society for Heart and Lung Transplantation (ISHLT) proposed the concept of Bronchiolitis Obliterans syndrome (BOS), where pulmonary function test changes become a surrogate of the histological presence of BO.[9] The BOS classification was updated in 2002.[10] Transplant centers across the world adopted the BOS system as the tool to recognize lung allograft dysfunction allowing comparison of long-term results.

Currently, lung transplantation is an accepted alternative of care for patients with chronic lung diseases. In 2006, the registry of the ISHLT reported results on 21,265 lung transplant recipients and 3,154 heart-lung recipients transplanted over 24 years across the world from over 200 transplant centers.[11] The main indications for lung transplantation are Chronic Obstructive Pulmonary Disease, Idiopathic Pulmonary Fibrosis, Cystic Fibrosis, Alpha-1 Antitrypsin Deficiency and other causes of end-stage of lung disease. The main indications for heart-lung transplant continue to be pulmonary hypertension (idiopathic or associated with congenital heart disease).

*Corresponding Author: Cesar A. Keller—Medical Director, Lung Transplant Program, Mayo Clinic Transplant Center, 4205 Belfort Road, Suite 1100, Jacksonville FL 32216, USA. Email: keller.cesar@mayo.edu

*Chronic Allograft Failure: Natural History, Pathogenesis, Diagnosis and Management,* edited by Nasimul Ahsan. ©2008 Landes Bioscience.

Despite the improvement in surgical technical aspects, better understanding and management of primary graft dysfunction and the use of immunosuppressive therapy with triple drug combinations (typically cyclosporine or tacrolimus associated with azathioprine or mycophenolate and steroids) with or without induction therapy with polyclonal anti-lymphocytic agents or Interleukin-2 receptor antagonists, the fact remains that regardless of which protocol is used, 40 to 50% of patients will require treatment for acute rejection during the first year after transplant.

Survival rates after lung transplantation have improved to 87% at 3 months, 78% at 1 year, 61% at 3 years and 49% at 5 years, but lag behind survival rates by other solid organ recipients, likely reflecting the effects of continuous antigenic and environmental challenges that the lung graft receives by the exposure to the environment through respiration, subjecting the graft to alloimmune-dependent and independent factors, producing recurrent airway and air-space injury.

Despite experience gained over 20 years of worldwide performance of lung transplants, BOS continues to be the most common cause of morbidity and mortality among lung transplant recipients surviving over one year. The most recent statistics reveal that over 40% of patients surviving transplantation for 5 years will subsequently develop BOS. BOS and chronic graft failure account for 44% of deaths occurring 1 to 3 years after lung transplantation.[11]

This chapter reviews the pathogenesis and diagnosis of bronchiolitis obliterans and its functional surrogate, bronchiolitis obliterans syndrome.

## Pathogenesis

The precise pathogenesis has not been fully elucidated and there is considerable debate regarding the etiology of BO. While chronic rejection is classified pathologically as either vascular or airway rejection, chronic vascular rejection involving the pulmonary vasculature and leading to pulmonary artery atherosclerosis is much less common.

Bronchiolitis obliterans is typified by a submucosal lymphocytic infiltrate and disruption of the epithelium of the small airways. Fibromyxoid granulation follows, which progressively obliterates the lumen, leading to fibrosis and irreversible partial or complete obstruction of bronchioli. There may be wide variability in epithelial changes, fibrosis and inflammation in patients with advanced BOS as analyzed in explants of BOS patients undergoing re-transplantation.[12]

Animal models have shown the importance of the integrity of the airway epithelium in preventing obliterative airways disease and airway epithelial cells have also been shown to have a high apoptosis index (>1%) in transplanted airway epithelial cells not treated with immunosuppressants. Also, murine gene expression profiling in models of obliterative airway disease have revealed the predicted response of airway injury leading to an innate and then adaptive immune response with both cell-mediated and humoral reactions leading to eventual airway epithelial loss.[13-15] The histopathologic features of BO imply that recurrent injury and secondary inflammatory process of epithelial cells and sub-epithelial structures in small airways lead to fibroproliferation and obliteration of airways.[16]

While the pathogenesis is complicated, likely involving multiple components of the immune system, emphasis in research has been placed on certain components.

### The Lymphocyte as a Major Contributor

Bronchiolitis obliterans is associated with an increase in reactive lymphocyte activity directed against donor-specific HLA antigens as assessed by primed lymphocyte testing (PLT) and oligoclonal CD4+ T-cell expansion in peripheral blood.[17,18] Zheng et al followed 29 patients over 3 years, 17 of which developed BOS stage 1. They showed a decreased BAL CD4+ count and increased BAL CD8+ in all posttransplant patients but an exaggerated CD3+ and CD8+ airway

wall infiltrate over time in those that developed BOS, suggesting that CD8+ cells may escape immunosuppression in patients that develop BO.[19] Leukotriene B4 (LTB4) is a lipid mediator derived from arachidonic acid. It has potent chemotactic activity for monocytes and granulocytes through its receptor BLT1. Medoff et al also demonstrated potent chemotactic activity for in vitro-activated CD4+ and CD8+ lymphocytes. Furthermore, BLT1 induced deficiency in a murine model resulted in decreased mortality and inflammation, reduced collagen deposition and reduced recruitment of lymphocytes. The expression of BLT1 is associated with airway obliteration in this model.[20]

### The Fibroblast as a Contributor

It has been hypothesized that airway epithelial cells can undergo epithelial mesenchymal transformation (EMT) resulting in transformation to fibroblasts. Markers of EMT such as matrix metalloproteinase zymographic activity have been demonstrated in epithelial cell cultures of transplant recipients and fibroblast markers such as S100A4 have stained positive in the epithelium of biopsy specimens.[21] On the other hand, Brocker et al demonstrated significant recipient-derived fibroblasts in BO lesions, showing a role for circulating fibroblast precursor cells, probably of recipient bone-marrow origin.[22]

### Humoral Immunity

A progressive increase in anti-HLA class 1 antibodies in patients with BOS is associated with a progressive decline in pulmonary function and graft—host HLA mismatches, particularly at the HLA-A locus are associated with increased risk for BO.[23,24] C4d, a stable component of complement activation is increasingly deposited in the bronchial wall of patients with BOS.[25]

### Possible Role for Endothelin 1 (ET1)

Suppression of endothelin 1 (ET1) has resulted in decreased proliferation of smooth muscle cells, delayed epithelial necrosis and attenuated recruitment of interleukin 1 and interleukin 2 immunoreactive cells in a rat model. Endothelin contributes to stenosis in a rat tracheal mode and may have a pro-inflammatory role in BO.[26] ET1 secretion is enhanced as a result of ischemia-reperfusion injury, which may be a risk factor for the development of BO.

## Diagnostic Criteria

The diagnosis of bronchiolitis obliterans follows two different approaches in clinical practice. The diagnosis may be strongly suspected after excluding other causes of functional decline and based on clinical diagnosis, the so-called bronchiolitis obliterans syndrome (BOS) or confirmed histologically, referred to as bronchiolitis obliterans (BO).

## Clinical Diagnosis

### Physical Exam

Symptoms of BO are usually nonspecific and more indolent than in acute rejection. Patients usually show progressive decline in spirometry with an obstructive pattern and may have nonspecific upper-respiratory tract symptoms early in the course, associated with dyspnea. Physical exam is rarely helpful in the early stages of decline but in advanced BO, ronchi, pops and squeaks may be heard.

### Pulmonary Function Testing

The Bronchiolitis Obliterans syndrome has proven to be the best surrogate marker for BO and while many other markers have been proposed, BOS is predictive of graft and patient survival.

See Table 1 for the Classification of Bronchiolitis Obliterans syndrome (BOS).[10,27]

Baseline $FEV_1$ is defined as the average of the best two $FEV_1$ values after transplant. Decline in $FEV_1$ is determined by two measurements measured at least 3 weeks apart. BOS-0p has been proposed as a

*Table 1. Classification of bronchiolitis obliterans syndrome*

| BOS Stage | PFT Criteria | Biopsy |
|---|---|---|
| 0: No abnormality | FEV1 > 80% baseline | a: negative or |
| BOS-0p | 80 < FEV1 < 90 or | b: positive for |
| | FEF$_{25-75\%}$ < 75% baseline | bronchiolits |
| 1: Mild | 65% < FEV1 < 80% baseline | a or b |
| 2: Moderate | 50% < FEV1 < 65% baseline | a or b |
| 3: Severe | FEV1 < 50% baseline | a or b |

potential stage for the later development of BOS and was added to the original staging system in 2001. BOS-Op is defined as a 10 to 19% decrease in and/or FEV$_1$ 25% or greater decrease in FEF$_{25-75\%}$ from baseline.[28] Occasionally there may be a slight improvement in lung function and BOS score over time. The worst score assigned to a patient should remain in effect.

Decline in forced expiratory flow between 25 and 75% has been deemed a more sensitive marker for BO than decline in FEV$_1$ in most double lung and heart-lung recipients. Patterson et al measured decline in FEF$_{25-75\%}$ and FEV$_1$ at a point where the values were <70% baseline and <30% baseline, comparing the ratios of both to that of the baseline values and concluded that FEF$_{25-75\%}$ is a more sensitive marker for BOS than FEV$_1$ (p <0.01). All patients in this study were heart-lung or double lung recipients.[29] A second study found similar results and did not specifically evaluate the value of FEF$_{25-75\%}$ in the single lung transplant group.[30] A recent study evaluating 197 single lung recipients retrospectively, showed that 81% of patients who met BOS-0p FEV$_1$ criteria died within 3 years. Using this predictive criterion, FEV$_1$ was superior to FEF$_{25-75\%}$ in single lung recipients. The specificity and positive predictive value curves were better for obstructive versus underlying restrictive disease in this group.[31] Daily home spirometry may detect decline earlier than scheduled clinical spirometry and is a valuable tool in the diagnosis.[32]

Two studies have suggested that measurements of the slope of alveolar plateau for He and N during single breath washouts as a marker for heterogeneous ventilation may be a more sensitive marker than conventional pulmonary function testing still, including mid-expiratory flow rates.[28-30,33]

Methacholine challenge is not widely used to diagnose bronchial hyper-responsiveness as a predictor for the later development of BO. Stanbrook et al challenged 60 patients at 3 months posttransplant and determined that there was a significantly increased risk for those with a positive challenge (p <0.006). Mean time to development of BOS after a positive challenge was 16.9 months, indicating potentially useful predictive value for this test.[34]

### Radiographic Diagnosis

Early in the course, chest X-ray is usually normal. The hyperlucent lung caused by BO has been evaluated by scintigraphy, showing matched ventilation perfusion defect and delayed washout, angiography which typically shows a diminished pulmonary artery and decreased peripheral vasculature and CT which may show decreased vascularity, diminished mean attenuation values, lung volume loss, loss in integrity of airways and bronchiectasis.[35]

In advanced disease bronchiectasis may be visible both on chest X-ray and CT. CT of the chest is more useful in early diagnosis and probably the most useful radiographic test. Several findings have been described on inspiration on CT, including bronchial wall thickening, mosaic perfusion pattern and bronchial dilatation but these findings lack sensitivity.[36-39] Knollman et al studied 52 patients using high resolution electron beam (HREB) CT after lung transplant. Mean lung attenuation was correlated with decline in lung function and was lower in patients who developed BOS both in inspiration and expiration (p <0.0001). In patients with BOS, parenchymal attenuation was less homogenous in patients with BOS, with a sensitivity of 78% (7 of 9 patients) and specificity of 85% (17 of 20 patients).[40] Mean lung attenuation measurements by CT has previously been suggested as a predictor for the development of BOS 1 year later with sensitivity of 69%, specificity of 71% and accuracy 84%[41]. Several studies have suggested air-trapping on expiration by high resolution CT (HRCT) to be the most sensitive radiographic marker for BO.[42-44] Reproducibility of distribution of air-trapping has been found to be worse in patients with BO following 3 sequential CT scans performed within an hour.[45] In contradiction however, one study evaluating air-trapping by CT has shown a sensitivity of only 50%, 44% and 64% for scans obtained before the clinical appearance of BOS, after the diagnosis and later in the course of disease respectively, possibly rendering this tool for diagnosis less useful than originally suggested.[46] Magnetic resonance imaging using 3HeMRI to detect ventilation defects appears to be more sensitive than CT but the clinical application is yet to be determined.[47]

Newer quantitative ventilation scintigrams have been proposed as a sensitive marker for progression of BO in a pilot study. The average contribution to total FEV$_1$ of the native lung in single lung recipients differs significantly according to underlying diagnosis at 9%, 38% and 27% for patients with obstructive, vascular and restrictive underlying disease respectively. There appears to be good correlation with decline in FEV$_1$ and ventilation by scintigraphy in patients with vascular or restrictive disease but not with obstructive pulmonary disease. Scintigraphy may be a useful adjunct to diagnosing early BOS in single lung transplant recipients with underlying vascular or restrictive disease.[48,49]

CT remains the most useful radiographic adjunct in the diagnosis of BO but there are no pathognomonic findings and results should be interpreted within clinical context.

### Role for Bronchoscopy and Open Lung Biopsy

While the sensitivity for transbronchial biopsy to diagnose BO may be as low as 38%, the value lies more in the ability to exclude infection and acute rejection which left untreated may be a risk factor for the later development of BO.[50,51] We perform surveillance bronchoscopy with transbronchial biopsy at our institution but others have argued that there is similar outcome and no greater risk in a program which performs bronchoscopy based only on clinical indication.[52,53] The definitive histological diagnosis of BO usually requires open lung biopsy. Furthermore, open or thoracoscopic lung biopsy may provide a new and unsuspected diagnosis in as many as one third of patients with suspected BO and respiratory decline.[54] Certain broncho-alveolar lavage (BAL) characteristics have been described in patients with BO. Clinically, the role of BAL remains primarily to exclude infection which appears to be a major risk-contributor. There are no pathognomonic findings and many of the described changes reflect inflammation which may be a marker for infection or nonspecific inflammation. Increased neutrophils and indirect neutrophil markers such as matrix-metalloproteinase 9 and chemokines such as interleukin 8 have been mentioned as early BAL markers while RANTES (regulated on activation: normal T-cells expressed and secreted) is expressed both in induced sputum and BAL at higher levels in BO.[55-57] One longitudinal study showed only interleukin 12 in BAL to be predictive of the later development of BOS.[58] There is an increase in BAL CD8+ T-cells and decrease in CD4+ cells accompanied by an increase in CD8+ and CD3+ airway wall cells with time in all lung transplant recipients but these changes are more pronounced in patients with BOS.[19]

The value of a bronchoscopic study including BAL and trans-bronchial biopsies in the diagnosis of BO therefore, is to exclude other potential causes of functional decline, rather than establishing histological diagnosis. Open lung biopsy may shed light on new diagnosis but adds little to the sensitivity of BOS as the surrogate marker for BO.

### Exhaled Nitric Oxide

Animal models have shown potent up-regulation of inducible nitric oxide synthase (iNOS) in airway epithelium and fibroblasts in models of BO. NO may play a dual role in promoting fibroblast proliferation and causing airway epithelial destruction.[59] A study of 16 patients with BOS found exhaled nitric oxide (FeNO) to be elevated in those with subsequent decline in lung function compared to those with a more stable course.[60] Verleden et al studied 32 lung transplant recipients, 13 of which developed BOS. The mean FeNO was $24.3 \pm 13.2$ parts per billion (ppb) in patients with BOS and $11.4 \pm 4.9$ ppb in stable patients at 2 years post-transplant (p = 0.0054), suggesting FeNO as a sensitive and accurate early marker for BO. The native lung does not appear to contribute significantly to elevated FeNO in single lung transplant recipients.[61,62]

## Histological Diagnosis

The histological diagnosis of BO is difficult to establish, due to a patchy distribution of disease in early stages. Five to ten transbronchial biopsies obtained during bronchosocopy, will usually contain enough alveolated tissue to establish presence or absence of acute rejection. Biopsies displaying acute rejection will show the characteristic perivascular mononuclear cell infiltrates in alveolated parenchyma.[8] Transbrochial biopsies however, will typically contain few if any distal bronchioles to evaluate BO and although occasionally the histological confirmation of suspected BOS can be achieved, the low yield of this technique makes this approach unreliable.[63] Histological confirmation of BOS therefore will occur in those few cases where an open lung biopsy will be clinically indicated, or via postmortem specimens. In lung transplantation, the histological term of bronchiolitis obliterans refers to the hyalinized fibrous plaques present in the sub-mucosa of small airways, which may lead to partial of complete luminal obliteration. The deposition of scar tissue may be eccentric or concentric and may be associated with destruction of the smooth muscle wall.[94] The presence of accompanying mononuclear cell infiltrates will characterize the "active" classification of BO and the absence of mononuclear infiltrates will characterize the "inactive" classification. Typically, histological samples for BO evaluation will be studied with Hematoxilin and Eosin (H&E) stained samples in addition to the use of Elastin Van Gieson (EVG stains). For histological examples of BO please refer to Figures 1-5.

## Predisposing Factors

The factors which may predispose to the development of BO are listed in Table 2. Infectious agents that have been more frequently mentioned in studies are listed in the table. Episodes of asymptomatic acute rejection have provided some of the rationale for surveillance biopsies but, as noted above, this approach is not taken at all transplant centers.

***Table 2. Predisposing factors for bronchiolits obliterans***

1. Acute rejection
2. Primary graft dysfunction
3. Lymphocytic bronchitis/bronchiolitis
4. Gastro-esophageal reflux (GERD) / bile acid aspiration
5. Infection
   - Bacterial
     - Chlamydia pneumoniae
   - Viral
     - CMV
     - Human Herpes Virus 6
     - Respiratory Syncitial virus
     - Influenza
     - Parainfluenza
     - Adenovirus
   - Fungal
     - Scedosporium Apiospermum
6. Donor-recipient mismatching
   - HLA mismatching
   - Donor-recipient gender mismatching
   - Positive response to PRA (panel reactive antibody)

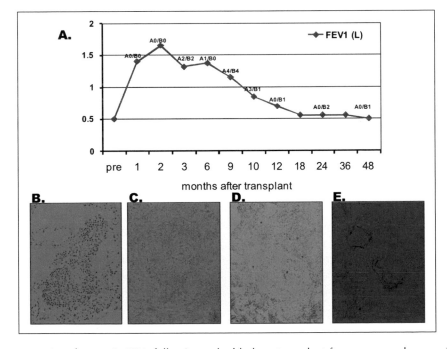

Figure 1. Figure shows progressive changes in $FEV_1$ following a double lung transplant for severe emphysema in a 52 year old female (A). Biopsy gradings obtained at each point in time are included. An open lung biopsy was obtained 9 months after transplant during an episode of severe clinical decline. Histological samples showed: B) Acute cellular rejection, severe, with a perivascular mononuclear cell infiltrate. C) Active bronchiolitis obliterans. The lumen of this bronchiole is significantly narrowed by inflamed scar tissue (D,E). Bronchiolitis obliterans and chronic vascular rejection (accelerated vascular stenosis). Lumina of both the bronchiole and the pulmonary artery are completely obliterated by scar tissue. The bronchiole can be recognized by the presence of a single elastic layer (top), whereas the pulmonary artery is exposed by its double elastica (bottom). B-D) Hematoxylin and eosin stain. E) Elastic van Gieson stain.

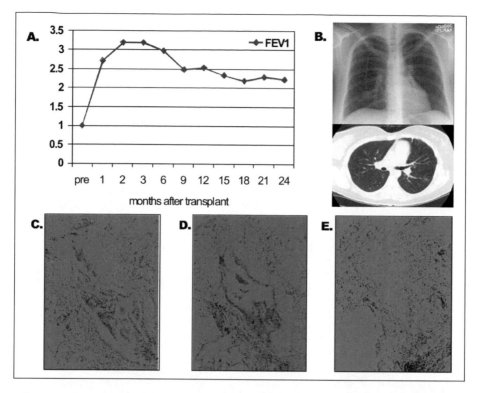

Figure 2. Figure shows changes in FEV$_1$ following a double lung transplant in a 26 year old female with a variant case of Cystic Fibrosis (A). She developed decline on FEV1 6-9 months after transplant. She had documented Gastro-Esophageal Reflux Disease and underwent fundoplication. Chest radiograph and CT chest were unrevealing (B). Histological samples showing BO, obtained with transbronchial biopsies are shown (C-E): Bronchiolitis Obliterans. Step sections of the same bronchiole show different degrees of obliteration. Hematoxylin and eosin stain.

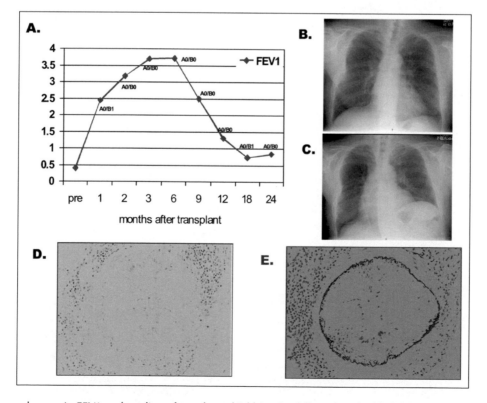

Figure 3. Figure shows changes in FEV1 and grading of transbronchial biopsies following a double lung transplantation on a 56 year old male with severe emphysema (A). Nine months after transplant the patient developed severe herpes zoster infection in the left neck, producing left phrenic nerve neuropathy, resulting in a paralyzed left hemidiaphragm, followed by decline in FEV1. Chest radiograph postdouble lung transplant is shown (B), paralyzed left hemidiaphragm post herpetic infection is shown (C). An open lung biopsy of the left lung was obtained at the time of performing surgical plication of the left diaphragm. Histological samples (D,E) showed concentric bronchiolitis obliterans. The lumen of this bronchiole is obliterated by concentric submucosal scar tissue. D) Hematoxylin and eosin stain. E) Elastic van Gieson stain.

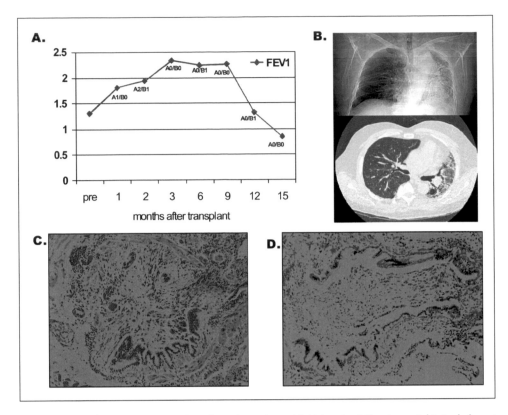

Figure 4. Figure shows progressive changes in FEV$_1$ and grading of transbronchial biopsies following a right single lung transplant for severe idiopathic pulmonary fibrosis in a 61 year old male (A). One year after transplant, patient developed a syndrome consistent with viral upper respiratory infection followed by severe decline in lung function. Chest radiographs and CT scans showed no infiltrates in transplanted lung and severe pulmonary fibrosis in native lung (B). Histology from open lung biopsies obtained are shown (C,D): Bronchiolitis obliterans. Eccentric scar tissue obliterates the lumen of this bronchiole. C) Hematoxylin and eosin stain. D) Elastic van Gieson stain.

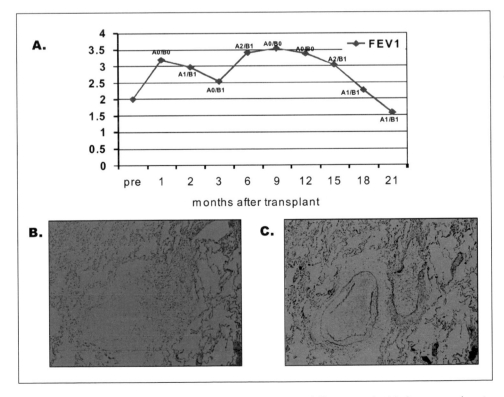

Figure 5. Figure shows changes in FEV$_1$ and grading of transbronchial biopsies following a double lung transplant in a 66 year old male with severe idiopathic pulmonary fibrosis. He developed progressive decline in lung function 15 months after transplant (A). Histology from open lung biopsy are shown (B,C): This bronchiole, which is completely obliterated by scar tissue, can only be recognized by the presence of smooth muscle and elastic fibers in the wall. The accompanying pulmonary artery shows pulmonary hypertension with medial hypertrophy. B) Hematoxylin and eosin stain. C) Elastic van Gieson stain.

## Acute Rejection

Acute rejection is the most widely accepted risk factor for BO and BOS. Systematic review of published literature supports the concept that BO arises from alloimmunologic injury marked by clinically diagnosed episodes of acute rejection.[64,65] Three or more episodes of mild to severe acute rejection (≥A2) has been cited as a risk factor for the subsequent development of bronchiolitis obliterans.[7,66,67] A retrospective study over 7 years and involving 228 patients, however, has shown minimal rejection (A1) to be a distinct risk factor for BOS with a comparable risk to that of A2 rejection, with possible new therapeutic implications. Even a solitary episode of A1 rejection may increase the risk for BO.[68,69]

## Lymphocytic Bronchitis/Bronchiolitis

Lymphocytic inflammation of airways seen in bronchial biopsies of lung transplant recipients may represent acute rejection after having excluded the confounding possibility of coexistent infection. This finding has been associated with BO onset and the strength of the relationship appears to grow over time.[70-72]

## Primary Graft Dysfunction

There is conflicting data regarding primary graft dysfunction (PGD), previously referred to as severe acute lung injury, as a risk factor for the development of BO. One study of 115 patients, 41 of whom developed BO suggested PGD as a risk factor for BO. While short-term survival of less than 30 days is clearly adversely affected by PGD and risk of death extends beyond the first transplant year, this complication was not clearly defined as an independent risk factor for the development of BO in a review of 5,262 patients from the United Network for Organ sharing/International Society of Heart and Lung Transplantation registry between 1994 and 2000. BOS may potentially have a relationship with PGD but in this review, primary graft dysfunction appeared to be an independent risk factor for mortality.[73-75] BOS may be under-reported in the ISHLT registry and so this data should be interpreted with caution.[76]

## Acid and Non-Acid Reflux

Non-immune insults on the airways may play a significant role in the pathogenesis of BO. D'Ovidio has demonstrated that freedom from BO is significantly shortened in patients with elevated BALF bile acids and acid reflux appears to adversely affect the $FEV_1$.[77] At the Duke Lung Transplant Program, 128 lung transplant recipients were evaluated for reflux disease by esophageal probe and 43 of 93 patients with abnormal pH study underwent surgical fundoplication. After the procedure 16 patients had improved BOS scores and 13 subsequently no longer met the criteria for BOS. Overall survival was improved in patients who had normal pH studies or who had Nissen fundoplication.[78, 79] Recently, a case has been made for safely being able to perform Nissen fundoplication in patients with end-stage lung disease awaiting transplantation.[80]

## Infection

Infection, particularly with respiratory viruses, may lead to rapid decline in lung function. A study of 100 lung transplant recipients with respiratory tract infections had nasopharyngeal and throat swabs taken. Fifty had respiratory tract infections (RTI) and a definite viral etiology was confirmed in 33 of 50 with RTI infections. An episode of acute rejection was defined in 16% (n = 8) in the infected group vs. 0 of 50 in the control group and a decline in FEV1 of greater than 20% was also defined in 18% of the infected group at 3 months vs. 0% in the control group (p = 0.006). Viruses identified included rhinovirus (9), corona virus (8), RSV (6), influenza A (5) and human metapneumovirus (1).[81]

Lower respiratory tract infections with viruses have been cited as a distinct risk factor for BOS and death.[82] In one study, human herpes virus 6 (HHV 6) was found in a surprisingly large number of patients (20 of 88) and was associated with risk for developing BOS > stage 1 and death.[83] In a study of 115 lung transplant recipients, 41 developed BOS. Two or more episodes of acute rejection were predictors of BOS onset, whereas CMV infection and a history of ischemia reperfusion injury were strongly associated not with onset but with development and progression of BOS.[73] Studies have suggested an association with CMV infection and ocurrence of BO.[84] Acute rejection often precedes the development of CMV pneumonia as a result of enhanced immunosuppression to treat the former and one study suggests that CMV pneumonia is not a risk factor for subsequent BOS.[85] While a direct relationship between CMV infection and bronchiolitis obliterans has not been clearly demonstrated in studies, prophylaxis has significantly improved survival and CMV donor positive, recipient negative status has been recognized as an independent risk factor for death.[86] In a consecutive series of 68 lung transplant recipients, CMV prophylaxis with CMV hyperimmune globulin and ganciclovir resulted in improved survival and a markedly decreased three year BOS diagnosis in the treatment group (n = 38).[87]

## Donor-Recipient Mismatching

Small studies have shown the presence of anti-HLA class 1 antibodies, detected by PRA-STAT ELISA to be predictive of the later development of BOS in humans and have been shown to induce growth factors, apoptosis and proliferation of airway epithelial cells and serve as a chemo-attractant for inflammatory cells in animal models.[88,89,23] Furthermore, in a study analyzing sera from 90 recipients, 9 patients developed de novo HLA antibodies posttransplant specific to donor antigens. A positive posttransplant flow-PRA was significantly associated with BOS grades 1, 2 or 3 (hazards ratio 3.19). Four of these patients died with BO and from this study it would appear that BO patients with positive flow-PRA may have a worse prognosis than other BO patients. A single study following 152 patients determined donor-recipient mismatch at the HLA-A locus to be significant for the development of BO.[24,90,91] An intriguing but small retrospective study evaluating 98 recipients for gender matching showed improvement in overall survival and BO free period for gender-mismatched donor and recipient pairs. Male donor, male recipient pairs appeared to have the worst survival probability. The Registry of International Society of Heart and Lung Transplantation adult lung transplant report of 2004 on the other hand cited female-donor to female recipient as a significantly protective combination in survival.[76,92]

# Prognosis

Overall survival rates from the Registry of the International Society for Heart and Lung Transplantation for all lung transplant recipients from January 1994 to June 2003 were 86% at 3 months, 76% at one year, 60% at 3 years, 49% at 5 years and 24% at 10 years. While mortality from 0 to 30 days posttransplant is overwhelmingly due to acute rejection (63%) and from 31 days to 1 year due to infection (37%), bronchiolitis obliterans and infection feature as leading causes of death from 1 to 3 years posttransplant (26.5% and 24.6% respectively). Between 3 and 5 years and after 5 years, bronchiolitis is easily the predominant cause of death (28.9% and 27.1%) respectively.

Airway obstruction easily outstrips the development of coronary vasculopathy by 5 years posttransplantation and 44% of recipients between April 1994 and June 2004 with 5 years follow-up had developed BOS.[93]

# References

1. Hardy JD, Webb WR, Dalton ML et al. Lung Homotransplantation in man. Jama 1963; 186:1065-74.
2. Veith FJ. Lung transplantation. Surg Clin North Am 1978; 58(2):357-64.
3. Reitz BA, Wallwork JL, Hunt SA et al. Heart-lung transplantation: successful therapy for patients with pulmonary vascular disease. N Engl J Med 1982; 306(10):557-64.
4. Unilateral lung transplantation for pulmonary fibrosis. Toronto Lung Transplant Group. N Engl J Med 1986; 314(18):1140-45.
5. Burke CM, Theodore J, Dawkins KD et al. Posttransplant obliterative bronchiolitis and other late lung sequelae in human heart-lung transplantation. Chest 1984; 86(6):824-29.
6. Paul LC. Chronic rejection of organ allografts: magnitude of the problem. Transplant Proc 1993; 25(2):2024-25.
7. Keller CA, Cagle PT, Brown RW et al. Bronchiolitis obliterans in recipients of single, double and heart-lung transplantation. Chest 1995; 107(4):973-80.
8. Yousem SA, Berry GJ, Cagle PT et al. Revision of the 1990 working formulation for the classification of pulmonary allograft rejection: Lung Rejection Study Group. J Heart Lung Transplant 1996; 15(1 Pt 1):1-15.
9. Cooper JD, Billingham M, Egan T et al. A working formulation for the standardization of nomenclature and for clinical staging of chronic dysfunction in lung allografts. International Society for Heart and Lung Transplantation. J Heart Lung Transplant 1993; 12(5):713-16.
10. Estenne M, Maurer JR, Boehler A et al. Bronchiolitis obliterans syndrome 2001: an update of the diagnostic criteria. J Heart Lung Transplant 2002; 21(3):297-310.
11. Trulock EP, Edwards LB, Taylor DO et al. Registry of the International Society for Heart and Lung Transplantation: twenty-third official adult lung and heart-lung transplantation report—2006. J Heart Lung Transplant 2006; 25(8):880-92.
12. Martinu T, Howell DN, Davis RD et al. Pathologic correlates of bronchiolitis obliterans syndrome in pulmonary retransplant recipients. Chest 2006; 129(4):1016-23.
13. Lande JD, Dalheimer SL, Mueller DL et al. Gene expression profiling in murine obliterative airway disease. Am J Transplant 2005; 5(9):2170-84.
14. Adams BF, Brazelton T, Berry GJ et al. The role of respiratory epithelium in a rat model of obliterative airway disease. Transplantation 2000; 69(4):661-64.
15. Alho HS, Salminen US, Maasilta PK et al. Epithelial apoptosis in experimental obliterative airway disease after lung transplantation. J Heart Lung Transplant 2003; 22(9):1014-22.
16. Nicod LP. Mechanisms of airway obliteration after lung transplantation. Proc Am Thorac Soc 2006; 3(5):444-49.
17. Duncan SR, Leonard C, Theodore J et al. Oligoclonal CD4(+) T-cell expansions in lung transplant recipients with obliterative bronchiolitis. Am J Respir Crit Care Med 2002; 165(10):1439-44.
18. SivaSai KS, Smith MA, Poindexter NJ et al. Indirect recognition of donor HLA class I peptides in lung transplant recipients with bronchiolitis obliterans syndrome. Transplantation 1999; 67(8):1094-98.
19. Zheng L, Orsida B, Whitford H et al. Longitudinal comparisons of lymphocytes and subtypes between airway wall and bronchoalveolar lavage after human lung transplantation. Transplantation 2005; 80(2):185-92.
20. Medoff BD, Seung E, Wain JC et al. BLT1-mediated T-cell trafficking is critical for rejection and obliterative bronchiolitis after lung transplantation. J Exp Med 2005; 202(1):97-110.
21. Ward C, Forrest IA, Murphy DM et al. Phenotype of airway epithelial cells suggests epithelial to mesenchymal cell transition in clinically stable lung transplant recipients. Thorax 2005; 60(10):865-71.
22. Brocker V, Langer F, Fellous TG et al. Fibroblasts of recipient origin contribute to bronchiolitis obliterans in human lung transplants. Am J Respir Crit Care Med 2006; 173(11):1276-82.
23. Jaramillo A, Smith MA, Phelan D et al. Development of ELISA-detected anti-HLA antibodies precedes the development of bronchiolitis obliterans syndrome and correlates with progressive decline in pulmonary function after lung transplantation. Transplantation 1999; 67(8):1155-61.
24. Sundaresan S, Mohanakumar T, Smith MA et al. HLA-A locus mismatches and development of antibodies to HLA after lung transplantation correlate with the development of bronchiolitis obliterans syndrome. Transplantation 1998; 65(5):648-53.
25. Magro CM, Pope Harman A, Klinger D et al. Use of C4d as a diagnostic adjunct in lung allograft biopsies. Am J Transplant 2003; 3(9):1143-54.
26. Tikkanen JM, Koskinen PK, Lemstrom KB. Role of endogenous endothelin-1 in transplant obliterative airway disease in the rat. Am J Transplant 2004; 4(5):713-20.
27. Cooper JD, Billingham M, Egan T et al. A working formulation for the standardization of nomenclature and for clinical staging of chronic dysfunction in lung allografts. International Society for Heart and Lung Transplantation. J Heart Lung Transplant 1993; 12(5):713-16.
28. Estenne M, Maurer JR, Boehler A et al. Bronchiolitis obliterans syndrome 2001: an update of the diagnostic criteria. J Heart Lung Transplant 2002; 21(3):297-310.
29. Patterson GM, Wilson S, Whang JL et al. Physiologic definitions of obliterative bronchiolitis in heart-lung and double lung transplantation: a comparison of the forced expiratory flow between 25% and 75% of the forced vital capacity and forced expiratory volume in one second. J Heart Lung Transplant 1996; 15(2):175-81.
30. Reynaud-Gaubert M, Thomas P, Badier M et al. Early detection of airway involvement in obliterative bronchiolitis after lung transplantation. Functional and bronchoalveolar lavage cell findings. Am J Respir Crit Care Med 2000; 161(6):1924-29.
31. Lama VN, Murray S, Mumford JA et al. Prognostic value of bronchiolitis obliterans syndrome stage 0-p in single-lung transplant recipients. Am J Respir Crit Care Med 2005; 172(3):379-83.
32. Finkelstein SM, Snyder M, Stibbe CE et al. Staging of bronchiolitis obliterans syndrome using home spirometry (see comment). Chest 1999; 116(1):120-26.
33. Estenne M, Van Muylem A, Knoop C et al. Detection of obliterative bronchiolitis after lung transplantation by indexes of ventilation distribution. Am J Respir Crit Care Med 2000; 162(3 Pt 1):1047-51.
34. Stanbrook MB, Kesten S. Bronchial hyperreactivity after lung transplantation predicts early bronchiolitis obliterans. Am J Respir Crit Care Med 1999; 160(6):2034-39.
35. Miravitlles M, Alvarez-Castells A, Vidal R et al. Scintigraphy, angiography and computed tomography in unilateral hyperlucent lung due to obliterative bronchiolitis. Respiration 1994; 61(6):324-29.
36. Morrish WF, Herman SJ, Weisbrod GL et al. Bronchiolitis obliterans after lung transplantation: findings at chest radiography and high-resolution CT. The Toronto Lung Transplant Group. Radiology 1991; 179(2):487-90.
37. Lentz D, Bergin CJ, Berry GJ et al. Diagnosis of bronchiolitis obliterans in heart-lung transplantation patients: importance of bronchial dilatation on CT. AJR Am J Roentgenol 1992; 159(3):463-67.
38. Loubeyre P, Revel D, Delignette A et al. Bronchiectasis detected with thin-section CT as a predictor of chronic lung allograft rejection (see comment). Radiology 1995; 194(1):213-16.
39. Ikonen T, Kivisaari L, Harjula AL et al. Value of high-resolution computed tomography in routine evaluation of lung transplantation recipients during development of bronchiolitis obliterans syndrome. J Heart Lung Transplant 1996; 15(6):587-95.
40. Knollmann FD, Kapell S, Lehmkuhl H et al. Dynamic high-resolution electron-beam CT scanning for the diagnosis of bronchiolitis obliterans syndrome after lung transplantation. Chest 2004; 126(2):447-56.
41. Knollmann FD, Ewert R, Wundrich T et al. Bronchiolitis obliterans syndrome in lung transplant recipients: use of spirometrically gated CT. Radiology 2002; 225(3):655-62.
42. Leung AN, Fisher K, Valentine V et al. Bronchiolitis obliterans after lung transplantation: detection using expiratory HRCT. Chest 1998; 113(2):365-70.
43. Bankier AA, Van Muylem A, Knoop C et al. Bronchiolitis obliterans syndrome in heart-lung transplant recipients: diagnosis with expiratory CT. Radiology 2001; 218(2):533-39.
44. Siegel MJ, Bhalla S, Gutierrez FR et al. Postlung transplantation bronchiolitis obliterans syndrome: usefulness of expiratory thin-section CT for diagnosis. Radiology 2001; 220(2):455-62.
45. Bankier AA, Van Muylem A, Scillia P et al. Air trapping in heart-lung transplant recipients: variability of anatomic distribution and extent at sequential expiratory thin-section CT. Radiology 2003; 229(3):737-42.
46. Konen E, Gutierrez C, Chaparro C et al. Bronchiolitis obliterans syndrome in lung transplant recipients: can thin-section CT findings predict disease before its clinical appearance? Radiology 2004; 231(2):467-73.
47. Gast KK, Viallon M, Eberle B et al. MRI in lung transplant recipients using hyperpolarized 3He: comparison with CT. J Magn Reson Imaging 2002; 15(3):268-74.
48. Ouwens JP, van der Bij W, van der Mark TW et al. The value of ventilation scintigraphy after single lung transplantation. J Heart Lung Transplant 2004; 23(1):115-21.
49. Johansson A, Moonen M, Enocson A et al. Detection of chronic rejection by quantitative ventilation scintigrams in lung-transplanted patients: a pilot study. Clinical Physiology and Functional Imaging 2005; 25(3):183-87.
50. Cagle PT, Brown RW, Frost A et al. Diagnosis of chronic lung transplant rejection by transbronchial biopsy. Mod Pathol 1995; 8(2):137-42.

51. Higenbottam T, Stewart S, Penketh A et al. Transbronchial lung biopsy for the diagnosis of rejection in heart-lung transplant patients. Transplantation 1988; 46(4):532-39.
52. heValentine VG, Taylor DE, Dhillon GS et al. Success of lung transplantation without surveillance bronchoscopy. J Heart Lung Transplant 2002; 21(3):319-26.
53. Swanson SJ, Mentzer SJ, Reilly JJ et al. Surveillance transbronchial lung biopsies: implication for survival after lung transplantation. J Thorac Cardiovasc Surg 2000; 119(1):27-37.
54. Weill D, McGiffin DC, Zorn GL Jr et al. The utility of open lung biopsy following lung transplantation. J Heart Lung Transplant 2000; 19(9):852-57.
55. Mamessier E, Milhe F, Badier M et al. Comparison of induced sputum and bronchoalveolar lavage in lung transplant recipients. J Heart Lung Transplant 2006; 25(5):523-32.
56. Hubner RH, Meffert S, Mundt U et al. Matrix metalloproteinase-9 in bronchiolitis obliterans syndrome after lung transplantation. Eur Respir J 2005; 25(3):494-501.
57. Riise GC, Andersson BA, Kjellstrom C et al. Persistent high BAL fluid granulocyte activation marker levels as early indicators of bronchiolitis obliterans after lung transplant. Eur Respir J 1999; 14(5):1123-30.
58. Meloni F, Vitulo P, Cascina A et al. Bronchoalveolar lavage cytokine profile in a cohort of lung transplant recipients: a predictive role of interleukin-12 with respect to onset of bronchiolitis obliterans syndrome. J Heart Lung Transplant 2004; 23(9):1053-60.
59. Romanska HM, Ikonen TS, Bishop AE, et al. Up-regulation of inducible nitric oxide synthase in fibroblasts parallels the onset and progression of fibrosis in an experimental model of posttransplant obliterative airway disease. J Pathol 2000; 191(1):71-77.
60. Brugiere O, Thabut G, Mal H et al. Exhaled NO may predict the decline in lung function in bronchiolitis obliterans syndrome. Eur Respir J 2005; 25(5):813-19.
61. Verleden GM, Dupont LJ, Delcroix M et al. Exhaled nitric oxide after lung transplantation: impact of the native lung. Eur Respir J 2003; 21(3):429-32.
62. Verleden GM, Dupont LJ, Van Raemdonck DE et al. Lung Transplant G. Accuracy of exhaled nitric oxide measurements for the diagnosis of bronchiolitis obliterans syndrome after lung transplantation. Transplantation 2004; 78(5):730-33.
63. Yousem SA, Paradis I, Griffith BP. Can transbronchial biopsy aid in the diagnosis of bronchiolitis obliterans in lung transplant recipients? Transplantation 1994; 57(1):151-53.
64. Sharples LD, McNeil K, Stewart S et al. Risk factors for bronchiolitis obliterans: a systematic review of recent publications. J Heart Lung Transplant 2002; 21(2):271-81.
65. Scott AI, Sharples LD, Stewart S. Bronchiolitis obliterans syndrome: risk factors and therapeutic strategies. Drugs 2005; 65(6):761-71.
66. Heng D, Sharples LD, McNeil K et al. Bronchiolitis obliterans syndrome: incidence, natural history, prognosis and risk factors. J Heart Lung Transplant 1998; 17(12):1255-63.
67. Bando K, Paradis IL, Similo S et al. Obliterative bronchiolitis after lung and heart-lung transplantation. An analysis of risk factors and management. J Thorac Cardiovasc Surg 1995; 110(1):4-13; discussion 4.
68. Khalifah AP, Hachem RR, Chakinala MM et al. Minimal acute rejection after lung transplantation: a risk for bronchiolitis obliterans syndrome. Am J Transplant 2005; 5(8):2022-30.
69. Hachem RR, Khalifah AP, Chakinala MM et al. The significance of a single episode of minimal acute rejection after lung transplantation. Transplantation 2005; 80(10):1406-13.
70. El-Gamel A, Sim E, Hasleton P et al. Transforming growth factor beta (TGF-beta) and obliterative bronchiolitis following pulmonary transplantation. J Heart Lung Transplant 1999; 18(9):828-37.
71. Girgis RE, Tu I, Berry GJ et al. Risk factors for the development of obliterative bronchiolitis after lung transplantation. J Heart Lung Transplant 1996; 15(12):1200-8.
72. Husain AN, Siddiqui MT, Holmes EW et al. Analysis of risk factors for the development of bronchiolitis obliterans syndrome. Am J Respir Crit Care Med 1999; 159(3):829-33.
73. Fiser SM, Tribble CG, Long SM et al. Ischemia-reperfusion injury after lung transplantation increases risk of late bronchiolitis obliterans syndrome. Annals of Thoracic Surgery 2002; 73(4):1041-7; discussion 7-8.
74. Christie JD, Kotloff RM, Ahya VN et al. The effect of primary graft dysfunction on survival after lung transplantation. Am J Respir Crit Care Med 2005; 171(11):1312-16.
75. Fisher AJ, Wardle J, Dark JH et al. Non-immune acute graft injury after lung transplantation and the risk of subsequent bronchiolitis obliterans syndrome (BOS). J Heart Lung Transplant 2002; 21(11):1206-12.
76. Boucek MM, Edwards LB, Keck BM et al. Registry for the International Society for Heart and Lung Transplantation: seventh official pediatric report—2004. J Heart Lung Transplant 2004; 23(8):933-47.
77. D'Ovidio F, Mura M, Tsang M et al. Bile acid aspiration and the development of bronchiolitis obliterans after lung transplantation. J Thorac Cardiovasc Surg 2005; 129(5):1144-52.
78. Davis RD Jr, Lau CL, Eubanks S et al. Improved lung allograft function after fundoplication in patients with gastroesophageal reflux disease undergoing lung transplantation. J Thorac Cardiovasc Surg 2003; 125(3):533-42.
79. Hadjiliadis D, Duane Davis R, Steele MP et al. Gastroesophageal reflux disease in lung transplant recipients. Clin Transplant 2003; 17(4):363-68.
80. Linden PA, Gilbert RJ, Yeap BY et al. Laparoscopic fundoplication in patients with end-stage lung disease awaiting transplantation. J Thorac Cardiovasc Surg 2006; 131(2):438-46.
81. Kumar D, Erdman D, Keshavjee S et al. Clinical impact of community-acquired respiratory viruses on bronchiolitis obliterans after lung transplant. Am J Transplant 2005; 5(8):2031-36.
82. Khalifah AP, Hachem RR, Chakinala MM et al. Respiratory viral infections are a distinct risk for bronchiolitis obliterans syndrome and death. Am J Respir Crit Care Med 2004; 170(2):181-87.
83. Neurohr C, Huppmann P, Leuchte H et al. Human herpesvirus 6 in bronchalveolar lavage fluid after lung transplantation: a risk factor for bronchiolitis obliterans syndrome? Am J Transplant 2005; 5(12):2982-91.
84. Cerrina J, Le Roy Ladurie F, Herve PH et al. Role of CMV pneumonia in the development of obliterative bronchiolitis in heart-lung and double-lung transplant recipients. Transpl Int 1992; 5 (Suppl 1):S242-45.
85. Tamm M, Aboyoun CL, Chhajed PN et al. Treated cytomegalovirus pneumonia is not associated with bronchiolitis obliterans syndrome. Am J Respir Crit Care Med 2004; 170(10):1120-23.
86. Ettinger NA, Bailey TC, Trulock EP et al. Cytomegalovirus infection and pneumonitis. Impact after isolated lung transplantation. Washington University Lung Transplant Group. Am Rev Respir Dis 1993; 147(4):1017-23.
87. Ruttmann E, Geltner C, Bucher B et al. Combined CMV prophylaxis improves outcome and reduces the risk for bronchiolitis obliterans syndrome (BOS) after lung transplantation. Transplantation 2006; 81(10):1415-20.
88. Reznik SI, Jaramillo A, Zhang L et al. Anti-HLA antibody binding to hla class I molecules induces proliferation of airway epithelial cells: a potential mechanism for bronchiolitis obliterans syndrome. J Thorac Cardiovasc Surg 2000; 119(1):39-45.
89. Maruyama T, Jaramillo A, Narayanan K et al. Induction of obliterative airway disease by anti-HLA class I antibodies. Am J Transplant 2005; 5(9):2126-34.
90. Schulman LL, Weinberg AD, McGregor CC et al. Influence of donor and recipient HLA locus mismatching on development of obliterative bronchiolitis after lung transplantation. Am J Respir Crit Care Med 2001; 163(2):437-42.
91. Palmer SM, Davis RD, Hadjiliadis D et al. Development of an antibody specific to major histocompatibility antigens detectable by flow cytometry after lung transplant is associated with bronchiolitis obliterans syndrome. Transplantation 2002; 74(6):799-804.
92. Roberts DH, Wain JC, Chang Y et al. Donor-recipient gender mismatch in lung transplantation: impact on obliterative bronchiolitis and survival. J Heart Lung Transplant 2004; 23(11):1252-59.
93. Trulock EP, Edwards LB, Taylor DO et al. Registry of the International Society for Heart and Lung Transplantation: twenty-second official adult lung and heart-lung transplant report—2005. J Heart Lung Transplant 2005; 24(8):956-67.
94. Khoor A, Yousem SA. Pathology of lung transplantation. In: Leslie KO, Wick MR, eds. Practical Pulmonary Pathology. A diagnostic approach, Philadelphia, Churchill Livingstone, Elsevier Inc, 2005:401-422.

# CHAPTER 15

# Treatment of Chronic Graft Failure after Lung Transplantation

## Francisco G. Alvarez and Cesar A. Keller*

## Abstract

Since lung transplantation became a reality 25 years ago, improvements in lung preservation, surgical techniques and post-operative management have improved the 1-year patient survival to almost 80%. Beyond the first year, though, bronchiolitis obliterans (BO), considered a form of chronic allograft rejection, has become a major threat to survival since it affects up to 50 to 60% of patients who survive five years after transplantation and more than half of these patients will eventually succumb to it within the next few years. Treatment for BO includes augmented immunosuppression with boluses of parenteral corticosteroids in addition to adjusting the immunosuppression regimen either by increasing the dose of the drugs being use or switching to other immunosuppressant s within the same class. The addition of other drugs like methotrexate, or newer agents like sirolimus have also being used. Aerosolized cyclosporine is a promising therapeutic alternative but its role in preventing or treating BO remains to be defined. Other strategies include using cytolytic therapy against T-lymphocytes, total lymphoid irradiation and photopheresis. To date none of these strategies have proven conclusively effective in preventing or treating an established BO. The major problem with the available data is that it is spread across too many transplant centers in the form of small, non controlled, non randomized studies which make interpretation of results very difficult. Large multicenter and well designed trials will be essential to advance our understanding of the complex pathophysiology of this condition and to find an effective therapy against BO.

## Introduction

Therapy for BOS is in general ineffective. This fact has highlighted the need for early detection with the hope that early modification of immunosuppressive therapy may slow down or prevent the development of BOS. This was the main reason that prompted the re-examination of the BOS diagnostic criteria by the International Society for Heart and Lung Transplantation (ISHLT) in 2001.[1]

Unfortunately, although several risks factors for BOS have been recognized, the predictive value of each one of those factors is not well understood. Furthermore, there are well known risks of increasing immunosuppression in those patients felt to be at a higher risk for developing BOS. Clearly more information is needed in this regard but the accumulated experience suggests that certain preventive measures should be strongly considered:

1. Based on growing evidence that even asymptomatic episodes of A1 rejection can increase the risks for BO,[2,3] there is an increasing consensus that such episodes should be treated more aggressively.

2. More aggressive prevention, prophylaxis and/or treatment against those infections also related to BO has been advocated by others.[4] Some reports have found an association between the development of BO and nonCMV viral infections;[5,6] bacterial infections;[7] and specially CMV infection, which has been implicated as a risk factors in several studies.[8-11]

3. Early detection and treatment of gastroesophageal reflux disease (GERD), even with fundoplication surgery, also seems to be effective in preventing or slowing the development of BOS.[12,13]

Bronchiolitis obliterans is a complex syndrome that involves both alloimmune and nonalloimmune mechanisms. Because the predominance of any of these mechanisms may vary in different patients it has been suggested that the treatment of BOS should be individualized accordingly.[14]

Currently BOS is usually treated by augmenting immunosuppression by changing medications within therapeutic classes, by the addition of new drugs or by applying nonmedicinal immune-modulating therapies (see Table 1). Unfortunately these therapeutic interventions are based mostly on small case series and rather anecdotal reports without enough statistical power to allow firm recommendations in most cases.

## Immunosuppressive Therapy

### Enhanced Immunosuppression

The rationale for augmented immunosuppression in patients with bronchiolitis obliterans is based on the realization that although BO is a complex and heterogeneous process there is ample evidence that alloimmune injury, defined histologically as acute rejection, is one of the more important risk factors for the development of BOS. Several studies have shown that recurrent episodes of acute rejection,[15] a single episode of A2 rejection;[7] or multiple or even one single episode of an A1 rejection[3,16] are associated with an increased risk of BO, so, at least in theory, in a subset of patients at risk or with established BO further suppression of the immune system should lead to prevention, stabilization or reversal of the BO process.

Multiple reports using different drugs have been published to date. Most are single centers trials with small number of patients, rather limited follow-up and without a control group. Some of these trials evaluate the effectiveness of the drugs to prevent or delay the development of BO while other focus on the treatment of an already established BO.

### High Dose of Corticosteroids

Very few studies have been reported using corticosteroids as only treatment for BOS/BO since they are usually administered in combination

*Corresponding Author: Cesar A. Keller—Medical Director, Lung Transplant Program, Mayo Clinic Transplant Center, 4205 Belfort Road, Suite 1100, Jacksonville FL 32216, USA. Email: keller.cesar@mayo.edu

*Chronic Allograft Failure: Natural History, Pathogenesis, Diagnosis and Management*, edited by Nasimul Ahsan. ©2008 Landes Bioscience.

## Table 1. Different therapeutic modalities that have been used in the treatment of BO

**I. Immnunosuppressive Therapy**

1. Enhanced immunosuppression
   a. High dose corticosteroids
   b. Cytolytic therapy
   c. Anti interleukin 2 receptor antibodies
2. Switch from cyclosporine to tacrolimus and/or from azathioprine to mycophenolate
3. Alternate immunosupression: sirolimus, everolimus, methotrexate
4. Inhaled immunosuppressive agents
5. Other
   a. Photopheresis
   b. Total lymphoid irradiation

**II. Non Immunosuppressive Therapy**

1. Macrolides
2. Gastroesophageal reflux management
3. Statins

**III. Management of Associated Infections**

1. Cytomegalovirus infections
2. Other viral infections
3. Atypical infections

**IV. Other**

1. Retransplantation

with other drugs. Although one of the first steps in treating a patient with suspected BO is the administration of pulse steroids there is not firm data that this approach is actually effective in altering the natural progression of BOS. Two studies, including only seven patients, reported favorable response to corticosteroids alone,[17,18] although later three of them had a relapse of the BOS. The steroid most commonly used in the United States is methylprednisolone at a dose of 10 mg/kg or up to 1 gram daily for three days.

There are few reports evaluating the efficacy of inhaled corticosteroids. One small randomized study[19] showed that nebulized budesonide in patients with recurrent episodes of acute rejection was successful in preventing BO but this was a very small study with only 11 patients. Another study in 14 patients with lymphocytic bronchiolitis[20] showed significant improvement on the $FEV_1$ and exhaled nitric oxide. The only prospective, randomized double-blind study of inhaled steroids in patients with BOS found no benefit in the treatment group.[21]

In summary, there is not enough data to determine the actual effectiveness of inhaled or systemic corticosteroids in patients with BOS.

## Cytolytic Therapy

Three cytolytic drugs, capable of producing a profound depletion of T-lymphocytes, have been used either as induction agents, as treatment for refractory rejection or as treatment for an established BO/BOS: Monoclonal antibody against CD3 lymphocytes (OKT3), antithymocyte globulin (ATG) and antilymphocyte globulin (ALG).

### Induction Therapy

According to the last ISHLT registry report,[22] 41 to 47% of patients transplanted received induction therapy, mostly with ATG, ALG or with a interleukin-2 receptor antagonist (IL-2R). Less than 5% received OKT3. Induction therapy with either ATG or ALG was superior to induction with IL-2R or no induction at all in reducing both the percentage of patients treated for rejection the first year post transplantation and, among recipients treated for rejection during the first year, in reducing the average number of rejection episodes. Other studies have shown similar results.[23,24] In a small, retrospective study comparing induction therapy with OKT3 versus ALG Ross et al[25]

found similar incidence of BOS but the time of onset was longer in the OKT3 group.

Unfortunately, according to the IHSLT registry,[22] induction therapy did not change freedom from BOS, although it had a small positive effect in survival over 10 years.

### Established BO

Three retrospective studies suggest that it is possible to arrest or slow down the decline in lung function in patients with stable BOS/BO.[26-28] The largest of these was the study by Date et al[26] who reported 64 courses of cytolytic therapy (ALG, ATGAM or OKT3) in 48 patients and showed a fall in the decline of the $FEV_1$. There was no difference between the different agents used.

Kesten et al[27] reported the use of ATG in 15 patients with BOS. Pulmonary function improved in 2 patients, remained stable in 5 and continued to deteriorate in 8. In another study, Snell[28] used ATGAM in 10 patients with BOS and reported a significant decrease in the rate of the decline in the $FEV_1$ over a mean follow-up period of 310 months in 9 out of the 10 patients.

In general, the use of cytolytic therapy has shown only marginal results at best and they are associated with significant side effects like opportunistic infections, cytokine release syndrome, hypotension, etc. To this date no firm data exists that this therapeutic modality alters the progression of BOS or improves survival.

### Interleukin 2 Receptor Antibodies

There are few studies using interleukin 2 receptors antagonists (IL-2R) in lung transplant recipients and their role in the prevention or treatment of BO is unknown. Slebos et al[29] compared 17 patients that received induction with basiliximab with 34 patients that received induction with anti-thymocyte globulin and found that although survival was similar in both groups there was a beneficial trend in freedom of BOS in the IL-2R treated group. Analysis of bronchoalveolar lavage also showed a significant decrease in the number of lymphocytes in the baxiliximab group. In another retrospective report Garrity et al[30] found that induction therapy with daclizumab resulted in a significant decrease in the incidence of grade 2 or greater acute rejection compared to 34 historical controls but it is not known if this will translate into a decreased incidence of BOS or improved survival. In contrast, Brock[31] found no difference in the rates of acute rejection between three different induction regimens with OKT3, ATG or daclizumab.

One study specifically evaluated the use of anti IL-2R monoclonal antibody in patients with established BOS.[32] In this small, noncontrolled trial five patients with the diagnosis of BOS were treated with a six month course of daclizumab (1 mg/kg IV) every 30 days and high dose steroids. Lung function stabilized in four patients for at least 7 months. Side effects included fungal infection in 4 patients and was the cause of death in one.

The latest IHSLT report[22] indicates that induction therapy with IL-2R antagonists does not significantly alter the freedom from BOS and, in fact, it seems to be less effective than induction with ALG or ATG in reducing the percentage of patients treated for acute rejection during the first year post transplantation. Until larger studies become available the effectiveness of IL-2R in BOS will remain a matter of speculation.

## Switching Cyclosporin to Tacrolimus and/or Azathioprine to Mycophenolate

### Tacrolimus

#### Preventive Treatment

One of the first reports that compared tacrolimus with cyclosporine as the primary calcineurin inhibitor after transplantation was published by Keenan et al.[33] In this report, which included 67 patients treated with cyclosporine and 66 patients treated with tacrolimus, obliterative bronchiolitis developed in 38% of those treated with cyclosporine versus 21.7% in the tacrolimus group (P = 0.025), although the one-year and two-year survival rates were similar in the two groups. After a follow-up of seven years[34] the survival for the group treated with cyclosporine was 39%, compared with 50% for the tacrolimus group.

In a trial including 74 patients published by Zuckermann et al,[35] with a subsequent follow-up at 36 months,[36] comparing the efficacy of the combination of tacrolimus or cyclosporine in combination with mycophenolate reported similar survival for both groups, although there was a significantly higher incidence of BOS in the cyclosporine group (41% versus 10% in the tacrolimus group, p = <0.01).

In another study, Reichenspurner[37] compared three different immunosuppressive protocols: CsA plus AZA (n = 34); Tac plus AZA with induction with ATG (n = 30); and Tac with MMF (n = 12) and found that freedom from acute rejection, incidence of rejection per 100 patient days was similar but survival was significantly better in the tacrolimus-azathioprine group. No comparison was made with the tacrolimus-mycophenolate group because this group had a much shorter follow-up time

### Treatment of Established BO

Several studies (see Table 2) have shown that switching cyclosporine to tacrolimus slows down and sometimes partially reverse, the progression of an established BOS,[38-46] although at least one study[44] showed no benefit in 11 patients with BOS. Unfortunately, with one exception,[45] these series are small, many of them lack an appropriate control group, the follow-up varies between 6 and 24 months and, most importantly, none of them showed conclusive evidence that this therapeutic maneuver prolongs survival. Furthermore, some of these reports describe patients that were also treated with other modalities like, total lymphoid irradiation,[40] or addition of other drugs like mycophenolate mofetil,[42] so it is difficult to determine to what extent the beneficial effects were due to tacrolimus itself. There is not enough data regarding the optimal tacrolimus dose. The available literature also suggests that tacrolimus is more beneficial at early stages of BOS. In fact, once bronchiolitis obliterans reaches stage III it seems that the addition of tacrolimus may be less effective,[38,41] although stabilization of lung function still may be achievable. Figures 1 and 2 show the effect of switching two patients with diagnosis of BOS from CsA to Tac. The first patient (Fig. 1) responded well with stabilization of pulmonary function for more than 30 months (methotrexate 5 mg/week was also added). The second patient (Fig. 2) shows a patient with progressive deterioration despite the use of Tac.

In the largest study to date,[45] Sarahrudi et al retrospectively analyzed the data from 244 patients from 13 European, Australian and Canadian lung transplant centers converted from cyclosporine to tacrolimus. Of these, 110 patients were switched because of recurrent-ongoing rejection while 134 had stage 1 to 3 bronchiolitis obliterans syndrome. In the BOS group it was reported that the use of tacrolimus resulted in a significant reduction in the number of acute rejections and a marked reduction in the rate of $FEV_1$ decline.

The study had some limitations: It was retrospective; 19 of the BOS patients also received cytolytic therapy; the diagnosis of episodes of acute rejection was made by biopsy only in 18% pre tacrolimus therapy and in 6% post tacrolimus initiation. Furthermore, no details were given about tacrolimus dose. Although as a group stabilization in $FEV_1$ was reported, there is not detailed data about what percentage of patients did not respond to therapy. Finally, there was no report on survival.

Although some studies have reported a modest increase in serum creatinine levels, the use of tacrolimus has been generally well tolerated without significant increase in the incidence of infections.

Clearly, more studies are needed to determine the effectiveness of tacrolimus in established bronchiolitis obliterans, but the current evidence suggests that tacrolimus may be relatively effective in preventing or delaying the development of BOS and, or stabilizing lung function after it has been established. To date there is not definitive data that the use of tacrolimus improves survival.

### Mycophenolate Mofetil

#### Preventive Treatment

Several small studies[47-49] suggest that mycophenolate is more effective than azathioprine in preventing episodes of acute rejection, but whether this will translate into a lower incidence of BO is a matter of speculation. One of these studies[48] reported a non significant decreased prevalence of BOS after 12 months of follow-up in the MMF group. In contrast, in a prospective, randomized study in 81 consecutive patients treated either with cyclosporine and steroids plus azathioprine or mycophenolate (dose of 1 gm twice a day),[50] found no difference in the number of episodes of acute rejection grade A2 or higher between both groups after 6 months of follow-up.

In the largest study to date, involving 315 patients (159 patients were treated with mycophenolate, 156 with azathioprine), which was also a prospective, randomized, open-label, multicenter trial, McNeil et al[51] compared the effects of MMF versus azathioprine in combination with induction therapy, cyclosporine and corticosteroids on the incidence of BOS After three years post lung transplantation there was no significant difference in the incidence of acute rejection, BOS or in survival between groups. The MMF dose was 1.5 gm twice a day for the first three months and 1 gm twice a day for the rest of the study.

***Table 2. Review of studies reporting the use of tacrolimus in patients that have been diagnosed with BOS***

| Report | N† | Tacrolimus Dose§ | Tacrolimus Blood Level Target | No. of Responders* | Follow-Up (Months) |
|---|---|---|---|---|---|
| Ross et al | 10 | 0.025-0.05 | 10-15 ng/mL (IMX assay) | 10 | 15 ± 3.2 |
| Cairn et al | 27 | 0.1 | 7-12 ng/mL (Abbott MIE) | 19 | Spirometry 12 Survival 48 |
| Kesten et al | 12 | Not specified | 10—20 ng/ml(IM$_x$ Abbot) | 7 | 6 months |
| Revell et all | 11 | Not specified | 8—13 ng/mL¶ | 8 | 12 |
| Sarahrudi et al | 11 | Not specified | 11.7 ± 2.9 ng/mL¶ | 0 | 12.7 ± 9.7 |
| Roman et al | 12 | Not specified | 5—20 ng/mL (Abbot MEIA) | 9 | 18.6 |
| Verleden et al | 10 | Not specified | 9.1 ± 0.9¶ | 7 | 6 |
| Sarahrudi et al | 134 | Not specified | Not specified | Not specified | 18.4 |
| Mentzer et al | 9 | 0.075-0.15 mg/kg | 10—40 ng/ml (IMx assay) | Not specified | 5.6 ± 2.8 |
| Knoop et al | 5 | 0.2 mg/kg/day | 5 to 20 ng/ml | 4 | 3 months |

† Number of patients that completed the stated follow-up period or had enough data for analysis. Only includes patients with diagnosis of BO/BOS.
* Response defined as reduction in decline, stabilization or improvement in $FEV_1$ and/or $FEF_{25-75\%}$.
§ in mg/kg/day in divided doses.
¶ No method identified.

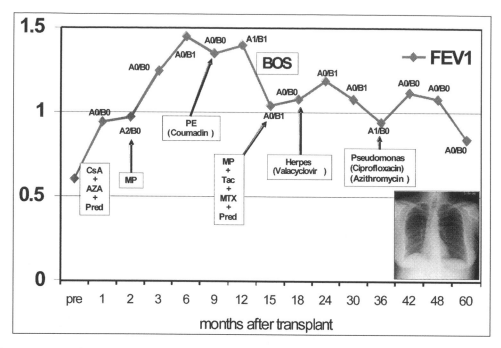

Figure 1. Medical management of BOS: Figure shows changes in FEV1 as well as lung biopsy results following a right single lung transplant performed on a 61 years old female with end stage emphysema. Initial immunosuppression included cyclosporine (CsA) plus Azathioprine (AZA) and prednisone (Pred). She received treatment for mild acute rejection (A2) 2 months post transplant with boluses of methylprednisolone (MP). Her lung function peaked at 6 months posttransplant. She required treatment for an episode of pulmonary embolism (PE) with anticoagulation 9 months post transplant. She develops BOS-2 15 months posttransplant. Received treatment with MPS, switching CsA to tacrolimus (Tac) and adding methotrexate (MTX) and subsequently treatment with Valacyclovir for herpes stomatitis. Lung function improved to subsequently decline again secondary to pseudomonas tracheobronchitis treated with Ciprofloxacin. Azithromycin was added. Lung function again improved transiently with eventual decline again (BOS-3). Available options to be considered include photopheresis and retransplantation.

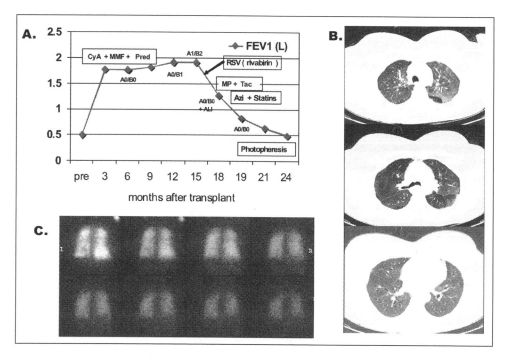

Figure 2. Refractory BOS: Figure shows changes in FEV1 and biopsy results on a 52 years old African American female recipient of a double lung transplant for pulmonary fibrosis A). Her lung function improved progressively after lung transplantation with a protocol of immunosuppression including CsA + Pred + Mycophenolate Mofetil (MMF). Lung function peaked 15 months post-transplant. An infection with Respiratory Syncytial Virus (RSV) produced acute bronchiolitis. She was treated with inhaled Rivabirin. Following RSV infection she showed progressive and relentless drop in FEV1, developing BOS-3. Her chest CT performed during expiration showed a mosaic pattern suggestive of bronchiolitis B) and her ventilation lung scan showed retention of Xenon during exhalation consistent with airflow obstruction C). Her BOS has been refractory to trials with MP, switching CsA to Tac and addition of Azithromycin and statins. Currently patient receives a trial with photopheresis. Repeat lung transplantation is not an option due to development of advanced coronary artery disease.

## Combination Therapy with Tacrolimus and Mycophenolate

In a prospective, randomized study Treede et al[52] compared either tacrolimus or cyclosporine in combination with mycophenolate mofetil. All patients also received steroids and induction therapy with ATG and found that although the number of acute rejection episodes per 100 patient days was significantly lower in the Tac-MMF group freedom from acute rejection and survival at 6 and 12 months were similar in both groups.

Two reports by Zuckerman et al comparing Tac or CsA with MMF were described in the tacrolimus section and found similar survival between these groups.[35,36] In contrast, Izbicki et al[53] compared 8 patients treated with a combination of CsA, AZA and prednisone with 21 patients treated with Tac, MMF and prednisone and reported a better 1 year survival in the Tac/MMF/prednisone group, although the small number of patients make this data difficult to interpret, specially when there was no difference between the study groups in the number of acute rejection episodes or incidence of BOS. Another small study by Gerbase[54] reported a decrease in the incidence of acute rejections and BOS in 12 patients treated with MMF and/or tacrolimus early after lung transplantation when compared to other 12 patients switched from a regimen of CsA-AZA to MMF and/or tacrolimus in the second year post transplantation.

### Established BOS

Three studies have described the use of MMF in patients with established BO. Speich et al[55] reported a case report of a 20 year-old patient with BO with stabilization of the $FEV_1$ after initiation of MMF. Whyte et al[56] also reported stabilization pulmonary function in seven of thirteen patients with BO treated with MMF. Roman et al[42] studied 12 patients with BO under immunosuppression with CsA/AZA/Prednisone. In all patients CsA was changed to tacrolimus, while in 5 patients AZA was changed to MMF. After a mean follow-up of 18.6 months there was a significant decrease in the number of acute rejection episodes and the $FEV_1$ values improved or stabilized in nine patients. It is impossible to determine with the data reported if MMF had a beneficial effects in these patients.

Interestingly, although the two largest prospective trials comparing mycophenolate with azathioprine[50,51] have shown no significant difference in the incidence of acute rejection, survival or the incidence of BOS, data presented in the last ISHLT report[22] suggests that in recipients older than 60 year-old the use of mycophenolate, regardless of the calcineurin inhibitor used, is associated with a decreased in the incidence of acute rejection. Still no firm data exists that indicates that the use of mycophenolate is associated with improved survival or decreased incidence of BO.

### Alternate Immunosuppression

#### Sirolimus/Everolimus

There is very little data regarding the use of sirolimus (SRL) as a therapy for BOS. Two small studies by Cahill et al[57] and Hernández et al[58] treated 12 and 11 patients respectively with BOS in combination with a calcineurin inhibitor and steroids. Cahill reported that as a group there were no significant changes in the $FEV_1$ or the $FEF_{25-75\%}$. Although some of the patients with the faster decline in spirometric parameters either stabilized or improved with SRL, no details were provided. Side effects included anemia, worsening renal function and malignancies in 2 patients. Hernández reported stabilization of lung function in 8 of 11 patients with BOS. Sirolimus was discontinued in two patients due to side effects.

In a multicenter, randomized, double-blind clinical trial,[59] 213 BOS-free maintenance patients received an oral derivative of SRL, everolimus (3 mg/day) or azathioprine (AZA, 1-3 mg/kg/day) in combination with cyclosporine and corticosteroids. Incidence of efficacy failure (defined as a decline in $FEV_1$ >15%) at 12 months was

significantly lower in the everolimus group, but became similar at 24 months. Although at 12 months, the everolimus group had significantly reduced incidence of decline in $FEV_1$ >15%, BOS and acute rejection, after 24 months, only incidence of acute rejection remained significantly less in the everolimus group. Serious adverse events and high serum creatinine values were more common with everolimus. The authors suggested that patients kept on prolonged maintenance treatment with everolimus may benefit from replacing AZA with everolimus 3 months after lung transplantation.

To date there is no firm evidence that SRL is effective in BOS. Our own experience has been disappointing and significant side effects, including myelosuppression and infections have been observed. Although SRL is well known as a renal sparing agent, the available evidence suggests that when used in combination with a calcineurin, even at low doses, significant deterioration of renal function still may occur.

### Methotrexate

To date there are three reports using methotrexate (MTx) for treatment of BO[60,61] or refractory rejection.[62] The two trials in patient with BO included 15 patients, of which it seems that 12 showed stabilization of lung function after a follow-up of at least 6 months. As with most studies in BO after lung transplantation these are small, nonrandomized and without control groups. Cahill[62] described the successful use of MTx in 12 patients with steroid-resistant acute rejection, but when comparing these 12 patients with a similar historical control group of 63 patients the authors found that the incidence of BO in the MTx group was 33% versus 23% in the historical control group, although this could be due to the fact that these patients with refractory acute rejection were at a higher risk for developing BO in the first place. There was not a significant difference in survival between these two groups. All these studies reported significant side effects.

In summary there is not sufficient data regarding the efficacy of methotrexate for treatment of BO after lung transplantation. Rather anecdotal reports suggest that some patients may respond to the drug but due to the few patients treated it is not possible to determine if the alleged benefit was due to the drug itself or just part of the natural history of BO. Our experience with MTx has been mixed although some patients seem to have stabilized after the initiation of this drug (see Fig. 1) Larger, multicenter, randomized studies are needed to determine the role of MTx in the treatment of BO, as well as the optimal dosage and which patients are most likely to benefit from this potentially dangerous therapy.

### Inhaled Immunosuppressive Therapy

#### Inhaled Cyclosporine

One of the most exciting developments in the area of lung transplantation is the use of inhaled cyclosporine. The concept of delivering a high dose of an immunosuppressant directly into the lungs is certainly attractive and has significant potential advantages that have been demonstrated with the use of other inhaled medications like bronchodilators, tobramycin, etc.

Studies in animal lung transplant models showed that administration of inhaled cyclosporine achieved 10 to 100-fold higher concentrations of cyclosporine in the lung than in heart or kidney tissue and suggested a benefit in decreasing proinflammatory cytokine production and early allograft rejection with low systemic exposure.[63,64]

In a series of several trials the group from the University of Pittsburgh reported that the addition of aerosolized cyclosporine to a standard immunosuppressive regimen to patients with either chronic or refractory acute rejection, resulted in improvement in rejection histology, reduction in proinflammatory cytokine production, dose-dependent improvement in pulmonary function stabilization of pulmonary function and overall improved survival compared with contemporary and historical controls from a transplant registry.[65-69]

In another small study, this time using prophylactic inhaled cyclosporine, Corcoran et al[70] reported that patients depositing more than 5 mg of the drug in the periphery of their lungs showed improved lung function on average, while the less than 5 mg deposition and placebo groups showed a decline in $FEV_1$ and more episodes of rejection.

In an open label, non randomized study Iacono et al[71] used inhaled cyclosporine in 39 patients with biopsy-proven BO (in 13 of these patients the drug was started before the diagnosis of BO was made) and compared this group with two control groups treated with standard immunosuppression, one composed of 51 patients from the University of Pittsburgh and 100 patients from the Novartis Lung Transplant Database, Stanford, CA, USA (a registry that includes patients from 12 major transplant centers around the world). Aerosol cyclosporine was administered for a mean of $69.9 \pm 66.3$ weeks. The primary goal of the study was to determine if inhaled cyclosporine improved survival after the diagnosis of bronchiolitis obliterans.

Using a multivariate survival analysis the authors reported that aerosol cyclosporine was associated with improved survival for the entire study group compared to both control groups. When only the subgroup of 26 patients that received aerosol cyclosporine after the diagnosis of bronchiolitis obliterans was made was used in the analysis there was a "survival advantage approaching significance" compared to the Pittsburgh control group and a significant one when compared to the multicentric group.

Although these open-label trials were promising the interpretation of the data was seriously limited by the lack of an adequate control group.

These results prompted a randomized, double blind, placebo controlled trial on inhaled cyclosporine initiated six weeks after transplantation in addition to a standard regimen of immunosuppression.[72] Although the power analysis stipulating a 33% difference in the frequency of acute rejection suggested the enrolment of 136 patients only 58 were enrolled due to a variety of factors. These patients were randomized to receive aerosol cyclosporine (n = 28) or aerosol placebo (n = 30). The study was closed two years after he last subject had been enrolled and all outcomes variables were followed until either the death of the patient or the end of the study.

The mean duration of treatment was $400 \pm 57$ days for patients receiving inhaled cyclosporine and $431 \pm 50$ days for the placebo group. Only 46% of the treatment group patients and 43% of the placebo patients completed the two year inhalation period of the study.

The primary end point of the study was the frequency of histologic acute rejection. Secondary end points included overall survival and chronic rejection-free survival. The rates of acute rejection grade 2 or higher, the primary end point of the study, showed no significant difference between the study and control groups (0.44 to 0.46 episodes per patient per year). Surprisingly, given the above result, the chronic rejection-free survival and the overall survival were significantly better in the treatment group. A hazard analysis of survival free of the bronchiolitis obliterans syndrome showed a total of 10 events in the inhaled cyclosporine group versus 20 events in the control group (p = 0.01). There were three deaths in the treatment group (11%) versus 14 deaths (47%) in the placebo group (p = 0.005). Indeed, multivariate survival regression analysis showed that the relative risk of death in the inhaled cyclosporine group compared to the placebo group was 0.20. No significant effects for CMV status, transplant type or HLA mismatch were found. The risk of infection or development of cancer was not increased in the treatment group. Side effects from inhaled cyclosporine were described as mild or moderate and transient.

It is unclear why the incidence of acute rejection episodes was similar in both groups. The authors speculated that chronic rejection presents in the airways as bronchiolitis obliterans while acute rejection presents as a vasculitic process and most of the aerosol cyclosporine concentrates at the airway level with a much lower vascular concentration. This

may be in contradiction, though, with a previous study from the same group that showed that inhaled cyclosporine was effective in reversing refractory acute rejection.[67]

Based on the above results a pharmaceutical company requested approval for inhaled cyclosporine to the Food and Drug Administration, which, to this date, still under review.

As promising as these results are, there has been criticism regarding the methodology of the study. Verdelen and Dupont[73] pointed out the much higher than previously reported prevalence of bronchiolitis obliterans syndrome when compared with the data collected by the ISHLT, specially in the placebo group, may explain the reported benefit of inhaled cyclosporine.

Some members of the FDA committee evaluating the potential approval of aerosolized cyclosporine[74] also raised concerns regarding other aspects of the methodology, including:

1. Poor randomization which did not stratified by single versus double lung transplant, or other baseline donor/recipient characteristics known to influence long term survival.
2. There was a wide range of dosing practices and problems with compliance with the protocol-specified regimen.
3. Only two of the 14 deaths in the placebo group were due to bronchiolitis obliterans. The predominant causes of death were infection and sepsis, which is consistent with the information on causes of death over time post transplant from registry data.
4. Information on the $FEV_1$ before enrollment is incomplete. Only 15 out of 26 patients in the aerosolized cyclosporine group and 17 out of 30 patients in the placebo control group have documented values of $FEV_1$ pre-enrollment.

The FDA also had concerns about insufficient safety data for the use of aerosolized cyclosporine in humans and suggested that larger, probably multicenter trials should be conducted. The available evidence is hard to ignore, though and several prominent members of the lung transplant community feel that it is very unlikely any institutional review board will approve a placebo-control trial for aerosol cyclosporine given the data already available.[75]

In summary, although the data about the use of inhaled cyclosporine in lung transplant recipients is far from complete, the available evidence is strong enough to justify larger trials with enough power to prove conclusively if this therapeutic modality is as effective as it seems.

## Other Immunosuppressive Therapy

### Photopheresis

Photopheresis is a combination of leukopheresis and the administration of the photosensitive drug 8-methoxypsoralen, followed by photoirradiation with long-wavelength ultraviolet A. The potential mechanism of action is unknown but it is speculated that the irradiation with ultraviolet A activates the 8-methoxypsoralen within the DNA of the leukocyte leading to the proliferative arrest of the activated T-cells.

Four single center reports, with a total of only 30 patients, have studied the effectiveness of photopheresis as treatment for BO.[76-79] It seems that stabilization of pulmonary function was achieved in 17 of these patients. From the scant data available apparently at least six of the nonresponders were on BOS stages 2 or 3, suggesting photopheresis may be more effective in early stages of BOS. Although photopheresis appears to be safe, based on the experience on other patients since there is insufficient data in lung transplant patients, it is worth noticing that there were three cases of malignancies reported in this small population (two lung cancers[79] and one posttransplant lymphoproliferative disease in the allograft[76]). It is impossible with the data available to determine if this is related to the use of photopheresis.

Photopheresis requires special equipment and can be cumbersome. O'Hagan[76] reported the process taking 6 hours and performed on two consecutive days twice a month until stabilization of pulmonary function and every 4 to 6 weeks thereafter. The fact that some patients deteriorated after photopheresis was discontinued suggests that continuous treatment may be needed. This could represent a major obstacle in patients living far away from centers with the capability to perform the procedure.

In summary there is not enough data to support the notion that photopheresis is effective in arresting the progression of BO, but given its apparent safety and the results of few small studies it is worth it to explore its potential in larger, randomized trials.

### Total Lymphoid Irradiation

There is no strong evidence that total lymphoid irradiation (TLI) is effective as treatment for BO. Valentine et al[80] reported the use of TLI for refractory acute rejection in six patients. One patient died two months later from ARDS. The incidence of acute rejection decreased significantly in the other five but there is no evidence that TLI had any effect in the subsequent development of BO.

Three other studies,[81-83] with a total of 60 patients, reported a significant reduction in the decline of the $FEV_1$ in at least 31 of them (no details on individual response were offered in the study of 12 patients by Chacon et al[81]). In the other two studies[82,83] 17 out of 48 patients could not complete the therapy due to side effects or relentless progression of BO. It is obvious that larger, multicenter and randomized studies are needed to better assess the efficacy of TLI in the treatment of BO.

## Non Immunosuppressive Therapy

### Macrolides

There has been a great interest in the use of macrolides for BOS since Gerhardt el al[84] reported significant improvement in the $FEV_1$ (mean improvement of 0.63 L) in five of six patients with BOS after treatment with azithromycin at a dose of 250 mg three times a week for a mean of 13.7 weeks. The study had several limitations, including lack of controls, small sample size and short follow-up. Furthermore, four out of the six patients had documented pseudomonas aeruginosa pulmonary or sinus colonization before starting azithromycin, so it could be argued that some of the benefits seen could be attributable to the antimicrobial effects of the drug.

In another study, Yates et al[85] reported a mean increase in the $FEV_1$ of 110 ml (range, −70 to 730 ml) between baseline and three months after initiation of azithromycin in a retrospective study of 20 patients with established BOS receiving azithromycin on alternate days with a mean follow-up of 6.25 months. This is only an 8% increase from the baseline mean $FEV_1$ of 1.44 L, according to the individual patients data reported but the improvement was sustained beyond three months in 12 of 17 patients. Although clearly some patients seemed to have benefited from this therapy it is obvious that not everybody did. Other study[86] found no improvement in lung function in 11 patients with BOS who were followed for 10 months.

The mechanism by which azithromycin could improve lung function in patients with BOS has not be determined. Macrolides have important anti-inflammatory properties[87,88] and several studies have reported beneficial effects in diffuse panbronchiolitis;[89] cystic fibrosis;[90,91] and asthma.[92] Other studies have shown that macrolides produce a reduction in inflammatory mediators such as interleukin 8, interleukin 6, tumor necrosis factor and interleukin 1β.[93,94] All these mediators have been implicated in the development of BOS in humans.[95-98] Gerhardt et al[84] also suggested that macrolides may also help as a promotility agent in patients with GERD, a well known risk factor for BOS.

In summary, the current evidence suggests that some patients with BOS may benefit from the addition of macrolides to their treatment. Large, prospective randomized trials are needed to better define the role of macrolides in the treatment of BOS.

### Treatment of Gastroesophageal Reflux

There is increasing evidence that GERD not only is common in both, patients with advanced lung disease listed for lung transplantation,[99] as well as in lung transplant recipients,[100-103] but that may also be a predisposing factor for BO.[104,105] The reason is not clear but it may be related to repetitive and prolonged contact time with aspirated gastric content in patients with denervated lungs, abnormal mucociliary and an impaired cough reflex.

The transplant group from Duke University has published several reports[12,13,106] indicating that treatment of reflux with fundoplication is not only safe but can improve lung function and increase the freedom from BOS period, although there was no difference in the incidence or severity of acute rejection. In fact, their success has moved this group to perform the operation in any patient that presents a positive esophageal pH study regardless of the presence, or not, of symptoms; or in any patient that meet BOS criteria and a positive esophageal pH test. Our experience in this regard has been limited but our results tend to corroborate the Duke University group data (see Fig. 3). Several questions remain with regard to the management of GERD in lung transplant recipients such as the role of medications like proton pump inhibitors in patients with GERD after transplantation; how much reflux is needed to justify this invasive procedure; best timing of the operation etc. In the mean time it seems reasonable to aggressively screen and treat any lung transplant patient for GERD, especially in the settings of declining lung function. Fundoplication should be strongly considered in these patients, especially if maximal medical treatment is ineffective.

### Statins

Statins were reported to have beneficial effects in the outcome of heart transplant patients beyond their know effects over blood lipids levels.[107,108] In the lung transplant literature only one trial has been published by the group of the University of Pittsburgh.[109] In this study, outcomes of 39 patients that received statins for hyperlipidemia was compared with 161 contemporaneous controls who did not received these drugs. Although the incidence of BO was similar for both groups when only those patients that received statins in the first year post transplant are analyzed (n = 15) it was found that none of these patients had developed BO, whereas the cumulative incidence of BO was 37% by 6 years after transplantation in the nonstatin group.

Although limited by the small number of patients and the retrospective nature of the study the results are provocative not only because of the data regarding the incidence of BOS but because several other parameters recorded (i.e., number of episodes of acute rejection, cellularity of the BAL, histopathology, etc) tend to confirm the main observation of the study. Larger, prospective, randomized studies are needed to confirm this initial data.

## Management of Associated Infections

### Cytomegalovirus

Considerable controversy exists on whether CMV infections may or may not be related to BOS. Given the unquestionable association between CMV infection and increased mortality in lung transplant recipients CMV prophylaxis and treatment of infections are indicated. Evidence-based guidelines for treatment of CMV infection have recently been published.[110] These guidelines suggest that all lung transplant recipients should be considered for CMV prophylaxis. Prophylaxis should consist of valgancyclovir 900 mg daily for at least 100 days posttransplant and consideration to longer prophylaxis (180 days) may be warranted. Combination prophylaxis with CMV IVIG should be considered. Patients should be monitored every 2 weeks for CMV viremia for the first 6 months and then continued monthly. Breakthrough CMV infection and disease should be treated with intravenous gancyclovir 5 mg/kg twice a day for up to 3 weeks, treatment should continue until the viral load is below detection limits of a sensitive

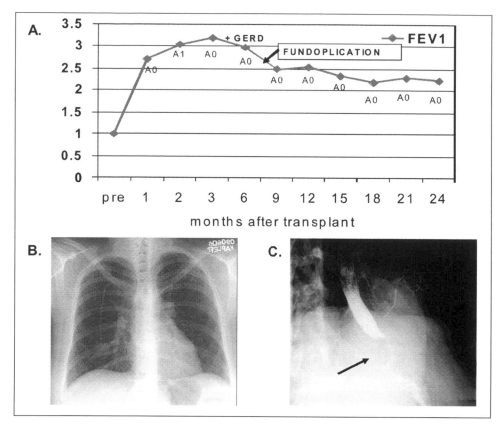

Figure 3. Surgical management of BOS: Figure shows progressive changes in FEV1 following double lung transplantation in a 26 years old female who was recipient of a double lung transplant due to end-stage lung disease associated with Shwachman Diamond syndrome (A). Her lung function and chest radiograph (B) were normal post double lung transplantation. She developed progressive decline in FEV1 consistent with BOS 1 and work-up revealed significant gastro-esophageal-reflux disease (GERD), despite medical therapy. She underwent a laparoscopic esophageal Nissen fundoplication (C). Her FEV1 has stabilized following surgery.

monitoring assay (PCR). Addition of Foscarnet should be considered for recurrent CMV infections, poor clinical responses or development of resistance to gancyclovir.

### Other Viral Respiratory Tract Infections

There is increasing evidence linking infections with community acquired viral respiratory tract infections like rhinovirus, corona virus, RSV, influenza A and Para influenza, with BOS (see Fig. 2). There no published data regarding avoidance of BOS by treating such infections. Influenza virus may be treated with neuraminidase inhibitors and other agents. Aerosolized Ribavirin, intravenous immunoglobulin, RSV immunoglobulin and palivizumab have been reported as viable therapies for RSV virus and some cases of parainfluenza virus infections, but its effectiveness as preventing subsequent development of BOS is unknown.[111] There is no standard therapy for other viruses.

The role of influenza vaccination in solid organ transplant recipients is controversial because variable data about antibody responses to the vaccine. Aggressive vaccination of close contacts and household members should be considered to reduce the chances of infection in lung transplant recipients[112]

Human Herpes-Virus-6 has also been linked to development of BOS but there are no controlled trials to assess the role of acyclovir or gancyclovir on treatment and prophylaxis of HHV-6 and its possible role on preventing BOS[113]

### Atypical Infections

There is one report linking *Chlamydia pneumoniae* donor seropositivity and recipient seronegativity independently associated with development of BOS, which raise a possible explanation for improvement in some of the patients with BOS treated with macrolide antibiotics.[114]

Although infection or colonization of airways of patients diagnosed with BOS is common and the standard of practice usually involves medical treatment of bronchial infections guided y broncho-alveolar lavage cultures and protected brush cultures results, there is no available data proving that the treatment of such infections will necessarily alter the outcome of BOS

### Other

#### Retransplantation

For those patients that do not respond to medical treatment the only option left is retransplantation, but this is still a controversial point given the historically poor results of retransplantation.[115] In a report of 15 patients Brugière et al[116] reported survival of 60%, 53% and 45% at 1 year, 2 years and 5 years respectively. Interestingly, infection in the retained graft was the cause of death in 4 of the six patients that died from infection which suggests that replacement of the primary graft should be the preferred procedure. Strueber et al[117] reported a 1 and 5 years survival of 78% and 62% respectively in 37 patients with BOS. The young age of this group (mean age 36 years) may have been a factor in these excellent results.

In summary, retransplantation is a reasonable option in well selected patients with BOS.

#### Mayo Clinic at Jacksonville Protocol for Treatment of Bronchiolitis Obliterans after Lung Transplantation

The Figure 4 shows our general protocol for treatment of BO. This is meant to be a general guideline only and treatment for each patient should be individualized according to the circumstances. The first step is to identify other potential causes of declining lung function like infections, airways problems, etc. Once those are ruled out and the diagnosis of BOS

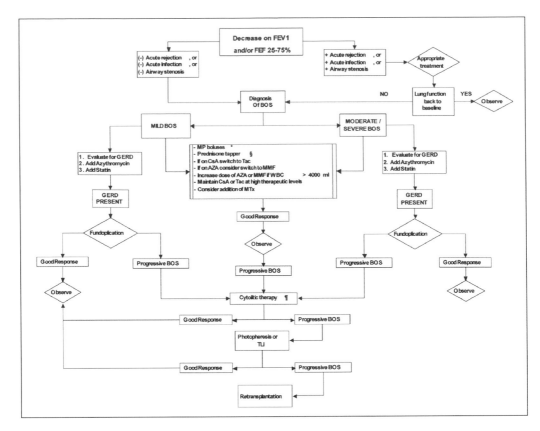

Figure 4. Mayo Clinic Jacksonville protocol for management of BOS (see text for detailed explanation). Abbreviations: AzA: azathioprine; CsA: cyclosporine; GERD: gastroesophageal reflux; MMF: mycophenolate; MP: methylprednisolone; Tac: tacrolimus; WBC: white cell count. *Methylprednisolone bolus: for mild BOS: 2.5 mg/kg/day for 3 days; For severe BOS: 10 mg/kg/day for 3 days. §Prednisone taper: for Mild BOS: 0.5 mg/kg/day, then taper down to 0.2 mg/kg/day over 4 weeks; For severe BOS: 1 mg/kg/day and taper down to 0.2 mg/kg/day over 4 weeks.

is established all patients undergo an evaluation to rule out GERD, even if they are asymptomatic. If positive, patients will be sent for fundoplication surgery. In addition, patients receive a bolus of methylprednisolone, 2.5 mg/kg for mild BOS and 10 mg/kg for moderate severe BOS. In both cases the bolus is administered daily for three consecutive days and it is followed by a prednisone taper starting with 0.5 mg/kg/day for mild BOS and 1 mg/kg/day for moderate or severe BOS. The prednisone dose is lowered over a period of four weeks down to 0.2 mg/kg/day.

Patients being treated with CsA or AzA are switched to Tac or MMF respectively; and for those already on these drugs we try to maximize the dosage if tolerated. We also consider at this point the addition of MTx, usually 5-10 mg per week. All patients are started on azhitromycin, 250 mg three times a week and a statin if they are not on any of these drugs already. If lung function stabilizes or improved they are followed closely; if lung function continues to decline we try cytolytic agents, usually thymoglobulin (ATG-rabbit) at a dose of 1.5 mg/kg infusion over 6 hours once a day for three consecutive days with close follow-up of lymphocyte count.

If the patient continues to deteriorate we proceed with a trial of photopheresis twice a week for 6 weeks. Total lymphoid irradiation is also an alternative at this point. For well selected patients retransplantation is our final option.

## References

1. Estenne M, Maurer JR, Boehler A et al. Bronchiolitis obliterans syndrome 2001: an update of the diagnostic criteria. J Heart Lung Transplant 2002; 21:297-310.
2. Haloun A, Despins P. Transplantation pulmonaire dans la mucoviscidose. Rev Prat 2003; 53:167-70.
3. Hopkins PM, Aboyoun CL, Chhajed PN et al. Association of minimal rejection in lung transplant recipients with obliterative bronchiolitis.[see comment]. Am J Respir Crit Care Med 2004; 170:1022-26.
4. Khalifah AP, Hachem RR, Chakinala MM et al. Respiratory viral infections are a distinct risk for bronchiolitis obliterans syndrome and death. Am J Respir Crit Care Med 2004; 170:181-87.
5. Billings JL, Hertz MI, Savik K et al. Respiratory viruses and chronic rejection in lung transplant recipients. J Heart Lung Transplant 2002; 21:559-66.
6. Hodges TN, Torres FP, Zamora MR et al. Treatment of respiratory syncitial viral and parainfluenza lower respiratory tract infections in lung transplant patients. J Heart Lung Transplant 2001; 20:170.
7. Girgis RE, Tu I, Berry GJ et al. Risk factors for the development of obliterative bronchiolitis after lung transplantation. J Heart Lung Transplant 1996; 15:1200-8.
8. Keenan RJ, Lega ME, Dummer JS et al. Cytomegalovirus serologic status and postoperative infection correlated with risk of developing chronic rejection after pulmonary transplantation. Transplantation 1991; 51:433-38.
9. Keller CA, Cagle PT, Brown RW et al. Bronchiolitis obliterans in recipients of single, double and heart-lung transplantation. Chest 1995; 107:973-80.
10. Kroshus TJ, Kshettry VR, Savik K et al. Risk factors for the development of bronchiolitis obliterans syndrome after lung transplantation. J Thorac Cardiovasc Surg 1997; 114:195-202.
11. Smith MA, Sundaresan S, Mohanakumar T et al. Effect of development of antibodies to HLA and cytomegalovirus mismatch on lung transplantation survival and development of bronchiolitis obliterans syndrome. J Thorac Cardiovasc Surg 1998; 116:812-20.
12. Cantu E 3rd, Appel JZ 3rd, Hartwig MG et al. J. Maxwell Chamberlain Memorial Paper. Early fundoplication prevents chronic allograft dysfunction in patients with gastroesophageal reflux disease. Ann Thorac Surg 2004; 78:1142-1151; discussion 1142-51.
13. Lau CL, Palmer SM, Howell DN et al. Laparoscopic antireflux surgery in the lung transplant population. Surg Endosc 2002; 16:1674-78.
14. Boehler A, Estenne M. Posttransplant bronchiolitis obliterans. Eur Respir J 2003; 22:1007-18.
15. Heng D, Sharples LD, McNeil K et al. Bronchiolitis obliterans syndrome: incidence, natural history, prognosis and risk factors. J Heart Lung Transplant 1998; 17:1255-63.

16. Hachem RR, Khalifah AP, Chakinala MM et al. The significance of a single episode of minimal acute rejection after lung transplantation. Transplantation 2005; 80:1406-13.

17. Paradis I, Yousem S, Griffith B. Airway obstruction and bronchiolitis obliterans after lung transplantation. Clin Chest Med 1993; 14:751-63.

18. Paradis IL, Duncan SR, Dauber JH et al. Effect of augmented immunosuppression on human chronic allograft rejection Am Rev Respir Dis 1992; 145:A705.

19. Takao M, Higenbottam TW, Audley T et al. Effects of inhaled nebulized steroids (budesonide) on acute and chronic lung function in heart-lung transplant patients. Transplant Proc 1995; 27:1284-85.

20. De Soyza A, Fisher AJ, Small T et al. Inhaled corticosteroids and the treatment of lymphocytic bronchiolitis following lung transplantation. Am J Respir Crit Care Med 2001; 164:1209-12.

21. Whitford H, Walters EH, Levvey B et al. Addition of inhaled corticosteroids to systemic immunosuppression after lung transplantation: a double-blind, placebo-controlled trial.[see comment]. Transplantation 2002; 73:1793-99.

22. Trulock EP, Edwards LB, Taylor DO et al. Registry of the International Society for Heart and lung Transplantation: Twenty-third official adult lung and heart-lung transplantation report-2006. J Heart Lung Transplant 2006; 25:880-92.

23. Griffith BP, Hardesty RL, Armitage JM et al. Acute rejection of lung allografts with various immunosuppressive protocols. Ann Thorac Surg 1992; 54:846-51.

24. Palmer SM, Miralles AP, Lawrence CM et al. Rabbit antithymocyte globulin decreases acute rejection after lung transplantation: results of a randomized, prospective study. Chest 1999; 116:127-33.

25. Ross DJ, Jordan SC, Nathan SD et al. Delayed development of obliterative bronchiolitis syndrome with OKT3 after unilateral lung transplantation. A plea for multicenter immunosuppressive trials.[see comment]. Chest 1996; 109:870-73.

26. Date H, Lynch JP, Sundaresan S et al. The impact of cytolytic therapy on bronchiolitis obliterans syndrome. J Heart Lung Transplant 1998; 17:869-75.

27. Kesten S, Rajagopalan N, Maurer J. Cytolytic therapy for the treatment of bronchiolitis obliterans syndrome following lung transplantation. Transplantation 1996; 61:427-30.

28. Snell GI, Esmore DS, Williams TJ. Cytolytic therapy for the bronchiolitis obliterans syndrome complicating lung transplantation.[see comment]. Chest 1996; 109:874-78.

29. Slebos DJ, Kauffman HF, Koeter GH et al. Airway cellular response to two different immunosuppressive regimens in lung transplant recipients. Clin Transplant 2005; 19:243-49.

30. Garrity ER Jr, Villanueva J, Bhorade SM et al. Low rate of acute lung allograft rejection after the use of daclizumab, an interleukin 2 receptor antibody. Transplantation 2001; 71:773-77.

31. Brock MV, Borja MC, Ferber L et al. Induction therapy in lung transplantation: a prospective, controlled clinical trial comparing OKT3, anti-thymocyte globulin and daclizumab. J Heart Lung Transplant 2001; 20:1282-90.

32. Ding IB, Baumgartner RA, Schwaiblmair M et al. Administration of anti-interleukin-2Ralpha monoclonal antibody in bronchiolitis obliterans syndrome after lung transplantation. Transplantation 2003; 75:1767-69.

33. Keenan RJ, Konishi H, Kawai A et al. Clinical trial of tacrolimus versus cyclosporine in lung transplantation.[see comment]. Ann Thorac Surg 1995; 60:580-584; discussion 584-85.

34. McCurry KR, Zaldonis DB, Keenan RJ et al. Long-term follow-up of a prospective, randomized trial of tacrolimus versus cyclosporin in human lung transplantation. Am J Transplant 2002; 2002:159.

35. Zuckermann A, Reichenspurner H, Birsan T et al. Cyclosporine A versus tacrolimus in combination with mycophenolate mofetil and steroids as primary immunosuppression after lung transplantation: one-year results of a 2-center prospective randomized trial.[see comment]. J Thorac Cardiovasc Surg 2003; 125:891-900.

36. Zuckermann A, Reichenspurner H, Jaksch P et al. Long term follow-up of a prospective randomized trial comparing tacrolimus versus cyclosporine in combination with MMF after lung transplantation. J Heart Lung Transplant 2003; 22:S76-77.

37. Reichenspurner H, Kur F, Treede H et al. Optimization of an immunosuppressive protocol after lung transplantation. Transplantation 1999; 68:67-71.

38. Cairn J, Yek T, Banner NR et al. Time-related changes in pulmonary function after conversion to tacrolimus in bronchiolitis obliterans syndrome. J Heart Lung Transplant 2003; 22:50-57.

39. Kesten S, Chaparro C, Scavuzzo M et al. Tacrolimus as rescue therapy for bronchiolitis obliterans syndrome. J Heart Lung Transplant 1997; 16:905-12.

40. Mentzer RM Jr, Jahania MS, Lasley RD. Tacrolimus as a rescue immunosuppressant after heart and lung transplantation. The U.S. Multicenter FK506 Study Group. Transplantation 1998; 65:109-13.

41. Revell MP, Lewis ME, Llewellyn-Jones CG et al. Conservation of small-airway function by tacrolimus/cyclosporine conversion in the management of bronchiolitis obliterans following lung transplantation. J Heart Lung Transplant 2000; 19:1219-23.

42. Roman A, Bravo C, Monforte V et al. Preliminary results of rescue therapy with tacrolimus and mycophenolate mofetil in lung transplanted patients with bronchiolitis obliterans. Transplant Proc 2002; 34:146-47.

43. Ross DJ, Lewis MI, Kramer M et al. FK 506 'rescue' immunosuppression for obliterative bronchiolitis after lung transplantation. Chest 1997; 112:1175-79.

44. Sarahrudi K, Carretta A, Wisser W et al. The value of switching from cyclosporine to tacrolimus in the treatment of refractory acute rejection and obliterative bronchiolitis after lung transplantation. Transpl Int 2002; 15:24-28.

45. Sarahrudi K, Estenne M, Corris P et al. International experience with conversion from cyclosporine to tacrolimus for acute and chronic lung allograft rejection. J Thorac Cardiovasc Surg 2004; 127:1126-32.

46. Verleden GM, Dupont LJ, Van Raemdonck D et al. Effect of switching from cyclosporine to tacrolimus on exhaled nitric oxide and pulmonary function in patients with chronic rejection after lung transplantation. J Heart Lung Transplant 2003; 22:908-13.

47. Zuckermann A, Klepetko W, Birsan T et al. Comparison between mycophenolate mofetil and azathioprine-based immunosuppression in clinical lung transplantation. J Heart Lung Transplant 1999; 18:432-40.

48. Ross DJ, Waters PF, Levine M et al. Mycophenolate mofetil versus azathioprine immunosuppressive regimens after lung transplantation: preliminary experience. J Heart Lung Transplant 1998; 17:768-74

49. O'Hair DP, Cantu E, McGregor C et al. Preliminary experience with mycophenolate mofetil used after lung transplantation. J Heart Lung Transplant 1998; 17:864-68.

50. Palmer SM, Baz MA, Sanders L et al. Results of a randomized, prospective, multicenter trial of mycophenolate mofetil versus azathioprine in the prevention of acute allograft rejection. Transplantation 2001; 71:1772-76.

51. McNeil K, Glanville AR, Wahlers T et al. Comparison of mycophenolate mofetil and azathioprine for prevention of bronchiolitis obliterans syndrome in de novo lung transplant recipients.[see comment]. Transplantation 2006; 81:998-1003.

52. Treede H, Klepetko W, Reichenspurner H et al. Tacrolimus versus cyclosporine after lung transplantation: a prospective, open, randomized two-center trial comparing two different immunosuppressive protocols. J Heart Lung Transplant 2001; 20:511-17.

53. Izbicki G, Shitrit D, Aravot D et al. Improved survival after lung transplantation in patients treated with tacrolimus/mycophenolate mofetil as compared with cyclosporin/azathioprine Transplant Proc 2002; 34:3258-59.

54. Gerbase MW, Spiliopoulos A, Fathi M et al. Low doses of mycophenolate mofetil with low doses of tacrolimus prevent acute rejection and long-term function loss after lung transplantation. Transplant Proc 2001; 33:2146-47.

55. Speich R, Boehler A, Thurnheer R et al. Salvage therapy with mycophenolate mofetil for lung transplant bronchiolitis obliterans: importance of dosage. Transplantation 1997; 64:533-35.

56. Whyte RI, Rossi SJ, Mulligan MS et al. Mycophenolate mofetil for obliterative bronchiolitis syndrome after lung transplantation. Ann Thorac Surg 1997; 64:945-48.

57. Cahill BC, Somerville KT, Crompton JA et al. Early experience with sirolimus in lung transplant recipients with chronic allograft rejection. J Heart Lung Transplant 2003; 22:169-76.

58. Hernandez RL, Gil PU, Gallo CG et al. Rapamycin in lung transplantation. Transplant Proc 2005; 37:3999-4000.

59. Snell GI, Valentine VG, Vitulo P et al. Everolimus versus azathioprine in maintenance lung transplant recipients: an international, randomized, double-blind clinical trial. Am J Transplant 2006; 6:169-77.

60. Boettcher H, Costard-Jäckle A, Möller F et al. Methotrexate rescue therapy in lung transplantation. Transplant Proc 2002; 34:3255-57.

61. Dusmet M, Maurer J, Winton T et al. Methotrexate can halt the progression of bronchiolitis obliterans syndrome in lung transplant recipients. J Heart Lung Transplant 1996; 15:948-54.

62. Cahill BC, O'Rourke MK, Strasburg KA et al. Methotrexate for lung transplant recipients with steroid-resistant acute rejection. J Heart Lung Transplant 1996; 15:1130-37.

63. Zenati M, Duncan AJ, Burckart GJ et al. Immunosuppression with aerosolized cyclosporine for prevention of lung rejection in a rat model. Eur J Cardiothorac Surg 1991; 5:266-71.

64. Keenan RJ, Duncan AJ, Yousem SA et al. Improved immunosuppression with aerosolized cyclosporine in experimental pulmonary transplantation. Transplantation 1992; 53:20-25.

65. Iacono A, Dauber J, Keenan R et al. Interleukin 6 and interferon-gamma gene expression in lung transplant recipients with refractory acute cellular rejection: implications for monitoring and inhibition by treatment with aerosolized cyclosporine. Transplantation 1997; 64:263-69.

66. Iacono AT, Keenan RJ, Duncan SR et al. Aerosolized cyclosporine in lung recipients with refractory chronic rejection. Am J Respir Crit Care Med 1996; 153:1451-55.

67. Iacono AT, Smaldone GC, Keenan RJ et al. Dose-related reversal of acute lung rejection by aerosolized cyclosporine. Am J Respir Crit Care Med 1997; 155:1690-98.

68. Keenan RJ, Iacono A, Dauber JH et al. Treatment of refractory acute allograft rejection with aerosolized cyclosporine in lung transplant recipients. J Thorac Cardiovasc Surg 1997; 113:335-40.

69. Keenan RJ, Zeevi A, Iacono AT et al. Efficacy of inhaled cyclosporine in lung transplant recipients with refractory rejection: correlation of intragraft cytokine gene expression with pulmonary function and histologic characteristics. Surgery 1995; 118:385-91.

70. Corcoran TE, Smaldone GC, Dauber JH et al. Preservation of post transplant lung function with aerosol cyclosporin. Eur Respir J 2004; 23:378-83.

71. Iacono AT, Corcoran TE, Griffith BP et al. Aerosol cyclosporin therapy in lung transplant recipients with bronchiolitis obliterans. Eur Respir J 2004; 23:384-90.

72. Iacono AT, Johnson BA, Grgurich WF et al. A randomized trial of inhaled cyclosporine in lung-transplant recipients. N Engl J Med 2006; 354:141-50.

73. Verleden GM, Dupont LJ. Inhaled cyclosporine in lung transplantation.[comment]. N Engl J Med 2006; 354:1752-1753; author reply 1752-53.

74. PulminiqTM (cyclosporine, USP) Inhalation Solution (CyIS) Briefing Document for the Pulmonary Advisory Committee meeting 2005 (obtained from FDA website www.fda.gov).

75. Department of Health and Human Services United States Food and Drug Administration center for Drug Evaluation and Research Pulmonary-Allergy Drugs Advisory Committee, 2005. (obtained from FDA website www.fda.gov).

76. O'Hagan AR, Stillwell PC, Arroliga A et al. Photopheresis in the treatment of refractory bronchiolitis obliterans complicating lung transplantation. Chest 1999; 115:1459-62.

77. Salerno CT, Park SJ, Kreykes NS et al. Adjuvant treatment of refractory lung transplant rejection with extracorporeal photopheresis. J Thorac Cardiovasc Surg 1999; 117:1063-69.

78. Slovis BS, Loyd JE, King LE. Photopheresis for chronic rejection of lung allografts. N Engl J Med 1995; 332:962.

79. Villanueva J, Bhorade SM, Robinson JA et al. Extracorporeal photopheresis for the treatment of lung allograft rejection. Ann Transplant 2000; 5:44-47.

80. Valentine VG, Robbins MC, Wehner JH et al. Total lymphoid irradtiation for refractory acute rejection in heart-lung and lung allografts. Chest 1996; 109(65):692-99.

81. Chacon RA, Corris PA, Dark JH et al. Tests of airway function in detecting and monitoring treatment of obliterative bronchiolitis after lung transplantation. J Heart Lung Transplant 2000; 19:263-69.

82. Diamond DA, Michalski JM, Lynch JP et al. Efficacy of total lymphoid irradiation for chronic allograft rejection following bilateral lung transplantation. Int J Radiat Oncol Biol Phys 1998; 41:795-800.

83. Fisher AJ, Rutherford RM, Bozzino J et al. The safety and efficacy of total lymphoid irradiation in progressive bronchiolitis obliterans syndrome after lung transplantation.[see comment]. Am J Transplant 2005; 5:537-43.

84. Gerhardt SG, McDyer JF, Girgis RE et al. Maintenance azithromycin therapy for bronchiolitis obliterans syndrome: results of a pilot study. Am J Respir Crit Care Med 2003; 168:121-25.

85. Yates B, Murphy DM, Forrest IA et al. Azithromycin reverses airflow obstruction in established bronchiolitis obliterans syndrome.[see comment]. Am J Respir Crit Care Med 2005; 172:772-75.

86. Shitrit D, Bendayan D, Gidon S et al. Long-term azithromycin use for treatment of bronchiolitis obliterans syndrome in lung transplant recipients. J Heart Lung Transplant 2005; 24:1440-43.

87. Culic O, Erakovic V, Parnham MJ et al. Anti-inflammatory effects of macrolide antibiotics. Eur J Pharmacol 2001; 429:209-29.

88. Culic O, Erakovic V, Cepelak I et al. Azithromycin modulates neutrophil function and circulating inflammatory mediators in healthy human subjects. Eur J Pharmacol 2002; 450:277-89.

89. Kudoh S, Azuma A, Yamamoto M et al. Improvement of survival in patients with diffuse panbronchiolitis treated with low-dose erythromycin. Am J Respir Crit Care Med1998; 157:1829-32.

90. Equi A, Balfour-Lynn IM, Bush A et al. Long-term azithromycin in children with cystic fibrosis: a randomised, placebo-controlled crossover trial. Lancet 2002; 360:978-84.

91. Wolter J, Seeney S, Bell S et al. Effect of long-term treatment with azithromycin on disease parameters in cystic fibrosis: a randomised trial. Thorax 2002; 57:212-16.

92. Zeiger RS, Schatz M, Sperling W et al. Efficacy of toleandomycin in outpatients with severe, corticosteroid-dependent asthma. J Allergy Clin Immunol 1980; 66:438-46.

93. Ianaro A, Ialenti A, Maffia P et al. Anti-inflammatory activity of macrolide antibiotics. Journal of Pharmacoloy and Experimental Therapy 2000; 292:156-63.

94. Suzuki H, Asada Y, Ikeda K et al. Inhibitory effect of erythromycin on IL-8 secretion from exudative cells in the nasal discharge of patients with chronic sinusitis. Laryngoscope 1999:407-10.

95. DiGiovine B, Lynch JP 3rd, Martinez FJ et al. Bronchoalveolar lavage neutrophilia is associated with obliterative bronchiolitis after lung transplantation: role of IL-8. J Immunol 1996; 157:4194-202.

96. Elssner A, Jaumann F, Dobmann S et al. Elevated levels of interleukin-8 and transforming growth factor-beta in bronchoalveolar lavage fluid from patients with bronchiolitis obliterans syndrome: proinflammatory role of bronchial epithelial cells. Munich Lung Transplant Group.[see comment]. Transplantation 2000; 70:362-67.

97. Scholma J, Slebos D-J, Boezen HM et al. Eosinophilic granulocytes and interleukin-6 level in bronchoalveolar lavage fluid are associated with the development of obliterative bronchiolitis after lung transplantation. Am J Respir Crit Care Med 2000; 162:2221-25.

98. Smith C, Jaramillo A, Lu KC et al. Neutralization of tumor necrosis factor-alpha or interleukin-1 prevents obliterative airway disease in HLA-A2 transgenic murine tracheal allografts. J Heart Lung Transplant 2001; 20:166-67.

99. Linden PA, Gilbert RJ, Yeap BY et al. Laparoscopic fundoplication in patients with end-stage lung disease awaiting transplantation. J Thorac Cardiovasc Surg 2006; 131:438-46.

100. Benden C, Aurora P, Curry J et al. High prevalence of gastroesophageal reflux in children after lung transplantation. Pediatr Pulmonol 2005; 40:68-71.

101. Hadjiliadis D, Duane Davis R, Steele MP et al. Gastroesophageal reflux disease in lung transplant recipients. Clin Transplant 2003; 17:363-68.

102. Reid KR, McKenzie FN, Menkis AH et al. Importance of chronic aspiration in recipients of heart-lung transplants. Lancet 1990; 336:206-8.

103. Young LR, Hadjiliadis D, Davis RD et al. Lung transplantation exacerbates gastroesophageal reflux disease. Chest 2003; 124:1689-93.

104. Hartwig MG, Appel JZ, Li B et al. Chronic aspiration of gastric fluid accelerates pulmonary allograft dysfunction in a rat model of lung transplantation. J Thorac Cardiovas Surg 2006; 131:209-17.

105. Palmer SM, Miralles AP, Howell DN et al. Gastroesophageal reflux as a reversible cause of allograft dysfunction after lung transplantation. Chest 2000; 118:1214-17.

106. Davis RD, Lau CL, Eubanks S et al. Improved lung allograft function after fundoplication in patients with gastroesophageal reflux disease undergoing lung transplantation. J Thorac Cardiovas Surg 2003; 125:533-42.

107. Wenke K, Meiser B, Thiery J et al. Simvastatin reduces graft vessel disease and mortality after heart transplantation. Circulation 1997; 96:1398-1402.

108. Kobashigawa JA, Katznelson S, Johnson JA et al. Effect of pravastatin on outcomes after cardiac transplantation. New Engl J Med 1995; 333:621-27.

109. Johnson BA, Iacono AT, Zeevi A et al. Statin use is associated with improved function and survival of lung allografts. Am J Respir Crit Care Med 2003; 167:1271-78.

110. Zamora MR, Davis RD, Leonard CT. Management of cytomegalovirus infection in lung transplant recipients: evidence-based recommendations. Transplantation 2005; 80:157-63.

111. Kumar D, Erdman D, Keshavjee S et al. Clinical impact of community-acquired respiratory viruses on bronchiolitis obliterans after lung transplant. Am J Transplant 2005; 5:2031-36.

112. Garantziotis S, Howell DN, McAdams HP et al. Influenza pneumonia in lung transplant recipients: clinical features and association with bronchiolitis obliterans syndrome.[see comment]. Chest 2001; 119:1277-80.

113. Neurohr C, Huppmann P, Leuchte H et al. Human herpesvirus 6 in bronchoalveolar lavage fluid after lung transplantation: a risk factor for bronchiolitis obliterans syndrome? Am J Transplant 2005; 5:2982-91.

114. Kotsimbos TC, Snell GI, Levvey B et al. Chlamydia pneumoniae serology in donors and recipients and the risk of bronchiolitis obliterans syndrome after lung transplantation. Transplantation 2005; 79:269-75.

115. Novick RJ, Stitt LW, Al-Kattan K et al. Pulmonary retransplantation: predictors of graft function and survival in 230 patients. Pulmonary Retransplant Registry. Ann Thorac Surg 1998; 65:227-34.

116. Brugiere O, Thabut G, Castier Y et al. Lung retransplantation for bronchiolitis obliterans syndrome: long-term follow-up in a series of 15 recipients.[see comment]. Chest 2003; 123:1832-37.

117. Strueber M, Fischer S, Gottlieb J et al. Long-term outcome after pulmonary retransplantation. J Thorac Cardiovasc Surg 2006; 132:407-12.

# CHAPTER 16

# Liver Transplantation—An Overview

Tiffany E. Kaiser, E. Steve Woodle and Guy W. Neff*

## Abstract

Liver transplantation has offered thousands of patients a new lease on life. The improvements in survivals are attributed to the various treatment modalities before and after liver transplantation. Certain diseases such as chronic Hepatitis C virus (HCV) infection is an epidemic that is currently the number one indication for liver transplantation while hepatitis B virus (HBV) infection has less of a mortality problem following liver transplantation since the advent of hepatitis B immune globulin and oral antivirals. The impact of HCV in liver transplantation is well known; however therapeutic interventions are less standardized and often depend upon institutional protocol. Over the past few decades we have see continual improvement in patient and graft survival; however the issue of recurrent disease remains a problem. The aim of immunosuppression management post liver transplant has shifted from that of controlling rejection to one of minimizing drug related toxicities. Current treatment strategies involve early corticosteroid withdrawal and calcinurin inhibitor minimization. This chapter will provide a comprehensive review of the literature and address many issues and complications with transplantation in patients suffering from chronic liver disease from a variety of disease entities.

## Introduction

Since the first successful orthotopic liver transplantation (OLT) by Thomas Starzl in 1967, surgical and medical improvements have lead to improved global survival rates. Despite advances in peri-operative management and decreased patient mortality, significant organ shortages and the increasing number of patients dying on the transplant waiting list have forced transplant clinicians to seek alternative organ sources. Transplant centers have investigated new techniques such as living donor split livers or the use of extended criteria donor (ECD) organ in hopes of abating this problem.

Early OLT outcomes were suboptimal. Liver transplantation was associated with a high rate of mortality, stemming from organ rejection and infection as well as the morbidity associated with steroid dependence and disease recurrence. However, in the early 1980s, the situation drastically improved with the introduction of a novel class of drugs, the calcineurin inhibitors (CNIs). Newer anti-metabolites, such as mycophenolate mofetil, shortly followed in the 1990s and provided additional defense against acute rejection and allowed for decreased exposure to CNIs. Innovation in both immunosuppression combined with a nearly two decades of additional experience with liver transplantation fueled a new era in OLT in which acute rejection became almost obsolete. Today the transplant clinician is challenged with optimizing long-term patient outcomes and controlling patterns of disease recurrence.

This chapter is intended to review the current state of liver transplantation. Patient selection, current organ allocation and an outline of current strategies to improve donor shortages will be addressed. Pre and post-operative variables associated with patent and graft survival, as well as disease recurrence will be outlined. Finally, a summary of current immunosuppression strategies and their associated toxicities will be reviewed.

## Pretransplant

### Cirrhosis and Liver Transplantation

Cirrhosis and chronic liver disease together were the 12th most common cause of death in the United States in 2004, accounting for 26,549 deaths (9.0 per 100,000 persons).[1] Traditionally complications of cirrhosis: ascites, spontaneous bacterial peritonitis, hepatic encephalopathy, portal hypertension, variceal bleeding and hepatorenal syndrome, are treated with standard medical management and/or palliative procedures. As liver disease progresses such options lose their effectiveness and without a liver transplant, patients will succumb to their disease.[1,2] The incidence of cirrhosis and progression to end-stage liver disease (ESLD) necessitating transplantation will increase to more than 500% its current rate as a result of chronic HCV and non-alcoholic fatty liver disease (NAFLD).[3]

### Donor Organ Shortage

OLT is the therapy of choice for cirrhosis. Unfortunately, the demand for organs far outweighs the supply. Today over 17,000 patients are waiting for the average 6,000 livers transplanted per year. Annually, the number of patients medically suited to undergo OLT continues to rise, while the pool of deceased donor organs remains insufficient; thus, the supply is not meeting the demand, as summarized in Figure 1.

To combat the growing shortage of donor organs, physicians and organ procurement agencies are expanding the donor pool through four mechanisms. The first is to increase the number of patients (and their families) who give consent for organ donation and thus extend the number of cadaveric livers. Second is improved management of the deceased donor through targeted pharmacotherapy and mechanical support allowing for an increased number of acceptable organs. The third mechanism is adaptation of the ECD, which allows for transplantation of organs that were previously considered unacceptable for transplantation. Transplanting surgeons consent recipients for these extended criteria donors and adapt their peri-operative and immunosuppressive pharmacotherapy to reduce the incidence and severity of ischemia reperfusion injury. Today, these extended criteria allografts are being successfully used for transplantation with graft and patient

*Corresponding Author: Guy W. Neff—Associate Professor of Medicine, Medical Director, Hepatic Transplantation, 231 Albert Sabin Way, ML 0595, University of Cincinnati Medical Center, Cincinnati, Ohio 45267, USA. Email: guy.neff@uc.edu

*Chronic Allograft Failure: Natural History, Pathogenesis, Diagnosis and Management*, edited by Nasimul Ahsan. ©2008 Landes Bioscience.

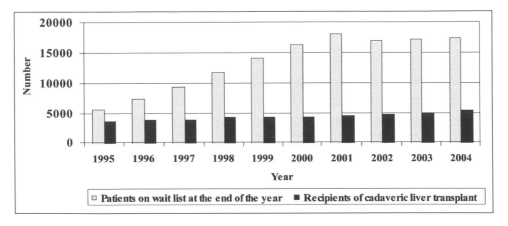

Figure 1. Organ supply versus demand 1995 to 2004.

survival that mirrors that of conventional donor recipients.[4,5] Finally, advances in medical practice such as living-donor liver transplantation and split-liver transplantation show promise in alleviating some of the deceased donor wait list demand.[6]

In the wake of strategies to increase the number of organs, the increased recipient pool has lead to challenges in terms of patient selection and effective organ allocation. The Model for End Stage liver disease (MELD) was introduced in the last decade to provide objective criteria for the grading of liver disease. This scoring system was created to estimate the chance of survival for patients based on both disease etiology and severity.[7,8] This prospective scoring system utilizes patient laboratory values for serum bilirubin, serum creatinine and international normalized ratio (INR) to predict survival.[7,8] The MELD scoring system is currently used by UNOS in prioritizing allocation of organs for OLT. Although designed to help prioritize patients, this system, does not recognize distinctions in "donor organ quality" such as; age, gender, fat content and heart beating versus nonheart beating status which ultimately influence outcomes of the OLT.[9]

## Posttransplant

### Patient Survival

Improvements in patient selection, surgical techniques and immunosuppressive regimens have drastically improved patient survival post OLT. Additionally, improved outcomes may be explained due to our increasing experience with factors associated with rejection risk and graft outcomes. Such factors include patient and donor variables, donor-recipient interactions, as well as intra, peri and post operative variables. All of these have culminated into an era of improved patient outcomes.

Survival following OLT can most appropriately be linked with ESLD etiology. Overall survival at one year post OLT ranges from 80-90%.[10] Figure 2 further divides patient survival at 3 months, 1, 3 and 5 years based on ESLD etiology. OLT recipients transplanted for either cholestatic or metabolic disease have the highest rate of survival at 3 and 5 years post OLT, while those with malignant neoplasms have the poorest.[10]

### Graft Survival

Graft survival rates have improved significantly with the advent of newer immunosuppressant agents. According to the OPTN/SRTR Annual Report reflecting patients transplanted through 2003, overall adjusted graft survival rates were 82.2%, 73.1% and 66.7% at 1, 3 and 5 years respectively post transplant.[10] As previously mentioned the introduction of CNIs revolutionized solid organ transplantation and reduced the incidence of liver acute rejection, by nearly 20%.[11,12] Mycophenolate mofetil, introduced in the early 1990's, has become a potent addition to most CNI sparing or avoidance protocols, aimed at

reducing the incidence and severity of CNI induced renal dysfunction. The anti-proliferative and angiogenesis effects of sirolimus have also been employed in patients who are transplanted for or who develop malignancies post transplantation. While, newer monoclonal antibodies directed against specific cell targets, such as the IL-2 receptor, allow for a further reduction in acute rejection rates. As a consequence, acute rejection rates have reached an unprecedented low and the current arsenal of immunosuppressants has allowed for the ability to tailor immunosuppresive regimens to the individual patient in order to maximize long-term patient and allograft survival.

### Causes of Graft Loss

Today, it is reasonable to expect that most liver allografts will serve their recipients throughout their respective life span. In fact, it is speculated that a significant number of patients die secondary to other comorbidities ((cardiovascular, endocrine etc.) with a functioning liver allograft. However, liver allograft failure does occur and continues to be evaluated as an area for improvement. When it occurs, acute cellular rejection typically occurs within the first 3 months post OLT and thereafter becomes less common. Due to improved immunosuppressive agents the incidence of acute rejection has declined, however; the reported incidence remains highly variable (20%-70%) likely due to different definitions and variations in nomenclature.[11,13] Chronic rejection in liver allografts has a relatively low incidence (<10%) but a rapid onset with a progressive course.[12,14]

Acute and chronic rejections remain important complications of liver transplantation; however; recurrent disease is the most common reason for intrinsic allograft failure. According to a review published by Kotlyar et al the overall recurrence rate of chronic hepatitis C (HCV) is greater than 90%, with a 5 year patient and graft survival rate of 70% and 57% respectively.[15] Table 1 illustrates the overall disease recurrence rates and 5 year patient and graft survivals according to ESLD etiology.[15]

With expansion of the donor pool and larger number of patients being listed with significant other comorbidities optimal management post OLT has yet to be achieved. Thus research to improve prevention as well as obtaining a better knowledge and treatment of disease recurrence will likely lengthen both the duration and quality of life for OLT recipients.

### Outcomes Following OLT

#### Hepatitis C Virus (HCV)

HCV is currently the leading cause of ESLD in the United States affecting 4 million.[16,17] As a result, HCV has become the most common indication for OLT. Current UNOS data reports that 42% of OLTs performed are for the treatment of ESLD secondary to chronic HCV infection.[18] HCV recurrence following OLT is universal and the natural

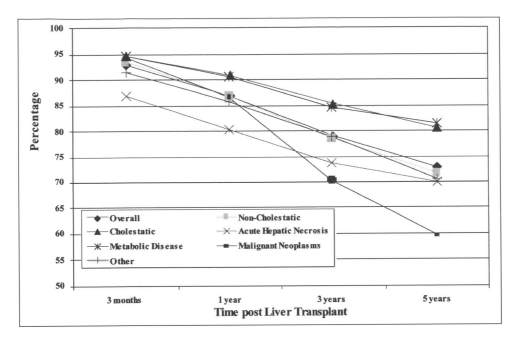

Figure 2. Relationship between patient survival post-OLT and primary causes of ESLD.

history of HCV is quite variable and difficult to predict.[19] Patient and graft survival tend to vary according to transplant centers, yet there is no dispute that fibrosis develops at a faster rate in the transplanted patients with HCV, as 20 to 30% of patients develop cirrhosis within 5 years of transplantation.[18,20,21] The rapid progression onto cirrhosis and the subsequent decrease in graft survival has lead many centers to adjust their post OLT treatment protocols to include antiviral therapy. [17,22-24]

Therapy with interferon-alpha monotherapy has not proven to be efficacious in the treatment of HCV post OLT.[23] However, therapies that include combination therapy (interferon and ribavirin) have shown limited yet improved success, with initial virological response and sustained viral load suppression seen anywhere from 10 to 20%.[25] Studies with pegylated interferon monotherapy have been published in the transplant setting with modest success.[26] Pegylated alfa-interferons combined with ribavirin remain the current standard of care for treatment of HCV post OLT. The benefits of this drug (pegylated interferon) are based on the prolonged duration of effect. Interferon is linked to a large inert, water-soluble molecule, polyethylene glycol (PEG) which allows for a longer half-life and increased trough levels without the peak levels attributable for some of the toxicity seen. The half-life for interferon is about 2-3 hours, while increased to approximately 54 hours when pegylated. This prolonged half-life allows for constant and steady levels of interferon within the serum.[27] Thus, patients are allowed once a week dosing instead of thrice weekly and appears to maintain a similar

safety profile as the short acting interferons.[28,29] The future is bright for patients suffering from HCV as there are presently more than 60 drugs submitted to the FDA for experimental approval. All in all, the benefits of pegylated interferons, combined with improved immuneosuppressant medications enhanced graft survival has been demonstrated.[21] Lastly, retransplantation in HCV transplant recipients continues to challenge transplant centers. The variance in opinions is best explained by experience and recipient morbidity status.[21,30-32]

**Hepatitis B Virus (HBV)**

HBV was long considered a poor reason for OLT due to suboptimal outcomes. In fact, not until the introduction of hepatitis b immunglobulin (HBIg) and nucleoside/tide agents, did this patient population overcome the burden of disease recurrence, often with subsequent organ failure and death. The literature regarding post OLT prophylaxis and/or suppression of HBV DNA tends to be center dependent throughout the world.[33-39] Fortunately, newer antiviral agents became available resulting in decreased mortality amongst our HBV transplant recipient population.[40] Most centers use some form of HBIG therapy and/or combination nucleotide analogue, lamivudine or adefovir.[39,41-48] The most difficult question to solve and agree upon is not only the use of HBIg but rather the length of therapy.[36,49] Cost effectiveness and utilization of HBIG has plagued patients throughout the world. Adequate treatment regimens to better determine the effective use of HBV antiviral therapy, with or without HBIG, need to be developed.

*Table 1. Disease recurrence post OLT*

| Etiology of Disease | Overall Recurrence Rate % | 5 Year Patient Survival % (CI) | 5 Year Graft Survival % (CI) |
|---|---|---|---|
| Hepatitis C | >90 | 70 (67-72) | 57 (54-59) |
| Hepatitis B | <5 with prophylaxis | 79 (74-83) | 68 (61-75) |
| Hepatocellular carcinoma | 8-15 | 52 (35-67) | 46 (31-60) |
| Primary biliary cirrhosis | 11-23 | 86 (83-89) | 73 (71-76) |
| Primary sclerosing cholangitis | 9-47 | 86 (83-89) | 73 (71-76) |
| Autoimmune hepatitis | 16-46 | 77 (71-82) | 68 (63-75) |
| Alcohol-induced Cirrhosis | <5 | 72 (68-76) | 65 (61-68) |
| Nonalcoholic steatohepatitis | 11-38 | 73 (63-77) | 66 (61-70) |

CI = 95% confidence interval. Adapted from Forman et al.[15]

In a cost analysis of HBIg versus oral anti-viral treatment modalities, HBIg was found to be extremely expensive and not devoid of drug related side effects.[50] Despite these economic concerns, centers throughout the world continue to use HBIg for prophylaxis and continue it for the lifetime of the OLT recipient. Discussions pertaining to drug discontinuation are sparse while the most common concerns regarding HBIg therapy are related to the duration of therapy.[36,49]

### Alcoholic Liver Disease (ALD)

OLT as a result of ALD has remained a controversial topic over the last two decades.[51] Long term survival in patients suffering from ALD appears to be similar to other indications for OLT.[52] The literature suggests alcohol use ranges from 8 to 22% for the first post-OLT year with cumulative rates reaching 30 to 40% by five years following transplantation. Recent work by DiMartini et al shows that after five years of follow-up, 22% of OLT recipients reported use of any alcohol by the first year and 42% had a drink by 5 years. By 5 years, 26% of patients had returned to binge drinking. In a univariate model, predictors of alcohol use included pre-OLT length of sobriety, a diagnosis of alcohol dependence, a history of other substance use and prior alcohol rehabilitation.[53]

Yet, most importantly, clinical interviews following OLT are essential to monitoring alcohol and substance use to detect and prevent recidivism. Treatment research is underway but has relied on traditional strategies that are not always applicable to OLT recipients. Further areas to improve clinical care include improving health behaviors, specifically, smoking cessation.[52]

### Hepatocellular Carcinoma (HCC)

Aggressive pretransplant therapies have changed the course of success for patients suffering from HCC while awaiting OLT.[54] Most importantly, adjuvant therapies for HCC while waiting for OLT provides an improvement in life expectancy.[55] The present UNOS requirements for OLT listing suggest that transplant centers follow the Milan and/or University at San Francisco (UCSF) criterions as summarized in Table 2. Arguments are ongoing as to which criteria are superior and which should be employed when listing a patient for OLT. An international report confirms the American findings in that the MELD scoring system has benefited patients suffering from HCC.[56] Some reports suggest the allocation based upon specific donor characteristics would promote less optimal grafts for this patient population.[57] All in all, OLT recipients with HCC continue to remain at risk of recurrence and require vigilant serial imaging and follow up post OLT.[58,59]

### *Immunosuppressive Therapy*

Initial immunosuppressive regimens in solid organ transplantation included corticosteroids and azathioprine; however graft survival rates were suboptimal. In the early 1980s a new era began, with the introduction of a novel drug class, calcineurin inhibitors (CNIs), cyclosporine and tacrolimus. Advances in immunosuppressive therapy have played an important role in liver transplantation allowing for improved graft rejection rates and survival.[60] Today, these compounds remain the backbone of immunosuppressive regimes. The use of CNIs was reported in 97% of patients discharged from the hospital after OLT in

2003-2004.[10] At 1 year post OLT, the use of CNIs remained prevalent with 93% maintained on CNIs (84% TAC, 9% CyA).[10] Additionally, patterns of change are emerging in terms of corticosteroid use post OLT from the traditional long term use to withdraw or elimination early in the postoperative period. Of the approximately 80% OLT recipients discharged on corticosteroids, only 49% are still using at 1 year and 33% at 2 years post.[10] Other agents included in maintenance immunosuppressive regimens are antimetabolites and sirolimus. In 2003-2004 the overall use of antimetabolites was reported in 58% of those discharged from the hospital post OLT and in 55% at 1 year post (52% mycophenolate the mofetil and 3% azathioprine). On the other hand, the use of sirolimus after OLT has been limited in the early post OLT period, but appears to be on the rise as the length of time post OLT increases. For example, only 5% of OLT recipients were discharged on sirolimus, while 12% were reported on the drug at 1 year post OLT.[10]

Due to the success of OLT, with excellent patient and graft survival, the focus of post OLT drug management has shifted from that of controlling rejection to one of minimizing drug toxicities.[61-64] Over the past few years, many strategies have been developed to minimize drug related adverse events ultimately as a means to improve morbidity and mortality. For years corticosteroids have been integral to the success of solid organ transplant, however, with the use of CNIs along with the development of newer agents such as mycophenolate mofetil (MMF) the reliance on corticosteroids has been reduced.[62] Corticosteroids adverse event profile and their effect on HCV and the liver allograft are the main objectives for corticosteroid elimination and/or minimization. Numerous studies have demonstrated the benefits of corticosteroid free immunosuppressive strategies in OLT.[62]

Currently, much interest has focused on renal insufficiency post OLT, likely due to the sentinel articles by Ojo et al and Gonwa et al in which the incidence of renal failure post nonrenal transplant was reported.[65,66] Chronic renal failure (CRF) developed at 5 years post transplant in 21% of recipients of a nonrenal organ and 18.1% in OLT recipients.[65] CNI therapy, a key component of immunosuppressive regimens have been implicated as a principal cause of post OLT renal dysfunction. This issue of post OLT renal dysfunction will likely intensify due to the fact that the current allocation system (MELD) prioritizes candidates with elevated serum creatinines, in addition we are transplanting older and sicker patients with worse baseline kidney function. As a means to address this ongoing problem, different renal sparing strategies have been developed to try to eliminate and/or reduce CNI from treatment protocols including: (1) antibody induction (2) CNI avoidance or minimization (3) mycophenolate mofetil and (4) conversion from CNI to sirolimus.

### Antibody Induction

Traditionally, in contrast to other types of solid organ transplants antibody induction in OLT is uncommon. The liver's capacity to regenerate following an immunologic insult and the immunoprotection provided by various mechanisms, including Kuffer cell immunoabsorption, traditionally made antibody induction obsolete. Recently however, the use of antibody induction in OLT appears to be on the rise. Overall antibody use for induction was 7% in 1997, compared to 21% in 2004.[10]

**Table 2. Milan and University of California at San Francisco (UCSF) staging criteria for hepatocellular carcinoma**[N]

| Staging | Single Tumor Maximum Diameter | Multiple Tumors Maximum Number | Largest Tumor Size | Total Tumor Size |
|---|---|---|---|---|
| Milan | ≤5.0 cm | 3 | ≤3.0 cm | NA |
| UCSF | ≤6.5 cm | 3 | ≤4.5 cm | ≤8.0 cm |

NA, not applicable

Increased utilization is likely due to the development and experience with new drugs (i.e., anti-thymocyte globulin and alemtuzamab) as well as novel immunosuppressive strategies employed as a means to delay exposure to the nephrotoxic CNIs and/or to help facilitate the removal and dose reduction of other agents such as corticosteroids.[67,68] Tchervenkov et al reported that anti-thymoctye globulin induction allows delayed CNI initiation without compromising patient and graft survival and secondly allowing for greater renal recovery in patients with baseline renal impairment (pretransplant serum creatinine ≥1.5 mg/dL).[67] Others report their experience with alemtuzumab induction.[69] Tzakis and colleges demonstrated significant improvement in renal function in the group that received alemtuzuamab induction combined with low dose tacrolimus (TAC) compared to controls which received the standard TAC dose plus corticosteroids.[69] In the first two month post OLT the incidence of acute rejection was significantly lower with the alemtuzumab group (15%) compared to control (6%); however the overall incidence of acute cellular rejection episodes was similar between groups (46% vs 55%). To date it is unclear what effect antibody induction has on long term renal function, additional prospective randomized trials with longer follow up are necessary.

### CNI Avoidance

Despite the recognition of chronic renal impairment resulting from the nephrotoxic effects of the CNIs, these agents remain an essential component of the immunosuppressive regimens post OLT. The latest data from the OPTN/SRTR annual report shows that greater than 95% of OLT recipients are discharged and remain on CNIs (either tacrolimus (majority) or cyclosporine) at one year follow up.[10] CNI avoidance protocols in OLT have not been achieved routinely; however, pilot trials have begun to delineate the limitations and promises of such approaches. CNI-sparing protocols appear to be much more promising in balancing the early need for rejection minimization while tapering doses and minimizing long-term toxicity.[70,71]

### Mycophenolate Mofetil (MMF)

MMF was approved by the FDA in 1995 and is an immunosuppressant with a side effect profile different from CNI and most importantly one that's non nephrotoxic, thus can be considered an attractive CNI replacement in patients with compromised renal function. A study published in 2005 by Reich et al reported encouraging results in that glomerular filtration rate (GFR) improved in greater than 15% of patients when MMF was added and the CNI dose was either reduced or eliminated.[72] However, the serious drawback to CNI withdrawal is lowered immunological safety as demonstrated in numerous studies reporting increased allograft rejection rates.[72-74] Even though this renal sparing option may be beneficial, assessment of risk for rejection should be done prior to CNI withdrawal. Perhaps, this strategy should be utilized in those where the risk for rejection is likely lower, for example; those with long term follow up, minimal prior rejection history and minimal CNI requirement prior to withdrawal.

### Sirolimus Conversion

Sirolimus, an mTOR agent has been proposed as a replacement for CNIs. The early overwhelming promise was temporized by medication related side effects and mortalities.[75,76] The primary Achilles' heal of sirolimus is the difficulty with acquiring real time drug levels that allow for appropriate daily dose adjustments. Therapeutic target serum levels with combination CNIs and sirolimus treatment regimes vary among disease entities and medical centers. Overall the clear-cut serum levels to achieve when combining a CNI with sirolimus and the prolonged turn around time for sirolimus serum levels make the use of this combination therapy very complex. Consequently drug related complications can ensue and often do.[77,78]

Protocols utilizing sirolimus conversion typically convert patients from CNI to sirolimus once renal insufficiency develops post OLT.

The optimal time point post OLT for sirolimus conversion has yet to be determined and varies among medical centers. The benefit of conversion is improvement in glomerular filtration rate.

Sirolimus's role in OLT recipients to preserve renal function continues to evolve. Most studies are reporting similar findings; that renal function stabilizes in a subset of patients, however, overall efficacy is limited by the significant number of patients' inability to tolerate the drug and those whose renal function continues to deteriorate despite its use. Additional larger, prospective randomized controlled trials are necessary to determine sirolimus's precise utility.

The introduction of new drugs and growing experience with their use has permitted the development of novel drug treatment strategies aimed to minimize rejection, improve graft function, as well as minimize toxicities. Additionally, with the multitude of different drug classes, each with a unique mechanism of action: monoclonal and polyclonal antibodies, CNI's, antimetabolites and mTOR inhibitors, as well as pharmacokinetic, pharmacodynamic and pharmacogenetic drug therapy monitoring patient immunosuppression management post OLT can be individualized to better address individual patient's needs, ultimately to minimize drug related toxicities.

## Discussion

The atmosphere for liver transplantation has evolved over the past half century culminating into a procedure with a great deal of success mostly through improvements in immune suppression protocols and the enhanced recognition as well as treatment of recurrent disease. The future for liver transplantation is bright. Medical centers will continue to face challenges with organ shortages and promoting improvements in the present organ allocation system; however novel strategies will continue to be developed to address such issues. Indications for OLT will likely change over the ensuing decade with the rise in cases of HCC and NAFLD. Fortunately, many disease entities will see continued improvement in survival outcomes, reduced drug related toxicities and advances in therapies aimed as preventing disease recurrence.

## References

1. National Center for Health Statistics. National Vital Statistics Report. Chronic liver disease/cirrhosis. Accessed December 28, 2006 at www.cdc.gov/nchs/fastats/liverdis.htm
2. Heidelbaugh JJ SM. Cirrhosis and Chronic Liver Failure: Part II. Complications and Treatment. American Family Physician 2006; 74(5):767-76.
3. Davis G. Chronic hepatitis C and liver transplantation. Rev Gastroenterol Disord 2004; 4(1):7-17.
4. Busuttil RW TK. The Utility of Marginal Donors in Liver Transplantation. Liver Transplantation 2003; 9(7):651-63.
5. Tector AJ MR, Chestovich P, Vianna R et al. Use of Extended Criteria Livers Decreases Wait Time for Liver Transplantation Without Adversely Impacting Posttransplant Survival. Annals of Surgery 2006; 244(3):439-50.
6. Trotter JF. Expanding the donor pool for liver transplantation. Curr Gastroenterol Rep 2000; 2(1):46-54.
7. Malinchoc M KP, Gordon FD, Peine CJ et al. A Model to Predict Poor Survival in Patients Undergoing Transjugular Intrahepatic Portosystemic Shunts. Hepatology 2000; 31(4):864-71.
8. Kamath PS WR, Malinchoc M, Kremers W et al. A Model to Predict Survival in Patients With End-Stage Liver Disease. Hepatology 2001; 33(2):464-70.
9. Lucianetti A GM, Bertani A. Liver transplantation in children weighting less that 6 kg: the Bergamo experience. Transplant Proc 2005; 37(2):1143-45.
10. 2005 Annual Report of the U.S. Organ Procurement and Transplantation Network and the Scientific Registry of Transplant Recipients: Transplant Data 1995-2004. Department of Health and Human Services, Health Resources and Services Administration, Healthcare Systems Bureau, Division of Transplantation, Rockville MD; United Network for Organ Sharing, Richmond, VA; University of Renal Research and Education Association, Ann Arbor, MI.
11. Horoldt BS BM, Gunson BK, Bramhall SR et al. Does the Banff Rejection Activity Index Predict Outcomes in Patients with Early Acute Cellular Rejection Following Liver Transplantation? Liver Transplantation 2006; 12:1144-51.

12. Matinlauri IH NM, Hockerstedt KA, Isoniemi HM. Changes in Liver Graft Rejections Over Time. Transplantation Proceedings 2006; 38:2663-66.
13. Neuberger J. Incidence, timing and risk factors for acute and chronic rejection. Liver Transplant Surgery 1999; 5 (suppl 1):S30.
14. Gouw AS VDHM, Van Den Berg AP, SSlooff MJ et al. The significance of Parenchymal changes of acute cellular rejecion in predicting chronic liver graft rejections. Transplantation 2002; 73(2):243-47.
15. Kotlyar DS CM, Reddy KR. Recurrence of Diseases Following Orthotopic Liver Transplantation. American Journal Gastroenterology 2006; 101:1370-78.
16. Younossi Z. Chronic Hepatitis C: a clinical overview. Cleveland Clinic Journal of Medicine 1997; 64(5):25-68.
17. Bizollon T DC, Baulieux J, Trepo C. Treatment of recurrent hepatitis C following liver transplantation. Current Gastroenterology Reports 1999; 1(1):15-19.
18. Charlton M SE, Wiesner R, Everhart J et al. Predictors of patient and graft survival following liver transplantation for hepatitis C. Hepatology 1998; 28(3):823-30.
19. D Mutimer GD, C Barrett, L Grellier et al. Lamivudine without HBIg for prevention of graft reinfectio by hepatitis B: Long-Term Follow-Up. Transplantation 2000; 70(5):809-15.
20. MW Fried YK, GA Smallwood, M Cong et al. Hepatitis G Virus Co-Infection in Liver Transplantation Recipients With Chronic Hepatitis C and Nonviral Chronic liver Disease. Hepatology 1997; 25(5):1271-75.
21. Charlton M. Recurrence of Hepatitis C Infection: Where Are We Now? Liver Transplantation 2005; 11(11 Suppl 1):S57-62.
22. S Roayaie MH, S Emre, TM Fishbein et al. Comparison of Surgical Outcomes for Hepatocellular Carcinoma in Patients with Hepatitis B Versus Hepatits C: A Western Experience. Annuals of Surgical Oncology 2000; 7(10):764-70.
23. EJ Gane SL, SM Riordan, BC Portmann et al. A Randomized Study Comparing Ribavirin and Interferon Alfa Monotherapy for Hepatitis C Recurrence After Liver Transplantation. Hepatology 1998; 27(5):1403-7.
24. V Mazzaferro MB, M Pasquali, E Regalia et al. Preoperative serum levels of wild-type and hepatitis B e antigen-negative hepatitis B virus (HBV) and graft infection after liver transplantation for HBV-related hepatocellular carcinoma. Journal of Viral Hepatitis 1997; 4(4):235-42.
25. Hoofnagle JH DA. The Treatment of Chronic Viral Hepatitis. New England Journal of Medicine 1997; 336(5):347-56.
26. GW Neff MM, CB O'Brien, S Nishida et al. Treatment of Established Recurrent Hepatitis C in Liver-Transplant Recipients with Pegylated Interferon-alfa-2b and Ribavirin Therapy. Transplantation 2004; 78(9):1303-7.
27. NE Algranati SS, M Modi. A branched chain methoxy 40kDa polyethylene glycol (PEG) moiety optimized the pharmacolinetics (PK) of Peginterferon 2a (PEG-INF) and may explain its enhanced efficacy in chronic hepatitis C. Hepatology 1999; 30(Suppl:702A):Abstract.
28. S Zeuzem VF, J Rasenack, EJ Heathcote et al. Peginterferon Alfa-2a In Patients with Chronic Hepatitis C. The New England Journal of Medicine 2000; 343(23):1666-72.
29. P Glue JF, R Rouzier-Panis, C Raffanel et al. Pegylated interferon-alpha2b: pharmacokinetics, pharmacodynamics, safety and preliminary efficacy data. Hepatitis C Intervention Therapy Group. Clinical Pharmacology & Therapeutics 2000; 68(5):556-67.
30. GW Neff COB, J Nery, NJ Shire et al. Factors that Identify Survival After Liver Retransplantation for Allograft Failure Caused by Recurrent Hepatis C Infection. Liver Transplantation 2004; 10(12):1497-503.
31. Rosen H. Retransplantation for Hepatitis C: Implication of Different Policies. Liver Transplantation 2000; 6(6 Suppl 2):S41-46.
32. McCashland T. Retransplantation for Recurrent Hepatitis C: Positive Aspects. Liver Transplantation 2003; 9(11):S67-72.
33. Z Ben-Ari EM, R Tur-Kaspa. Experience with lamivudine therapy for hepatitis B virus infection before and after liver transplantation and review of the literature. Journal of Internal Medicine 2003; 253(5):544-52.
34. S Fagiuoli VM, M Pompili, S Gianni et al. Liver transplanation: the Italian experience. Digestive & Liver Disease 2002; 34(9):640-48.
35. JR Nery CN-A, KR Reddy, R Cirocco et al. Use of liver grafts from donors positive for antihepatitis b-core antibody (ANTI-HBc) in the era of prophylaxis with hepatitis-b immunoglobulin and lamivudine. Transplantation 2003; 75(8):1179-86.
36. CM Lo SF, CH Liu, CL Lai et al. Prophylaxis and Treatment of Recurrent Hepatitis B After Liver Transplantation. Transplanation 2003; 75 (3 Suppl):S41-44.
37. S Zheng JW, W Wang, D Huang et al. Lamivudine as prophylaxis against hepatitis B virus reinfection following orthotopic liver transplantation. Chung-Hua i Hsueh Tsa Chih [Chinese Medical Journal] 2002; 82(7):445-48.

38. HE Vargas FD, J Rakela. A concise update on the status of liver transplantation for hepatitis B virus: the challenges in 2002. Liver Transplantation 2002; 8(1):2-9.
39. AB van Nunen RdM, RA Heijtink, AC Vossen et al. Passive immunization of chronic hepatitis B patients on lamivudine therapy: a feasible issue? Journal of Viral Hepatitis 2002; 9(3):221-28.
40. Neff GW OBC, Nery J, Shire N et al. Outcomes in Liver Transplant recipients with hepatitis B virus: resistance and recurrence patterns from a large transplant center over the last decade. Liver Transplantation 2004; 10(11):1372.
41. Grellier L MD, Ahmed M, Brown D et al. Lamivudine prophylaxis against reinfection in liver transplantation for hepatitis B cirrhosis. The Lancet 1996; 348(9036):1212-15.
42. Markowitz JS MP, Conrad AJ, Markmann JF et al. Prophylaxis Against Hepatitis B Recurrence Following Liver Transplantation Using Combination Lamivudine and Hepatitis B Immune Globulin. Hepatology 1998; 28(2):585-89.
43. Dodson SF dM, Bonham CA, Geller DA et al. Lamivudine After Hepatitis B Immune Globulin Is Effective in Preventing Hepatitis B Recurrence After Liver Transplantation. Liver Transplantation 2000; 6(4):434-39.
44. Kruger M. European hepatitis B immunoglobulin trials: prevention of recurrent hepatitis B after liver transplantation. Clinical Transplantation 2000; 14(Suppl 2):14-19.
45. Perrillo R. Antiviral therapy to prevent and treat hepatitis B virus infection in hepatic allografts. Clinical Transplantation 2000; 14(Suppl 2):25-28.
46. Perrillo RP KM, Sievers T, Lake JR. Posttransplantation: emerging and future therapies. Seminars in Liver Disease 2000; 20(Suppl 1):13-17.
47. Seehofer D RN, Steinmuller T, Muller AR et al. Combination prophylaxis with hepatitis B immunoglobulin and lamivudine after liver transplantation minimizes HBV recurrence rates unless evolution of pre-transplant lamivudine resistance. Zeitschrift fur Gastroenterologie 2002; 40(9):795-99.
48. Lok A. Prevention of Recurrent Hepatitis B PostLiver Transplantation. Liver Transplantation 2002; 8(10 Suppl 1):S67-73.
49. Villamil F. Hepatitis B: Progress in the Last 15 Years. Liver Transplantation 2002; 8(10 Suppl 1):S59-66.
50. McGory R. Pharmacoeconomic analysis of HBV liver transplant therapies. Clinical Transplantation 2000; 14(Suppl 2):29-38.
51. Lucey MR BT. Alcoholic liver disease: to transplant or not to transplant? Alcohol Alcohol 1992; 27(2):103-8.
52. DiMartini A WR, Fireman M. Liver transplantation in patients with alcohol and other substance use disorders. Psychiatr Clin North Am 2002; 25(1):195-209.
53. DiMartin A DN, Dew MA, Javed L et al. Alcohol Consumption Patterns and Predictors of Use Following Liver Transplantation for Alcoholic Liver Disease. Liver Transplantation 2006; 12(5):813-20.
54. DiBisceglie A. Pretransplant Treatments for Hepatocellular Carcinoma: Do They Improve Outcomes? Liver Transplantation 2005; 11(11 Suppl 1):S10-13.
55. Gores G. Hepatocellular Carcinoma: Gardening strategies and bridges to Transplantation. Liver Transplantation 2003; 9(2):199-200.
56. Ravaioli M GG, Ballardini G, Cavrini G et al. Liver Transplantation with the Meld System: A Prospective Study from a Single European Center. American Journal of Transplanation 2006; 6(7):1572-77.
57. Ravaioli M GG, Ercolani G, Cescon M et al. Liver Allocation for Hepatocellular Carcinoma: A European Center Policy in the preMELD Era. Clinical Transplantation 2006; 81(4):525-30.
58. Shah SA GP, Gallinger S, Cattral MS et al. Factors Associated with Early Recurrence after Resection for Hepatocellular Carcinoma and Outcomes. Journal American College of Surgery 2006; 202(2):275-83.
59. Said A LM. Liver transplantation: and update. Current Opinion Gastroenterology 2006; 22(3):272-78.
60. Cattral MS LL, Levy GA. Immunosuppression in Liver Transplantation. Seminars in Liver Disease 2000; 20(4):523-31.
61. Starzl T. Back to the Future. Transplantation 2005; 79(9):1009-14.
62. O'Grady J. Corticosteroid-Free Strategies in Liver Transplantation. Drugs 2006; 66(14):1853-62.
63. Regazzi MB AM, Rinaldi M. New Strategies in Immunosupression. Transplant Proceedings 2005; 37:2675-78.
64. Hirose R VF. Immunosuppression: Today, Tomorrow and Withdrawal. Seminars in Liver Disease 2006; 26(3):201-10.
65. Ojo AO HP, Port FK, Wolfe RA et al. Chronic Renal Failure after Transplantation of a Nonrenal Organ. The New England Journal of Medicine 2003; 349(10):931-40.
66. Gonwa TA MM, Melton LB, Hays SR et al. End-Stage Renal Disease (ESRD) After Orthotopic Liver Transplantation (OLTX) Using Calcineurin-Based Immunotherapy. Transplanation 2001; 72(12):1934-39.

67. Tchervenkov JI TG, Cantarovich M, Barkun JS et al. The Impact of Thymoglobulin on Renal Function and Calcineurin Inhibitor Initiation in Recipients of Orthotopic Liver Transplant: A Retrospective Analysis of 298 Consecutive Patients. Transplant Proceedings 2004; 36:1747-52.

68. Bogetti D SH, Jarzembowski TM, Manzelli A et al. Thymoglobulin induction protects liver allografts from ischemia/reperfusion injury. Clinical Transplantation 2005; 19:507-11.

69. Tzakis AG TP, Kato T, Nishida S et al. Preliminary experience with Alemtuzumab (Campath-1H) and Low-Dose Tacrolimus Immunosuppression in Adult Liver Transplant Recipients. Transplanation 2004; 77(8):1209-14.

70. McAlister VC PK, Malatjalian DA, Colohan S et al. Orthotopic Liver Transplantation Using Low-Dose Tacrolimus and Sirolimus. Liver Transplantation 2001; 7(8):701-8.

71. Fung J KD, Kadry Z, Patel-Tom K et al. Immunosuppression in Liver Transplantation Beyond Calcineurin Inhibitors. Liver Transplantation 2005; 11(3):267-80.

72. Reich DJ CP, Hodge EE. Mycophenolate Mofetil for Renal Dysfunction in Liver Transplant Recipients on Cyclosporine or Tacrolimus: Randomized, Prospective, Multicenter Pilot Study Results. Transplanation 2005; 80(1):18-25.

73. Schlitt HJ BA, Boker KH, Schmidt HH et al. Replacement of calcineurin inhibitors with mycophenolate mofetil in liver-transplant patients with renal dysfunction: a randomised controlled study. The Lancet 2001; 357:587-91.

74. Hong MK AP, Jones RM, Vaughan RB et al. Predictors of Improvement in renal function after calcineurin inhibitor withdrawal for postliver transplant renal dysfunction. Clinical Transplantation 2005; 19:193-98.

75. Fung J MA. Rapamycin: Friend, Foe or Misunderstood? Liver Transplantation 2003; 9(5):469-72.

76. Trotter J. Sirolimus in Liver Transplantation. Transplant Proceedings 2003; 35(Suppl 3A):193S-200S.

77. Neff GW MM, Tzakis AG. Ten Years of Sirolimus Therapy in Orthotopic Liver Transplant Recipients. Transplantation Proceedings 2003; 35(Suppl 3A):209S-16S.

78. Montalbano M NG, Yamaskiki N, Meyer D et al. A Retrospective Review of Liver Transplant Patients Treated with Sirolimus from a Single Center: An Analysis of Sirolimus-Related Complications. Transplantation 2004; 78(2):264-68.

# Hepatic Allograft Loss:
## Pathogenesis, Diagnosis and Management

### Mohammad Ali*

## Introduction

Liver transplantation is the established therapeutic modality for the treatment of both acute and chronic end stage liver disease. After successful transplantation 85% recipients usually survive one year, 69% for five years and 61% for ten years. Well functioning liver allograft is the prime issue on which survival and most of the morbidity of the recipients is centered. Several conditions that affect the out come of the graft include donor and recipient operative factors, quality and quantity of the graft, graft procurement, preservation, infections, rejections, immunological destruction and recurrent diseases. It is difficult to differentiate the conditions clinically and by laboratory investigations in certain circumstances. Graft dysfunction causes multiple problems for the recipient, which includes physical sufferings, financial involvement and even mortality. It needs multidisciplinary management strategies that include intervention and surgical procedures. The quality of compromised graft decreases gradually and finally develops the irreversible stage of graft failure. It may require retransplantation in 20% of patients.[1] One and five year's survival after retransplantation is 55% and 47% respectively.[2] Every attempt should be taken to prevent primary graft loss, which will minimize retransplantation and improve over all patient survival.

## Donor Status Affecting the Graft Function

Donor condition and liver quality have profound influence on good graft function. Technical variants of hepatic allograft are considered as risk factor for graft survival.

### Graft Quality

There are three types of hepatic allograft—standard, nonstandard and marginal. The use of marginal graft has become more common due to sharp increase in recipients in respect to shortage of donors. Marginal graft function depends on multiple factors. Advanced age (>75 yrs), prolonged hospitalization in intensive care unit, massive doses of inotropes, high sodium concentration, hypoxic and septic complications dictates the graft function.[3,4] Besides these, viral serologies, hepatic steatosis, cold ischemia time (CIT) and warm ischemia time (WIT) also have profound effect on graft function. Donor hypernatraemia is supposed to have relation with post transplant mortality.[5-9] Steatosis has been shown to have definite effects in the prognosis of liver transplant. A graft of 25% macrosteatosis causes 12.5% primary nonfunction (PNF) and 62.5% delayed nonfunction.[10] Donor liver biopsy is mandatory to define the degree of steatosis before retrieval. Cold ischemia time is an important factor in the overall outcome. It varies from 12 to 14 hours during deceased donor liver transplant

(DDLT) in different centers. It should be minimized as far as possible by organizing the retrieval technique and policy.

### Primary Nonfunction

Primary graft nonfunction (PNF) is a life threatening early complication after liver transplantation. Hepatic allograft fails to function after implantation. It occurs in 5-10% of recipients.[11] Pathologically it is due to microcirculatory injury leading to loss of centrilobular architecture, lymphocytic infiltration and central cholestasis. It has multifactorial etiology, such as prolonged cold ischemic injury, macrosteatosis and extreme donor age.[12,13] Many conditions may simulate with the features of PNF. It is diagnosed by exclusion of hyper acute rejection, hepatic arterial thrombosis and portal vein thrombosis. Poor liver function is reflected by reduction or absence of bile production, high transaminases, increased lactate level, coagulopathy, severe encephalopathy and altered renal function.[14-16] Doppler study and ancillary examinations are needed to exclude vascular impairment. All the supportive measures including liver dialysis with molecular adsorbent recirculating system (MARS) is essential for hepatic support. Mortality is the usual out come unless retransplantation is done.

### Fatty Donor Liver (Steatosis)

Steatosis of the donor liver has significant impact on allograft function and patient survival after orthotopic liver transplant (OLT). The standard regime is to avoid fatty liver or to use micro steatotic liver graft (<25%). But due to extreme donor organ shortage various grades of steatotic grafts are often used. Initial poor graft function, primary nonfunction (PNF),[18] allograft infarction, biliary complications are more with macro steatotic liver grafts(>30%). All these ailments may lead to graft failure which may need retransplantation in 6% of the recipients.[19]

## Graft Size Mismatch

Size of the allograft is an important factor for outcome in both Deceased Donor Liver Transplant (DDLT) and Living Donor Liver Transplant (LDLT). The Graft recipient weight ratio (GRWR) is the ratio of standard graft size to recipient's body weight; which is 0.8-1% and it is 40-50% of the standard liver volume (SLV). Both the small for size and large for size graft have impact on overall graft function.

### Small for Size Graft

Small for size graft is a condition when GRWR <0.8%. The graft suffers portal hyperperfussion and insufficient venous outflow leading to arterial hypoperfussion with reduced liver function. The regeneration

*Mohammad Ali—Professor & Head, Dept. of Hepato-Biliary-Pancreatic Surgery & Liver Transplantation, Room 911, BIRDEM Hospital, 122 Kazi Nazrul Islam Avenue, Shahbag, Dhaka-1000, Bangladesh. Email: hbbirdem@yahoo.com

capacity of the graft also reduced. It can produce multiple functional problems. These conditions are together called small for-size syndrome (SFSS).[20,21] It has two variety of clinical presentations– the small for size dysfunction (SFSD) and small for size nonfunction (SFSNF). In SFSD there is mild to moderate alteration of liver function. SFSNF is the more severe clinical condition when there is hyper bilirubinaemia (>100 micromole /L), encephalopathy (Grade 3 or 4), elevated Alanine aminotransferase (ALT)/Aspartate aminotransferase (AST), ascitis, pulmonary and renal failure. Small for size graft problem is diagnosed by exclusion of arterial or portal vein occlusion, outflow congestion, biliary leakage, rejections and infectious complications. The critical state of SFSS is seen in split liver implantation in DDLT, especially during implantation of one liver to two adult recipients and during LDLT when lateral segments (seg II and III) are implanted in an adult recipient.[22] Graft dysfunction should be evaluated by imaging studies. Contrast enhanced CT is needed for evaluation of hepatic parenchymal abnormalities such as congestion, fatty changes, infarction, abscess, intrahepatic biliary changes and extrahepatic abnormalities.[23]

Liver attenuation indices (LAI) in Houns field units(HU), derived from differences between mean hepatic and splenic attenuation. It is calculated on unenhanced CT images obtained on 10th posttransplant day of LDLT. CT hypo attenuation reflects histological graft dysfunction. One year graft survival is less when predicted LAI <5.[24]

### Large for Size Graft

Size mismatch between graft and recipient frequently occur in pediatrics liver transplants. Large for size graft is considered when GRWR >3%. It may cause graft compression leading to compromise of hepatic vascular flow, thrombotic complications, vessel kinking and low portal flow. Technical difficulties in vascular anastomosis may occur due to disproportion of vessel diameter.[25,26] Appropriate size graft should always be considered. In difficult circumstances delayed abdominal wound closure, mesh closure or creation of hernia with closure of skin only is safe to protect the large graft.

### Hepatic Ischemia-Reperfusion Injury

Allograft suffers a period of ischemia during liver transplantation. The condition is aggravated on restoring the blood supply called ischemia-reperfusion (I-R) injury. The event in the graft after the onset of reperfusion includes endothelial swelling, vasoconstriction, leukocyte entrapment and platelet aggregation within the sinusoids.[27-31] All these conditions result in failure of micro-circulation. There is decrease in nitric oxide (NO) and increase in endothelin (ET) production leading to narrowing of sinusoidal lumen and decreased leukocyte activities, which cause cellular injury and necroptosis. This reperfusion injury, may cause primary non function and delayed poor functions of the graft. Multiple factors are involved in I-R injury. Donor factors such as ischemia caused by hypotension and steatotic liver grafts[32,33] are more susceptible to I-R injuries than normal grafts. The operative factors include reduced portal venous or hepatic arterial inflow and the length of cold and warm ischemia time. Cold ischemic storage and reperfusion of liver causes ultra structural changes which become more prominent with increase of storage time. I-R injuries are common in marginal grafts and graft from non heart beating donors. Preservation injury is diagnosed when there is increase of Gamma- glutamyl tranpeptidase(GGT) but normal serum bilirubin level. Donor selection, technical expertise and reduction of both cold and warm ischemia time can reduce the I-R injury significantly.

### Allograft Retrieval Injury

The hepatic allograft retrieval should be well planned and organized. There may be injury to vessels due to failure of recognition of aberrant or replaced hepatic artery. Such injury causes inadequate perfusion of the segment of the graft and ischemic injury of biliary system. Apart from the arterial injury there may be portal vein and hepatic vein injury.

Maintenance of proper outflow of the graft is mandatory especially during retrieval of graft from living donor. Preservation of drainage of anterior sector (seg V and VIII) is important to avoid the outflow congestion of the sector. Middle hepatic vein might need to be incorporated in the right lobe graft for its good function. All precautions are maintained to prevent contusion, laceration or subcapsular haematoma formation of the graft. Repair, reconstruction and preparation of injured graft causes prolonged back table preparation. It has deleterious effects on the overall graft function after its implantation.

### Effects of Warm and Cold Ischemia on Graft Function

Ischemia of hepatic allograft causes swelling of the sinusoidal cells which subsequently narrows the sinusoidal lumen and disturbs the microcirculatory units of liver. It causes outflow block phenomenon of liver and leads to its dysfunction. Allograft procurement procedure passes through two stages of probable ischemic injury. One is warm ischemia during dissection and cannulation. Another one is cold ischemia during perfusion, preparation and implantation of the graft to the recipient. Cold ischemia is more tolerable than warm ischemia. Cold ischemia during first 8 hours doesn't cause significant injury but warm ischemia causes earlier ischemic injuries. The intensity of lesions with 4 hours period of warm ischemia is equal to intensity of lesions with 16 hours period of cold ischemia.[34] Chance of cold and warm ischemia is more in DDLT than LDLT. Efforts are to be taken to minimize the injuries.

## Vascular Problems of Liver Graft

Vascular complications in the early and late periods following liver transplant causes significant morbidity and mortality. It leads to altered hepatic functions, biliary complications, necrosis, abscess formation and finally allograft failure. Vascular evaluation is essential during harvesting and preparation of the allograft for identification of anomalies. This ensures the adequate arterial flow and venous drainage of all the segments of the graft. Anatomical consideration is critical for preparation of reduced size, split and living donor liver grafts. Delineation of vasculature of recipient, specially the hepatic artery and portal vein is mandatory for planned vascular anastomosis.

### Hepatic Arterial Thrombosis (HAT)

Thrombosis of hepatic artery is the most severe and frequent vascular complication of liver transplantation.[35] This is noted in 3-14% of adults and in 26% of children.[36,37] Most of the incidence is noted in the first 2 months of transplant.[38] HAT manifests as acute massive hepatic necrosis, grossly elevated hepatic enzymes, biliary leakage, biliary stricture and episodes of sepsis within the infracted liver.[39,40] HAT causes a high mortality of up to 75%. Factors contributing to the development of HAT includes technical difficulties, prolonged cold-ischemia, pediatric recipient due to small diameter of hepatic artery.[41] ABO incompatible graft, use of vascular conduits, or retransplantation. Increased thrombin activity, elevated levels of plasminogen activator and increased hematocrit level is considered as risk factors.[42-44] The clinical presentations of HAT are variable. It causes dramatic increase in transaminases, decrease in bile production with its thin colour, persistent elevation of prothrombin time and bilirubin. Duplex sonography is the most standard method of evaluation of HAT (sensitivity >90%). In the absence of Doppler flow to the hepatic artery, with clinical and laboratory presentation suggestive of HAT. Immediate reoperation and revascularization by thrombectomy or the use of conduit is essential. Management of HAT after 2 weeks is usually done after evaluation by magnetic resonance angiography, 3-dimensional Gadolinium enhanced MR angiography,[45] 3- D multislice CT angiography[46] or by conventional angiography. Delayed onset HAT is usually treated by transcatheter thrombolytic therapy, installation of urokinase to hepatic artery, percutaneous balloon angioplasty and vascular stenting. Surgical revascularization is the option when all these means are exhausted.[47] In the presence of severe hepatic impairment with persistently elevated bilirubin and diffuse biliary stricture, retransplantation is usually the last resort.

### Hepatic Arterial Stenosis

Hepatic arterial stenosis is noted in 11% of allografts. It causes significant morbidity and mortality. It is associated with biliary complications in 60% of cases. The diagnosis requires noninvasive investigation followed by angiography. Immediate surgical revascularization is mandatory. Percutaneous balloon angioplasty can be performed with 80-100% success rate but there is chance of restenosis in 30-60%.[48] About 50% of patient with hepatic artery stenosis and biliary stricture can be treated successfully without re-operation.

### Portal Vein Thrombosis

Portal vein thrombosis (PVT) is more common in children than adult recipients. It is noted in 11.2% cases. It is commonly noted in children with biliary atresia. Predisposing factor is poor inflow due to preexisting partial portal vein thrombosis or disproportion of vessel diameter as in hypoplasia, kinking and redundance of portal vein. Other factors like high haematocrit, hypovolumia, over correction of coagulation and compression of the graft as in large-for-size graft also contribute to PVT. Hemodynamic instability, evidence of graft dysfunction and massive ascitis is the usual presentation of PVT. It requires Doppler ultrasound and MR Portography to delineate the extent of thrombus. PVT is treated initially by infusing urokinase directly into the portal vein.[49] Surgical intervention in the form of PV reconstruction, grafting and shunt procedures are required for those in which conservative measures are unsatisfactory.

### Portal Vein Stenosis

Portal vein stenosis is noted in about 3% of liver transplant.[50,51] It usually follows after difficult portal vein anastomosis and in those requiring inter-position grafts. Portal vein stenosis usually manifest as graft dysfunction with portal hypertension. It is treated by percutaneous balloon angioplasty and finally surgical reconstruction to salvage the allograft.

### Hepatic Vein Occlusion

Hepatic vein occlusion is a rare complication usually associated with hypercoagulable state. It may also occur if any anastomosis of hepatic vein or its graft is stenosed. It is more common in right lobe graft in LDLT and patients with pre operative Budd-chiari syndrome. Venous congestion of the segment V and VIII of right lobe graft may occur when the middle hepatic vein is not incorporated to the graft. The segment VI and VII may become congested when the dominant right inferior accessory vein is ligated. Any hepatic vein >5 mm diameter needs reconstruction. Hepatic venous congestion is detected by Doppler ultrasonography initially, followed by dual phase computerized tomography as zone of variable attenuation.[52] Partial hepatic venous occlusion is well tolerated but complete obstruction is catastrophic. It is treated percutaneously by balloon angioplasty and metallic stenting.[53] Complete hepatic vein obstruction needs retransplantation.

### IVC Occlusion

Inferior venacaval stenosis or occlusion is noted in less than 2% of patients. These patients present with hepatic venous outflow obstruction and truncal edema. Long time anticoagulation, percutaneous balloon angioplasty and stenting are the main management strategy.

### Hepatic Arterial Steal Syndrome

Arterial steal syndrome (ASS) is the arterial hypo perfusion of the graft caused by shifting of blood flow to the splenic or gastro duodenal artery. It causes the biliary system to suffer from ischemic changes. It is noted in 5.9% of cases. It is presented with elevated liver enzymes, impaired graft function and cholastasis. It is treated by splenectomy or coil embolization of splenic or gastro duodenal artery. If untreated a serious complication may occur in 30% of patients. Banding of the splenic artery may be performed prophylactic ally to prevent the development of arterial steal syndrome.[54]

## Biliary Problems of Liver Graft

Biliary complications causing allograft dysfunction or failure continue to remain an important problem after OLT. The incidence is 5-8% and mortality of about 10%.[55] Biliary problems are greater in adult to adult LDLT than pediatric LDLT and cadaveric transplantation. Biliary anastomosis is very sensitive to ischemic injury, HAT, stenosis of hepatic artery and arterial steal syndrome. Factors involved in the generation of biliary complications are secondary to technical failure, prolonged graft cold preservation, severe acute rejections, ABO incompatibility and donor duct ischemia. Most biliary complications occur within 1st three months of transplant. In general leaks occur early and stricture develops later. Bile leaks were equally frequent in both choledochocholedochostomy (CC) over a T-tube and Roux-en-Y choledochojejunostomy with internal stent (C-RY). Strictures are more common after C-RY type of reconstruction.[56]

The biliary problems include biliary leaks, dehiscence of biliary anastomosis, biliary tree necrosis, intrahepatic biloma and liver abscess. These lesions may progress to sectoral or diffuse biliary stricture, sepsis and graft loss. Multiple aetiological factors are involved such as small duct size, two or more biliary anastomosis, excessive periductal collection of donor liver and inadvertent ligation of small bile duct in the graft.[57] Two types of stricture are noted in OLT. One is anastomotic stricture and other one is non-anastomotic stricture. Anastomotic stricture occurs due to combination of surgical technique and local ischemia. Non-anastomotic strictures are multiple and secondary to ischemia.

Most of the biliary complications present as cholangitis, abdominal pain, jaundice, pruritus or altered liver function. A prompt diagnosis is essential to ensure long term graft survival and to minimize morbidity and mortality. Large study revealed about 50% patients initially treated for rejection but subsequently proved to have biliary complications.

The laboratory diagnosis of biliary complications includes elevated alkaline phosphatase, GGT, white cell count and altered coagulation profile. Other evaluation include abdominal ultrasound, computerized tomography, magnetic resonance imaging (MRI), CT angiography, Radio nuclide scanning by Hydroxy Imino Diacetic Acid (HIDA), Doppler ultrasound of hepatic arterial flow, T-tube cholangiography, endoscopic retrograde cholangiography (ERC) and percutaneous transhepatic cholangiography (PTC). Liver biopsy is often needed to exclude acute rejection, viral hepatitis or recurrence of primary disease.

Appropriate therapy for bile leak depends on location of bile leak and evaluation of hepatic arterial problem. HAT or stenosis should be corrected first prior to management of bile leak. Conservative and interventional approaches are applied first for management of all the biliary problems. The procedures include percutaneous image guided drainage, endoscopic or percutaneous dilatation, internal or interno-external stenting and trans-jejunal loop dilatation. If the endoscopic or percutaneous trans-hepatic approaches fails than CC may be converted to C-RY or redo of the hepaticojejunostomy. Retransplantation is considered during poor graft function, intractable cholangitis, untreatable biliary complications and complete obstruction of hepatic arterial flow leading to poor quality of life.

## Infectious Complications

Sepsis is a major cause of morbidity and mortality after OLT, noted is about 23% patients.[58] The commonly encountered infections in the 1st month is due to nosocomial and 2nd to 6th months is by the opportunistic pathogens. Relapsed or reactivation of infections occur about six months and after wards due to community acquired infections.[59] The causative factors are post-operative bleeding, hepatic arterial thrombosis, biliary leakage, necrosis and abscess in the graft. Immunosuppressants also contribute greatly for initiation and maintenance of all the infectious complications. Bacterial infection occurs in 55%, fungal in 22% and viral in 22% of recipients. These infections affect the graft function in various ways.

Bacterial infections are mainly caused by gram negative and gram positive organisms, usually noted in first two weeks. This coincides with the highest period of immunosuppressive therapy. Mostly noted in the liver, biliary tree, peritoneal cavity and laparotomy wound.[60] It is related to technical difficulties, biliary obstruction, bile leak and hepatic arterial stenosis or thrombosis.

Fungal infections mostly occur in first 8 weeks, caused by Candida species in 80% and Aspargillous infections in 15% cases.[61] Risk factors include high steroid doses, broad spectrum antibiotics, immunosuppressant, prolonged operative time, redo orthotopic liver transplantation (OLTX) and transplant in fulminant hepatic failure (FHF).

Cytomegalovirus (CMV) is the most common viral pathogen after OLTX. Mostly noted in 3 to 8 weeks time with maximum in 5th week. CMV is associated with about four fold increased in the relative risk of mortality within 1year of liver transplants.[62] The potential sources of CMV after liver transplantations are the donated allograft, blood products transfused and reactivation of endogenous virus. Epstein-Barr Virus (EBV) has been associated with development of posttransplant lympho proliferative disorder (PTLD) in 2.7% of the recipients. PTLD usually occurs in 6 or more months after transplantation and involves the transplanted graft. Extranodal Epstein—Barr virus is responsible for various illness especially infectious mononucleosis and lymphoproliferative diseases. Herpes simplex virus causes ulcers in the lips, mouth and hepatitis in 0.4% cases.

All these bacterial, viral, fungal infection can affect the graft function directly or indirectly. Prophylactic measures and adequate therapy should be the standard protocol.

## Fibrosis or Cirrhosis of the Graft

Graft loss due to fibrosis or cirrhosis is caused by secondary biliary cirrhosis, ischemic type biliary strictures, portal fibrosis or denovo hepatitis.[63,55] Median time between transplantation and graft loss due to fibrosis or cirrhosis is about 2.2 yrs.[64] Early evaluation and management of primary pathology minimizes the graft loss.

## ABO Incompatible Grafts

Liver transplantation across ABO incompatible (ABO-I) donors is a controversial issue. ABO-I liver graft is used as life- saving procedure in emergency situation when the ABO compatible graft is not available as in extreme donor shortage. The risks of ABO-I transplantation include hyperacute rejection, hepatic artery thrombosis and biliary complications. The grafts are complicated by wide spread hemorrhagic necrosis, intraorgan thrombosis and deposition of antibody and complement. This syndrome is called as "single–organ disseminated intravascular coagulation".[65] Graft survival for 1 yr is 66%, 2 yrs is 30% and 5 yrs is 20% for ABO-I liver transplants compared to 76% 2 yrs survival for ABO compatible donors. The global 5 years graft survival is not more than 20%.[66-68] Antilymphocytic globulin, plasmaphoresis, splenectomy, soluble antigen and ABO immunoadsorvents have been used prophylactically. The success rate is variable and associated with high incidence of sepsis.[69,70] New immunosuppressive protocol for adult ABO-I transplant without splenectomy is now suggested.[71]

## Summary

Allograft dysfunction is the most important complication of liver transplantation. Graft dysfunction may occur at the early and late posttransplant period. Improved result is noted due to advancement in surgical technique, immunosuppressive protocols and post-operative care. However, graft dysfunction and failure still remains a problem. It has important consequences because the recipient will either die or need a new liver graft for survival. A graft failure also implies an additional impact on the extremely limited donor pool. All out measures for selection of suitable donors and recipient, refinement of surgical technique, immunological modulation should be the prime consideration. Early diagnosis of graft dysfunction, interventions, prevention and control of sepsis has great contribution for maintaining the good graft functions and offer survival benefit to the recipient.

## References

1. Duran FG, Cerca dillo RA, Santosh et al. Late orthotopic liver transplantation: Indications and survival, Transplantation Proc 1998; 30:1876-77.
2. Mark mann JF, Markowitz JS, Yersiz H et al. Long term survival after retransplantation of the liver. Annals of surgery 1997; 226(4):408-20.
3. Ploeg RJ, D'Allessandro AM, Knechtle SJ et al. Risk factors for primary dysfunction after liver transplantation a multivariate analysis. Transplantation 1993; 55(4):807-13.
4. Deschenes M, Belle SH, Krom RA et al. Early allograft dysfunction and liver transplantation. Transplantation 1998; 66(3):302-10.
5. Totsuka E, Fung U, Hakamada K et al. Analysis of clinical variables of donors and recipients with respect to short-term graft outcome in human liver transplantation. Transplant Proc 2004; 36:2215.
6. Totsuka E, Dodson F, Urakami A et al. Influence of high serum sodium levels on early postoperative graft function in human liver transplantation: effect of correction of donor hypernatremia. Liver Transpl Surg 1999; 5:421-28.
7. Jawan B, Goto S, Lai CY et al. The effect of hypernatremia on liver allografts in rats. Anesth Analg 2002; 95:1169-72.
8. Vanda Walker SG. The effects of donor sodium levels on recipient liver graft function, J Transpl Coord 1998; 8:205-8.
9. Angelico M, Gridelli B, Strazzabosco M. A.I.S.F. Commission on Liver Transplantation. Practice of adult liver transplantation in Italy. Recommendations of the Italian Association for the Study of the Liver(AISF). Dig Dis Dis 2005; 37:461-67.
10. Zamboni F, Franchello A, David E et al. Effect of macrovescicular steatosis and other donor and recipient characteristics on the outcome of liver transplantation. Clin Transplant 2001; 15:53-57.
11. Vertemati M, Savatella G, Minola E et al. Morphometric analysis of primary graft nonfunction in liver transplantation, Hepatology 2005; 46:451-59.
12. Marubayashi S, Asahara T, Dohi K. Ischemia-reperfusion injury of liver and therapeutic intervention. Hepatol Res 2000; 16:233-53.
13. Zamboni F, Frenchello A, David E et al. Effect of macrovascular steatosis and other donor and recipient characteristics on the outcome of liver transplantation. Clin Transpla 2001; 15:53-57.
14. Marino IR, Doyle HR, Aldrighetti R et al. Efect of donor age and sex on outcome of liver transplantation Hepatology. 1995; 22:1754-62.
15. Pruim J, Van Woerden WF, Knol E. Donor data in liver grafts with primary non function- a preliminary analysis by the European liver registry. Transplant Proc 1989; 21:2383-84.
16. Snover DC. Liver transplantation In Sale GEed. The pathology of organ transplantation.Oxford. Butter worth Publishers; 1990; 103-9.
17. Adam R. Reynes, M. Johann. M. Marina I, Ascar Cinglal et al. The outcome of steatotic grafts in liver transplantation, Transplant Proc 1991; 23:1538-40.
18. Trevasani F, Colantoni A, Cavaceni P et al. The use of donor fatty liver for liver transplantation: a challenge or a guagmire? J Hepatol 1996; 24:114-21.
19. Deborah Verran, Taras Kusyk, Dorothy Painter et al. Clinical Experience Gained from the use of 120 steatotic donor liver for orthotopic liver transplantation. Liver Transpl; 2003; 9(5):500-5.
20. Kiuchi T, Tanaka K, Ito et al. Small for size graft in living donor liver transplantation-how far should we go? Liver transpl 2003; 9:S29-S35.
21. Kiuchi T, Kasahara M, Uryuharn K et al. Impact of graft size mismatching on graft prognosis in liver transplantation from living donors. Transplantation 1999; 67:321-27.
22. Felix Dahm, Panco Georgiev, Pierre Alain Clavien. Small-for-size syndrome after partial liver transplantation-Definition, Mechanisms of diseases and clinical implications. Am J Transplantation 2005; 5:2605-10.
23. Quiroga S, Sebastia MC, Margaritc et al. Complications of orthotopic liver transplantation spectrum of findings with helical CT. Radiographic 2001; 21:1085-102.

24. Jai Young Cho, Kyung-Suk Suh, Hae Won Hee et al. Hypoattenuation in unenhanced CT reflects histological graft dysfunction and predicts 1- year mortality after living donor liver transplantation. Liver Transpl 2006; 12:1403-11.

25. Cheng YF, Chen CL, Huang TL et al. Risk factors intraoperative portal vein thrombosis in pediatric living donor liver transplantation. Clin Transplant 2004; 18:390-94.

26. Kiuchi T, Kashara M, Urauhara K et al. Impact of graft size mismatching on graft prognosis in liver transplantation from living donors. Transplantation 1999; 67:321-27.

27. Vollmar B, Glasz J, Lei Derer R et al. Hepatic microcirculatory perfusion failure is a determinant for liver function in warm ischaemia-reperfussion. Am J Pathol 1994; 145:1421-23.

28. Marzi I, Takei Y, Rüker M et al. Endothelin-I is involved in hepatic sinusoidal vasoconstriction after ischemia and reperfusion. Transplant Int 1994; 7:S503-6.

29. Jaeschke H, Farhood A, Smith CW. Neutrophils contribute to ischemia reperfusion injury in rat liver in vivo. FASEB J 1990; 4:3355-59.

30. Yadan SS, Howell DN, GAO W et al. L- selection and ICAM-I medicine reperfusion injury and neutrophil adhesion in the warm ischemic mouse liver. Am J Physiol 1998; 275:G1341-52.

31. Cywes R, Rackham MA, Tietze L et al. Role of platelets in hepatic allograft preservation injury in the rat. Hepatology 1993; 18:635-47.

32. Adam R, Reynes M, Johann M et al. The outcome of steatotic grafts in liver transplantation. Transplant proc 1991; 23:1538-40.

33. Hui A-M, Kawasaki S, Makuuchi M et al. Liver injury following normothermic ischemia in steatotic rat liver. Hepatology 1995; 20:1287-93.

34. Vukovic R, Simic M, Tasic M. Analysis of ischemic lesions of the liver after various periods of warm and cold ischemia. Med Pregl 1996; 49(7-8):263-67.

35. Langnas A, Marujo W, Stratta R et al. Vascular complications after orthotopic liver Transplantation. Am J Surg 1991; 161:82-83.

36. Wozney P, Zajko AB, Bron KM et al. Vascular complications after liver transplantation: a 5-year experience, AJR 1986; 147:657-63.

37. Lallier M, St-Vil D, Dubois J et al. Vascular complications after pediatric liver transplantation. J Pediatr Surg 1995; 30:1122-26.

38. Holbert BL, Campbell WL, Skolnick ML. Evaluation of the transplanted liver and post operative complications. Radiol Clin North Am 1995; 33:521-40.

39. Mazzaferro V, Esquivel CO, Makowaka I et al. Hepatic artery thrombosis after pediatric liver transplantation, a medical or surgical event? Transplantation 1989; 47:971-77.

40. Merion RM, Burtch GD, Ham JM et al. The hepatic artery in liver transplantation. Transplantation 1989; 48:438-43.

41. Broniszezak D, Szymezak M, Kaminski A et al. Complications after pediatric liver transplantation from the living donors. Transplantation Proceedings 2006; 38:1456-58.

42. Langnas A, Marujo W, Stratta R et al. Vascular complications after orthotopic liver Transplantation. Am J Surg 1991; 161:82-83.

43. Eid A, Lyass S, Venturero M et al. Vascular complications post orthotopic liver transplantation. Transplant Proc 1999; 31:1903-4.

44. Schhuetze S, Linenberger M. Acquired protein S deficiency with multiple thrombotic complications after orthotopic liver transplant. Transplantation 1999; 67:1366-69.

45. David B, Stafford- Johnson MD, Brian H et al. Vascular complications of liver transplantation: Evaluation with Gadolininium enhanced MR Angiography Radiology 1998; 207:153-60.

46. Winter TC, Freeny PC, Nghiem HV et al. Hepatic arterial anatomy in transplantation candidates: evaluation with three dimensional CT arteriography. Radiology 1995; 195:363-70.

47. Sakamoto Y, Harihara Y, Nakatsuka T et al. Rescue of liver grafts from hepatic artery occlusion In living- related liver transplantation. Br J Surg 1999; 86:886-89.

48. Vignali C, Cioni R, Petruzzi P et al. Role of interventional radiology in the management of vascular complication after liver transplantation. Transplant Proc 2004; 36:552-54.

49. Yankes J, Uglietta J, Grant J et al. Percutaneous transhepatic recanalization and thrombolysis of the sup. mesenteric vein. Am J Roentgenol 1988; 151:289-90.

50. Bechstein WO, Blumhardt G, Ringe B et al. Surgical complications in 200 consecutive liver transplants. Transplant Proc 1987; 19:3830-31.

51. Lerut J, Jzakias AG, Bron K et al. Complications of venous reconstruction in human orthotopic liver transplantation. Ann Surg 1987; 2005:404-14.

52. Kim B, Kim T, Kim J et al. Hepatic venous congestion after living donor liver transplantation with right lobe graft: two phage CT findings. Radiology 2004; 232:173-80.

53. Sze D, Semba C, Razavi M et al. Endovascular treatment of hepatic venous outflow obstruction after piggyback technique of liver transplantation. Transplantation 1999; 68:446-49.

54. Natascha C, Nussler, Utz Settacher et al. Diagnosis and treatment of arterial Steal Syndromes in liver transplant recipients. Liver Transpl 2003; 9(6):596-602.

55. Lemmer ER, Spearman CW, Krieg JE et al. The management of biliary complications following orthotopic liver transplantation. S Afr J Surg 1997; 35:77-81.

56. Franklin Greif MD, Oscar L, Poronsther MD et al. The incidence, timing and management of biliary tract complications after orthotopic liver transplantation. Ann Surgery 1994; 219(1):40-45.

57. Hariharay, Makuuchi M, Takayama T et al. A simple method to avoid a biliary complication after living-related liver transplantation. Transplant Proc 1998; 30(7):3199.

58. Nemes B, Sarvary E, Sotonyi P et al. Factors in Association with sepsis after liver transplantation: The Hungarian experience. Transplantation Proc 2005; 37:2227-28.

59. Fishman JA, Rubin RH. Infection in organ-transplant recipients, N Engl J Med 1998; 338(24):1741-51.

60. Dummer S, Kusne S, Liver transplantation and related infections. Semin Respir Infect 1993; 8:191-98.

61. Winston DJ, Emmanouilides C, Bussutil RW. Infections in liver transplant recipients. Clin Infect Dis 1995; 21:1077-91.

62. Falagas MF, Syndman DR, Grittifh J et al. Effect of cytomegalovirus infection status on first-year mortality rates among orthotopic liver transplant recipients, The Boston center for Liver Transplantation CMVIG study Group. Ann Intern Med 1997; 126(4):225-79.

63. Batts KP, Ludwig J, Chronic Hepatitis An update on terminology and reporting. Am J Surg pathol 1995; 19:1409-17.

64. Egbert Sieders, Paul M.J.G. Peeters, Dlisabeth M. TenVergert et al. Graft loss after pediatric liver transplantation. Ann of Surg 2002; 235(1):125-32.

65. Demetris AJ, Jaffe R, Tzakis A et al. Antibody-mediated rejection of human orthotopic liver allograft. A study of liver transplantation across ABO blood group barriers. Am J Pathol 1988; 132:489-502.

66. Gordon RD, Fung JJ, Markus B et al. The antibody crossmatch in liver transplantation. Surgery 1986; 100:705-15.

67. Farges O, Nocci kalil A, Samuel D et al. The use of ABO-incompatible grafts in liver transplantation: a life-saving procedure in highly selected patients. Transplantation 1995; 59:1124-33.

68. Gugenheim J, Samuel D, Reynes M et al. Liver transplantation across ABO blood group barriers. Lancet 1990;336:519-23.

69. Cacciarelli TV, So SK, Lim J et al. A reassessment of ABO incompatibility in pediatric liver transplantation. Transplantation 1995; 60:757-68.

70. Tokunaga y, Tanaka K, Fujita S et al. Living related liver transplantation across ABO blood groups with FK506 and OKT3. Transpl Int 1993; 6:313-18.

71. Roberto Troisi, Lucien Noens, Roberto Montalti et al. ABO- mismatch adult living donor liver transplantation using antigen-specific immunoadsorption and quadruple immunosuppression without splenectomy. Liver Transpl 2006; 12:1412-17.

# CHAPTER 18

# Chronic Allograft Dysfunction—Liver

Susan Lerner, Pauline Chen and Paul Martin*

## Abstract

In the United States, more than six thousand patients undergo liver transplantation (LT) annually with generally excellent outcomes reflected in patient survival of 88% at one year and 80% at three years and graft survival of 83% and 74%.[1] Advances in immunosuppression have made acute cellular rejection a negligible threat to hepatic graft survival in compliant patients and generally long-term threats to graft survival reflect recurrent disease rather than rejection. Chronic rejection although relatively infrequent in LT recipients remains an important cause of graft loss and may reflect noncompliance. However, other threats to graft viability have been recognized as the practice of liver transplant has evolved. Use of nonheart beating donors results in frequent non-anastomotic stricturing and higher graft failure (relative risk 1.85) with the need for retransplantation.[2] Older donor grafts are a factor in more severe recurrence of HCV. Interferon therapy to treat HCV recurrence has been implicated in profound graft dysfunction reminiscent of chronic rejection. With longer term follow-up recurrence of nonviral disease has become more obvious. Recurrent cholestatic liver disease, most notably primary sclerosing cholangitis, can lead to graft loss. As the significance of non-alcoholic fatty liver disease as a cause of decompensated cirrhosis and hepatocellular carcinoma has become more fully appreciated, recurrent hepatic steatosis has now entered into the differential of graft dysfunction as has de novo hepatic steatosis.

As time from LT increases, the differential evolves with early (i.e., first three postoperative months) graft dysfunction due to impaired graft function due to a suboptimal donor, technical issues such as hepatic artery thrombosis or anastomotic biliary stricture, acute cellular rejection or opportunistic infection such as cytomegalovirus infection. Beyond the initial three months, the differential of graft dysfunction increasingly reflects recurrent disease, although technical problems most notably unrecognized hepatic artery thrombosis as well as anastomotic strictures can present with cholangitis or cholestatic liver biochemistries. Acute rejection beyond three to six months postLT often reflects inadequate immunosuppression. Hepatic allograft dysfunction can occur early or late and the definitions can vary as to the time point that separates those two categories.

Monitoring hepatic allograft dysfunction is an intergral part of the longterm management of the liver transplant recipient. As time from transplant increases, the frequency and intensity of follow-up diminishes although late graft dysfunction may be the first clue to important processes such as chronic rejection.or hepatic artery thrombosis. For both of these processes retransplantation may be required. Retransplants now account for approximately 8% to 10% of all transplants performed in the United States per year.[1]

Hepatic allograft dysfunction is indicated by increasing or persistent elevation of serum levels of alanine aminotransferase, alkaline phosphatase, or bilirubin. Depending on the biochemical pattern, the initial evaluation may be with one of the three key diagnostic studies: ultrasonography with Doppler assessment of the hepatic vasculature, cholangiography, or liver biopsy.[3] Of the three, liver biopsy usually plays the most important role in elucidating the cause of late allograft dysfunction. Important causes of late graft dysfunction often are impossible to distinguish from each other in the absence of a biopsy. Decisions regarding the treatment of late allograft dysfunction can be particularly challenging since elevations in hepatic enzymes are nonspecific. The choice of major alterations in a patient's immunosuppressive regimen or even retransplantation thus often rests on histologic findings.[3]

## Role of Liver Biopsy

Liver biopsy is invaluable in determining the etiology of liver allograft dysfunction. While most late causes of liver allograft injury are first detected because of abnormalities in routinely monitored liver tests, the clinical, serological and histopathological features often overlap, notably making it particularly difficult to differentiate recurrent disease from rejection.

Tissue obtained through biopsy should be assessed for adequacy; at least six to eight portal tracts should be available in the sample. Clinicians should then correlate any pathologic findings with a thorough systematic examination of the patient, keeping in mind the original disease, the immunosuppression used, liver tests, viral serology and immunology and radiology findings.[4] The involvement of an experienced liver pathologist familiar with the spectrum of normal and abnormal findings in hepatic allografts is crucial.

## Causes of Chronic Graft Loss

Late hepatic allograft dysfunction may result from a variety of insults, including rejection (acute and chronic), vascular stenosis/thrombosis, de novo or recurrent infection, biliary complications, recurrent disease, autoimmune hepatitis and drug toxicity. Other possible causes of late allograft dysfunction include posttransplant lymphoproliferative disease or the recurrence of hepatocellular carcinoma.[3] In a Pittsburgh study of medium-term survivors, the causes for allograft dysfunction included viral hepatitis (33%, of which 2/3 were recurrent viral and 1/3 de novo viral), rejection (22%), recurrent nonviral disease (14%) and obstructive cholangiopathy (6%). Of note, there was poor correlation between the symptoms, laboratory findings and histological changes.[5]

When determining the cause of allograft dysfunction, clinicians should take certain factors into consideration: the original disease, donor factors such as prolonged cold ischemia time or the use of a

*Corresponding Author: Paul Martin—Professor of Medicine, Recanati/Miller Transplantation Institute, The Mount Sinai Medical Center, One Gustave L. Levy Place, Box 1104, New York, NY 10029-6574, USA. Email: paul.martin@mountsinai.org

*Chronic Allograft Failure: Natural History, Pathogenesis, Diagnosis and Management*, edited by Nasimul Ahsan. ©2008 Landes Bioscience.

*Table 1. Causes of early versus late allograft dysfunction*

| Early | Late |
|-------|------|
| Rejection—Acute | Rejection—Acute or Chronic |
| Vascular Complications | Vascular Complications |
| Biliary Complications | Biliary Complications |
| Primary nonfunction | Recurrent Disease |
| Infections | Infections |
| | De novo autoimmune hepatitis |
| | Neoplasms |
| | Drug hepatotoxicity |

nonheart beating donor, operative events such as a difficult arterial anastamosis or posttransplant complications including episodes of rejection or biliary disease and immunosuppressive therapy including any changes in that therapy.

## Acute Rejection

Acute cellular rejection typically occurs within the first six to twelve weeks following liver transplantation; thereafter, the incidence decreases. Patients will have elevated liver enzymes, but definitive diagnosis is determined by liver biopsy and pathological exam. Of those patients who receive such a diagnosis, 80-90% will respond to high-dose intravenous corticosteroids.

Late acute rejection episodes occur more than 30 days after liver transplantation and may develop in 15-20% of recipients. Compared to acute rejection, LAR appears to be a more omnious form of rejection. Factors predisposing patients to LAR include underlying liver disease, a reduction in immunosuppression, or patient noncompliance. These episodes may be only be recognized at a late stage and are less responsive to corticosteroid therapy. One study showed that as few as 50% of patients diagnosed with late acute rejection responded to steroid treatment. Patients who are at higher risk for developing late acute rejection are those less than 30 years of age and those with autoimmune disease or acute fulminant failure as their primary diagnosis.[6]

Late acute rejection has slightly different features than typical acute rejection. The most common histologic characteristics are (1) predominantly mononuclear portal inflammation containing lymphocytes, neutrophils and eosinophils; (2) venous subendothelial inflammation of portal or central veins or perivenular inflammation; and (3) inflammatory bile duct damage.[7]

## Chronic Rejection

Chronic rejection has a more complex clinical and pathological picture than acute rejection. Although the incidence over the last decade has been reduced to 3%-5% of all transplants what it had been previously, chronic rejection remains an important cause of liver allograft failure.[7] In chronic rejection, extensive fibrosis develops over the course of months to years.[8] While chronic rejection has been documented in liver grafts as early as a month after transplantation, it does not generally occur until at least six months after transplant.

*Table 2. Workup of late allograft dysfunction*

Laboratory Tests
Review History of Recipient Disease, donor factors, postsurgical course, immuno suppressive history, rejection episodes
Ultrasound with Doppler to assess biliary vascular system
    Hepatic artery thrombosis—arteriography and/or retransplantation
    Bile duct dilatation—cholangiogram—endoscopic or radiologic intervention and/or retransplantation
Liver Biopsy

The etiology of chronic rejection is likely multifactorial. Episodes of acute rejection are not necessarily portents of chronic rejection, but chronic rejection is often preceded by one or more episodes of acute rejection, usually steroid-resistant. Acute rejection episodes, even when completely reversed by immunosuppressive treatment, increase the likelihood that chronic rejection will develop.[8] The roles of histocompatibility differences and sex-mismatching remain controversial.[7] Cytomegalovirus infection may also be a risk factor; CMV infection is associated with the secretion of chronic fibroblast factors which may promote the fibrosis associated with chronic rejection.[9]

Chronic rejection is a more insidious process and carries a worse prognosis than acute rejection; it more frequently results in graft failure and loss.[10] Although the overall incidence of chronic rejection has been decreasing, chronic rejection resulting in graft failure still occurs in about 3% to 4% of all liver transplant recipients.[11]

Because the clinical features of chronic rejection are similar to those of acute rejection or recurrent disease, a percutaneous biopsy is imperative when attempting to make the diagnosis. In chronic rejection, portal tracts and perivenular regions are primarily affected. While cytokeratin staining can also be used to facilitate the diagnosis,[7] the minimum diagnostic criteria are: (1) biliary epithelial senescence changes affecting a majority of the bile ducts with or without bile duct loss; or (2) foam cell obliterative arteriopathy; or (3) bile duct loss affecting > 50% of the portal tracts.[4] Any pathologic changes can be further divided into early and late stages.

Demetris has postulated that the pathogenesis of chronic rejection reflects the presence of a conventional basement membrane I in biliary cells that consists of an extracellular matrix which includes Type IV collagen, fibronectin and laminin unique among hepatic cells. This extracellular matrix is critical for initiating the lymphocyte signaling pathway and many of the molecules expressed on the bile duct epithelium likely end up recruiting or stimulating inflammatory cells in various ways.[12] Vierling posits that the biliary injury is the final phase of a T-cell mediated nonsuppurative destructive cholangitis, which is marked inflammatory injury caused by the local release of cytokines and other effector molecules and by the ischemic damage caused by injury to the peribiliary capillary plexus and obstruction of the more proximal arteries.[13]

In contrast with chronic rejection in other solid organ allografts, chronic rejection is reversible to some extent in liver transplant recipients. It is therefore crucial that chronic rejection be recognized as early as possible.[14] Steroids have little or no impact on chronic rejection, but tacrolimus seems not only to decrease the incidence of chronic rejection with regular use but also to aid in tempering its progression. In those patients with chronic rejection who are not on tacrolimus as part of their routine immunosuppression, converting to a tacrolimus-based immunosuppresion may be a useful salvage therapy. Those patients with chronic rejection already taking tacrolimus may respond to an increased dose although its side effects, including drug-induced diabetes, renal toxicity and neurotoxicity may limit the ability to increase the dose.

Finally, adding another immunosuppressive agent to a patient's maintenance regimen can occasionally help to slow the course of chronic rejection. The addition of sirolimus or mycophenalate mofetil to the immunosuppressive regimen has proven effective in many cases.

## Vascular Complications

Hepatic artery complications are the most common vascular complications, occurring in 5-12% of adult liver recipients. (3) These complications, which include stenosis or thrombosis, usually occur early in the postoperative course but can also be recognized in up to 2.8% of patients as far out as four weeks after transplant. While established risk factors are small arterial size, the use of extension grafts and prolonged ischemia time, a recent study implicates CMV donor seropositivity as a potential cause of late hepatic artery thrombosis (HAT).[15]

The time differentiation between early and late HAT presentation is generally 30 days. The presentation of delayed HAT is often much more variable: cholangitis with or without strictures or abscesses, bile leaks, or simply biochemical dysfunction.[15] Biopsy findings of early HAT show centrilobular coagulative necrosis or ballooning degeneration, infarcts and increased apoptosis with minimal inflammation. Biopsy findings of late HAT may show bile duct necrosis, hepatic abscesses, or bile ductular reactions.[16]

The clinical sequelae of HAT range from asymptomatic graft dysfunction, cholangitis due to biliary stricturing, to hepatic abscess formation or even graft failure although the latter is more typical in early HAT. Unfortunately most patients with HAT will require retransplantation.[17] Fibrinolysis during angiography, surgical thrombectomy and immediate vascular reconstruction have all been described, alone or in combination with varying success rates and in at least some instances, patients have been able to avoid retransplantation.[17]

Portal vein thrombosis and stenosis can also lead to late graft failure typically presenting with refractory ascites, recurrent variceal bleeding and encephalopathy. Risk factors include a hypercoaguable state, prior portocaval shunting and the use of interposition grafts. Biopsy findings in these patients may show portal fibrosis, occasional sclerosis of portal vein branches, hepatocyte atrophy and sinusoidal dilatation and, more rarely, infarcts.[16] While portal vein vascular complications are very rare, thrombosis and stenosis, particularly if they occur early posttransplant, can result in graft loss if revascularization does not occur. Balloon dilatation and stenting can occasionally delay the need for retransplantation.

Thrombosis of the hepatic veins or vena cava is a much more unusual complication, unless the patient had originally presented with Budd-Chiari syndrome and had an underlying coagulopathy. The use of radiologic balloon dilatation or surgical repair can occasionally obviate the need for retransplantation.

## Biliary Complications

Biliary complications occur in up to 20% of patients after liver transplantation. In all cases, HAT must first be excluded. Biliary complications include leaks, anastomotic strictures and non-anastomotic strictures. Leaks often occur early after surgery or at the time of T-tube removal; and in this setting, they can usually be quickly identified and treated. Anastomotic strictures can be treated endoscopically or percutaneously with balloon dilatation and stenting.

Surgical conversion to a hepaticojejunostomy may be required if prior non-operative interventions fail. Biliary reconstructive surgery is usually indicated for patients with anastomotic strictures and early allograft dysfunction; the role of surgical reconstruction is less clear in those with later presentations of biliary complication especially. with non-anastomotic strictures and unfavorable allograft histology (presence of severe fibrosis and lack of biliary features).

Liver biopsy is particularly important in evaluating patients with delayed allograft dysfunction and suspected biliary complications to exclude recurrent viral hepatitis, primary biliary cirrhosis, or primary sclerosing cholangitis before attributing graft dysfunction exclusively to a biliary complications.[18]

Non-anastomotic strictures unrelated to hepatic artery thrombosis are associated with numerous risk factors, including cytomegalovirus, prolonged cold ischemia time and use of grafts from nonheart beating donors. Some of these strictures may be treatable by biliary stenting, but the majority remain refractory to treatment. Retransplantation is often necessary.

## Impact of Donor Factors

As liver transplantation has evolved, the criteria for acceptable donor organs have been relaxed. The most notable of this is in the use of nonheart beating organs or organs that are procured as a result of donation after cardiac death (DCD). These donors are usually individuals with devastating irreversible neurologic injuries who do not meet formal brain death criteria.[19] Despite ethical controversies surrounding this type of organ donation, the contribution of organs from DCD has grown rapidly and now represents 5% of all deceased donors in liver transplantation.[20] Several centers have reported varying experiences with the use of DCD's with some noting no difference between brain-dead and DCD donors and others reporting increased rates of primary nonfunction, biliary complications and hepatic artery stenosis. A high incidence of intrahepatic ischemic type biliary strictures has been reported. Many of these patients demonstrated progressive stricturing and beading of the intrahepatic biliary system. These complications were responsible for a significant degree of morbidity resulting in re-operation, multiple endoscopic and percutaneous biliary interventions, retransplantation and death.[21]

An analysis from the national UNOS database between 1993 and 2001 performed by Abt et al revealed that of 27,000 patients analyzed 144 were recipients of livers from DCD. Recipients of a DCD graft were found to have a 30% increase in the risk of graft failure when compared with a brain-dead donor allograft. Primary nonfunction accounted for some of this difference. Prolonged cold ischemic time and a recipient on life support before transplantation were predictors of early graft failure among recipients of allografts from DCD. Each additional hour of cold ischemic time increased the risk of graft failure by 17%. Among DCD grafts with a cold ischemic time greater than 8 hours, the incidence of graft failure within 60 days of transplantation was 30.4% and increased to 58.3% when cold ischemic time exceeded 12 hours. Conversely, when cold ischemic time was less than 8 hours, only 10.8% of DCD grafts failed within 60 days.[20] Additional investigation is indicated to better define which DCD organs are optimal for hepatic transplantation, the acceptable limits of donor warm ischemic time and which recipients are best served with the use of these organs.

## Recurrence of Original Disease

The most common etiology of late hepatic allograft dysfunction is recurrence of the original disease. Chronic hepatitis B is a frequent cause of advanced liver disease; the disease affects 1.25 million people in the United States and more than 300 million people worldwide. Liver transplantation for HBV was historically limited by high rates of HBV reinfection and decreased patient survival. In its recurrent form, hepatitis B infection was marked by rapidly progressive hepatic deterioration and extremely high mortality rates. However, with the use of both hepatitis B immunoglobulin (HBIG) and antiviral agents, there has been a significant reduction in HBV recurrent disease. Patient and graft survival in patients with hepatitis B is now equivalent to that seen in those with other indications for orthotopic liver transplantation.[22-25]

Patients treated long-term with antiviral agents can occasionally develop recurrent disease due to treatment induced mutations in the viral genome. The histopathologic presentation of hepatitis B infection in liver allografts is similar to that seen in the native liver. An atypical pattern of recurrent HBV, fibrosing cholestatic hepatitis (FCH), occurs in a very small number of patients. These patients present with a severe cholestatic syndrome, which may clinically resemble acute or chronic rejection. On pathologic exam, this rapidly progressing fibrosis demonstrates bridging and confluent necrosis; clinically, it can lead to early graft failure in a small number of patients. This more aggressive course of HBV infection after transplant probably results from enhanced viral replication and attenuated host response in these patients. FCH is only seen in immunosuppressed patients.

The diagnosis of recurrence is made by demonstrating the presence of HBV surface antigen or DNA in the recipient's serum or the presence of HBV surface antigen or core antigen in liver tissue.[26]

Chronic hepatitis C affects more than 170 million people worldwide and can lead to cirrhosis and liver cancer. In most liver transplant programs, hepatitis C is the leading indication for liver transplantation; unfortunately the hepatic allograft is universally reinfected after transplantation. Recurrent Hepatitis C may even lead to graft failure requiring retransplantation.[27-32]

In the posttransplant setting, HCV is frequently more rapidly progressive than in nontransplanted patients and can lead to graft cirrhosis in 8-30% of patients within 5-7 years of transplantation.[33] Unlike hepatitis B, no therapy has been conclusively shown to alter recurrence or progression. A number of factors, however, have been identified with more aggressive recurrence of HCV: the pretransplant viral load, the use of corticosteroids or antilymphocyte therapy, the presence of genotype 1a, the use of organs from older donors, prolonged warm ischemic time and the development of biliary strictures. In addition, it appears that early recurrence is associated with greater severity and poorer prognosis.[3,34]

There are two histopathological patterns of severe chronic HCV: (1) aggressive hepatitis with prominent interface activity and (2) fibrosing cholestatic hepatitis. The predominant features of HCV include mononuclear portal inflammation, often arranged into nodular aggregates, necroinflammatory and ductular-type interface activity and mild macrovesicular steatosis. Features of fibrosing cholestatic hepatitis include centrilobular hepatocyte swelling and degeneration, cholestasis, hepatocyte apoptosis, portal expansion because of a ductular reaction, fibrosis and a mild mixed portal inflammation. Fibrosing cholestatic hepatitis is associated with massive HCV replication.[4]

A liver biopsy is required when differentiating between rejection and recurrent hepatitis C; it is also imperative in determining treatment. Unnecessary augmentation of immunosuppression can accelerate fibrogenesis in chronic HCV or trigger cholestatic hepatitis and untreated acute rejection can progress to chronic rejection.

Treatment with interferon and ribavirin therapy in patients with HCV may be beneficial in up to 25% of recipients with recurrent HCV.[4] As with nontransplant patients the best results reported to date have been achieved with the combination of pegylated interferon and ribavirin, with sustained responses seen in 26-45% of patients.[35] Treatment of recurrent HCV is more challenging in the transplant setting with important issues of tolerability in patients often with some degree of renal impairment.

Two antiviral strategies have been proposed. Proponents of pre-emptive or early posttransplant treatment argue that HCV recurrence should be treated before significant graft damage has occurred. Opponents of this early strategy argue that preemptive treatment increases the risk of graft rejection and does not distinguish patients who truly need treatment from those in whom the therapy and its attendant complications are unnecessary. These opponents argue instead for waiting and targeting treatment at those patients with clinically significant graft reinfection.

A major concern is reports of severe graft dysfunction apparently induced by therapy in at least some cases after apparently a successful antiviral response.[33] A recent report describes a form of graft dysfunction with features of de novo autoimmune hepatitis (AIH) in those patients receiving PEG-IFN alfa-2b and Ribavirin for HCV recurrence. In this series 17% of patients experienced unexplained liver test abnormalities in spite of HCV-RNA clearance in most cases. Graft rejection, anastomotic complications and concomitant infections were excluded as possible etiologies in all of these cases. The laboratory and histological characteristics (antibody appearance, severe interface hepatitis with plasmacellular infiltration and rosettes) of these cases led the investigators to postulate that the patients had developed de novo AIH, which arose from the vigorous, drug-induced immune response.[36]

Autoimmune hepatitis affects predominantly middle-aged women and typically responds to immunosuppressive therapy. When AIH leads to chronic liver failure, it is an indication for orthotopic liver transplantation. About 20-30% of patients undergoing liver transplanta-

tion for AIH develop features of recurrent disease, half of them within the first year posttransplant. However, few of these patients go on to require retransplantation and there appears to be no decrease in graft or patient survival in patients with recurrent AIH.

AIH recurrence should be suspected in transplant recipients with abnormal LFT results in the absence of histological features typical of acute rejection on liver biopsy.[37] The minimum diagnostic criteria for recurrence are (1) interface hepatitis with portal lymphocytic infiltrates; (2) significant titers of antinuclear antibodies, smooth muscle antibodies, or antibodies to liver kidney microsome type 1; (3) hyper-gammaglobulinemia; and (4) exclusion of virus-induced or drug-related hepatitis and late acute or chronic rejection. Early pathologic changes include lobular hepatitis with hepatocyte rosetting, all of which can evolve into a chronic phase characterized by lymphoplasmacytic portal inflammation with prominent interface activity. Plasmacytic infiltrates characterize AIH but are not diagnostic requisites.[4]

De novo AIH should also be considered in the differential of late graft dysfunction. The first report of de novo AIH described the characteristic form of the disease in initially unaffected hepatic allografts at a median period of two years postliver transplant. Further reports have confirmed the existence of this diagnosis. In addition to the pathognomonic changes usually associated with AIH, this condition is characterized by a biochemical hepatitis, circulating autoantibodies, elevated immunoglobulins, inflammatory infiltration in the graft with interface hepatitis and an indication for transplantation other than AIH. Most studies suggest that the condition is relatively uncommon, occurring in less than 1% of liver allograft recipients and more commonly in children than adults. There is often a good response to treatment with additional immunosuppression with corticosteroids, but in some cases there is a relentless progression to cirrhosis and graft failure.[38]

Primary Biliary Cirrhosis (PBC) is characterized as chronic, progressive, destructive, nonpurulent inflammation of small- and middle-sized bile ducts. The etiology remains unknown but is likely secondary to disordered immune regulation. Antimitochondrial antibodies are present in 95% of patients. Patients with symptomatic PBC have, without transplantation, a median survival time of five to 10 years and decompensated patients between three and five years. There is no medical therapy and the only curative treatment is liver transplantation.

Liver transplantation significantly improves the prognosis of these patients. Recurrence can occur, however, usually three years or more after transplantation. Typically antimitochondrial antibodies persist post-OLT and their presence has no diagnostic significance in evaluating graft dysfunction. Pathologic findings in grafts affected by recurrent PBC are nearly identical to those seen in the native livers. Pathognomonic for PBC is a noninfectious granulomatous cholangitis in the presence of antimicrobial antibodies and absence of other causes such as infectious and biliary strictures.[4] Despite its similarity with the primary disease, recurrent PBC may be difficult to diagnose by histology since it often mimics chronic rejection, biliary obstruction, graft-versus-host diseases, ischemic lesions and drug-induced graft injury.[33]

Multiple recent studies have looked at the impact of PBC recurrence on graft survival, analyzing patients with long-term follow-up of up to 15 years after liver transplantation. Although protocol liver biopsies reveal histological recurrence, very few patients develop graft dysfunction.[33] In fact, there are no significant differences in patient or graft survival for patients with or without recurrent PBC. Of note, recurrent disease may be more frequent and earlier in patients on tacrolimus-based immunosuppression compared to patients on cyclosporine-based regimens although this observation has not been universally reproduced. Rapid weaning of immunosuppression may also predispose patients to recurrent disease.[39]

Primary Sclerosing Cholangitis (PSC) is a chronic progressive cholestatic liver disorder that has no effective medical therapy. PSC's pathogenic mechanism is unclear; histologically, there is inflammation and fibrosis of intrahepatic and/or extrahepatic bile ducts. Some have postulated that disordered immune regulation may be factor in the development of this disease.

A diagnosis of PSC is based on clinical, biochemical, cholangiographic and histological criteria. Cholestatic liver indices and perinuclear antineutrophil cytoplasmic antibodies are typical. Irregular ductal strictures, beading, diverticular outpouching and pruning are seen on cholangiography. Biopsies reveal a fibrous cholangitis that consists of concentric rings of fibrous tissue with edema and inflammatory cells around interlobular bile ducts. There are also fibro-obliterative lesions that have replaced ducts.[39] Patients with PSC do quite well with transplantation in the absence of a complicating cholangiocarcinoma but seem to have an increased predisposition to biliary strictures and possibly chronic rejection. Unfortunately, PSC has also been shown to recur in long-term studies. Like PBC, PSC generally recurs late, after three years posttransplantation. The initial finding is a persistent elevation in serum alkaline phospatase. Careful exclusion of other conditions associated with intrahepatic and extrahepatic biliary strictures is mandatory—ischemic strictures, chronic rejection, ABO incompatibility between donor and recipient and viral and bacterial infections.

Microscopic findings in recurrent PSC are identical to those described in native diseased livers and in other causes of biliary strictures. Subtle findings that suggest low-grade strictures include mild portal edema; mild nonspecific acute and chronic "pericholangitis" often accompanied by a very mild type of ductular reaction; sinusoidal clusters of neutrophils; and centrilobular hepatocanalicular cholestasis. More significant strictures usually cause lamellar periductal edema, increased portal tract ductal profiles and/or concentric periductal fibrosis.[4]

Recurrent PSC is a slowly progressive process that leads eventually to biliary fibrosis, although need for retransplantation is uncommon.[40] The overall survival rates for recurrent and nonrecurrent groups are similar.[41]

Other liver diseases that have been shown to recur after liver transplantation include alcoholic cirrhosis and nonalcoholic steatohepatitis. Nonalcoholic fatty liver disease is an increasingly recognized liver failure resulting in LT. It is typically associated with obesity, type II Diabetes Mellitus and hyperlipidemia. De novo development of NAFLD after LT is also being recognized as a cause of graft failure. There appears to be a significant association with weight gain after transplant and its development.[42]

## Infectious Causes of Graft Failure

Infections after liver transplantation can also lead to allograft dysfunction. These types of infections tend to occur early after transplant and are largely related to immunosuppression.

Approximately 50% of patients experience a bacterial infection related to intravascular lines, wound infections, biliary tract leaks and urinary tract infections. Bacterial infections can be associated with cholestasis, which itself can be precipitated by some of the antibiotics used to treat the infections. Histologically, there is bile ductular proliferation with prominent bile plugging; these changes generally resolve as the bacterial infection is eradicated.[3]

Fungal infections tend to occur in the first six months after liver transplantation and are often associated with a high mortality rate. The organisms implicated in these infections tend to be primarily Candida specics, followed by Aspergillus. Invasive fungal disease should be considered in all liver recipients with unexplained allograft dysfunction.[3] The propensity for fungal infections is influenced by surgical factors, including technical complexity, prolonged operation time, greater transfusion requirements and bleeding complications requiring re-operation. Increased length of post-OLT ICU stay is also associated with higher risk.[16]

Cytomegalovirus infections are also common post liver transplant. Seronegative recipients of seropositive donors have the highest risk of developing symptomatic CMV disease. Of those who develop CMV disease, approximately 50% are documented to have CMV hepatitis. In addition to hepatitis, CMV can cause fever, leucopenia, thrombocytopenia, atypical lymphocytosis, pneumonitis and retinitis. The diagnosis is made using polymerase chain reaction. Histologically, typical nuclear inclusion bodies or microabscesses are noted on liver biopsy.[3]

Cytomegalovirus may be a cofactor in other complications following liver transplantation and can be associated with bacterial and fungal infections, acute and chronic rejection and hepatic artery thrombosis. CMV infections may also contribute to the development of allograft rejection by up-regulating MHC Class I and II molecules via the release of interferons. In terms of CMV's relationship with Hepatitis C, it appears that short-term CMV viremia does not appear to predispose patients to more severe hepatitis C recurrence.[15]

Other viruses that can cause allograft function include herpes simplex and the Epstein-Barr virus. Herpes simplex infections are most typically diagnosed in the first three weeks post liver transplant and can involve the oral and genital mucosa, the eye, the esophagus and the liver. Herpes hepatitis can lead to serious allograft dysfunction and may even lead to fulminant hepatic failure and death with a predominat lymphocyte infiltration on liver biopsy.[16]

The most common manifestations of EBV infection in liver transplant recipients include hepatitis and posttransplant lymphoproliferative disorders (PTLD). EBV hepatitis usually occurs four to six months after transplantation and may be associated with over-immunosuppression and the use of antilymphocyte therapy. EBV may also be associated with PTLD and B-cell lymphoproliferative disorders. PTLD involves the liver in particular; but while it may present initially as hepatic dysfunction, PTLD rarely leads to massive hepatic necrosis.[3] It is more frequent in pediatric recipients reflecting their lack of prior EBV infection before they became immunosuppressed.

## Posttransplant Malignancies

Since adoption of the Model for End-stage Liver Disease (MELD) system for organ allocation in February of 2002, patients with hepatocellular carcinoma (HCC) meeting acceptable criteria have been afforded priority for cadaver liver allografts. As a result there has been nearly a 3-fold increase in the number of patients transplanted with HCC. Recent studies demonstrate a recurrence in 8-11% of patients. In addition, not surprisingly, recurrence of HCC after transplantation shortens survival. Most recurrences present within 2 years, however up to 10 % of patients can be diagnosed up to 4 years after transplant and therefore surveillance must be life-long. Therapeutic options for recurrent HCC are limited. Hepatic artery chemoembolization and systemic chemotherapy are of little value in the postoperative setting. Evidence seems to support that patients with recurrence amenable to surgical treatment should be resected or ablated.[43]

In most studies, the incidence of de novo malignancies in the OLT population is 3%-15%, roughly twice that in the general population. Development of a malignancy in the setting of solid organ transplantation is multifactorial and depends upon an individual's predisposition to malignancy, pretransplantation disease states, recipient viral status and the use and intensity of immunosuppression regimens. The risk is cumulative and increases with time from transplantation.[16] However, these malignancies rarely involve the new allograft.

## Drug Hepatotoxicity

Hepatic allograft dysfunction can also occur to drug hepatoxicity; therefore, a detailed drug-use history including the use of over the counter medications and herbal products is essential in all liver transplant recipients presenting with late hepatic allograft dysfunction. Both of the calcineurin inhibitors are associated with liver dysfunction; changes are usually mild and nonspecific.[44] Azathioprine can also result in graft damage, manifesting as hepatitis, cholestasis, nodular regenerative hyperplasia, or veno-occlusive disease.[45-46]

Drug-related hepatic injury is conventionally divided into three categories: cytotoxic (hepatocellular), cholestatic and mixed patterns. Hepatic injury may be accompanied by systemic manifestations of hypersensitivity such as fever, rash and eosinophilia. Biochemical parameters frequently show hyperbilirubinemia associated with elevations in aminotransferase and alkaline phosphatase levels. The histologic picture caused by drug hepatoxicity can be very nonspecific, so a liver biopsy may not be very useful in diagnosis, although it may help exclude other causes of hepatic allograft dysfunction.

If a drug is believed to be the cause of hepatic dysfunction, the patient should immediately discontinue the offending drug. Fortunately, most drug-induced hepatitis subsequently reverses. Occasionally, however, drug-induced hepatic injury can result in fulminant hepatic failure with complete destruction of the allograft.[3]

# References

1. Based on OPTN data as of November 2006. Health Resources and Services Administration contract 231-00-0115.
2. Merion RM, Pelletier SJ, Goodrich N et al. Donation after cardiac death as a strategy to increase deceased donor liver availability. Ann Surg 2006; 244(4):555-62.
3. Wiesner RH, Narayanan Menon KV. Late hepatic allograft dysfunction. Liver Transpl 2001; 7(11):S60-73.
4. Banff Working Group. Liver biopsy interpretation for causes of late liver allograft dysfunction. Hepatology 2006; 4(2):489-501.
5. Pappo O, Ramos H, Starzl TE et al. Structural integrity and identification of causes of liver allograft dysfunction occurring more than 5 years after transplantation. Am J Surg Pathol 1995; 19:192-206.
6. Mor E, Gonwa TA, Husberg BS et al. Late-onset acute rejection in orthotopic liver transplantation—associated risk factors and outcome. Transplantation 1992; 54:821-24.
7. Inomata Y, Tanaka K. Pathogenesis and treatment of bile duct loss after liver transplantation. J Hepatobiliary Pancreat Surg 2001; 8:316-22.
8. Libby P, Pober JS. Chronic rejection. Immunity 2001; 14:387-97.
9. Gao L, Zheng S. Cytomegalovirus and chronic allograft rejection in liver transplantation. World J Gastroenterol 2004; 10(13):1857-61.
10. Hayry P. Chronic rejection: an update on the mechanisms. Transplatn Proc 1998; 30:3993-95.
11. Neuberger J. Incidence, timing and risk factors for acute and chronic rejection. Liver Transpl Surg (Suppl) 1999:S30-36.
12. Demetris AJ. Immune cholangitis: liver allograft rejection and graft-versus-host disease. Mayo Clinic Proc 1998; 73:367-69.
13. Vierling, JM. Immunology of acute and chronic hepatic allograft rejection. Liver Transpl Surg 1999; 5:S1-20.
14. Sebagh M, Blakolmer K, Falissard B et al. Accuracy of bile duct changes for the diagnosis of chronic liver allograft rejection: reliability of the 1999 Banff schema. Hepatology 2002; 35:117-25.
15. Silva MA, Jambulingam PS, Gunson BK et al. Hepatic artery thrombosis following orthotopic liver transplantation: a 10-year experience from a single centre in the United Kingdom. Liver Transpl 2006; 12:146-51.
16. Washington K. Update on postliver transplantation infections, malignancies and surgical complications. Adv Anat Pathol 2005; 12(4):221-26.
17. Stange B, Glanemann M, Nuessler N et al. Hepatic artery thrombosis after adult liver transplantation. Liver Transpl 2003; 9:612-20.
18. Sutcliffe R, Maguire D, Mroz A et al. Bile duct strictures after adult liver transplantation: a role for biliary reconstructive surgery? Liver Transpl 2004; 10(7):928-34.
19. Abt PL, Desai NM, Crawford MD et al. Survival following liver transplantation from nonheart beating donors. Ann Surg 2004, 239(1):87-92.
20. Abt PL et al. Donation after cardiac death. J Am Coll Surg 2006, 203(2):208-25.
21. Abt P, Crawford M, Desai N et al. Transplantation 2003, 75(10):1659-63.
22. Anselmo D, Ghobrial R, Jung L et al. New era of liver transplantation for hepatitis B: a 17-year single-center experience. Ann Surg 2002; 235:611-20.
23. Ishitani M, McGory R, Dickson R et al. Successful retransplantation for recurrent posttransplant hepatitis B virus infection in the primary allograft. Transplant Proc 1996; 28:1714-16.
24. Ishitani M, McGory R, Dickson R et al. Retransplantation of patients with severe posttransplant hepatitis B in the first allograft. Transplantation 1997; 64:410-14.
25. Roche B, Samuel D, Feray C et al. Retransplantation of the liver for recurrent hepatitis virus infection—the Paul Brousse experience. Liver Transpl Surg 1999; 5:166-74.
26. Thung, SN. Histologic findings in recurrent HBV. Liver Transpl 2006; 12:S50-53.
27. Rosen H. Retransplantation for hepatitis C: implications of different policies. Liver Transpl 2000; 6:S41-46.
28. Ghobrial R. Retransplantation for recurrent hepatitis C. Liver Transpl 2002; 8:S38-43.
29. Berenguer M, Prieto M, Palau A et al. Severe recurrent hepatitis C after liver retransplantation for hepatitis C virus—related graft cirrhosis. Liver Transpl 2003; 9:228-35.
30. Ghobrial R, Farmer D, Baquerizo A et al. Orthotopic liver transplantation for hepatitis C: outcome, effect of immunosuppression and causes of retransplantation during an 8-year single-center experience. Ann Surg 1999; 229:824-83.
31. Ghobrial R, Colquhoun S, Rosen H et al. Retransplantation for recurrent hepatitis C following tacrolimus or cyclosporine immunosuppression. Transplant Proc 1998; 30:1740-41.
32. Rosen H, Martin P. Hepatitis C infection in patients undergoing liver transplantation. Transplantation 1998; 66:1612-16.
33. Jacob DA, Neumann UP, Bahra M et al. Long-term follow-up after recurrence of primary biliary cirrhosis after liver transplantation in 100 patients. Clin Transplant 2006; 20:211-20.
34. Carmiel-Haggai M, Fiel MI, Gaddipati HC et al. Recurrent hepatitis C after retransplantation: factors affecting graft and patient outcome. Liver Transpl 2005; 11(12):1567-73.
35. Berenguer M, Palau A, Fernandez A et al. Efficacy, predictors of response and potential risks associated with antiviral therapy in liver transplant recipients with recurrent hepatitis C. Liver Transpl 2006; 12:1067-76.
36. Berardi S, Lodato F, Gramenzi A et al. High incidence of allograft dysfunction in liver transplant patients treated with Peg-Interferon alfa-2b and Ribavirin for hepatitis C recurrence: possible de novo autoimmune hepatitis? Gut 2006; 6.
37. Molmenti EP, Netto GJ, Murray NG et al. Incidence and recurrence of autoimmune/alloimmune hepatitis in liver transplant recipients. Liver Transpl 2002; 8(6):519-26.
38. Heneghan MA, Portmann BC, Norris SM et al. Graft dysfunction mimicking autoimmune hepatitis following liver transplantation in adults. Hepatology 2001; 34:464-70.
39. Faust, TW. Recurrent primary biliary cirrhosis, primary sclerosing cholangitis and autoimmune hepatitis after transplantation. Liver Transpl 2001; 7(11, S1):S99-108.
40. Graziadei I, Wiesner R, Batts K et al. Recurrence of primary sclerosing cholangitis following liver transplantation. Hepatology 1999; 29:1050-56.
41. Khettry U, Keaveny A, Goldar-Najafi A et al. Liver transplantation for primary sclerosing cholangitis: a long-term clinicopathologic study. Hum Pathol 2003; 34(11):1127-36.
42. Seo S, Maganti K, Khehra M et al. De novo nonalcoholic fatty liver disease after liver transplantation. Liver Transpl 2007; 13(6):788-90.
43. Roayaie S, Schwartz JD, Sung MW et al. Recurrence of hepatocellular carcinoma after liver transplant: patterns and prognosis. Liver Transpl 2004; 10(4):534-40.
44. Kassianides C, Nussenblatt R, Palestine AG et al. Liver injury from cyclosporine A. Dig Dis Sci 1990; 35:693-97.
45. Gane E, Portmann B, Saxena R et al. Nodular regenerative hyperplasia of the liver graft after liver transplantation. Hepatology 1994; 20:88-94.
46. Sterneck M, Wiesner RH, Ascher N et al. Azathioprine hepatotoxicity after liver transplantation. Hepatology 1991; 14:806-10.

# Liver Graft Loss Due to Vascular Complications

Barbara Stange, Matthias Glanemann and Natascha C. Nüssler*

## Abstract

Vascular complications occur in about 10% of patients undergoing orthotopic liver transplantation. Depending on the involved vessels and the time point after liver transplantation, the clinical course of these patients may vary considerably, ranging from complete absence of symptoms to acute or chronic graft failure.

In this chapter we describe the pathophysiology, clinical picture and therapeutic options of the different types of vascular complications after OLT. Hepatic artery complications are the most frequent vascular problems after OLT, whereas complications involving the portal vein or the vena cava are less frequent.

Patients with vascular complications that are associated with reduced but still existing blood flow, such as stenosis, kinking or steal syndromes, may benefit from surgical or interventional therapies which aim at re-establishing a normal blood flow in the respective vessel. The best results can be achieved in patients with short duration of impaired graft perfusion. Re-establishment of normal blood flow is much more difficult to achieve in patients with complete occlusion of arterial or venous vessels of liver grafts. In these patients, therapy is often focused on the secondary symptoms of impaired graft perfusion, such as biliary destruction or portal hypertension.

## Introduction

Orthotopic liver transplantation (OLT) has become an accepted therapy for acute and chronic end-stage liver disease with 5-year survival rates up to 80% in experienced transplant centers.[1] However, the postoperative course of liver transplant recipients can still be affected by a number of complications. Among these, vascular complications, which mostly involve the hepatic artery or portal vein rather than the vena cava or the liver veins, are associated with high morbidity including acute and chronic graft loss.[2,3]

This chapter reports on vascular complications following orthotopic liver transplantation subsequently resulting in liver graft loss. Herein, we focus on the incidence, clinical presentation and therapeutic options of arterial, portal-venous and venous complications as well as their consequences on immediate and long term graft function.

## Arterial Complications

Arterial complications following orthotopic liver transplantation are the most frequent vascular complications.[4] In this context, hepatic artery thrombosis (HAT), hepatic artery stenosis (HAS) and hepatic artery aneurysms (HAA) have been frequently reported.[2,5,6] Arterial steal syndromes, which may have a significant impact of posttransplant liver graft function due to decreased arterial blood supply, are also being discussed.[7]

### Hepatic Artery Thrombosis (HAT)

HAT is the most frequent arterial complication occurring in 2.5% to 9% of adult liver transplant recipients (Fig. 1).[5] Ischemia of the liver graft due to HAT results in graft morbidity or graft loss with retransplantation rates up to 80%, or even death.[8]

Predisposing factors for HAT mostly stem from donor and graft features such as donor age greater than 60 years, the necessity for back table reconstruction of the hepatic artery, or early rejection episodes.[9,10] In some studies, cold ischemia time, usage of blood products, retransplantation, or Roux-en-Y biliary reconstruction were also addressed as risk factors for HAT.[10,12] Moreover, CMV-seronegative patients receiving a seropositive allograft were reported to be at risk for early HAT.[10,11,12] In our own patient population we observed that the incidence of HAT was increased in patients with arterial anastomosis of the graft artery with the supracoeliac aorta (5.7-fold increase), whereas arterial reconstruction of accessory right or accessory left arteries was not associated with an increased risk of HAT.[2,5]

Clinical presentation of HAT may vary considerably and is dependent on the time of onset. Early HAT (within the first 30 days after OLT) mostly presents with initial nonfunction or severe dysfunction of the liver graft and has a mortality rate of up to 30%. Of the remaining, almost every third patient has to undergo retransplantation.[5] In contrast, in patients with late HAT (beyond 30 days post-OLT) biliary tract complications are the predominant clinical feature.[13] Although the mortality rate in late HAT is usually low, the incidence of retransplantation due to chronic graft loss can be as high as 60%.[13]

Due to the high risk of graft loss, early diagnosis and subsequent therapy is absolutely mandatory. Doppler ultrasonography should be routinely performed after surgery and in case of suspicious arterial findings angiography should follow. (Fig. 1) The possibility of angioplasty and fibrinolysis, or stent placement should be discussed once early HAT has been diagnosed.[12,14] If arterial patency is not achieved, surgical thrombectomy may be an alternative treatment as graft saving modality. As reported in literature, graft loss can be avoided using this treatment policy, however with variable success rates.[2,15,16,17]

The management of late HAT differs significantly from the management of acute HAT. Patients with late HAT may present with a variety of symptoms including increase of serum transaminase levels, cholestasis or severe septic complications as well as liver abscess or biliary tract destruction.[5,13] (Fig. 2) Therefore, treatment modalities can vary from surveillance or antibiotic therapy to interventional abscess drainage, endoscopic treatment with internal/external drainage, hepaticojejunostomy, or partial hepatectomy.[5,18,19,20] Although many patients with late HAT remain clinically stable for many years with conservative treatment and experience an acceptable quality of life, most of them

*Corresponding Author: Natascha C. Nüssler—Department of General-, Visceral- & Transplantation Surgery, Charité, Campus Virchow-Klinikum, Universitätsmedizin Berlin, Augustenburger Platz 1, 13353 Berlin, Germany. Email: natascha.nuessler@charite.de

*Chronic Allograft Failure: Natural History, Pathogenesis, Diagnosis and Management*, edited by Nasimul Ahsan. ©2008 Landes Bioscience.

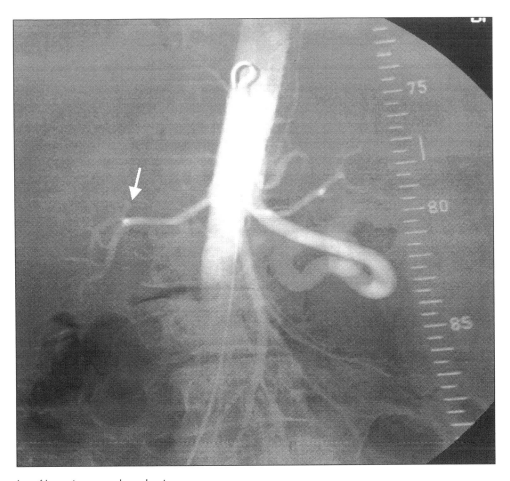

Figure 1. Angiography of hepatic artery thrombosis.

ultimately require retransplantation.[5] Due to organ shortage and high mortality related to retransplantation, maximum efforts to rescue liver grafts with HAT should therefore be carried out.

### *Hepatic Artery Stenosis*

Stenosis of the hepatic artery has a comparable incidence to HAT (3.2% in our population).[2,9,21] Stenoses are found at the anastomosis or in the graft artery and are situated extra- or intrahepatically. Stenoses in the anastomotic region are usually caused by technical errors. Postanastomotic stenoses result from twisting and kinking of graft arteries, which may be caused by technical problems or damage of the vascular wall.[21]

Depending on onset and localization of the stenosis, clinical symptoms vary and are remarkably similar to HAT. Patients may be asymptomatic or show only a slight increase of transaminases or cholestatic enzyme levels. On the other hand, some patients suffer from cholangitis or biliary duct destruction, which may lead to chronic graft loss during the later follow-up.

Therapy of HAS consists of balloon dilatation and/or stenting as well as surgical resection and direct arterial reconstruction or saphenous vein interposition.[2,22-24] Untreated HAS will lead to HAT in more than two thirds of the patients within 6 months.[23] Ischemic biliary complications of HAS should be treated endoscopically as

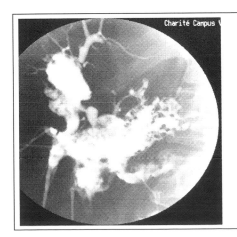
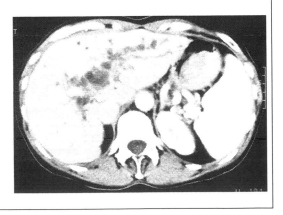

Figure 2. Biliary tract deconstruction after HAT in ERC and CT. Reproduced with permission from Strange et al, Liver Trans 2003; 9(6):612-620.[5]

long as possible. However, retransplantation due to ongoing biliary tract destruction may become necessary in up to 30% of the patients.[2,21]

### Hepatic Artery Aneurysms

With an incidence of less than 1%, hepatic artery aneurysms (HAA) are uncommon vascular complications after OLT, but are associated with a mortality rate up to 69%.[25] Most of the affected patients have acute, life-threatening gastrointestinal or intra-abdominal bleeding.[6]

Intrahepatic HAAs may develop following fine needle biopsy or percutaneous transhepatic cholangiography. Extrahepatic HAAs are usually localized at the site of the arterial anastomosis. In most of these patients local infections due to insufficient hepaticojejunostomy, bile leaks or bowel perforation may have led to the development of HAA.[25,26]

Emergency therapy of ruptured extrahepatic HAAs often consists of resection of the aneurysm and ligation of the hepatic artery. As a consequence of hepatic artery ligation, long-term complications may arise including graft loss, ischemic biliary lesions such as bile duct necrosis or liver abscess. Therefore, the surgical approach should include the resection of the aneurysm with simultaneous reconstruction of the artery (for example with an autologous venous graft, or an arterial allograft) in order to avoid ischemic biliary complications. If reconstruction is impossible, simultaneous hepaticojejunostomy is advisable. In general, surgery should be considered even in asymptomatic patients because of the high morbidity and mortality of HAA after OLT.[2,6]

Surgical repair of intra-hepatic HAAs is not possible and retransplantation is usually required. Superselective arterial embolisation may be used as a bridging procedure and in rare cases and in absence of infection also as therapeutic option alone.[25,26]

Stenting of aneurysms has been reported, however in patients with an infectious cause of HAA surgery should be preferred, since surgery may also allow treatment of the infectious focus.[27,28]

### Arterial Steal Syndromes (ASS)

Arterial steal syndromes after OLT are characterized by arterial hypoperfusion of the graft caused by shifting of blood flow into the splenic or gastroduodenal artery, which is observed in up to 6% of patients undergoing OLT.[7,29,30] Several mechanisms leading to ASS,

like stenosis of the graft artery or splanchnic hyperemia and hyperperfusion of the spleen in patients with end-stage liver disease, have been identified. Angiographic findings such as an enlarged splenic artery and early perfusion of the splenic artery have been reported to be indicative for ASS, however, reliable criteria for the diagnosis of steal syndromes have not been defined.[30] (Fig. 4) Untreated ASS may lead to serious complications in more than 30% of patients with a re-OLT rate of up to 23%.[7]

Patients may present with elevated liver enzyme levels, impaired graft function, or ischemic biliary tract destruction. In the majority of patients, duplex ultrasound examination of the graft does not reveal pathological findings and an angiography is therefore mandatory for the diagnosis of ASS. An enlarged splenic artery (> 4mm or 150% of hepatic artery diameter) as well as dynamic findings indicating relative hypoperfusion of the graft are indications for the diagnosis of ASS (Fig. 3). Other alterations of the hepatic artery, immunologic, toxic or infectious reasons for graft dysfunction have to be ruled out.[7]

Since the vast majority of ASS involve the splenic artery, the goal of any treatment modality is to reduce the blood flow steal into the splenic artery and to increase blood flow into the graft artery. This can be achieved by either splenectomy, ligation, banding, or coil embolisation of the splenic artery.

All procedures result in significant improvement of liver graft perfusion, but differ significantly in morbidity and mortality. Splenectomy in transplant recipients is associated with a significant risk for complications like portal vein thrombosis or overwhelming postsplenectomy infection.[33] Ligation of the splenic artery may lead to splenic infarction or sepsis in few cases, whereas banding (artificial stenosis with nonadsorbable suture close to the origin of the splenic artery) of the splenic artery is not associated with severe complications. However recurrence of ASS has been observed after banding of the splenic artery.

Efforts should be made to identify patients at risk already before OLT. These patients should undergo prophylactic banding or ligation of the splenic artery during OLT in order to prevent the development of ASS after OLT.[7]

If the diagnosis of ASS is made after OLT, coiling of the splenic artery should be the procedure of choice. Embolization of the splenic

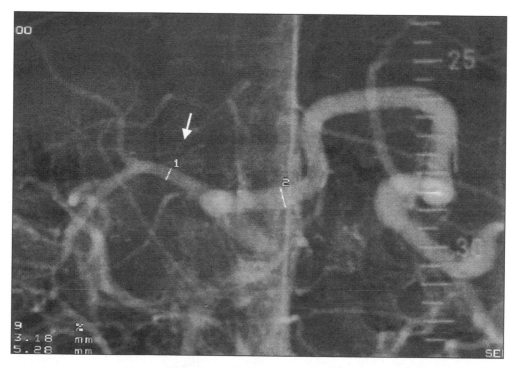

Figure 3. Angiography of lienalis steal syndrome.

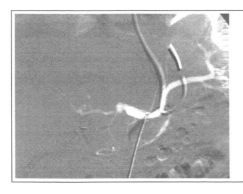
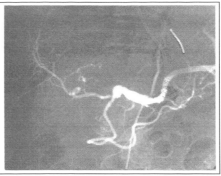

Figure 4. Lienalis steal syndrome before and after coil embolisation.

artery is a safe therapeutic option without the need for surgery, however coils must be placed in the central part of the artery to avoid complications of the spleen (Fig. 4).[7,29]

Therapy of gastroduodenal steal syndrome should also consist of ligation or transarterial embolisation of the gastroduodenal artery.[31,32]

Because of the high incidence of complications, all patients with ASS should preferably be treated by banding or ligation of the stealing artery or coil embolisation.

## Venous Complications

Venous complications following orthotopic liver transplantation are rare events. Less than 5% of adult liver transplant recipients experience this kind of complication. In this context, complications involving the inferior vena cava and the liver veins occur even less frequent than portal-venous complications.[3]

### Complications of the Portal Vein

Portal vein complications, such as thrombosis (PVT) or stenosis (PVS), are relatively rare after liver transplantation, affecting approximately 1 to 2.7% of transplant recipients. Interestingly, more than a third of the patients with portal complications also have arterial abnormalities.[3,8,34]

Portal-venous complications may be caused by technical problems, such as differences in the diameter of the recipient and donor portal vein, use of cryopreserved grafts, excessively long vessel stumps, thrombotic occlusion of the veno-venous bypass, or portosystemic shunt surgery prior to transplantation.[36,37] In addition, morphological alterations of the vessel wall, such as recanalization after thrombotic occlusion, modification of the standard end-to-end anastomosis and concomitant splenectomy are associated with an increased risk of portal vein complications after transplantation.[3]

Clinical symptoms of PVT depend on the time of onset and are classified as early (first week postoperatively) or late (first week onwards)

occlusions.[8] Acute PVT may result in graft failure often requiring immediate retransplantation, if thrombectomy and reconstruction of the portal anastomosis are unsuccessfull.[3,34] In case of late PVT, typical signs of portal hypertension including ascites, splenomegaly, or esophageal varices are more common. In this situation, portosystemic shunt operation and TIPS application have been successful.[3,8] However, TIPS deposition is limited to individual cases with intra- but not extrahepatic occlusion of the portal vein, since a patent portal venous branch is required for successful shunting.

Clinical symptoms are usually less pronounced in patients with stenosis of the portal vein than in patients with complete occlusion of the portal vein. Only patients with a stenosis grade of more than 80% develop symptoms.[3]

PVS may be treated surgically (renewal of the anastomosis) or interventionally, the latter being the preferred choice. Percutaneous angioplasties as well as intravascular stent placement are effective and safe treatment modalities with good mid-term patency.[38-42] (Fig. 5) Interesting to note, more than 50% of patients with portal vein complications require only symptomatic treatment, including sclerotherapy of esophageal varices or no treatment at all.[3]

### Complications of the Inferior Vena Cava

This kind of vascular complication in liver transplant recipients is a rare but serious problem reaching a mortality rate up to 24%.[43-45] Obstruction or stenosis of the inferior vena cava have been reported with an incidence of 1%-14,8% depending on the kind of vena cava anastomosis.[43,46,47]

Actually, two main types of vena cava anastomosis were performed: (i) the traditional technique described by Starzl in 1963, with resection of the retrohepatic part of the recipients inferior vena cava using a veno-venous bypass followed by an end-to-end cavocavostomy and (ii) a technique with preservation of the vena cava, first described by Calne in 1968 as piggy-back technique and now used with several variations.[48-50]

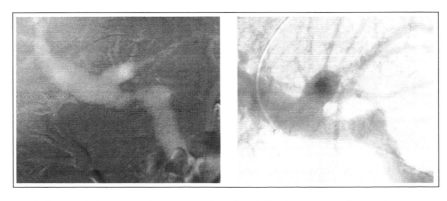

Figure 5. Portal vein stenosis before and after angioplasty. Reproduced with kind permission from: Glanemann M et al. Transpl Int 2001; 14(1):48-51;[39] ©2001 Blackwell Publishing.

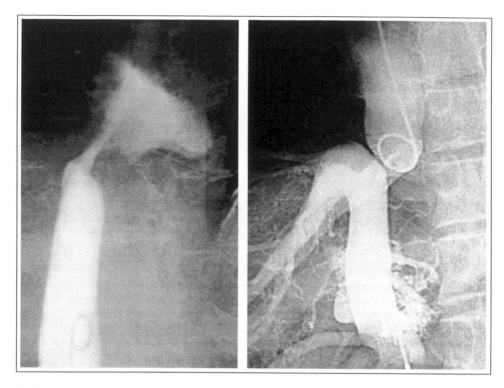

Figure 6. Stenosis of inferior vena cava. Reproduced from Glanemann et al, World J Surg 2002; 26(3):342-347,[43] with kind permission from Springer Science and Business Media.

In the past, the traditional technique was preferred and several studies showed a low incidence of about 1,7% of caval complications like stenosis or occlusion.[43,46]

Actually, the vena cava preserving technique gained more and more acceptance. Some studies show similar complication rates of the vena cava anastomosis compared to the traditional technique (<1%-1,5%),[44,51] whereas other authors report on higher complication rates between 4,6-14,8% when using the cava preserving technique.[45-47]

Technical failure (too small anastomosis or kinking due to a too long segment of the donor retrohepatic cava) or compression of the vena cava by a large organ is the underlying cause in one third of the patients with cava stenosis and recurrence of Budd Chiari syndrome in one fourth of the patients.[3]

The clinical symptoms include ascites, detoriating liver function, hepatomegaly, renal insufficiency, peripheral edema and other unspecific symptoms.[43] Occlusion of the vena cava during the early course after liver transplantation is often associated with severe graft dysfunction and surgical revision of the vena cava anastomosis may be difficult. Our own results have been discouraging, since most patients did not recover and had to be retransplanted.[43,45] In contrast, patients with stenosis of the vena cava could be treated successfully with balloon angioplasty or stent application (Fig. 6).[34,43-45] In some patients surgical treatment of severe ascites, such as Denver-shunt application, might already be sufficient. However, the retransplantation rate due to complications of the vena cava can be as high as 40%.[43,45]

### Complications of the Liver Veins

Due to the persistent shortage of cadaveric livers, living donor liver transplantation (LDLT) has emerged as a strong alternative. In this procedure, the graft hepatic veins are either anastomosed end-to-end to the stump of the recipient's hepatic veins or in end-to-side fashion directly to the inferior vena cava. This surgical technique can induce mechanical or functional stenosis of the hepatic veins, which may cause venous outflow obstruction of the transplanted liver resulting in graft loss. In addition, partial liver grafts usually grow considerably after transplantation and as consequence the liver veins are twisted, which might compromise the hepatic venous outflow. The incidence of hepatic vein stenoses in LDLT has been reported to be about 2%-8,6%.[52,53]

Clinical symptoms of liver vein stenosis may consist of refractory ascites, elevated liver enzymes, enlarged liver, or impaired graft function.

Hepatic venous outflow obstruction immediately after transplantation is a surgical emergency and usually requires re-operation. In late onset of hepatic venous outflow obstruction, surgical correction may be difficult and therefore interventional procedures are preferred in this situation. Balloon angioplasty via the transjugular or transhepatic route seems to be an effective treatment option with good results and good mid-term patency.[52-54] Failed angioplasty may be treated by stent placement, surgical revision or retransplantation.[55,56]

### References
1 Consensus conference on indication of liver transplantation. Hepatology 1994; 20(suppl):1-68.
2 Settmacher U, Stange B, Haase R et al. Arterial complications after liver transplantation. Transpl Int 2000; 13(5):372-78.
3 Settmacher U, Nussler NC, Glanemann M et al. Venous complications after orthotopic liver transplantation. Clin Transplant 2000; 14(3):235-41.
4 Langnas AN, Marujo W, Stratta RJ et al. Vascular complications after orthotopic liver transplantation. Am J Surg 1991; 161(1):76-82; discussion 82-83.
5 Stange BJ, Glanemann M, Nuessler NC et al. Hepatic artery thrombosis after adult liver transplantation. Liver Transpl 2003; 9(6):612-20.
6 Stange B, Settmacher U, Glanemann M et al. Aneurysms of the hepatic artery after liver transplantation. Transplant Proc 2000; 32(3):533-34.
7 Nussler NC, Settmacher U, Haase R et al. Diagnosis and treatment of arterial steal syndromes in liver transplant recipients. Liver Transpl 2003; 9(6):596-602.
8 Sanchez-Bueno F, Hernandez Q, Ramirez P et al. Vascular complications in a series of 300 orthotopic liver transplants. Transplant Proc 1999; 31(6):2409-10.
9 Vivarelli M, Cucchetti A, La Barba G et al. Ischemic arterial complications after liver transplantation in the adult: multivariate analysis of risk factors. Arch Surg 2004; 139(10):1069-74.

10  Oh CK, Pelletier SJ, Sawyer RG et al. Uni- and multi-variate analysis of risk factors for early and late hepatic artery thrombosis after liver transplantation. Transplantation 2001; 71(6):767-72.

11  Madalosso C, de Souza NF, lstrupDM et al. Cytomegalovirus and its association with hepatic artery thrombosis after liver transplantation. Transplantation 1998; 66(3):294-97.

12  Silva MA, Jambulingam PS, Gunson BK et al. Hepatic artery thrombosis following orthotopic liver transplantation: a 10-year experience from a single centre in the United Kingdom. Liver Transpl 2006; 12(1):146-51.

13  Bhattacharjya S, Gunson BK, Mirza DF et al. Delayed hepatic artery thrombosis in adult orthotopic liver transplantation-a 12-year experience. Transplantation 2001; 71(11):1592-96.

14  Cotroneo AR, Di Stasi C, Cina A et al. Stent placement in four patients with hepatic artery stenosis or thrombosis after liver transplantation. J Vasc Interv Radiol 2002; 13(6):619-23.

15  Turrion VS, Alvira LG, Jimenez M et al. Incidence and results of arterial complications in liver transplantation: experience in a series of 400 transplants. Transplant Proc 2002; 34(1):292-93.

16  Tian MG, Tso WK, Lo CM et al. Treatment of hepatic artery thrombosis after orthotopic liver transplantation. Asian J Surg 2004; 27(3):213-17.

17  Sheiner PA, Varmer CV, Guarrera JV. Selective revascularization of hepatic artery thromboses after liver transplantation improves patient and graft survival. Transplantation 1997; 64(9):1295-99.

18  Verdonk RC, Buis CI, Porte RJ et al. Biliary complications after liver transplantation: a review. Scand J Gastroenterol Suppl 2006; 243:89-101.

19  Zhou G, Cai W, Li H et al. Experiences relating to management of biliary tract complications following liver transplantation in 96 cases. Chin Med J (Engl) 2002; 115(10):1533-37.

20  Guckelberger O, Stange B, Glanemann M et al. Hepatic resection in liver transplant recipients: single center experience and review of the literature. Am J Transplant 2005; 5(10):2403-9.

21  Jain A, Costa G, Marsh W et al. Thrombotic and nonthrombotic hepatic artery complications in adults and children following primary liver transplantation with long-term follow-up in 1000 consecutive patients. Transpl Int 2006; 19(1):27-37.

22  Ueno T, Jones G, Martin A et al. Clinical outcomes from hepatic artery stenting in liver transplantation. Liver Transpl 2006; 12(3):422-27.

23  Saad WE, Davies MG, Sahler L et al. Hepatic artery stenosis in liver transplant recipients: primary treatment with percutaneous transluminal angioplasty. J Vasc Interv Radiol 2005; 16(6):795-805.

24  Vignali C, Bargellini I, Cioni R et al. Diagnosis and treatment of hepatic artery stenosis after orthotopic liver transplantation. Transplant Proc 2004; 36(9):2771-73.

25  Marshall MM, Muiesan P, Srinivasan P et al. Hepatic artery pseudoaneurysms following liver transplantation: incidence, presenting features and management. Clin Radiol 2001; 56(7):579-87. Erratum in: Clin Radiol 2001; 56(9):785.

26  Leelaudomlipi S, Bramhall SR, Gunson BK et al. Hepatic-artery aneurysm in adult liver transplantation. Transpl Int 2003; 16(4):257-61.

27  Maleux G, Pirenne J, Aerts R et al. Case report: hepatic artery pseudoaneurysm after liver transplantation: definitive treatment with a stent-graft after failed coil embolisation. Br J Radiol 2005; 78(929):453-56.

28  Muraoka N, Uematsu H, Kinoshita K et al. Covered coronary stent graft in the treatment of hepatic artery pseudoaneurysm after liver transplantation. J Vasc Interv Radiol 2005; 16(2 Pt 1):300-2.

29  Langer R, Langer M, Schulz A et al. The splenic steal syndrome and the gastroduodenal stel syndrome in patients before and after liver transplantation. Aktuelle Radiol 1992; 2(2):55-58.

30  Geissler I, Lamesch P, Witzigmann H et al. Splenohepatic arterial steal syndrome in liver transplantation: clinical features and management. Transpl Int 2002; 15(2-3):139-41.

31  Vogl TJ, Pegios W, Balzer JO et al. Arterial steal syndrome in patients after liver transplantation: transarterial embolization of the splenic and gastroduodenal arteries. Rofo 2001; 173(10):908-13.

32  Nishida S, Kadono J, DeFaria W et al. Gastroduodenal artery steal syndrome during liver transplantation: intraoperative diagnosis with Doppler ultrasound and management. Transpl Int 2005; 18(3):350-53.

33  Troisi R, Heese UJ, Decruyenaere J et al. Functional life-threatening disorders and splenectomy following liver transplantation. Clin Transpl 1999; 13:380-88.

34  Karatzas T, Lykaki-Karatzas E, Webb M et al. Vascular complications, treatment and outcome following orthotopic liver transplantation. Transplant Proc 1997; 29(7):2853-55.

35  Cavallari A, Vivarelli M, Bellusci R et al. Treatment of vascular complications following liver transplantation: multidisciplinary approach. Hepatogastroenterology 2001; 48(37):179-83.

36  Millis JM, Seaman DS, Piper JB et al. Portal vein thrombosis and stenosis in pediatric liver transplantation. Transplantation 1996; 62(6):748-54.

37  Kuang AA, Renz JF, Ferrell LD et al. Failure patterns of cryopreserved vein grafts in liver transplantation. Transplantation 1996; 62(6):742-47.

38  Park KB, Choo SW, Do YS et al. Related Articles, Percutaneous angioplasty of portal vein stenosis that complicates liver transplantation: the mid-term therapeutic results. Korean J Radiol 2005; 6(3):161-66.

39  Glanemann M, Settmacher U, Langrehr JM et al. Portal vein angioplasty using a transjugular, intrahepatic approach for treatment of extrahepatic portal vein stenosis after liver transplantation. Transpl Int 2001; 14(1):48-51.

40  Perkins JD. Percutaneous transhepatic balloon dilation for portal venous stenosis. Liver Transpl 2006; 12(2):321-22.

41  Wang JF, Zhai RY, Wei BJ et al. Percutaneous intravascular stents for treatment of portal venous stenosis after liver transplantation: midterm results. Transplant Proc 2006; 38(5):1461-62.

42  Shan H, Xiao XS, Huang MS et al. Portal venous stent placement for treatment of portal hypertension caused by benign main portal vein stenosis. World J Gastroenterol 2005; 11(21):3315-18.

43  Glanemann M, Settmacher U, Langrehr JM et al. Results of end-to-end cavocavostomy during adult liver transplantation. World J Surg 2002; 26(3):342-47.

44  Navarro F, Le Moine MC, Fabre JM et al. Specific vascular complications of orthotopic liver transplantation with preservation of the retrohepatic vena cava: review of 1361 cases. Transplantation 1999; 68(5):646-50.

45  Cescon M, Grazi GL, Varotti G et al. Venous outflow reconstructions with the piggyback technique in liver transplantation: a single-center experience of 431 cases. Transpl Int 2005; 18(3):318-25.

46  Lerut JP, Molle G, Donataccio M et al. Cavocaval liver transplantation without venovenous bypass and without temporary portocaval shunting: the ideal technique for adult liver grafting? Transpl Int 1997; 10(3):171-79.

47  Hesse UJ, Defreyne L, Pattyn P et al. Hepato-venous outflow complications following orthotopic liver transplantation with various techniques for hepato-venous reconstruction in adults and children. Transpl Int 1996; 9(1):182-84.

48  Starzl TE, Marchioro TL, Von Kaula KN et al. Homotransplantation of the liver in humans. Surg Gynecol Obstet 1963; 117:659-76.

49  Calne RY, Williams R. Liver transplantation in man. I. Observations on technique and organization in five cases. Br Med J 1968; 4(5630):535-40.

50  Belghiti J, Panis Y, Sauvanet A et al. A new technique of side to side caval anastomosis during orthotopic hepatic transplantation without inferior vena caval occlusion. Surg Gynecol Obstet 1992; 175(3):270-72.

51  Parrilla P, Sanchez-Bueno F, Figueras J et al. Analysis of the complications of the piggy-back technique in 1,112 liver transplants. Transplantation 1999; 67(9):1214-17.

52  Karakayali H, Boyvat F, Coscun M et al. Venous complications after orthotopic liver transplantation. Transplant Proc 2006; 38(2):604-6.

53  Ko GY, Sung KB, Yoon HK et al. Endovascular treatment of hepatic venous outflow obstruction after living-donor liver transplantation. Vasc Interv Radiol 2002; 13(6):591-99.

54  Kubo T, Shibata T, Itho K et al. Outcome of percutaneous transhepatic venoplasty for hepatic venous outflow obstruction after living donor liver transplantation. Radiology 2006; 239:285-90.

55  Mazariegos GV, Garrido V, Jaskowski-Phillips S et al. Management of hepatic venous obstruction after split-liver transplantation. Pediatr Transplant 2000; 4(4):322-27.

56  Huang TL, Chen TY, Tsang LL et al. Hepatic venous stenosis in partial liver graft transplantation detected by color Doppler ultrasound before and after radiological interventional management. Transplant Proc 2004; 36(8):2342-43.

# CHAPTER 20

# Late Allograft Failure: Liver

## Jeffrey S. Crippin*

## Abstract

Dysfunction and subsequent loss of a liver allograft can have dire consequences for the recipient. Acute and chronic rejection, an ongoing risk for the lifetime of the allograft in the vast majority of liver transplant recipients, is a minor problem, except when levels of immunosuppression fall. The more common complications leading to late loss of the allograft in liver recipients are recurrence of the disease that lead to the transplant in the first place and biliary strictures related to different etiologies. This manuscript provides an overview of the scope of the problem and different management options for the transplant physician.

## Introduction

Loss of the liver allograft follows a different course than the loss of kidney, heart and lung allografts. While the latter are frequently lost due to allograft rejection or complications of immunosuppression, the transplanted liver is often lost due to recurrence of the disease that ultimately led to the transplant in the first place. Hepatitis C, the most common reason for liver transplantation worldwide, infects the transplanted liver without exception, though the severity of the disease and its progression varies from patient to patient. The challenge to the transplant team is controlling and, hopefully, eliminating these disease processes, with ongoing excellent function of the allograft. This chapter will examine the diseases responsible for late graft loss in liver transplant recipients, as well as management options (Table 1).

## Acute and Chronic Rejection

Although late allograft loss due to acute or chronic rejection is less common in liver transplant recipients than in recipients of other solid organ transplants, the assumption should not be that this does not occur. Bouts of **acute cellular rejection** (ACR) are uncommon years after a transplant. Documented ACR is usually related to inadequate levels of immunosuppression, often related to noncompliance or an event leading to lower drug levels. The latter may be precipitated by something as simple as a viral gastroenteritis, with emesis of immunosuppressive medications or decreased absorption related to diarrhea. The administration of other medications affecting metabolism of the immunosuppressive agent may lower drug levels, as well. Isoniazid, rifampin and phenytoin can lead to lower levels of calcineurin inhibitors in the face of maintenance doses, due to an increase in activity of the cytochrome P450 system.

Treatment of ACR, regardless of its timing, is the same. Pulse doses of corticosteroids followed by a slow taper and an increase in trough levels of calcineurin inhibitors, TOR inhibitors (sirolimus), or antiproliferative agents (azathioprine and mycophenolate mofetil) is effective. ACR refractory to this therapy, may lead to the need for polyclonal antibody therapy directed at T-cell receptors (rabbit anti-thymocyte globulin). Unfortunately, ACR occurring after the first post-operative month has been associated with a significant rate of allograft loss. One study showed allograft failure in 27% of cases of ACR following the first 30 days.[1] Noncompliance was thought to be the overriding factor. In spite of these findings, retransplantation for allograft failure due to rejection is uncommon, particularly in the modern era.[2]

**Chronic rejection** is characterized by bile duct loss and obliterative vasculopathy, routinely occurring within 6 months of the transplant. Chronic rejection now occurs in 5% of liver transplants, down from an incidence of 15-20% seen in the early years of the field.[3,4] Studies from the 1990's showed the risk for the development of chronic rejection is increased in patients with refractory episodes of acute cellular rejection, multiple episodes of rejection,[5] late onset ACR, male to female gender

### Table 1. Treatment of late allograft dysfunction

| Etiology | Treatment Considerations |
|---|---|
| Acute cellular rejection | Corticosteroids |
| | Higher calcineurin inhibitor levels |
| | Addition of another agent |
| | Polyclonal anti-thymocyte globulin |
| | Retransplantation |
| Chronic rejection | Change in calcineurin inhibitor |
| | Addition of another agent |
| | Retransplantation |
| Hepatitis C | Minimize immunosuppression |
| | Pegylated interferon/ribavirin |
| | Retransplantation |
| Hepatitis B | Addition or change in antiviral agent |
| | Retransplantation |
| Autoimmune hepatitis | Corticosteroids |
| | Anti-proliferative agent |
| | Retransplantation |
| Primary biliary cirrhosis | Ursodeoxycholic acid |
| | Change in immunosuppression? |
| | Retransplantation |
| Primary sclerosing cholangitis | Consider ursodeoxycholic acid |
| | Change in immunosuppression? |
| | Maintain stricture patency |
| | Retransplantation |
| Biliary strictures | Maintain stricture patency |
| | Optimize hepatic arterial flow |
| | Antibiotics for cholangitis |
| | Retransplantation |

*Jeffrey S. Crippin—Liver Transplantation, Division of Gastroenterology, Washington University School of Medicine, 660 South Euclid, Campus Box 8124, St. Louis, Missouri 63110, USA. Email: jcrippin@wustl.edu

*Chronic Allograft Failure: Natural History, Pathogenesis, Diagnosis and Management*, edited by Nasimul Ahsan. ©2008 Landes Bioscience.

mismatch,[6] nonwhite race,[7] a pretransplant diagnosis of autoimmune hepatitis or autoimmune biliary disease,[8] and possibly a role for CMV infection, though studies are conflicting on the final issue.[9,10] Patients present with cholestasis and the diagnosis is established by liver biopsy. Biliary strictures should be ruled out with cholangiography, either by magnetic resonance imaging or directly with endoscopic or percutaneous examination. Patency of the hepatic artery, a common cause of biliary ischemia, strictures and cholestasis should be ruled out, as well.

The treatment of chronic rejection is centered on manipulation of immunsuppression. Both allograft and patient survival improve when cyclosporine A is changed to tacrolimus.[11] Making the diagnosis early was crucial in this study, as survival was dependent on whether or not the change in medication occurred when the bilirubin was less than 10 mg/dL. If chronic rejection cannot be reversed with changes in immunosuppression, cholestasis worsens and patients suffer from the ravages of secondary biliary cirrhosis. Pruritus, fatigue and fat malabsorption can result. Retransplantation can be the only option, however, survival rates for retransplantation are lower than for first time transplants and there is an increased risk of developing chronic rejection in the second liver allograft.[3]

## Hepatitis C

Infection of the liver allograft with hepatitis C occurs in essentially all patients transplanted for cirrhosis secondary to hepatitis C. Progression of the disease proceeds at variable rates, depending on a number of factors, though the process is poorly understood.[12] Disease progression follows an accelerated course, compared to the progression of liver injury in the native liver.[13] In the allograft, disease progression is affected by corticosteroid boluses,[14,15] the use of monoclonal anti-lymphocyte antibody therapy for refractory rejection (more of a historical problem due to less frequent use),[14,15] donor age greater than 50 years,[16,17] and a high pretransplant viral load[18] (Table 2). By five years following the transplant, cirrhosis of the allograft is present in up to 20-40% of cases. Though long term studies are not available, this percentage likely increases with time. The development of cirrhosis starts a course of rapid deterioration, with decompensation (the development of complications of end stage liver disease, such as, ascites, variceal bleeding and hepatic encephalopathy) occurring in 40% within 1 year and 60% within 3 years, compared to less than 5% and 10%, respectively, in patients with cirrhosis secondary to hepatitis C of the native liver.[19,20] Once decompensation occurs, survival rates drop precipitously, with a three year survival rate of <10%.[19]

Treatment of hepatitis C in the allograft creates a dilemma, due to the absence of controlled studies and the variable natural history. No specific optimal immunosuppressive strategy has been defined, though most centers tend to rapidly taper corticosteroids to low doses or off within 6-12 months of the transplant. High dose steroids are avoided, thus, conservative approaches to findings suggestive of ACR are often used. Equivocal biopsy findings often lead to a period of several days of observation, followed by a second biopsy to verify the previous findings. If ACR is present, higher levels of calcineurin inhibitors or the addition of an anti-proliferative agent, such as mycophenolate mofetil or azathioprine, are often used as first line therapy.

### Table 2. Factors associated with late graft failure due to hepatitis C

High pretransplant viral load
Treatment of acute rejection with corticosteroids
Treatment of acute rejection with monoclonal anti-lymphocyte
   antibody therapy
Donor age greater than 50 years
Cytomegalovirus infection

### Table 3. Treatment of hepatitis C following liver transplantation: Factors for consideration

History of treatment pretransplant
  • Multiple treatment trial failures
  • Unchanged viral load vs. relapse
  • Poor patient tolerance
Factors affecting dosing
  • Pretreatment anemia
  • Pretreatment leucopenia
  • Pretreatment thrombocytopenia
  • Renal insufficiency
Hepatic histology
  • Minimal fibrosis
  • Bridging fibrosis or cirrhosis

The use of antiviral agents are also fraught with problems. The timing of therapy is an often debated subject. Some advocate treatment early in the course of the disease, prior to the development of hepatic fibrosis. Others favor waiting for fibrotic changes, due to the overall poor response rate and poor patient tolerance. Interferon alfa (standard and pegylated forms) and ribavirin, associated with a 50-60% sustained virologic response rate in nontransplant patients, leads to a cure in only 20-25% of cases, usually due to poor patient tolerance and the need for dose reductions.[21,22] Anemia, neutropenia and thrombocytopenia may reach intolerable levels, necessitating dose reduction, leading to decreased efficacy. Anemia can effectively be treated with supplemental erythropoietin at a dose of 40,000 u subcutaneously weekly. Neutropenia, usually defined as an absolute neutrophil count less than 500-1000/mm³, can be treated with, filgrastim, a human granulocyte colony-stimulating factor, used once to thrice weekly, based on the patient's response. Dose reduction of pegylated interferon and ribavirin is another option, however, dose reduction potentially decreases efficacy. In the face of all of these issues, cases must be considered on an individual basis. Factors to assist in the decision process include previous history of treatment, patient tolerance, presence of anemia/leukopenia/thrombocytopenia, baseline renal function and hepatic histology (Table 3).

Treatment of the hepatitis C transplant recipient with allograft failure is usually no different from the treatment of other patients with cirrhosis. Portal hypertensive bleeding is managed endoscopically and with decompressive procedures such as a transjugular intrahepatic portosystemic shunt (TIPS). Ascites is treated with a sodium restricted diet and diuretics, though renal insufficiency related to pretransplant azotemia or to calcineurin inhibitor nephrotoxicity, may limit their effectiveness. Encephalopathy can be controlled with the use of lactulose and non-absorbable antibiotics.

End stage liver disease secondary to chronic hepatitis C may lead to consideration of retransplantation. However, survival rates following a second transplant for hepatitis C are clearly lower than first transplants with a one year patient survival rate of 57% in one study.[23] Lower survival rates are seen with pretransplant bilirubin levels greater than 10 mg/dL and a serum creatinine greater than 2.0 mg/dL.[24] The approach to retransplantation varies from center to center. Some centers use extended criteria donors or donors after cardiac death, in an attempt to retain optimal donor organs for first time recipients. Other centers choose to not retransplant hepatitis C patients at all.

## Hepatitis B

In the early 1990's, patient and allograft survival following liver transplantation for hepatitis B were so poor, many transplant centers considered hepatitis B a contraindication to transplantation. However, the use of high dose hepatitis B immunoglobulin (HBIG) as immunoprophylaxis, drastically improved survival rates.[25] HBIG binds circulating hepatitis B surface antigen, preventing it from infecting the liver allograft. The addition of a nucleoside or nucleotide analogue has further improved survival rates, with a marked decrease in the rate of allograft failure.[26]

The development and availability of additional antiviral agents has markedly simplified the management of hepatitis B in the transplant and nontransplant patient. Suppression of viral replication decreases hepatic necroinflammatory activity and fibrosis. Ultimately, this has lead to a decrease in the number of liver transplants being performed for this disease.

Most patients transplanted for hepatitis B are maintained on long term intravenous or intramuscular HBIG and an antiviral agent. HBIG is dosed to maintain a hepatitis B surface antibody titer of at least 500, though some patients maintain hepatitis B surface antigen negativity with lower titers. Hepatitis B surface antigen positivity in the serum is usually associated with failure to maintain the surface antibody titer at the appropriate level, often due to noncompliance. Once hepatitis B surface antigen positivity has been documented, additional doses of HBIG can be administered with the hope of increasing the antibody titer to adequate levels to neutralize antigenemia. However, persistent viremia means long term immunoprophylaxis is no longer needed.

The challenge in the control of the hepatitis B infection is centered on the development of resistant strains of the virus. Lamivudine, the first nucleoside analogue used to treat hepatitis B, leads to the development of resistant strains in nearly 70% of nontransplant patients treated for four years.[27] The development of a resistant strain may have biochemical and clinical implications. An increase in the level of HBV-DNA while on treatment suggests a resistant strain has formed. With time, this will usually lead to elevated liver biochemistries and clinical symptoms, such as fatigue, myalgias and jaundice, all associated with progressive liver disease. With the availability of additional antiviral agents, hepatitis B can be controlled. Patients with lamivudine resistant strains may benefit from the addition of or a change to the antiviral agent, adefovir dipivoxil.[28] The addition of adefovir allows lamivudine to continue to suppress wild type virus, while adefovir suppresses the lamivudine resistant strain or strains. Adefovir is also an excellent first line drug following the transplant. Entecavir is another nucleoside analogue with excellent potency. It should not be used in patients with lamivudine resistance, due to cross resistance seen between lamivudine and entecair. Tenofovir and telbivudine are other agents available for the management of hepatitis B. Dose reduction may be needed due to renal insufficiency. Entecavir, tenofovir and telbivudine have not been studied in liver transplant recipients, however, no adverse reactions are known with commonly used immunosuppressive agents.

If the hepatitis B infection cannot be controlled, the disease will progress, eventually resulting in cirrhosis and its complications. Management of these complications are no different from the management of the patient with hepatitis C listed in the section above. Retransplantation is another potential option. Prior to the widespread use of HBIG and antiviral agents, retransplantation in patients with allograft failure secondary to hepatitis B usually had a dismal outcome.[29] However, the use of high dose HBIG in combination with an antiviral agent has markedly improved outcomes following retransplantation, though published data on the topic is lacking.

## Autoimmune Disease

In general, the incidence of recurrent autoimmune disease in the liver allograft is low, presumably related to the effect of immunosuppressive agents on the disease process. However, patients transplanted for autoimmune disease can develop recurrent disease in the allograft with evidence of allograft dysfunction and failure.

**Autoimmune hepatitis** accounts for approximately 5% of liver transplants performed each year. Histologic evidence of autoimmune hepatitis in the allograft develops in 16-46% of cases, based on interface hepatitis and a lymphoplasmacytic portal infiltrate.[30,31] The timing of recurrence varies and is usually related to decreasing levels of immunosuppressants. One study showed a mean time of recurrence at 2.5 years following the transplant.[31] Risk factors for recurrent disease include HLA DR3 positive recipients receiving DR3 negative allografts. The diagnosis is established histologically, since autoantibody formation may be affected by immunosuppression. If the disease is not controlled, progression to cirrhosis is likely, complicated by portal hypertension and jaundice. The presence of disease recurrence routinely leads to an adjustment in immunosuppression. If a patient is on a calcineurin inhibitor only, addition of a lymphocyte inhibitor, such as azathioprine or mycophenolate mofetil, often leads to improved biochemical and histologic changes, though most experience is anecdotal. Increased doses of corticosteroids may also decrease hepatic inflammation, though persistently high corticosteroid doses are not preferred due to the probability of steroid induced side effects.

De novo autoimmune hepatitis in the liver allograft can develop in patients without a previous history of the disease. Reported cases have been in association with a pretransplant diagnosis of primary biliary cirrhosis,[32] though similar findings have been seen in patients transplanted for other diseases, albeit unreported. Histologically, findings similar to those seen in nontransplant patients are present. Treatment is similar to that used in nontransplant patients, i.e., corticosteroids and a lymphocyte inhibitor. If the addition of these agents to a calcineurin inhibitor is unsuccessful at controlling the disease, consideration can be given to changing the calcineurin inhibitor. Corticosteroids are tapered off over the course of several months. Increasing the trough level of the calcineurin inhibitor may also lower abnormal liver biochemistries, though the chance of worsening azotemia, hypertension and diabetes makes this option less attractive. In patients with disease refractory to this medical regimen, the use of polyclonal antibodies to T-cell receptors has been useful in slowing disease progression in selected cases.

**Primary biliary cirrhosis** is marked by bile ductular damage leading to cirrhosis in the native liver. Similar changes can occur in the allograft, albeit only up to 20% of cases, occurring at 3-6 years following the transplant.[33,34] Other causes of allograft dysfunction should be ruled out, such as acute and chronic rejection, biliary obstruction and drug hepatotoxicity. Antimitochondrial antibody titers, the key autoantibody used for diagnosis, do not necessarily correlate with disease recurrence or severity. Progression of the disease is slow. There are no reported cases of allograft failure secondary to PBC, however, long term survivors with recurrent disease are certainly at risk. The use of bile acid therapy is common once disease recurrence has been documented. Ursodeoxycholic acid, 13-15 mg/kg/day, the standard dose used in nontransplant patients, has been associated with prolonged survival in the nontransplant population. Whether similar results are seen with transplant recipients has not been studied. Another approach, based on the autoimmune nature of the disease, is manipulation of the immunosuppression regimen. Anecdotal experience has shown improvement in liver biochemistries with the addition of mycophenolate mofetil, though controlled studies have not been performed, due to the small number of patients afflicted.

**Primary sclerosing cholangitis,** a disease marked by fibrotic stricturing of the intra- and extrahepatic bile ducts, ultimately resulting in a secondary biliary cirrhosis, appears in the allograft in 10-20% of cases.[35,36] The diagnosis is established cholangiographically. Classic histologic changes of fibrous cholangitis and fibro-obliterative ductal lesions are helpful, though inconsistent. Other causes of biliary strictures should be ruled out, such as ischemic strictures secondary to hepatic artery thrombosis or stenosis, HIV cholangiopathy, or strictures secondary to prolonged cold ischemia. Management varies from case to case. Manipulation of immunosuppression may help,

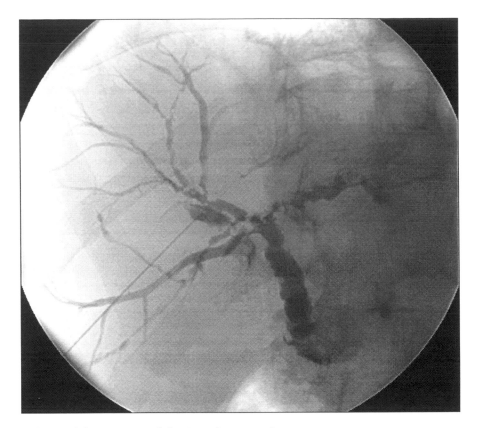

Figure 1. Cholangiogram showing biliary strictures following a liver transplant.

in spite of multiple negative studies in nontransplant patients. A change in immunosuppression may lead to an improvement in liver biochemistries, however, improvement of biliary strictures, once established, is unlikely. Stent placement, either percutaneously or endoscopically, can maintain patency of tight strictures, allowing adequate bile flow and preventing episodes of bacterial cholangitis. Bile acid therapy has no known effect on the natural history of the disease in the nontransplant patient, though it is frequently used in both transplant and nontransplant patients, due to improvement in liver biochemistries. Long term studies in patients with PSC in the allograft are lacking. Worsening strictures may lead to recurrent bouts of bacterial cholangitis necessitating the use of antibiotics on a regular basis. Refractory cholangitis or the development of cirrhosis are potential indications for retransplantation.

## Biliary Strictures

The biliary tree of the allograft receives its sole blood supply from hepatic arterial flow. In the event of a hepatic arterial thrombosis or stenosis, the lack of blood flow to the biliary tree can lead to catastrophic results, with stricturing similar to that seen in primary sclerosing cholangitis (Fig. 1). Similar cholangiographic findings can occur in the face of a transplant across ABO blood groups,[37] prolonged cold ischemia time (>10-12 hours),[38] and with livers from donors after cardiac death.[39]

Once an ischemic event has occurred, improving blood flow does not reverse the injury. Although efforts to revascularize are common in the setting of an acute hepatic artery thrombosis in the early days after the transplant, the presence of collateral flow makes these efforts less necessary when the event occurs later. Stricture dilation via endoscopic or percutaneous approaches with placement of stents to allow drainage will minimize the damage caused by stricture formation. The timing of placement and replacement of stents will vary from patient to patient and center to center. Many patients can be maintained for years with routine follow up. However, as the strictures worsen with time, the complications of biliary cirrhosis begin to appear and stent placement is of little use.

Stricture formation, leading to impaired bile flow, may result in cholangitis and even abscess formation. Cholangitis can be treated with broad spectrum antibiotics, such as ciprofloxacin. If a hepatic abscess is present, percutaneous drainage may be needed with an externally draining tube. Retransplantation remains a possibility for appropriate candidates with refractory cholangitis or complications of cirrhosis.

## References

1. Anand A, Hubscher S, Gunson B et al. Timing, significance and prognosis of late acute allograft rejection. Transplantation 1995; 60:1098-1103.
2. Jain A, Reyes J, Kashyap R et al. Long-term survival after liver transplantation in 4,000 consecutive patients at a single center. Ann Surg 2000; 232:490-500.
3. Neuberger J. Incidence, timing and risk factors for acute and chronic rejection. Liver Transpl Surg 1999; 5:S30-36.
4. Demetris AJ, Murase N, Lee RG et al. Chronic rejection. A general overview of histopathology and pathophysiology with emphasis on liver, heart and intestinal allografts. Ann Transplant 1997; 2:27-44.
5. Demetris A, Adams D, Bellamy C et al. Update of the International Banff Schema for Liver Allograft Rejection working recommendations for the histopathologic stages and report of chronic rejection. Hepatology 2000; 31:792-99.
6. Candinas D, Gunson BK, Nightingale P et al. Sex mismatch as a risk factor for chronic rejection of liver allografts. Lancet 1995; 346:1117-21.
7. Freese DK, Snover DC, Sharp HL et al. Chronic rejection after liver transplantation: a study of clinical histopathological and immunological features. Hepatology 1991; 13:882-91.
8. Hayashi M, Keeffe EB, Krams SM et al. Allograft rejection after liver transplantation for autoimmune liver diseases. Liver Transpl Surg 1998; 4:208-14.
9. Manez R, White LT, Linden P et al. The influence of HLA matching on cytomegalovirus hepatitis and chronic rejection after liver transplantation. Transplantation 1993; 55:1067-71.
10. Paya CV, Wiesner RH, Hermans PE et al. Lack of association between cytomegalovirus infection, HLA matching and the vanishing bile duct syndrome after liver transplantation. Hepatology 1992; 16:66-70.

11. Sher LS, Cosenza CA, Michel J et al. Efficacy of tacrolimus as rescue therapy for chronic rejection in orthotopic liver transplantation ; a report of the US Multicenter Liver Study Group. Transplantation 1997; 64:258-63.

12. Van Vlierberghe H, Troisi R, Colle I et al. Hepatitis C infection-related liver disease: patterns of recurrence and outcome in cadaveric and living-donor liver transplantation in adults. Transplantation 2004; 77:210-14.

13. Gane EJ, Portmann BC, Naoumov NV et al. Long-term outcome of hepatitis C infection after liver transplantation. N Engl J Med 1996; 334:815-20.

14. Neumann UP, Berg T, Bahra M et al. Long-term outcome of liver transplants for chronic hepatitis C: a 10-year follow-up. Transplantation 2004; 77:226-31.

15. Charlton M and Seaberg E. Impact of immunosuppression and acute rejection on recurrence of hepatitis C: results of the National Institute of Diabetes and Digestive and Kidney Diseases Liver Transplantation Database. Liver Transpl Surg 1999; 5:S107-14.

16. Berenguer M, Prieto M, San Juan F et al. Contribution of donor age to the recent decrease in patient survival among HCV-infected liver transplant recipients. Hepatology 2002; 36:202-10.

17. Firpi RJ, Abdelmalek MF, Soldevila-Pico C et al. One-year protocol liver biopsy can stratify fibrosis progression in liver transplant recipients with recurrent hepatitis C. Liver Transpl 2004; 10:1240-47.

18. Charlton M, Seaberg E, Wiesner R et al. Predictors of patient and graft survival following liver transplantation for hepatitis C. Hepatology 1998; 28:823-30.

19. Berenguer M, Prieto M, Rayon JM et al. Natural history of clinically compensated HCV-related cirrhosis following liver transplantation. Hepatology 2000; 32:852-58.

20. Fattovich G, Giustina G, Degos F et al. Morbidity and mortality in compensated cirrhosis type C: a retrospective follow-up study of 384 patients. Gastroenterology 1997; 112:463-72.

21. Chalasani N, Manzarbeitia C, Ferenci P et al. Peginterferon alfa-2a for hepatitis C after liver transplantation: two randomized, controlled trials. Hepatology 2005; 41:289-98.

22. Samuel D, Bizollon T, Feray C. Interferon alfa-2b plus ribavirin in patients with chronic hepatitis C after liver transplantation: a randomized study. Gastroenterology 2003; 124:642-50.

23. Rosen HR, Martin P. Hepatitis C infection in patients undergoing liver retransplantation. Transplantation 1998; 66:1612-16.

24. Rosen HR, Madden JP, Martin P. A model to predict survival following liver retransplantation. Hepatology 1999; 29:365-70.

25. Samuel D, Muller R, Alexander G et al. Liver transplantation in European patients with the hepatitis B surface antigen. N Engl J Med 1993; 329:1842-47.

26. Angus PW, McCaughan GW, Gane EJ et al. Combination low-dose hepatitis B immune globulin and lamivudine provided effective prophylaxis against posttransplantation hepatitis B. Liver Transpl 2000; 6:429-33.

27. Leung N, Lai CL, Chang TT et al. Extended lamivudine treatment in patients with chronic hepatitis B enhances hepatitis B e antigen seroconversion rates: results after three years of therapy. Hepatology 2001; 33:1527-32.

28. Schiff ER, Lai CL, Hadziyannas S et al. Adefovir dipivoxil therapy for lamivudine-resistant hepatitis B in pre and posttransplantation patients. Hepatology 2003; 38:1419-27.

29. Crippin J, Foster B, Carlen S et al. Retransplantation in hepatitis B-a multicenter experience. Transplantation 1994; 57:823-26.

30. Molmenti EP, Netto GJ, Murray NG et al. Incidence and recurrence of autoimmune/alloimmune hepatitis in liver transplant recipients. Liver Transpl 2002; 8:519-26.

31. Duclos-Vallee JC, Sebagh M, Rifai K et al. A 10 year follow up study of patients transplanted for autoimmune hepatitis: histological recurrence precedes clinical and biochemical recurrence. Gut 2003; 52:893-97.

32. Jones DEJ, James OFW, Portmann B et al. Development of autoimmune hepatitis following liver transplantation for primary biliary cirrhosis. Hepatology 1999; 30:53-57.

33. Sanchez EQ, Levy MF, Goldstein RM et al. The changing clinical presentation of recurrent primary biliary cirrhosis after liver transplantation. Transplantation 2003; 76:1583-88.

34. Sylvestre PB, Batts KP, Burgart LJ et al. Recurrence of primary biliary cirrhosis after liver transplantation: histologic estimate of incidence and natural history. Liver Transpl 2003; 9:1086-93.

35. Graziadei IW, Wiesner RH, Batts KP et al. Recurrence of primary sclerosing cholangitis following liver transplantation. Hepatology 1999; 29:1050-56.

36. Goss JA, Shackleton CR, Farmer DG et al. Orthotopic liver transplantation for primary sclerosing cholangitis. A 12-year single center experience. Ann Surg 1997; 225:472-81.

37. Sanchez-Urdazpal L, Batts KP, Gores GJ et al. Increased bile duct complications in liver transplantation across the ABO barrier. Ann Surg 1993; 218:152-58.

38. Sanchez-Urdazpal L, Gores GJ, Ward EM et al. Ischemic-type biliary complications after orthotopic liver transplantation. Hepatology 1992; 16:49-53.

39. Abt P, Crawford M, Desai N et al. Liver transplantation from controlled nonheart-beating donors: an increased incidence of biliary complications. Transplantation 2003; 75:1659-63.

# CHAPTER 21

# Liver Allograft Failure
# Due to Recurrent Disease—Pathology

**Urmila Khettry\* and Atoussa Goldar-Najafi**

## Abstract

L iver transplantation (LT) is an acceptable mode of therapy for
end-stage liver diseases of varying etiology. With the exception
of certain disorders of genetic and toxic etiology, most other
diseases can recur in the liver allograft. The recurrent diseases can share
clinicopathological features of other graft disorders such as acute and
chronic rejection and ischemic injury. Recurrent hepatitis C is a major
problem in the post-LT period. The salient clinicopathological features
of recurrent diseases are discussed.

## Introduction

Many chronic or fulminant liver diseases of varying etiology result
in hepatic failure. For patients with hepatic failure, liver transplantation
(LT) offers the only chance of recovery followed by the possibility of
leading a relatively normal life. However, in a majority of cases, LT only
removes the target organ i.e., the diseased native liver and replaces it
with another target organ, albeit non-diseased, without eradicating the
underlying cause of that disease.

The underlying etiologies of hepatic failure can be: 1. *Infectious,*
notably, the viral hepatitides like chronic hepatitis C (HCV) and
hepatitis B (HBV), 2. *Dysregulated immune mediated disorders,* such
as, autoimmune hepatitis (AIH), primary biliary cirrhosis (PBC)
and primary sclerosing cholangitis (PSC), 3. *Toxic-metabolic diseases,*
particularly, resulting from the excessive use of alcohol, drug-induced
injury and non-alcoholic steatohepatitis (NASH), 4. *Hereditary
disorders* like hemochromatosis, alpha-1-antitrypsin deficiency and
numerous others resulting from congenital errors of metabolism, 5.
*Neoplastic diseases,* 6. *Vascular disorders,* like the Budd-Chiari syndrome and 7. *Others.*

With the exception of the inborn errors of metabolism and
drug-induced hepatic failure, most other liver disorders necessitating
LT may recur in the transplanted organ. These recurrent diseases often
pose a diagnostic problem for the pathologist and clinician alike. The
presence of other transplant-related issues, donor liver's characteristics
and on going immunosuppression may alter the pathological findings.
In addition, because the LT recipients are closely monitored post-LT,
the slightest abnormalities in liver function tests may prompt a biopsy
of the allograft with the possibility of detecting the recurrence of the
original disease in an early possibly non-diagnostic stage. Both these
factors can make a definitive diagnosis on a single specimen a rather
challenging proposition for the pathologist and every effort should
be made to correlate histolological findings with clinical, serological,
radiological and molecular parameters, as needed, as well as observation
of features evolve over time.

## Infectious Diseases

Of all infectious processes resulting in liver failure and necessitating LT, chronic infection with hepatotropic viruses, HCV and HBV,
are most significant. It must be noted that re-infection with the virus
does not justify the usage of the term recurrent hepatitis. For the latter,
hepatitic findings must be demonstrated on allograft biopsies.

### Hepatitis C

HCV is the leading indication for LT worldwide, representing
30%-42% of all liver transplants and a further increase predicted for
the future.[1-5] Graft re-infection following transplantation is universal
and can be detected as early as 3 weeks post-LT.[3]

Histologically, the earliest findings in a typical case are scattered
acidophil bodies and prominent kupffer cells.[6] Lobular disarray and
mild portal and lobular chronic inflammation may be seen as well.
Moderate to severe steatosis is seen in some cases of infection with
genotype 3.[7] The mild lobular hepatitis seen early in the post-LT course
evolves over time with features of chronic hepatitis (Fig. 1) commonly
associated with HCV, i.e., portal chronic inflammation with tight
lymphocytic aggregates, minimal biliary epithelial changes, interface
hepatitis, varying degrees of steatosis and continued lobular hepatitis.
In some cases progressive fibrosis (Fig. 2) develops leading to cirrhosis. The estimated median duration to graft cirrhosis is 10 years with
cumulative risks at 5 and 7 years being 10% and 44%, respectively.[4]
Post-LT disease progression is significantly more rapid than HCV in
a nontransplant setting.[1] Furthermore, an increase in recurrent HCV
related progressive fibrosis has been noted recently.[8,9] Older donor age,
genotype 1B, changes in immunosuppression, lack of azathioprine
and early withdrawl of steroids have been reportedly associated with
fibrosis progression.[4,8-12] Since the histologic disease may not correlate
with viremia or biochemical markers of liver disease, protocol biopsies
have been advocated by some investigators. [4,8,13]

The so called fibrosing cholestatic type (FCH) of recurrent HCV
is seen in a minority, about 4%-9%, of post-LT patients. These patients
present with progressive jaundice and massive increase in viral counts.[1]
Graft failure is inevitable in most cases. Histologically, hepatocytic ballooning and degeneration, cholestasis, brisk ductular reaction, variable
portal fibrosis and mixed portal reaction, usually mild are seen. Ischemic
cholangiopathy and biliary obstruction must be excluded clinically.

An aggressive form of recurrent HCV other than the FCH can also
be seen.[1] In our study on 92 post-LT HCV patients from 1999-2003,
66.2% of patients developed recurrence with either "typical" or "autoimmune hepatitis-like" (AIH-like) features (Fig. 3).[14] AIH-like cases
had high grade plasma cell rich inflammation on allograft biopsies

\*Corresponding Author: Urmila Khettry— Department of Pathology, Lahey Clinic Medical Center, 41 Mall Road, Burlington, MA 01805, USA.
Email: urmila.khettry@lahey.org

*Chronic Allograft Failure: Natural History, Pathogenesis, Diagnosis and Management,* edited by Nasimul Ahsan. ©2008 Landes Bioscience.

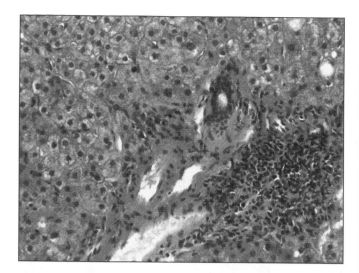

Figure 1. Typical recurrent HCV in an allograft with mild to moderate portal and periportal chronic inflammation and portal lymphoid aggregate (original magnification 25X, hematoxylin and eosin stain).

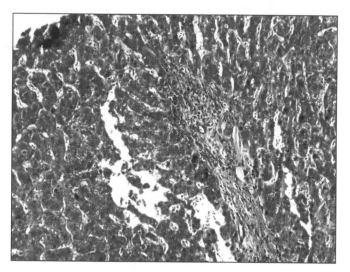

Figure 2. Recurrent HCV with progressive fibrosis in an allograft. Part of an early fibrous septa is depicted in this photomicrograph (original magnification 25X, trichrome stain).

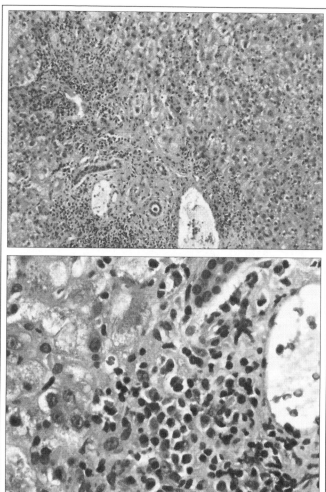

Figure 3. AIH-like recurrent HCV in an allograft with moderate to severe portal, periportal and lobular lymphoplasmacytic inflammation. No significant bile duct injury or venous endothelitis is seen (original magnification upperframe 25X, lowerframe x40, hematoxylin and eosin stain).

with increased incidence of central venulitis. On a single biopsy acute cellular rejection (ACR) was at times difficult to exclude. Progressive fibrosis (stage ≥2) was seen in 39.3% with an increased frequency in the AIH-like group. Native liver inflammation grade ≥2 with plasmacytic periseptitis (Fig. 4) correlated with post-LT progressive fibrosis.

### Hepatitis B (HBV)

Prior to the availability of anti-HBV therapy, the post-LT outcome of HBV patients was dismal.[4,15] Histologically, severe lobular hepatitis progressing rapidly to chronic hepatitis and cirrhosis were seen resulting in graft failure and death usually within 2 years following the transplant. The post-LT course was particularly poor in patients positive for HBV-DNA and or HBV e-antigen before transplantation. A small subset of patients developed the FCH type of recurrent HBV with high viremia. The allograft biopsies from these patients stained strongly with immunostains for HBV surface antigen and depicted features described earlier under the HCV section.

Currently, anti-HBV therapy is highly effective. Overall, fewer HBV patients are undergoing LT and those requiring transplant have an outcome comparable to that seen in other liver allograft recipients.[4,15]

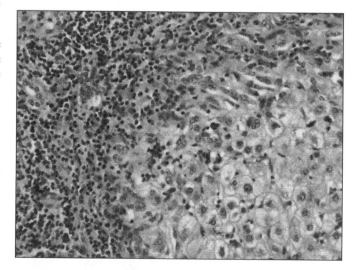

Figure 4. Native liver from an HCV patient depicting periseptal plasma cell-rich chronic inflammation (plasmacytic periseptitis). Post-LT this patient developed progressive AIH-like recurrent HCV (original magnification 25X, hematoxylin and eosin stain).

## Dysregulated Immune-Mediated Diseases

LT is commonly performed for advanced stage liver disorders resulting from dysregulated immunity. There are three diseases that fall under this category: AIH, PBC and PSC. The overall post-LT course for these diseases is generally considered good. Although, it was initially believed that LT might be "curative" for these disorders based on the premise that "self-antigen(s)" and not "allo-antigen(s)" are the target of immune-attack, long-term follow-up has not supported that hypothesis.[16-28] The criteria to optimally diagnose recurrence of these diseases must be strict but still allow the flexibility of observing the development over time on sequential biopsies and take into account other causes of allograft dysfunction.

### Autoimmune Hepatitis

The concept of AIH recurring in an allograft may not be scientifically accurate, still such a designation for the process that resembles the pre-LT disease has been accepted for clinical use and is found helpful in tailoring proper therapy.

The incidence of recurrent AIH is highly variable, 17%-82% and the estimated risk increases over time from 8% during the first year to 68% by 5 years post-LT. The presence of HLA DR3 or DR4 in the recipient and high grade inflammation, grade ≥3, in the native liver have been reported as risk factors/predictors associated with recurrence.[17,20]

The diagnosis of AIH in a nontransplant setting requires a combination of clinical, serological and histological correlation by utilizing the International Autoimmune Hepatitis Group Scoring System.[17] However, this scoring system is of a limited value in post-LT patients where immunosuppressive regimens and other forms of graft dysfunction may modify the serologic and clinicopathological features. Before a diagnosis of recurrent AIH can be made with certainty, other causes of hepatitis should be excluded and the histological features should be correlated with clinical and serological findings.

The earliest histological findings consist of a mild lobular hepatitis with prominent plasma cells.[20] Scattered acidophil bodies often surrounded by lymphoplasmacytic cells can also be seen.[20] These features may mimic hepatitis caused by Epstein Barr virus or certain drugs. In sequential biopsies, lymphoplasmacytic infiltrate, usually brisk, concentrates in the portal and periportal location. A biopsy with these findings may mimic ACR, particularly if the infiltrate is dense and obscures the portal structures. Response, albeit partial, to increased immunosuppressive therapy may further add to the differential diagnostic problem. Presence of plasma cells, interface activity, lack of bile duct injury and absent portal venous endotheliitis favor recurrent AIH. Prominent central venulitis (Fig. 5) with prominent plasma cells is also seen frequently in recurrent AIH along with bridging necrosis. Although, central venulitis without prominent plasma cells can be seen in ACR, bridging necrosis is not a feature associated with ACR. Increased risk of ACR and chronic rejection (CR) have been reported in post-LT AIH patients, however, difficulties in differential diagnosis between recurrent AIH and ACR may have, perhaps in a limited way, contributed to this reported association.

Sequential biopsies may be required for accurate diagnosis and are warranted because recurrent AIH can progress with fibrosis and result in liver failure. In our series of 12 patients, 5 had initial histologically confirmed recurrent AIH between 35-280 days post-LT. Two of the 5 developed progressive fibrosis leading to cirrhosis by 5 and 12 years.

### Primary Biliary Cirrhosis

Recurrent PBC was first described in 1982 based on the pathological findings of "florid duct lesion" (FDL) on allograft biopsies in patients transplanted for PBC.[23] This was followed by reports from several centers refuting the existence of this phenomenon.[1,24] However, over the past decade, based on the results of long-term follow-up from many institutions, a consensus support for recurrent PBC has emerged.[23-25]

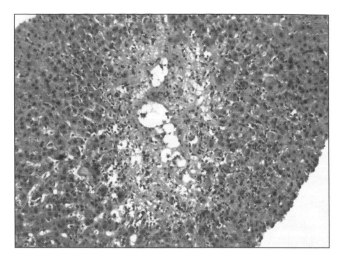

Figure 5. Severe lymphoplasmacytic central venulitis in a patient with recurrent AIH. The photomicrograph shows centrilobular hepatocytic degeneration and necrosis with prominent lymphoplasmacytic inflammation (original magnification 25X, hematoxylin and eosin stain).

The criteria for the diagnosis of PBC in a nontransplant setting are well-defined, however, the recurrence in an allograft is difficult to diagnose.[24] The confounding factors that make the diagnosis of PBC recurrence difficult include: possible association of portal granulomas in ACR within 1 year post-LT, overlapping clinical and histologic features with ACR and CR, persistence of anti mitochondrial antibodies following transplantation and focal/segmental occurrence of the diagnostic FDLs on allograft biopsies.

The gold standard for the diagnosis of recurrent PBC remains histological demonstration of a FDL in at least one allograft biopsy (Fig. 6).[25] In the absence of a diagnostic lesion on an allograft biopsy, moderate to severe lymphocytic cholangitis with biliary epithelial injury and overall a "biliary gestalt" in the proper clinical setting should raise the possibility of recurrent PBC.[1] Plasma-cell rich periportal hepatitis has been reported as an early marker predictive of PBC recurrence.[26] Immunohistochmical evidence for PBC recurrence in the form of apical staining of biliary epithelium with anti PDH-E2 monoclonal antibodies has been reported,[27] however, the sensitivity and specificity of this marker has not been fully elucidated. A minority of patients with recurrence may develop progressive fibrosis. A post-LT syndrome similar to AIH may be seen in some patients.[25]

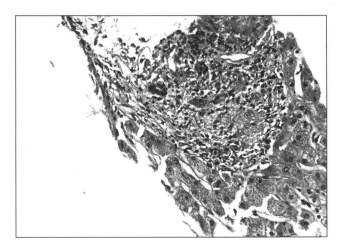

Figure 6. Fragmented liver allograft biopsy from a case of recurrent PBC with a poorly formed florid duct lesion (original magnification 25X, hematoxylin and eosin stain).

In our long-term study involving 43 patients, recurrence with FDL seen in at least 1 post-LT biopsy, was diagnosed in 18.6%.[25] Another group (11.65%) had features more consistent with an AIH-like disorder, termed autoimmune liver disease-not otherwise specified (AILD-NOS). Both these groups had portal plasmacytosis in their earlier biopsies, progressive fibrosis in approximately half of the cases and similar post-LT survival rates. In our opinion, both these post-LT disorders were recurrent autoimmune phenomenon, recurring in the "true" form in the former group and with altered clinicopathological expression in the latter, possibly related to allo-antigens and immunosuppression.

The reported incidence of recurrent PBC is 8%-30% and may be influenced by tacrolimus-based immunosuppression.[1,23,24] Long-term follow-up with sequential histological evaluation is critical in adequately managing all post-LT disorders that might arise in these patients.

### Primary Sclerosing Cholangitis

Doubts and controversies were raised initially regarding the diagnosis of recurrent PSC in liver allografts, but now it is an accepted phenomenon.[1,28-31] The challenge in PSC cases is further complicated by the very nature of the disease process. The characteristic histological finding of fibrous obliterative paucicellular cholangitis is often absent on liver biopsy specimens due to the preferential involvement of larger ducts in most cases. In transplant patients, the diagnosis of recurrent PSC poses even more difficulties because the fibrotic stricturing process can be mimicked by ischemic cholangiopathy, certain infections and reflux cholangiopathy related to Roux loop biliary reconstruction usually employed in these patients. Therefore, the diagnosis of recurrent PSC requires typical cholangiographic features along with distinct biochemical and histological characteristics. Exclusion of other stricturing condition is mandated which include complications resulting from hepatic artery thrombosis, prolonged cold ischemia time, ABO-incompatibility, biliary infections and CR.[28] It is also recommended that a stricturing process in a PSC patient within the post-LT day 90 should not be diagnosed as recurrent PSC.[28]

In a typical case of cholangiographically suspected recurrent PSC, the histological features are those of varying grades of chronic biliary tract obstruction.[29,30] Mild portal edema with mild acute and chronic pericholangitis and peripheral ductular reaction are seen early, while further progression results in periductal lamellar edema and fibrosis with ductopenia (Fig. 7). The inflammatory infiltrate accompanying these biliary changes is generally sparse and mixed with a few neutrophils. Canalicular and intrahepatocytic cholestasis may be seen as well.

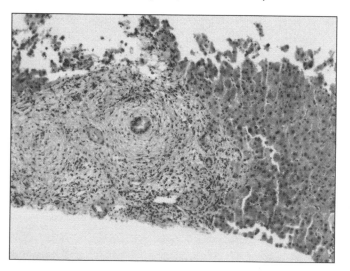

Figure 7. Recurrent PSC in an allograft with periductal lamellar edema, fibrosis and mild chronic inflammation (original magnification 25X, hematoxylin and eosin stain).

Progressive fibrosis may be seen in a minority of patients with ductopenia and biliary fibroobliterative lesions. However, fibroobliterative lesions of interlobular bile ducts are not easily discernible on an allograft biopsy and may also be seen in cases of ischemic cholangiopathy. The periductal changes and ductular reaction are helpful in excluding early CR with paucity of ducts.

PSC recurs in 6%-40% of patients.[30] The post-LT long-term outcome of PSC patients is good with an overall 5-year survival over 80%.[31] In this series, the mean time for the radiological diagnosis was 421 days (range 92-1,275 days) post-LT and for the appearance of histological features 1,380 days (range 420-3,240 days).

Our experience with 42 long-term post-LT PSC patients yielded 6 with a post-LT graft disorder best characterized as recurrent PSC and another 12 with a disorder identical to the AILD-NOS group we had observed earlier with our PBC patients.[25] Progressive fibrosis with cirrhosis was seen in one-third of patients in both groups. Overall, the post-LT histopathological findings in our patients were notable for the inconsistent and incomplete presence of features usually associated with a diagnosis of PSC and its differential diagnostic counterparts such as CR and ischemic cholangiopathy. Evaluation of sequential biopsies with clinical and radiological correlation was absolutely necessary to segregate patients in the different categories. No definite risk factor for PSC recurrence has been defined thus far.

## Toxic and Metabolic Diseases

### Alcoholic Liver Disease

Since alcohol is an exogenous toxin, theoretically, liver transplantation should provide a cure for the end-stage liver disease resulting from its abuse. Although, recidivism or relapse of alcoholism post-LT is a feared complication, it is uncommon for LT recipients to return to alcohol abuse.[32] Some patients with alcoholic liver disease also have HCV. Recurrent HCV, particularly with genotype 3, may present with severe steatosis.[7] Exclusion of alcohol abuse in these cases is of utmost importance because this HCV genotype is amenable to treatment.

### Non Alcoholic Steatohepatitis (NASH)

There are now several reports describing pathologically confirmed recurrence of NASH in liver allografts.[33,34] Furthermore, a significant number of patients with cryptogenic cirrhosis have NASH as the underlying etiology of their end-stage liver disease and can develop recurrence with steatosis and fibrosis in their allografts.[35,36] The appearance and progression of NASH in the allograft may be further hastened by the immunosuppressive therapy, which in most cases tends to be diabeticogenic.

## Hereditary Disorders

LT is an effective and curative treatment for numerous hereditary disorders that may or may not be associated with liver disease in the patient. For genetic diseases such as Wilson disease, alpha-1-antitrypsin deficiency, hereditary amyloidosis and others, the transplanted liver corrects the defect by providing the normal unmutated enzyme or protein.[37-43] As a result, post-LT recurrence of these diseases is not an issue. However, some other hereditary diseases that are associated with liver disease and hepatic failure but have their primary site of defect at a site other than the liver may not be cured following LT.

### Hemochromatosis

Although there are many genetic disorders that can result in hepatic iron-overload, in practice, the most frequently encountered form is the *HFE*-related hemochromatosis. Still, end-stage *HFE*-related hemochromatosis is an uncommon indication for LT. Since the primary iron absorption defect is in the small intestine, theoretically, the new transplanted liver should not be able to reverse the underlying genetic abnormality and therefore, the defective iron absorption continues following LT. Although, some studies have reported post-LT recurrent iron accumulation in patients with iron overload in native livers, others have failed to observe either biochemical or histological evidence of post-LT iron overload in these patients.[44] These differences can be explained by a variety of reasons: (1) underlying genetic heterogeneity of the iron overload disorder; (2) varying grades of pre-LT iron-depletion; (3) wide differences in perioperative usage of blood products; (4) poor post-LT survival of these patients due to infectious and cardiac complications; and (5) varying, but generally short post-LT follow-up of patients in most studies.[45-48]

### Porphyrias

Of all known porphyrias, erythropoietic protoporphyria (EPP) may involve the liver. The underlying cause of EPP is the deficiency of enzyme ferrochelatase and as a result protoporphyrin accumulates in the bone marrow. Some patients with EPP may develop chronic liver disease due to the toxic effects of protoporphyrin on the hepatobiliary structures necessitating LT. Recurrence of EPP in the transplanted liver has been well-documented.[49,50] Recurrent EPP is diagnosed based on the presence of birefringent pigment on liver allograft biopsy. In one study, 11 of 17 patients had such deposits on liver allograft biopsies and 3 required reLT for recurrent EPP.[50]

## Malignant Tumor

End-stage chronic liver diseases may be associated with the development of carcinoma in the liver. While, carcinoma arising in most cirrhotic livers is hepatocellular in origin, the malignancy complicating long-standing primary sclerosing cholangitis is usually cholangiocarcinoma (CC). The tumors in these cases may be known prior to LT or may be discovered incidentally by the pathologist during the evaluation of the native liver. Because of the poor post-LT survival associated with CC and high stage (stage ≥3) hepatocellular carcinoma (HCC), transplantation is seldom performed in a patient known to harbor one of these tumors.[51] However, because of the scoring advantage given to patients with low-stage (stage ≤2) HCC in the current MELD era, there is an increase in HCC patients undergoing LT.[52] Despite the use of Milan criteria at most centers, HCC recurrence rate of 17% at 4 years has been reported.[53]

Recurrent HCC may be seen within the transplanted liver or at distant sites.[52] The histopathological diagnosis of recurrent HCC is usually not a challenge for the pathologist. However, it is prudent to remember that radiologically detected tumor in a few of these patients could originate in other sites and may not represent recurrent HCC. Therefore, every attempt should be made to examine the original native liver tumor slides and if needed, employ ancillary tests such as immunohistochemistry to confirm the origin of the post-LT tumor.

A careful examination of the native liver to determine the accurate pathological staging and other characteristics of HCC is the most important contribution that a pathologist can make in the post-LT management of these patients. Native liver HCC features associated with increased recurrence and/or poor survival include: macroscopic vascular invasion, poor histologic grade, tumor cumulative size of 8.0 cm and tumor expression of p53 and Ki-67.[53-55] The size and number of HCC are the key factors in assigning them the favorable stage in the MELD scoring system. Understaging by pre-operative radiological studies occurs in about 21% of patients and patients found to be substantially outside of the Milan criteria have a significant risk (50%-60%) of recurrence at 4 years.[56] It is important for the pathologist to know about the history of any pre-LT ablative therapy because the mass detected on the gross examination of the native liver might be larger in size than that documented on pre therapy radiologic studies due to expansile necrosis and added fibrosis. Also, pathologically confirmed 100% tumor necrosis in these cases is associated with very low recurrence rate.[57]

## Vascular Disorders

Thrombotic disorders, particularly chronic Budd-Chiari syndrome is an uncommon indication for LT and can recur in the allograft.[58] In most instances, the patient must remain on optimally controlled anticoagulation in the post-LT period. Any recurrence of thrombosis involving the vascular system during the early post-operative period following live donor liver transplantation can have catastrophic results with infarction of varying portions of the partial graft.[59]

## Conclusion

LT is a valid therapeutic option for liver failure from varying etiologies. Recurrence of the original disease is seen in some patients. Histological confirmation of diagnosis may require sequential biopsies. In recent years, a more rapid progression with fibrosis and cirrhosis has been noted with HCV recurrence.

## References

1. Banff Working Group. Liver biopsy interpretation for causes of late liver allograft dysfunction. Hepatology 2006; 44:489-501.
2. Aboulioud MS, Escobar F, Douzdijian V et al. Recurrent disease after liver transplantation. Transplant Proc 2001; 33:2716-19.
3. Montablano M, Neff GW. An update in liver transplantation in patients with hepatitis B and hepatitis C. Minerva Gastroenterologica E Dietologica 2005; 51:109-26.
4. Berenguer M. Recurrent allograft disease: viral hepatitis. Acta Gastro-Enterologica Belgica 2005; 68:337-46.
5. Charlton M. Recurrence of hepatitis C infection: where are we now? Liver Transpl 2005; 11:S57-62.
6. Khettry U, Robiou C, Jenkins R et al. Recurrent hepatitis C in liver allografts: early histologic indicators and correlation with HCV RNA in liver tissue. Int J Surg Pathol 1998; 6:197-204.
7. Gordon FD, Pomfret EA, Pomposelli JJ et al. Severe steatosis as the initial histologic manifestation of recurrent hepatitis C genotype 3. Hum Pathol 2004; 35:636-38.
8. Berenguer M, Ferrell L, Watson J et al. HCV related fibrosis progression following liver transplantation: increase in recent years. J Hepatol 2000; 32:673-84.
9. Alonso O, Loinaz C, Abradelo M et al. Changes in the incidence and severity of recuurent hepatitis C after liver transplantation over 1990-1999. Transplant Proc 2003; 35:1836-37.
10. Hunt J, Gordon FD, Lewis WD et al. Histologic recurrence and progression of hepatitis C after orthotopic liver transplantation: influence of immunosuppressive regimens. Liver Transpl 2001; 7:1056-63.
11. Neumann UP, Berg T, Bahra M et al. Long-term outcome of liver transplants for chronic hepatitis C: a 10 year follow-up. Transplantation 2004; 77:226-31.
12. Gordon FD, Poterucha JJ, Germer J et al. Relationship between hepatitis C genotype and severity of recurrent hepatitis C after liver transplantation. Transplantation 1997; 63:1419-22.
13. Firpi RJ, Abdelmalek MF, Soldevila-Pico C et al. One-year protocol liver biopsy can stratify fibrosis progression in liver transplant recipients with recurrent hepatitis C infection. Liver Transplant 2004; 10:1240-47.
14. Khettry U, Huang W-Y, Simpson MA et al. Patterns of recurrent hepatitis C after liver transplantation in a recent cohort of patients. Hum Pathol 2006; in press.
15. Wang ZF, Zhu ZJ, Shen ZY. Advances in prophylaxis and treatment of recurrent hepatitis B after liver transplantation. Hepatobiliary Pancreat Dis Int 2005; 4:509-14.
16. Duclos-Vallee JC. Recurrence of autoimmune hepatitis, primary biliary cirrhosis and primary sclerosing cholangitis after liver transplantation. Acta Gastroenterol Belg 2005; 68:331-36.
17. Gonzalez-Koch A, Czaja AJ, Carpenter HA et al. Recurrent autoimmune hepatitis after orthotopic liver transplantation. Liver Transpl 2001; 7:302-10.
18. Ratziu V, Samuel D, Sebagh M et al. Long-term follow-up after liver transplantation for autoimmune hepatitis: evidence of recurrence of primary disease. J Hepatol 1999; 30:131-41.
19. Vogel A, Heinrich E, Bahr MJ et al. Long-term outcome of liver transplantation for autoimmune hepatitis. Clin Tranaplant 2004; 18:62-69.
20. Ayata G, Gordon FD, Lewis D et al. Liver Transplantation for autoimmune hepatitis: a long-term pathologic study. Hepatology 2000; 32:185-92.
21. Prados E, Cuervas-Mons V, de la Mata M et al. Outcome of autoimmune hepatitis after liver transplantation. Transplantation 1998; 66(12):1645-50.

22. Neuberger J. Transplantation for autoimmune hepatitis. Seminars in Liver Disease 2002; 22(4):379-86.

23. MacQuillan GC, Neuberger J. Liver transplantation for primary biliary cirrhosis. Clin Liver Dis 2003; 7:941-56.

24. Neuberger J. Liver transplantation for primary biliary cirrhosis. Autoimmunity Reviews 2003; 2:1-7.

25. Khettry U, Anand N, Faul PN et al. Liver transplantation for primary biliary cirrhosis: a long-term pathologic study. Liver Transpl 2003; 9:87-96.

26. Sebagh M, Farges O, Dubel L et al. Histological features predictive of recurrence of primary biliary cirrhosis after liver transplantation. Transplantation 1998; 65:1328-33.

27. Van de Water J, Gerson LB, Ferrell LD et al. Immunohistochemical evidence of disease recurrence after liver transplantation for primary biliary cirrhosis. Hepatology 1996; 24:1079-84.

28. Wiesner RH. Liver transplantation for primary sclerosing cholangitis: timing, impact of inflammatory bowel disease and recurrence of disease. Best Pract Res Clin Gastroenterol 2001; 15:667-80.

29. Khettry U, Keaveny A, Goldar-Najafi A et al. Liver transplantation for primary sclerosing cholangitis: a long-term clinicopathologic study. Hum Pathol 2003; 34:1127-36.

30. Bjoro K, Brandsaeter B, Foss A et al. Liver transplantation in sclerosing cholangitis. Semin Liver Dis 2006; 26:69-79.

31. Graziadei IW, Wiesner RH, Batts KP et al. Recurrence of primary sclerosing cholangitis following liver transplantation. Hepatology 1999; 29:1050-56.

32. Goldar-Najafi A, Gordon FD, Lewis WD et al. Liver transplantation for alcoholic liver disease with or without hepatitis C. Int J Surg Pathol 2002; 10:115-22.

33. Burke A, Lucey MR. Non-alcoholic fatty liver disease, non-alcoholic steatohepatitis and orthotopic liver transplantation. Am J Transpl 2004; 4:686-93.

34. Kim WR, Poterucha JJ, Porayko MK et al. Recurrence of non-alcoholic steatohepatitis following liver transplantation. Transplantation 1996; 62:1802-5.

35. Ayata G, Gordon FD, Lewis WD et al. Crytogenic cirrhosis: clinicopathologic findings at and after liver transplantation. Hum Pathol 2002; 33:1098-104.

36. Ong J, Younossi ZM, Reddy V et al. Crytogenic cirrhosis and posttransplantation non-alcoholic fatty liver disease. Liver Transpl 2001; 7:797-801.

37. Wang XH, Cheng F, Zhang F et al. Copper metabolism after living related liver transplantation for Wilson's disease. World J Gastroenterol 2003; 9:2836-38.

38. Wang XH, Zhang F, Li XC et al. Eighteen living related liver transplants for Wilson's disease: A single-center. Transplant Proc 2004; 36:2243-45.

39. Asonuma K, Inomata Y, Kasahara M et al. Living related liver transplantation from heterozygote genetic carriers to children with Wilson's disease. Pediatr Transplant 1993; 3:201-5.

40. Goss JA, Stribling R, Martin P. Adult liver transplantation for metabolic liver disease. Clin Liver Dis 1998; 2:187-210.

41. Kumar KS, Lefkowitch J, Russo MW et al. Successful sequential liver and stem cell transplantation for hepatic failure due to primary AL amyloidosis. Gastroenterology 2002; 122:2026-31.

42. Burdelski M, Ullrich K. Liver transplantation in metabolic disorders: summary of the general discussion. Eur J Pediatr 1999; 158:S95-96.

43. Kayler LK, Merion RM, Lee S et al. Long-term survival after liver transplantation in children with metabolic disorders. Pediatr Transplant 2002; 6:295-300.

44. Brandhagen DJ, Alvarez W, Therneau TM et al. Iron overload in cirrhosis-HFE genotypes and outcome after liver transplantation. Hepatology 2000; 31(2):456-60.

45. Parolin MB, Batts KP, Weisner RH et al. Liver allograft iron accumulation in patients with and without pretransplantation hepatic hemosiderosis. Liver Transpl 2002; 8:331-39.

46. Crawford DHG, Fletcher LM, Hubscher SG et al. Patients and graft survival after liver transplantation for hereditary hemochromatosis: implications for pathogenesis. Hepatology 2004; 39:1655-62.

47. Whittington CA, Kowdley KV. Review article: haemochromatosis. Aliment Pharmacol Ther 2002; 16:1963-75.

48. Kowdley KV, Brandhagen DJ, Gish RG et al. Survival after liver transplantation in patients with hepatic iron overload: the national hemochromatosis transplant registry. Gastroenterology 2005; 129:494-503.

49. McGuire BM, Bonkovsky HL, Carithers RL et al. Liver transplantation for erythropoietic protoporphyria liver disease. Liver Transpl 2005; 11:1590-96.

50. Meerman L, Haagsma EB, Gouv ASH et al. Long-term follow-up after liver transplantation for erythropoietic protoporphyria. Eur J Gastroenterol Hepatol 1999; 11:431-38.

51. Nissen NN, Cavazzoni E, Tran TT et al. Emerging role of transplantation for primary liver cancers. Cancer J 2004; 10:88-96.

52. Khettry U, Azabdaftari G, Simpson MA et al. Impact of model for end-stage liver disease (MELD) scoring system on pathological findings at and after liver transplantation. Liver Transpl 2006; 12:958-65.

53. Merli M, Nicolini G, Gentili F et al. Predictive factors of outcome after liver transplantation in patients with cirrhosis and hepatocellular carcinoma. Transplant Proc 2005; 37:2535-40.

54. Guzman G, Alagiozian-Angelova V, Layden-Almer JE et al. p53, Ki 67 and serum alpha feto-protein as predictors of hepatocellular carcinoma recurrence in liver transplants patients. Mod Pathol 2005; 18:1498-503.

55. Zavaglia C, De Carlis L, Alberti AB et al. predictors of long-term survival after liver transplantation for hepatocellular carcinoma. Am J Gastroenterol 2005; 100:2708-16.

56. Roberts JP. Tumor surveillance-what can and should be done? Screening for recurrence of hepatocellular carcinoma after liver transplantation. Liver Transpl 2005; 11:S45-46.

57. Sotiropoulos GC, Malago M, Molmenti EP et al. Disease course after liver transplantation for hepatocellular carcinoma in patients with complete tumor necrosis in liver explants after performance of bridging treatments. Eur J Med Res 2005; 10:539-42.

58. Halff G, Todo S, Tzakis AG et al. Liver transplantation for the Budd-Chiari syndrome. Ann Surg 1990; 211:43-49.

59. Pantanowitz L, Pomfret EA, Pomposelli JJ et al. Pathologic analysis of right lobe graft failure in adult-to-adult donor liver transplantation. Int J Surg Pathol 2003; 11:283-94.

# CHAPTER 22

# Chronic Allograft Enteropathy

**Gonzalo P. Rodriguez-Laiz and Kishore R. Iyer***

## Abstract

Current immunosuppression has made small bowel transplantation the standard of care for patients with short bowel syndrome who face complications of total parenteral nutrition (TPN). With standardization and refinement of operative techniques and improvements in rates and outcomes of acute cellular rejection, attention is rightfully turning to late outcomes and quality of life of long-term survivors. Despite seemingly adequate maintenance therapy, chronic rejection affects the long-term outcome post intestinal transplantation and can lead to graft loss with attendant morbidity and even mortality. Pari passu with this, comes the issue of retransplantation.

The findings of chronic rejection are subtle in mucosal (endoscopic) biopsies and a clinical presentation of worsening nutritional status in a patient who was previously doing well, coupled with a high index of suspicion, can precede the pathologic findings of this entity. The diagnosis is more reliably made on full-thickness biopsies, generally obtained after exploratory laparotomy for obstructive symptoms, or on explant of the whole allograft.

The final consequences of chronic rejection are graft loss and resumption of TPN and/or retransplantation. In this chapter, we will discuss chronic rejection in small bowel transplantation, from the clinician's perspective and will try to show the many hurdles faced during the evaluation and treatment of this complex issue.

## Introduction

Small bowel transplantation is the standard of care for patients with intestinal failure, who develop TPN-related complications, regardless of the etiology. These complications are not benign and can range from lack of vascular access, most commonly due to recurrent line sepsis and thrombosis, to TPN-associated cholestasis and liver failure. This paradigm recognizes that for the majority of patients with intestinal failure, TPN is life saving and allows long survival with a good quality of life; it is only the subset of patients with potentially life-threatening complications of TPN who are suitable candidates for intestinal transplantation.

Despite earlier claims favoring combined liver and intestinal transplantation over the use of isolated intestinal grafts,[1,2] the data actually supports the concept of isolated intestinal transplantation prior to the development of liver failure,[3,4] or even as rescue therapy once cholestasis and hepatic fibrosis (or even early cirrhosis) have already ensued.[5] The theoretical advantage of simultaneous liver replacement conferring a protective immunological benefit on intestinal graft survival has not translated into actual survival advantage for patients receiving combined liver-intestinal transplants as opposed to those receiving isolated intestinal transplants.[3,6]

In spite of significant improvement in immunosuppressive therapy, rejection remains the leading cause for graft loss amongst intestinal transplant recipients. Acute rejection is more commonly seen as early as the second post-transplant week and can result in graft loss within the first two months after transplantation.[7] Chronic rejection, on the other hand, may also start early, but generally is responsible for graft loss beyond the first year mark and most commonly over two years post transplant.[7,8] It must be noted that in intestinal transplantation, the consequences of uncontrolled rejection and over-immunosuppression may often be indistinguishable from uncontrolled sepsis, as one is confronted with an increasingly immunosuppressed patient who has an intestinal allograft that is at least patchily devoid of a mucosal barrier which allows unrestricted translocation of infectious organisms in an exquisitely vulnerable patient.

## Epidemiology

It is hard to accurately quantify the magnitude of the problem, since the denominator is not very large. In a large series of 172 intestinal transplant recipients, Parizhskaya et al found 15 cases (8%) of chronic rejection by pathologic examination of the resected allograft, full thickness biopsies, or autopsy specimens.[9]

There have been several well written papers describing the different findings, both clinical and pathological, that affect the intestinal allograft, with a good correlation between the presenting symptoms and the gross and microscopic findings, in human subjects[9,10] and in experimental animal models.[11-13]

## Pathophysiology

There is no single proven mechanism that leads to the development of chronic rejection in intestinal allografts. It is not even clearly established that this process is a true or pure immunologic event, although several studies have shown an increased rate of chronic rejection in experimental models without the use of immunosuppressive therapy.[14] Therefore, the term chronic allograft enteropathy seems better suited to adequately describe several changes that affect the transplanted intestine and ultimately render it unable to adequately support the nutritional needs of the recipient.

## Clinical Presentation

The symptoms that are widely recognized as associated with the diagnosis of chronic allograft enteropathy include:
- Persistent diarrhea
- Intestinal pseudo-obstruction
- Nonhealing mucosal ulcers
- Late development of feeding intolerance
- Late malnutrition requiring TPN
- Gastrointestinal obstruction secondary to enteric strictures

*Corresponding Author: Kishore R. Iyer—Adult and Pediatric, Intestinal Transplant and Rehabilitation, Recanati/Miller Transplantation Institute, The Mount Sinai Medical Center, One Gustave L. Levy Place, New York, New York 10029, USA. Email: kishore.iyer@msnyuhealth.org

*Chronic Allograft Failure: Natural History, Pathogenesis, Diagnosis and Management*, edited by Nasimul Ahsan. ©2008 Landes Bioscience.

These symptoms can occur in isolation or combined and in some instances, there might be a previous history of acute cellular rejection. However, that is not a prerequisite and intestinal grafts can progress to the development of chronic enteropathy without having suffered any prior clinical episode of acute cellular or humoral rejection. Notwithstanding the above, two of the risk factors that have been clearly associated with the development of chronic rejection are the occurrence of an acute rejection episode within 30 days of transplantation and a higher number of acute rejection episodes.[9]

Early symptoms suggestive of chronic intestinal graft dysfunction may be subtle and only regular surveillance and close follow-up of these patients can help reveal these clinical findings. In addition, it is important to recognize that late intestinal graft dysfunction can also be the result of many other unrelated pathologies, both infectious (viral syndromes, bacterial enteritis and parasitic diseases) and non-infectious (atherosclerotic disease, thrombotic events, mechanical obstruction due to adhesions, etc). However, the majority of late intestinal graft losses can be attributed to what has been termed chronic rejection.

The typical patient with chronic allograft enteropathy will show some evidence of weight loss and increased appetite, when there is no significant peritoneal scarring and no loss of motility. In other instances, native proximal gastrointestinal distension (gastric and duodenal) as a consequence of the impaired intestinal motility may be the first ominous sign of impending chronic enteropathy, as demonstrated in acute and chronic rejection in experimental models.[15,16]

The typical pathologic changes of chronic allograft enteropathy are confined to submucosal and muscular planes—hence a mucosal biopsy is insufficient to demonstrate the typical findings associated with this pathology and a full thickness biopsy may be indicated. In other occasions, suspected adhesions causing obstructive symptoms force exploratory laparotomy and dilated or shrunken loops of small bowel, as well as generalized fibrosis and scarring, are good indicators of the gross diagnosis.

## Pathology

The gross pathological findings have been described as features compromising the full thickness of the graft and thus can be demonstrated on resected specimens (partial or total enterectomies) obtained at laparotomy or on postmortem examination. The most striking gross feature has been termed sclerosing peritonitis and has been seen in both humans,[17] as well as research animal models.[10] In the first of these two reports, the patients underwent exploratory laparotomy for distal ileal obstruction and operative findings included serositis, dense fibrous adhesions and contraction of the mesentery. These findings could have otherwise been attributed to a generalized peritonitis, which could be infectious or postperforation in origin; however, they were only present in the grafts and none of the native intestinal segments had any evidence of inflammatory or fibrous changes, thus demonstrating the immunologic aggression towards the "foreign" viscera. Similar findings were noted in the second report, where the post-mortem examination in the sacrificed animals showed evidence of shortening of the mesentery, with a fibrotic appearance and a slightly white surface. The lymph nodes contained within this mesentery were enlarged, whereas the native bowel did not present any macroscopic changes or any abnormality.

In our own experience, the bowel grafts that fail as a consequence of chronic allograft enteropathy and are resected, uniformly show shrunken small bowel loops, with absence of motility and some degree of peritoneal fibrosis, as seen in Figure 1. The bowel loops also tend to be clumped and are firm to the touch. The subsequent en-bloc resection is generally a tedious and slow process, since the chronic inflammatory response fixes the graft to the peritoneal wall and adjacent viscera and extreme care has to be taken to complete the enterectomy without injuring any adjacent organ and key retroperitoneal structures.

Unfortunately, these findings are just the macroscopic proof of the damage and offer no help in an attempt to make a preoperative diagnosis of the disease. Furthermore, it is conceivable that the ability to make an early diagnosis might assist in attempts to initiate aggressive therapy to prevent graft loss due to chronic allograft enteropathy.

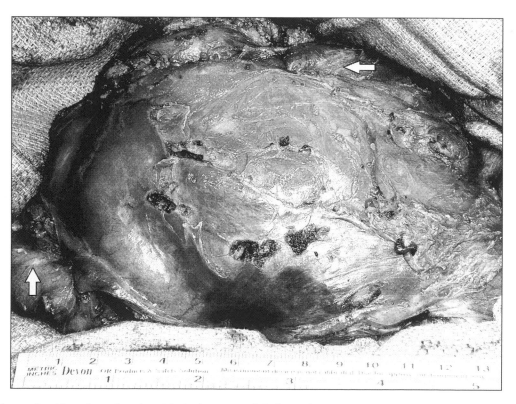

Figure 1. Sclerosing peritonitis at time of explant. Vertical arrow on left shows the native sigmoid colon. Horizontal arrow in upper right shows the right hepatic edge.

However, there are some microscopic findings seen on routine endoscopic biopsies, which could be regarded as early or premature signs of associated or upcoming chronic graft enteropathy and were described by Parizhskaya et al in a report of the evolution of morphologic nonspecific changes observed in serial biopsies taken from patients who ultimately had chronic rejection of their grafts.[9] These findings are:

- Mild fibrosis of the lamina propria.
- Focal loss of the crypts of Lieberkuhn.
- Granulation tissue replacement of the lamina propria.
- Thickened, flattened, or blunt villi.

Late findings are those associated with vascular damage, which is most notorious on the mesenteric side of the graft and thus necessitates a full thickness biopsy. This is usually obtained during a partial small bowel resection (typically in a case of chronic rejection that manifests as small bowel obstruction) or in an explanted graft.

The findings during exploratory laparotomy are those that one might expect when the diagnosis of chronic allograft enteropathy is suspected. The bowels appear hypomotile and can be dilated, but most commonly are normal or smaller in caliber, with a densely fibrotic surface and often, a rubbery consistency. There may be focal areas of strictures, which are usually firm and not suitable for stricturoplasty. Lysis of adhesions can be performed, but offers no help since the problem is not obstruction due to adhesions, but rather the hampered intestinal motility, as demonstrated by Pernthaler et al[18] and subsequently by Klaus et al in a small bowel rat model.[16]

The chronically rejected allograft bowel is often a densely matted phlegmon that may encase important retroperitoneal structures such as the ureters in the fibrous process. The resected specimen often does not permit identification of individual bowel loops. The pathognomonic microscopic features listed below are confirmatory for the diagnosis, but are not always present; the gross appearance of the explant with poorly-distinguished "pan-fibrosis" suffices to make the diagnosis.

The microscopic findings distinctive for chronic rejection in small bowel transplantation have been described in several reports.[19,20] The most commonly found features have generally a late appearance (more than three months post transplant) and may include (see Figures 2-6):

- Loss of villous architecture.
- Obliterative arteriopathy involving medium-sized vessels
- Eccentric subendothelial thickening.
- Loss of crypts, that may also show regenerative and/or metaplastic changes.
- Nonhealing mucosal ulcers.

These changes may show a slow progression, ultimately leading to complete loss of the functional capacity of the small bowel graft. Thus, the transplanted intestine becomes progressively unable to support the nutritional needs of the individual.

There is a huge paucity of data on the mechanisms of human intestinal transplant rejection—while an immune basis is suspected, the relative role of cellular and humoral immune processes remains to be clarified.

Clinical outcomes following chronic rejection of the intestinal allograft are limited to small case series and essentially reflect the poor outcome of intestinal retransplantation. More recent series confined to small numbers of patients, suggest some improvement in outcome with 9 of 12 pediatric patients being alive a year following retransplantation at Pittsburgh (Mazariegos, personal communication).

## Conclusions and Outlook for the Future

In the decade or so since the advent of clinical human intestinal transplantation, considerable progress has been achieved resulting in improved early survival among patients being transplanted for complications derived from TPN. There is a trend towards earlier transplantation of patients before they manifest a progression of their TPN-related complications, with an expectation of further improvement in the overall outcome.

As early survival improves and earlier survivors are now going well past the 5-year mark, there is an increasing recognition of late graft loss due to chronic rejection of the small bowel allograft. The absence of pathognomonic microscopic changes in all cases of chronic allograft enteropathy has led to a shift towards defining chronic rejection as a clinical syndrome of poor late graft function, accompanied by exten-

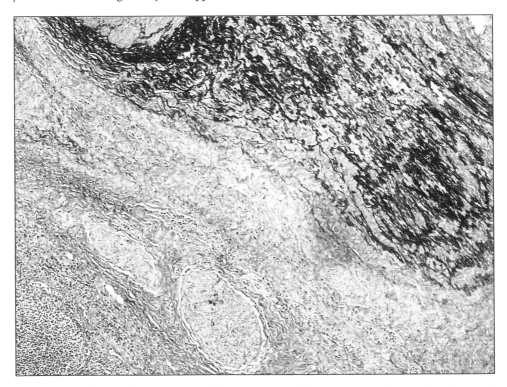

Figure 2. Chronic inflammatory infiltrate, along with neural hyperplasia and fibrosis, characteristic of chronic rejection. Elastin, 20X.

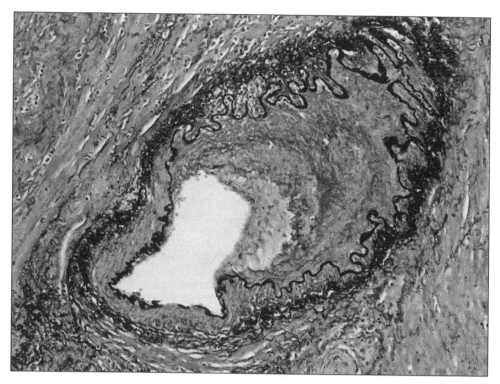

Figure 3. Eccentric intimal proliferation and thickening in a medium size artery. Elastin, 40X.

sive and significant fibrosis in the bowel and often, but not invariably, distinctive vascular changes in the medium sized vessels.

The recent advances in immunosuppression, with the addition of antibody induction in most cases, along with improvements in the early outcomes after intestinal transplantation, will surely expose us more often to chronic allograft enteropathy. Along with this will be the opportunity to reduce the approximately 10% late graft loss it currently causes in intestinal transplantation.

*Acknowledgement*

The authors wish to acknowledge the invaluable help provided by Dr. Raffaella Morotti, Associate Professor of Pathology at the Lillian and Henry M. Stratton-Hans Popper Department of Pathology, in supplying the microscopic illustrations for this chapter.

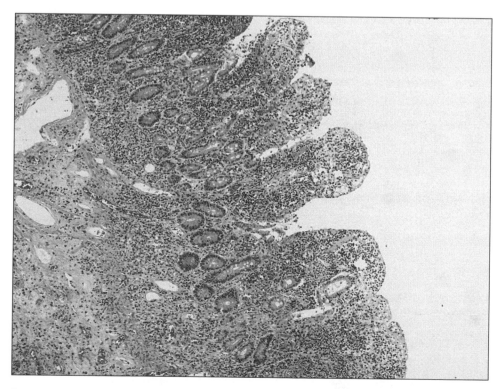

Figure 4. Chronic inflammatory infiltrate in the lamina propria, with loss of surface epithelium, blunting of the villi and crypt loss. H&E, 10X.

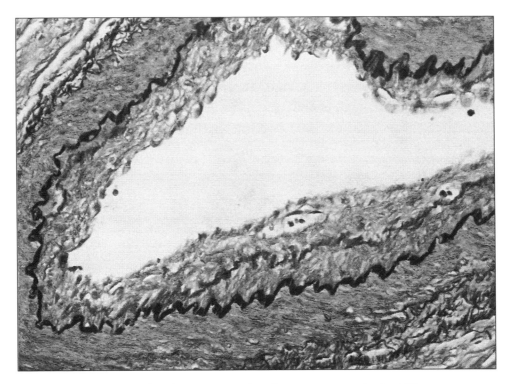

Figure 5. Intimal hyperplasia and narrowing of the lumen in a medium/large size artery. Elastin, 40X.

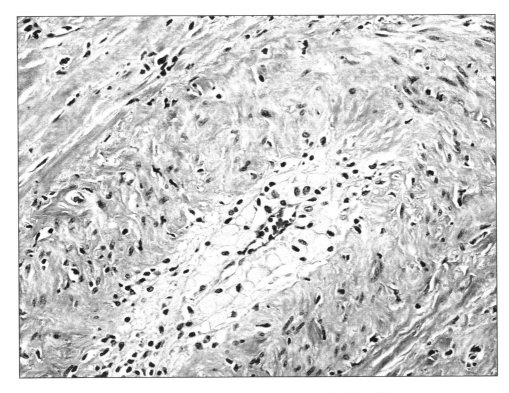

Figure 6. Foamy macrophage infiltration in the intima of a small submucosal vessel. H&E, 20X.

## References

1. Goulet O, Jan D, Lacaille F et al. Intestinal transplantation in children: preliminary experience in Paris. JPEN J Parenter Enteral Nutr 1999; 23(5 Suppl):S121-25.
2. Abu-Elmagd K, Reyes J, Bond G et al. Clinical intestinal transplantation: a decade of experience at a single center. Ann Surg 2001; 234(3):404-16; discussion 416-17.
3. Pomfret EA, Fryer JP, Sima CS et al. Liver and Intestine Transplantation in the United States, 1996-2005. Am J Transplant 2007; 7:1376-89.
4. Nishida S, Levi D, Kato T et al. Ninety-five cases of intestinal transplantation at the University of Miami. J Gastrointest Surg 2002; 6(2):233-39.
5. Sudan DL, Kaufman SS, Shaw BW Jr et al. Isolated intestinal transplantation for intestinal failure. Am J Gastroenterol 2000; 95(6):1506-15.
6. Grant D, Abu-Elmagd K, Reyes J et al. 2003 report of the intestine transplant registry: a new era has dawned. Ann Surg 2005; 241(4):607-13.
7. Noguchi Si S, Reyes J, Mazariegos GV et al. Pediatric intestinal transplantation: the resected allograft. Pediatr Dev Pathol 2002; 5(1):3-21.
8. Iyer KR, Srinath C, Horslen S et al. Late graft loss and long-term outcome after isolated intestinal transplantation in children. J Pediatr Surg 2002; 37(2):151-54.

9. Parizhskaya M, Redondo C, Demetris A et al. Chronic rejection of small bowel grafts: pediatric and adult study of risk factors and morphologic progression. Pediatr Dev Pathol 2003; 6(3):240-50.

10. Klaus A, Margreiter R, Pernthaler H. Diffuse mesenterial sclerosis: a characteristic feature of chronic small-bowel allograft rejection. Virchows Arch 2003; 442(1):48-55.

11. Kuusanmaki P, Lauronen J, Paavonen T et al. How to diagnose chronic rejection. A study in porcine intestinal allografts. Scand J Immunol 1997; 46(5):514-19.

12. Orloff SL, Yin Q, Corless CL et al. A rat small bowel transplant model of chronic rejection: histopathologic characteristics. Transplantation 1999; 68(6):766-79.

13. Ma H, Wang J, Wang J et al. Features of chronic allograft rejection on rat small intestine transplantation. Pediatr Transplant 2007; 11(2):165-72.

14. Langrehr JM, Banner B, Lee KK et al. Clinical course, morphology and treatment of chronically rejecting small bowel allografts. Transplantation 1993; 55(2):242-50.

15. Sugitani A, Bauer AJ, Reynolds JC et al. The effect of small bowel transplantation on the morphology and physiology of intestinal muscle: a comparison of autografts versus allografts in dogs. Transplantation 1997; 63(2):186-94.

16. Klaus A, Klima G, Margreiter R et al. Myoelectric activity during chronic small bowel allograft rejection in rats. Dig Dis Sci 2002; 47(11):2506-11.

17. Macedo C, Sindhi R, Mazariegos GV et al. Sclerosing peritonitis after intestinal transplantation in children. Pediatr Transplant 2005; 9(2):187-91.

18. Pernthaler H, Kreczy A, Plattner R et al. Myoelectric activity during small bowel allograft rejection. Dig Dis Sci 1994; 39(6):1216-21.

19. Lee RG, Nakamura K, Tsamandas AC et al. Pathology of human intestinal transplantation. Gastroenterology 1996; 110(6):1820-34.

20. Demetris AJ, Murase N, Lee RG et al. Chronic rejection. A general overview of histopathology and pathophysiology with emphasis on liver, heart and intestinal allografts. Ann Transplant 1997; 2(2):27-44.

# CHAPTER 23

# Renal Allograft Survival:
## Epidemiologic Considerations

### Titte R. Srinivas and Herwig-Ulf Meier-Kriesche*

## Introduction

Kidney transplantation is the treatment of choice for patients with end-stage renal disease. From initial pioneering experiences 50 years ago, kidney transplantation has become a clinical reality with over a 100,000 patients living with a functioning kidney transplant at the end of 2004 in the United states alone.[1] The growing epidemic of chronic renal disease progressing to ESRD worldwide along with the pandemic of Type 2 diabetes will likely see an increasing number of patients receiving a kidney transplant in the coming decades.

We review the trends in graft and patient survival that have been observed in kidney transplantation over the recent era and integrate this with what is known of the pathophysiology of allograft failure. We will also review the relevant long-term outcomes with the commonly used immunosuppressive regimens in kidney transplantation given their differing impact on graft survival. The relevance of past experiences with immunosuppression and graft survival in the context of the emerging epidemic of polyoma virus nephropathy and the increasing recognition of humoral rejection will also be discussed.

## The Patient Survival Advantage Conferred by Kidney Transplantation

In the initial experiences, renal transplantation was largely viewed as a modality that conferred benefits such as a better quality of life to the patient and a cost saving measure that reduced medical expenditure. Clear cut demonstration of a survival advantage conferred by transplantation was not a straightforward matter. First, randomized trials comparing dialysis versus transplantation were neither feasible nor ethical. Second, the transplant recipient represents a healthier, wealthier and younger subset of the larger dialysis population that includes those patients not deemed suitable transplant candidate.

Wolfe et al conducted an elegant analysis that compared transplanted patients with those that were waitlisted but not transplanted.[2] As depicted in Figure 1, despite the increased short-term risk of death after renal transplantation, cadaveric transplantation conferred a significant long-term survival benefit. These beneficial effects of transplantation were similar in men and women, most marked in diabetics and extended across all other causes of ESRD. Blacks and Native Americans benefited somewhat less that Asians and Whites but the overall benefits extended across all racial subgroups. As could be expected, those in the 20-39 age group benefited the most improvements were nevertheless notable in those over the age of 70.[2] This seminal study transformed the view of kidney transplantation from a procedure that actually prolonged life as opposed to one that merely enhanced its quality.

## Causes of Graft Loss

In the current era, it is reasonable to expect that most kidney allografts will serve their recipients through their life span. Indeed death with preserved kidney function is probably the ultimate goal for all kidney transplant recipients.[3] The predominant causes of kidney transplant failure are summarized in Table 1.

Despite a survival advantage over dialysis, excess mortality in the kidney transplant population is a significant problem and over half of these patients die with a functioning kidney transplant. Approximately 7 percent of the functioning grafts in the United States and Canada can be expected to fail each year.[3] The predominant cause of loss of kidney allograft function remains rejection (graft injury due to alloimune responses to the allograft). Factors contributing to ongoing alloimmune responses include breakdown in immunosuppression as a result of patient noncompliance, or therapeutic decisions to minimize exposure to complications of immunosuppressive drugs [3]. Most kidneys that fail before the death of the patient exhibit a chronic progressive decline in function unaccompanied by any specific pathologic findings. Previously designated as chronic rejection, the umbrella term of chronic allograft nephropathy (CAN) is now preferred. CAN encompasses a variety of features such as fibrointimal vascular sclerosis, glomerulosclerosis, tubular atrophy and interstitial fibrosis with a variable contribution from calcineurin inhibitor nephrotoxicity. A number of these changes overlap with changes associated with ageing and corresponding age related changes in the donor kidney. Both immunologic and non-immunologic factors have been implicated in the pathogenesis of CAN. Of the immunologic risk factors,

*Table 1. Causes of kidney transplant failure*

| | |
|---|---|
| Death with Function | 40-45% |
| Failure of the Transplant Kidney | 55-60% |
| Chronic Allograft Nephropathy (Chronic Transplant Glomerulopathy 5%) | 30% |
| Recurrent or de novo disease (Including BK Virus Nephropathy: 1-10%) | 10% |
| Miscellaneous and mixed picture (Unknown, multifactorial, end-stage renal disease from medical illness) | 10% |
| Technical and thrombosis | 2% |
| Outright rejection | 5% |

Adapted from: Halloran PF et al. In: Weir MR, ed. Medical Management of Kidney Transplantation. 2005;[3] with copyright permission from Lippincott Williams & Wilkins.

*Corresponding Author: Herwig-Ulf Meier-Kriesche—University of Florida, 1600 SW Archer Road, Box 100224, Gainesville FL 32610-0224, USA. Email: meierhu@medicine.ufl.edu

*Chronic Allograft Failure: Natural History, Pathogenesis, Diagnosis and Management*, edited by Nasimul Ahsan. ©2008 Landes Bioscience.

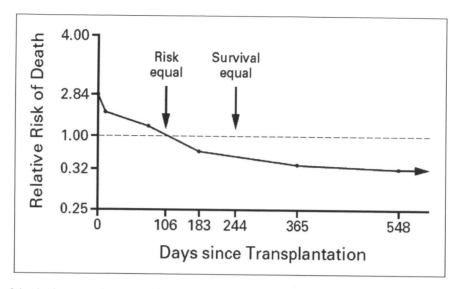

Figure 1. Relative risk of death after transplantation. The reference group comprised waitlisted patients who were not transplanted. From: Wolfe RA et al. N Engl J Med 1999; 341(23):1725-1730;[2] with copyright permission of the Massachusetts Medical Society.

acute rejection is most accepted as the harbinger of future CAN. Acute rejection episodes that are not accompanied by return of renal function to baseline and those that occur in the late posttransplant period and recurrent episodes of acute rejection put patients at the highest risk for CAN. Evidence for immunologic factors in the genesis of CAN is further underscored by the finding that half-life of grafts decreases with increasing HLA mismatch. A growing body of evidence points towards humoral immunity as an important factor underlying the development of CAN. Histologic CAN is strongly associated with C4d staining in allograft biopsies and the presence of anti-HLA antibodies; amelioration of the humoral response may forestall loss of renal function in some cases.[4,5]

Non-immunologic factors include brain death in the donor, ischemia-reperfusion injury, calcineurin inhibitor toxicity, hypertension, diabetes mellitus (posttransplant or preexisting), hyperlipidemia and CMV infection.[6] Polyoma virus nephropathy also may contribute to attrition of graft function although its exact contributions are uncertain.[7]

## Graft Survival

The introduction of azathioprine (AZA) in 1962 heralded the clinical era of renal transplantation with one year graft survival rates between 45 and 50 percent. By the 70's the superior graft survival in living donor transplantation as opposed to cadaveric transplantation became apparent. Further strides in renal transplantation were made with the introduction of cyclosporine (CsA) and the more widespread use of antibody induction in the 80's and better methods to treat acute rejection such as OKT3 (monoclonal anti-CD3). The 90's saw the introduction of newer immunosuppressants such as mycophenolate mofetil (MMF) and tacrolimus (TAC); acute rejection rates reached an unprecedented low. According to the OPTN/SRTR Annual Report reflecting patients transplanted through 2004, overall adjusted graft survival rates for recipients of deceased donor non-ECD kidneys were 91% at one year and 69 percent at 5 years. As a comparison, 1 and 5-year graft survival rates in living donor kidney recipients were 97% and 80 percent respectively (Fig. 2).[1]

In addition to survival rates at discrete time points after transplantation, another widely used and quoted parameter is the graft half-life.[8] The conditioned half-life is defined as the time to loss of fifty percent of allografts that have survived the first year after transplantation.[8] This parameter has been advocated as one that largely measures factors impacting graft survival after the first year posttransplantation into the long-term. Gjertson et al had stated in an early publication that despite improvements in short term graft survival, graft half-life remained remarkably stable at around 8 years.[9]

Hariharan et al. using SRTR data reported that the projected conditioned half-life of allografts improved from 7.9 to 13.8 years for the cadaveric donor transplant and 12.7 to 21.6 years for the living donor transplant in the period spanning 1998 to 1995.[10] These estimates reflected a gain of about 6 years in the projected half lives. This projected improvement in graft survival was noted in an era with little new drug development, modest decrease in acute rejection rates but increasing one-year graft survival rates. Advances in both the ability to use the immunosuppressive armamentarium and the improved delivery of overall care to the transplanted patient were deemed responsible for these projected benefits.[10] However these were projected half-lives based on projections conditioned on 1 year graft survival. Subsequent analysis of real half-lives from the SRTR database in actually showed far more modest improvement in long-term graft survival despite an impressive decrease in acute rejection rates.[11] Indeed, the improvements in the recent era are largely restricted to those subjects receiving retransplants. First transplants showed a cumulative increase in graft survival of less than 6 months. The paradox presented in the current era in renal transplantation is that despite remarkable decreases in acute rejection rates, long-term graft survival has not shown a parallel improvement. This was demonstrated in an analysis of SRTR data pertaining to all adult transplant recipients transplanted between 1995 and 2000.[11] Significant decreases in acute rejection rates were noted in the first 6 months, first year and the second year after transplantation.[11] This impressive decline in acute rejection rates over the study period did not however translate into improved long-term allograft survival. Furthermore, a disturbing statistically significant worsening of death-censored graft survival was noted. Moreover, a disturbing trend was noted in the clinical behavior of the acute rejection episodes. A greater proportion of acute rejection episodes were associated with a lack of restoration of renal function to pre-rejection stable baseline values. Lack of return to baseline level of renal function after treatment of an acute rejection episode was associated with an incremental increase in the relative hazard for death-censored graft survival.[11]

It is well recognized that rejection episodes that are accompanied by return to renal function to baseline after treatment are less likely to be associated with chronic progressive renal dysfunction compared to those that do not. In an earlier study of USRDS data, pertaining to the recent era in transplantation, a greater proportion of patients were noted to have more severe, treatment resistant behavior associated with the acute rejection episodes.[12] The teleological explanation that can be offered to explain this trend is that

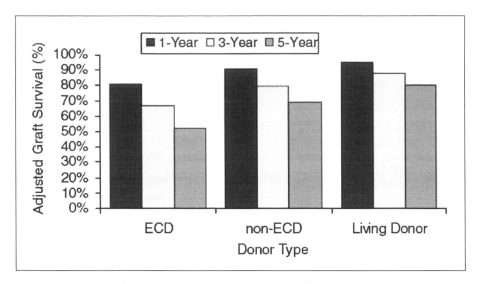

Figure 2. Adjusted 1-year, 3-year and 5-year kidney graft survival, by donor type, for transplants received 1998-2003. From: Cohen DJ et al. Am J Transplant 2006; 6(5 Pt 2):1153-1169;[1] with copyright permission of Blackwell Publishing.

the more potent immunosuppression in use in transplantation in the current era probably selects out the expression of more severe episodes of acute rejection while suppressing the more treatment responsive rejections noted under less potent immunosuppressive regimens of the past.

Taken together, the need to effect a significant shift from the earliest paradigm in transplantation, i.e., early immunologic success portends excellent long-term outcome is probably appropriate. Unfortunately, in the aggregate, early events such as acute rejection and early markers such as renal function, fall short as markers of long-term success.

## Death with a Functioning Allograft

As more elderly patients receive a kidney transplant, death with a functioning kidney transplant is assuming an increasingly prominent position as a cause of late allograft loss. The three most important potentially modifiable causes of death with a functioning transplant are heart disease, infection and malignancy.

The immunosuppressed state that is inevitable with transplantation and the toxicities of individual immunosuppressive medications interact with recipient factors and the ambient level of allograft function in modifying the expression of each of the principal etiologies of death with a functioning allograft. When graft loss is analyzed in a manner independent of patient death (a parameter called death-censored graft loss), a better estimate of effects of risk factors on graft dysfunction and ultimately graft loss can be obtained. However, estimates of rates of death censored graft loss may be influenced in a biased manner by risk factors that influence both patient death and attrition of graft function. As an example, diabetes mellitus and hypertension are both risk factors for patient death and renal insufficiency by synergistic effects in mediating cardiovascular disease.

Extrapolations from clinical practice in the general population are used to develop treatment strategies that influence cardiovascular disease after renal transplantation. These measures include, smoking cessation, reduction of body weight in the obese, glycemic control and aggressive measures against hypertension and hyperlipidemia. While specific evidence supporting the efficacy of individual measures based on results of randomized trials in the transplant population is not readily available, results of the ALERT trial which assessed the effects of lipid lowering in renal transplant recipients on cardiovascular events lend credence to this overall approach. In this trial, the use of fluvastatin in renal transplant recipients reduced somewhat the incidence of cardiac deaths and nonfatal MI without reducing overall cardiovascular mortality or the need for coronary intervention.[13]

## Risk Factors for Patient and Graft Survival

### *Waiting Time and Premptive Transplantation*

An analysis of the USRDS data showed that increasing waiting times for a transplant (on dialysis) adversely impacted both graft and patient survival. This study was conducted using data pertaining to 73,103 primary adult renal transplants registered at the United States Renal Data System Registry from 1988 to 1997 for the primary endpoints of death with functioning graft and death-censored graft loss. A longer waiting time on dialysis emerged a significant risk factor for both death-censored graft survival and patient death with functioning graft after renal transplantation (P <0.001 for each). Relative to pre-emptive transplants, increasing waiting times incrementally increased both mortality risk and risk for death censored graft survival after transplantation.[14]

In a retrospective cohort study of 8,481 patients, pre-emptive transplantation was associated with a 52 percent reduction in the relative hazard for graft failure in the first year after transplantation.[15] These risk reductions were 82 percent in the second year post-transplantation and 86 percent in the subsequent year compared with transplantation occurring after the start of dialysis. Interestingly, increasing duration on dialysis prior to transplantation significantly increased the relative odds for acute rejection within the first 6 months posttransplantation.[15]

In an effort to minimize donor related confounding factors a paired kidney study involving 2405 pairs of kidneys was performed.[16] Each kidney of the pair was transplanted into recipients with differing times on ESRD. Six-antigen matched kidneys were excluded from this study as a disproportionate number of these kidneys, through a national sharing program are transplanted pre-emptively, bypassing the waiting time requirement. Five and ten-year unadjusted graft survival and death censored graft survival were significantly inferior in those subjects with more than 24 months of dialysis time versus those who incurred less than 6 months on dialysis (Fig. 3). These trends remained significant after adjustment for multiple significant confounders that influence waiting time such as high panel reactive antibody, advanced age and African-American race. Indeed part of the advantage of living-donor compared with deceased donor transplantation may be explained by the effect of waiting time. This effect of waiting time is dominant enough that a recipient of a deceased donor kidney with an ESRD time less than 6 months obtains graft survival roughly equivalent to living donor transplant recipients who wait for their transplant on dialysis for more than 2 years.[16]

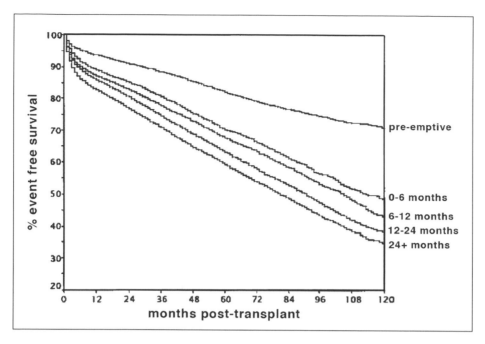

Figure 3. Unadjusted Graft Survival in recipients of cadaveric renal transplants by length of dialysis treatment prior to transplant (n = 56,587). From: Meier-Kriesche HU, Kaplan B. Transplantation 2002; 74(10):1377-1381;[16] with copyright permission of Lippincott Williams & Wilkins.

A follow-up study utilized a nontraditional endpoint, death occurring after graft loss (DAGL). In an analysis involving 78,564 primary renal transplants reported to the USRDS between 1988 and 1998, dialysis for more than two years was associated with a greater than two fold relative hazard for DAGL; transplant time was not associated with risk of DAGL.[17]

In summary, increasing time accrued with ESRD is a strong independent risk factor that adversely affects both graft and patient survival in a dose-dependent fashion. Importantly, this risk factor is potentially the most modifiable in the practical clinical setting and underscores the need to refer all patients with ESRD for transplantation in a timely manner and pre-emptive transplantation before the start of dialysis wherever possible.

### *Cardiovascular Disease Is Determined by the Level of Allograft Function*

Decreased renal function, even of modest severity, is increasingly implicated as a risk factor for cardiovascular death.[18] In renal transplant recipients as well, cardiovascular disease is the most significant cause of death.

The effects of varying levels of allograft function on cardiovascular death after renal transplantation have been demonstrated. In an analysis of 58,900 recipients of a primary renal transplant reported to the USRDS between 1988 and 1998 and who had at least 1 year of graft survival were evaluated for the relationship between the primary study endpoint of cardiovascular death beyond 1 year of transplantation and the level of renal function.[19] Serum creatinine values at 1 year after transplantation were independently associated with the risk for cardiovascular death; serum creatinine values greater than 1.5 mg/dL were associated with a significant and progressive increase in the risk for cardiovascular death (Fig. 4).[19] This risk of cardiovascular death was significantly higher when patients who lost allograft function were included in the analysis. An association between worsening renal function and infectious death was noted. However the level of renal function was not associated with malignancy related death.[19]

The survival advantage conferred by renal transplantation is likely related to abrogation of cardiovascular risk. In a study, first-kidney-transplant patients reported to the USRDS from 1995 to 2000 were

evaluated for the primary endpoint of cardiovascular death by transplant vintage.[20] These subjects were then compared to all 66813 adult kidney waitlisted patients reported in the same time period. Cardiovascular death rates peaked during the first 3 months following transplantation probably reflecting perioperative events and decreased subsequently with increasing transplant vintage. This salutary effect of successful renal transplantation on cardiovascular death was demonstrated in both living and deceased donor transplant recipients. Furthermore, risk for cardiovascular death was successfully reduced by successful renal transplantation even in patients with end-stage renal disease secondary to diabetes; a group known to be at the highest risk for cardiovascular events.

Thus, successful renal transplantation abrogates the cardiovascular risk attendant to renal dysfunction. As a corollary, the level of renal function accrued after transplantation is the primary determinant of the posttransplantation risk of cardiovascular risk. This latter point underscores the need to ensure the most optimal level of renal function that can be achieved and maintained in the renal transplant recipient.

### *Hypertension and Kidney Allograft Survival*

Hypertension in the transplant recipient is multifactorial. Contributors to its genesis and maintenance include, pre-existing hypertension, renal parenchymal hypertension, calcineurin inhibitors, corticosteroids and volume overload.

Progression of glomerular disease in native kidneys occurs with uncontrolled hypertension; control of blood pressure attenuates progression to renal failure in subjects with glomerular disease. In kidney transplant recipients, higher levels of systolic blood pressure either during the first year posttransplant or throughout the posttransplant period correlate with lower graft survival.[21]

Several retrospective studies shave demonstrated associations between high levels of blood pressure and cardiovascular risk or patient survival after renal transplantation. Higher levels of systolic blood pressure after transplantation correlate with progressively inferior patient survival. Well controlled hypertension leads to regression of left ventricular hypertrophy and improvement in carotid atherosclerosis after renal transplantation.

Thus both retrospective studies and prospective evidence point towards hypertension as an important factor that impacts graft and

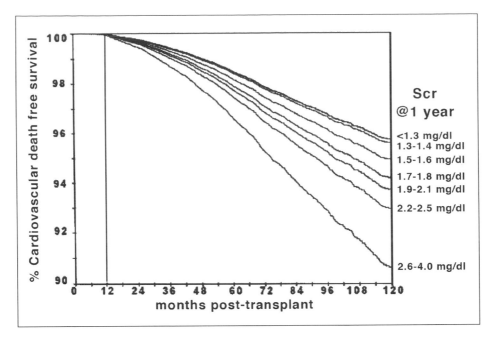

Figure 4. Cardiovascular death by serum creatinine at 1 year posttransplantation. From: Meier-Kriesche HU et al. Transplantation 2003; 75(8):1291-1295;[19] with copyright permission of Lippincott Williams & Wilkins.

patient survival after kidney transplantation. As a general rule, in the absence of specific clinical trials in the transplant population, blood pressure targets advocated in the nontransplant population are recommended in the effort to mitigate cardiovascular risk and retard progression of renal disease in the renal transplant recipient.

### HLA Matching

With improved immunosuppression available today HLA matching has taken on less significance. Notably, excellent results have been noted in living donor transplants using unrelated donors where HLA mismatch is more the rule than the exception. However, it is clear from registry data that the fewer HLA mismatches, the better the long-term graft survival. This holds particularly true for a donor-recipient pair with zero mismatches.[22] The effect of HLA matching is not merely one of decreased acute rejection as the effect of HLA mismatches persists after correcting for acute rejection in multivariate analyses. Certain studies have found a particularly negative effect of mismatch at the DR locus.[23] This relationship of HLA matching and long term graft survival is one of the salient lines of evidence for an immunologic basis of CAN.

### Donor Source

Kidneys from living donors when compared to cadaveric donors are associated superior graft survival. This advantage of living donation persists after correction for ischemic time and delayed graft function remains one of the enduring factors associated with good outcomes. Not only does a living donor kidney contribute to superior long-term graft survival, there also appears to be mortality benefit, that is not fully understood associated with a live donor source as compared to a cadaveric kidney. The advantage of living donor kidneys holds equally true for unrelated donors and appears to be in large part non-immunologically mediated (at least non-HLA mediated). Most likely the advantage of living donor kidneys reflects a combination of kidneys of excellent quality transplanted with minimal trauma into the most compliant patient.

### Donor Age

The older the donor kidney (above the age of 18), the shorter the expected graft life.[24] This effect persists after correction for recipient characteristics which could produce similar effects. The reason for this poorer outcome is in all likelihood due to diminished functional reserve of the older kidney along with age related involutive changes that further impair the kidney's ability to adapt to injurious stressors. One interesting hypothesis is that the older kidney itself may be on a path to programmed senescence posttransplantation which would be most manifest in an older donor kidney.[25]

### Delayed Graft Function

Delayed graft function (DGF) has impacts negatively on long term graft survival. DGF has been related to an increased incidence of acute rejection and the association of DGF and poor outcomes may be due to the increased incidence of acute rejection.[26] Other single center reports report no adverse impact of DGF on graft survival.

An analysis of USRDS data from 37,216 primary cadaveric renal transplants from 1985 to 1992 showed that DGF was independently associated with significant reductions in both long and short-term graft survival.[27] Similar results were reported in an analysis of UNOS data from 27096 primary cadaveric renal transplants between 1994 and 1997.[28] While large databases provide data pertaining to graft survival in large numbers of patients, they are unfortunately lacking in the detailed patient level data that is the inherent advantage of a well performed albeit small single center study. In such a study, DGF's adverse effects on graft survival were suggested as largely explicable by inferior renal function at 1 year.[29]

Overall, the impact of DGF on graft survival remains a controversial area and will likely evolve with advances in Immunosuppression and organ preservation. This area is of immense clinical importance and will likely gain additional prominence with the larger numbers of ECD and DCD transplants that are being performed each year.

### Acute Rejection

Acute rejection remains the single greatest risk factor for chronic allograft nephropathy (CAN) and graft loss.[15] While acute rejection rates are decreasing, the association of acute rejection and chronic allograft failure is increasing over the last 10 years.[14] The histologic grade of rejection, severity of renal functional impairment at the time of diagnosed rejection, timing of rejection episode and completeness of response to anti-rejection treatment have all been shown to have prognostic significance. These issues have also been discussed in the section on trends in graft survival in this chapter.

## Recipient Age

Elderly patients contribute to the greatest growth in the ESRD population. Successful transplantation improves survival and the quality of life in the elderly. As may be expected, death with a functioning graft is the predominant cause of graft loss. As may also be expected cardiovascular events, infection and malignancies constitute the principal causes of death. Acute rejection is relatively infrequent in the elderly transplant recipient. Intensification of immunosuppression in the elderly worsens infectious complications without decreasing acute rejection or improved graft survival. Despite this immunologic advantage death censored graft loss worsens with increasing age of the recipient. This was shown in an analysis of USRDS data from 59,509 white patients examining recipient age as an independent factor impacting graft loss due o chronic attrition of renal function.[30] While death with functioning graft was expectedly more common with increasing recipient age, death censored graft loss also worsened with increasing age. Compared to the age group of 18-59 years, the recipient aged 50-64 accrued a 29% higher relative risk of graft loss while those 65 years or greater incurred a 67% higher relative risk for graft loss.[30] This effect of increasing age on incrementally increasing the risk for death censored graft loss was independent of traditional determinants of graft survival such as DGF and acute rejection. Effects also were independent of the type of immunosuppressive regimen employed. While rejection is relatively infrequent in the elderly, return to baseline function after treatment of acute rejection episodes may be suboptimal. In an analysis of USRDS data from 40,289 white primary renal transplant recipients over the period 1998 to 1997, both increasing donor and recipient age were independent risk factors for graft survival.[31] The best graft survival was in the situation where a kidney from a young donor was transplanted into a young recipient and the worst graft survival was exhibited when kidneys from older donors were transplanted into older donors. It thus appears that donor and recipient age have a synergistic detrimental impact on graft survival. Whether this reflects an intrinsically senescent kidney transplanted into a senescent biologic milieu is unclear.

Death rates due to infection increase linearly with increasing recipient age in waitlisted elderly ESRD patients.[32] In the transplanted elderly patient, infectious mortality rises exponentially.[32] Overall mortality and cardiovascular mortality rise with increasing age in both the transplanted and waitlisted elderly.[32] However, the magnitude of this increased mortality is not greater in transplanted patients.[32] The slope for cardiovascular mortality is halved with successful transplantation in the elderly.[32] Malignancy related death showed a slight increase in the elderly transplanted patients perhaps reflecting the additive effects of pharmacologic Immunosuppression to the effects of senescence on the immune system.[32]

## Recipient Size

Obesity (body mass index; BMI >30 kg/sq m.) constitutes an important risk factor for cardiovascular disease, hypertension and diabetes mellitus in the general population. Obesity may also predispose to progressive renal insufficiency as it has been accompanied by the development of proteinuria and focal and segmental glomerulosclerosis, both of which are attributed to hyperfiltration in the obese. Obese dialysis patients may have a survival advantage. In the transplanted population, obese patients have been shown to be at increased risk for death. The effects of obesity (BMI) on long term graft and patient survival were examined a study of 51,927 patients reported to the USRDS between 1988 and 1997.[33] Increased mortality was noted at extremes (both low and high) of BMI.[33] This U-shaped relationship resembles that seen in the general population and also described the relationship between obesity and cardiovascular and infectious mortality. Furthermore, increasing BMI was associated with worsening death-censored graft survival.[33]

Transplantation does confer a survival advantage to the obese dialysis patient.[34] This survival advantage holds true for both living donor and cadaveric transplantation as has been shown for the entire waitlisted population.[34]

## Recipient Race

From the early days of transplantation, black patients experienced inferior graft and patient survival rates. Explanations have included heightened rejection risk, differing HLA-polymorphisms, differing pharmacokinetics of immunosuppressants, hypertension, noncompliance and socioeconomic status. However, the difference between African-Americans and Caucasians is becoming narrower with newer immunosuppressive regimens and differential dosing of the agents (see section on immunosuppressants).[35,36] Despite the need for increasing immunosuppression to maintain freedom from acute rejection, African-American transplant patients are at decreased risk for infectious death. This represents a potential modifiable factor wherein increased immunosuppression may be delivered relatively safely to confer protection against ejection risk and possibly obviate some of the racial differences in graft survival that were seen in the past.

## Extended Criteria Donors (ECD; Marginal Donors)

ECDs are defined by UNOS as those donors who, based on certain clinical characteristics have been demonstrated to increase risk of 10-year allograft loss by 70 percent. Deceased donor characteristics that define ECD kidneys include age ≥60 years, or age 50-59 years and two of the following conditions: cerebrovascular accident as the cause of death, pre-existing hypertension, or terminal serum creatinine greater than 1.5 mg/dl. In a study of 122,175 patients on the UNOS transplant waitlist between 1992 to 1997, the effect of an ECD transplant on survival of recipients was examined relative to those that remained on the waitlist.[37] On an average, the ECD recipient lived 5 years longer than the waitlisted patient whereas the recipient of an ideal cadaveric kidney accrued a 13-year survival benefit. Diabetics obtained the greatest proportional survival benefit and those with hypertensive renal disease incurred the greatest absolute gains in life-years. Transplantation of an ECD kidney did not increase mortality risk in any subgroup examined. UNOS has implemented policies that allow consenting patients to opt for both an ECD kidney and the ideal kidney. Those patients who decline listing for an ECD kidney could potentially incur the increased mortality attendant to waiting time. It is conceivable that the younger nondiabetic waitlisted patient could wait longer on the list than an older diabetic recipient who would gain the mortality benefit conferred by transplantation.[38]

## Immunosuppressive Therapy

As noted previously, the introduction of azathioprine and then cyclosporine ushered in the clinical era of transplantation and the widespread improvement in early graft survival respectively. Subsequent years saw the use of cyclosporine use in combination with corticosteroids with or without azathioprine and the introduction of anti-T-cell antibodies to treat rejection and the use of antibody induction therapy. In the early 1990s rejection rates in the first year were still around thirty to forty percent with triple therapy using cyclosporine, azathioprine and corticosteroids. Mycophenolate mofetil (MMF) was then introduced in 1995. Each of the three Phase III trials of MMF in renal transplantation clearly established the superiority of MMF over the control group azathioprine or placebo for the primary endpoint of acute rejection. However, the trials were not adequately powered to allow analysis of differences in other measures of clinical interest such as graft and patient survival. Also, Phase III trials are designed and conducted with the intent to show the superiority of one drug over the other over the period of the study, immunosuppressive agents may not be used in practice in the same manner as they were in these very trials. As an example, MMF was initially approved based on its combination with cyclosporine and corticosteroids for use in renal transplantation. However, despite the absence of a definitive prospective randomized trial, clinicians were using increasingly, the combination of MMF with tacrolimus over cyclosporine an MMF. Therefore, in many respects, comparison between regimens as they are used in clinical practice is of relevance. This is particularly of relevance as the introduction of the m-TOR inhibitor, sirolimus and its use in combination with calcineurin inhibitors widened the choice of both the calcineurin inhibitor and antiproliferative available to transplant physicians.

Fortunately, questions unanswered by the pivotal trials and their subsequent analyses have been answered to some extent further studies utilizing a large transplant database that is maintained by the United Network for Organ sharing (UNOS) and distributed, enriched with social security and Center for Medicare and Medicaid Services (CMS) data by the United States Renal Data System (USRDS). The UNOS data is also distributed also to the Scientific Registry of Renal Transplant Recipients (SRTR), a unique database that contains all Organ Procurement and Transplant Network (OPTN) data since 1987.[39,40] Each of these large databases contains data on patients receiving MMF as part of their maintenance immunosuppressive regimen, dating back to 1992.[35,41-45] Analyses of data from the USRDS or the SRTR afford the advantages that derive from data obtained from large numbers of patients across multiple centers. Large databases such as the USRDS and the SRTR also provide data that reflects use of a drug in the relatively uncontrolled setting of clinical practice as opposed to the constrained environment of a clinical trial. In addition endpoints like graft loss and patient death are reported with remarkable accuracy in these databases, when both CMS and Social Security Death Master File data are integrated.[39] However, by the same token the databases have many limitations as many biases govern the choice of immunosuppressive agents used in practice. Importantly the USRDS and SRTR databases do not provide any dosing or drug concentration data and for that reason, any associations derived in analyses, can only be extended to the pattern of clinical use of particular drug combinations during the pertinent historic timeframes analyzed. Even more importantly, it should be noted that retrospective analyses provide evidence of associations between risk factors (e.g., treatment regimen or recipient race) and outcomes (e.g., time to graft loss or decline in renal function) and do yield valuable measures of the strength and significance of such associations.[46] However, proof of causality relationships between a risk factor and outcome should necessarily include among other lines of evidence, prospective experimental evidence.[47] Discussed below are the contributions of various immunosuppressant agents and regimens to outcomes in renal transplantation.

## Mycophenolate Mofetil (MMF)

MMF's impact on long-term graft loss in kidney transplantation was addressed in a USRDS database analysis where data from 66,744 renal transplant recipients reported to the USRDS between 1988 and 1997 were analyzed.[41] This analysis was restricted to patients receiving a solitary kidney transplant in that time period. The cumulative risk for chronic allograft failure, the primary study end-point, defined as graft loss beyond 6 months after transplantation censored for death, rejection and other reasons for graft loss with a discrete attributable cause increased in the azathioprine treated patients as opposed to those on MMF. 4-year death censored graft loss was significantly better in the MMF group as opposed to the AZA group (85.6% vs. 81.9%) and 4 year patient survival in the MMF group was significantly superior to that in the AZA group ( 91.4% vs. 89.8, p = 0.002). These effects of MMF persisted after correcting for acute rejection (AR).[45]

Among African-American patients a more striking reduction in acute rejection rates was observed in patients on MMF (20.5%) versus AZA (32.8%) within the first 6 months posttransplantation.[35] MMF therapy was associated with clinically important and significant risk

reductions in death with functioning graft, death-censored graft loss and chronic allograft failure, particularly in the African-American. Importantly, the beneficial effects of MMF on acute rejection and graft loss were not accompanied by an increase in the risk of death with a functioning graft. MMF may thus exhibit a wider therapeutic index than AZA (paralleling theoretical expectations based on the relatively specific and potent immunosuppressive properties of MMF).[48,49]

Subsequent analyses of USRDS and SRTR data have shown that MMF is associated with fewer episodes of late AR and greater stability of graft function compared to AZA. The effects on preventing late AR are particularly prominent in African-Americans and may account for the superior graft survival observed in this population with MMF.[43-45] Some of the benefit of MMF in preserving long-term renal allograft function may thus arise from prevention of late episodes of acute rejection especially in high risk populations such as African-Americans.

### MMF in Combination with Tacrolimus

In recent years however, cyclosporine has been largely replaced in most centers by tacrolimus.[50] This shift reflects a gradual change in practice patterns based on empiric observations and expectation of less nephrotoxicity with tacrolimus that preceded actual published results of formal studies evaluating this combination. Gonwa et al have reported on the three year follow-up of their experience with two hundred twenty-three recipients of first cadaveric kidney allografts who were randomized to receive tacrolimus with MMF, tacrolimus with azathioprine, or cyclosporine (Neoral) with MMF.[51] At 1 year the lowest rate of steroid resistant acute rejection was lowest in the tacrolimus MMF group without significant differences in overall incidence of acute rejection. Furthermore patients randomized to either treatment arm containing tacrolimus exhibited lower serum creatinine concentrations in comparison with those receiving cyclosporine. Given these results, it is quite conceivable that long-term graft function could be expected to be superior when MMF is used in combination with tacrolimus as opposed to cyclosporine.

### MMF Compared to Sirolimus in Cyclosporine Based Regimens

Recent years have shown an increasing use of MMF or sirolimus with calcineurin inhibitors with very few centers using azathioprine in newly transplanted recipients.[50]Thus a more current and relevant comparison in the recent era is provided by a study in which MMF was compared to sirolimus, in calcineurin inhibitor based regimens. In one such study, data pertaining to 23,016 primary recipients reported to the Scientific Registry of Transplant Recipients between 1998 and 2003 were analyzed.[52] The regimen combining cyclosporine and sirolimus was associated with significantly lower graft survival (74.6% vs. 79.3% at 4 years, p = 0.002) and death-censored graft survival (83.7% vs. 87.2%, p = 0.003) compared to cyclosporine and MMF. In multivariate analyses, the cyclosporine-sirolimus combination was associated with a significantly increased risk for graft loss, death-censored graft loss and decline in renal function (HR = 1.22, p = 0.002; HR = 1.22, p = 0.018 and HR = 1.25, p <0.001, respectively). Furthermore, regimens combining MMF with cyclosporine or tacrolimus were associated with better serial creatinine clearances than regimens combining sirolimus with cyclosporine (also see below). It should also be noted that while sirolimus was initially approved for use with cyclosporine, concerns of the nephrotoxicity of this regimen have prompted a gradual shift away from this combination toward it's use with Tacrolimus or MMF.[53]

### MMF Versus Sirolimus with Tacrolimus

The perception of the lower nephrotoxicity of tacrolimus along with the renal function concerns about the cyclosporine/sirolimus combination have in recent years led to increasing use of tacrolimus in combination with sirolimus.[54-57] A relatively recent analysis of SRTR data showed that recipients treated with tacrolimus and MMF exhibit better graft survival than those receiving cyclosporine and MMF or tacrolimus with sirolimus.[58] A statistically significant difference was demonstrated between the tacrolimus and sirolimus regimen versus the tacrolimus and MMF arm. Progressive separation of the survival curves over time was observed, with a clinically significant difference at 3 years after transplantation (Tacrolimus + Rapamycin, 80.3% versus Tacrolimus + MMF, 85.9%, p <0.001). This difference in graft survival was similar in magnitude to the previously shown difference between Cyclosporine + Rapamycin and Cyclosporine + MMF, in both the univariate analysis and the multivariate analysis.[52] These results are in consonance with the results of a prospective study reported by Mendez et al.[46,59]

### MMF with Sirolimus

There has been increasing interest in avoiding the use of calcineurin inhibitors (CNI) in kidney transplantation in order to avoid their nephrotoxicity. Excellent short term results have been reported with sirolimus used in combination with MMF and corticosteroids in kidney transplantation.[60,61]

Outcomes over a longer follow-up period with the sirolimus (SRL) and mycophenolate mofetil (MMF) combination regimen (SRL/MMF) in solitary kidney transplant recipients transplanted between 2000 and 2005 reported to the Scientific Registry of Renal Transplant Recipients were evaluated recently.[62] Of this cohort, 3.5 percent received SRL/MMF (n = 2040). Six month acute rejection rates were higher with SRL/MMF (SRL/MMF:16.0% vs. other regimens: 11.2%, p <0.001). Overall graft survival was significantly lower on SRL/MMF (Fig. 5). SRL/MMF was associated with twice the hazard for graft loss (AHR = 2.0, 95% C.I. 1.8, 2.2) relative to TAC/MMF. In analyses restricted to patients who remained on the discharge regimen at 6 months posttransplant, conditional graft survival in deceased donor transplants was significantly lower with SRL/MMF

compared to patients on TAC/MMF or CsA/MMF regimens at 5 years posttransplant (64%, 78%, 78% respectively, p = 0.001) (Fig. 6) and across all patient subgroups. Corroborative evidence comes from the Symphony study wherein standard-dose CsA based regimens are being compared to low-dose CsA, TAC or SRL in combination with MMF, daclizumab and corticosteroids in renal transplantation.[63] In the reported one year results of this study, biopsy-proven acute rejection at 6 months in SRL/MMF patients was 33 percent versus 11 percent with TAC/MMF (p <0.01) and 22 percent with CsA/MMF. With regard to allograft function, calculated glomerular filtration rate was 57.3 mL/min with SRL/MMF versus 65.4 mL/min with TAC/MMF (p <0.0001). Lastly, one-year graft survival was significantly inferior in SRL/MMF patients (TAC/MMF: 94 percent; SRL/MMF: 89 percent; p = 0.017).[64] Notably, targeted SRL concentrations in the Symphony study may be lower than those reported previously by Flechner et al,[65,66] but the higher acute rejection rates and the worse graft survival certainly mirror findings from the US retrospective data. These results likely reflect the poor protection against acute rejection afforded by this regimen as also poor tolerability of the regimen.

### Discussion

Kidney transplantation is the renal replacement modality of choice for the ESRD patient. Advances in immunosuppression, medical management of transplant recipients, management of immunosuppression and advances in overall medical care have made kidney transplantation a clinical reality.

Remarkable strides have been made in both patient and graft survival over the last fifty years. However, a disturbing trend of lack of improvement in long-term success rates of kidney transplantation is becoming apparent. The earlier expectation of freedom from acute rejection in the early posttransplant period has not been translated into improved long-term outcomes. Many factors could account for this disconnect.

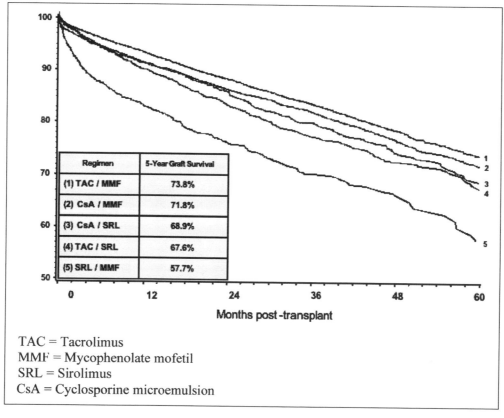

| Regimen | 5-Year Graft Survival |
|---|---|
| (1) TAC / MMF | 73.8% |
| (2) CsA / MMF | 71.8% |
| (3) CsA / SRL | 68.9% |
| (4) TAC / SRL | 67.6% |
| (5) SRL / MMF | 57.7% |

TAC = Tacrolimus
MMF = Mycophenolate mofetil
SRL = Sirolimus
CsA = Cyclosporine microemulsion

Figure 5. Overall graft survival by immunosuppressant regimen for deceased donor transplant recipients. From: Srinivas TR et al. Am J Transplant 2006; in press;[62] with copyright permission of Blackwell Publishing.

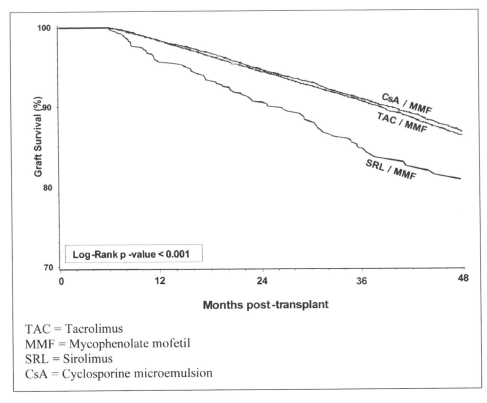

Figure 6. Kaplan-Meier plot of overall graft survival in deceased donor transplant recipients by immunosuppressant regimen for patients on treatment at six months. From: Srinivas TR et al. Am J Transplant 2006; in press;[62] with copyright permission of Blackwell Publishing.

Transplantation is now being performed in patients with multiple comorbidities who are also much older. In order to maximize utilization of cadaveric kidneys, both NHBD and ECD kidneys are being transplanted increasingly. These clinical situations are intrinsically fraught with risk for DGF and subsequent adverse clinical events such as acute rejection which are then compounded by the burden of comorbidity.

Potent immunosuppression may also be associated with a change in the clinical expression of rejection. More treatment sensitive rejections may not express themselves at the time period immediately after transplantation when immunosuppression is at it most intense but present themselves in the late posttransplant period where both the prescribed intensity of immunosuppression and the frequency of clinical and laboratory follow-up decline. Such rejections as do express early in the posttransplant period may in fact represent breakthrough through maximal immunosuppression and thus have an intrinsic predisposition to treatment resistance and as a corollary, suboptimal restoration of graft function.

More potent immunosuppression also carries an additional price in the incidence of infectious complications. The burden of opportunistic infection such as cytomegalovirus infections is well recognized and transplant physicians have become well versed in the prophylaxis and treatment of such infections. However, of great concern in the transplant community is the growing epidemic of polyoma virus nephropathy.[7] In the absence of specific treatment, strategies rely on prevention by serial screening for the virus by PCR.[67] Such screening is not universal across programs and the exact details of interpretation are still a work in progress. Furthermore, the mainstay of management, minimization of immunosuppression, is fraught with the risk for acute rejection and the consequent need to intensify immunosuppression.

While not specifically discussed in this chapter, corticosteroid avoidance is of concern to patients and physicians alike in the effort to minimize the cardiovascular, musculoskeletal, metabolic and cosmetic side effects of these agents. Excellent early results have been reported with these protocols.[68-70] The growing tendency to offer these protocols to patients is clearly evident as reflected in SRTR data.[71] However, the impact of such protocols on late declines in graft survival may be difficult to assess for many years as has been the experience in the past.[72] Certainly, findings such as increasing interstitial fibrosis in protocol biopsies of patients who undergo steroid elimination suggest the need for caution.[73]

Calcineurin inhibitor nephrotoxicity is well recognized. By the same token, remarkable strides in renal allograft survival have been realized in the era of calcineurin inhibition. While studies of serial protocol biopsies do show a higher cumulative incidence of CNI related changes, these changes were necessarily reported in grafts that had survived to inclusion in the study. In of themselves such studies do not test the hypothesis that elimination of CNIs and the use of alternative CNI-free regimens will necessarily translate into durable long-term graft survival. As discussed smaller single center studies in highly selected patients showed promise of the CNI-free regimen of SRL with MMF. However, registry data and the larger Symphony study show that graft survival is inferior with this combination. Thus, CNI-free immunosuppression using the currently available agents is not yet a universally applicable strategy.

Evidence for or against the use of an immunosuppressant should necessarily be derived from well designed adequately powered prospective clinical trials. Such trials are difficult to perform and have in the past been designed to test one agent against another. However, extrapolation from the efficacy of a single agent to the clinical behavior of the same agent when it is part of a different regimen is not a straightforward extrapolation as has been highlighted by the experience with m-TOR inhibitors in combination with both CNIs and MMF. Also of note is that the regimen that is most widely used in kidney transplantation, TAC + MMF + Corticosteroids came into widespread use based on clinical intuition and practice based observations long before the first clinical trials that showed its efficacy.

Registry data have been criticized as not yielding prospective data or proof for cause and effect. However, they are and continue to be a valuable reflection of outcomes as they relate to the use of immunosuppression in clinical practice. They thus should be taken into consideration with all other lines of evidence in the understanding of outcomes and the improvement of practice.

In conclusion, attainment and maintenance of improvements in graft and patient survival in renal transplantation will need an integrative approach which combines knowledge gained from clinical trials, outcomes data generated from large registries and accruing clinical experience.

# References

1. Cohen DJ, St Martin L, Christensen LL et al. Kidney and pancreas transplantation in the United States, 1995-2004. Am J Transplant 2006; 6(5 Pt 2):1153-69.
2. Wolfe RA, Ashby VB, Milford EL et al. Comparison of mortality in all patients on dialysis, patients on dialysis awaiting transplantation and recipients of a first cadaveric transplant. N Engl J Med 1999; 341(23):1725-30.
3. Halloran PF, Gourishankar S, Vongwiwatan A et al. Approaching the Renal transplant with Deteriorating Function: Progressive Loss of Renal Function is not Inevitable. In: Weir MR, editor. Medical Management of Kidney Transplantation. Philadelphia: Lippincott Williams & Wilkins, 2005.
4. Mauiyyedi S, Pelle PD, Saidman S et al. Chronic humoral rejection: identification of antibody-mediated chronic renal allograft rejection by C4d deposits in peritubular capillaries. J Am Soc Nephrol 2001; 12(3):574-82.
5. Theruvath TP, Saidman SL, Mauiyyedi S et al. Control of antidonor antibody production with tacrolimus and mycophenolate mofetil in renal allograft recipients with chronic rejection. Transplantation 2001; 72(1):77-83.
6. Womer KL, Vella JP, Sayegh MH. Chronic allograft dysfunction: mechanisms and new approaches to therapy. Semin Nephrol 2000; 20(2):126-47.
7. Randhawa P, Brennan DC. BK virus infection in transplant recipients: an overview and update. Am J Transplant 2006; 6(9):2000-5.
8. Opelz G, Mickey MR, Terasaki PI. Calculations on long-term graft and patient survival in human kidney transplantation. Transplant Proc 1977; 9(1):27-30.
9. Gjertson DW. Survival trends in long-term first cadaver-donor kidney transplants. Clin Transpl 1991; 225-35.
10. Hariharan S, Johnson CP, Bresnahan BA et al. Improved graft survival after renal transplantation in the United States, 1988 to 1996. N Engl J Med 2000; 342(9):605-12.
11. Meier-Kriesche HU, Schold JD, Kaplan B. Long-term renal allograft survival: have we made significant progress or is it time to rethink our analytic and therapeutic strategies? Am J Transplant 2004; 4(8):1289-95.
12. Meier-Kriesche HU, Ojo AO, Hanson JA et al. Increased impact of acute rejection on chronic allograft failure in recent era. Transplantation 2000; 70(7):1098-100.
13. Holdaas H, Fellstrom B, Jardine AG et al. Effect of fluvastatin on cardiac outcomes in renal transplant recipients: a multicentre, randomised, placebo-controlled trial. Lancet 2003; 361(9374):2024-31.
14. Meier-Kriesche HU, Port FK, Ojo AO et al. Effect of waiting time on renal transplant outcome. Kidney Int 2000; 58(3):1311-17.
15. Mange KC, Joffe MM, Feldman HI. Effect of the use or nonuse of long-term dialysis on the subsequent survival of renal transplants from living donors. N Engl J Med 2001; 344(10):726-31.
16. Meier-Kriesche HU, Kaplan B. Waiting time on dialysis as the strongest modifiable risk factor for renal transplant outcomes: a paired donor kidney analysis. Transplantation 2002; 74(10):1377-81.
17. Kaplan B, Meier-Kriesche HU. Death after graft loss: an important late study endpoint in kidney transplantation. Am J Transplant 2002; 2(10):970-74.
18. Muntner P, He J, Hamm L et al. Renal insufficiency and subsequent death resulting from cardiovascular disease in the United States. J Am Soc Nephrol 2002; 13(3):745-53.
19. Meier-Kriesche HU, Baliga R, Kaplan B. Decreased renal function is a strong risk factor for cardiovascular death after renal transplantation. Transplantation 2003; 75(8):1291-95.
20. Meier-Kriesche HU, Schold JD, Srinivas TR et al. Kidney transplantation halts cardiovascular disease progression in patients with end-stage renal disease. Am J Transplant 2004; 4(10):1662-68.
21. Mange KC, Cizman B, Joffe M et al. Arterial hypertension and renal allograft survival. JAMA 2000; 283(5):633-38.
22. Held PJ, Kahan BD, Hunsicker LG et al. The impact of HLA mismatches on the survival of first cadaveric kidney transplants. N Engl J Med 1994; 331(12):765-70.
23. Vereerstraeten P, Abramowicz D, De Pauw L et al. Experience with the Wujciak-Opelz allocation system in a single center: an increase in HLA-DR mismatching and in early occurring acute rejection episodes. Transpl Int 1998; 11(5):378-81.
24. Takemoto S, Terasaki PI. Donor age and recipient age. Clin Transpl 1988; 345-56.
25. Halloran PF, Melk A, Barth C. Rethinking chronic allograft nephropathy: the concept of accelerated senescence. J Am Soc Nephrol 1999; 10(1):167-81.
26. Troppmann C, Gillingham KJ, Benedetti E et al. Delayed graft function, acute rejection and outcome after cadaver renal transplantation. The multivariate analysis. Transplantation 1995; 59(7):962-68.
27. Ojo AO, Wolfe RA, Held PJ et al. Delayed graft function: risk factors and implications for renal allograft survival. Transplantation 1997; 63(7):968-74.
28. Shoskes DA, Cecka JM. Deleterious effects of delayed graft function in cadaveric renal transplant recipients independent of acute rejection. Transplantation 1998; 66(12):1697-701.
29. Boom H, Mallat MJ, de Fijter JW et al. Delayed graft function influences renal function, but not survival. Kidney Int 2000; 58(2):859-66.
30. Meier-Kriesche HU, Ojo AO, Cibrik DM et al. Relationship of recipient age and development of chronic allograft failure. Transplantation 2000; 70(2):306-10.
31. Meier-Kriesche HU, Cibrik DM, Ojo AO et al. Interaction between donor and recipient age in determining the risk of chronic renal allograft failure. J Am Geriatr Soc 2002; 50(1):14-17.
32. Meier-Kriesche HU, Ojo AO, Hanson JA et al. Exponentially increased risk of infectious death in older renal transplant recipients. Kidney Int 2001; 59(4):1539-43.
33. Meier-Kriesche HU, Vaghela M, Thambuganipalle R et al. The effect of body mass index on long-term renal allograft survival. Transplantation 1999; 68(9):1294-97.
34. Glanton CW, Kao TC, Cruess D et al. Impact of renal transplantation on survival in end-stage renal disease patients with elevated body mass index. Kidney Int 2003; 63(2):647-53.
35. Meier-Kriesche HU, Ojo AO, Leichtman AB et al. Effect of mycophenolate mofetil on long-term outcomes in African american renal transplant recipients. J Am Soc Nephrol 2000; 11(12):2366-70.
36. Neylan JF. Immunosuppressive therapy in high-risk transplant patients: dose-dependent efficacy of mycophenolate mofetil in African-American renal allograft recipients. US Renal Transplant Mycophenolate Mofetil Study Group. Transplantation 1997; 64(9):1277-82.
37. Ojo AO, Hanson JA, Meier-Kriesche H et al. Survival in recipients of marginal cadaveric donor kidneys compared with other recipients and wait-listed transplant candidates. J Am Soc Nephrol 2001; 12(3):589-97.
38. Merion RM, Ashby VB, Wolfe RA et al. Deceased-donor characteristics and the survival benefit of kidney transplantation. JAMA 2005; 294(21):2726-33.
39. Dickinson DM, Bryant PC, Williams MC et al. Transplant data: sources, collection and caveats. Am J Transplant 2004; 4 Suppl 9:13-26.
40. Merion RM. 2004 SRTR Report on the State of Transplantation. Am J Transplant 2005; 5(4 Pt 2):841-42.
41. Meier-Kriesche H, Ojo AO, Arndorfer JA et al. Mycophenolate mofetil decreases the risk for chronic renal allograft failure. Transplant Proc 2001; 33(1-2):1005-06.
42. Meier-Kriesche HU, Ojo AO, Leichtman AB et al. Interaction of mycophenolate mofetil and HLA matching on renal allograft survival. Transplantation 2001; 71(3):398-401.
43. Meier-Kriesche HU, Steffen BJ, Hochberg AM et al. Long-term use of mycophenolate mofetil is associated with a reduction in the incidence and risk of late rejection. Am J Transplant 2003; 3(1):68-73.
44. Meier-Kriesche HU, Steffen BJ, Hochberg AM et al. Mycophenolate mofetil versus azathioprine therapy is associated with a significant protection against long-term renal allograft function deterioration. Transplantation 2003; 75(8):1341-46.
45. Ojo AO, Meier-Kriesche HU, Hanson JA et al. Mycophenolate mofetil reduces late renal allograft loss independent of acute rejection. Transplantation 2000; 69(11):2405-9.
46. Kaplan B, Schold J, Meier-Kriesche HU. Overview of large database analysis in renal transplantation. Am J Transplant 2003; 3(9):1052-56.
47. Hill AB. Statistical Evidence and Inference. In: Hill AB, editor. Principles of Medical Statistics. London: The Lancet Limited, 1971; 309-23.
48. Allison AC, Eugui EM. Mycophenolate mofetil and its mechanisms of action. Immunopharmacology 2000; 47(2-3):85-118.

49. Morris RE, Wang J, Blum JR et al. Immunosuppressive effects of the morpholinoethyl ester of mycophenolic acid (RS-61443) in rat and nonhuman primate recipients of heart allografts. Transplant Proc 1991; 23(2 Suppl 2):19-25.
50. Kaufman DB, Shapiro R, Lucey MR et al. Immunosuppression: practice and trends. Am J Transplant 2004; 4 Suppl 9:38-53.
51. Gonwa T, Johnson C, Ahsan N et al. Randomized trial of tacrolimus + mycophenolate mofetil or azathioprine versus cyclosporine + mycophenolate mofetil after cadaveric kidney transplantation: results at three years. Transplantation 2003; 75(12):2048-53.
52. Meier-Kriesche HU, Steffen BJ, Chu AH et al. Sirolimus with neoral versus mycophenolate mofetil with neoral is associated with decreased renal allograft survival. Am J Transplant 2004; 4(12):2058-66.
53. Andoh TF, Lindsley J, Franceschini N et al. Synergistic effects of cyclosporine and rapamycin in a chronic nephrotoxicity model. Transplantation 1996; 62(3):311-16.
54. MacDonald AS. A worldwide, phase III, randomized, controlled, safety and efficacy study of a sirolimus/cyclosporine regimen for prevention of acute rejection in recipients of primary mismatched renal allografts. Transplantation 2001; 71(2):271-80.
55. Formica RN Jr, Lorber KM, Friedman AL et al. Sirolimus-based immunosuppression with reduce dose cyclosporine or tacrolimus after renal transplantation. Transplant Proc 2003; 35(3 Suppl):95S-98S.
56. MacDonald A. Improving tolerability of immunosuppressive regimens. Transplantation 2001; 72(12 Suppl):S105-12.
57. MacDonald AS. Rapamycin in combination with cyclosporine or tacrolimus in liver, pancreas and kidney transplantation. Transplant Proc 2003; 35(3 Suppl):201S-8S.
58. Meier-Kriesche HU, Schold JD, Srinivas TR et al. Rapamycin in combination with Tacrolimus is associated with worse renal allograft survival compared to Mycophenolate Mofetil combined with Tacrolimus. Am J Transplant In Press 2005.
59. Mendez R, Gonwa T, Yang HC et al. A prospective randomized trial of tacrolimus in combination with sirolimus or mycophenolate mofetil in kidney transplantation; Results at one year. Transplantation In Press 2005.
60. Flechner SM, Goldfarb D, Modlin C et al. Kidney transplantation without calcineurin inhibitor drugs: a prospective, randomized trial of sirolimus versus cyclosporine. Transplantation 2002; 74(8):1070-76.
61. Larson TS, Dean PG, Stegall MD et al. Complete avoidance of calcineurin inhibitors in renal transplantation: a randomized trial comparing sirolimus and tacrolimus. Am J Transplant 2006; 6(3):514-22.
62. Srinivas TR, Schold JD, Eagan AA et al. Mycophenolate Mofetil/Sirolimus Compared to other Common Immunosuppressive Regimens In Kidney Transplantation. Am J Transplant 2006; In Press.
63. Ekberg H, Tedesco-Silva H, Demirbas A, et al Symphony comparing standard immunosuppression to low-dose cyclosporine, tacrolimus or sirolimus in combination with MMF, daclizumab and corticosteroids in renal transplanation. Am J Transplant 2006.
64. Ekberg H, Tedesco-Silva H, Demirbas A, et al Symphony comparing standard immunosuppression to low-dose cyclosporine, tacrolimus or sirolimus in combination with MMF, daclizumab and corticosteroids in renal transplanation. Am J Transplant 2006.
65. Flechner SM, Goldfarb D, Modlin C et al. Kidney transplantation without calcineurin inhibitor drugs: a prospective, randomized trial of sirolimus versus cyclosporine. Transplantation 2002; 74(8): 1070-76.
66. Flechner SM, Kurian SM, Solez K et al. De novo kidney transplantation without use of calcineurin inhibitors preserves renal structure and function at two years. Am J Transplant 2004; 4(11):1776-85.
67. Brennan DC, Agha I, Bohl DL et al. Incidence of BK with tacrolimus versus cyclosporine and impact of pre-emptive immunosuppression reduction. Am J Transplant 2005; 5(3):582-94.
68. Kumar MS, Xiao SG, Fyfe B et al. Steroid avoidance in renal transplantation using basiliximab induction, cyclosporine-based immunosuppression and protocol biopsies. Clin Transplant 2005; 19(1):61-69.
69. Kaufman DB, Leventhal JR, Koffron AJ et al. A prospective study of rapid corticosteroid elimination in simultaneous pancreas-kidney transplantation: comparison of two maintenance immunosuppression protocols: tacrolimus/mycophenolate mofetil versus tacrolimus/sirolimus. Transplantation 2002; 73(2):169-77.
70. Khwaja K, Asolati M, Harmon J et al. Outcome at 3 years with a prednisone-free maintenance regimen: a single-center experience with 349 kidney transplant recipients. Am J Transplant 2004; 4(6):980-87.
71. Meier-Kriesche HU, Li S, Gruessner RW et al. Immunosuppression: evolution in practice and trends, 1994-2004. Am J Transplant 2006; 6(5 Pt 2):1111-31.
72. Sinclair NR. Low-dose steroid therapy in cyclosporine-treated renal transplant recipients with well-functioning grafts. The Canadian Multicentre Transplant Study Group. CMAJ 1992; 147(5):645-57.
73. Laftavi MR, Stephan R, Stefanick B et al. Randomized prospective trial of early steroid withdrawal compared with low-dose steroids in renal transplant recipients using serial protocol biopsies to assess efficacy and safety. Surgery 2005; 137(3):364-71.

# Predictive Parameters of Renal Graft Failure

Paola Romagnani*

## Abstract

The incidence of end stage renal disease (ESRD) is increasing at a faster rate than the availability of kidney donors, but unfortunately the improvement in short-term graft survival rates has not been followed by substantial amelioration in long-term outcome. Almost half of cadaveric allografts will still be lost within 10 years after transplantation, chronic allograft nephropathy (CAN) being the most frequent cause of graft failure over time. Furthermore, the increasing demand of organ for transplantation creates an urgent need for optimizing the outcome of transplantation by achieving long-term graft acceptance with normal organ function. A series of variables concerning donor, recipient and the peritransplant period influence long-term graft outcome. Donor age, organ size and quality, time on dialysis are the most relevant factors related to the donor and the recipient. Cold ischemia and delayed graft function are the most relevant peritransplant factors. After the transplant, only measurement of renal allograft function and proteinuria are applied in clinical practice as early risk markers. However, the very low incidence of CAN in recipients without previous acute rejection (AR) and the relevant role still played by HLA matching, suggests that immunologic factors are the main determinants of long-term kidney transplant outcome. Several laboratory tests have been proposed to better define the immunological status of the recipient and its possible response to different degrees of immunosuppression. In particular, measurement of serum and urinary cytokines might enter clinical practice in the next future and allow to identify high risk patients, helping to individualize immunosuppressive therapies.

## Introduction

Long-term renal allograft survival has not paralleled improvements made in the past three decades in short-term survival.[1] Almost half of all cadaveric allografts will still be lost within 10 years after transplantation. The predominant causes of late renal allograft loss are chronic allograft nephropathy (CAN) and recipient mortality, the latter often from cardiovascular disease.[2] Furthermore, the increasing demand of organ for transplantation creates an urgent need for optimizing the outcome of transplantation by achieving long-term graft acceptance with normal organ function.[3] To identify transplant recipients at risk for late renal allograft loss, who may benefit from preventive and therapeutic strategies at an early stage after transplantation, a large variety of predictive clinical markers have been proposed.[4,5] Some of these markers are related to recipient and donor selection,[5] and are taken into account before transplantation to allocate the grafts.[6] After the transplant, only measurement of renal allograft function,[7,8] and proteinuria,[9] are applied in clinical practice as early risk markers.[5] Unfortunately, all these clinical parameters display a limited predictive value. More importantly,

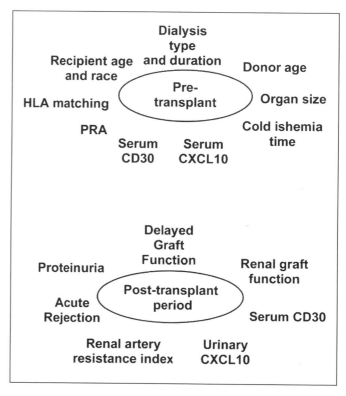

Figure 1. Predictive parameters of graft failure and their relationship with different phases of transplantation.

these markers do not have predictive ability on the immunological status of the recipient and its possible response to different degrees of immunosuppression. Thus, a pressing need exists for the identification of better early risk parameters.[4,5] In the recent years, novel serologic and urinary markers of graft failure have been proposed. These novel markers might indeed be used to individualize immunosuppressive therapies. In this chapter, we will analyse all known risk factors as well as novel immunological markers which might enter clinical practice in the next future (Fig. 1).

## Currently Used Predictors of Graft Outcome

### Donor Age

The shortage of cadaver donors for kidney transplantation has prompted many centers to use kidneys from older donors.[10] Kidneys from older donors exhibit a series of changes characterized by glomerular,

*Paola Romagnani—Department of Clinical Pathophysiology, Excellence Center for Research, Transfer and High Education DENOTHE, University of Florence, Viale Pieraccini 6, 50139, Florence, Italy. Email: p.romagnani@dfc.unifi.it

*Chronic Allograft Failure: Natural History, Pathogenesis, Diagnosis and Management*, edited by Nasimul Ahsan. ©2008 Landes Bioscience.

vascular and tubular senescence. These changes may be aggravated by atherosclerosis, hypertension, or diabetes, which are highly prevalent in older individuals. On this basis, the use of older donors was questioned. In a study on United Network for Organ Sharing (UNOS) data of 35,621 adult cadaveric renal transplant recipients receiving kidneys from adult donors between 1988 and 1994 and followed for a minimum of 1 year after transplantation, increased odds of graft failure were seen with increasing donor age, previous transplantation and elevated panel-reactive antibody (PRA). Furthermore, kidneys from older donors were associated with poorer graft survival rates with African-American recipients, but no detrimental effects when used for older, Hispanic, Asian, or female recipients were observed.[10] The kidneys from older donors had poorer graft survival than the kidneys from younger donors when transplanted into recipients of repeated transplants, although the impact of repeated transplant and donor age on graft survival appear to be independent, suggesting that kidneys from donors over the age of 55 overall have reduced functional reserve, which has an adverse effect on long-term function.[10] Thus, attempts should be made to better estimate functional reserve among the older age group, but age alone should not be the sole factor for exclusion of a potential donor.[11,12] Indeed, growing evidence suggests that the long-term survival of single or dual kidney grafts from donors older than 60 years of age is excellent, provided that the grafts are evaluated histologically before implantation.[13,14] Based on biopsy findings, up to 35% and 44% of kidneys from deceased donors older than 60 years may be adequate for a single or dual transplant, respectively.[14,15] Using these kidneys should increase the current organ supply by another 25-30% and limit the negative impact of donor age on graft survival.

### Recipient Age and Race

Not only the age of the donor, but also the age of the recipient is increasing in recent years. The UNOS data show that the results are worse for recipients above age 50 years. The main cause of graft failure is death with a functioning graft. As expected, the older the age, the higher the risk of death. On the other hand, the risk of graft failure caused by acute or chronic rejection tends to decrease with age.[16,17]

Single center studies and registry data have had conflicting results regarding the impact of recipient age on CAN. Initially, it was shown that increased recipient age is an independent risk factor for the development of CAN, but only in Caucasians.[18] The outcome differences between ethnic groups after kidney transplantation have led to the characterization of African Americans as having high immune risk and multicenter clinical trials have reported better outcomes when African Americans receive higher doses of immunosuppression, suggesting pharmacokinetic and pharmacodynamic differences.[19] However, a recent analysis of the UNOS data obtained from cadaveric kidney transplants between 1995 and 2000 data does not support the hypothesis that advanced recipient age is a risk factor for CAN even in caucasians.[20]

### Organ Size

There have been several reports suggesting that kidneys from small donors may be at increased risk for late graft failure if they are transplanted into large recipients. Overall, poorer graft survival was seen in pediatric donor transplants, where progressive increase in donor age was associated with improved graft survival when the donors were 6-11 years, whereas progressive increase in donor weight was associated with improved graft survival when the donors were 0-5 years.[21] These results obtained in pediatric patients were recently confirmed by a large analysis of all first cadaver kidney transplantations performed during the years 1994 to 1999 in the United States.[22] In this study, large recipients who received kidneys from small donors made up 1.5% of the population and had a 43% increased risk of late graft failure compared with medium-size recipients who received kidneys from medium-size donors. Medium-size recipients who received kidneys from small donors made up 12.0% of the population and had a 16%

increased risk of late graft failure. Disparities in recipient and donor size had similar adverse affects on mortality. However, effects of recipient obesity and donor gender on late graft survival were no longer statistically significant after the effects of donor and recipient body size were taken into account, suggesting that the relative size of the donor and recipient should possibly be taken into account when choosing kidneys for transplantation.[22] Furthermore, the graft/recipient mass ratio has a significant impact on filtration rate and proteinuria. This last effect is noticeable even after a short interval, particularly when the adaptive response of the graft to the recipient mass is expected to be considerable.[23] In conclusion, kidney volume strongly correlates with function in living kidney donors and is an independent determinant of post-transplant graft outcome. These findings suggest that transplantation of larger kidneys confers an outcome advantage and larger kidneys should be preferred when selecting from otherwise similar living donors.[24]

Very recently, glomerular size was also proposed as a possible risk factor for long-term graft failure. Indeed, patients showing large glomerular volume, high Banff chronic score and poor early renal function in stable grafts showed a poor graft outcome.[25]

### Dialysis Type and Duration

Near doubling of death rates among wait listed patients argues against the routine strategy of placing end stage renal disease (ESRD) patients on maintenance dialysis before the possibility of transplantation is considered. This leads to the question of whether transplantation of these patients prior to the initiation of dialysis (pre-emptive transplantation) would improve outcome significantly. Results of previous studies examining this issue have been conflicting.[26,27] However, several subsequent large analyses have demonstrated that pre-emptive transplantation leads to substantial improvements in patient and graft survival when compared to transplantation after a period of dialysis therapy.[28,29] The adverse effects of dialysis therapy on post-transplant outcomes are duration-dependent. Thus, increasing periods of exposure to dialysis prior to transplantation are associated with proportionately greater increases in the risk of graft loss and patient death after transplantation. A retrospective analysis from the United States Renal Data Systems (USRDS) database found that being on dialysis for 0 to 6 months prior to transplantation conferred a 17% increased risk for death-censored graft loss as compared with pre-emptive transplantation.[30] Since this risk increased linearly with incremental time on dialysis, being on maintenance dialysis for 36 months before transplantation resulted in the risk of graft loss being 68% higher than that observed with pre-emptive transplantation. Similar conclusions were reached by other authors.[31] Thus, although dialysis initiation at the appropriate time should not be deferred in favour of waiting for a kidney, transplantation of suitable patients at the earliest time offers the best outcomes for patient and graft survival. An objective estimate of this risk was obtained from the results of a retrospective analysis that examined 10-years adjusted graft survivals for deceased and living donor transplants as a function of the duration of pretransplant dialysis.[32] This study demonstrated that 10-year graft survival in cadaveric kidney transplant recipients for those transplanted pre-emptively *vs.* patients on dialysis for more than 2 years was 69 and 39% respectively, while for living donor transplant was 75 and 49%, respectively. In this context, living donor transplants performed after prolonged dialysis exposure yields results similar to deceased donor transplants performed within 6 months of dialysis.

The role of pretransplant dialysis modality in affecting post-transplant outcomes is controversial. Some studies have reported lower delayed graft function (DGF) rates with its attendant long-term benefits for patients treated with peritoneal dialysis, while others have shown worse outcomes related to excess graft thrombosis and infections for these patients.[33-35] A large study performed on 22,776 patients in the US concluded that transplantation in peritoneal dialysis patients is more frequently associated with early, but not late, graft failure.[36] DGF was less common in peritoneal dialysis patients but this potential benefit appears to be offset by other factors which are associated with early graft loss. These factors seem to be represented by association between prior peritoneal dialysis treatment with increased risk of allograft thrombosis.[34,35]

## Retransplant

Nowadays, kidney retransplantation, although considered to be a high-risk procedure, is increasingly performed worldwide. Second graft survival rates seem to be 10% lower than that of first grafts, although recently the survival difference has been reduced to 1%.[37] Rejection and early graft loss may be more common, especially in highly sensitized patients.[38,39] However, other reports found no difference in terms of acute rejection (AR) episodes, as well as 5-year and 8-year patient and graft survivals.[39,40] An additional study observed no significant difference in survival rates in different age groups.[41] In conclusions, the increased risk of graft loss observed in some studies in retransplanted patients is probably linked to highly sensitized patients rather than to retransplantation per se.

## Cold Ischemia Time (CIT)

Long CIT, with or without delayed graft function (DGF), has been associated with reduced graft survival.[42] These authors found that the risk for graft failure censored for death was significantly influenced by cold CIT, as well as by donor serum creatinine and their time-interaction variables. This means, for example, that an increased donor creatinine in kidneys from donors with different ages represents comparable damage. It also means that the deleterious effect of increased CIT affects living and cadaveric donor kidneys to the same extent.

Roodnat et al showed that the longer the CIT, the higher the graft failure risk.[43] These risks are highest in the directly postoperative phase and decrease in time to approach slowly. One year after transplantation, the difference in risks has disappeared. This means that the risk of a long CIT predominantly concerns the immediate postoperative phase, whereas a small deleterious effect is exerted for a longer period after transplantation. The influence of increased CIT on graft function has been studied in a few univariate[44,45] and multivariate analyses.[18,46-54] A significant effect on graft function or on the graft failure risk was shown by some of them.[18,45-47,51-53] Time dependency of the CIT was studied by Smits et al who concluded that there is a permanent detrimental effect of cold ischemia time on the risk for overall graft failure (no time dependency). By contrast, Roodnat et al did find a time interaction of cold ischemia time.

## Delayed Graft Function (DGF)

The reported frequency of delayed function of cadaveric kidney transplants greatly varies worldwide, from 2% to 50%.[55-62] Such variability mainly results from differences in the rates reported by different national and international transplantation registries, whether heart-beating or nonheart-beating donors were included, as well as the ambiguity in definition of the event. The Organ Procurement and Transplantation Network database[57] in the United States shows that 50% of patients with DGF start to recover renal function by day 10 after transplantation, whereas 33% regain function by day 10-20 and 10-15% do so subsequently. The rate of primary nonfunction is 2-15%.[57] Usually, the outcome is better after living donor transplantation than after cadaver donor transplantation. As expected, in kidneys from living donors, the frequency of DGF is 4-10%, average 5%.[46,57] These experimental results lend support to the finding that, in kidney-transplant recipients, DGF is a negative risk factor for long-term renal allograft survival.[63] Multivariate analysis confirmed this disorder as an independent predictor of graft loss and showed a relative risk of graft loss

2.9 times greater for delayed than for immediate kidney function.[64] Its importance on long-term graft outcome is further supported by findings that in cadaver transplants from 1994 to 1998 in the United States, the half-life of kidneys with no DGF was 11.5 years, compared with 7.2 years for those with delayed function.[65] Also, duration is a significant prognostic factor. In 126 kidney-transplant recipients grouped according to function recovery after surgery, 5-year graft survival was 90% in those with immediate graft function, 84% in those who recovered within the first week and 50% in those with temporary nonfunction for longer than 1 week.[66] Similarly, others have found that DGF of longer than 6 days strongly decreased long-term survival of transplanted kidneys.[67] Additionally, the analysis from the USRDS,[56] involving more than 37,000 primary cadaveric renal transplants, revealed that DGF was independently predictive of 5-year graft loss and its presence with early AR further reduced the rate of 5-year graft survival.

One of the most controversial issues is whether DGF is harmful in the absence of rejection. This interaction can be difficult to dissect because kidneys with this condition have a higher frequency of acute rejection (AR) and biopsy is not always done during the period of delayed function, so rejection may be underdiagnosed. Nevertheless, in a multicentre study of 57,000 first cadaveric transplants reported to the UNOS registry,[68] DGF had a strong effect independent of rejection. In the presence of rejection, its effect was even stronger, with graft half-life decreasing from 9.4 to 6.2 years.[68] In another study,[69] at 10-years after transplantation, the actuarial kidney graft survival was 64% in patients with no history of DGF or rejection episodes, 44% in those with DGF, 36% in those with history of rejection and 15% in those with both risk factors, further suggesting a potential additive negative effect on graft outcome. However, not all studies agree with the conclusion that delayed function is a strong predictor of long-term outcome of the graft. Indeed, some investigators have found that it is associated with increased graft loss in the first 6-12 months but not in the subsequent post-transplantation period.[70,71] In a cohort of 3,800 cadaveric renal transplants, patients whose function recovered well by 1 month after DGF had significantly lower 4-year graft survival compared with those with no history of this condition.[72] However, patients with delayed function but good graft function at 6 months showed similar long-term graft survival to those with no such history.

Although these discrepancies are difficult to reconcile, the controversial definition of both long-term outcome and DGF, as well as differences in the medication management that affects tubular cell proliferation or in the techniques used to monitor for rejection could explain, at least partly, some of the apparent variability of findings linking DGF to graft survival.

## Renal Graft Function

A natural candidate for a surrogate marker for graft loss that has been proposed is renal function (serum creatinine or calculated GFR levels). Using data from the USRDS, renal function along with the change in renal function demonstrated a high relative risk for ultimate graft survival and graft loss.[7] However, while renal function is a strong risk factor and highly correlated with graft failure, the utility of renal function as a predictive tool for graft loss is limited. Simliar conclusions were achieved by using the percentage of change in inverse serum creatinine (Delta1/Cr).[68]

## Proteinuria

Proteinuria (especially low-grade proteinuria defined as proteinuria <1 g/day) is observed in many patients within the first month following transplantation. Proteinuria 1 year after transplantation was associated with poor renal outcome in renal transplantation in several studies.[9,73-76] The risk of graft loss associated with proteinuria seemed linearly dependent upon its quantity.[9] Accordingly, low-grade proteinuria is often referred as 'subclinical' or 'negligible' and only persistent proteinuria >0.5-1.0 g/day for at least 3-6 months is considered significant according to the American Society of Transplantation.[77] However, a more recent study demonstrated that early low-grade proteinuria (even <0.5 g/day) and short-term increase in proteinuria are independent powerful predictors of graft loss in renal transplantation.[78] Proteinuria was associated with donor age, cardiovascular cause of donor death, prolonged cold and warm ischemia times and AR episodes.[78]

These findings suggest that proteinuria is the consequence of many factors including pretransplant renal lesions, ischemia-reperfusion injury and immunologic aggression. Accordingly, it was previously reported that microalbuminuria was more frequent,[79] and others found that proteinuria was more abundant in patients with a history of AR.[76] However, short-term reduction in proteinuria was associated with improved long-term graft survival, probably because it blocks the proteinuria-induced progression of renal function decline.[79] Altogether, these findings suggest that early proteinuria may thus reflect the existence of renal lesions fixed to a great extent.

## Acute Rejection (AR)

Patients with a history of AR episodes are more likely to develop chronic dysfunction or rejection than those without such episodes. In a Minnesota study, for example, 20% of living donor recipients and 36% of cadaver donor recipients who sustained an AR episode within the first 60 days post-transplant subsequently developed chronic rejection;[80] in comparison, patients with no history of rejection had less than a 1% chance of eventually developing chronic disease. This very low incidence of CAN in recipients without previous AR suggests that immunologic factors are the main determinants of long-term kidney transplant outcome.[81] These and other studies suggest that the benefit of living related transplantation results from the fact that a living related graft progresses from acute to CAN at a slower rate than a cadaveric graft.[81,82] Furthermore, a cadaveric graft that is free of AR 3 months after transplantation has an equal likelihood of functioning at 5 years as that of a graft from a living related donor.[82] The association between subclinical rejection and outcome has yielded contradictory results.[82-84] Presence of subclinical rejection with CAN was associated with old donors, percentage of PRA and presence of AR before protocol biopsy.[84] Treatment of subclinical rejection with corticosteroids leads to better histologic and functional outcomes in renal transplant recipients.[85] A recent study suggests that glomerular enlargement, an adaptation mechanism that occurs in the first year after transplant, is impaired in patients with subclinical rejection. Moreover, impaired glomerular enlargement is associated with progression of CAN,[86] thus explaining the association between the occurrence of subclinical rejection and graft failure.

## The Renal Artery Resistance Index (RI)

The renal artery resistance index (RI), assessed by Doppler ultrasonography, was recently identified as a new risk marker for late renal allograft loss. In a first study, the renal segmental arterial resistance index (the percentage reduction of the end-diastolic flow as compared with the systolic flow) was measured by Doppler ultrasonography in 601 patients at least three months after transplantation. A renal arterial resistance index of 80 or higher measured at least three months after transplantation is associated with poor subsequent allograft performance and death.[87] This finding was questioned because RI in that study was not measured at predetermined time points and ultrasonography is operator-dependent.[87] A further study investigated the predictive value of renal vascular resistance (RVR), a less operator-dependent method as assessed by mean arterial pressure divided by renal blood flow, for the prediction of recipient mortality and death-censored graft loss.[88] RVR was compared to commonly used risk markers such as creatinine clearance, serum creatinine and proteinuria in 793 first-time cadaveric renal transplant recipients at predetermined time points after transplantation. This study showed that RVR is a prominent risk marker for recipient mortality and death-censored graft loss.[88] However, the predictive

value of RVR for recipient mortality owed mainly to the impact of mean arterial blood pressure. In contrast, RVR constituted more than the sum of its components for death-censored graft loss, but showed less predictive value than serum creatinine in univariate analysis. As the assessment of RVR is expensive and time-consuming, the authors concluded that RVR holds no clinical merit for the follow-up of renal transplant recipients.[88]

## HLA Matching

Historically, registry analyses have demonstrated a benefit of HLA matching for renal transplantation. Before 1985, the 1-year loss rate for HLA-matched grafts in the United States was 17% compared with 42% for those with six A, B and DR mismatched antigens (a 25% difference).[89] In 1990, this difference decreased to 17%.[90] In 1995, the loss rate was 10% for HLA-matched and 18% for HLA-mismatched transplants.[90] In 2001, only 7% of HLA-matched and 12% of HLA-mismatched transplants were lost—a fourfold decrease from 1985.[91] Although the loss rate declined markedly in the period since 1985, the relative rate of loss remains approximately twice as high for mismatched transplants.

Half-life estimates, based on the rate of graft loss after the first year, indicate that HLA-matched grafts function 50% longer than those with mismatches.[92-93] Ten-year survival for zero mismatched transplants performed between 1979 and 1984 was 41% compared with 25% for those with five to six HLA mismatches.[90] Interestingly, an identical 16% difference in the estimated 10-year survival for zero and five to six mismatched transplants was projected for transplants performed between 1995 and 2000 (63% versus 47%, respectively).

After adjusting for other confounding factors, the hazard ratio of graft loss for HLA-mismatched kidneys compared with those with zero A, B and DR mismatches was 1.38. After censoring death with function, the hazard ratio for HLA-mismatched transplants was 1.55.[93] Rejection is another measure of outcome used to assess the efficacy of HLA matching.

A comparison of paired kidneys, in which one kidney was transplanted in an HLA-matched recipient and the other in a mismatched recipient, found that those receiving an HLA-matched transplant had a 6% lower incidence of rejection (13% versus 19%).[93] Accordingly, DR mismatches are associated with a increased risk of AR and early graft loss.[94-96] There is some evidence that HLA-A and HLA-B antigen mismatches are associated with late graft loss.[97,98] Thus, although more limited than previously thought, HLA matching influences graft outcome in renal transplantation.[99]

## Pretransplant Determination of Anti-HLA Antibodies

Anti-HLA antibodies can be induced in potential kidney transplant recipients after exposure to foreign HLA antigens in the form of blood transfusions, pregnancies, or previously rejected transplants. Hyperacute graft rejection, the direct consequence of donor-reactive anti-HLA antibodies, has become rare after introduction of the lymphocytotoxic crossmatch test[100] in which recipient serum is reacted against donor lymphocytes prior to transplantation. However, even if the crossmatch test is negative, patients with highly reactive PRA before transplantation are at an increased risk of rejection, which might be explained by a generally more reactive immune system or donor-specific antibodies that go undetected in the complement-dependent lymphocytotoxicity assay (CDC) due to the test's insufficient sensitivity. Recipients with high PRA have been shown to benefit from potent immunosuppression and HLA-matched kidneys.[101] Two different classes of anti-HLA antibodies can occur in sera of kidney graft recipients: antibodies directed against HLA class I antigens, expressed on nucleated cells and platelets; and antibodies against HLA class II antigens, expressed on B lymphocytes, activated T lymphocytes and cells of the macrophage lineage, including endothelial cells. Although extensively studied, the individual contribution of anti-HLA class I and class II antibodies to kidney graft rejection was unresolved until recently, mainly for technical reasons. Anti-HLA antibodies are routinely assessed in the CDC assay, in which a patient's serum is incubated with a panel of human lymphocytes obtained from different blood donors. Although testing against resting T lymphocytes for the detection of anti-HLA class I antibodies and against B cells for the detection of anti-HLA class II antibodies, has been used to allow a differentiation of anti-class I from anti-class II antibodies, the results have been less than satisfactory. Moreover, CDC has been criticized for not being able to detect noncomplement-binding, low affinity, or low-titer antibodies. Recently sensitive flow cytometry and ELISA techniques utilize solubilized class I or class II HLA antigens fixed onto microtiter plates or beads were developed.[102-105] These techniques allow precise differentiation between the two HLA antibody classes. However, the clinical relevance in kidney transplantation of antibodies detected by these sensitive techniques has been a matter of debate. More recent results performed using the ELISA technology suggest that presensitization of kidney recipients against *either* HLA class I or class II antigens was of no clinical consequence in first graft recipients, even in the presence of HLA mismatches. Only sensitization against *both* HLA class I and class II resulted in significantly increased rejection of HLA-mismatched grafts.[106] Among retransplant recipients, presensitization against HLA class I in the absence of anti-class II reactivity appeared to be harmful, suggesting a qualitative difference between anti-HLA class I antibodies produced in response to blood transfusions or pregnancies and antibodies produced as a result of graft rejection. Still, the vast majority of recipients do not possess any pretransplant anti-HLA antibodies.

## New Predictive Markers of Graft Outcome and Their Potential Applications

### Pretransplant Serum CD30 Levels

The CD30 molecule, a member of the tumor necrosis factor/nerve growth factor receptor superfamily, is a relatively large 120-kd glycoprotein that is preferentially expressed on type 2 T helper cells (Th2).[107] A soluble form of CD30 (sCD30) is released into the bloodstream after activation of CD30+ T-cells.[108] Recipients with a high pretransplant serum sCD30 content of $\geq 100$ U/mL had a significantly lower graft survival rate than recipients with a low sCD30 of <100 U/mL.[109,110] The sCD30 effect was evident in all important subgroups of patients: in first-graft recipients as well as in recipients of regrafts; in presensitized patients with a PRA of $\geq$ 5%; as well as in nonsensitized patients with a PRA of <5%. Using sCD30, 22% of the patients were categorized as high risk.[82] Recipients with a high pretransplant sCD30 required significantly more rejection treatment during the first post-transplant year and continued to lose grafts at a higher rate during the 5-year follow-up period,[110] suggesting that pretransplant sCD30 predicts not only the risk of acute but also chronic rejection. Except for tacrolimus, sirolimus and anti-IL-2R antibodies, which were not examined, none of the commonly used immunosuppressants were able to abrogate the effect of high pretransplant sCD30.[110] Recent in vitro data further support the involvement of CD30+ T-cells in the alloimmunization process.[111] When considered in combination with PRA, the predictive value of sCD30 increased. At the end of the third year there was a 17% difference in graft survival between CD30-negative/PRA-negative and CD30-positive/PRA-positive patients. It has been shown that the influence of HLA matching on graft survival is stronger in patients with preformed lymphocytotoxic antibodies than in nonsensitized patients.[112] Importantly, in an analysis of recipients without PRA, a strong HLA matching effect became apparent in nonsensitized recipients with high sCD30.[113] In other words, nonsensitized patients with high pretransplant

sCD30 appear to benefit greatly from an HLA well-matched kidney. Accordingly, in a pilot study of 56 kidney graft recipients, a nondecreasing plasma sCD30 content during the first 3 to 5 posttransplant days is a good predictor of impending graft rejection.[114]

### CXCR3-Binding Chemokines

Growing evidence, suggests that CXCL10 is critical in promoting and amplifying host alloresponses responsible for AR. In CXCL10- or CXCR3-gene deficient mice, cardiac transplants are not acutely rejected and undergo permanent engraftment.[115,116] Accordingly, neutralization of CXCL10 with monoclonal antibodies prolongs the allograft survival in both cardiac and small bowel allograft rejection.[116,117] Furthermore, the intragraft expression of CXCL10 has been reported in association with the rejection of renal,[118] lung[119] and cardiac allografts.[120,121] In a study evaluating the role of chemokines in renal transplant, induction of CXCR3 and CXCL10 mRNA was associated with AR.[118] Thus, the importance of CXCL10-CXCR3 interactions in the pathogenesis of graft failure appears to be clearly demonstrated in multiorgan models. Recent evidence indicates that CXCR3 and CXCL10 are also highly expressed in conjunction with the development of transplant vasculopathy in cardiac allografts[122] and that glomerular infiltration by CXCR3+ T-cells is associated with a transplant glomerulopathy and with the development of CAN in biopsies of kidney transplants.[123] In addition to its potent effects on immune responses,[124-130] CXCL10 also alters functions of vascular endothelial cells and pericytes,[130-135] thus promoting the development of CAN. These dual functions of CXCL10 on immune and vascular cells can account for its role in the pathogenesis of both AR and CAN, making this chemokine a novel target for future therapeutic strategies aimed at preventing transplant rejection. For all these reasons, measurement of pretransplant serum levels of CXCL10 were investigated to verify its possible predictive power of recipient's risk of graft rejection and transplant failure.[136,137] CCL22 was chosen as a control chemokine because its serum levels are increased in several immune-mediated inflammatory disorders,[138] but targeting the receptor for CCL22, CCR4, has no effect on graft survival.[139] Patients with normally functioning grafts showed significantly lower pretransplant serum levels of CXCL10 compared to patients who experienced graft failure. Such a difference was even more significant in patients who lost their graft within 2 years after transplantation. Pretransplant serum CCL22 levels did not correlate with the graft outcome.[136]

Life time analysis was performed after the assignment of patients to four groups according to the 0-25th (<64 pg/mL), 25th-50th (>64 and <97 pg/mL), 50th-75th (>97 and <157 pg/mL) and 75th-100th (>157 pg/mL) percentiles of CXCL10 serum levels. This analysis showed significantly lower, death-censored, 5-yrs survival rates of the grafts with increasing quartiles of CXCL10 serum levels. As expected, life-time analysis performed in relation to CCL22 serum levels revealed no difference in graft survival rates.[136]

Given the critical role of CXCL10 in the pathogenesis of AR, the relationship between pretransplant serum levels of CXCL10 and the occurrence of AR was then analysed. The frequency of AR episodes in the first month after transplant significantly increased in relation to increasing pretransplant serum CXCL10 levels. Patients who underwent rejection episodes within the first month after transplantation showed higher serum CXCL10 levels as compared with nonrejectors. In particular, patients with serum CXCL10 levels above the 75th percentile (>157 pg/mL) showed a nearly 2-fold greater frequency of rejection than patients whose serum CXCL10 levels were below the 75th percentile. Rejection episodes were not only more frequent, but also more severe, in patients showing high pretransplant serum levels of CXCL10. Indeed, among patients experiencing early rejection episodes, those who required antihuman thymocyte globulin treatment because of its severity, were characterized by significantly higher pretransplant serum levels of CXCL10 compared to patients with less severe rejection episodes. Furthermore, among kidney recipients, high pretransplant serum levels of CXCL10 were significantly and inversely correlated with the numbers of days elapsed until the first AR episode. Accordingly, CXCR3 binding chemokines were found to be increased in the urine of patients with AR.[140] The urinary levels of CXCL10 had a higher sensitivity and predictive value than serum creatinine in terms of monitoring the response to antirejection therapy[140] and urinary levels of CXCL9 and CXCL10 were sensitive and specific predictor for AR, mirroring response to antirejection therapy.[141] Recently, we also demonstrated that patients developing CAN had significantly higher pretransplant serum concentrations of CXCL10 than patients with normally functioning grafts.[136,137] Multivariate analyses indicated that high serum levels of CXCL10 were a significant risk factor for acute graft rejection and graft failure.[136] This effect was independent of other well known risk factors, such as recipient age and gender, number of HLA-A, -B and -DR mismatches, primary disease, type of immunosuppression, PRA, number of transplants, cold ischemia time, donor age and gender and even sCD30 levels, suggesting that, at least in this cohort of patients, measurement of pretransplant serum CXCL10 levels has a higher predictive power than sCD30.

Taken together, these results indicate that extremely high pretransplant serum levels of CXCL10 predict the risk for the development of AR and CAN with subsequent transplant failure. More recently, Matz et al demonstrated that high urinary levels of CXCL10 in the first days after transplant is predictive of AR, as well as of short and long-term graft function.[142] Thus, measurement of CXCR3-binding chemokines in serum or urine may be useful to select those patients requiring more aggressive immunosuppressive regimens.

## Conclusions

In spite of the continuous progress in immunosuppressive and supportive therapy, a number of factors still interfere with the complete success of renal transplantation. Current evidence suggests that a careful selection of the recipient and the donor and evaluation of some clinical variables acting mainly in the peritransplant period may help to reduce the recipient's risk for late renal allograft loss. However, the very low incidence of CAN in recipients without previous AR suggests that immunologic factors are the main determinants of long-term kidney transplant outcome. Recent evidence suggest that measurement of some cytokines before or immediately after transplantation can provide help in predicting graft outcome and in subsequent individualization and optimization of immunosuppression and make it reasonable to propose that frequencies of AR, CAN and subsequent graft failure should not be judged as absolute, but rather be evaluated in relation to the single patient's risk for each event.

## References

1. Meier-Kriesche HU, Schold JD, Srinivas TR et al. Lack of improvement in renal allograft survival despite a marked decrease in acute rejection rates over the most recent era. Am J Transplant 2004; 4:378-83.
2. Kreis HA, Ponticelli C. Causes of late renal allograft loss: chronic allograft dysfunction, death and other factors. Transplantation 2001; 71:5-9.
3. Lechler RI, Sykes M, Thomson AW et al. Organ transplantation—how much of the promise has been realized? Nat Med 2005; 11:605-13.
4. Marsden PA. Predicting outcomes after renal transplantation—new tools and old tools. N Engl J Med 2003; 349:182-84.
5. Lachenbruch PA, Rosenberg AS, Bonvini E et al. Biomarkers and surrogate endpoints in renal transplantation: present status and considerations for clinical trial design. Am J Transplant 2004; 4:451-57.
6. Susal C, Pelzl S, Simon T et al. Advances in pre- and post-transplant immunologic testing in kidney transplantation. Transplant Proc 2004; 36:29-34.
7. Kaplan B, Schold J, Meier-Kriesche HU. Poor predictive value of serum creatinine for renal allograft loss. Am J Transplant 2003; 3:1560-65.
8. Kasiske BL andany MA, Danielson B. A thirty percent chronic decline in inverse serum creatinine is an excellent predictor of late renal allograft failure. Am J Kidney Dis 2002; 39:762-68.
9. Roodnat JI, Mulder PG, Rischen-Vos J et al. Proteinuria after renal transplantation affects not only graft survival but also patient survival. Transplantation 2001; 72:438-44.
10. Hariharan S, McBride MA, Bennett LE et al. Risk factors for renal allograft survival from older cadaver donors. Transplantation 1997; 64:1748-54.
11. Alexander JW, Bennett LE, Breen TJ. Effect of donor age on outcome of kidney transplantation. A two-year analysis of transplants reported to the United Network for Organ Sharing Registry. Transplantation 1994; 57:871-76.
12. Basar H, Soran A, Shapiro R et al. Renal transplantation in recipients over the age of 60: the impact of donor age. Transplantation 1999; 67:1191-93.
13. Remuzzi G, Cravedi P, Perna A et al. Long-term outcome of renal transplantation from older donors. N Engl J Med 2006; 354:343-52.
14. Raspollini MR, Messerini L, Taddei GL. Long-term outcome of renal transplantation from older donors. N Engl J Med 2006; 354:2071-72.
15. Ruggenenti P, Perico N, Remuzzi G. Ways to boost kidney transplant viability: a real need for the best use of older donors. Am J Transplant 2006; 6:2543-47.
16. Cecka JM. The UNOS renal transplant registry. In: Cecka JM, Terasaki PI, eds. Clinical Transplants 2000. Los Angeles: UCLA Tissue Typing Laboratories, 2001:1-18.
17. Ponticelli C. Renal transplantation 2004: where do we stand today? Nephrol Dial Transplant 2004; 19:2937-47.
18. Meier-Kriesche HU. Relationship of recipient age and development of chronic allograft failure. Transplantation 2000; 70:306-10.
19. Light JA. Kidney transplants in African Americans and non-African Americans: equivalent outcomes with living but not deceased donors. Transplant Proc 2005; 37:699-700.
20. Keith DS, Cantarovich M, Paraskevas S et al. Recipient age and risk of chronic allograft nephropathy in primary deceased donor kidney transplant. Transpl Int 2006; 19:649-56.
21. Bresnahan BA, McBride MA, Cherikh WS et al. Risk factors for renal allograft survival from pediatric cadaver donors: an analysis of united network for organ sharing data. Transplantation 2001; 72:256-61.
22. Kasiske BL, Snyder JJ, Gilbertson D. Inadequate donor size in cadaver kidney transplantation. J Am Soc Nephrol 2002; 13:2152-59.
23. Giral M, Nguyen JM, Karam G et al. Impact of graft mass on the clinical outcome of kidney transplants. J Am Soc Nephrol 2005; 16:261-68.
24. Poggio ED, Hila S, Stephany B et al. Donor kidney volume and outcomes following live donor kidney transplantation. Am J Transplant 2006; 6:616-24.
25. Azevedo F, Alperovich G, Moreso F et al. Glomerular size in early protocol biopsies is associated with graft outcome. Am J Transplant 2005; 5:2877-82.
26. John AG, Rao M, Jacob CK. Pre-emptive live-related renal transplantation. Transplantation 1998; 66:204-9.
27. Papalois VE, Moss A, Gillingham KJ et al. Pre-emptive transplants for patients with renal failure: an argument against waiting until dialysis. Transplantation 2000; 70:625-31.
28. Kasiske BL, Snyder JJ, Matas AJ et al. Pre-emptive kidney transplantation: the advantage and the advantaged. J Am Soc Nephrol 2002; 13:1358-64.
29. Mange KC, Joffe MM, Feldman HI. Effect of the use or non-use of long-term dialysis on the subsequent survival of renal transplants from living donors. N Engl J Med 2001; 344:726-31.
30. Meier-Kriesche HU, Port FK et al. Effect of waiting time on renal transplant outcome. Kidney Int 2000; 58:1311-17.
31. Goldfarb-Rumyantzev A, Hurdle JF, Scandling J et al. Duration of end-stage renal disease and kidney transplant outcome. Nephrol Dial Transplant 2005; 20:167-75.
32. Meier-Kriesche HU, Kaplan B. Waiting time on dialysis as the strongest modifiable risk factor for renal transplant outcomes: a paired donor kidney analysis. Transplantation 2002; 74:1377-81.
33. Van Biesen W, Vanholder R, Van Loo A et al. Peritoneal dialysis favorably influences early graft function after renal transplantation compared to hemodialysis. Transplantation 2000; 69:508-14.
34. Ojo AO, Hanson JA, Wolfe RA et al. Dialysis modality and the risk of allograft thrombosis in adult renal transplant recipients. Kidney Int 1999; 55:1952-60.
35. Murphy BG, Hill CM, Middleton D et al. Increased renal allograft thrombosis in CAPD patients. Nephrol Dial Transplant 1994; 9:1166-69.
36. Snyder JJ, Kasiske BL, Gilbertson DT, et al. A comparison of transplant outcomes in peritoneal and hemodialysis patients. Kidney Int 2002; 62:1423-30
37. Hirata M, Terasaki P.I. Renal retransplantation. Clin Transpl 1994; 10:419-33.

38. Stratta RJ, Oh CS, Sollinger HW et al. Kidney retransplantation in the cyclosporine era. Transplantation 1988; 45:40-45.

39. Mouquet H, Benalia H, Chartier-Kastler E et al. Renal retransplantation in adults. Comparative prognostic study. Prog Urol 1999; 9:239-43.

40. Delmonico FL, Tolkoff-Rubin N, Auchincloss H Jr et al. Second renal transplantations Ethical issues clarified by outcome; outcome enhanced by a reliable crossmatch. Arch Surg 1994; 129:354-60.

41. Pour-Reza-Gholi F, Nafar M, Saeedinia A et al. Kidney retransplantation in comparison with first kidney transplantation. Transplant Proc 2005; 37:2962-64.

42. Troppmann C, Gillingham KJ, Gruessner RW et al. Delayed graft function in the absence of rejection has no long-term impact. A study of cadaver kidney recipients with good graft function at 1 year after transplantation. Transplantation 1996; 61:1331-7.

43. Roodnat JI, Mulder PG, Van Riemsdijk IC et al. Ischemia times and donor serum creatinine in relation to renal graft failure. Transplantation 2003; 75:799-804.

44. Terasaki P, Cecka JM, Gjerdson DWJ et al. High survival rates of kidney transplants from spousal and living unrelated donors. N Engl J Med 1995; 333:333-36.

45. Cecka JM. The UNOS scientific renal transplant registry. Clin Transpl 1992;1-18.

46. Smits JMA, Houwelingen van HC, Meester de J et al. Permanent detrimental effect of non-immunological factors on long-term renal graft survival. Transplantation 2000; 70:317-23.

47. Chertow GM, Milford EL, MacKenzie HF et al. Antigen independent determinants of cadaveric kidney transplant failure. JAMA 1996; 276:1732-36.

48. Connolly JK, Dyer PA, Martin S et al. Importance of minimizing HLA-DR mismatch and cold preservation time in cadaveric renal transplantation. Transplantation 1996; 61:709-14.

49. Madrenas J, Newman S, McGregor JR et al. An alternative approach for statistical analysis of kidney transplant data: Multivariate analysis of single center experience. Am J Kidney Dis 1988; 12:524-30.

50. Heaf JG, Ladefoged J. Hyperfiltration, creatinine clearance and chronic graft loss. Clin Transpl 1998; 12:11-18.

51. Gjertson JW. A multifactor analysis of kidney graft outcomes at one and five years post-transplantation: 1996 UNOS update. Clin Transpl 1996; 343-60.

52. Thorogood J, Houwelingen JC, Persijn GG et al. Prognostic indices to predict survival of first and second renal allografts. Transplantation 1991; 52:831-36.

53. Held PJ, Kahan BD, Hunsicker LG et al. The impact of HLA mismatches on the survival of first cadaveric kidney transplants. N Engl J Med 1994; 331:765-70.

54. Peters TG, Shaver TR, Ames JE et al. Cold ischemia and outcome in 17,937 cadaveric kidney transplants. Transplantation 1995; 59:191-96.

55. Perico N, Cattaneo D, Sayegh MH et al. Delayed graft function in kidney transplantation. Lancet 2004; 364:1814-27.

56. Ojo AO, Wolfe RA, Held P et al. Delayed graft function: risk factors and implications for renal allograft survival. Transplantation 1997; 63:968-74.

57. The Organ Procurement and Transplantation Network (http://www.OPTN.org) (accessed 2004).

58. Jacobs SC, Cho E, Foster C et al. Laparoscopic donor nephrectomy: the University of Maryland 6-year experience. J Urol 2004; 171:47-51.

59. The Canadian Multicentre Transplant Study Group. A randomized clinical trial of cyclosporine in cadaveric renal transplantation. Analysis at three years. N Engl J Med 1986; 314:1219-25.

60. Gjertson DW. Impact of delayed graft function and acute rejection on kidney graft survival. In: Cecka JM, Terasaki PI, eds. Clinical Transplants. Los Angeles: The Regents of the University of California, 2000:467-79.

61. Koning OHJ, Ploeg RJ, Van Bockel JH et al. Risk factors for delayed graft function in cadaveric kidney transplantation. Transplantation 1997; 63:1620-28.

62. Senel FM, Karakayali H, Moray G et al. Delayed graft function: predictive factors and impact on outcome in living-related kidney transplantations. Ren Fail 1998; 20:589-95.

63. Whjttaker JR, Veith FJ, Soberman R et al. The fate of the renal transplant with delayed function. Surg Gynecol Obstet 1973; 136:919-22.

64. Halloran PF, Aprile MA, Farewell V et al. Early function as the principle correlate of graft survival: a multivariate analysis of 200 cadaveric renal transplants treated with a protocol incorporating antilymphocyte globulin and cyclosporine. Transplantation 1988; 46:223-28.

65. Halloran PF, Hunsicker LG. Delayed graft function: state of the art, 2000. Summit Meeting, Scottsdale, Arizona, USA. Am J Transplant 2001; 1:115-20.

66. Yokoyama I, Uchida K, Kobayashi T et al. Effect of prolonged delayed graft function on long-term graft outcome in cadaveric kidney transplantation. Clin Transplant 1994; 8:101-6.

67. Giral-Classe M, Hourmant M, Cantarovich D et al. Delayed graft function of more than six days strongly decreases long-term survival of transplanted kidneys. Kidney Int 1998; 54:972-78.

68. Shoskes DA, Cecka M. Deleterious effects of delayed graft function in cadaveric renal transplant recipients independent of acute rejection. Transplantation 1998; 66:1697-701.

69. Troppman C, Gruessner AC, Gillingham KJ et al. Impact of delayed graft function on long-term graft survival after solid organ transplantation. Transplant Proc 1999; 31:1290-92.

70. Prommool S, Jhangri GS, Cockfield SM et al. Time dependency of factors affecting renal allograft survival. J Am Soc Nephrol 2000; 11:565-73.

71. Woo YM, Jardine AG, Clark AF et al. Early graft function and patient survival following cadaveric renal transplantation. Kidney Int 1999; 55:692-99.

72. Sanfilippo F, Vaughn W, Spees EK et al. The detrimental effects of delayed graft function in cadaver donor renal transplantation. Transplantation 1984; 38:643-48.

73. Yidiz A, Erkoc R, Sever MS et al. The prognosis importance of severity and type of proteinuria of post-transplant proteinuria. Clin Transpl 1999; 13:241-44.

74. Massy ZA, Guijarro C, Wiederkehr MR et al. Chronic allograft rejection: immunologic and non-immunologic risk factors. Kidney Int 1996; 49:518-24.

75. Hohage H, Kleyer U, Bruckner D et al. Influence of proteinuria on long-term transplant survival in kidney transplant recipients. Nephron 1997; 75:160-65.

76. Fernandez-Fresnedo G, Plaza JJ, Sanchez-Plumed J et al. Proteinuria: a new marker of long-term graft and patient survival in kidney transplantation. Nephrol Dial Transplant 2004; 19:47-51.

77. Kasiske BL, Vasquez MA, Harmon WE et al. Recommendations for the outpatient surveillance of renal transplant recipients. American Society of Transplantation. J Am Soc Nephrol 2000; 11:1-86.

78. Halimi JM, Laouad I, Buchler M et al. Early low-grade proteinuria: causes, short-term evolution and long-term consequences in renal transplantation. Am J Transplant 2005; 5:2281-88.

79. Halimi JM, Al-Najjar A, Buchler M et al. Microalbuminuria in renal transplant recipients: influence of history of acute rejection episodes and sodium intake. Transplant Proc 2002; 34:801-2.

80. Basadonna GP, Matas AJ, Gillingham KJ et al. Early versus late acute renal allograft rejection: impact on chronic rejection. Transplantation 1993; 55:993-95.

81. Humar A, Hassoun A, Kandaswamy R et al. Immunologic factors: the major risk for decreased long-term renal allograft survival. Transplantation 1999; 68:1842-46.

82. Knight RJ, Burrows L, Bodian C. The influence of acute rejection on long-term renal allograft survival: a comparison of living and cadaveric donor transplantation. Transplantation 2001; 72:69-76.

83. Roberts IS, Reddy S, Russell C et al. Subclinical rejection and borderline changes in early protocol biopsy specimens after renal transplantation. Transplantation 2004; 77:1194-98.

84. Moreso F, Ibernon M, Goma M et al. Subclinical rejection associated with chronic allograft nephropathy in protocol biopsies as a risk factor for late graft loss. Am J Transplant 2006; 6:747-52.

85. Rush D, Nickerson P, Gough J et al. Beneficial effects of treatment of early subclinical rejection: a randomized study. J Am Soc Nephrol 1998; 9:2129-34.

86. Ibernon M, Goma M, Moreso F et al. Subclinical rejection impairs glomerular adaptation after renal transplantation. Kidney Int 2006; 70:557-61.

87. Giraudeau B, Halimi J-M, Jay SJ et al. Renal Arterial Resistance Index. NEJM 2003; 349:1573-74.

88. de Vries AP, van Son WJ, van der Heide JJ et al. The predictive value of renal vascular resistance for late renal allograft loss. Am J Transplant 2006; 6:364-70.

89. Mickey MR. HLA matching in transplants from cadaver donors. In: Terasaki PI, ed. Clinical Kidney Transplants 1985. Los Angeles: UCLA Tissue Typing Laboratory, 1985:45.

90. Cecka JM. The UNOS Scientific Renal Transplant Registry—ten years of kidney transplants. Clin Transpl 1997;1-14.

91. 2002 OPTN/SRTR Annual Report 1992-2001, HHS/HRSA/OSP/DOT, UNOS, URREA, Rockville MD, Richmond VA, Ann Arbor MI 2002.

92. Takiff H, Cook DJ, Himaya NS et al. Dominant effect of histocompatibility on ten-year kidney transplant survival. Transplantation 1988; 45:410-15.

93. Takemoto SK, Terasaki PI, Gjertson DW et al. Twelve years' experience with national sharing of HLA-matched cadaveric kidneys for transplantation. New Engl J Med 2000; 343:1078-84.

94. Pirsch JD, Ploeg RJ, Gange S et al. Determinants of graft survival after renal transplantation. Transplantation 1996; 61:1581-86.

95. Barocci S, Valente U, Gusmano R et al. HLA matching in pediatric recipients of a first kidney graft. A single center analysis. Transplantation 1996; 61:151-54.

96. Connolly JK, Dyer PA, Martin S et al. Importance of minimizing HLA-DR mismatch and cold preservation time in cadaveric renal transplantation. Transplantation 1996; 61:709-14.

97. Thorogood J, Persijn GG, Schreuder GM et al. The effect of HLA matching on kidney graft survival in separate post-transplantation intervals. Transplantation 1990; 50:146-50.

98. Zantvoort FA, D'Amaro J, Persijn GG et al. The impact of HLA-A matching on long-term survival of renal allografts. Transplantation 1996; 61:841-44.

99. Takemoto S, Port FK, Claas FH et al. HLA matching for kidney transplantation. Hum Immunol 2004; 65:1489-505.

100. Patel R, Terasaki PI. Significance of the crossmatch test in kidney transplantation. N Engl J Med 1969; 280:735-39.

101. Shenton BK. The detection and relevance of pretransplant antibodies. Transplant Immunol 1994; 2:135-37.

102. Kao KJ, Scornik JC, Small SJ. Enzyme-linked immunoassay for anti-HLA antibodies: an alternative to panel studies by lymphocytotoxicity. Transplantation 1993; 55:192-96.

103. Müller-Steinhardt M, Fricke L, Kirchner H et al. Monitoring of anti-HLA class I and II-antibodies by flow cytometry in patients after first cadaveric kidney transplantation. Clin Transplant 2000; 14:85-89.

104. Lobashevsky AL, Senkbeil RW, Shoaf J et al. Specificity of preformed alloantibodies causing B-cell positive flow crossmatch in renal transplantation. Clin Transplant 2000; 14:533-42.

105. Christiaans MH, Nieman F, van Hooff JP et al. Detection of HLA class I and II antibodies by ELISA and complement-dependent cytotoxicity before and after transplantation. Transplantation 2000; 69:917-27.

106. Süsal C, Opelz G. Kidney graft failure and presensitization against HLA class I and class II antigens. Transplantation 2002; 73:1269-73.

107. D'Elios MM, Romagnani P, Scaletti C et al. In vivo CD30 expression in human diseases with predominant activation of Th2-like T-cells. J Leukoc Biol 1997; 61:539-44.

108. Romagnani S, Del Prete G, Maggi E et al. CD30 and type 2 T helper (Th2) responses. J Leukoc Biol 1995; 57:726-30.

109. Pelzl S, Opelz G, Wiesel M et al. Soluble CD30 as a predictor of kidney graft outcome. Transplantation 2002; 73:3-6.

110. Süsal C, Pelzl S, Döhler B et al. Identification of highly responsive kidney transplant recipients using pretransplant soluble CD30. J Am Soc Nephrol 2002; 13:1650-56.

111. Chan KW, Hopke CD, Krams SM et al. CD30 expression identifies the predominant proliferating T lymphocyte population in human alloimmune responses. J Immunol 2002; 169:1784-91.

112. Opelz G, Wujciak T, Mytilineos J et al. HLA compatibility and organ transplant survival. Rev Immunogenet 1999; 1:334-42.

113. Süsal C, Pelzl S, Opelz G. Strong HLA matching effect in nonsensitized kidney recipients with high pretransplant soluble CD30. Transplantation 2003; 76:1231-32.

114. Pelzl S, Opelz G, Daniel V et al. Evaluation of post-transplantation soluble CD30 for diagnosis of acute renal allograft rejection. Transplantation 2003; 75:421-23.

115. Hancock W, Lu B, Gao W et al. Requirement of the chemokine receptor CXCR3 for acute allograft rejection. J Exp Med 2000; 192:1515-19.

116. Hancock WW, Gao W, Csizmadia V et al. Donor-derived IP-10 initiates development of acute allograft rejection. J Exp Med 2001; 193:975-80.

117. Zhang Z, Kaptanoglu L, Haddad W et al. Donor T-Cell Activation Initiates Small Bowel Allograft Rejection Through an IFN-γ-Inducible Protein-10-Dependent Mechanism. J Immunol 2002; 168:3205-12.

118. Segerer S, Cui Y, Eitner F et al. Expression of chemokines and chemokine receptors during human renal transplant rejection. Am J Kidney Dis 2001; 37:518-53.

119. Agostini C, Calabrese F, Rea F et al. CXCR3 and its ligand CXCL10 are expressed by inflammatory cells infiltrating lung allografts and mediate chemotaxis of T-cells at sites of rejection. Am J Pathol 2001; 158:1703-11.

120. Melter M, Exeni A, Reinders ME et al. Expression of the chemokine receptor CXCR3 and its ligand IP-10 during human cardiac allograft rejection. Circulation 2001; 104:2558-64.

121. Fahmy NM, Yamani MH, Starling RC et al. Chemokine and chemokine receptor gene expression indicates acute rejection of human cardiac transplants. Transplantation 2003; 75:72-78.

122. Zhao DX, Hu Y, Miller GG et al. Differential Expression of the IFN-{gamma}-Inducible CXCR3-Binding Chemokines, IFN-Inducible Protein 10, Monokine Induced by IFN and IFN-Inducible T-Cell {alpha} Chemoattractant in Human Cardiac Allografts: Association with Cardiac Allograft Vasculopathy and Acute Rejection. J Immunol 2002; 169:1556-60.

123. Akalin E, Dikman S, Murphy B et al. Glomerular Infiltration by CXCR3+ ICOS+ Activated T-Cells in Chronic Allograft Nephropathy with Transplant Glomerulopathy. Am J Transplant 2003; 3:1116-20.

124. Rossi D, Zlotnik A. The biology of chemokines and their receptors. Annu Rev Immunol 2000; 217-42.

125. Bonecchi R, Bianchi G, Bordignon PP et al. Differential expression of chemokine receptors and chemotactic responsiveness of type 1 T helper cells (Th1s) and Th2s. J Exp Med 1998; 187:129-34.

126. Loetscher M, Gerber B, Loetscher P et al. Chemokine receptor specific for IP10 and Mig: structure, function and expression in activated T-lymphocytes. J Exp Med 1996; 184:963-69.

127. Romagnani P, Annunziato F, Lazzeri E et al. Interferon-inducible protein 10, monokine induced by interferon gamma and interferon-inducible T-cell alpha chemoattractant are produced by thymic epithelial cells and attract T-cell receptor (TCR) alphabeta+ CD8+ single-positive T-cells, TCRgammadelta+ T-cells and natural killer-type cells in human thymus. Blood 2001; 97:601-7.

128. Romagnani P, Maggi L, Mazzinghi B et al. CXCR3-mediated opposite effects of CXCL10 and CXCL4 on TH1 or TH2 cytokine production. J Allergy Clin Immunol 2005; 116:1372-79.

129. Lazzeri E, Romagnani P. CXCR3-binding chemokines: novel multifunctional therapeutic targets. Curr Drug Targets Immune Endocr Metabol Disord 2005; 5:109-18.

130. Lasagni L, Francalanci M, Annunziato F et al. An alternatively spliced variant of CXCR3 mediates the inhibition of endothelial cell growth induced by IP-10, Mig and I-TAC and acts as functional receptor for platelet factor 4. J Exp Med 2003; 197:1537-49.

131. Wang X, Yue TL, Ohlstein EH et al. Interferon-inducible protein-10 involves vascular smooth muscle cell migration, proliferation and inflammatory response. J Biol Chem 1996; 271:24286-93.

132. Romagnani P, Lasagni L, Annunziato F et al. CXC chemokines: the regulatory link between inflammation and angiogenesis. Trends Immunol 2004; 25:201-9.

133. Romagnani P, Beltrame C, Annunziato F et al. Role for interactions between IP-10/Mig and CXCR3 in proliferative glomerulonephritis. J Am Soc Nephrol 1999; 10:2518-26.

134. Romagnani P, Annunziato F, Lasagni L et al. Cell cycle-dependent expression of CXC chemokine receptor 3 by endothelial cells mediates angiostatic activity. J Clin Invest 2001; 107:53-63.

135. Bonacchi A, Romagnani P, Romanelli RG et al. Signal transduction by the chemokine receptor CXCR3: activation of Ras/ERK, Src and phosphatidylinositol 3-kinase/Akt controls cell migration and proliferation in human vascular pericytes. J Biol Chem 2001; 276:9945-54.

136. Rotondi M, Rosati A, Buonamano A et al. High pretransplant serum levels of CXCL10/IP-10 are related to increased risk of renal allograft failure. Am J Transplant 2004; 4:1466-74.

137. Lazzeri E, Rotondi M, Mazzinghi B et al. High CXCL10 expression in rejected kidneys and predictive role of pretransplant serum CXCL10 for acute rejection and chronic allograft nephropathy. Transplantation 2005; 79:1215-20.

138. Romagnani P, Rotondi M, Lazzeri E et al. Expression of IP-10/CXCL10 and MIG/CXCL9 in the thyroid and increased levels of IP-10/CXCL10 in the serum of patients with recent-onset Graves' disease. Am J Pathol 2002; 161:195-206.

139. Hancock W, Wang L, Ye Q et al. Chemokines and their receptors as markers of allograft rejection and targets for immunosuppression. Curr Opin Immunol 2003; 15:479-86.

140. Hu H, Aizenstein BD, Puchalski A et al. Elevation of CXCR3-binding chemokines in urine indicates acute renal-allograft dysfunction. Am J Transplant 2004; 4:432-37.

141. Hauser IA, Spiegler S, Kiss E et al. Prediction of acute renal allograft rejection by urinary monokine induced by IFN-gamma (MIG). J Am Soc Nephrol 2005; 16:1849-58.

142. Matz M, Beyer J, Wunsch D et al. Early post-transplant urinary IP-10 expression after kidney transplantation is predictive of short- and long-term graft function. Kidney Int 2006; 69:1683-90.

# CHAPTER 25

# Clinico-Pathological Correlations of Chronic Allograft Nephropathy

Jeremy R. Chapman*

## Abstract

Despite, or perhaps because of, common usage "CAN" is a poorly defined term. At Westmead we combine both pathology and physiology to arrive at the following definition: "Progressive graft dysfunction accompanied by chronic interstitial fibrosis, tubular atrophy, vascular occlusive changes and glomerulosclerosis". Chronic allograft nephropathy (CAN) is the cause of a majority of renal transplant failure in most countries where losses due to acute rejection have abated. Pathologists have decided to abandon the term and replace it with a specific pathological description "IFTA" or interstitial fibrosis and tubular atrophy—simply a pathology description of the appearance of a kidney biopsy.

Long term protocol biopsy series have demonstrated the combined impacts of immune and non-immune influences on the pathological appearances. CNI nephrotoxicity, sub-clinical and chronic humoral rejection remain realistic targets for changed therapeutic strategies. Clinical strategies for both monitoring and investigating CAN are sadly flawed, with serum creatinine, hypertension and proteinuria occurring very late and badly underestimating graft histology. New non-interventional investigational tools are needed to replace the current gold standard: protocol biopsy.

## Introduction

Our pleasure at conquering the problem of graft loss from acute rejection has been tempered by the sad realisation that the long term prognosis of grafts has not been improved.[1,2] Death from vascular, malignant or infective disease and chronic failure of the kidney transplant have now become our most significant targets for improving the quality and quantity of life after transplantation.[1,3]

During the 1980s and early 1990s there was clearly a failure to understand the phenomenon then called "chronic rejection". The Banff meeting of pathologists coined the term "chronic allograft nephropathy" and by doing so disconnected the appearance of chronic graft damage from any presumption of it being solely due to an immune pathophysiological mechanism. In common parlance CAN has become little more than short hand for progressive loss of kidney function through graft fibrosis, leading to graft failure and the need for dialysis. New thinking amongst the histopathologists has now lead to abandonment of Chronic Allograft Nephropathy and replacement with "IFTA" or interstitial fibrosis and tubular atrophy. This has been designed to be taken simply as a pathological description of the appearance of the kidney biopsy.[4] It will not be the rallying cry that "CAN" has been, but one can understand the reason for change because "CAN" has taken on a life of its own, disconnected from underlying clinical realities and permitting abrogation of responsibility for prevention and treatment in the individual patient. Over the past few years longitudinal histological studies have helped in the understanding of the limited responses to injury that the kidney possesses leading to development of strategies for both prevention and treatment.

## The Clinicians' View of Chronic Allograft Nephropathy

Intercept and slope have been introduced to describe the functional deterioration of renal allografts.[5] Intercept describes the initial functional capacity of a kidney and slope describes the rate of deterioration after transplantation. One can thus contrast two kidneys both of which are losing Glomerular Filtration Rate (GFR) at 5 ml/min/year, one of which starts at an intercept of 100ml/min while the other from an elderly hypertensive donor starts at only 30 ml/min. The first graft will last for nearly twenty years before the recipient finally returns to dialysis, while the second graft will appear to fail rapidly over only 4 years. A naive clinician will miss the fact that the first graft is being damaged using crude clinical measures such as serum creatinine, but the second graft will be identified as a concern quite rapidly despite the fact that the underlying rate of damage is identical in the two examples. The renal functional reserve disguises the underlying histological damage and serum creatinine is an insensitive indicator.[6] Clinical reliance on serum creatinine has been justified through management of the acute posttransplant course, where it has provided a sensitive and specific indicator for both acute rejection and other reversible causes of renal damage. This fact has mislead us as we have translated this monitoring model to long term management systems.

Assessment of the pathological damage been undertaken in series of routine protocol biopsies both in the early posttransplant period and in a few transplant programs also for chronic transplant monitoring. This chapter reviews our current understanding of the pathophysiological correlates of chronic allograft nephropathy.

## The Pathologists' View of Chronic Allograft Nephropathy

The "Banff" meetings have provided the focus for classifying renal allograft histology[7-10] and was responsible for introducing "Chronic Allograft Nephropathy" into our lexicon in the early 1990s. CAN is defined by the presence of tubular atrophy and interstitial fibrosis of the allograft but does not imply aetiology in the way assumed by the previous term "chronic rejection".[7] Underlying pathophysiology can

*Jeremy R. Chapman—Department of Renal Medicine, Westmead Hospital, WESTMEAD, NSW 2145, Australia.
Email: jeremy_chapman@wsahs.nsw.gov.au

*Chronic Allograft Failure: Natural History, Pathogenesis, Diagnosis and Management*, edited by Nasimul Ahsan. ©2008 Landes Bioscience.

be detected histologically in order to assign a presumed aetiology in 60% of chronic allograft biopsies.[11] Nodular arteriolar hyalinosis and a striped pattern of fibrosis associated with calcineurin inhibitor toxicity; disruption of the elastica, inflammatory cells in the fibrotic intima and proliferation of myofibroblasts in the intima, with or without deposition of the complement component C4d, implying chronic immune mediated rejection.

The Banff 97 system[7] identified the proportion of interstitial fibrosis and tubular atrophy which are generally widespread even in the small tissue biopsy samples from 18G renal allograft core biopsies. CAN grades were based on the proportion of the cortical area of the biopsy affected by chronic interstitial fibrosis and tubular atrophy, with Grade I from 6–25% of the area of the biopsy, Grade II being 26–50% and Grade III being more than 50%. The CAN grade and scoring of the chronic Banff qualifiers for interstitial fibrosis (ci) and tubular atrophy (ct) were not directly linked, providing a limitation of the system.

The recent rethinking of use of the phrase "CAN" has become necessary because it has been hijacked as a catch phrase for progressive allograft dysfunction and not as a pathological entity. The Banff group has determined that there should be reversion to a focus on the components of histology that reflect the aetiology. The revised Banff Schema[4] uses the phrase IFTA to replace CAN. However, more than 800 publications using the term CAN can't all be wrong, so the Banff 97 classification remains a reality in the available studies and literature and will continue to influence thinking for many years to come.

In a report which accumulated 2,127 protocol biopsies there was one graft loss and three direct interventions for bleeding,[12] reinforcing the view that it is safer to biopsy a kidney and gain an understanding of subclinical events than to rely on flawed and failed non-invasive monitoring. The protocol biopsy provides the rigorous scientific basis for understanding the pathophysiological and also the gold standard for development of both new diagnostic and therapeutic options.

## What Does Longitudinal Histology Reveal?

Our group has published two different series of protocol biopsies addressing the pathophysiological causes and correlates of CAN from biopsies at implantation and at 3 months and 12 months. We have then been able to follow the patients for up to 15 years to understand the impact of early damage mediated through histological changes.[13,14]

A second series of biopsies was taken came from 120 simultaneous pancreas kidney (SPK) transplant recipients biopsied annually for ten years to yield approximately 1,000 biopsies.[15-19]

Our first series included kidneys from both deceased and living donors of all ages, while the kidney pancreas transplant donors were all under 45years old with normal implantation biopsies and thus capable of demonstrating the evolution of long term graft fibrosis[15] and the natural history of CAN unentangled by preceding donor disease. Many other transplant programs have reported long term protocol biopsy studies providing reinforcement of hypotheses and refinement of understanding of the clinicopathological correlations.[20-26]

## What Are the Predictors of Histology in the First Year?

Both chronic and acute changes are seen on protocol biopsies 3 to 12 months after transplantation.[13,14] At three months only 22% of grafts were normal, 56% had CAN grade I and 24% CAN grade II or III. Even by three months interstitial fibrosis and tubular atrophy dominate and were correlated with the age of the donor, delayed graft function and episodes of vascular rejection. Chronic vascular damage in the three month biopsy was due to donor vascular disease and correlated with total cold ischaemic time and prior vascular rejection. Not surprisingly, 12 month histology reflects both chronic changes on the 3 month biopsy and the continuing acute inflammatory changes still present at 3 months. Subclinical rejection is the presence of acute interstitial infiltrates and tubulitis, but without deteriorating serum creatinine. Presence of subclinical rejection on the 3 month protocol biopsy, predicted interstitial fibrosis and CAN on the 12 month biopsy. In other words a subclinical infiltrate is fibrosing and damaging the graft, but it is not evident because it is within the functional capacity of the kidney and does not impact immediately on graft function. Figure 1 demonstrates the significant predictors of CAN in the 12 month biopsy. Graft survival was predicted, in this series by development of proteinuria and by histological changes seen on the three month protocol biopsy scored as ci (chronic interstitial fibrosis) and cv (chronic vascular damage) Figure 2 shows the actuarial graft survival based upon the degree of interstitial fibrosis seen within the first year. Cyclosporine exposure, hypertension and donor and recipient age did not correlate with outcomes in this relatively small series. Similar findings have been

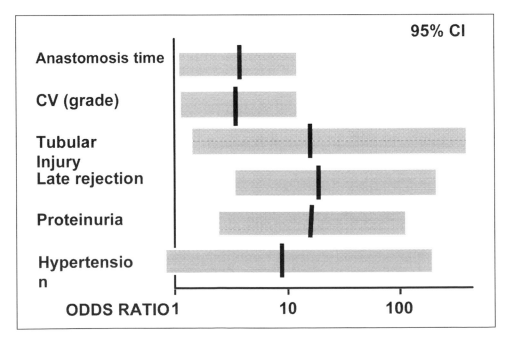

Figure 1. Predictors of Chronic Allograft Nephropathy at three months.[13]

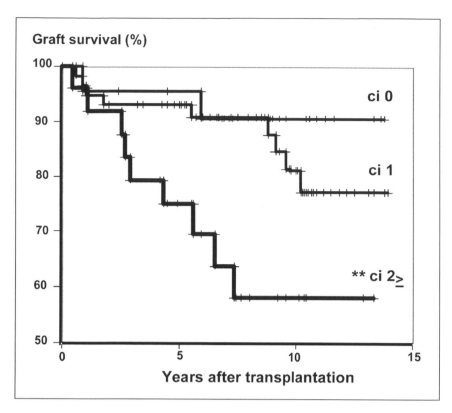

Figure 2. Correlation between Graft survival and chronic interstitial fibrosis at three months after transplantation.[13]

made by others, reflecting lack of change in short term creatinine values despite progression of chronic histological lesions[21] and correlation between chronic histological changes on six month protocol biopsy and graft survival after only three years of follow up.[22]

## What Are the Predictors of Long Term Histology?

If early pathology can predict outcome, it would be important to identify the linkages between early damage and subsequent graft failure since this knowledge could suggest therapeutic options and help us understand how a damaged graft ultimately fails.

Our second protocol biopsy series was part of our long term study of simultaneous pancreas and kidney transplantation in patients with insulin dependent diabetes mellitus.[15] In this study we measured GFR and preformed an oral glucose tolerance test and renal transplant biopsy at 0 3 and 12 months and then annually for ten years. Survivals at one year were: patient (96%), kidney (94%) and pancreas (85%) and at 10 years were patient (80%), kidney (77%) and pancreas (67%).[15] Measured GFR in the survivors was 61 ml/min at one year and 50 ml/min between 6 and 10 years but the pathological changes were severe.

## Chronic Interstitial Fibrosis and Tubular Atrophy

Biopsies at the time of transplantation in the relatively young donors selected for SPK transplants were essentially normal, but there was a dramatic increase in interstitial fibrosis and tubular atrophy in the first year.[17] Around two thirds of the total burden of fibrosis seen at the ten year mark had already occurred in the first year and was associated with acute tubular necrosis, acute rejection and untreated subclinical rejection. Subclinical rejection was an important cause of fibrosis initially but declined with time after transplantation to be replaced by calcineurin inhibitor toxicity as the principle late cause of the IFTA. Renal function correlated loosely with the amount of fibrosis, with a measured GFR of 65 ml/min in patients with normal biopsies, to 59ml/min and 44 ml/min with mild and moderate fibrosis respectively. The three most prominent and important patterns of graft fibrosis are Calcineurin Inhibitor (CNI) Nephrotoxicity, untreated sub-clinical rejection and

chronic humoral or antibody mediated rejection. If we are to impact on long term graft survival we will need to prevent progression of CAN/IFTA through early recognition and treatment of these different causative patterns of histological damage, especially CNI nephrotoxicity, as well as cellular and humoral rejection processes.

## Calcineurin Inhibitor Nephrotoxicity and Arteriolar Hyalinosis

While absence of histological signs of rejection has been used to assist in the diagnosis of acute CNI nephrotoxicity, nodular arteriolar hyalinosis (ah), striped interstitial fibrosis (if) and tubular microcalcification (Ca) are the classical hallmarks of chronic CNI nephrotoxicity. In our series these signs of CNI nephrotoxicity were seen in 100% (ah), 88%(if) and 79%(Ca) by ten years.[18] Approximately half of the kidneys had at least two hallmarks at five years but this was essentially a universal finding by ten years. 5mg/kg/day of Cyclosporine was associated with CNI toxicity within 5 years.

While striped fibrosis and nodular arteriolar hyalinosis are widely accepted as due to CNI nephrotoxicity, diffuse arteriolar hyalinosis and tubular calcification may be due to diabetes and hypertension. In our series, arteriolar hyalinosis correlated with cyclosporine trough levels over 200 ng/ml at 3 months, but not with glucose tolerance or glycosylated haemoglobin. Hypertension was seen before the appearance of arteriolar hyalinosis and then occurred equally in hypertensive and normotensive patients. In the only randomized controlled study comparing Tacrolimus and Cyclosporine there were no histological differences in the CNI's in the two year protocol biopsies.[20]

## Subclinical Rejection

Sub-clinical rejection (SCR) is only really accepted by those who have examined a protocol biopsy taken from a stable patients and found severe interstitial cellular infiltrates and tubulitis. Since it can be eliminated by heavy levels of immunosuppression SCR may not be seen or be relatively rare in recent studies.[24] However, quite severe interstitial inflammation and tubulitis may be seen in protocol biopsies

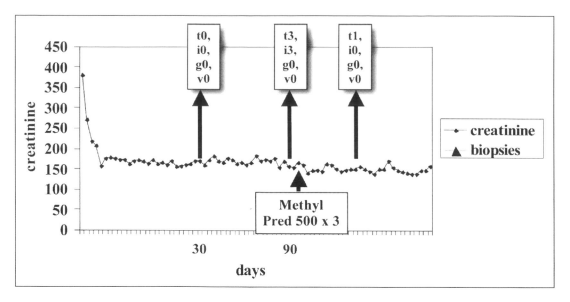

Figure 3. Serum creatinine and acute Banff biopsy qualifiers in a patient's 1 and 3 month protocol biopsies.

(Fig. 3), but with declining frequency and severity after transplantation. Immunosuppression impacts substantially, being less common with tacrolimus than cyclosporine and also mycophenolate mofetil than azathioprine.[16]

SCR causes interstitial fibrosis and tubular atrophy, identified by the correlation of SCR at three months with subsequent CAN/IFTA in both of our series.[14,16] In those studies we demonstrated that increased intensity of immunosuppression reduced the incidence of SCR and prevented subsequent IFTA with the most recent protocol of Tacrolimus and Mycophenolate Mofetil effectively abolishing subclinical rejection by three months and substantially reducing the incidence of CAN/IFTA on the one year biopsy.[16]

One of the more recent discoveries has been that chronic inflammation restricted to areas of interstitial fibrosis, ignored in the Banff schema, correlated with increased fibrosis both in our series and in those of others. Interestingly interstitial fibrosis without cellular infiltration carries a much less severe prognosis than when it is associated with cellular infiltration as demonstrated in three protocol biopsy studies.[25-27] It remains to be seen if treatment of the inflammatory process in these late cases with interstitial fibrosis will improve long term graft survival.[28]

## Glomerular Damage

Transplant glomerulopathy is characterised by enlarged glomeruli, mesangial matrix expansion, changes in mesangial cells and splitting of the glomerular basement membrane.[29] It is almost certainly accounted for by Chronic Antibody Mediated Rejection in which transplant glomerulopathy, donor specific antibodies and deposition of C4d have been demonstrated.[28,30] It is important to test this hypothesis with a well designed study of early intervention based either on detection of donor specific antibody or presence of C4d on protocol biopsy, but not by detection of glomerulopathy since this will almost certainly be too late to impact on long term outcomes.

Chronic and progressive glomerulosclerosis of the transplanted kidney is reminiscent of secondary Focal Sclerosis in the native kidney. Early glomerular damage is associated both with cold ischaemia and with episodes of acute CNI nephrotoxicity. Late and progressive glomerulosclerosis becomes most evident by 5 years and results from both preceding tubulointerstitial nephritis and arteriolar hyalinosis which presumably creates glomerular ischaemia. In our SPK series GFR fell in parallel with the increase in glomerulosclerosis but underestimated the degree of glomerulosclerosis, with a mean result of 59 ml/min with no

glomerulosclerosis, 56 ml/min with 1-20% sclerosed glomeruli and a remarkably preserved 52 ml/min despite more than 20% of glomeruli lost to fibrosis. Serum creatinine is not only a poor measure of GFR, but measured GFR is also a poor indicator of glomerular loss implying that only protocol biopsy can be used as a gold standard for both clinical management and scientific enquiry.

## Conclusions

The clinicopathologoical correlates of Chronic Allograft Nephropathy are both immune and non-immune mediated damage. Modern immunosuppression has reduced damage due to acute severe or vascular rejection, but undiagnosed and thus untreated sub-clinical rejection, as well as chronic antibody mediated rejection remain significant contributors to the long term outcomes of renal transplantation. "IFTA" also results from chronic CNI nephrotoxicity which creates damage through both arteriolar hyalinosis and interstitial fibrosis. Both immune and non-immune mechanisms add to pre-existing donor disease and ischaemia-reperfusion injury, as well as the later events such as infection with BK virus to explain the histology of each individual graft. The end game, exemplified by declining GFR and eventually rising serum creatinine, results from the impact of both interstitial fibrosis and arteriolar hyalinosis independently and together causing glomerular sclerosis. If we continue to simply monitor serum creatinine, proteinuria and hypertension to detect patients who may have histological CAN we will continue to experience the late graft failure rates that we see today.

## References

1. ANZDATA Registry report 2004. In: McDonald S, Excell L, eds. Australia and New Zealand Dialysis and Transplant Registry, Adelaide, South Australia.
2. Meier-Kriesche HU, Schold JD, Kaplan B. Long term renal allograft survival: have we made significant progress or is it time to rethink our analytic and therapeutic strategies? Am J Transplant 2004; 4:1289-95.
3. US Renal Data System, USRDS 2004 Annual Report: Atlas of end-stage renal disease in the United States, Bethesda, MD. National Institutes of Health, National Institute of Diabetes and Digestive and Kidney Diseases 2004.
4. Solez K, Colvin RB, Racusen LC et al. Banff '05 Meeting Report: Differential diagnosis of chronic allograft injury and elimination of chronic allograft nephropathy ('CAN'). Am J Transplant 2007; 7:518-26.
5. Hunsicker LG, Bennett LE. Acute rejection reduces creatinine clearance at 6 months following renal transplantation but does not affect subsequent slope of Ccr. Transplantation 1999; 67:S83.
6. Kassirer JP. Clinical evaluation of kidney function—glomerular function. N Engl J Med 1971; 285(7):385-89.

7. Racusen LC, Solez K, Colvin RB et al. The Banff 97 working classification of renal allograft pathology. Kidney Int 1999; 55:713-23.

8. Solez K. International standardization of criteria for histologic diagnosis of chronic rejection in renal allografts. Clin Transplant 1994; 8:345-50.

9. Solez K, Axelsen RA, Benediktsson H et al. International standardization of criteria for the histologic diagnosis of renal allograft rejection: the Banff working classification of kidney transplant pathology. Kidney Int 1993; 44:411-22.

10. Dean DE, Kamath S, Peddi VR et al. A blinded retrospective analysis of renal allograft pathology using the Banff schema: implications for clinical management. Transplantation 1999; 68:642-45.

11. Colvin RB. Chronic allograft nephropathy. N Engl J Med 2003; 349:2288-90.

12. Furness PN, Philpott CM, Chorbadjian MT et al. Protocol biopsy of the stable renal transplant: a multicenter study of methods and complication rates. Transplantation 2003; 76:969-73.

13. Kuypers DR, Chapman JR, O'Connell PJ et al. Predictors of renal transplant histology at three months. Transplantation 1999; 67:1222-27.

14. Nankivell BJ, Fenton-Lee CA, Kuypers DR et al. Effect of histological damage on long-term kidney transplant outcome. Transplantation 2001; 71:515-23.

15. Nankivell BJ, Borrows RJ, Fung CL-S et al. The natural history of chronic allograft nephropathy. N Engl J Med 2003; 349:2326.

16. Nankivell BJ, Borrows RJ, Fung CL-S et al. Natural history, risk factors and impact of subclinical rejection in kidney transplantation. Transplantation 2004; 78:242-49.

17. Nankivell BJ, Borrows RJ, Fung CL-S et al. Delta analysis of posttransplantation tubulointerstitial damage. Transplantation 2004; 78:434-41.

18. Nankivell BJ, Borrows RJ, Fung CL-S et al. Calcineurin inhibitor nephrotoxicity: longitudinal assessment by protocol histology. Transplantation 2004; 78:557-65.

19. Nankivell BJ, Borrows RJ, Fung CL-S et al. Evolution and pathophysiology of renal-transplant glomerulosclerosis. Transplantation 2004; 78:461-68.

20. Solez K, Vincenti F, Filo R. Histopathological findings from 2-year protocol biopsies from a US multicenter kidney transplant trial comparing tacrolimus versus cyclosporine: a report of the FK506 kidney transplant study group. Transplantation 1998; 66:1736-40.

21. Seron D, Moreso F, Fulladosa X et al. Reliability of chronic allograft nephropathy diagnosis in sequential protocol biopsies. Kidney Int. 2002; 61(2):727-33.

22. Dimeny E, Wahlberg J, Larsson E et al. Can histopathological findings in early renal allograft biopsies identify patients at risk for chronic vascular rejection? Clin Transplant 1995; (2):79-84.

23. Seron D, Moreso F, Ramon JM et al. Protocol renal allograft biopsies and the design of clinical trials aimed to prevent or treat chronic allograft nephropathy. Transplantation. 2000; 69(9):1849-55.

24. Rush D for the FKC008 Study Group. Abstract World Transplant Congress, Boston 2006.

25. Shishido S, Asanuma H, Nakai H et al. The impact of repeated subclinical acute rejection on the progression of chronic allograft nephropathy. J Am Soc Nephrol 2003 (4):1046-52.

26. Cosio FG, Grande JP, Wadei H et al. Predicting subsequent decline in kidney allograft function from early surveillance biopsies. Am J Transplant 2005; (10):2464-72.

27. Moreso F, Ibernon M, Goma M et al. Subclinical rejection associated with chronic allograft nephropathy in protocol biopsies as a risk factor for late graft loss. Am J Transplant 2006; (4):747-52.

28. Rush D, Nickerson P, Gough J et al. Beneficial effects of treatment of early subclinical rejection: a randomized study. J Am Soc Nephrol 1998; (11):2129-34.

29. Hamburger J, Crosnier J, Dormont J. Observations in patients with a well tolerated homotransplant kidney: possibility of a new secondary disease. Ann N Y Acad Sci 1964; 120:558-77.

30. Mauiyyedi S, Pelle PD, Saidman S et al. Chronic humoral rejection: identification of antibody-mediated chronic renal allograft rejection by C4d deposits in peritubular capillaries. J Am Soc Nephrol 2001; 12:574-82.

# CHAPTER 26

# Recurrent Glomerular Disease in the Allograft:
## Risk Factors and Management

Hani M. Wadei,* Xochiquetzal J. Geiger and Martin L. Mai

## Abstract

In contrast to the major improvement in immunologically mediated allograft loss, little advances have been made in the area of recurrent glomerular disease which currently stands as the third leading cause of renal allograft loss. This chapter will focus on the current status of recurrent glomerular disease in kidney transplant recipients. The epidemiology, risk factors and implications of recurrent disease on graft survival will be discussed collectively. This will be followed by individual review of recurrent glomerulo-nephritides with special attention to predictors and management of recurrent disease. When appropriate, prevention of disease recurrence will also be discussed.

### Epidemiology of Recurrent Glomerulonephritis in the Allograft

Glomerulonephritis (GN) is the cause of end stage renal disease (ESRD) in 30% of kidney transplant recipients.[1] In these patients, the risk of recurrent disease ranges between 2 and 20%.[2-6] This wide variability in the prevalence of disease recurrence among reports is attributed to: (i) the different follow-up periods, (ii) inability to differentiate between recurrent and de-novo GN in absence of native kidney biopsy and (iii) different biopsy practices among centers. For example, unless a biopsy is performed, a slowly deteriorating graft function and/or evidence of proteinuria can be easily blamed on chronic allograft nephropathy (CAN), underestimating the true prevalence of recurrent diseases.

### Onset

Early recurrence is common in focal segmental glomerulosclerosis (FSGS) and hemolytic-uremic syndrome (HUS) (hours to days) while recurrent IgAN tends to be more delayed with increased risk of recurrence with longer duration of follow-up. In a study that included almost 5000 transplant recipients, the mean time to recurrence was 475, 594, 664 and 846 days for FSGS, membrano-proliferative GN (MPGN), membranous GN (MGN) and IgAN, respectively.[5]

## Predictors of Allograft Loss from Recurrent Disease

An important predictor of allograft loss from recurrent disease is the type of GN with FSGS, HUS and MPGN carry the worst prognosis.[4-6] After a mean of 5.4 years follow-up period, the relative risk of allograft loss from recurrent disease was 5.4, 2.4 and 2.2 for HUS, MPGN and FSGS, respectively. Figure 1 depicts the expected time from transplant to allograft loss from recurrent disease in 1505 primary renal allograft recipients with biopsy-proven glomerular disease. As shown, recurrent FSGS and MPGN induce rapid loss of the allograft while allograft loss from recurrent IgAN is mostly a late event.

Recurrent disease is more common in males,[5,7] young age group.[5] and retransplants.[5] Living donor (LRD) transplantation, high panel reactive antibody (PRA) and better human leukocyte antigen (HLA) matching have been reported as risk factors. Male gender have more than two fold increased risk of allograft loss than their female counterparts (relative risk RR 2.24, P = 0.003).[4,5] Also for each 10% increase in PRA, there is 10% increased risk of allograft loss from recurrent disease.[4] Although allograft loss is more common in the pediatric age group,[3] among adult kidney transplant recipients, age does not seem to alter the risk of allograft loss from recurrent disease.[4] Previous allograft loss from recurrent disease is a major risk factor for subsequent loss from recurrence. This is well documented in a study by Briggs et al who showed that allograft loss from recurrent disease increases from 3% for primary renal transplants to 48% for secondary transplants.[6] The effect of LRD transplantation and the degree of HLA matching on risk of recurrent disease is controversial. In the Renal Allograft Disease Registry (RADR) report that included 5000 transplant recipients, risk of recurrence is comparable between living and decease donor (DD) transplantation.[5] Similar risk of recurrence is also reported in a study by Odorico et al that involved 364 kidney transplant recipients with native GN: LRD and living unrelated (LURD)/DD kidney transplantation (16.8% versus 12.8%, p = NS).[8] In this report, the degree of HLA matching did not impact recurrence and allograft loss was similar in both groups: 56.5% for LRD and 48.3% for LURD/DD transplant (P = NS). In the Australian registry, LRD transplant do not have an increase risk of allograft loss from recurrent disease compared to DD transplants.[4] Andresdottir et al reported a much worse allograft survival in recipients of HLA-identical transplants. Five, 10 and 20 years allograft survival was 88%, 70% and 63%, respectively compared to 100% in the group who did not have glomerular cause of renal failure (33 patients 1968-1996). There were additional 11 cases with proteinuria and allograft dysfunction, 9 of whom had biopsy proven recurrent disease, raising the overall risk of allograft loss or dysfunction from recurrent disease to 45% at 12 years.[9] The poor allograft survival in this subgroup has raised concerns that disease recurrence offsets the benefit of HLA-identical transplants. The impact of recurrent disease on patient survival was also studied extensively with no observed effect of disease recurrence or allograft loss from recurrent disease on 10-years and 20 patients' survival.[9,10] The risk of recurrence is also higher in identical twins not receiving immunosuppressive agents.[5,11] However, the 4-year actuarial risk of recurrence, graft and patient survivals are comparable between those who did and those who did not receive maintenance steroids.[10] Despite the short follow-up period

*Corresponding Author: Hani M. Wadei—Department of Transplantation Medicine, Mayo Clinic, 4205 Belfort Road, Suite 1100, Jacksonville, Fl 32216, USA. Email: wadei.hani@mayo.edu

*Chronic Allograft Failure: Natural History, Pathogenesis, Diagnosis and Management*, edited by Nasimul Ahsan. ©2008 Landes Bioscience.

**Table 1.  Summary of the risk of recurrence and allograft loss from recurrent disease in 5 major publications that spans the period from 1967 to 1996**

|  | Hariharan et al[284] | Hariharan et al[7] | Briganti et al[4] | Briggs et al[6] | Odorico et al[8] |
|---|---|---|---|---|---|
| Studied period | 1987-1996 | 1984-1994 | 1988-1997 | 1980-1991 | 1967-1994 |
| Publication year | 1999 | 1998 | 2002 | 1999 | 1996 |
| Country | USA | USA | Australia | Europe | USA |
| Mean follow-up period (years) | 5.4 | 7.3 | NA | NA | 14.1 |
| Risk of recurrent disease (%) | 3.4 | 6.3 | NA | NA | 12.2 |
| Short term risk of allograft loss (%) | NA | 4.5 (1 y) | 3.7 (5 y) | 3 (2 y) | 5 (5 y) |
| Long term risk of allograft loss (%) | NA | 15 (7 y) | 8.4 (10 y) |  | 8 (10 y) |

in the steroid free group, these results are encouraging and are similar to historically reported risk of allograft loss in those maintained on steroids with the same duration of follow-up.

### Effect of Recurrence on Allograft Loss

Does recurrent disease affect allograft survival? With the remarkable improvement in allograft survival for both LRD and DD kidney transplantation, the contribution of recurrent disease to allograft loss is expected to increase. In fact, the experience from the University of Minnesota suggests that the incidence of allograft failure due to recurrent MPGN has doubled from 3.5% (1988-1994) to 7.2% (1995-2003) while rates of death censored allograft failure remained the same.[12] The risk of allograft loss from recurrent disease at 5 years from transplant ranges around 4%, with increasing threat with longer follow-up. Briganti et al studied the risk of allograft loss from recurrent disease in 1505 kidney transplant recipients with biopsy proven GN.[4] The risk of allograft loss from recurrent disease increased with the duration of follow-up. At 1, 5 and 10 years the risk of allograft loss was 0.6%, 3.7% and 8.4%, respectively. At 10-year, recurrent disease was the third most common cause of allograft failure following chronic rejection and death with functioning graft.[4] Despite the impact of recurrent disease, 10-years allograft survival was similar in recipients with and without GN as a cause of ESRD mainly due to lower incidence of death with functioning graft in the GN group (14.1% versus 25.2%, P < 0.001) (Fig. 2). So it appears that although recurrent disease negatively impacts allograft survival, its overall effect on allograft loss in a given transplant population is less futile. The risk of recurrent GN and allograft loss from recurrent disease in 5 major publications that covers the period from 1967 to 1996 is summarized in Table 1.

### Clinical Manifestation and Differential Diagnosis

In the majority of cases recurrent disease carries the same pathological and clinical presentations as the primary lesion.[2] Clinically manifesting recurrent disease is hallmarked by proteinuria, hematuria and deterioration of allograft function. In this regard, worsening or new onset proteinuria has been correlated to allograft glomerular pathology[13] and is an indication to proceed with allograft biopsy. Recurrent GN needs to be differentiated from other causes of progressive decline in allograft function and other glomerular diseases affecting the allograft. The commonest causes of declining function include acute rejection, CAN and polyoma-virus associated nephropathy (discussed in Chapters 37, 38). The distinction between these can be easily made on histopathological evaluation of the allograft together with determination of BK viral load in blood and urine. Other glomerular lesions that might affect the allograft include de novo glomerular disease and transplant glomerulopathy (TGN). In the case of de novo disease, a native kidney disease other than GN

or a glomerular lesion other than the incident one needs to be the cause of renal failure. In many cases, biopsy evidence of the original disease is lacking and this determination is hard to make. TGN is easier to separate from recurrent disease due to the lack of immune complex deposition in the former. However, this differentiation hinges on immunofluorescent (IF) and electron microscopic (EM) assessment of the specimen, which if not done TGN will be easily confused with MPGN.

### Prevention

To prevent recurrence the original factors that induced the native kidney disease has to be changed. Since our understanding of these factors or milieu is limited, little advances have been made in this field with few exceptions. Plasmapheresis have been tried to prevent recurrence of FSGS, MPGN type I, antiGBM disease and HUS with variable success rates. Delaying transplantation to ensure complete resolution of the immunological insult may be helpful in anti-GBM but not in other types of GN in reducing recurrence risk. By aiming to eliminate antigenic stimulation, bilateral native nephrectomy was previously advocated to decrease the recurrence risk.[14] The role of bilateral native nephrectomy was examined in a large series by Odorico et al.[17] Rates of recurrence were compared between patients with (n = 61) and without (n = 303) bilateral native nephrectomies. After a 10 years follow-up, there was a significant increase recurrence risk in nephrectomized patients (42% versus 19.4%, P < 0.02) mainly due to increase in recurrence risk in those with FSGS who undergo nephrectomy.

### IgA Nephropathy and Henoch-Schönlein Purpura

Since its initial description by Berger et al[20] a recurrence rate ranging from 25-60% have been reported with IgAN.[15-19] As previously discussed, this may be an underestimation of the true incidence of recurrent IgAN.

### Incidence, Etiology, Clinical and Pathological Presentation

Recurrent IgAN usually manifests at a relatively late time point from transplantation. In a study that included 106 kidney transplants with IgAN, 37 (35%) of whom experienced recurrent disease, the mean interval between engraftment and onset of any urinary abnormality is 2.8 years (range 2 months to 17.5 years) with almost 10% of the cases manifesting more than 10 years after transplantation (Fig. 3)[19] Therefore, time to biopsy is critical in determining recurrence risk of IgAN with more cases identified when biopsies are obtained at a later time point from the transplant. In fact, Odum and colleagues identifies the duration of follow-up to be the only predictor of recurrence in 27 recipients with recurrent IgAN followed for 3 to 183 months after transplantation.[15] When long follow up period and use of protocol biopsies are

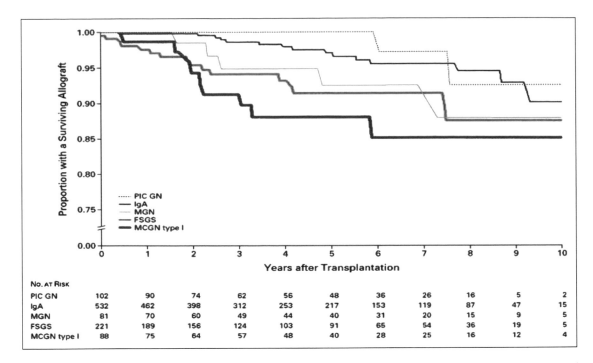

Figure 1. Time from transplant to allograft loss varies by the type of recurrent glomerular disease with the majority of losses from recurrent FSGS and MPGN occur early while allograft loss from recurrent IgAN tends to be rather a late event. PIC GN denotes pauci-immune crescentic glomerulonephritis, IgA: IgA nephropathy, MGN: membranous glomerulonephropathy, FSGS: focal segmental glomerulosclerosis and MCGN: mesangiocapillary glomerulonephritis. From: Briganti E et al. N Engl J Med 2002; 347:103;[4] ©2002 with permission of the Massachusetts Medical Society. All rights reserved.

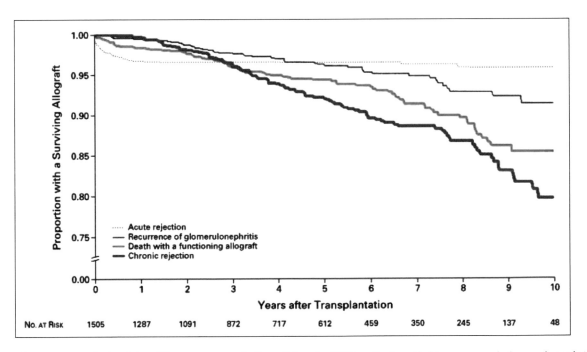

Figure 2. Kaplan–Meier analysis of the different causes of allograft loss in 1505 patients with biopsy-proved glomerulonephritis who received a primary renal transplant in Australia from 1988 through 1997. Recurrent disease was the third most common cause of allograft loss after 10 years of follow-up. From: Briganti E et al. N Engl J Med 2002; 347:103;[4] ©2002 with permission of the Massachusetts Medical Society. All rights reserved.

incorporated, the risk of recurrent IgAN is as high as 80%.[18] However, clinically apparent recurrent IgAN within the first year of transplantation is not uncommon with a third of the patients with recurrent IgAN may clinically manifest at this early time point.[20] Other relative predictors of recurrence include male recipients, age <30 years[18,19,21] and shorter time course to ESRD [median 5 yr (0-25 yr) for the recurrent group versus 10 yr (0-37 yr) for the nonrecurrent group, P = 0.015].[22] IgA structure,

angiotensin-converting enzyme genotype and duration on dialysis are not associated with increased risk of recurrence.[22,23,15]

While the beneficial effects of mycophenolate mofetil (MMF) in IgAN has been reported, the natural history of recurrent IgAN remains unaltered by the type of current immunosuppressive agents.[11,19,24-26] In a retrospective study, Chandrakantan and colleagues reported a similar recurrence risk (follow-up period 3 years) of IgAN in patient treated

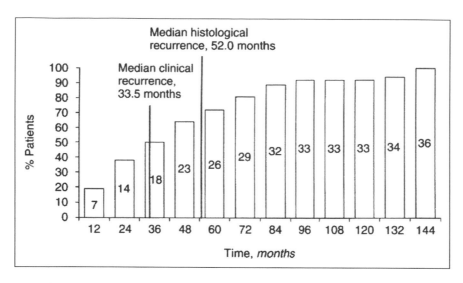

Figure 3. Time of appearance of proteinuria and/or hematuria in patients with recurrence of IgA glomerulonephritis in 106 recipients. The numbers in the columns refer to the number of patients with clinical recurrence. Note that there is still a risk of IgAN recurrence even 10 years after transplantation. Adapted by permission from Macmillan Publishers Ltd: Ponticelli C et al. Kidney Int 2001, 60:1948;[19] ©2001.

with either azathioprine (AZA, n = 61) or MMF (n = 91); AZA 10% versus MMF 8.1% (p = 0.76). Changing from AZA to MMF did not modify the disease progression and overall allograft survival rate remained similar in both groups.[20]

Transmission of pre-existing IgAN to the allograft is well documented.[27-29] Using implantation time biopsy, Suzuki et al found mesangial IgA deposition in 16% (82/510) of the allografts.[30] These clinically insignificant mesangial deposits disappeared over time on repeat biopsies.[27] In another study involving 24% (83/342) allograft with pre-existing IgAN, a gradual disappearance of mesangial deposits was observed on follow-up biopsies (at 1, 3 and 6 months posttransplant) without any clinical sequeals.[27] On the contrary, both higher recurrence rate (38% vs 9%, p = 0.04) and poor allograft survival are noted in recipients with primary diagnosis of IgAN who received an allograft with pre-existing IgA deposition.[29]

Various HLA antigens (HLA B12, B35 and DR4) which were previously associated with increased recurrence risk do not seem to be currently of predictive value.[21,22,31] Conflicting reports exist about the association of LRD kidney transplantation and recurrent IgAN (associated,[16,22,32-34] not-associated[15,17,19,20,24,31,35]). Others have found similar recurrence risk between LRD and DD transplantation with increased risk of recurrence only in those who received HLA identical LRD (RR 2.7, p = 0.001) compared to HLA identical DD.[21] The overall allograft survival is similar between HLA identical and those with ≥1 HLA mismatches (p = 0.5) indicating that in IgAN recipients, the benefit of zero mismatched transplant might be altered or even lost.[21] Interestingly, pooled data from multiple reports indicate that: (i) long-term LRD allograft survival (5 and 10 years) in IgAN recipients is similar if not better than the LRD allograft in non-IgAN recipients,[31,32] (ii) the cumulative graft survival in IgAN recipients is not reduced in LRD[17,20,22,33] and (iii) evidence of accelerated graft loss is lacking in recurrent IgAN in LRD transplant and therefore, suggesting that LRD transplantation should be performed in patients with IgAN.

Allograft loss from recurrent IgAN is uncommon in the first 3 years after transplantation. In one series that included 37 recurrent IgAN cases, the median time to allograft loss is 4.3 years [19]; however, the prevalence of functional impairment and eventually allograft loss increases with longer follow-up.[20,35,36] For that reason, the 5 years allograft survival is similar between cases with or without recurrent IgAN; however, at 10 years worse allograft survival is evident in recurrent IgAN.[16]

Predictors of poor allograft outcome include: previous allograft loss from recurrent IgAN, elevated serum creatinine (SCr) at time of diagnosis, systolic hypertension, proteinuria more than 1 gram/day and crescentic transformation of IgAN.[18,19] Of those, the relative risk (RR) of allograft loss from recurrent IgAN is 12.9 for each log-unit increase in SCr and 4.9 for proteinuria >1 gram/day.[19] Compared to mesangial IgAN or FSGS, the presence of crescentic IgAN on the original biopsy is also associated with poor allograft survival (5-year graft survival 71% versus 100%; p = 0.03).[34]

Clinical presentations of recurrent IgAN are variable and seem to correlate with histological findings.[15,37,38] Crescentic recurrent IgAN is rare and is universally associated with significant allograft dysfunction with rapid rise in SCr and high risk of allograft loss.[15] Pathological findings (Fig. 4) are characterized by more chronicity and relative lack of acute inflammatory infiltrates, probably as a reflection of the effect of the immunosuppression medications.[38,39]

### Treatment

Management of recurrent IgAN is challenging. As previously mentioned, modulating the immunosuppression medications does not seem to change the clinical course of the disease.[20] Blocking the renin-angiotensin system (RAS) with either an angiotensin-converting enzyme inhibitor (ACE-I) or and angiotensin-I receptor blocker (ARB) allows significant reduction in proteinuria[40] and improves graft survival.[19,41] This beneficial effect of angiotensin blocking agents is not observed when these drugs are started after histological diagnosis of recurrent IgAN is made.[20]

To summarize, recurrent IgAN is an indolent disease and its negative impact on allograft survival becomes apparent with longer duration of follow-up. Living donation is justifiable in IgAN patients and despite minimally increased risk of recurrence the overall allograft survival is comparable to deceased donor transplantation. Treatment is often frustrating but preemptive blocking the RAS is advisable.

HSP represents the systemic variant of IgAN and recurrent HSP usually manifests as isolated renal allograft involvement indistinguishable from recurrent IgAN. Extrarenal manifestations of HSP such as rash, arthralgia, gastrointestinal bleeding and abdominal pain are uncommon but have been noted. Meulders et al reported the actuarial 5 years risks for HSP recurrence and for allograft loss due to recurrence to be 35% and 11%, respectively.[42] A tendency towards high recurrence with shorter duration of the original disease was noted; however, delaying transplantation for more than a year from starting dialysis has not been shown to alter recurrence risk.[42]

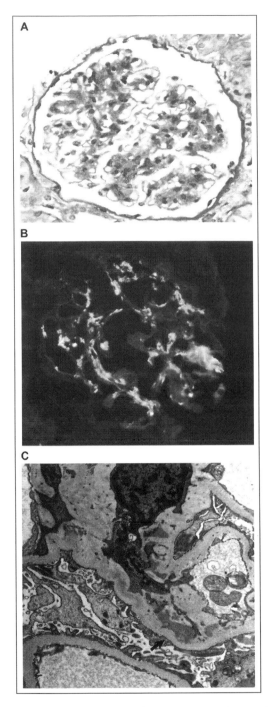

Figure 4. Recurrent IgAN. A) Light microscopic picture showing moderate mesangial expansion with both increased matrix and mesangial cells. The glomerular basement membrane is unremarkable. (PAS, × 400). B) Immunofluorescence picture showing characteristic dominant or codominant mesangial staining with IgA. (anti-IgA immunofluorescence, × 400). C) Electron microscopic picture showing immune complex deposits present within the mesangium, primarily underlying the paramesangial glomerular basement membrane (arrow). There are no glomerular basement membrane deposits. (TEM, × 5800).

## Focal Segmental Glomerulosclerosis (FSGS)

Among all glomerulonephropathy, FSGS (Fig. 5) is the leading cause of nephrotic syndrome in the United State[43] and the disease has a high chance of recurrence that approaches 30% in primary allografts. Once a first transplant fails due to recurrent disease, the risk of recurrence in the second transplant approaches 80-100%.[44] Secondary FSGS usually does not recur unless the etiological stimulus for FSGS persists after transplantation.

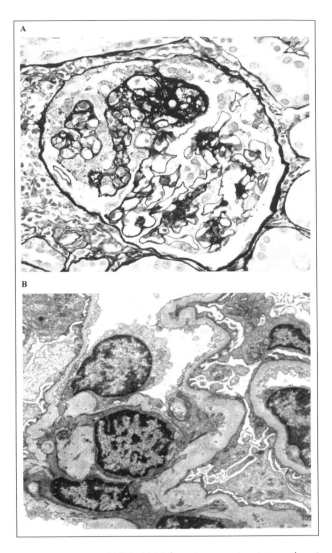

Figure 5. Recurrent FSGS. A) Light microscopic picture showing segmental sclerosis with tuft adhesion to adjacent Bowman's capsule. A few capillary loop foam cells are also seen. The uninvolved portion of the glomerulus shows minimal histologic changes. (Jones' silver stain, × 400). B) Electron microscopic picture showing mild mesangial matrix expansion with extensive foot process effacement. No immune complex deposits are present (TEM, × 4700).

### Incidence, Etiology, Clinical and Pathological Presentation

Recurrent FSGS manifests relatively early with 80% of affected patients manifest in the first post-transplant year.[6,45] In a study that included 71 patients with FSGS, 36% of whom had biopsy proven recurrent disease, the mean duration to recurrence was 7.5 months (range: 0.5 to 44 months).[46] Pardon et al used proteinuria (>3g/day) rather than biopsy findings as a diagnostic criteria and observed recurrent disease in 6 of 12 cases (50%) within the first 24 hours posttransplantation.[47] In another study, 11 of the 12 recipients who experienced FSGS recurrence had significant proteinuria within the first month after transplant.[48]

Predictors of recurrent FSGS include; young recipients' age, Caucasian race, short interval between FSGS diagnosis and ESRD, mesangial hypercellularity on native kidney biopsy and a history of prior allograft loss to recurrent disease.[45,46,49,50] Recipients' gender and duration on dialysis appear to have no impact on recurrence.[44,50] The effect of HLA matching on recurrent FSGS is debatable. Zimmerman reported a recurrence risk of 82% in recipients of 4 or more antigen matched kidneys compared to 53% with less matched kidneys and he cautioned against using HLA-identical transplants in patients with FSGS.[51] Baum and colleagues analyzed the North American Pediatric Renal Transplant Cooperative Study (NAPRTCS) database and identified 752 FSGS pediatric transplant recipients.[52] There were more allograft failures attributed to recurrent disease in the living donor group (26.7% versus 15.9%, P = 0.06) while the 5-years allograft survival was similar in recipients of living or DD kidneys (69% versus 60%, P = NS) suggesting that the beneficial effect of living donation on allograft survival is offset in patients with recurrent FSGS.[52] On the contrary, Abott and colleagues found no association between living donation and allograft loss from FSGS recurrence.[53] A recent analysis of the United States Renal Data System (USRDS) database indicates that FSGS patients receiving a zero mismatched LRD transplant have the lowest annually adjusted death censored allograft loss rates (ADGL of 10.5 grafts per 1000 per year) compared to recipients of mismatched living and DD kidneys (ADGL of 36.5 and 63.2, respectively).[54] Based on the current literature, LRD kidney transplantation is justifiable in patients with FSGS, however, due to the high risk of recurrence and risk of allograft loss from recurrent disease, proper counseling of the donor recipient pair is advisable.[55] Although induction therapy with either anti-thymocyte globulin or anti-IL2 receptor antibodies has been associated with increase incidence of recurrence, this has not been reproduced in subsequent reports.[47,56,57] Neither steroid withdrawal nor avoidance[10,58] and posttransplant immunosuppression with either azathioprine or cyclosporine A (CyA) does appear to affect risk of recurrence,[45,48] however, mounted doses of steroids and/or CyA might be needed to maintain remission of recurrent FSGS.[50,59] Anecdotal reports indicate an adverse effect of rapamune conversion with development of recurrent FSGS and subsequent improvement after switching back to CyA.[60] Donor factors, with the exception of old donors' age, do not seem to affect recurrence risk.[47]

Allograft loss is an early event occurring 4-28 months from transplant[45] Abott and colleagues examined predictors of allograft loss after disease recurrence and identified Caucasian recipients' race, Caucasian recipients of African American donor kidneys and young recipients age to be linked to increased risk of allograft loss after disease recurrence.[53] The immediate onset of proteinuria after transplantation is a negative prognostic factor with 55% of the patients who developed proteinuria within 2 days of transplantation returned to dialysis in less than 24 months while 7 of the 9 with late onset proteinuria (>2 days) had stable allograft function after mean follow-up of 44 months.[44] Apart from allograft loss, recurrent FSGS is associated with increased incidence of primary nonfunction, delayed graft function and acute rejection.[47,52,61,62] Recurrent FSGS has a detrimental effect on allograft survival with a relative risk of allograft loss two folds higher in those with recurrent disease (RR, 2.25).[5] Single center studies also indicate a 50% allograft loss rate from recurrent disease and this number has been remarkably constant over the last decade.[46] However, patient survival is not affected by recurrent disease.[52,53] Since the main presentation of recurrent FSGS is proteinuria, some centers advocate adequate control of pretransplant proteinuria either medically using ACEI or through surgical nephrectomy. However, nephrectomy has been shown to increase the risk of recurrent glomerular disease irrespective of etiologies, therefore, pretransplant nephrectomy is not widely accepted.[8,63]

## Pathogenesis of Recurrent FSGS

### Permeability Factor

It has long been suspected that a circulating factor survives over time in patients with FSGS and induces posttransplant recurrence of the disease. This hypothesis is supported by (1) infusion of sera from patients with recurrent FSGS into rats induces proteinuria,[64]

(2) transplacental transmission of such circulating factor induces proteinuria in the fetus[65] and (3) the injection of the supernatant of plasma from FSGS patients into rats induces rapid (within 2 minutes) and dose dependent proteinuria.[66-68]

This presumptive permeability factor (molecular mass between 30-50 kd[66]) is heat labile, slightly anionic and sensitive to proteases[68] and its biological activity can be blocked by adding normal plasma or normal plasma components, CyA and indomethacin.[69-71] The circulating factor is active in highly diluted plasma[68,72] and may share structural similarity with immunoglobulins.[67,68,73]

Despite the inability to isolate this factor, its biological activity (perm-selectivity) can be evaluated. Savin et al developed an in vitro bioassay that evaluates the volume changes of isolated rat glomeruli upon exposure to FSGS serum and gives an indirect measure of permeability activity ($P_{alb}$) in which $P_{alb}$ of 1 indicates maximal effects. Following its description, high $P_{alb}$ was associated with high risk of recurrence: practically 100% with values greater than 0.69.[74] The cumulative risk of recurrence was 17% in those with $P_{alb}$ < 0.50 and 86% in those with values higher than 0.5.[74] Plasmapheresis decreased the $P_{alb}$ and proteinuria levels and the eluted plasmapheresis fluid injected in rats induced FSGS. However, the ability of the $P_{alb}$ to predict recurrent disease is inconsistent among reports.[72] Indeed, recurrence still occurrs in patients with very low $P_{alb}$, with a frequency of FSGS recurrence of 31% in those with values <0.1.[74]

Recently, accumulating evidence indicates that the loss of a naturally occurring inhibitor rather than the presence of a circulating factor is responsible for the increased perm-selectivity in FSGS patients. In a unique experiment, Coward and colleagues induced redistribution of the slit diaphragm proteins 48 hours after exposure to nephrotic serum, with nephrin, podocin and CD2AP being retained within the cytosol.[75] Interestingly, the addition of normal serum reversed the delocalization of nephrin, strongly supporting the possibility of a defective or missing protective factor in nephrotic plasma.[75] Moreover, co-incubation with nephrotic urine reversed the increased permselectivity activity induced by FSGS serum while normal urine had no such inhibitory effect.

## Role of Podocin Mutation

Genetic studies identify podocin mutation to be responsible for 50% of familial autosomal recessive FSGS and almost 10 to 30% of sporadic cases of FSGS.[76,77] Multicenter studies demonstrate that patients with two pathogenic NPHS2 mutations (homozygous or compound heterozygous) bear a very low risk of recurrent FSGS (<10%), whereas those with only one mutation (heterozygous) run a risk comparable to nonfamilial FSGS patients.[77] Weber studied 338 patients from 272 families with steroid-resistant nephrotic syndrome and identified podocin mutation in 43% (35 of 81 families) of cases of familial FSGS compared to 10% (n = 18 of 172) in those with sporadic disease. Homozygous and compound heterozygous mutations were more common in familial than in sporadic cases who tended to have more isolated heterozygous mutations. Of the 32 cases with homozygous/compound heterozygous mutations who received a kidney transplant, none experienced early FSGS recurrence and only one had recurrent disease 2 years after transplantation.[77] The incidence of recurrence was much higher in those with heterozygous mutations.[77] Similar findings are reported by Bertelli et al.[78] The clinical outcome of recurrence, response to plasmapheresis and immunosuppressive therapy are comparable between those with or without podocin mutation, suggesting a common mechanism of recurrence in recipients of both groups.

Current knowledge indicates that patients with podocin mutations do not seem to have detectable anti-podocin antibodies.[77-79] The fact that patients with two NPHS2 pathogenic mutations have early disease onset and low chance of recurrence points to podocin mutation as the main pathogenic factor. However in patients with single NPHS2 mutation, additional environmental or unidentified genetic factors are needed to precipitate FSGS, these factors are not corrected by transplant and leads to early disease recurrence in 66-75% of cases. This theory is supported by the detectable $P_{alb}$ activity in the majority of patients with NPHS2 mutation who subsequently recurred after transplant.[79,80] Hence, molecular defects of podocin might be part of a multifactorial disease in which plasma factors might also be involved.[78]

### Prevention

Gohh et al subjected 10 FSGS patients with high risk of recurrence (due to history of rapid progression to end stage renal failure, history of previous allograft loss to recurrent FSGS or elevated to $P_{alb}$ to 8) to prophylactic pretransplant plasmapheresis. After 8-42 months follow-up period, seven of the 10 had no evidence of recurrence including 3 of the 6 who previously lost their allograft from recurrent disease.[81] Two of those who experienced recurrent disease lost their allograft while the third patient had significant allograft dysfunction and nephrotic range proteinuria despite posttransplant plasmapheresis. Although these results are compatible with a favorable response of pretransplant plasmapheresis in preventing disease recurrence there is a lack of randomized studies.

Pretransplantations analysis of podocin mutations to predict disease recurrence is not widely available and has not been studied. However, parents of children with homozygous/compound heterozygous mutation are obligate carriers of the disease and should be cautiously considered for kidney donation.

### Treatment

## Plasmapheresis

Despite the lack of a clear pathophysiological mechanism of action, current literature indicates that plasmapheresis reduces proteinuria, reverses foot process effacement, decreases permeability factor ($P_{alb}$) activity and induces complete remission in 50 to 75% of cases.[46,74,82-84] However, the response to plasmapheresis is variable from one patient to another. Relapses after cessation of successful treatment is not uncommon and prolonged or chronic plasmapheresis therapy may be required to maintain remission in selected patient population.[85]

Davenport et al systematically reanalyzed 12 publications of plasmapheresis use in recurrent FSGS spanning the period from 1992 to 2000.[84] Of the 44 identified cases, 32 (73%) responded to plasmapheresis. The number of plasmapheresis treatments was similar in responders and nonresponders, however, therapy was initiated earlier in responders with a median interval from diagnosis of recurrence to plasmapheresis of 10 days compared to 19.5 days in nonresponders (p = 0.34). The presence of glomerulosclerosis on pretreatment biopsy predicted treatment failure. Relapse after successful treatment occurred in 10 cases. Similarly, in another report three patients with early diagnosis who were immediately initiated on plasmapheresis therapy had reversal of epithelial foot process effacement and remission of proteinuria.[46] From the above studies, early and aggressive institution of plasmapheresis is recommended in attempt to reverse disease recurrence. Plasmapheresis also improves the overall graft outcome. Deegens reported the outcome of 13 patients with recurrent FSGS treated with plasmapheresis.[86] Complete remission was achieved in 7 (54%) while the remaining 4 (31%) had partial remission. Five-year graft survival was better in the plasmapheresis group compared to 11 historical controls who did not receive plasmapheresis (85% vs 30%, P = 0.02). Similar results were also reported when Protein A column was used.[85,87]

## Medical Management

The role of cyclophosphamide in FSGS recurrence is less clear.[88,85] Kershaw treated 3 patients with recurrent FSGS with cyclophosphamide with resolution of proteinuria and preservation of allograft function.[88] It remains unclear whether the combination of cyclophosphamide and plasmapheresis is superior to plasmapheresis alone. Dall'Amico treated 13 patients with either plasmapheresis alone or in association with cyclophosphamide and compared the allograft outcome to 5 untreated cases. Four of the five (80%) cases who did not receive treatment lost their grafts. The allograft survival was slightly better in those who received combined therapy compared to those who received plasmapheresis alone [4 of 11 (36%) and 1 of 2 (50%)] but a firm conclusion is hard to make due to the limited number of cases.[44] Although FSGS recurs despite using conventional dose CyA, high dose CyA induces disease remission either alone or in combination with plasmapheresis[45,50,89-92] and it has been shown to decrease the $P_{alb}$.[93,94] Salomon et al reported their experience with high dose intravenous CyA in 17 recipients with recurrent FSGS, 4 of whom received concomitant plasmapheresis therapy. CyA was administered for 6 to 27 days (mean, 19.5 days). Initial response was achieved in 14 (82%) and 11 (65%) had prolonged remission 6 months to 9 years posttherapy. Actuarial 1 and 5 years allograft survival was 92% and 75%, respectively.[92] In a similar report, 13 out of 16 patients (81%) responded to high CyA with relapses upon dose reduction.[90] CyA dose reduction or switching to tacrolimus (TAC) was successful once long lasting remission was achieved.[90] However, histological evidence of calcineurin inhibitor (CIN) nephrotoxicity was apparent in 4 of the 9 (44%) patients who had posttreatment biopsy. In another report of resistant recurrent FSGS, the combination of antiCD20 monoclonal antibody (rituximab), plasmapheresis and cyclophosphamide induced a long lasting disease remission.[95] Finally, therapy with ACE-I and/or ARB should also be considered to decrease proteinuria and to treat hypertension.

In summary, recurrent FSGS has a significant impact on allograft survival. Although, the underlying pathophysiology is poorly understood, the current evidence suggests interaction between genetic predispositions and factor/s that alter podocyte physiology. Early diagnosis and vigilant monitoring for worsening proteinuria is warranted while initiation of plasmapheresis provides the best hope in reversing disease recurrence. In patients with FSGS, LRD transplant is not contraindicated as long as the risk and consequences of recurrence is fully explained.

## Systemic Lupus Erythromatosis (SLE)

### Incidence, Etiology, Clinical and Pathological Presentation

Recurrent lupus nephritis is rare and affects only 1-4% of renal allograft.[96,97] However, with more dedicated biopsy practices, the prevalence of recurrent lupus nephritis is documented to be much higher.[98-101] Goral and colleagues performed IF and EM assessment for all biopsy specimens obtained on recipients with history of lupus nephritis (LN) and reported disease recurrence in 15 of 50 patients (30%).[101] Eight of the 15 cases (53%) had evidence of mesangial LN that would have been missed if only light microscopy (LM) has been used. With less stringent pathological assessment, Stone et al identified 9 cases (8.5%) of recurrent LN in a series of 106 kidney transplant recipients transplanted between 1984 and 1996 (mean follow-up 62.6 months) and similar findings were also reported by Moroni and colleagues [3 of 33 cases (9%) and mean follow-up 91 months]. The highest reported risk of recurrent LN is by Nyberg and colleagues who performed surveillance biopsies and used dedicated histopathological assessments on biopsies obtained from patients with lupus nephritis. Seven of the sixteen (44%)

studied patients had evidence of disease recurrence. The prevalence of recurrent LN in surveillance biopsies was 25%.[102]

A common practice is to ensure disease quiescence both clinically and serologically for a period of 6 month to 1 year prior to transplantation.[101,103] Although a longer period on dialysis has been advocated, studies that examined risk factors for disease recurrence showed that recurrence risk is not related to pretransplant duration of SLE, duration on dialysis, pre-emptive transplantation, age at transplant, gender, type of transplant (LRD vs. DD), number of HLA matching or SLE flares before or after transplantation.[99,101] Maintenance steroids might be needed in patients with extra-renal manifestations of SLE, however, the risk of recurrence is similar in steroid treated versus steroid spared recipients.[10]

In the above mentioned studies, the mean duration from transplantation to diagnosis of recurrent LN is 3.1 to 4.3 years, however the diagnosis was made as early as 5 days and as late as 15.3 years from transplantation.[99,101] Clinically, most recurrent lupus cases presents with hematuria or proteinuria while elevated serum creatinine tends to be a late finding.[99,101] Interestingly, recurrent disease is rarely associated with systemic manifestations or serological evidence of SLE.[99,101,104] Pathologically, World Health Organization (WHO) class II LN (mesangial LN) is the most common finding and compromises 40-53% of cases while focal proliferative and membranous LN (class III and V) are diagnosed in the remainder of patients. Diffuse proliferative LN (class IV) is rare but carries a worse prognosis with high risk of allograft loss.[99] Pathological class transformation is not uncommon.[101,102] Whether this pathological transformation is a reflection of early diagnosis or the effect of the immunosuppression medications is currently unknown. Retransplantation after kidney loss from recurrent LN is successful in the majority of cases.[99]

Apart from disease recurrence, SLE per se might impacts allograft survival. Two studies reported poorer 3 and 10 years allograft survival in recipients with LN, mainly due to increased incidence of rejection and thrombosis in these patients. In a recent study, long term patient and allograft survival in 33 Caucasian renal transplant recipients with mainly proliferative LN was compared to 70 well matched controls. The actuarial 15 years patient survival and graft survival rates were 80% and 69% in patients with SLE and 83% and 67% in the matched controls, respectively (p = NS).[98] Thrombotic complications were more common in SLE patients but allograft loss from thrombosis was similar between the two groups. Other posttransplant complications including delayed graft function, rejection episodes and severe infections were similar between the two groups. The discrepancy in allograft survival among the above published studies can be related to different immunosuppression regimens used. Indeed, Zara et al reported worse allograft survival in SLE recipients in the preCyA era.[105]

## Treatment

Optimum treatment strategy is yet to be established. The promising results of mycophenolate mofetil in inducing and maintaining remission in native kidney LN make it an attractive option in treating disease recurrence. Most recently, the rituximab was given at 4 weekly infusions (375 mg/m² each) in association with oral steroids induced complete disease remission in 5 patients and partial remission in another 3 patients with proliferative LN (4 with focal proliferative and 6 with diffuse proliferative).[106] Similar favorable response to rituximab is observed in 2 of 3 patients with diffuse proliferative LN.[107] Moreover, the response to rituximab is sustained for more than 2 years in 6 patients with refractory LN.[108]

In summary, recurrent LN is not uncommon and its incidence increases by using more dedicated practices aiming at early diagnosis. However, allograft loss from recurrent disease is limited to few cases primarily due to the mild histological features. Apart from ensuring disease quiescence, no specific precautions need to be undertaken to prevent recurrent disease. Once occurs, mounting baseline immunosuppression with or without rituximab is indicated.

# Hemolytic Uremic Syndrome (HUS)

## Incidence, Etiology, Clinical and Pathological Presentation

Ferraris and colleagues reported their experience with 66 renal transplants conducted on 62 pediatric patients with Classical HUS followed for an average of 7 years.[109] Following the initial episode of HUS, none of the cases developed recurrent HUS either before or after transplantation despite posttransplant use of CyA in the majority of patients. Patient's survival was similar between HUS cases and controls while the actuarial 10 years allograft survival was better in those in HUS compared to those with other GN (79% compared to 58%, P < 0.001).[109] Similar findings were reported by Artz and colleagues who identified only one (6%) case of recurrent HUS among 18 recipients with Classical HUS pretransplantation.[110] One-year allograft survival was 78% which was not different from matched controls (p = 0.78).[110]

Atypical HUS accounts for almost 10% of all HUS cases. For atypical HUS, most series reported a posttransplantation recurrence risk around 65%.[110-114] The risk of medication induced recurrent atypical HUS is negligible unless re-exposure to the precipitating drug has occurred.[112] Cases with no identifiable triggering event (idiopathic) have a risk of recurrence around 50%.[115] While those with documented complement regulator factor H (CFH) or C3b-cleaving enzyme factor I (FI) mutations have the highest risk (80-100% compared to 55% in those without mutations).[115-117] On the other hand kidney graft outcome is favorable in patients with membrane complement protein (MCP) mutations as the kidney is the main site of synthesis of this protein and kidney transplantation alone corrects the underlying factor deficiency. This fact is further supported by lack of recurrence up to 10 years posttransplant in 2 patients with these mutations.[117] Additional 8 patients with MCP mutation had the recurrence rate of 12.5%.[118,119] Table 2 summarizes the risk of recurrence in the different forms of atypical HUS.

Potential risk factors for recurrence were examined by Ducloux and colleagues who reviewed ten studies comprising 159 grafts in 127 patients. Overall, a low risk of recurrence (27.8%) was reported and was probably due to inclusion of both pediatric and adult HUS cases.[120] Older age at onset of HUS (16.9 ± 7.6 years vs. 9.9 ± 6.5 years; p < 0.02) and shorter mean interval between HUS and ESRD (0.79 ± 0.39 years vs. 2.78 ± 2.47 years; p < 0.01) were reportedly associated with HUS recurrence.[120] Shorter mean interval between HUS and transplantation tended to relate to increased recurrence risk (2.5 ± 2.7 years vs. 6.0 ± 6.4 years; P < 0.01), but this association lost its significance on multivariate analysis.[120] It is still controversial whether grafts from LRDs or the use of CINs are related to higher risk of HUS recurrence.[111-113,120] Indeed, recurrent HUS still manifests despite CIN free immunosuppression protocols.[121] In the case of LRDs, a recent report found no relationship between the type of transplant and disease recurrence.[110,112] However, caution should be undertaken in selecting LRDs especially in the case of familial HUS as HUS occurrence in prior kidney donors has been described.[111,122,123] Potential LRDs to HUS recipients need to be tested for CFH mutation with the understanding that some donors who tested negative for the mutation still developed HUS after kidney donation, probably due to the affection of other untested mutations.[115,123]

Most recurrent cases manifest in the first month posttransplant with few exceptions of delayed recurrence of up to 10 years.[111-113,120] Rapidly progressive graft dysfunction is the most common presentation. Systemic manifestation of thrombotic microangiopathy with thrombocytopenia, Coomb's negative hemolytic anemia and schistocytes is common but not universal.[124] Neurological deficits and fever are rare. The diagnosis hinges on histological demonstration of evidence of microthrombi in the allograft. Similar lesions can also be seen in anti-

***Table 2.*** *Risk of HUS recurrence in the allograft depends on the etiology of native kidney HUS*

| Etiology of HUS | Posttransplant risk of recurrence |
| --- | --- |
| **Classical HUS** | **Negligible** |
| Atypical HUS | |
| Complement mutations | |
| CFH mutation | 80-100% |
| FI mutation | 80-100% |
| MCP mutation | 12.5% |
| Idiopathic | 50% |
| Medication induced | Negligible |

body mediated rejection (AMR), anti-phospholipid antibody (APLA) syndrome and CIN related HUS. The presence of peri-tubular capillary staining for C4d in the case of AMR and serological evidence of APLA differentiates these diseases from recurrent HUS.

Recurrent HUS carries a poor outcome with allograft failure that approaches 100% in most series. The median time from diagnosis of recurrence to allograft loss is usually very short (3 to 8 days).[110,113] In a meta-analysis that included 127 patients in 10 reports, the one-year graft survival was 76.6% in patients without recurrence and 33.3% in patients with recurrence (p < 0.001).[120] Other studies reported 4-year actuarial allograft survival of only 10-20%.[112,114]

### Prevention

Although there is no effective way of preventing recurrent HUS, guidelines published by the European Society of Pediatric Nephrology (ESPN) recommends testing for mutations of the complement regulatory proteins in patients with Atypical HUS. These guidelines are available at http://espn.cardiff.ac.uk/guidelines.htm. This approach will better determine the risk of disease recurrence and guide future consideration for living donation. A reasonable approach is to first test for serum C3 levels with the understanding that normal C3 levels does not necessarily exclude complement dysfunction. Decreased CH50 activity can be found in some but not all cases with CFH or MCP mutations. Measurement of CFH in serum is useful to identify those with CFH mutation associated with reduced CFH levels. Finally, genetic testing looking for mutations in CFH and MCP genes will identify those with mutations associated with dysfunctional proteins. Search for FI mutations should be performed in patients with reduced FI serum levels.[125]

### Treatment

Treatments proposed for patients with native kidney HUS, namely plasma exchange, plasma infusion, antiplatelet agents and coticosteroids have been tried in recurrent HUS with limited response. Colon and colleagues reported nearly 100% allograft loss in 8 recurrent HUS cases treated with plasma exchange or plasma infusion.[114] Similarly, Miller et al treated 7 recurrent HUS cases with either plasma exchange or infusion and reported allograft loss in 86% of the cases.[113] The correction of a precipitating event like treating concurrent cytomegalo-virus infection or discontinuation of an offending medication (CIN) along with plasma exchange might reverse disease recurrence.[113,126] Intravenous immunoglobulin infusion and rituximab were successfully used in two transplant recipients with recurrent HUS and may provide a rationale in desperate cases.[127,128] Intensification of plasma exchange in anticipation for transplantation or in the early posttransplant period has been tried with mixed results.[124,126,129] As the liver is the main source of CFH in the circulation, a logical alternative is simultaneous liver-kidney transplantation. This option is only indicated in the case of CFH mutation and requires genotyping testing prior to transplantation. Although, liver transplantation normalizes the CFH activity and may protect the transplanted kidney from recurrence of disease,[130,131] the results

of simultaneous liver-kidney transplantation in HUS are discouraging.[116] To circumvent this possibility, Saland and colleagues used pre and intra-operative plasma exchange to replenish the deficient factor at time of surgery with immediate hepatic function and absence of HUS recurrence 28 months after transplantation.[129] Other cases resulted in primary hepatic nonfunction, severe neurological deficits and fatal infections despite achieving normal factor H levels.[130,131]

In summary, recent advances have proved that HUS is a heterogeneous disorder. Understanding the etiological factors and the genetic background behind HUS is crucial in the management of these patients and may allow the clinician to predict the recurrence risk and tailor therapies accordingly. Atypical HUS, especially when associated with CFH or FI mutations, has a protracted course with high risk of allograft failure. In these cases, combined liver kidney transplantation is promising, but yet with poor outcomes. Ways to prevent complement activation in the immediate posttransplant period are needed and should be the subject of future research.

## Membranous Nephropathy

### Incidence, Etiology, Clinical and Pathological Presentation

More is now known about the incidence, etiology and clinical and pathological presentations of membranous nephropathy after kidney transplantation. The incidence of biopsy proven membranous nephropathy in all patients presenting for renal transplantation varies between 0.3-2.6%.[5,132-134] This represents 3-6% of the patients with any form of biopsy proven glomerulonephritis.[4,132,135] Because histological confirmation is available in only 47-81% of patients with a clinical diagnosis of glomerulonephritis, the true incidence of membranous nephropathy may actually be higher.[4-6,136,137]

The etiology of recurrent or de novo membranous nephropathy posttransplant is not known. Data from study of Heymann's nephritis in rats implicates megalin, dipeptidyl-peptidase IV and neutral endopeptidase in the formation of immune deposits.[138,139] Two of these proteins, dipeptidyl-peptidase IV and neutral endopeptidase are present on the human podocyte in contradistinction to megalin, which is found only on the rat podocyte.[140,141] Interesting human data describing severe antenatal membranous nephropathy associated with maternally derived neutral endopeptidase antibodies suggests a role of podocyte antigens in some forms of this disease.[142] More work in confirming putative antigens or identifying new proteins and mechanisms of disease continues and hopefully will focus on the transplant patient.[143-146]

The clinical hallmark of recurrent or de novo membranous nephropathy is proteinuria with or without renal dysfunction. Of 32 case reports, onset of nephrotic range proteinuria occurs in a mean of 10 months and 21 months for recurrent disease and de novo disease respectively.[147,148] This difference has not been confirmed by subsequent studies, where proteinuria due to membranous nephropathy develops at any time posttransplant.[132,149] Renal dysfunction is variable in both entities and an association with hypertension or edema is not described.[149]

Pathologically, recurrent and de novo membranous nephropathy cannot be differentiated in transplant kidneys and may occur with other pathology in the kidney.[150,151] However, mild to moderate mesangial hypercellularity, immune deposits containing complement, uneven distribution of immune complexes and mixed Ehrenreich and Churg stages of disease are found in recurrent and de novo disease and not in native kidney idiopathic disease.[151] Some of these differences have been confirmed in series of de novo disease, emphasizing the concomitant presence of current or past rejection changes.[152,153] Figure 6 presents a case with recurrent MN.

Onset of stage I-II disease develops at a mean of 18 months following transplant compared to the longer mean interval of 33 months for stage III-IV disease.[151] The findings of different membranous stages in the same biopsy and the change from segmental deposition to a diffuse pattern of immune deposits over time are consistent with progression of the membranous disease process.[151] In this series, concomitant transplant glomerulopathy is present in those patients with progression of renal dysfunction and not seen in patients with preserved renal function, with one exception.[151] This suggests a negative prognostic influence of the presence of transplant glomerulopathy associated with membranous nephropathy. An unusual variant of recurrent membranous nephropathy with immune complex spherule deposits of diameter similar to nuclear pores but immunoperoxidase and immunofluorescence staining negative for nuclear membrane was described in a single allograft recipient 14 months posttransplant. Interestingly, the pathology of the native kidney biopsy 20 years prior to the transplant revealed the same spherule deposits.[154] The source of the spherules remains elusive. Another atypical form of membranous nephropathy complicated by crescentic changes has been shown to recur in successive transplants.[155] This has been described once before, also in a patient with native crescentic membranous nephropathy.[156]

### Recurrent, de Novo and Pre-Existing Membranous Nephropathy Posttransplant and Associated Graft Loss

It is challenging to determine the true incidence of recurrent and de novo membranous nephropathy posttransplantation. Small series, missing native kidney biopsies and the lack of immunofluorescence and electron microscopy evaluation of transplant biopsies skew findings. Earlier studies/reviews of the literature report a risk of recurrent membranous nephropathy between 4-57%.[132,133,147] When compiling series with 10 or more patients of native kidney membranous nephropathy undergoing transplantation, the recurrence rate is 17%.[132] Two larger and more recent studies with 30 and 64 patients with native kidney membranous nephropathy treated with kidney transplantation establish the rate of recurrence at 20-29%.[8,150] The actuarial risk of recurrence in the study by Cosyns et al reached 29% at 3 years and remained there with 10 years follow-up.[150] According to similar studies/reviews, de novo membranous nephropathy develops in 0.4-9.3% of all patients presenting for transplant.[134,148,151-153,157,158] A compilation of available studies as of 1994 establishes the onset of de novo membranous nephropathy in 1.4% of all transplanted patients, making it numerically much more common than recurrent disease.[153] A large multi-center prospective study should help better define risk of recurrent and de novo membranous nephropathy posttransplant in the near future.[135] It should be noted that in the majority of these studies, the patients were biopsied for cause; therefore, those with sub-clinical disease were missed.[149] The actual number of recurrent and de novo disease posttransplant is likely higher. Although rare, donor kidneys with pre-existing membranous nephropathy have been transplanted verifying another source of this pathology in kidney transplants.[159,160] Biopsy documentation of resolution of the donor membranous changes is observed in one patient whereas clinical improvement (normalization of protein excretion or stable creatinine) is noted in two others.[159,160]

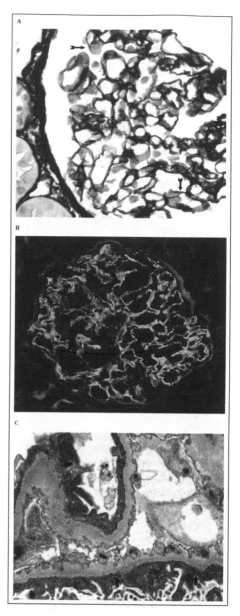

Figure 6. Recurrent membranous nephropathy. A) Light microscopic picture showing the glomerular basement membrane with rare, barely discernable pinholes without spike formation (arrows). The mesangium is unremarkable. (Jones' silver stain, × 400). B) Immunofluorescence picture showing a finely granular capillary loop staining with IgG without mesangial staining, characteristic of idiopathic membranous nephropathy. (anti-IgG immunofluorescence, × 400). C) Electron microscopic picture showing scattered small subepitheial immune complex-type deposits with minimal surrounding basement membrane reaction, as seen in early stage I membranous nephropathy (arrows). Overlying foot processes are moderately effaced. (TEM, × 5800).

Graft loss in those who develop recurrent or de novo disease is variable. Briganti et al reviewed 81 cases of native kidney membranous nephropathy and found overall graft loss of 40.1% at 10 years. This is not significantly different than the graft loss of 45.8% for other types of nonglomerular kidney disease. Graft loss from recurrent membranous is 12.5%, while 15.3% was blamed on chronic rejection.[4] Hariharan et al report a multi-center review of 4913 patients and show membranous nephropathy occurred in 16 with overall graft loss of 44% at 5 years and a kidney half-life of 1193 days. Again, the relative risk of graft loss was not significantly different than those without recurrent or de novo disease, possibly due to limited number of patients or follow-up.[5] Berger et al reach the same conclusion, recurrent or de novo membranous nephropathy did not affect graft survival.[134] When specifically looking at graft loss in those developing recurrent disease, Cosyns et al review 27 reported cases in the literature and show a graft loss of 72% at five years. The author's own series of 30 patients demonstrate graft loss of 38% at 5 years and 52% at 10 years versus 11% and 21% respectively in those transplant with membranous who had not developed recurrence.[150] Importantly, this difference is not statistically significant. In de novo membranous nephropathy, Schwarz et al found graft loss in 14 of 21 patients, with de novo disease responsible in only one. This did not differ statistically from 851 patients without de novo involvement, where 5 year graft loss was 28.6% in de novo disease versus 39.2% in those without (p value not given).[153] Freese et al found a relative risk of graft loss of 2.0 comparing de novo disease to those without (p = 0.03). This de novo group included 11 patients with membranous nephropathy, but unfortunately, also 9 with proliferative and 8 with a combination of both lesions pathologically making conclusions difficult at best.[161] Although membranous nephropathy after transplant carries no statistical increase risk of graft loss over those without the disease in most studies, there is a trend over time of increased graft loss.[4,150]

### Treatment

Risk factors for membranous nephropathy posttransplant and its course have been studied. Although a male predominance of membranous nephropathy after transplant has been described in some reports, in none is it statistically different.[132,149,150,152,153] Age, ethnicity, duration of dialysis, transfusions, pretransplant immunosuppression along with donor age, sex and ethnicity have all been evaluated in different reports and are associated with no increased risk of disease.[132,150,152,153] Curiously, the rapidity of the native disease increases risk of membranous after transplant in some studies,[162,163] but the opposite was found to be true in others making conclusions difficult.[132,150] HLA matching has been carefully studied. Although early reports implicated better matching as a risk factor, subsequent reports have not confirmed that finding.[134,147,151,153,157,164-166] HLA-DR3 is a risk factor for development of idiopathic membranous nephropathy, but current studies show at best a trend without statistically significant risk of this HLA protein and membranous nephropathy after transplant.[132,150,151,164,166] Interestingly, one study found HLA-DR4 in 46.7% of patients with de novo disease compared to 13.6 % in control, p < 0.005.[151] Another paper demonstrated donor HLA-A3 in 75% of recurrent membranous nephropathy versus 15% in controls at 3 years, p = 0.0035.[166] Larger studies will be needed to confirm these findings. Many studies have shown increased development of membranous nephropathy after transplant when the graft is from a living donor.[134,148,162,163,165,167,168] These findings have not been confirmed by subsequent reports.[132,137,147,150,153] Acute tubular necrosis posttransplant is not a danger for development of membranous nephropathy and neither are proteinuria or forms of rejection.[151] No particular immunosuppression reduces the probability of membranous posttransplantation.[9,132-134,137,147,150,153,169,170] Graft loss is not increased with rapid steroid taper protocols.[10] Not unexpectedly, nephrotic proteinuria, elevated creatinine (>1.3 mg/dl) and rejection with membranous nephropathy on biopsy increase the threat of graft loss.[147,152,153,157] Bilateral native nephrectomy pretransplant also increases the risk of graft loss compared to no nephrectomy.[8] While Heidet et al describe a high risk of recurrent de novo membranous nephropathy in subsequent transplants after loss of a graft with this disease; most studies have not shown a prohibitive risk of membranous nephropathy recurring after native disease.[149,164,171]

Because of the lack of prospective, randomized trials for the treatment of membranous nephropathy after kidney transplant, recommendations come only from case reports. Spontaneous remission has been described in idiopathic membranous nephropathy and in at least one case posttransplant.[133] Control of blood pressure and hyperlipidemia along with the use of angiotensin converting enzyme inhibitors is recommended in idiopathic disease, but has not been studied in disease after transplant.[149] Ideura et al have used LDL apheresis to successfully treat nephrosis associated with membranous nephropathy posttransplant refractory to pulse methylprednisolone with proteinuria falling from 3000 mg to 500 mg 5 months posttreatment.[172] No subsequent reports were found on literature review. Pulse or high dose corticosteroids have been used with success in three papers.[152,153,173] These findings were not supported by other studies.[133,134,153,157,172,174] Use of cyclophosphamide, CyA, tacrolimus or mycophenolate mofetil have shown no benefit or isolated success in treating this disease and like corticosteroids, cannot be enthusiastically recommended.[149,150,163,175,176] Rituxamab was used in one patient with recurrent membranous unresponsive to maximizing anti-lipid and angiotensin converting enzyme inhibitor therapy. The patient was given 375 mg/m[2] weekly times four with reduction in proteinuria from 16,000 mg before treatment to 500 mg 2 years post-infusion with stable graft function.[177] Studies are ongoing to determine the efficacy of this treatment on a larger scale. In the mean time, treatment choices are difficult given the paucity of favorable data and care must be demonstrated to avoid complications of immunosuppression.[132]

## Membranoproliferative Glomerulonephritis

Membranoproliferative glomerulonephritis (MPGN) has no pathognomonic clinical or laboratory characteristics, but histologically it is divided into three separate entities. MPGN 1 is defined by subendothelial deposits, MPGN 2 is characterized by electron dense deposits and MPGN 3, which may be a variant of MPGN 1, has subendothelial and subepithelial deposits. Each subtype will be reviewed separately.

## Membranoproliferative Glomerulonephritis 1

### Incidence, Etiology, Clinical and Pathological Presentation

MPGN 1 is an uncommon cause of end stage renal disease in patients undergoing transplantation.[178] Reports of the incidence of disease vary from 0.2 to 3%.[10,178-181] It is estimated to represent between 3.7 to 15% of disease due to glomerulonephritis.[165,178,179,182,183] One author noted a decrease in the incidence of MPGN 1 over the past few decades, perhaps from a favorable change in environmental factors such as decreased exposure to infections.[175]

The etiology of idiopathic MPGN 1 is unknown. Secondary causes of this disease in renal transplantation include systemic illnesses and various infections, particularly hepatitis C.[178,184,185] Of note, the close association of hepatitis C and MPGN 1 reported in the United States and Japan are not seen in other parts of the world.[178,186-191] MPGN 1 has been linked with other viral infections posttransplant such as Epstein Barr virus, cytomegalovirus, hepatitis G and hepatitis B.[192-195] Only idiopathic recurrent MPGN 1 will be discussed.

The clinical hallmarks of MPGN 1 are proteinuria and/or graft dysfunction.[2,165,171,179,180,196,197] In two studies, the proteinuria/graft dysfunction develops at a mean of 20-21 months after transplantation, but with a broad range of 1 to 72 months.[136,179] Another study reported a longer median time to recurrence of 4.7 years.[178] Although complement levels may be valuable markers of disease pretransplantation, they were not reliable posttransplant.[165,179,198,199] No clinical characteristic appears to identify patients at risk for recurrence.[179,200]

Posttransplantation, MPGN 1 must be distinguished from chronic transplant glomerulopathy (CTG). By light microscopy, these two processes are hardly distinguishable, except perhaps by the rare presence of crescents in more severe forms of MPGN 1.[165,196] However, they can be distinguished by immunofluorescent and electron microscopy.[196] MPGN 1 will demonstrate staining with C3 in irregular and granular distribution of the peripheral capillary loops with segmental and granular deposits of IgM and IgG in the capillary walls and variably in the mesangium, whereas there is no consistent pattern with CTG. Dense subendothelial deposits are seen on electron microscopy with MPGN 1 versus electron lucent zone of finely flocculent substance in the subendothelial area with CTG. These findings were confirmed in a biopsy study directly comparing 12 patients with MPGN 1 and 10 with CTG. C3 staining intensity was greater than IgM in the MPGN 1 patients and IgM greater than C3 staining in those with CTG.[196] These immunofluorescent characteristics of CTG confirm findings of other authors.[201,202] Therefore, immunofluorescence and electron microscopy are important tools in diagnosing recurrent MPGN 1. Figure 7 presents a case with recurrent MPGN secondary to cryoglobulinemic glomerulonephritis in a patient with HCV infection.

### Recurrent, de Novo and Pre-Existing MPGN 1 Posttransplant and Associated Graft Loss

In 1982, Cameron reviewed the literature and determined the rate of recurrence of MPGN 1 to be 30%.[182] In this review, the author felt the true incidence may be as low as 15% based on the results of two large case reports at the time and recognizing that the inappropriate labeling of chronic transplant glomerulopathy as MPGN 1 would falsely increase estimates, as has been suggested by others.[96,165,182,183,196,203-206] Reports of recurrent MPGN 1 may not reflect the true incidence for other reasons also. Often, biopsies are not available on many transplanted patients; thus the denominator or cases transplanted with MPGN 1 may be larger than thought which would decrease the actual recurrence rate.[179] Follow-up in many of these studies is short yet the incidence of recurrence increases with time; therefore, it has been postulated that the true incidence may be higher than published results.[179] Unrecognized hepatitis C infection causing MPGN 1 may increase reported recurrence of idiopathic MPGN 1 in the past. More recent reports show recurrence rates ranging from 3-65%.[2,3,136,137,165,171,175,178-181,183] A current cumulative review of 14 studies, including some older and newer publications, show recurrent MPGN 1 in 61 of 218 patients transplanted, or 28%.[207] Clearly more work is needed to better define this risk. In recipients of HLA-identical living related donor kidney transplant Andresdottir et al show recurrence in 3 of 4 patients posttransplantation, all within 2 years.[179] This difference was not statistically significant from other types of transplant and other reports have not shown an increased risk of any type of living related donor transplant compared to deceased donor transplant.[180] De novo MPGN 1 has been described in 3 patients and pre-existing disease in donor kidneys has also rarely been reported.[153,208-210] Thus, determining the source of the MPGN 1 may be confusing, particularly if native disease or the donor kidney was not examined with biopsy. Recurrence in successive transplant grafts is observed and in the largest series of 5 patients published by Andresdottir et al recurrent disease was noted in 4 patients or 80%.[2,179,211,212]

Cameron reports graft loss of 10% due to recurrence of MPGN 1, 14 of 140 grafts.[182] Graft loss of 10-30% has been shown or quoted by other authors.[2,3,136,165,175,213] A recent cumulative review of the literature concludes that graft loss from recurrence of MPGN 1 is 16%, developing in 34 of 218 recipients.[207] Determining graft loss by looking at case reports or small series is hard for several reasons. Native kidney biopsies are not always available to confirm initial diagnosis, thus diagnosis is based on clinical judgment and this prejudices data.[179] Papers do not always separate different types of MPGN and they may be reported as a group adversely affecting conclusions.[5,7,181] Small numbers make solid estimates of risk problematic. Large registry data of biopsy proven MPGN 1 native and graft disease may provide better insight of graft loss risk. Analysis of the ERA-EDTA registry from 1980-1991, showed 6% graft loss from disease recurrence of 564 MPGN 1 patients transplanted, with graft loss generally developing after 2 years.[6] Additional

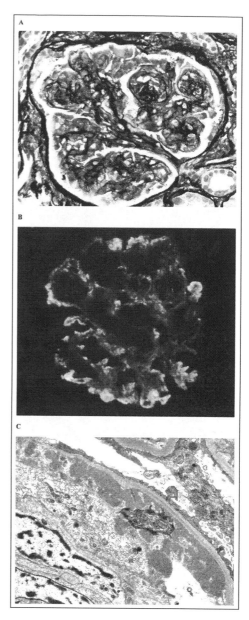

Figure 7. Recurrent MPGN. A) Light microscopic picture showing an enlarged glomerulus with mesangial expansion and endocapillary proliferation, imparting a lobular appearance. Segmental glomerular basement membrane splitting with subendothelial eosinophilic material corresponding to cryoglobulin deposits is present. (Jones' silver stain, × 400). B) Immunofluorescence picture showing segmental, irregular and chunky capillary loop and mesangial staining with IgM. Similar staining with C3 and lesser IgG is also present. (anti-IgM immunofluorescence, × 400). C) Electron microscopic picture showing large subendothelial deposits with mesangial interposition and glomerular basement reduplication, imparting a split appearance. The deposits have vague substructure, frequently seen with cryoglobulin deposits. (TEM, × 7400).

registry data from an interesting Australian paper reports graft loss due to recurrent MPGN 1 in 9 of 88 patients, an actuarial 10 year graft loss due to recurrent disease of 14.4%.[4] In the same study, actuarial 10 year graft loss of all causes was 44% while that due to chronic rejection was 23%. The adjusted hazard ratio for graft loss for those with MPGN 1 as primary disease was 2.91, p = 0.001.[4] These estimates of 6-14% graft loss are close to the original report of Cameron. In those patients who develop recurrent MPGN 1 posttransplant, graft loss occurs in 34 of 61 patients, or 56%.[207] Those who progress to graft loss do so a mean of 40 months after diagnosis of recurrent disease in one report.[179]

## *Risks and Treatment*

Risk of recurrent MPGN 1 has been assessed. Although age is not a factor in some reports, Little et al demonstrate an increased risk of recurrence in the pediatric population of age less than 9 years.[178-180] Risk of recurrent MPGN 1 is not associated with gender, rate of progression of original disease, interval from initiation of dialysis to transplant, immunosuppression, rejection, cold ischemia, donor age or sex, or decade of observation.[136,137,178-180] Although living related donation as a whole is not considered a hazard, there was increased risk of recurrence and graft loss in 4 HLA-identical living donor transplants described by Andresdottir et al.[178-180,214] In HLA-identical living donation, recurrent disease is the primary cause of death censored graft loss.[9] HLA-B8, DR3, SC01, GLO2 and appear to increase susceptibility to MPGN 1 in native kidney disease.[215] In two papers, HLA-B8 or HLA-B8, DR3 is tied to recurrence posttransplant; specifically HLA-B8 recipients receiving female living donor kidneys appear to be at increased risk in one report.[179,180] Persistently low C3 or newly developed decreases in C3 and recurrent MPGN 1 are described by Berthoux et al but these findings have not been confirmed by others, pre or posttransplant.[3,165,178,182,199,216,217] Severity of histology in native kidney disease appears to predict recurrence posttransplantation and graft success. Graft survival of 88% at 10 years in those without crescentic disease, < 20% interstitial fibrosis and less than 3+ mesangial proliferation in native kidneys was found compared to 8% survival in a group with nephrosis and one crescent on initial biopsy.[178] Mesangial proliferation and crescents at presentation are factors associated with disease recurrence in the univariate analysis while only crescentic change on initial presentation is significant on multivariate examination, with a hazard ratio of 6.88, p = 0.005.[178] This risk of crescentic native kidney presentation and recurrence has been found by others.[182,211,217] Native kidney nephrectomy does not alter risk of MPGN 1 with subsequent transplantation.[8] Some reports suggest a protective effect when using CyA immunosuppression, but this is not verified by others.[181,218,219] Rapid steroid taper does not appear to increase risk of recurrence in any form of MPGN.[10] Risk of recurrence after repeat transplant following graft loss from MPGN 1 is controversial and without definitive conclusion.[179,214]

Like native kidney MPGN 1 where one paper notes no progress in treatment success in the past 30 years, therapy for recurrent disease is difficult.[178] With no randomized, controlled studies available, case reports are the only references for treatment options.[3,171,175,183,200,217,218,220,221] Spontaneous remission of disease has been mentioned, but must be a rare event.[175] Dipyridamole has been used in native kidney disease and in reports of recurrent disease treatment, usually with other agents.[207,211,217,222] Cyclophosphamide has been used in four separate case reports with success.[198,207,217,222] Administered orally with high dose intravenous or oral corticosteroids and with or without plasmapheresis, 3 of 4 patients maintain graft function and 2 of these are in remission of heavy proteinuria.[198,207,217] One study of plasmapheresis and high dose oral corticosteroid in a single patient shows biopsy proven improvement histologically after 6 exchanges. Additional pheresis was held because of infectious complications leaving the patient with stable but marginal function and 6 grams of proteinuria.[213] Others have not seen biopsy improvement with 19 pheresis exchanges.[200] High dose mycophenolate mofetil (3 grams per day) started after biopsy proven recurrence results in remission of proteinuria in one patient with MPGN 1.[223] Combined liver-kidney transplant in 2 patients with hepatitis C and end stage renal disease due to associated MPGN 1 demonstrate no proteinuria (no clinical evidence of recurrent MPGN 1) and excellent kidney graft function with 35-45 months follow-up.[224] This success has been shown by others.[225] Undoubtedly, more work is needed before recommendations of treatment may be offered.

## Membranoproliferative Glomerulonephritis 2

### *Incidence, Etiology, Clinical and Pathological Presentation*

MPGN 2 is an uncommon cause of glomerulonephritis and end stage renal disease.[226,227] It represents 2.5% of the end stage renal disease population due to glomerulonephritis in an Australian registry.[165] Of those presenting for transplant, 0.98 to 4.6% have MPGN 2 and it is seen in 1.2 to 11% of biopsy proven glomerulonephritis.[4,6,136,226,227]

Alas, the etiology of recurrent idiopathic MPGN 2 is not known. Continuous dense ribbon-like glomerular basement membrane deposits staining for C3 characterize this type of MPGN. What causes these deposits is unclear.[226] C3 nephritic factor activity (C3NeF), which may be seen in the serum of 60% of patients with native kidney disease (and 30% of those with MPGN 1), is associated with hypocomplementemia.[203,228,229] Its role in deposit formation and disease is unclear.[228] Persistent C3NeF and hypocomplementemia posttransplant occurs, but not universally and dense deposit recurrence may develop without its presence.[229] Interestingly, a single case report of a patient with MPGN 2 transplanted post bilateral nephrectomy developed biopsy proven recurrence of dense deposits at one month associated with measurable C3NeF that eventually became undetectable at one year with no clinical evidence of disease.[228] Follow-up biopsy was not mentioned.

Clinically, the hallmark of recurrent MPGN 2 is proteinuria.[165] Proteinuria is usually subnephrotic and often develops in the first year posttransplant, but may occur as early as 3 weeks.[171,175,230,231] In one report, proteinuria was present in 5 of the 11 recurrences.[226] Biopsy evidence of disease may precede proteinuria and has been demonstrated at 12 days posttransplant.[226,232] Exacerbation of hypertension is seen less than 30% of the time and microscopic hematuria may be present with or without recurrence.[227] Although low C3 and the presence of C3NeF may be seen with recurrence, the correlation is poor with 37.5% of patients with recurrent MPGN 2 having normal C3 levels in one study.[226,227,229,231] Thus, complement is not a reliable clinical marker of the disease. Renal dysfunction is often present at diagnosis, but may be more a reflection of indication to biopsy (change in creatinine) than a reliable clinical marker of disease onset.[226,227] Slow decline in renal function is felt to be the norm but in one report graft failure due to recurrence developed as early as 13-39 months.[171,220,226,227]

Dense eosinophilic ribbon-like thickening of the glomerular basement membrane on light or electron microscopy characterizes the pathologic changes of MPGN 2 and morphologic variations are described in native disease.[233,234] Endocapillary proliferation may be seen preceding the dense deposits.[226,232] Droz et al show that the dense deposits precede the C3 staining which is confirmed by Andresdottir et al but these authors also found cases of globular C3 deposits only in the mesangium with no deposits in the glomerular basement membrane and no electron dense deposit seen on electron microscopy.[226,235] Histological changes vary between isolated dense deposits to marked mesangial proliferation to crescentic disease in one report. There appeared to be no correlation with histology of native kidney disease.[227] Neutrophils are seen in those with native disease as well as the transplant graft and they were not related to deposits in the basement membrane while appearing mainly in the biopsies characterized by crescents.[226] There is no evidence that MPGN 2 evolves to the other types of MPGN.[232] Other lesions such as rejection and calcineurin inhibitor toxicity are often found concurrently with MPGN 2 pathology.[226] This work emphasizes the importance of performing light, immunofluorescence and electron microscopy to correctly identify pathology in this population.[226,227,232,235]

## Recurrent, de Novo and Pre-Existing MPGN 2 Posttransplant and Associated Graft Loss

Recurrence of MPGN 2 appears to be 'the rule rather than the exception'; nonetheless, there is one report with recurrence as low as 18%.[136,165] In 1982, Cameron reviewed the reports of 61 transplants where biopsy evidence of recurrent disease was present in 42 of 48 specimens (88%), although clinical disease was present in only 24%.[182] Others have reported rates varying from 41% to 100%.[178,226,227,230,236,237] A summary of many of these studies published in 1999 showed recurrence in 75 of 128 (59%) recipients and in 54 of 55 (98%) of those with full microscopic examination.[226] Registry data has not separated the different types of MPGN nor did not comment on biopsy proven recurrence exclusive of graft loss and therefore, did not contribute to the current knowledge.[4,5] De novo and pre-existing idiopathic MPGN 2 must be rare, as no information was found in the literature.

Graft loss due to recurrence has varied from the extremes of 0-100% in review of case reports, often due to small sample size.[231,235,238] Cameron reviewed the literature as of 1982 and determined that 6 of 61 grafts were lost due to recurrent disease.[182] Andresdottir et al saw graft loss in 3 of 11 patients (27%), whereas review of the literature up to 1999 showed loss of 19 of 128 (15%) transplants.[226] The ERA-EDTA registry of 397 patients with MPGN 2 notes a graft failure rate from recurrence of 19%.[6] The NAPRTCS registry show kidney transplant loss due to MPGN 2 in 11 of 75 (15%) over a 5 year follow-up.[227] Graft loss due to recurrence or all causes is much higher in MPGN 2 than the general database. In contrast, a large Australian registry reports graft loss due to recurrence of 0% with 10 years follow-up in 18 recipients.[4] Thus, it is possible that histological recurrence may be followed by a benign clinical course at least in some patients.

### Risks and Treatment

Risk factors for the development of recurrence or graft loss have been reported. Male gender and heavy proteinuria may be risk factors for graft loss in the event of recurrence disease, but gender, age, ethnicity, duration of pretransplant dialysis, donor source and proteinuria with native disease have not been associated with increase rate of recurrence.[175,214,226,227,239] As mentioned earlier, neither complement nor C3NeF have been predictors of new disease posttransplant.[3,165,226,227,229-231,236] Crescentic disease on native kidney biopsy is not correlated with recurrence in one report, but Little et al did find crescents on initial biopsy to increase risk of recurrent MPGN 2 with a hazard ratio of 6.88.[178,227] It appears that crescentic changes on transplant biopsies predict worse graft survival.[226,227,239] Immunosuppression does not appear to alter risk and rapid steroid taper appears to be well tolerated.[10,165,226] Nephrectomy pretransplant did not statistically decrease rate of recurrence of MPGN, although numerically, no recurrence was noted in the nephrectomy group (0 of 9) versus 9 of 64 recurrent glomerulonephritis in the nonnephrectomy group of untyped MPGN.[8] Seven patients have received a second graft after losing the first to recurrent disease and graft loss has been found in only one, so retransplant can be considered in these patients.[2,199,226,238,239]

There is no established treatment for MPGN 2.[165,171,175,182,200,220] In the NAPRTCS series, altering immunosuppression by changing calcineurin inhibitors, increasing oral corticosteroids (or using pulse doses in 4 recipients) and use of plasmapheresis in 2 patients are discussed. Graft loss occurred in 2 of the 4 pulse steroid and both of the pheresis patients. Plasmapheresis has been used successfully in two individuals with recurrent MPGN 2.[240,241] Reports of treatment are lacking; therefore, no treatment recommendations will be offered.

## Membranoproliferative Glomerulonephritis 3

MPGN 3 is a rare disease. It is distinct from MPGN 1 by the presence of subepithelial deposits and large lucent areas in the glomerular basement membrane of unclear etiology.[242] It was present in only 2 of

1505 patients with glomerulonephritis undergoing transplantation in a large Australian registry.[4] Etiology of disease recurrence posttransplant is unknown. Three cases of recurrent MPGN 3 provide clinical information for review.[223,243,244] These patients develop recurrent disease characterized by heavy proteinuria and/or renal dysfunction within 21 months of transplant. Hypocomplementemia was noted in one recipient.[223] There is no pathological information distinguishing disease posttransplant from native disease. Of 2 patients with biopsy proven MPGN 3 followed for 10 years posttransplant, one patient developed graft loss.[4] Two of the case report patients suffered graft loss, one two months and the other 6 years after diagnosis.[243,244] The other has stable but abnormal graft function 24 months postrecurrence.[223] No risk factors for disease recurrence are reported. One patient underwent plasmapheresis without benefit and another was refractory to pulse steroids but experienced disease remission with mycophenolate mofetil at 3 grams daily.[223,244] No other information is available in the literature.

## Pauci-Immune Glomerulonephritis

### Incidence, Etiology, Clinical and Pathological Presentation

The pauci-immune glomerulonephritides are characterized by necrotizing small vessel vasculitis and little to no immune complex deposits usually accompanied by anti-neutrophil cytoplasmic antibody (ANCA).[245] Wegener's granulomatosis (WG), microscopic polyarteritis (MPA) and renal-limited vasculitis (RLV) are uncommon causes of glomerulonephritis in those presenting for transplant. However, under diagnosis in the general dialysis population is described where of 1277 subjects, 32 test positive for cANCA and 65 for pANCA.[246] Twenty-two percent with cANCA are also positive for PR3 and 31% of pANCA patients are MPO positive. Of the cANCA subjects, 5 new diagnoses of WG and 2 newly recognized RLV are made. Of the pANCA patients, 12 new MPA, 1 WG and 3 RLV are found. Thus, these diseases may be unrecognized in some of the reports. A single center review of 583 recipients note 7 patients (1.2%) with idiopathic rapidly progressive glomerulonephritis and 3 (0.5%) with WG.[137] The Australian registry of 1505 patients with end stage renal disease due to biopsy proven glomerulonephritis undergoing kidney transplantation show 70 with extracapillary and intracapillary glomerulonephritis (4.7%), 17 with MPA (1.1%) and 15 diagnosed as WG (1%).

The etiology of pauci-immune glomerulonephritis recurrence posttransplant is unknown. The role of ANCA in disease pathogenesis in unclear and recurrent disease may develop with or without ANCA.[247-250]

Clinically, recurrent disease may be heralded by renal dysfunction, hematuria or extra-renal manifestations of vasculitis (pulmonary hemorrhage, sinusitis, ureter necrosis, mononeuritis multiplex, arthritis, eye disease, bowel inflammation and rash).[249,251,252] In the largest review of the literature, recurrence developed between 4 and 89 months after transplant, although disease as early as 5 days has been confirmed.[249,250] The value of ANCA levels in diagnosing recurrence posttransplant has yet to be established.[249,251,252] Pathologically, there appear to be no differences in disease presentation when compared to native kidney reports.[247-249,252]

### Recurrent, de Novo and Pre-Existing Pauciimmune Glomerulonephritis Posttransplant and Associated Graft Loss

The small numbers in case reports have made determining the rate of recurrence difficult. Nachman et al pooled the work of the University North Carolina at Chapel Hill and University of Lund in Sweden along with the best studies of 25 published between 1970 and 1997 to better define recurrence posttransplant of pauci-immune glomerulonephritis.[249] Relapse of disease is defined as one of the following: rapid rise in creatinine with active urinary sediment; a kidney biopsy showing active necrosis or crescent development; hemoptysis, pulmonary hemorrhage or new lung nodules in the absence of infection; active vasculitis of the pulmonary or gastrointestinal tract; iritis or uveitis; new mononeuritis multiplex; or necrotizing vasculitis on biopsy of any tissue.[253] ANCA is not necessary for diagnosis. Of 127 patients with WG or MPA, 22 relapsed (17.3%) with an average time to recurrence of 31 months. Clinical information is available in 21 patients, 12 had kidney involvement whereas 10 had extra-renal disease only. Of the WG subjects, the disease specific recurrence is 20.4%, while relapse occurred in 15.7% of those with MPA or RLV. This difference is not statistically significant. The largest single center report of 33 patients with average follow-up between 51 and 74 months detailed relapse of disease in a single WG recipient involving the sinuses and nerves.[251] No kidney biopsy demonstrated evidence of disease. The relapse rate of 1 in 15 or 7% (CI 0-30%) is similar to that reported by Nachman et al. A study designed to report the success of a rapid steroid taper protocol versus lifelong corticosteroids followed 6 WG and 13 RLV from 26 to 92 months and showed no recurrence.[10] De novo and pre-existing pauci-immune glomerulonephritis have not been described. Retransplantation of those suffering graft losses from pauci-immune glomerulonephritis is reported, but not the risk of recurrence.[249]

Data from pooled studies, single center reports and a national registry describing graft loss from recurrent pauci-immune glomerulonephritis are reviewed. In the pooled studies, 2 of 16 subjects with clinical information lost grafts due to relapse of disease.[249] Long-term remission posttreatment occurred in 11 of these patients with one patient death and one patient with reversal of extra-renal disease but progressive kidney dysfunction. Of 33 transplant recipients in one single center study, none lost grafts to relapse and the 5 year overall graft survival for WG 51%, MPA 69% and RLV 63% compared to 56% in a control arm, representing no statistical significance.[251] Two other single center reports note no graft loss due to relapse in 9 WG patients and 20 with RLV.[10,137] The Australian registry shows graft loss in 2 of 102 subjects, or 2%.[4] The 10 year actuarial graft loss due to recurrence is 7.7%, with all cause graft loss of 53.7%. This is lower than reported 10 year graft loss due to recurrent MPGN 1, focal segmental glomerulosclerosis, membranous nephropathy and IgA nephropathy.

### Risks and Treatment

Pretransplant disease characteristics, disease subtype, donor type, ANCA type or titer, duration of follow-up, duration of dialysis support, pretransplant nephrectomy and the use of CyA or rapid steroid taper do not affect relapse of pauci-immune glomerulonephritis.[8,10,249,254,255] Since ANCA titer is not a factor in timing of transplant, the same authors suggest proceeding with surgery only after disease remission, although this sentiment is not universal.[256] It is recognized that no data support a benefit of waiting any defined time, although one author suggests a 6 month disease free interval before transplant if ANCA is positive.[3,249]

Reports of treatment vary widely with no randomized, controlled trials to guide therapy. After relapse, use of cyclophosphamide based treatment attained remission in 11 of 16 patients and has been recommended by several authors in what appears to be treatment of first choice.[3,165,171,175,220,247-250,254,257-260] Corticosteroid pulses, plasmapheresis and/or intravenous immunoglobulin have been selected by some, particularly for crescentic disease.[247,250,252] High dose mycophenolate mofetil has successfully induced remission in two patients, one biopsy proven.[223,261]

### Scleroderma/Systemic Sclerosis

Renal scleroderma crisis occurs in 10-25% of patients with various degrees of scleroderma and may lead to end stage renal disease.[262] Angiotensin converting enzyme inhibitors have had a significant favorable impact in avoiding or treating this complication of scleroderma.[262,263]

Nonetheless, end stage renal disease results in some and studies of transplant experience are available. The Australian registry notes 7 patients with biopsy proven scleroderma representing 0.2% of the total patients and 0.5% of those with biopsy proven glomerular disease.[4] Renal scleroderma crisis develops when endothelial cell injury of uncertain etiology leads to blood vessel changes in the kidney, producing hypertension, microangiopathy and renal ischemia.[262] Clinical features of presentation after transplantation may include accelerated hypertension and renal dysfunction.[263] Of note, improvement in systemic manifestations of scleroderma such as Raynaud's phenomenon, sclerodactyly and arthralgias has been reported posttransplant.[264] Pathological changes are similar to those in native kidney disease.[263] Recurrence rates have not been established, but graft loss due to relapse has been noted. The UNOS registry contains follow-up from 1987 to 1997 of 86 patients with scleroderma.[263] All cause graft loss is 44% with a 5 year graft survival of 47%. The reason for graft loss is determined in only 14 of 38 patients and 3 of these kidneys, or 21%, are lost due to recurrent scleroderma. The authors comment that this graft survival is equivalent to those transplanted with systemic lupus erythematosus. The ERA-EDTA registry demonstrates a 3 year graft survival of 44% in 28 patients undergoing transplant for scleroderma, with no mention of graft loss due to relapse.[265] The Australian registry showed no graft loss with 10 years follow-up in 7 recipients.[4] Paul et al have implicated the value of bilateral nephrectomy pretransplant in reducing relapse risk, but this has not been confirmed by others.[264] There appears to be no harm or advantage of using CyA based immunosuppression and the role of angiotensin converting enzyme inhibitors posttransplant has yet to be evaluated.[3,263]

## Anti-Glomerular Basement Membrane Disease

### Incidence, Etiology, Clinical and Pathological Presentation

Antibody directed at NC1 domain of the alpha-3 chain of type IV collagen is the source of anti-glomerular basement membrane disease (anti-GBM disease). In an Australian registry report, anti-GBM disease represented 2.9% of all biopsy proven glomerulonephritis presenting for transplantation.[4] Others note anti-GBM disease in 7.4% of 364 patients transplanted for glomerulonephritis and 0.8% to 1.4% of 804 and 583 recipients respectively with end stage renal disease of all causes.[8,10,137]

The etiology of recurrent anti-GBM disease posttransplant is linked to the presence of circulating anti-GBM antibody.[165,175] The stimulus for anti-GBM antibody production posttransplant in those with negative titers pretransplant is unknown. Anti-GBM antibody deposition in the glomerulus posttransplant is not always associated with clinical symptoms (or graft loss).[175] Current descriptions of clinical presentation are lacking due to low relapse rates.[8,10,137,251,266]

The pathological presentation of anti-GBM disease is IgG linear staining of glomerular basement membrane on immunofluorescence. Please note that similar staining has been described in allograft glomerulopathy, but this should easily be distinguished from recurrent anti-GBM disease by the absence of anti-GBM antibody in most cases.[201]

### Recurrent, de Novo and Pre-Existing Anti-GBM Disease Posttransplant and Associated Graft Loss

Many authors suggest that clinical recurrence of anti-GBM disease is rare if recipients are transplanted one year or more after anti-GBM antibody disappears.[4,5,165,171,175,182,183,221,266-268] In current literature, although one author shows relapse of anti-GBM in 2 of 27 patients (7%), many have reported no recurrences in small studies.[10,137,251,266] Most believe the current clinical practice results in relapse rates of 5% or less.[3,171,175,182,221] De novo presentation of disease (presumably in a patient without Alport's nephritis) is described by one paper without reference.[175] No report of pre-existing disease in donor kidneys was found.

Graft loss occurred in 3 of 8 patients in a report by Haubitz et al but none of these were graft failures from recurrent disease. Odorico et al show graft loss secondary to relapse in 1 of 27 patients, but other small studies have demonstrated none.[137,251,266] In the Australian report of 44 recipients with anti-GBM disease, graft loss from recurrent disease did not occur.[4] Deegens et al found the overall graft survival at 5 years to be 54%, which compared favorably to the control group graft survival of 56%.

### Risks and Treatment

Risk of relapse of anti-GBM disease is linked to anti-GBM antibody production. It has been suggested that antibody production in native disease is brief, but Lockwood et al have demonstrated persistent anti-GBM antibody titer up to 2 years after plasmapheresis treatment.[182,267] Care must be taken to confirm disappearance of antibody, even if there are no signs of clinical disease. Successful transplant in the presence of antibody has been reported, but others have shown more favorable outcome if anti-GBM antibody is undetectable at the time of transplant.[267,269] Many clinicians recommend avoiding transplant until the anti-GBM antibody is undetectable for 6-12 months.[4,5,171,175,182,183,221,251,266-268,270,271] There is no need for bilateral nephrectomy and no harm from rapid steroid taper.[8,10] There is no description of superiority of one form of immunosuppression to prevent recurrence. Treatment recommendations are sparse (perhaps from recent lack of experience associated with low relapse rates). Success has been noted with pulse corticosteroids, cyclophosphamide and plasmapheresis or cyclophophamide and immunoadsorption.[252,272]

## Alport's Syndrome and Anti-GBM Disease

### Incidence, Etiology, Clinical and Pathological Presentation

In certain parts of the world Alport's syndrome constitutes 1.8% of those with end stage renal disease.[273] The genetic abnormality producing the disease in patients rests with a mutation in the COL4A5 gene on the X chromosome resulting in abnormal alpha 5 chains or autosomal recessive mutations in the COL4A3 or A4 gene on chromosome 2 producing abnormal alpha 3 and 4 chains of type IV collagen.[274] These alterations in alpha-3, 4 and 5 chains of type IV collagen prevent formation of the alpha-3, 4, 5 protomer of collagen type IV, thus impairing the integrity of basement membrane. Inability to form this protomer because of abnormalities in any of its chains leaves no alpha-3, 4, or5 chains in the basement membrane, which over time produces disease through decomposition of collagen.[274] This process is unlikely to occur when transplantation for end stage renal disease is performed as the graft is immunologically unique; however, these patients are at risk for the development of anti-GBM disease. The alpha-3 chain is consider the Goodpasture's antigen while the alpha-3, 4, 5 protomer of collagen IV is the Goodpasture's protomer and these are the targets of antibody formation in any form of anti-GBM disease.[274] The immune system in patients with Alport's syndrome will be introduced to alpha-3, 4 and 5 chains in the glomerular basement membrane of the donor kidney and this rarely can produce anti-GBM antibody causing crescentic glomerulonephritis and loss of the graft.[275] Since many Alport's syndrome patients suffer the X linked form of the disease with the COL4A5 gene mutation producing no alpha-5 chains, the etiology of the anti-GBM disease is an antibody directed against the newly present donor alpha-5 chain of type IV collagen.[260,276,277] Anti-GBM antibody directed to the Goodpasture antigen, alpha-3 chain, is also seen and may be more prominent in the recessive forms of Alport's syndrome due to COL4A3 and A4 mutations. Onset of anti-GBM disease rarely includes systemic symptoms of vasculitis such as fever, arthralgias and rash; as the

hallmark of disease presentation are markers of direct kidney injury like proteinuria, hematuria and renal dysfunction.[276,278] It usually develops within the first year of transplant with a male preponderance, although late disease has been described.[278,279] There are no unique pathological characteristics of this disease in Alport's syndrome.[275-280] Please note that linear deposition of anti-GBM antibody has been reported without clinical disease.[281,282] Why some Alport's syndrome patients develop these anti-GBM antibodies and others do not is not fully understood. Some postulate that the generation of anti-GBM antibodies in Alport's will be strongest to alpha chain antigens never before seen by that immune system, as may happen with large mutations with absent alpha chain generation. On the other hand, Alport's characterized by lesser mutations that produce decreased but immunologically detectable alpha chains may induce weaker or absent anti-GBM antibody response and no clinical disease.[277,280]

### Rate of Anti-GBM Disease Development and Graft Loss

Three large series have described anti-GMB disease occurrence in Alport's syndrome posttransplant.[279,281,282] Of 35 transplants, Peten et al noted linear antibody deposits on kidney biopsy of 5 patients in the absence of crescentic change and with no adverse effect on renal function. Circulating anti-GBM antibody was detected in a single patient at 8 months and disappeared at 24 months without clinical sequelae.[281] Of 30 patients described by Gobel et al kidney biopsies were performed in 21 with no evidence of crescentic anti-GBM disease. Again, one recipient had transient antibody in the serum. Biopsy of this individual did show linear deposition of GBM antibody with no crescentic changes and no effect on renal function.[282] The largest study by Byrne et al evaluated 52 kidney transplants in those with Alport's syndrome with biopsy proven crescentic disease noted in 2 recipients. One of these recipients had no evidence of immune deposits and may represent pauci-immune glomerulonephritis.[279] These studies and other authors contend the rate of anti-GBM disease occurrence in Alport's syndrome post-transplant is 5% or less.[277,279,281-283] Graft loss in Alport's due to anti-GBM disease has also been reviewed. Of these three series totally 117 transplants, graft loss due to confirmed anti-GBM disease occurred in only one patient.[279,281,282] The Australian registry reported no graft loss due to glomerulonephritis in 77 Alport's syndrome recipients.[4] However, it is clear from small series reports that the development of crescentic anti-GBM disease carries a high risk of graft loss and is difficult to treat.[276,277] Graft survival overall is equal to control group.[279]

### Risks and Treatment

Alport's syndrome patients at risk for anti-GBM disease post-transplant are often men, have established sensori-neural hearing loss and usually have developed end stage renal disease before the age of 30.[277] This suggests a genetic predisposition for recurrence and Ding et al have established that large deletions of the COL4A5 gene are found in 46% of those complicated by anti-GBM disease posttransplant compared to its normal frequency of 16% in the general Alport's syndrome population.[277] Nonetheless, there is no established predictive value of these genetic mutations in determining risk of anti-GBM disease in individual patients, except graft failure due to anti-GBM disease.[276,277] Recurrence of anti-GBM disease in subsequent grafts is usually devastating. Treatment has included bolus corticosteroids, OKT3, ALG, ATG, cyclophophamide and plasmapheresis with mixed results.[276,278] Detecting anti-GBM antibody in the serum may be a marker of early disease.[275,276] Many assays are directed against the alpha-3 chain, but anti-GBM antibodies against the alpha-5 chain are common.[275,277,283] Better assays may improve diagnostic skill allowing early treatment which may favorably alter disease course.

# References

1. Cohen DJ, St. Martin L, Christensen LL et al. Kidney and Pancreas Transplantation in the United States, 1995-2004. Am J Transplant 2006; 6(5p2):1153-69.
2. Morzycka M, Croker BP Jr, Siegler HF et al. Evaluation of recurrent glomerulonephritis in kidney allografts. Am J Med 1982; 72(4):588-98.
3. Kotanko P, Pusey CD, Levy JB. Recurrent glomerulonephritis following renal transplantation. Transplantation 1997; 63(8):1045-52.
4. Briganti EM, Russ GR, McNeil JJ et al. Risk of renal allograft loss from recurrent glomerulonephritis. N Engl J Med 2002; 347(2):103-9.
5. Hariharan S, Adams MB, Brennan DC et al. Recurrent and de novo glomerular disease after renal transplantation: a report from Renal Allograft Disease Registry (RADR). Transplantation 1999; 68(5):635-41.
6. Briggs JD, Jones E. Recurrence of glomerulonephritis following renal transplantation. Scientific Advisory Board of the ERA-EDTA Registry. European Renal Association-European Dialysis and Transplant Association. Nephrol Dial Transplant 1999; 14(3):564-65.
7. Hariharan S, Peddi VR, Savin VJ et al. Recurrent and de novo renal diseases after renal transplantation: a report from the renal allograft disease registry. Am J Kidney Dis 1998; 31(6):928-31.
8. Odorico JS, Knechtle SJ, Rayhill SC et al. The influence of native nephrectomy on the incidence of recurrent disease following renal transplantation for primary glomerulonephritis. Transplantation 1996; 61(2):228-34.
9. Andresdottir MB, Hoitsma AJ, Assmann KJ et al. The impact of recurrent glomerulonephritis on graft survival in recipients of human histocompatibility leucocyte antigen-identical living related donor grafts. Transplantation 1999; 68(5):623-27.
10. Ibrahim H, Rogers T, Casingal V et al. Graft loss from recurrent glomerulonephritis is not increased with a rapid steroid discontinuation protocol. Transplantation 2006; 81(2):214-19.
11. Vongwiwatana A, Gourishankar S, Campbell PM et al. Peritubular Capillary Changes and C4d Deposits Are Associated with Transplant Glomerulopathy But Not IgA Nephropathy. Am J Transplant 2004; 4(1):124-29.
12. Sturdevant M, Garcia-Roca R, Nguyen T et al. Recurrent disease after kidney transplant—little progress in 2 decades. Transplantation 2006; 82(1 Suppl 2):167-68.
13. Myslak M, Amer H, Morales P et al. Interpreting posttransplant proteinuria in patients with proteinuria pretransplant. Am J Transplant 2006; 6(7):1660-65.
14. The 13th report of the human renal transplant registry. Transplant Proc 1977; 9(1):9-26.
15. Odum J, Peh CA, Clarkson AR et al. Recurrent mesangial IgA nephritis following renal transplantation. Nephrol Dial Transplant 1994; 9(3):309-12.
16. Wang AYM, Lai FM, Yu AW-Y et al. Recurrent IgA nephropathy in renal transplant allografts. Am J Kidney Dis 2001; 38(3):588-96.
17. Frohnert PP, Donadio JV Jr, Velosa JA et al. The fate of renal transplants in patients with IgA nephropathy. Clin Transplant 1997; 11(2):127-33.
18. Namba Y, Oka K, Moriyama T et al. Risk factors for graft loss in patients with recurrent IGA nephropathy after renal transplantation. Transplant Proc 2004; 36(5):1314-16.
19. Ponticelli C, Traversi L, Feliciani A et al. Kidney transplantation in patients with IgA mesangial glomerulonephritis. Kidney Int 2001; 60(5):1948-54.
20. Chandrakantan A, Ratanapanichkich P, Said M et al. Recurrent IgA nephropathy after renal transplantation despite immunosuppressive regimens with mycophenolate mofetil. Nephrol Dial Transplant 2005; 20(6):1214-21.
21. McDonald SP, Russ GR. Recurrence of IgA nephropathy among renal allograft recipients from living donors is greater among those with zero HLA mismatches. Transplantation 2006; 82(6):759-762.
22. Freese P, Svalander C, Norden G et al. Clinical risk factors for recurrence of IgA nephropathy. Clin Transplant 1999; 13(4):313-17.
23. Coppo R, Amore A, Cirina P et al. IgA serology in recurrent and nonrecurrent IgA nephropathy after renal transplantation. Nephrol Dial Transplant 1995; 10(12):2310-15.
24. Kessler M, Hiesse C, Hestin D et al. Recurrence of immunoglobulin a nephropathy after renal transplantation in the cyclosporine era. Am J Kidney Dis 1996; 28(1):99-104.
25. Choi MJ, Eustace JA, Gimenez LF et al. Mycophenolate mofetil treatment for primary glomerular diseases. Kidney Int 2002; 61(3):1098-14.
26. Nowack R, Birck R, van der Woude FJ. Mycophenolate mofetil for systemic vasculitis and IgA nephropathy. Lancet 1997; 349(9054):774.
27. Ji S, Liu M, Chen J et al. The fate of glomerular mesangial IgA deposition in the donated kidney after allograft transplantation. Clin Transplant 2004; 18(5):536-40.
28. Tolkoff-Rubin NE, Cosimi AB, Fuller T et al. IGA nephropathy in HLA-identical siblings. Transplantation 1978; 26(6):430-33.
29. Moriyama T, Nitta K, Suzuki K et al. Latent IgA deposition from donor kidney is the major risk factor for recurrent IgA nephropathy in renal transplantation. Clin Transplant 2005; 19 Suppl 14:41-48.
30. Suzuki K, Honda K, Tanabe K et al. Incidence of latent mesangial IgA deposition in renal allograft donors in Japan. Kidney Int 2003; 63(6):2286-94.
31. Kim YS, Moon JI, Jeong HJ et al. Live donor renal allograft in end-stage renal failure patients from immunoglobulin A nephropathy. Transplantation 2001; 71(2):233-38.
32. Andresdottir MB, Hoitsma AJ, Assmann KJ et al. Favorable outcome of renal transplantation in patients with IgA nephropathy. Clin Nephrol 2001; 56(4):279-88.
33. Choy BY, Chan TM, Lo SK et al. Renal transplantation in patients with primary immunoglobulin A nephropathy. Nephrol. Dial. Transplant 2003; 18(11):2399-404.
34. Soler MJ, Mir M, Rodriguez E et al. Recurrence of IgA Nephropathy and Henoch-Schonlein Purpura After Kidney Transplantation: Risk Factors and Graft Survival. Transplant Proc 2005; 37(9):3705-9.
35. Bumgardner GL, Amend WC, Ascher NL et al. Single-center long-term results of renal transplantation for IgA nephropathy. Transplantation 1998; 65(8):1053-60.
36. Ohmacht C, Kliem V, Burg M et al. Recurrent immunoglobulin A nephropathy after renal transplantation: a significant contributor to graft loss. Transplantation 1997; 64(10):1493-96.
37. Jeong HJ, Hong SW, Kim YS et al. Histologic factors associated with nephrotic-range proteinuria in recurrent IGA nephropathy. Transplant Proc 2003; 35(1):291.
38. Oka K, Imai E, Moriyama T et al. A clinicopathological study of IgA nephropathy in renal transplant recipients: beneficial effect of angiotensin-converting enzyme inhibitor. Nephrol Dial Transplant 2000; 15(5):689-95.
39. Toki K, Oka K, Kyo M et al. Clinicopathologic evaluation of IgA nephropathy in renal transplant recipients. Transplant Proc 2001; 33(1-2):1249-53.
40. Praga M, Gutierrez E, Gonzalez E et al. Treatment of IgA nephropathy with ACE inhibitors: a randomized and controlled trial. J Am Soc Nephrol 2003; 14(6):1578-83.
41. Courtney AE, McNamee PT, Nelson WE et al. Does angiotensin blockade influence graft outcome in renal transplant recipients with IgA nephropathy? Nephrol Dial Transplant 2006; 21(12):3550-54.
42. Meulders Q, Pirson Y, Cosyns JP et al. Course of Henoch-Schonlein nephritis after renal transplantation. Report on ten patients and review of the literature. Transplantation 1994; 58(11):1179-86.
43. Kitiyakara C, Kopp JB, Eggers P. Trends in the epidemiology of focal segmental glomerulosclerosis. Semin Nephrol 2003; 23(2):172-82.
44. Dall'Amico R, Ghiggeri G, Carraro M et al. Prediction and treatment of recurrent focal segmental glomerulosclerosis after renal transplantation in children. Am J Kidney Dis 1999; 34(6):1048-55.
45. Banfi G, Colturi C, Montagnino G et al. The recurrence of focal segmental glomerulosclerosis in kidney transplant patients treated with cyclosporine. Transplantation 1990; 50(4):594-96.
46. Artero M, Biava C, Amend W et al. Recurrent focal glomerulosclerosis: natural history and response to therapy. Am J Med 1992; 92(4):375-83.
47. Pardon A, Audard V, Caillard S et al. Risk factors and outcome of focal and segmental glomerulosclerosis recurrence in adult renal transplant recipients. Nephrol Dial Transplant 2006; 21(4):1053-59.
48. Wuhl E, Fydryk J, Wiesel M et al. Impact of recurrent nephrotic syndrome after renal transplantation in young patients. Pediatr Nephrol 1998; 12(7):529-33.
49. Senggutuvan P, Cameron JS, Hartley RB et al. Recurrence of focal segmental glomerulosclerosis in transplanted kidneys: analysis of incidence and risk factors in 59 allografts. Pediatr Nephrol 1990; 4(1):21-28.
50. Ingulli E, Tejani A. Incidence, treatment and outcome of recurrent focal segmental glomerulosclerosis posttransplantation in 42 allografts in children—a single-center experience. Transplantation 1991; 51(2):401-5.
51. Zimmerman CE. Renal transplantation for focal segmental glomerulosclerosis. Transplantation 1980; 29(2):172.
52. Baum MA, Stablein DM, Panzarino VM et al. Loss of living donor renal allograft survival advantage in children with focal segmental glomerulosclerosis. Kidney Int 2001; 59(1):328-33.
53. Abbott KC, Sawyers ES, Oliver JD 3rd et al. Graft loss due to recurrent focal segmental glomerulosclerosis in renal transplant recipients in the United States. Am J Kidney Dis 2001; 37(2):366-73.
54. Cibrik DM, Kaplan B, Campbell DA et al. Renal allograft survival in transplant recipients with focal segmental glomerulosclerosis. Am J Transplant 2003; 3(1):64-67.
55. First MR. Living-related donor transplants should be performed with caution in patients with focal segmental glomerulosclerosis. Pediatr Nephrol 1995; 9 Suppl:S40-42.

56. Raafat R, Travis LB, Kalia A et al. Role of transplant induction therapy on recurrence rate of focal segmental glomerulosclerosis. Pediatr Nephrol 2000; 14(3):189-94.

57. Hubsch H, Montane B, Abitbol C et al. Recurrent focal glomerulosclerosis in pediatric renal allografts: the Miami experience. Pediatr Nephrol 2005; 20(2):210-16.

58. Boardman R, Trofe J, Alloway R et al. Early steroid withdrawal does not increase risk for recurrent focal segmental glomerulosclerosis. Transplant Proc 2005; 37(2):817-18.

59. Deegens JK, Wetzels JF. Treatment of recurrent focal glomerulosclerosis after renal transplantation: is prednisone essential to maintain a sustained remission? Transplantation 2003; 75(7):1080-81.

60. Hocker B, Knuppel T, Waldherr R et al. Recurrence of proteinuria 10 years posttransplant in NPHS2-associated focal segmental glomerulosclerosis after conversion from cyclosporin A to sirolimus. Pediatr Nephrol 2006; 21(10):1476-79.

61. Dantal J, Baatard R, Hourmant M et al. Recurrent nephrotic syndrome following renal transplantation in patients with focal glomerulosclerosis. A one-center study of plasma exchange effects. Transplantation 1991; 52(5):827-31.

62. Kim EM, Striegel J, Kim Y et al. Recurrence of steroid-resistant nephrotic syndrome in kidney transplants is associated with increased acute renal failure and acute rejection. Kidney Int 1994; 45(5):1440-45.

63. Wetzels JF. New insights into the pathogenesis and the therapy of recurrent focal glomerulosclerosis. Am J Transplant 2005; 5(10):2594; author reply 2595.

64. Zimmerman SW. Increased urinary protein excretion in the rat produced by serum from a patient with recurrent focal glomerular sclerosis after renal transplantation. Clin Nephrol 1984; 22(1):32-38.

65. Kemper MJ, Wolf G, Muller-Wiefel DE. Transmission of Glomerular Permeability Factor from a Mother to Her Child. N Engl J Med 2001; 344(5):386-87.

66. Sharma M, Sharma R, McCarthy ET et al. "The FSGS factor:" enrichment and in vivo effect of activity from focal segmental glomerulosclerosis plasma. J Am Soc Nephrol 1999; 10(3):552-61.

67. Le Berre L, Godfrin Y, Lafond-Puyet L et al. Effect of plasma fractions from patients with focal and segmental glomerulosclerosis on rat proteinuria. Kidney Int 2000; 58(6):2502-11.

68. Sharma M, Sharma R, McCarthy ET et al. The Focal Segmental Glomerulosclerosis Permeability Factor: Biochemical Characteristics and Biological Effects. Exp Biol Med 2004; 229(1):85-98.

69. McCarthy ET, Sharma M. Indomethacin protects permeability barrier from focal segmental glomerulosclerosis serum. Kidney Int 2002; 61(2):534-41.

70. Sharma R, Sharma M, McCarthy ET et al. Components of normal serum block the focal segmental glomerulosclerosis factor activity in vitro. Kidney Int 2000; 58(5):1973-79.

71. Candiano G, Musante L, Carraro M et al. Apolipoproteins prevent glomerular albumin permeability induced in vitro by serum from patients with focal segmental glomerulosclerosis. J Am Soc Nephrol 2001; 12(1):143-50.

72. Godfrin Y, Dantal J, Perretto S et al. Study of the in vitro effect on glomerular albumin permselectivity of serum before and after renal transplantation in focal segmental glomerulosclerosis. Transplantation 1997; 64(12):1711-15.

73. Dantal J, Godfrin Y, Koll R et al. Antihuman immunoglobulin affinity immunoadsorption strongly decreases proteinuria in patients with relapsing nephrotic syndrome. J Am Soc Nephrol 1998; 9(9):1709-15.

74. Savin VJ, Sharma R, Sharma M et al. Circulating factor associated with increased glomerular permeability to albumin in recurrent focal segmental glomerulosclerosis. N Engl J Med 1996; 334(14):878-83.

75. Coward RJM, Foster RR, Patton D et al. Nephrotic Plasma Alters Slit Diaphragm-Dependent Signaling and Translocates Nephrin, Podocin and CD2 Associated Protein in Cultured Human Podocytes. J Am Soc Nephrol 2005; 16(3):629-37.

76. Karle SM, Uetz B, Ronner V et al. Novel mutations in NPHS2 detected in both familial and sporadic steroid-resistant nephrotic syndrome. J Am Soc Nephrol 2002; 13(2):388-93.

77. Weber S, Gribouval O, Esquivel EL et al. NPHS2 mutation analysis shows genetic heterogeneity of steroid-resistant nephrotic syndrome and low posttransplant recurrence. Kidney Int 2004; 66(2):571-79.

78. Bertelli R, Ginevri F, Caridi G et al. Recurrence of focal segmental glomerulosclerosis after renal transplantation in patients with mutations of podocin. Am J Kidney Dis 2003; 41(6):1314-21.

79. Becker-Cohen R, Bruschi M, Rinat C et al. Recurrent Nephrotic Syndrome in Homozygous Truncating NPHS2 Mutation Is Not Due to AntiPodocin Antibodies. Am J Transplant 2006; 15:15.

80. Carraro M, Caridi G, Bruschi M et al. Serum glomerular permeability activity in patients with podocin mutations (NPHS2) and steroid-resistant nephrotic syndrome. J Am Soc Nephrol 2002; 13(7):1946-52.

81. Gohh RY, Yango AF, Morrissey PE et al. Preemptive plasmapheresis and recurrence of FSGS in high-risk renal transplant recipients. Am J Transplant 2005; 5(12):2907-12.

82. Artero ML, Sharma R, Savin VJ et al. Plasmapheresis reduces proteinuria and serum capacity to injure glomeruli in patients with recurrent focal glomerulosclerosis. Am J Kidney Dis 1994; 23(4):574-81.

83. Pradhan M, Petro J, Palmer J et al. Early use of plasmapheresis for recurrent posttransplant FSGS. Pediatr Nephrol 2003; 18(9):934-38.

84. Davenport RD. Apheresis treatment of recurrent focal segmental glomerulosclerosis after kidney transplantation: re-analysis of published case-reports and case-series. J Clin Apher 2001; 16(4):175-78.

85. Belson A, Yorgin PD, Al-Uzri AY et al. Long-term plasmapheresis and protein A column treatment of recurrent FSGS. Pediatr Nephrol 2001; 16(12):985-89.

86. Deegens JK, Andresdottir MB, Croockewit S et al. Plasma exchange improves graft survival in patients with recurrent focal glomerulosclerosis after renal transplant. Transpl Int 2004; 17(3):151-57.

87. Dantal J, Bigot E, Bogers W et al. Effect of plasma protein adsorption on protein excretion in kidney-transplant recipients with recurrent nephrotic syndrome. N Engl J Med 1994; 330(1):7-14.

88. Kershaw DB, Sedman AB, Kelsch RC et al. Recurrent focal segmental glomerulosclerosis in pediatric renal transplant recipients: successful treatment with oral cyclophosphamide. Clin Transplant 1994; 8(6):546-49.

89. Vincenti F, Biava C, Tomlanovitch S et al. Inability of cyclosporine to completely prevent the recurrence of focal glomerulosclerosis after kidney transplantation. Transplantation 1989; 47(4):595-98.

90. Raafat RH, Kalia A, Travis LB et al. High-dose oral cyclosporin therapy for recurrent focal segmental glomerulosclerosis in children. Am J Kidney Dis 2004; 44(1):50-56.

91. Srivastava RN, Kalia A, Travis LB et al. Prompt remission of postrenal transplant nephrotic syndrome with high-dose cyclosporine. Pediatr Nephrol 1994; 8(1):94-95.

92. Salomon R, Gagnadoux MF, Niaudet P. Intravenous cyclosporine therapy in recurrent nephrotic syndrome after renal transplantation in children. Transplantation 2003; 75(6):810-14.

93. Sharma R, Savin VJ. Cyclosporine prevents the increase in glomerular albumin permeability caused by serum from patients with focal segmental glomerular sclerosis. Transplantation 1996; 61(3):381-83.

94. Sharma R, Sharma M, Ge X et al. Cyclosporine protects glomeruli from FSGS factor via an increase in glomerular cAMP. Transplantation 1996; 62(12):1916-20.

95. Hristea D, Hadaya K, Marangon N et al. Successful treatment of recurrent focal segmental glomerulosclerosis after kidney transplantation by plasmapheresis and rituximab. Transpl Int 2007; 20(1):102-5.

96. Ramos EL. Recurrent diseases in the renal allograft. J Am Soc Nephrol 1991; 2(2):109-21.

97. Mojcik CF, Klippel JH. End-stage renal disease and systemic lupus erythematosus. Am J Med 1996; 101(1):100-7.

98. Moroni G, Tantardini F, Gallelli B et al. The long-term prognosis of renal transplantation in patients with lupus nephritis. Am J Kidney Dis 2005; 45(5):903-11.

99. Stone JH, Millward CL, Olson JL et al. Frequency of recurrent lupus nephritis among ninety-seven renal transplant patients during the cyclosporine era. Arthritis Rheum 1998; 41(4):678-86.

100. Nyberg G, Karlberg I, Svalander C et al. Renal transplantation in patients with systemic lupus erythematosus: increased risk of early graft loss. Scand J Urol Nephrol 1990; 24(4):307-13.

101. Goral S, Ynares C, Shappell SB et al. Recurrent lupus nephritis in renal transplant recipients revisited: it is not rare. Transplantation 2003; 75(5):651-56.

102. Nyberg G, Blohme I, Persson H et al. Recurrence of SLE in transplanted kidneys: a follow-up transplant biopsy study. Nephrol Dial Transplant 1992; 7(11):1116-23.

103. Lochhead KM, Pirsch JD, D'Alessandro AM et al. Risk factors for renal allograft loss in patients with systemic lupus erythematosus. Kidney Int 1996; 49(2):512-17.

104. Azevedo LS, Romao JE Jr, Malheiros D et al. Renal transplantation in systemic lupus erythematosus. A case control study of 45 patients. Nephrol Dial Transplant 1998; 13(11):2894-98.

105. Zara CP, Lipkowitz GS, Perri N et al. Renal transplantation and end-stage lupus nephropathy in the cyclosporine and precyclosporine eras. Transplant Proc 1989; 21(1 Pt 2):1648-51.

106. Sfikakis PP, Boletis JN, Lionaki S et al. Remission of proliferative lupus nephritis following B-cell depletion therapy is preceded by down-regulation of the T-cell costimulatory molecule CD40 ligand: an open-label trial. Arthritis Rheum 2005; 52(2):501-13.

107. Leandro MJ, Edwards JC, Cambridge G et al. An open study of B lymphocyte depletion in systemic lupus erythematosus. Arthritis Rheum 2002; 46(10):2673-77.

108. Smith KG, Jones RB, Burns SM et al. Long-term comparison of rituximab treatment for refractory systemic lupus erythematosus and vasculitis: Remission, relapse and retreatment. Arthritis Rheum 2006; 54(9):2970-82.

109. Ferraris JR, Ramirez JA, Ruiz S et al. Shiga toxin-associated hemolytic uremic syndrome: absence of recurrence after renal transplantation. Pediatr Nephrol 2002; 17(10):809-14.

110. Artz MA, Steenbergen EJ, Hoitsma AJ et al. Renal transplantation in patients with hemolytic uremic syndrome: high rate of recurrence and increased incidence of acute rejections. Transplantation 2003; 76(5):821-26.

111. Kaplan BS, Papadimitriou M, Brezin JH et al. Renal transplantation in adults with autosomal recessive inheritance of hemolytic uremic syndrome. Am J Kidney Dis 1997; 30(6):760-65.

112. Lahlou A, Lang P, Charpentier B et al. Hemolytic uremic syndrome. Recurrence after renal transplantation. Groupe Cooperatif de l'Ile-de-France (GCIF). Medicine (Baltimore) 2000; 79(2):90-102.

113. Miller RB, Burke BA, Schmidt WJ et al. Recurrence of haemolytic-uraemic syndrome in renal transplants: a single-centre report. Nephrol Dial Transplant 1997; 12(7):1425-30.

114. Conlon PJ, Brennan DC, Pfaf WW et al. Renal transplantation in adults with thrombotic thrombocytopenic purpura/haemolytic-uraemic syndrome. Nephrol Dial Transplant 1996; 11(9):1810-14.

115. Bresin E, Daina E, Noris M et al. Outcome of Renal Transplantation in Patients with nonShiga Toxin-Associated Hemolytic Uremic Syndrome: Prognostic Significance of Genetic Background. Clin J Am Soc Nephrol 2006; 1(1):88-99.

116. Remuzzi G, Ruggenenti P, Colledan M et al. Hemolytic Uremic Syndrome: A Fatal Outcome after Kidney and Liver Transplantation Performed to Correct Factor H Gene Mutation. Am J Transplant 2005; 5(5):1146-50.

117. Caprioli J, Noris M, Brioschi S et al. Genetics of HUS: the impact of MCP, CFH and IF mutations on clinical presentation, response to treatment and outcome. Blood 2006; 108(4):1267-79.

118. Fremeaux-Bacchi V, Moulton EA, Kavanagh D et al. Genetic and functional analyses of membrane cofactor protein (CD46) mutations in atypical hemolytic uremic syndrome. J Am Soc Nephrol 2006; 17(7):2017-25.

119. Richards A, Kemp EJ, Liszewski MK et al. Mutations in human complement regulator, membrane cofactor protein (CD46), predispose to development of familial hemolytic uremic syndrome. Proc Natl Acad Sci USA 2003; 100(22):12966-71.

120. Ducloux D, Rebibou JM, Semhoun-Ducloux S et al. Recurrence of hemolytic-uremic syndrome in renal transplant recipients: a meta-analysis. Transplantation 1998; 65(10):1405-7.

121. Florman S, Benchimol C, Lieberman K et al. Fulminant recurrence of atypical hemolytic uremic syndrome during a calcineurin inhibitor-free immunosuppression regimen. Pediatr Transplant 2002; 6(4):352-55.

122. Bergstein J, Michael A Jr, Kellstrand C et al. Hemolytic-uremic syndrome in adult sisters. Transplantation 1974; 17(5):487-90.

123. Donne RL, Abbs I, Barany P et al. Recurrence of hemolytic uremic syndrome after live related renal transplantation associated with subsequent de novo disease in the donor. Am J Kidney Dis 2002; 40(6):E22.

124. Olie KH, Florquin S, Groothoff JW et al. Atypical relapse of hemolytic uremic syndrome after transplantation. Pediatr Nephrol 2004; 19(10):1173-76.

125. Noris M, Remuzzi G. Hemolytic Uremic Syndrome. J Am Soc Nephrol 2005; 16(4):1035-1050.

126. Olie KH, Goodship TH, Verlaak R et al. Posttransplantation cytomegalovirus-induced recurrence of atypical hemolytic uremic syndrome associated with a factor H mutation: successful treatment with intensive plasma exchanges and ganciclovir. Am J Kidney Dis 2005; 45(1):e12-15.

127. Yassa SK, Blessios G, Marinides G et al. Anti-CD20 monoclonal antibody (Rituximab) for life-threatening hemolytic-uremic syndrome. Clin Transplant 2005; 19(3):423-26.

128. Banerjee D, Kupin W, Roth D. Hemolytic uremic syndrome after multivisceral transplantation treated with intravenous immunoglobulin. J Nephrol 2003; 16(5):733-35.

129. Saland JM, Emre SH, Shneider BL et al. Favorable long-term outcome after liver-kidney transplant for recurrent hemolytic uremic syndrome associated with a factor H mutation. Am J Transplant 2006; 6(8):1948-52.

130. Remuzzi G, Ruggenenti P, Codazzi D et al. Combined kidney and liver transplantation for familial haemolytic uraemic syndrome. The Lancet 2002; 359(9318):1671-72.

131. Cheong HI, Lee BS, Kang HG et al. Attempted treatment of factor H deficiency by liver transplantation. Pediatr Nephrol 2004; 19(4):454-58.

132. Couchoud C, Pouteil-Noble C, Colon S et al. Recurrence of membranous nephropathy after renal transplantation. Incidence and risk factors in 1614 patients. Transplantation 1995; 59(9):1275-79.

133. Marcen R, Mampaso F, Teruel JL et al. Membranous nephropathy: recurrence after kidney transplantation. Nephrol Dial Transplant 1996; 11(6):1129-33.

134. Berger BE, Vincenti F, Biava C et al. De novo and recurrent membranous glomerulopathy following kidney transplantation. Transplantation 1983; 35(4):315-19.

135. Hariharan S, Savin VJ. Recurrent and de novo disease after renal transplantation: a report from the Renal Allograft Disease Registry. Pediatr Transplant 2004; 8(4):349-50.

136. O'Meara Y, Green A, Carmody M et al. Recurrent glomerulonephritis in renal transplants: fourteen years' experience. Nephrol Dial Transplant 1989; 4(8):730-34.

137. Schwarz A, Krause PH, Offermann G et al. Recurrent and de novo renal disease after kidney transplantation with or without cyclosporine A. Am J Kidney Dis 1991; 17(5):524-31.

138. Ronco P, Allegri L, Brianti E et al. Antigenic targets in epimembranous glomerulonephritis. Experimental data and potential application in human pathology. Appl Pathol 1989; 7(2):85-98.

139. Ronco PM, Ardaillou N, Verroust P et al. Pathophysiology of the podocyte: a target and a major player in glomerulonephritis. Adv Nephrol Necker Hosp 1994; 23:91-131.

140. Kerjaschki D, Horvat R, Binder S et al. Identification of a 400-kd protein in the brush borders of human kidney tubules that is similar to gp330, the nephritogenic antigen of rat Heymann nephritis. Am J Pathol 1987; 129(1):183-91.

141. Kerjaschki D, Neale TJ. Molecular mechanisms of glomerular injury in rat experimental membranous nephropathy (Heymann nephritis). J Am Soc Nephrol 1996; 7(12):2518-26.

142. Debiec H, Guigonis V, Mougenot B et al. Antenatal membranous glomerulonephritis due to antineutral endopeptidase antibodies. N Engl J Med 2002; 346(26):2053-60.

143. Debiec H, Guigonis V, Mougenot B et al. Antenatal membranous glomerulonephritis with vascular injury induced by antineutral endopeptidase antibodies: toward new concepts in the pathogenesis of glomerular diseases. J Am Soc Nephrol 2003; 14 Suppl 1:S27-32.

144. Debiec H, Nauta J, Coulet F et al. Role of truncating mutations in MME gene in fetomaternal alloimmunisation and antenatal glomerulopathies. Lancet 2004; 364(9441):1252-59.

145. Kerjaschki D. Pathomechanisms and molecular basis of membranous glomerulopathy. Lancet 2004; 364(9441):1194-96.

146. Ronco P, Debiec H. New insights into the pathogenesis of membranous glomerulonephritis. Curr Opin Nephrol Hypertens 2006; 15(3):258-63.

147. Josephson MA, Spargo B, Hollandsworth D et al. The recurrence of recurrent membranous glomerulopathy in a renal transplant recipient: case report and literature review. Am J Kidney Dis 1994; 24(5):873-78.

148. Davison AM, Johnston PA. Allograft membranous nephropathy. Nephrol Dial Transplant 1992; 7 Suppl 1:114-18.

149. Poduval RD, Josephson MA, Javaid B. Treatment of de novo and recurrent membranous nephropathy in renal transplant patients. Semin Nephrol 2003; 23(4):392-99.

150. Cosyns JP, Couchoud C, Pouteil-Noble C et al. Recurrence of membranous nephropathy after renal transplantation: probability, outcome and risk factors. Clin Nephrol 1998; 50(3):144-53.

151. Monga G, Mazzucco G, Basolo B et al. Membranous glomerulonephritis (MGN) in transplanted kidneys: morphologic investigation on 256 renal allografts. Mod Pathol 1993; 6(3):249-58.

152. Truong L, Gelfand J, D'Agati V et al. De novo membranous glomerulonephropathy in renal allografts: a report of ten cases and review of the literature. Am J Kidney Dis 1989; 14(2):131-44.

153. Schwarz A, Krause PH, Offermann G et al. Impact of de novo membranous glomerulonephritis on the clinical course after kidney transplantation. Transplantation 1994; 58(6):650-54.

154. Kowalewska J, Smith KD, Hudkins KL et al. Membranous glomerulopathy with spherules: an uncommon variant with obscure pathogenesis. Am J Kidney Dis 2006; 47(6):983-92.

155. Lazowski P, Sablay LB, Glicklich D. Recurrent crescentic membranous nephropathy in two successive renal transplants: association with choroidal effusions and retinal detachment. Am J Nephrol 1998; 18(2):146-50.

156. Hill GS, Robertson J, Grossman R et al. An unusual variant of membranous nephropathy with abundant crescent formation and recurrence in the transplanted kidney. Clin Nephrol 1978; 10(3):114-20.

157. Antignac C, Hinglais N, Gubler MC et al. De novo membranous glomerulonephritis in renal allografts in children. Clin Nephrol 1988; 30(1):1-7.

158. Honkanen E, Tornroth T, Pettersson E et al. Glomerulonephritis in renal allografts: results of 18 years of transplantations. Clin Nephrol 1984; 21(4):210-19.

159. Nakazawa K, Shimojo H, Komiyama Y et al. Preexisting membranous nephropathy in allograft kidney. Nephron 1999; 81(1):76-80.

160. Parker SM, Pullman JM, Khauli RB. Successful transplantation of a kidney with early membranous nephropathy. Urology 1995; 46(6):870-72.

161. Freese PM, Svalander CT, Molne J et al. Renal allograft glomerulopathy and the value of immunohistochemistry. Clin Nephrol 2004; 62(4):279-86.

162. Lieberthal W, Bernard DB, Donohoe JF et al. Rapid recurrence of membranous nephropathy in a related allograft. Clin Nephrol 1979; 12(5):222-228.

163. Obermiller LE, Hoy WE, Eversole M et al. Recurrent membranous glomerulonephritis in two renal transplants. Transplantation 1985; 40(1):100-2.

164. Heidet L, Gagnadoux ME, Beziau A et al. Recurrence of de novo membranous glomerulonephritis on renal grafts. Clin Nephrol 1994; 41(5):314-18.

165. Mathew TH. Recurrence of disease following renal transplantation. Am J Kidney Dis 1988; 12(2):85-96.

166. Andresdottir M, Hoitsma A, Wetzels J. Donor characteristics as risk factors for recurrence of membranous nephropathy after renal transplantation. Am J Transplant 2006; 6(s2):947.

167. Agarwal SK, Dash SC, Mehta SN et al. Recurrence of idiopathic membranous nephropathy in HLA-identical allograft. Nephron 1992; 60(3):366.

168. First MR, Vaidya PN, Maryniak RK et al. Proteinuria following transplantation. Correlation with histopathology and outcome. Transplantation 1984; 38(6):607-12.

169. Lal S, Luger A, Hashefi M et al. De novo membranous glomerulopathy in an renal transplant patient treated with FK 506. The first reprted case. Int J Artif Organs 1997; 20(7):379-82.

170. Montagnino G, Colturi C, Banfi G et al. Membranous nephropathy in cyclosporine-treated renal transplant recipients. Transplantation 1989; 47(4):725-27.

171. Choy BY, Chan TM, Lai KN. Recurrent glomerulonephritis after kidney transplantation. Am J Transplant 2006; 6(11):2535-42.

172. Ideura T, Hora K, Kaneko Y et al. Effect of low-density lipoprotein-apheresis on nephrotic syndrome due to membranous nephropathy in renal allograft: a case report. Transplant Proc 2000; 32(1):223-26.

173. Johnston PA, Goode NP, Aparicio SR et al. Membranous allograft nephropathy. Remission of nephrotic syndrome with pulsed methylprednisolone and high-dose alternate-day steroids. Transplantation 1993; 55(1):214-16.

174. Innes A, Woodrow G, Boyd SM et al. Recurrent membranous nephropathy in successive renal transplants. Nephrol Dial Transplant 1994; 9(3):323-25.

175. Dantal J, Giral M, Hoormant M et al. Glomerulonephritis recurrences after kidney transplantation. Curr Opin Nephrol Hypertens 1995; 4(2):146-54.

176. Harzallah K, Badid C, Fouque D et al. Efficacy of mycophenolate mofetil on recurrent glomerulonephritis after renal transplantation. Clin Nephrol 2003; 59(3):212-16.

177. Gallon L, Chhabra D. AntiCD20 Monoclonal Antibody (Rituximab) for the Treatment of Recurrent Idiopathic Membranous Nephropathy in a Renal Transplant Patient. Am J Transplant 2006; 6(12):3017-21.

178. Little MA, Dupont P, Campbell E et al. Severity of primary MPGN, rather than MPGN type, determines renal survival and posttransplantation recurrence risk. Kidney Int 2006; 69(3):504-11.

179. Andresdottir MB, Assmann KJ, Hoitsma AJ et al. Recurrence of type I membranoproliferative glomerulonephritis after renal transplantation: analysis of the incidence, risk factors and impact on graft survival. Transplantation 1997; 63(11):1628-33.

180. Karakayali FY, Ozdemir H, Kivrakdal S et al. Recurrent glomerular diseases after renal transplantation. Transplant Proc 2006; 38(2):470-72.

181. Shimizu T, Tanabe K, Oshima T et al. Recurrence of membranoproliferative glomerulonephritis in renal allografts. Transplant Proc 1998; 30(7):3910-13.

182. Cameron JS. Glomerulonephritis in renal transplants. Transplantation 1982; 34(5):237-45.

183. Ramos EL, Tisher CC. Recurrent diseases in the kidney transplant. Am J Kidney Dis 1994; 24(1):142-54.

184. Brunkhorst R, Kliem V, Koch KM. Recurrence of membranoproliferative glomerulonephritis after renal transplantation in a patient with chronic hepatitis C. Nephron 1996; 72(3):465-67.

185. Roth D, Cirocco R, Zucker K et al. De novo membranoproliferative glomerulonephritis in hepatitis C virus-infected renal allograft recipients. Transplantation 1995; 59(12):1676-82.

186. Cosio FG, Roche Z, Agarwal A et al. Prevalence of hepatitis C in patients with idiopathic glomerulopathies in native and transplant kidneys. Am J Kidney Dis 1996; 28(5):752-58.

187. Johnson RJ, Gretch DR, Yamabe H et al. Membranoproliferative glomerulonephritis associated with hepatitis C virus infection. N Engl J Med 1993; 328(7):465-70.

188. Madala ND, Naicker S, Singh B et al. The pathogenesis of membranoproliferative glomerulonephritis in KwaZulu-Natal, South Africa is unrelated to hepatitis C virus infection. Clin Nephrol 2003; 60(2):69-73.

189. Ohsawa I, Ohi H, Endo M et al. High prevalence of hepatitis C virus antibodies in older patients with membranoproliferative glomerulonephritis. Nephron 1999; 82(4):366-67.

190. Sabry AA, Sobh MA, Irving WL et al. A comprehensive study of the association between hepatitis C virus and glomerulopathy. Nephrol Dial Transplant 2002; 17(2):239-45.

191. Yamabe H, Johnson RJ, Gretch DR et al. Hepatitis C virus infection and membranoproliferative glomerulonephritis in Japan. J Am Soc Nephrol 1995; 6(2):220-23.

192. Andresdottir MB, Assmann KJ, Hilbrands LB et al. Primary Epstein-Barr virus infection and recurrent type I membranoproliferative glomerulonephritis after renal transplantation. Nephrol Dial Transplant 2000; 15(8):1235-37.

193. Berthoux P, Laurent B, Cecillon S et al. Membranoproliferative glomerulonephritis with subendothelial deposits (type 1) associated with hepatitis G virus infection in a renal transplant recipient. Am J Nephrol 1999; 19(4):513-18.

194. Birk PE, Chavers BM. Does cytomegalovirus cause glomerular injury in renal allograft recipients? J Am Soc Nephrol 1997; 8(11):1801-08.

195. Andresdottir MB, Assmann KJ, Hilbrands LB et al. Type I membranoproliferative glomerulonephritis in a renal allograft: A recurrence induced by a cytomegalovirus infection? Am J Kidney Dis 2000; 35(2):E6.

196. Andresdottir MB, Assmann KJ, Koene RA et al. Immunohistological and ultrastructural differences between recurrent type I membranoproliferative glomerulonephritis and chronic transplant glomerulopathy. Am J Kidney Dis 1998; 32(4):582-88.

197. Cheigh JS, Mouradian J, Susin M et al. Kidney transplant nephrotic syndrome: relationship between allograft histopathology and natural course. Kidney Int 1980; 18(3):358-65.

198. Muczynski KA. Plasmapheresis maintained renal function in an allograft with recurrent membranoproliferative glomerulonephritis type I. Am J Nephrol 1995; 15(5):446-49.

199. Curtis JJ, Wyatt RJ, Bhathena D et al. Renal transplantation for patients with type I and type II membranoproliferative glomerulonephritis: serial complement and nephritic factor measurements and the problem of recurrence of disease. Am J Med 1979; 66(2):216-25.

200. Dwyer K, Hill P, Murphy B. Early recurrence of type 1 membranoproliferative glomerulonephritis following cadaveric renal transplantation. Aust N Z J Med 2000; 30(1):103-4.

201. Habib R, Zurowska A, Hinglais N et al. A specific glomerular lesion of the graft: allograft glomerulopathy. Kidney Int Suppl 1993; 42: S104-11.

202. Maryniak RK, First MR, Weiss MA. Transplant glomerulopathy: evolution of morphologically distinct changes. Kidney Int 1985; 27(5):799-806.

203. Cameron JS, Turner DR, Heaton J et al. Idiopathic mesangiocapillary glomerulonephritis. Comparison of types I and II in children and adults and long-term prognosis. Am J Med 1983; 74(2):175-92.

204. Hamburger J, Crosnier J, Noel LH. Recurrent glomerulonephritis after renal transplantation. Annu Rev Med 1978; 29:67-72.

205. Cameron JS, Turner DR. Recurrent glomerulonephritis in allografted kidneys. Clin Nephrol 1977; 7(2):47-54.

206. Michielsen P. Recurrence of the original disease. Does this influence renal graft failure? Kidney Int Suppl 1995; 52:S79-84.

207. Lien YH, Scott K. Long-term cyclophosphamide treatment for recurrent type I membranoproliferative glomerulonephritis after transplantation. Am J Kidney Dis 2000; 35(3):539-543.

208. Brunt EM, Kissane JM, Cole BR et al. Transmission and resolution of type I membranoproliferative glomerulonephritis in recipients of cadaveric renal allografts. Transplantation 1988; 46(4):595-98.

209. Curschellas E, Landmann J, Durig M et al. Morphologic findings in "zero-hour" biopsies of renal transplants. Clin Nephrol 1991; 36(5):215-22.

210. Valenzuela R, Hamway SA, Deodhar SD et al. Histologic, ultrastructural and immunomicroscopic findings in 96 one hour human renal allograft biopsy specimens. Immunologic and clinical significance. Hum Pathol 1980; 11(2):187-95.

211. Glicklich D, Matas AJ, Sablay LB et al. Recurrent membranoproliferative glomerulonephritis type 1 in successive renal transplants. Am J Nephrol 1987; 7(2):143-49.

212. Zimmerman SW, Hyman LR, Uehling DT et al. Recurrent membranoproliferative glomerulonephritis with glomerular properdin deposition in allografts. Ann Intern Med 1974; 80(2):169-75.

213. Saxena R, Frankel WL, Sedmak DD et al. Recurrent type I membranoproliferative glomerulonephritis in a renal allograft: successful treatment with plasmapheresis. Am J Kidney Dis 2000; 35(4):749-52.

214. Little MA, Dupont P, Campbell E et al. Response to /`Differences between type I and II membranoproliferative glomerulonephritis/`. Kidney Int 2006; 70(8):1527-27.

215. Welch TR, Beischel L, Balakrishnan K et al. Major-histocompatibility-complex extended haplotypes in membranoproliferative glomerulonephritis. N Engl J Med 1986; 314(23):1476-81.

216. Berthoux FC, Ducret F, Colon S et al. Renal transplantation in mesangioproliferative glomerulonephritis (MPGN): relationship between the high frquency of recurrent glomerulonephritis and hypocomplementemia. Kidney Int Suppl 1975; (3):323-27.

217. Cahen R, Trolliet P, Dijoud F et al. Severe recurrence of type I membranoproliferative glomerulonephritis after transplantation: remission on steroids and cyclophosphamide. Transplant Proc 1995; 27(2):1746-47.

218. Ahsan N, Manning EC, Dabbs DJ et al. Recurrent type I membranoproliferative glomerulonephritis after renal transplantation and protective role of cyclosporine in acute crescentic transformation. Clin Transplant 1997; 11(1):9-14.

219. Tomlanovich S, Vincenti F, Amend W et al. Is cyclosporine effective in preventing recurrence of immune-mediated glomerular disease after renal transplantation? Transplant Proc 1988; 20(3 Suppl 4):285-88.

220. Chadban S. Glomerulonephritis recurrence in the renal graft. J Am Soc Nephrol 2001; 12(2):394-402.

221. Floege J. Recurrent glomerulonephritis following renal transplantation: an update. Nephrol Dial Transplant 2003; 18(7):1260-65.

222. Masutani K, Katafuchi R, Ikeda H et al. Recurrent nephrotic syndrome after living-related renal transplantation resistant to plasma exchange: report of two cases. Clin Transplant 2005; 19 Suppl 14:59-64.

223. Wu J, Jaar BG, Briggs WA et al. High-dose mycophenolate mofetil in the treatment of posttransplant glomerular disease in the allograft: a case series. Nephron Clin Pract 2004; 98(3):c61-66.

224. Cantarell MC, Charco R, Capdevila L et al. Outcome of hepatitis C virus-associated membranoproliferative glomerulonephritis after liver transplantation. Transplantation 1999; 68(8):1131-34.

225. Hiesse C, Samuel D, Bensadoun H et al. Combined liver and kidney transplantation in patients with chronic nephritis associated with end-stage liver disease. Nephrol Dial Transplant 1995; 10 Suppl 6:129-33.

226. Andresdottir MB, Assmann KJ, Hoitsma AJ et al. Renal transplantation in patients with dense deposit disease: morphological characteristics of recurrent disease and clinical outcome. Nephrol Dial Transplant 1999; 14(7):1723-31.

227. Braun MC, Stablein DM, Hamiwka LA et al. Recurrence of membranoproliferative glomerulonephritis type II in renal allografts: The North American Pediatric Renal Transplant Cooperative Study experience. J Am Soc Nephrol 2005; 16(7):2225-33.

228. Fremeaux-Bacchi V, Weiss L, Brun P et al. Selective disappearance of C3NeF IgG autoantibody in the plasma of a patient with membranoproliferative glomerulonephritis following renal transplantation. Nephrol Dial Transplant 1994; 9(7):811-14.

229. Leibowitch J, Halbwachs L, Wattel S et al. Recurrence of dense deposits in transplanted kidney: II. Serum complement and nephritic factor profiles. Kidney Int 197; 5(4):396-403.

230. Habib R, Antignac C, Hinglais N et al. Glomerular lesions in the transplanted kidney in children. Am J Kidney Dis 198; 0(3):198-207.

231. Turner DR, Cameron JS, Bewick M et al. Transplantation in mesangiocapillary glomerulonephritis with intramembranous dense "deposits": recurrence of disease. Kidney Int 1976; 9(5):439-48.

232. Aita K, Ito S, Tanabe K, Toma H et al. Early recurrence of dense deposit disease with marked endocapillary proliferation after renal transplantation. Pathol Int 2006; 56(2):101-9.

233. Joh K, Aizawa S, Matsuyama N et al. Morphologic variations of dense deposit disease: light and electron microscopic, immunohistochemical and clinical findings in 10 patients. Acta Pathol Jpn 1993; 43(10):552-65.

234. Silva F. Membranoprolifertive glomeulonephritis. Philadelphia: Lippincott-Raven, 1998.

235. Droz D, Nabarra B, Noel LH et al. Recurrence of dense deposits in transplanted kidneys: I. Sequential survey of the lesions. Kidney Int 1979; 15(4):386-95.

236. Bennett WM, Fassett RG, Walker RG et al. Mesangiocapillary glomerulonephritis type II (dense-deposit disease): clinical features of progressive disease. Am J Kidney Dis 1989; 13(6):469-76.

237. Muller T, Sikora P, Offner G et al. Recurrence of renal disease after kidney transplantation in children: 24 years of experience in a single center. Clin Nephrol 1998; 49(2):82-90.

238. Briner J. Glomerular lesions in renal allografts. Ergeb Inn Med Kinderheilkd 1982; 49:1-76.

239. Eddy A, Sibley R, Mauer SM et al. Renal allograft failure due to recurrent dense intramembranous deposit disease. Clin Nephrol 1984; 21(6):305-13.

240. Kurtz KA, Schlueter AJ. Management of membranoproliferative glomerulonephritis type II with plasmapheresis. J Clin Apher 2002; 7(3):135-37.

241. Oberkircher OR, Enama M, West JC et al. Regression of recurrent membranoproliferative glomerulonephritis type II in a transplanted kidney after plasmapheresis therapy. Transplant Proc 1988; 20(1 Suppl 1):418-23.

242. Strife CF, Jackson EC, McAdams AJ. Type III membranoproliferative glomerulonephritis: long-term clinical and morphologic evaluation. Clin Nephrol 1984; 21(6):323-34.

243. Morales JM, Martinez MA, Munoz de Bustillo E et al. Recurrent type III membranoproliferative glomerulonephritis after kidney transplantation. Transplantation 1997; 63(8):1186-88.

244. Ramesh Prasad GV, Shamy F, Zaltzman JS. Recurrence of type III membranoproliferative glomerulonephritis after renal transplantation. Clin Nephrol 2004; 61(1):80-81.

245. Jennette JC, Falk RJ, Andrassy K, et al. Nomenclature of systemic vasculitides. Proposal of an international consensus conference. Arthritis Rheum 1994; 37(2):187-92.

246. Weidemann S, Andrassy K, Ritz E. ANCA in haemodialysis patients. Nephrol Dial Transplant 1993; 8(9):839-45.

247. Lobbedez T, Comoz F, Renaudineau E et al. Recurrence of ANCA-positive glomerulonephritis immediately after renal transplantation. Am J Kidney Dis 2003; 42(4):E2-6.

248. Lowance DC, Vosatka K, Whelchel J, et al. Recurrent Wegener's granulomatosis. Am J Med 1992; 92(5):573-75.

249. Nachman PH, Segelmark M, Westman K, et al. Recurrent ANCA-associated small vessel vasculitis after transplantation: A pooled analysis. Kidney Int 1999; 56(4):1544-50.

250. Reaich D, Cooper N, Main J. Rapid catastrophic onset of Wegener's granulomatosis in a renal transplant. Nephron 1994; 67(3):354-57.

251. Deegens JK, Artz MA, Hoitsma AJ et al. Outcome of renal transplantation in patients with pauci-immune small vessel vasculitis or anti-GBM disease. Clin Nephrol 2003; 59(1):1-9.

252. Nyberg G, Akesson P, Norden G et al. Systemic vasculitis in a kidney transplant population. Transplantation 1997; 63(9):1273-77.

253. Nachman PH, Hogan SL, Jennette JC et al. Treatment response and relapse in antineutrophil cytoplasmic autoantibody-associated microscopic polyangiitis and glomerulonephritis. J Am Soc Nephrol 1996; 7(1):33-39.

254. Grotz W, Wanner C, Rother E et al. Clinical course of patients with antineutrophil cytoplasm antibody positive vasculitis after kidney transplantation. Nephron 1995; 69(3):234-36.

255. Allen A, Pusey C, Gaskin G. Outcome of renal replacement therapy in antineutrophil cytoplasmic antibody-associated systemic vasculitis. J Am Soc Nephrol 1998; 9(7):1258-63.

256. Schmitt WH, Haubitz M, Mistry N et al. Renal transplantation in Wegener's granulomatosis. Lancet 1993; 342(8875):860.

257. Clarke AE, Bitton A, Eappen R et al. Treatment of Wegener's granulomatosis after renal transplantation: is cyclosporine the preferred treatment? Transplantation 1990; 50(6):1047-51.

258. Curtis JJ, Diethelm AG, Herrera GA et al. Recurrence of Wegener's granulomatosis in a cadaver renal allograft. Transplantation 1983; 36(4):452-54.

259. Oberhuber G, Prior C, Bosmuller C et al. Early recurrence of Wegener's granulomatosis in a kidney allograft under cyclosporine treatment. Transpl Int 1988; 1(1):49-50.

260. Turney JH, Adu D, Michael J et al. Recurrent crescentic glomerulonephritis in renal transplant recipient treated with cyclosporin. Lancet 1986; 1(8489):1104.

261. Adams PL, Iskandar SS, Rohr MS. Biopsy-proven resolution of immune complex-mediated crescentic glomerulonephritis with mycophenolate mofetil therapy in an allograft. Am J Kidney Dis 1999; 33(3):552-54.

262. Steen VD. Scleroderma renal crisis. Rheum Dis Clin North Am 2003; 29(2):315-33.

263. Chang YJ, Spiera H. Renal transplantation in scleroderma. Medicine (Baltimore) 1999; 78(6):382-85.

264. Paul M, Bear RA, Sugar L. Renal transplantation in scleroderma. J Rheumatol 1984; 11(3):406-8.
265. Tsakiris D, Simpson HK, Jones EH, et al. Report on management of renale failure in Europe, XXVI, 1995. Rare diseases in renal replacement therapy in the ERA-EDTA Registry. Nephrol Dial Transplant 1996; 11 Suppl 7:4-20.
266. Haubitz M, Kliem V, Koch KM et al. Renal transplantation for patients with autoimmune diseases: single-center experience with 42 patients. Transplantation 1997; 63(9):1251-57.
267. Lockwood CM, Peters DK. Plasma exchange in glomerulonephritis and related vasculitides. Annu Rev Med 1980; 31:167-79.
268. Netzer KO, Merkel F, Weber M. Goodpasture syndrome and end-stage renal failure—to transplant or not to transplant? Nephrol Dial Transplant 1998; 13(6):1346-48.
269. Couser WG, Wallace A, Monaco AP et al. Successful renal transplantation in patients with circulating antibody to glomerular basement membrane: report of two cases. Clin Nephrol 1973; 1(6):381-88.
270. Beleil OM, Coburn JW, Shinaberger JH et al. Recurrent glomerulonephritis due to anti-glomerular basement membrane-antibodies in two successive allografts. Clin Nephrol 1973; 1(6):377-80.
271. Dixon FJ, McPhaul JJ, Lerner RA. The contribution of kidney transplantation o the study of glomerulonephritis—the recurrence of glomerulonephritis in renal transplants. Transplant Proc 1969; 1(1):194-96.
272. Khandelwal M, McCormick BB, Lajoie G et al. Recurrence of anti-GBM disease 8 years after renal transplantation. Nephrol Dial Transplant 2004; 19(2):491-94.
273. Persson U, Hertz JM, Wieslander J et al. Alport syndrome in southern Sweden. Clin Nephrol 2005; 64(2):85-90.
274. Hudson BG. The molecular basis of Goodpasture and Alport syndromes: beacons for the discovery of the collagen IV family. J Am Soc Nephrol 2004; 15(10):2514-27.
275. Turner AN, Rees AJ. Goodpasture's disease and Alport's syndromes. Annu Rev Med 1996; 47:377-86.
276. Browne G, Brown PA, Tomson CR et al. Retransplantation in Alport posttransplant anti-GBM disease. Kidney Int 2004; 65(2):675-81.
277. Ding J, Zhou J, Tryggvason K et al. COL4A5 deletions in three patients with Alport syndrome and posttransplant antiglomerular basement membrane nephritis. J Am Soc Nephrol 1994; 5(2):161-68.
278. Shah B, First MR, Mendoza NC et al. Alport's syndrome: risk of glomerulonephritis induced by anti-glomerular-basement-membrane antibody after renal transplantation. Nephron 1988; 50(1):34-38.
279. Byrne MC, Budisavljevic MN, Fan Z et al. Renal transplant in patients with Alport's syndrome. Am J Kidney Dis 2002; 39(4):769-75.
280. Brainwood D, Kashtan C, Gubler MC et al. Targets of alloantibodies in Alport anti-glomerular basement membrane disease after renal transplantation. Kidney Int 1998; 53(3):762-66.
281. Peten E, Pirson Y, Cosyns JP et al. Outcome of thirty patients with Alport's syndrome after renal transplantation. Transplantation 1991; 52(5):823-26.
282. Gobel J, Olbricht CJ, Offner G et al. Kidney transplantation in Alport's syndrome: long-term outcome and allograft anti-GBM nephritis. Clin Nephrol 1992; 38(6):299-304.
283. Kalluri R, van den Heuvel LP, Smeets HJ et al. A COL4A3 gene mutation and posttransplant anti-alpha 3(IV) collagen alloantibodies in Alport syndrome. Kidney Int 1995; 47(4):1199-204.
284. Hariharan S, Adams MB, Brennan DC et al. Recurrent and de novo glomerular disease after renal transplantation: a report from renal allograft disease registry. Transplant Proc 1999; 31(1-2):223-24.

# CHAPTER 27

# Pathology of Kidney Allograft Dysfunction

Bela Ivanyi*

## Abstract

The pathologic features, clinical correlations and differential diagnoses of the major causes of kidney allograft dysfunction are reviewed. Rejection is an inflammatory process of the recipient during which donor cells bearing alloantigens are destroyed; the rejection process is classified traditionally as hyperacute, acute or chronic. Hyperacute rejection is caused by high titers of preformed antibodies to donor endothelial cell antigens that lead to thrombosis in the renal vasculature and ischemic death of the graft. Acute cellular rejection is primarily mediated by contact-dependent cellular cytotoxicity. The indicator lesions are interstitial infiltrates composed predominantly of T-lymphocytes and tubulitis, with or without intimal arteritis. Acute humoral rejection is caused by donor-specific HLA class I antibodies. The morphological features include complement 4d positivity along the peritubular capillaries, neutrophils in the peritubular and glomerular capillaries, microthrombi in the microvessels and transmural arteritis, with or without fibrinoid necrosis. Chronic rejection is mediated by humoral and/or cellular alloimmune mechanisms and is characterized morphologically by obliterative intimal fibrosis in the arteries, double-contoured glomerular capillary loops and circumferentially multiplied and split basement membrane in the peritubular capillaries. Chronic allograft nephropathy is the result of a series of alloimmune and non-immune insults that cumulate and subsequently manifest in arterial intimal fibroelastosis, focal-global glomerular sclerosis, interstitial fibrosis and tubular atrophy. Calcineurin inhibitor toxicity may cause acute toxic tubulopathy and/or acute vascular toxicity or chronic hyaline arteriolopathy. Prolonged warm or cold ischemia prior to transplantation leads to posttransplant acute tubular necrosis. The indicator lesions are diffuse tubular damage (loss of brush border, focal epithelial necrosis, desquamation of cells into the tubular lumen, etc.,) and interstitial edema.

## Introduction

The gold standard for the assessment of structural abnormalities in the transplanted kidney is the morphological examination of a core-needle biopsy specimen. The biopsy is performed at the time of a graft dysfunction, when the etiology cannot be explored by clinical or non-invasive means, or at predetermined intervals, in order to recognize subclinical rejection, to identify drug toxicity or to determine the efficacy of new immunosuppressive drugs. The standard evaluation procedure (at least two cores whenever possible) involves light microscopic staining on serial sections (H&E, PAS, trichrome and methenamine silver) and immunostaining for complement (C) 4d.[1-3] The biopsy is regarded as adequate if 10 or more glomeruli with at least 2 arteries can be investigated by light microscopy (LM). For optimum interpretative power of the histologic changes, comparison with a time-zero biopsy is strongly advised.[1] Some pathologists also perform elastin staining, apply the full immunofluorescence (IF) panel (IgG, IgA, IgM, C3, C1q and fibrinogen) and investigate the upregulation of tubular HLA-DR antigen,[4,5] because such stainings sometimes furnish additional diagnostic information. Tissue sampling for electron microscopy (EM) is suggested if there is a clinical suspicion of de novo or recurrent glomerular disease. The ultrastructural examination of glomerular and peritubular capillaries facilitates a more definitive identification of glomerular and peritubular capillary lesions of chronic rejection,[6] but it is not routinely performed because of cost-benefit concerns.

This survey discusses the pathologic features (summarized in Tables 1 and 2), clinical correlations and differential diagnoses of the major causes of allograft dysfunction (for more details, see refs. 5 and 7). Chapter 38 in this book is dedicated to the topic of BK polyomavirus nephropathy.

## Transplant Rejection

Rejection is an inflammatory response of the recipient during which donor cells bearing antigens recognized as foreign (allogenic) are destroyed. The targeted antigens are the HLA class I and II antigens, the non HLA minor alloantigens and the ABO blood group antigens. The class I antigens are expressed by all renal parenchymal cells, but most intensely by the vascular endothelial cells. The class II antigens are displayed by capillary endothelial cells,[8] mesangial cells, proximal tubular epithelial cells and interstitial dendritic cells. Interferon-γ, produced by activated T-lymphocytes, enhances HLA antigen expression and particularly the expression of class II antigens in the tubules. Two types of alloantigen presentation exist. In the direct pathway, the CD8+ and CD4+ T-lymphocytes of the recipient recognize the allogenic molecules on the surface of the antigen-presenting cells in the graft. The CD8+ cells then differentiate into cytotoxic T-lymphocytes and cause the contact-dependent killing of the target cells. The CD4+ T-cells differentiate into $T_h1$ effector cells, which produce cytokines that induce a local delayed hypersensitivity reaction and stimulate B-lymphocytes and CD8+ T-lymphocytes. In the indirect pathway, the T-lymphocytes of the recipient recognize alloantigens after they are presented by the recipient's own antigen-presenting cells. $T_h1$ effector cells are generated and these mediate tissue injury via a local delayed hypersensitivity reaction and B-cell responses.[9,10]

Traditionally, three forms of rejection are distinguished: hyperacute, acute and chronic. Acute rejection is subclassified into cell-mediated and antibody-mediated types. Acute cellular rejection is primarily initiated by direct allorecognition and chronic rejection by indirect allorecognition. The pathology of rejection has an intrinsic focal nature.

*Bela Ivanyi—Department of Pathology, Allomas u. 2, H-6720 Szeged, Hungary. Email: ivanyi@patho.szote.u-szeged.hu

*Chronic Allograft Failure: Natural History, Pathogenesis, Diagnosis and Management*, edited by Nasimul Ahsan. ©2008 Landes Bioscience.

***Table 1. Lesions of acute rejection, acute calcineurin inhibitor (CNI) nephrotoxicity and posttransplant acute tubular necrosis (ATN)***

| Acute Rejection | Acute CNI Toxicity* | ATN |
|---|---|---|
| **Cellular** | **Acute afferent arteriolopathy** | Dilation and flattening of tubules |
| Interstitial infiltrates of activated lymphocytes | Endothelial swelling/vacuolization | Loss of brush border |
| Lymphocytic tubulitis; HLA-DR expression of tubules | Necrosis and early hyaline replacement of individual myocytes | Individual necrosis and desquamation of tubular epithelial cells |
| Lymphocytic intimal arteritis | | Enlarged regenerative nuclei |
| Transplant glomerulitis | | Interstitial edema |
| **Humoral** | **Tubulopathy** | |
| Fibrinoid necrosis of small arteries and/or afferent arterioles | Small, evenly distributed vacuoles mainly in the proximal straight tubules | |
| Glomerular microthrombi | | |
| Neutrophils in peritubular capillaries | *Rare manifestation: thrombotic microangiopathy | |
| Ischemic tubular damage | | |
| Complement 4d localized along peritubular capillaries | | |

## Hyperacute Rejection

This is produced by high titers of preformed antibodies targeting antigens on donor endothelial cells, such as blood group antigens or HLA class I antigens. The antibodies initiate complement fixation, platelet activation, the lysis of endothelial cells and activation of the clotting system, with thrombosis, circulation blockage and ischemic death of the grafted kidney.[11]

### Morphology

LM reveals swelling and lysis of the endothelial cells, denudation of the basement membrane, margination of the neutrophils and thrombi in the arteries, arterioles and glomeruli. The neutrophils are typically incorporated into the thrombi. The renal compartments undergo ischemic injury. Within 1 day, cortical and medullary infarcts develop. IF discloses strong staining for fibrin in the microvasculature and often in the interstitium. IgM or IgG, C3 and C4d may be observed in the walls of the arteries and arterioles and along the glomerular and peritubular capillaries.

### Clinical Correlation

Hyperacute rejection is extremely rare in consequence of the regular screening for donor-specific antibodies prior to transplantation. However, certain cases fail to be identified for some reason[12] and within minutes after the vascular anastomes have been established, the implanted kidney becomes cyanotic, edematous and flaccid and anuria develops. Hyperacute rejection is irreversible and the graft must be removed.

### Differential Diagnosis

Other conditions that can cause a primary graft failure are acute tubular necrosis (ATN), perfusion injury and major arterial occlusion. A common feature of all these is that IF reveals no immunoglobulin deposits along the endothelial surfaces.

***Table 2. Lesions of chronic rejection, chronic allograft nephropathy (CAN), chronic calcineurin inhibitor (CNI) nephrotoxicity and BK polyomavirus nephropathy***

| Chronic Rejection | CAN | Chronic CNI Toxicity | BK Polyoma Virus Nephropathy* |
|---|---|---|---|
| **Transplant arteriopathy** | **Intimal fibroelastosis in the arteries** | **Hyaline arteriolopathy** | |
| New-onset intimal fibrosis | | Myocytes are replaced | |
| Foam cells/mononuclears in the intima | | by beaded hyaline deposits that bulge into the adventitia | |
| Absent elastosis | | | |
| **Transplant glomerulopathy** | **No change or subendothelial hyalinosis in the arterioles** | **Nonspecific segmental or global glomerular sclerosis** | **Cytopathic changes** |
| Double-contoured glomerular capillaries | | | Enlarged tubular cells, karyomegaly, nuclear inclusions |
| **Transplant capillaropathy** | **Nonspecific segmental or global glomerular sclerosis** | **Striped interstitial fibrosis and tubular atrophy** | **Cytolytic changes** |
| At least 3 peritubular capillaries with 5 or more circumferential basement membrane layers on electron microscopy | | | Lysis of tubular epithelial cells, denudation of basement membrane |
| Complement 4d positivity | | | |
| **Interstitial fibrosis and tubular atrophy** | **Interstitial fibrosis and tubular atrophy** | | **Interstitial inflammatory infiltrates** |

* Adjunct immunohistochemical or electron microsopic studies confirm the diagnosis.

## Acute Cellular Rejection

This is mediated primarily by CD8[+] cytotoxic T-lymphocytes and, to a lesser extent, by local delayed hypersensitivity reaction induced by CD4[+] helper T-lymphocytes. Other effector mechanisms, such as antibody-dependent cellular cytotoxicity or local humoral responses, may also play a role in the evolution of alloimmune damage to the kidney.

### Morphology

The rejection process affects the interstitium, tubules, glomeruli and arteries, in varying combinations and with varying degrees of involvement. Three patterns are distinguished: tubulointerstitial, vascular and glomerular.

### Conventional Type of Tubulointerstitial Rejection

The majority of cases belong in this category and are characterized by infiltration of the interstitium by mononuclear leukocytes, accompanied by tubulitis (Fig. 1), peritubular capillaritis and interstitial edema.[1] The infiltrates consist mainly of CD8[+] and CD4[+] T-lymphocytes and macrophages. Scattered B-lymphocytes, plasma cells, eosinophils, neutrophils and natural killer cells are minor components of the infiltrate. Tubulitis (infiltration of the tubular epithelium by lymphocytes) in non-atrophic tubules is a defining lesion of acute tubulointerstitial rejection[1,13] and is usually accompanied by acute tubular injury: epithelial cell lysis or apoptosis induced by tubular wall-localized cytotoxic T-lymphocytes, epithelial thinning and detachment, enlarged regenerative nuclei and mitotic figures. The features of peritubular capillaritis include endothelial cell hypertrophy, accumulation and margination of lymphocytes and monocytes in the capillary lumina,[13,14] migration of endothelium-adherent inflammatory cells to the interstitium and, on occasion, lysis or apoptosis of endothelial cells in the close vicinity of cytotoxic T-lymphocytes. IF reveals the upregulation of class II antigens on tubular epithelial cells.

### Other Types of Tubulointerstitial Rejection

These occur infrequently and are characterized by interstitial infiltrates rich in B-lymphocytes[15] or plasma cells,[16,17] mild tubulitis and a worse outcome.

Vascular Rejection is diagnosed if infiltration of the intima of the arteries by T-lymphocytes and macrophages (Fig. 2) is noted.[18] The lesion is termed intimal arteritis (synonyms: endothelialitis or endarteritis) and may occur with or without intimal edema and fibrin deposition. Vascular rejection involves the large arteries more frequently than the interlobular arteries and arterioles.

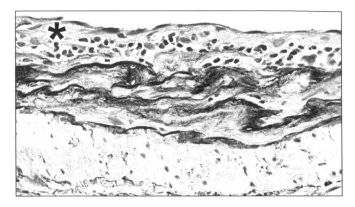

Figure 2. Acute cellular vascular rejection. Lymphocytes localize in the intima (asterisk) of the interlobular artery. Elastin staining.

In about 10% of the cases of acute cellular rejection, there is a marked infiltration of the glomerular tufts by T-lymphocytes and monocytes; this situation is denoted transplant glomerulitis.

### Clinical Correlation

Acute cellular rejection is the most common cause of a graft dysfunction in the first 3 months after transplantation. However, it may develop even years later. The cardinal sign is a sudden asymptomatic increase (>20%) in the serum creatinine level. With current immunosuppression regimens, symptoms such as pain, fever or oliguria are rare. Conventional tubulointerstitial rejection responds well to high-dose intravenous corticosteroid therapy. In contrast, the B-cell-rich or plasma cell-rich subtypes are usually steroid-resistant.[15-17] Vascular rejection can be reversed with anti-lymphocyte antibody preparations. Refractory cases have a poor outcome. A rare complication of acute cellular rejection is spontaneous rupture of the kidney allograft.[19]

### Differential Diagnosis

Acute tubulointerstitial rejection must be differentiated from BK polyoma virus nephropathy (rejection: lack of nuclear inclusion bodies in the tubular epithelial cells), acute bacterial tubulointerstitial nephritis (rejection: lymphocytic, but not neutrophilic[20] tubulitis), drug-induced acute interstitial nephritis (rejection: intense tubular HLA-DR staining) and posttransplant lymphoproliferative disease (rejection: predominance of T-lymphocytes, not B-lymphocytes). Intimal arteritis is pathognomic of acute cellular rejection. Transplant glomerulitis must be distinguished from recurrent or de novo proliferative glomerulonephritis. In the latter conditions, IF demonstrates glomerular immune deposits.

## Acute Antibody-Mediated Rejection

This form is mainly caused by donor-specific HLA class I antibodies which induce complement-mediated cytotoxic injury to the endothelial cells. Risk factors include a historically positive cross match, an increased lymphocytotoxic panel-reactive antibody titer, sensitization through previous renal transplantation, pregnancy or blood transfusions. The complement split product C4d binds covalently near the original antibody targets and resists rapid elimination. The immunohistochemical identification of C4d along the peritubular capillaries has been demonstrated to be strongly correlated with circulating donor-specific antibodies and acute humoral rejection and has therefore been used as a marker of humoral rejection.[21-23]

### Morphology

LM reveals the accumulation and margination of the neutrophils and later monocytes in the peritubular and glomerular capillaries, endothelial swelling/necrosis and microthrombi in the glomerular and peritubular capillaries and small arteries. Larger arteries may display

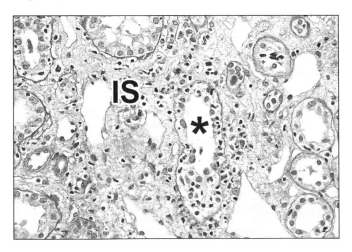

Figure 1. Acute cellular tubulointerstitial rejection. Lymphocytes infiltrate the interstitium (IS) and tubules. The latter phenomenon is termed tubulitis (asterisk). PAS staining.

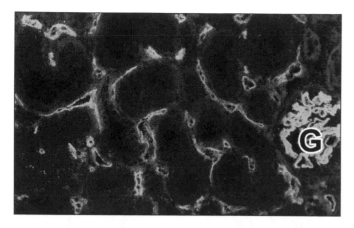

Figure 3. Acute humoral rejection by immunofluorescence. Bright linear staining of peritubular capillaries with complement 4d component. G—glomerulus.

transmural infiltrates of lymphocytes and neutrophils, with or without fibrinoid necrosis of the medial smooth muscle cells. There is little, if any, mononuclear inflammatory infiltration of the interstitium or tubulitis.[24] The tubules exhibit degenerative changes or ATN. There may be focal interstitial microhemorrhages and patchy cortical infarctions. IF reveals (Fig. 3) bright linear staining of the cortical and medullary peritubular capillaries with C4d.[25] In some cases, IgG or IgM and C3 can be observed together with fibrin along the glomerular and peritubular capillaries and in arterial fibrinoid necrosis. On EM, the peritubular capillaries display severe swelling of the endothelial cells, with rupture of the cell membrane (cytolysis), an increased rate of apoptosis, denudation of the basement membrane and intra- and extracapillary fibrin deposition.[26] For a definitive diagnosis of acute antibody-mediated rejection, the serologic evidence of circulating antibodies to donor HLA or other anti-donor endothelial antigens is required.[3]

### Clinical Correlation

Acute antibody-mediated rejection is a relatively infrequent condition (2-8%). It can arise at any time following transplantation, but most commonly in the first few weeks. After an initial period of good graft function, there is a sudden rise in the serum creatinine level, accompanied by a reduced urine output. Oligoanuria develops within days. The prognosis is poor. If an early diagnosis is achieved, therapeutic efforts may reverse the rejection process.

### Differential Diagnosis

Conditions that should be considered are thrombotic microangiopathy (TMA) secondary to calcineurin inhibitor (CNI) toxicity, anticardiolipin syndrome, viral infection due to cytomegalovirus or parvovirus B19 and recurrent hemolytic-uremic syndrome. C4d-positive peritubular capillaritis supports the diagnosis of acute antibody-mediated rejection.

## Chronic Rejection and Chronic Allograft Nephropathy

Chronic rejection involves ongoing, smoldering damage to the allograft, mediated predominantly by humoral alloimmune mechanisms. In addition to the chronic alloimmune damage to the renal parenchyma, the nephrons are depleted by a variety of alloantigen-independent factors, such as advanced donor age, ischemic injury to the graft during the implantation period, hypertension, CNI nephrotoxicity, infection, an increased ureteral pressure, etc., The alloimmune and non-immune mechanisms cumulate and, in association with additional mechanisms of injury, such as internal architectural disruption of the kidney, chronic cortical ischemia, persistent chronic inflammation, replicative senescence, epithelial-to-mesenchymal transition, etc., lead

to progressive nonspecific interstitial fibrosis and tubular atrophy.[27,28] Since chronic rejection, CNI nephrotoxicity, hypertensive vascular disease and chronic infection/reflux cannot always be distinguished in the biopsy specimen, the Banff classification of kidney transplant pathology[1] has coined the term chronic allograft nephropathy (CAN) to emphasize the multifactorial nature of the chronic kidney allograft injury. At the moment, there is no therapeutic consequence via which to distinguish chronic rejection from CAN. The usual diagnostic label is either CAN, with or without features of chronic rejection, or chronic rejection/CAN.

### Morphology

The interstitium displays a varying degree of fibrosis and the tubules in the fibrotic areas are atrophic (Fig. 4). The marker lesions of chronic rejection are transplant arteriopathy, transplant glomerulopathy and transplant capillaropathy, present either alone or in combination (Table 2). Transplant arteriopathy is diagnosed if the narrowing of medium-sized and large arteries by new-onset intimal fibrosis is observed (Fig. 5). The lesion may be associated with foam cells/mononuclears in the intima, breaks in the internal elastic lamina and the formation of neo-media. Elastosis is absent. Transplant glomerulopathy is characterized by double-contoured capillary loops

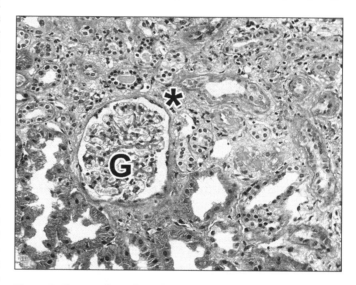

Figure 4. Chronic allograft nephropathy. At the asterisk, the interstitium displays fibrosis (blue) and the tubules lack segment-specific features indicating atrophy. G—glomerulus. Trichrome staining.

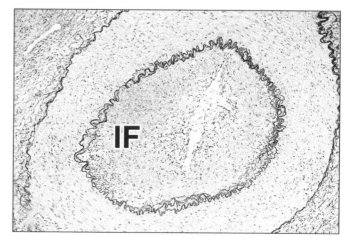

Figure 5. Chronic rejection, transplant arteriopathy. The arcuate artery is severely narrowed by intimal fibrosis (IF). Elastosis is absent. Elastin staining.

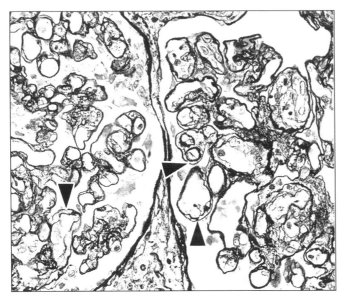

Figure 6. Chronic rejection, transplant glomerulopathy. Double-contoured capillary loops (arrowheads). Silver impregnation.

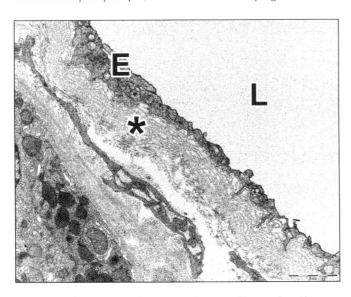

Figure 7. Chronic rejection, transplant capillaropathy. Electron microscopy reveals the multiplied and split basement membrane of the peritubular capillary profile (asterisk). L—lumen, E—endothelial layer.

(Fig. 6), with or without increase in the mesangial matrix, mesangial hypercellularity and mesangial interposition. On EM, the glomerular basement membrane is reduplicated or multiplied and this accounts for the phenomenon of doublecontours. Transplant capillaropathy is identified ultrastructurally (Fig. 7): the capillaries display a serrated contour and the basement membranes are circumferentially multiplied and split (5 or more layers in at least 3 profiles).[29] Chronic rejection may be accompanied by features of active cellular and/or humoral rejection. CAN is diagnosed if the lesions of chronic rejection cannot be verified (Table 2). Chronic rejection/CAN may coexist with CNI nephrotoxicity or recurrent glomerulonephritis.

### Clinical Correlation

Chronic rejection/CAN is the major cause of graft loss after 1 year. It is manifested by an insidious, progressive decline in the glomerular filtration rate, frequently accompanied by proteinuria (often in the nephrotic range) and hypertension.

### Differential Diagnosis

The main conditions to be differentiated are chronic CNI nephrotoxicity, de novo or recurrent glomerulonephritis and BK polyomavirus nephropathy. Hyaline arteriolopathy indicates CNI nephrotoxicity. Glomerulonephritis can be readily detected if the full immunohistochemical panel is used and embedding for EM is carried out. Nuclear inclusion bodies in the tubular cells, tubular epithelial injury, a varying degree of interstitial inflammation and tubular atrophy suggest polyomavirus nephropathy. Ancillary techniques, such as immunohistochemistry on paraffin sections with antibody to the SV40 large T-antigen or electron microscopy of infected tubular epithelial cells (virions 40 nm in diameter, often arranged in paracrystalloid arrays), confirm the diagnosis of polyomavirus nephropathy.

## The Banff Classification of Kidney Transplant Pathology

A scheme for the nomenclature and classification of renal allograft pathology was created by a group of experts in Banff, Canada in 1991 to standardize renal allograft biopsy interpretation and reporting and has subsequently been regularly updated. The diagnostic categories include normal, antibody-mediated rejection, "suspicious" for acute cellular rejection, acute/active cellular rejection, chronic/sclerosing allograft nephropathy and other changes not considered to be due to rejection. Vascular, glomerular, interstitial and tubular lesions of acute rejection and CAN are defined and scored 0 to 3 +.[1] The scoring codes are designated $v$, $g$, $i$ or $t$ for acute vascular, glomerular, interstitial or tubular changes and $cv$, $cg$, $ci$ or $ct$ for the corresponding chronic changes, respectively. The degree of arteriolar hyaline thickening ($ah$) is also defined and scored. The scores provide the basis for arbitrary grades of mild, moderate and severe.

## Calcineurin Inhibitor (CNI) Toxicity

CNIs (cyclosporine and tacrolimus) can cause acute or chronic nephrotoxicity; the lesions are identical. The pathogenetic pathways are complex and beyond the scope of the present paper.

## Acute Nephrotoxicity

This may appear as functional toxicity, a delayed recovery from post-transplant ATN, toxic tubulopathy or vascular toxicity.[5,7,30]

## Toxic Tubulopathy

### Morphology

Isometric tubular vacuolization is observed focally, predominantly affecting the proximal straight tubules (Table 1). The vacuoles are small, clear and evenly distributed throughout the cytoplasm. Ultrastructurally, they correspond to dilations of the endoplasmic reticulum. Focal tubular calcification may also be seen.

### Clinical Correlation

The patients present with an acute allograft dysfunction, the condition usually being associated with an elevated serum drug level. Following lowering of the dose of the drug, toxic tubulopathy is reversible.

### Differential Diagnosis

The features of this tubulopathy are indistinguishable from those of radiocontrast nephrotoxicity or osmotic nephrosis, conditions to be considered while making the diagnosis.

## Vascular Toxicity

CNIs may have a direct toxic effect on endothelial cells, with or without platelet aggregation. Two types of toxicity have been identified: acute arteriolopathy and TMA.

## Acute Arteriolopathy

### Morphology

The lesions are confined to the afferent arteriolar profiles: there is swelling and vacuolization of the endothelial cells, vacuolization, necrosis/apoptosis and possibly drop-out of individual myocytes and early replacement of the damaged myocytes by rounded plasma protein insudates (hyalinization). Acute arteriolopathy and toxic tubulopathy can coexist.

### Clinical Correlation

The patients present with an acute allograft dysfunction and the serum drug level is usually elevated. The clinical outcome varies. Some patients recover their renal function after the dose of CNI has been reduced, while others develop irreversible renal damage. In clinical practice, the diagnosis of arteriolopathy frequently leads to the cessation of CNI administration.

## TMA

### Morphology

The main feature is the formation of thrombi in the lumina of the arterioles and glomerular capillaries.

### Clinical Correlation

CNI-induced TMA is rare. The clinical picture resembles the hemolytic-uremic syndrome. The diagnosis prompts withdrawal of the drug because, if the lesions are widespread and associated with extensive arterial mural necrosis and luminal thrombosis, graft loss develops.

### Differential Diagnosis

CNI-induced TMA cannot be differentiated from other forms of TMA by morphology alone. The possibility of acute humoral rejection (C4d positivity), bacterial or viral infection, recurrent hemolytic-uremic syndrome and anticardiolipin antibody-induced TMA should therefore be considered. Pronounced arterial intimal changes are not typical of CNI-induced TMA.

## Chronic Nephrotoxicity

### Morphology

The indicator lesion is hyaline arteriolopathy, which affects the afferent arterioles and the distal portions of the small interlobular arteries. The necrotized/apoptotic smooth muscle cells are replaced by beaded hyaline deposits, situated on the outer aspect of the media, that bulge into the adventitia. The insudates are positive for IgM and C3 and the staining displays a pearl necklace pattern. In advanced cases, the entire wall is replaced by the hyaline material (Fig. 8) and the lumen is severely narrowed. The interstitium shows striped fibrosis and corresponding tubular atrophy. The glomerular changes include compensatory hypertrophy, mesangial matrix expansion, capillary collapse and focal-segmental and/or focal-global sclerosis.

### Clinical Correlation

Chronic CNI nephrotoxicity occurs several months after transplantation; the incidence increases with time. There is a slow, insidious rise in the serum creatinine level. Some patients have an elevated serum level of cyclosporine or tacrolimus. The kidney damage is irreversible. The appropriate management is a dose reduction or discontinuation of cyclosporine or tacrolimus, with the administration of an alternative immunosuppressive agent.

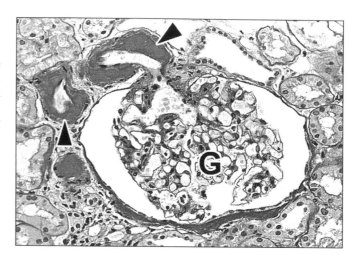

Figure 8. Chronic calcineurin inhibitor toxicity, hyaline arteriolopathy. The smooth muscle cells in the afferent arteriolar profiles (arrowheads) are replaced by homogeneous substance termed hyaline material. G—glomerulus. PAS staining.

### Differential Diagnosis

Hyaline arteriolopathy must be differentiated from the arteriolar hyalinosis seen in aging, hypertension or diabetes. In the latter three conditions, the hyalinosis is mainly subendothelial and rarely extends into the adventitia and necrosis of the myocytes is not observed. Focal-segmental glomerulosclerosis seems to be a result of hyperfiltration injury in response to the loss of functioning nephrons rather than a direct form of toxicity to the podocytes and/or endothelial cells and should be distinguished from recurrent or de novo idiopathic focal sclerosis. The presence of hyaline arteriolopathy, striped fibrosis and tubular atrophy with or without isometric vacuolization of the proximal straight tubules and the absence of the full-blown nephrotic syndrome argue against idiopathic focal sclerosis.

## Posttransplant ATN

Prolonged warm or cold ischemia causes ATN in the allograft.

### Morphology

The features include dilation and flattening of the tubular profiles, the loss of brush borders, individual necrosis of the tubular epithelial cells, cytoplasmic basophilia and enlarged regenerative nuclei. The desquamation of tubular epithelial cells into the tubular lumen may be observed. The tubular changes are accompanied by interstitial edema and sparse lymphocytic/mononuclear infiltrates, without tubulitis.[5]

### Clinical Correlation

The condition is a common complication of cadaveric renal transplantation. It manifests as oligoanuria in the immediate postoperative period, which resolves spontaneously over days and sometimes weeks. The recipients are maintained on dialysis until the urine output has been restored and the renal function has been recovered. Because of potential exaggeration of the ischemic tubular damage, CNIs are introduced only when the allograft function has been established.

### Differential Diagnosis

A renal biopsy is often necessary to differentiate from acute rejection. In ATN, tubular damage and interstitial edema are the leading features; marked interstitial inflammation, tubulitis or intimal arteritis indicates rejection.

# References

1. Solez K, Axelsen RA, Benediktsson H et al. International standardization of criteria for the histologic diagnosis of renal allograft rejection: the Banff working classification of kidney transplant pathology. Kidney Int 1993; 44:411-22.

2. Racusen LC, Solez K, Colvin RB et al. The Banff 97 working classification of renal allograft pathology. Kidney Int 1999; 55:713-23.

3. Racusen LC, Colvin RB, Solez K et al. Antibody-mediated rejection criteria—an addition to the Banff'97 classification of renal allograft rejection. Am J Transplant 2003; 3:1-7.

4. Mihatsch MJ, Nickeleit V, Gudat F. Morphologic criteria of chronic renal allograft rejection. Transplant Proc 1999; 31:1295-97.

5. D'Agati VD, Jenette JCh, Silva FG. Pathology of renal transplantation. In: D'Agati VD, Jenette JCh, Silva FG, eds. Nonneoplastic Kidney Diseases. Atlas of Nontumor Pathology. Silver Spring: ARP Press, 2005:667-708.

6. Ivanyi B. Transplant capillaropathy and transplant glomerulopathy: ultrastructural markers of chronic renal allograft rejection. Nephrol Dial Transplant 2003; 18:655-60.

7. Colvin RB. Renal transplant pathology. In: Jenette JCh, Olson JL, Schwartz MM, Silva FG, eds. Heptinstall's Pathology of the Kidney, 5th edition. Philadelphia: Lippincott-Raven, 1998:1409-540.

8. Muczynski KA, Ekle DM, Coder DM et al. Normal human kidney HLA-DR-expressing renal microvascular endothelial cells: characterization, isolation and regulation of MHC class II expression. J Am Soc Nephrol 2003; 14:1336-48.

9. Le Moine A, Goldman M, Abramowicz D. Multiple pathways to allograft rejection. Transplantation 2002; 73:1373-81.

10. Joosten SA, Sijpkens YWJ, van Kooten C et al. Chronic renal allograft rejection: pathophysiologic considerations. Kidney Int 2005; 68:1-13.

11. Shimizu A, Colvin RB. Pathological features of antibody-mediated rejection. Curr Drug Targets—Cardiovasc Haematol Disorders 2005; 5:199-214.

12. Terasaki PI, Cai J. Humoral theory of transplantation: further evidence. Curr Op Immunol 2005; 17:541-45.

13. Robertson H, Kirby JA. Posttransplant renal tubulitis: the recruitment, differentiation and persistence of intra-epithelial T-cells. Am J Transplant 2003; 3:3-10.

14. Ivanyi B, Hansen HE, Olsen TS. Postcapillary venule-like formation of peritubular capillaries in acute renal allograft rejection. Arch Pathol Lab Med 1992; 116:1062-67.

15. Sarwal M, Chua M-S, Kabham N et al. Molecular heterogeneity in acute renal allograft rejection identified by DNA microarray profiling. N Eng J Med 2003; 349:125-38.

16. Douglas C, Nadasdy T, Wing-Hong A et al. Plasma-cell rich acute renal allograft rejection. Transplantation 1999; 68:791-97.

17. Desvaux D, Le Gouvello S, Pastural M et al. Acute renal allograft rejections with major interstitial oedema and plasma cell-rich infiltrates: high γ-interferon expression and poor clinical outcome. Nephrol Dial Transplant 2004; 19:933-39.

18. Matheson PJ, Dittmer ID, Beaumont BW et al. The macrophage is the predominent inflammatory cell in renal allograft intimal arteritis. Transplantation 2005; 79:1658-62.

19. Szenohradszky P, Smehak G, Szederkenyi E et al. Renal allograft rupture: a clinicopathologic study of 37 nephrectomy cases in a series of 628 consecutive renal transplants. Transplant Proc 1999; 31:2107-11.

20. Fonseca LE, Shapiro R, Randhawa PS. Occurrence of urinary tract infection in patients with renal allograft biopsies showing neutrophilic tubulitis. Mod Pathol 2003; 16:281-85.

21. Mauiyyedi S, Crespo M, Collins AB et al. Acute humoral rejection in kidney transplantation: II. Morphology, immunopathology and pathologic classification. J Am Soc Nephrol 2002; 13:779-87.

22. Böhmig GA, Exner M, Habicht A et al. Capillary C4d deposition in kidney allografts: a specific marker of alloantibody-dependent graft injury. J Am Soc Nephrol 2002; 13:1091-99.

23. Feucht HE, Mihatsch MJ. Diagnostic value of C4d in renal biopsies. Curr Op Nephrol Hypertension 2005; 14:592-98.

24. Trpkov K, Campbell P, Pazderka F et al. Pathologic features of acute renal allograft rejection associated with donor-specific antibody. Analysis using the Banff grading schema. Transplantation 1996; 61:1586-92.

25. Nadasdy GyM, Bott C, Cowden D et al. Comparative study for the detection of peritubular capillary C4d deposition in human renal allografts using different methodologies. Hum Pathol 2005; 36:1178-85.

26. Liptak P, Kemeny E, Morvay Z et al. Peritubular capillary damage in acute humoral rejection: an ultrastructural study on human renal allografts. Am J Transplant 2005; 5:2870-76.

27. Nankivell BJ, Chapman JR. Chronic allograft nephropathy: current concepts and future directions. Transplantation 2006; 81:643-54.

28. Nangaku M. Chronic hypoxia and tubulointerstitial injury: a final common pathway to end-stage renal failure. J Am Soc Nephrol 2006; 17:17-25.

29. Ivanyi B, Fahmy H, Brown H et al. Peritubular capillaries in chronic renal allograft rejection: a quantitative ultrastructural study. Hum Pathol 2000; 31:1129-38.

30. Mihatsch MJ, Ryffel B, Gudat F et al. Cyclosporine nephropathy. In: Tisher CC, Brenner BM, eds. Renal Pathology with Clinical and Functional Correlations. Philadelphia: J.B. Lippincott Co., 1994:1555-86.

31. Liptak P, Ivanyi B. Primer: histopathology of calcineurin-inhibitor toxicity in renal allografts. Nature Clinical Practice Nephrol 2006; 2:398-404.

# Islet of Langerhans:
## Cellular Structure and Physiology

Amanda Jabin Gustafsson and Md. Shahidul Islam*

## Abstract

Islets of Langerhans, named after their discoverer Paul Langerhans, constitute a unique endocrine organ of critical importance in the metabolism of nutrients and energy homeostasis. Individual islets consist of three major types of electrically excitable cells, namely β-cells that secrete insulin, α-cells that secrete glucagon and δ-cells that secrete somatostatin. Islets develop from the gut endoderm and a set of transcription factors including PDX1, PAX4 and PAX6, play important roles in determination of the cell types and their functions. These microorgans are coordinated by neural and hormonal networks and secret hormones in an oscillatory manner. Nutrient metabolism and incretin hormones trigger insulin secretion. Important cellular components and messengers that determine insulin secretion include glucose transporters, glucokinase, ATP/ADP-ratio, mitochondrial metabolism, ATP-sensitive potassium channels, cAMP and $Ca^{2+}$. Because of their importance in the pathogenesis of diabetes and increased interest in islet transplantation during recent years, islets of Langerhans continue to be a field of extensive research throughout the world.

## Introduction

The islets of Langerhans, named after the German pathologist Paul Langerhans, constitute a critical organ unique in that it is split into about a million units hidden in the pancreas, thus contributing to the enormous difficulty in harvesting them. It was in 1869 that Paul Langerhans, then a 22-year old physician, studied anatomy and histology of the pancreas in great details. In his dissertation he described small, clearer areas in the pancreas which stained differently from the rest of the pancreas. Langerhans' speculation was that these structures were lymphatic tissue. Others thought that these structures could be embryonic remnants. It was not earlier than 24 years later that the structures were named "islets of Langerhans" by the French histologist Edouard Laguesse, who also suggested that the structures constituted the endocrine part of the pancreas with a possibility to produce a hormone with glucose-lowering effect.[1] It is unclear why nature has chosen to locate islets in the pancreas. The location is however advantageous, since the hormones are secreted directly into the portal vein enabling direct control of the hepatic function. Furthermore, it is speculated that a vascular system that allows the exocrine pancreas tissue to be nourished by endocrine hormones may have had importance during some stages of evolution explaining location of islets amidst the acinar lobules.[2] The islets have a pivotal role in regulation of glucose homeostasis in the body. The blood glucose-lowering hormone insulin is antagonized by glucagon and together they make a fine-tuning system that ensures that the glucose levels in the blood are kept in a narrow interval irrespective of food intake or starving situation. Impaired function or destruction of the insulin secreting cells in the islets underlies pathogenesis of different forms of diabetes, which is a major health problem throughout the world. Islet transplantation for the cure of type 1 diabetes appears more and more a reality and extensive research is going on in this field at the moment. The goal of this chapter is to give a bird's eye view of anatomy and physiology of the islets of Langerhans with an emphasis on clinical implications.

## Histological Features of the Islets of Langerhans

In humans, islets of Langerhans constitute spherical or ellipsoid clusters of cells with a diameter between ~50-250 μm (Fig. 1).[3] The number of islets in a given pancreas increases with a decreasing diameter of the islets.[4] Most of the islets in the pancreas are of small diameter, i.e., ~50-100 μm. However, medium sized islets with a diameter of ~100-200 μm contribute most to the total islet volume at all ages with the only exception of the newborn, where it is the opposite.[4] Some islets in diabetics can be very large, up to ~350 μm in diameter, because of oedema and deposition of amyloid.[3]

It is generally stated in textbooks that the islets of Langerhans constitute 2-3% of the pancreas and that the pancreas has one million islets. However, the total number of islets, in the pancreas, depends on age, BMI, size of the pancreas and conditions such as pregnancy.[5] The distribution of the islets in the pancreas also varies to some extent. Thus the concentration of islets in the tail is significantly higher than in the head and body of the pancreas.[6] Experience with islet transplantation during recent years has yielded information about the number of islets that can be recovered from one human pancreas. For clinical transplantation recovery of more than 300 000 islets per human pancreas is considered to be successful. However only 50-60% of the isolations are successful at the best isolating centers.[7] By using improved protocols for isolation of islets, on the average 349 000 islet equivalents can be recovered from one pancreas.[8] Each islet is surrounded by a collagen capsule. Characterization of different collagens in the islet-exocrine interface shows that collagen I, IV, V and VI are present and that collagen VI is the major component in the extracellular matrix.[7]

The islets consist of cords of polyhedral cells that are in close proximity to fenestrated capillaries and are lined by basement membrane on their free sides. Each islet contains from a few number of cells to several thousands cells. By using immunohistochemical techniques and specialized staining procedures, one can identify three major types of cells, namely the α-, β- and δ-cells that are irregularly mixed and

*Corresponding Author: Md. Shahidul Islam—Research Center, Department of Clinical Research and Education, Karolinska Institutet, Stockholm South General Hospital, S-118 83 Stockholm, Sweden. Email: shaisl@ki.se

*Chronic Allograft Failure: Natural History, Pathogenesis, Diagnosis and Management*, edited by Nasimul Ahsan. ©2008 Landes Bioscience.

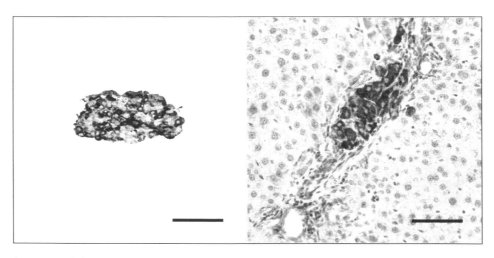

Figure 1. The figure shows a single human islet isolated for transplantation (left) and a rat islet syngeneically transplanted intraportally into the liver (right). The islets were stained for insulin (brown) and counterstained with hematoxylin. Scale bars represent 100 micrometer. Courtesy of Joey Lau, Uppsala University, Sweden. A color version of this figure is available online at www.Eurekah.com.

scattered throughout the islet.[9] In addition there are other minor cell types, namely the pancreatic polypeptide-secreting (PP)-cells and the dendritic cells. Most of the cells in the human islet are insulin-secreting β-cells (64%). Among the remaining are 26% glucagon-secreting α-cells, 8% δ-cells and 0.3% PP-cells. δ-cells secrete somatostatin and possibly gastrin. In each islet there are 5-20 dendritic cells which express class II antigen with phagocytosis capacity.[10] The β-cells secrete islet amyloid polypeptide (IAPP) in addition to insulin. However all β-cells probably do not secrete IAPP since only 54% of β-cells stain for IAPP.[3]

Different islet cells also have their characteristic appearances of the granules as seen under electron microscope. The α-cells possess granules that are 200-250 nm in diameter, have an inner rounded core and a less dense peripheral halo. The secretory granules in the β-cells have a diameter of 300-350 nm, are mainly found in the cell pole facing the blood capillaries and consist of a dense polymorphous core, sometimes several, surrounded with a spherical membranous sac. The granules in δ-cells are numerous with a diameter around 300 nm and are gathered at the vascular side of the cell. The PP-cells have granules that are irregularly shaped, oval or round and much smaller, 150 nm in diameter, than granules of α- and β-cells.[11]

In spite of the fact that the islets are structurally separated, they seem to have a mechanism to coordinate their work. Individual β-cells in an islet are able to communicate with each other through paracrine mechanisms via secretory products or via a local vascular system within the islet. Electrophysiological studies show electrical synchronization between different β-cells through gap junctions. β-cells also communicate with non β-cells via gap junctions. The gap junctions are made of connexin36 and permit passages of small molecules of up to 900 D between cells.[2] The functional importance of connexin36 of β-cells is evident from the fact that in connexin36 knockout mice, there is impaired oscillation of insulin secretion.[12]

## Development and Regeneration of Islet Cells

Islets of Langerhans arise from gut tube endoderm.[13] A hierarchy of transcription factors such as PDX1, FoxM1, Ngn3, IA1 and a MafB-MafA switch and many growth factors, such as activins, TGF-βs, fibroblast growth factors, epidermal growth factors, hedgehogs and wnts play important roles in pancreas development.[14-16] Activin and bone morphogenic factors induce progenitor cells that express the transcription factor PDX1 to develop endocrine cells from pancreatic duct epithelial cells.[17] Coordination of specific transcription and growth factors is essential for generation of the

different cell types in the islets during embryonic development. Notch signalling controls differentiation of progenitor cells to mature endocrine cells.[18] Identification of important transcription factors for the islet cells during their different stages of development has given important information about pancreas development as well as about several types of diabetes.[13] PDX1 is crucial for the development of the pancreas and for the function of differentiated β-cells. It controls transcription of several genes involved in glucose sensing and insulin synthesis. Among other transcription factors that are important for development of islet cells are PAX4 and PAX6. Deletion of Pax4 causes complete loss of β- and δ-cells, but increase of the number of α-cells.[19] Pax6 knockout causes decrease of all endocrine cell types and total absence of α-cells. Knockout of Pax4 and Pax6 leads to total loss of endocrine cells in the pancreas. Disruption of the Pdx1 gene in human results in agenesis of the pancreas. In human, mutations of Pdx1 gene are associated with development of maturity onset diabetes of the young type 4 (MODY4) and also predispose to the development of type 2 diabetes.[20]

Mature β-cells have a slow turnover process which involves replication of already existing β-cells, differentiation from precursor cells in the ductal epithelium and neogenesis. The individual β-cell mass can also increase by hypertrophy.[21] Remarkable examples of increase of the β-cell mass include pregnancy and obesity. Reduction of β-cells takes place through apoptosis.

## Vasculature and Innervation

The islets are richly vascularized by a capillary network connected with the exocrine pancreas and receive at least 10% of the pancreatic blood flow. The endothelial cells contain more fenestrae compared with corresponding acinar capillaries.[22] One to three arterioles enter the core of the islet and form fenestrated capillaries that reassemble again into venules either before or after leaving the islet. In this way a serial connected insulo-acinar portal system is formed with capacity to carry insulin in high concentrations.[2] The intra-islet blood flow is thought to provide a mechanism for paracrine interactions, although the structural basis for such interactions is poorly understood.

The islets are mainly innervated by the parasympathetic system, but are also innervated by sympathetic, peptidergic and non-peptidergic nerves.[2] The parasympathetic preganglionic origins are the dorsal motor nucleus of the vagus nerve and nucleus ambiguus. The nerve branches reach their ganglia in the pancreas tissue and demyelinated nerve fibres innervate the islet cells. It has been suggested that there is a close relationship between innervation of the islets and innervation of the acinar cells in the pancreas. The main part of the fibres enters the islets with the arterioles.[23] The parasympathetic ganglia form neuroinsular complexes with both α- and β-cells. Autonomic transmitters such as acetylcholine and noradrenalin affect the secretion of hormones from the islet cells. The parasympathetic system mainly increases insulin secretion, but glucagon and polypeptide secretion is also stimulated by acetylcholine. The sympathetic system inhibits glucose-induced insulin release and stimulates glucagon release in order to maintain or increase glycemia.

## Regulation of Insulin Secretion

In vitro studies of islets by electrophysiology or imaging techniques assume that β-cells have a resting state when they do not secrete insulin and a stimulated state when they do. However, in vivo under physiological conditions, large insulin secretion occurs even under fasting states and secretion increases after food intake. About 75% of total insulin secretion into the portal vein in human occurs in the form of dramatic oscillations with an interpulse interval of about five minutes.[24,25] The pulsatile pattern of insulin secretion which has many physiological advantages is lost in individuals with type 2 diabetes and their first degree relatives. Liver extracts about 80% of the insulin during the first passage. In the systemic circulation the pulsatile nature of insulin secretion is evidenced by small oscillations in insulin concentration. Regulation of insulin secretion occurs by regulation of the amplitude rather than frequency of insulin oscillation. Even islets transplanted into the liver secrete insulin in a pulsatile manner. It is unclear what signals synchronize insulin secretion from a large number of islets but neural networks are thought to play a role in this process.

After a mixed meal, there is an increase in the concentrations of nutrients including glucose, amino acids and free fatty acids in the plasma and the amplitude of insulin pulses increases. Molecular mechanisms that couple glucose stimulation to insulin secretion have been extensively studied. Glucose must be metabolized to be able to trigger insulin secretion and in this respect the enzyme glucokinase works as a glucose sensor. Several mutations in the glucokinase gene can lead to maturity onset diabetes of the young (MODY).[26] Metabolism of pyruvate in the mitochondria and mitochondrial ATP production are essential for glucose-stimulated insulin secretion. In addition to ATP, several other factors generated from mitochondria potentiate insulin secretion. Mutations or deletions in mitochondrial DNA result in some uncommon forms of diabetes.

Cytoplasmic ATP/ADP ratio acts as intracellular messenger that couples nutrient metabolism to electrical activity of β-cells. In this respect, the ATP-sensitive potassium channels ($K_{ATP}$ channels) act as sensors of cellular metabolism. These channels are the targets for sulfonylurea drugs. $K_{ATP}$ channels of β-cells consist of two subunits, the channel subunit KIR6.2 and the sulfonylurea receptor SUR1. Activating mutations in the genes that encode KIR6.2 and SUR1 cause permanent neonatal diabetes. Inactivating mutations usually of genes that encode SUR1 cause familial hyperinsulinemic hypoglycaemia of infancy.

An increase in the cytoplasmic ATP/ADP ratio leads to closure of $K_{ATP}$ channels which leads to depolarization of the plasma membrane. This in turn leads to activation of the L-type voltage gated $Ca^{2+}$ channels and influx of $Ca^{2+}$. This leads to $Ca^{2+}$ induced $Ca^{2+}$ release from the endoplasmic reticulum.[27] An increase in the cytoplasmic free $Ca^{2+}$ concentration is an essential trigger for insulin exocytosis. In addition to nutrients, insulin secretion is also regulated by neurotransmitters and incretin hormones secreted from the gut. Glucagon-like peptide 1 (GLP-1) is an important incretin hormone that not only increases insulin secretion, but also increases somatostatin secretion and inhibits glucagon secretion. Furthermore, it promotes β-cell survival and proliferation. These actions of GLP-1 are mediated by cAMP as well as $Ca^{2+}$ and other signalling pathways. Incretin and incretin mimetics are already being used as drugs for treatment of type 2 diabetes.[28,29]

## Future Perspectives

In spite of enormous importance of islets in health and disease, there are so far no imaging or isotopic techniques available for visualization or quantification of islets in living individuals. In this regard recent attempts to image islets by positron emission tomography or magnetic resonance imaging using magnetic nanoparticles are interesting approaches.[30,31] Future research should be directed to enable novel discoveries in this aspect.

### *Acknowledgements*

This work was supported in part by the Swedish Research Council Grant K2006-72X-20159-01-3, funds from Karolinska Institutet, Swedish Medical Society, Novo Nordisk Foundation, Stiftelsen Irma och Arvid Larsson-Rösts minne, Svenska diabetesstiftelsen and Stockholms Läns Landsting. A.J.G. is supported by the Karolinska Institutets MD PhD program.

## References

1. Fossati P. Edouard Laguesse at Lille in 1893 created the term "endocrine" and opened the endocrinology era. Hist Sci Med 2004; 38:433-39.
2. Weir GC, Bonner-Weir S. Islets of Langerhans: the puzzle of intraislet interactions and their relevance to diabetes. J Clin Invest 1990; 85:983-87.
3. Iki K, Pour PM. Distribution of Pancreatic Endocrine Cells, Including IAPP-expressing Cells in Nondiabetic and Type 2 Diabetic Cases. J.Histochem.Cytochem. 2007; 55(2):111-18.
4. Hellman B. Actual distribution of the number and volume of the islets of Langerhans in different size classes in non-diabetic humans of varying ages. Nature 1959; 184(Suppl 19):1498-99.
5. Kin T, Murdoch TB, Shapiro AM et al. Estimation of pancreas weight from donor variables. Cell Transplant 2006; 15:181-85.
6. Wittingen J, Frey CF. Islet concentration in the head, body, tail and uncinate process of the pancreas. Ann Surg 1974; 179:412-14.
7. Hughes SJ, Clark A, McShane P et al. Characterisation of collagen VI within the islet-exocrine interface of the human pancreas: implications for clinical islet isolation? Transplantation 2006; 81:423-26.
8. Lakey JR, Tsujimura T, Shapiro AM et al. Preservation of the human pancreas before islet isolation using a two-layer (UW solution-perfluorochemical) cold storage method. Transplantation 2002; 74:1809-11.
9. Cabrera O, Berman DM, Kenyon NS et al. The unique cytoarchitecture of human pancreatic islets has implications for islet cell function. Proc Natl Acad Sci USA 2006; 103:2334-39.
10. Leprini A, Valente U, Celada F et al. Morphology, cytochemical features and membrane phenotype of HLA-DR+ interstitial cells in the human pancreas. Pancreas 1987; 2:127-35.
11. Cavallero C, Spagnoli LG, Cavallero M. Ultrastructural study of the human pancreatic islets. Arch Histol Jpn 1974; 36:307-21.
12. Ravier MA, Guldenagel M, Charollais A et al. Loss of connexin36 channels alters beta-cell coupling, islet synchronization of glucose-induced Ca2+ and insulin oscillations and basal insulin release. Diabetes 2005; 54:1798-807.
13. Edlund H. Developmental biology of the pancreas. Diabetes 2001; 50 Suppl 1:S5-9.
14. Zhang H, Ackermann AM, Gusarova GA et al. The FoxM1 transcription factor is required to maintain pancreatic beta-cell mass. Mol Endocrinol 2006; 20:1853-66.
15. Nishimura W, Kondo T, Salameh T et al. A switch from MafB to MafA expression accompanies differentiation to pancreatic beta-cells. Dev Biol 2006; 293:526-39.
16. Mellitzer G, Bonne S, Luco RF et al. IA1 is NGN3-dependent and essential for differentiation of the endocrine pancreas. EMBO J 2006; 25:1344-52.
17. Hill DJ. Development of the endocrine pancreas. Rev Endocr Metab Disord 2005; 6:229-38.
18. Edlund H. Pancreatic organogenesis—developmental mechanisms and implications for therapy. Nat Rev Genet 2002; 3:524-32.
19. Sosa-Pineda B, Chowdhury K, Torres M et al. The Pax4 gene is essential for differentiation of insulin-producing beta cells in the mammalian pancreas. Nature 1997; 386:399-402.
20. Ahlgren U, Jonsson J, Jonsson L et al. Beta-cell-specific inactivation of the mouse Ipf1/Pdx1 gene results in loss of the beta-cell phenotype and maturity onset diabetes. Genes Dev 1998; 12:1763-68.
21. Bonner-Weir S. Islet growth and development in the adult. J Mol Endocrinol 2000; 24:297-302.
22. Henderson JR, Moss MC. A morphometric study of the endocrine and exocrine capillaries of the pancreas. Q J Exp Physiol 1985; 70:347-56.

23. Gilon P, Henquin JC. Mechanisms and physiological significance of the cholinergic control of pancreatic beta-cell function. Endocr Rev 2001; 22:565-604.

24. Porksen N, Nyholm B, Veldhuis JD et al. In humans at least 75% of insulin secretion arises from punctuated insulin secretory bursts. Am J Physiol 1997; 273:E908-14.

25. Song SH, McIntyre SS, Shah H et al. Direct measurement of pulsatile insulin secretion from the portal vein in human subjects. J Clin Endocrinol Metab 2000; 85:4491-99.

26. Vaxillaire M, Froguel P. Genetic basis of maturity-onset diabetes of the young. Endocrinol Metab Clin North Am 2006; 35:371-84.

27. Islam MS. The Ryanodine Receptor Calcium Channel of β-Cells: Molecular Regulation and Physiological Significance. Diabetes 2002; 51:1299-1309.

28. List JF, Habener JF. Glucagon-like peptide 1 agonists and the development and growth of pancreatic beta-cells. Am J Physiol Endocrinol Metab 2004; 286:E875-81.

29. Stonehouse AH, Holcombe JH, Kendall DM. Management of Type 2 diabetes: the role of incretin mimetics. Expert Opin Pharmacother 2006; 7:2095-105.

30. Tai JH, Foster P, Rosales A et al. Imaging islets labeled with magnetic nanoparticles at 1.5 tesla. Diabetes 2006; 55:2931-38.

31. Lu Y, Dang H, Middleton B et al. Long-Term Monitoring of Transplanted Islets Using Positron Emission Tomography Mol Ther 2006; 14(6):851-56.

# CHAPTER 29

# Islet Transplantation

Breay W. Paty and A.M. James Shapiro*

## Introduction

Islet transplantation restores endogenous insulin secretion in individuals with type 1 diabetes by infusing insulin-secreting islet cells, isolated from cadaveric pancreata, into the liver.[1] The immune mediated destruction of insulin-secreting beta cells within pancreatic islets that occurs in type 1 diabetes results in an almost complete loss of endogenous insulin secretion, making these individuals completely dependent on exogenous insulin for their survival.[2] The benefits of "intensive" glucose control have been clearly demonstrated in this group of patients.[3] However, the primary limitation of intensive insulin therapy is an increased risk of hypoglycemia.[4] This risk is particularly high among patients with long-standing diabetes and is often accompanied by increased glycemic lability which occurs when insulin doses do not precisely match physiologic requirements. A variety of therapeutic options have emerged to try to mimic "physiologic" insulin secretion, such as insulin analogues, continuous insulin infusion pumps and continuous glucose monitoring. For people at particularly high risk of severe hypoglycemia and glycemic lability, restoring endogenous insulin secretion by whole pancreas or pancreatic islet transplantation has several potential advantages over exogenous insulin therapy, including normal to near-normal blood glucose control, the possibility of insulin independence and of reduced secondary complications of diabetes. However, as with any transplant therapy, the potential benefits must be carefully weighed against the underlying risks of the procedure and of chronic immunosuppression.

Recent progress in clinical islet transplantation has significantly improved the 1-year rate of insulin independence from about 10% to approximately 70-80% at experienced centers.[5,6] However, to date there is very little long-term data regarding islet transplant function after five years. Most published studies involve islet transplantation in the absence of a kidney transplant (islet transplant alone [ITA]). There is some evidence suggesting that simultaneous islet kidney [SIK] transplantation or islet after kidney [IAK] transplantation may have comparable outcomes.[7,8] The following chapter describes the recent progress made in the field of islet transplantation including improvements in islet isolation, immunosuppression, metabolic and clinical outcomes. It also discusses the continuing challenges facing islet transplantation that must be overcome in order for this therapy to be more broadly applied in type 1 diabetes.

## History

Islet alotransplantation was first performed successfully in mice in the early 1970s[9,10] and soon progressed to clinical application in humans with the first human islet alotransplant in the modern era occurring at the University of Minnesota in 1977.[11] Larger pancreatic fragments were used instead of isolated islets, in order to try to maximize the total transplanted islet mass. However, despite considerable effort, these early attempts did not result in insulin independence and only led to reduced insulin requirements in a few individuals for limited periods of time.[12] Furthermore, the transplant procedure tended to be associated with considerable morbidity, including portal vein thrombosis and portal hypertension. The disappointing results of early attempts at islet allotransplantation were balanced by more encouraging results of islet autotransplantation.[13-15] This procedure involves isolating islets from nondiabetic individuals undergoing total pancreatectomy, usually for chronic, painful pancreatitis and re-infusing them into the liver in order to maintain some degree of endogenous insulin secretion. Because it involves transplantation of non-allogeneic tissue, islet autotransplantation eliminates both immune rejection and immunosuppressive toxicity as potential causes of islet graft loss. The liver was chosen as the site of transplantation because it appeared to be the most suitable; it is well vascularized and relatively easy to access via portal venous catheterization. Furthermore, experience with alternative sites, such as the spleen, met with little success and even led to life-threatening complications in some cases.[16] The overall insulin independence rate in one series of intrahepatic islet autotransplants was 34% between 2-10 years posttransplant.[15] However, when only recipients who received >300,000 islets were analyzed, the insulin independence rate was 74% at >2 years; this result was substantially greater than that observed at the time after islet allotransplantation. Subsequent metabolic studies of successful islet autotransplant recipients have shown that, although insulin and glucagon secretion are subnormal, islet secretory function is adequate to maintain prolonged insulin independence, given an adequate number of transplanted cells.[17,18] Progress continued throughout the 1980s in the areas of islet isolation and purification with a number of important developments, including the refinement of enzyme blends for more successful pancreas digestion,[19] the adoption of the Ricordi chamber for improved enzymatic and mechanical digestion of pancreatic tissue,[20] the use of controlled intraductal perfusion of digestive enzyme[21] and the adaptation of COBE cell processor for density purification of isolated islets.[22] Each of these advances led to improvements in islet yield and purity and consequently, to greater clinical success. Throughout this period, islet isolation and transplantation were successfully performed in larger animal models including dogs and pigs.[23-27] Efforts at human islet transplantation increased throughout the 1990s with gradual improvements in islet yields and clinical outcomes. Small clinical trials in Europe were able to achieve a 1-year insulin independence rate of approximately 50% using cyclosporine, mycophenolate mofetil and glucocorticoid immunosuppression.[28,29] However, combined data from the International

*Corresponding Author: A.M. James Shapiro—Director, Clinical Islet Transplant Program, University of Alberta. 2000 College Plaza, 8215—112th Street, Edmonton, AB, Canada T6G 2C8. Email: amjs@islet.ca

Islet Transplant Registry from this period[30] illustrate the challenges that continued to limit the success of islet transplantation, including suboptimal islet yields due to islet apoptosis during isolation, poor islet engraftment due to immediate nonspecific inflammatory responses after transplantation, persistent allo and autoimmune rejection of transplanted islets, direct toxic effects of immunosuppressive agents on islet survival and "exhaustion" of islet function due to persistent metabolic demands. The current era of clinical islet transplantation was ushered in with the publication of a series of seven type 1 diabetes patients who received islet transplants under a steroid-free immunosuppressive regimen, termed the "Edmonton Protocol".[31] The protocol was developed to prevent activation of the immune cascade by inhibiting T-cell activation, blocking IL-2 recruitment and the production of other cytokines and preventing "signal 3" proliferation and clonal expansion. Furthermore, it avoided the well-documented diabetogenic effects of glucocorticoids,[32,33] which likely played a major role in the limited success of previous islet transplant protocols. Another difference in approach was the standard use of two islet infusions from separate donors in order to ensure an adequate mass of transplanted cells. Early metabolic testing indicated near-normal glycemic control with mean hemoglobin A1c levels in the normal range and C-peptide levels equivalent to controls.[34] However, despite good glycemic control, only 21% of recipients had normal insulin responses to intravenous glucose stimulation while 56% of recipients had normal responses to arginine stimulation.[35] The results suggested that despite a total infused islet mass of 12,000-16,000 islets per kilogram body weight of each recipient, the total number of functioning islets after transplantation was likely to be substantially lower. Nevertheless, these results represented a major step forward in the clinical success of islet transplantation and stimulated renewed interest in this procedure as a viable means of restoring endogenous insulin secretion in selected individuals with type 1 diabetes.

## Islet Isolation and Transplantation

Improvements in clinical islet transplantation are attributable both to improvements in targeted immunosuppression and progress in islet isolation and purification. However, current islets yields are estimated to still be only 20-50% of total donor islet mass[36] and predictable yields of high-quality islets are still not routine. Islets for human transplantation are prepared using Good Manufacturing Practice (GMP)-grade facilities according to a standardized procedure.[6,37] The donor pancreas is distended by controlled ductal perfusion with the use of Liberase human islet enzyme (Roche). The pancreas is digested in a Ricordi chamber (Fig. 1) and purified on continuous Ficoll gradients on a cooled COBE apheresis system. The islets are then washed and resuspended in transplant medium for infusion into the recipient.

The factors contributing to successful islet isolation begin when a pancreas donor is identified. Pancreata from adult donors (>20 years), with high body mass index (BMI), no history of cardiac arrest or hypotension, minimally elevated blood glucose levels (<10 mmol/L), minimal cold ischemic time (<12 hours), low serum amylase or lipase levels, as well as the experience of the surgical retrieval team each can have a major impact on islet yield and the subsequent success of transplantation.[38-40] Islets are particularly susceptible to ischemic damage.[41,42] Reperfusion injury and the liberation of reactive oxygen species during the isolation process contribute to oxidative stress.[43-45] Consequently, the method of pancreas preservation during transport has been found to have a major impact on islet survival. Current pancreas preservation utilizes hypothermic, "two-layer method" (TLM) perfluorocarbon (PFC)/University of Wisconsin (UW) solution.[46-49] Perfluorocarbons have a high affinity for oxygen, but effectively release oxygen to tissues, thereby minimizing ischemia.

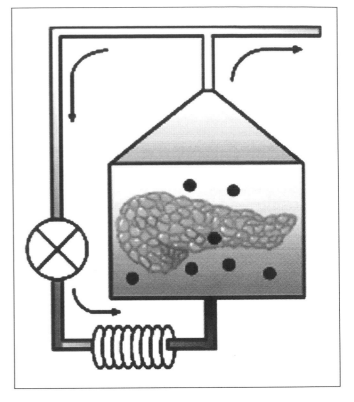

Figure 1. Schematic representation of the Ricordi chamber used for the mechanical and enzymatic digestion of pancreata during islet isolation. The pancreas is sectioned and placed within a steel (or acrylic) chamber along with steel marbles. Then the heated collagenase solution is pumped through the chamber while it is mechanically agitated. Samples are taken periodically to determine the degree of liberation of islets and when to stop the process.

In Vitro studies indicate that the oxygenated graft generates adenosine triphosphate (ATP), which drives the sodium-potassium ATP pump and helps maintain membrane integrity.[50] PFCs also appear to improve vascular endothelium and prevent or repair reperfusion injury.[51-53] Lower cost alternatives, such as Histidine-Tryptophan Ketoglutarate (HTK) solution have also been tried.[54] More recently, other methods, such as the use of oxygenated polymerized, stroma-free hemoglobin-pyridoxalated (Poly SFH-P), a hemoglobin-based $O_2$ carrier[55] may lead to less ischemic injury and further improve islet yields. Culturing islets for short periods (8-48 hrs) after isolation has become standard practice, since it allows immune preconditioning of recipients and does not appear to have a detrimental effect on transplant outcome.[56-59] Also, islet culturing allows the addition of agents such as nicotinamide and antioxidants to the purified islets, which may have cytoprotective effects.[60,61] Once islets are isolated and purified and an ABO-blood group-matched recipient is identified, the portal vein is cannulated under fluoroscopic guidance. Islets, suspended in solution, are then infused by gravity into the portal circulation and flow with the blood to lodge in the hepatic sinusoids (Fig. 2). Portal pressure is monitored periodically throughout the procedure to ensure that it does not rise significantly. To minimize the risk of bleeding, the catheter tract is plugged using a combination of coils and fibrinogen paste. Islet cell survival after transplantation has been estimated to be only 10-20%.[62] This low rate of survival is likely due to a number of factors including cellular hypoxia and the initiation of an inflammatory cascade including the so-called instant blood mediated inflammatory reaction (IBMIR), which may trigger the release of tissue factors detrimental to islet survival.[63,64]

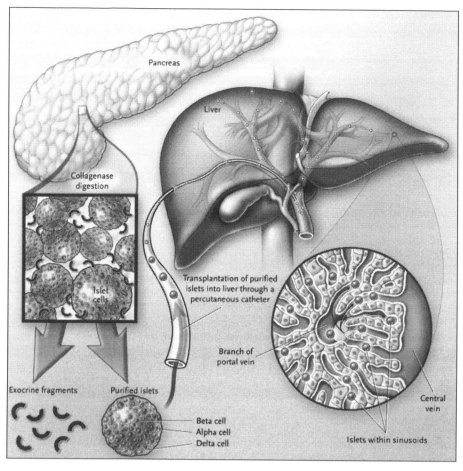

Figure 2. The four basic steps of islet transplantation: 1) collagenase digestion of the donor pancreas; 2) purification of the isolated islets using density centrifugation; 3) infusion of the purified islet preparation into the portal circulation; 4) implantation of the infused islets into the hepatic sinusoids. (Adapted wtih permission from R.P. Robertson. New Engl J Med 2004; 350(7):694-705;[1] ©2004 Massachusetts Medical Society).

## Current Immunosuppression

The dual nature of the immune response to transplanted islets, both alloimmune and autoimmune, is perhaps the most challenging hurdle to overcome.[65] To be effective, immunosuppressive regimens must have adequate potency to minimize both alloimmune rejection and autoimmune recurrence, while avoiding diabetogenic effects and direct toxic effects on beta cells. Current immunosuppression achieves this, at least in part, using a combination of sirolimus (SRL)/tacrolimus (TAC) in most cases.[6,31] Sirolimus, a macrolide that blocks T and B lymphocyte responses to IL-2 and other cytokines by interfering with downstream phosphorylation, thereby preventing T and B-cell recruitment and activation,[66] has a number of characteristics that make it uniquely suitable for maintenance or temporary immunosuppression in islet allograft recipients. It has efficacy in both auto and alloimmunity and it acts in synergy with other immunosuppressive agents, including ones that block costimulation.[67] When use in combination with tacrolimus, a potent calcineurin inhibitor, studies have demonstrated the SRL/TAC combination results in improved outcomes in a variety of animal transplant models.[68-70] Early human data also demonstrated improved graft survival with this combination.[71,72]

For induction, the monoclonal antibody IL-2 inhibitors have been found to be as effective and better tolerated than anti-thymocyte globulin[73,74] with a reduced incidence of acute rejection in renal transplantation.[75,76] However, a small human case series using basiliximab induction did not seem to show any benefit.[77] Similarly, the chimeric monoclonal antibody infliximab, showed no apparent benefit on human graft survival or function.[78]

The standard Edmonton Protocol consists of induction with daclizumab, a humanized IgG monoclonal antibody that inhibits interleukin-2 (IL-2) mediated lymphocyte activation, administered immediately pretransplant at a dose of 1 mg/kg, then repeated over 8 weeks posttransplant for a total of 5 doses. In cases where a second islet infusion is given beyond the 8-week induction window, the course of daclizumab is repeated. Sirolimus is given at a loading dose of 0.2 mg/kg orally immediately pretransplant, with maintenance dosing initially at 0.15 mg/kg/day adjusted to a 24-hour target serum trough level of 12-15 ng/ml for three months, then reduced to 7-12 ng/ml thereafter. Tacrolimus is initiated at 2 mg orally twice daily and adjusted to target 12-hour trough levels of 3-6 ng/ml.

Modifications of the Edmonton Protocol have also yielded a high rate of insulin independence, even after a single islet infusion. The University of Minnesota transplanted a series of eight women with type 1 diabetes who achieved insulin independence after a single islet infusion, with five recipients remaining insulin independent for at least one year.[79] Induction immunosuppression consisted of rabbit antithymocyte globulin (RATG) with single dose methylprednisolone (on day –2 only, 2 mg/kg), daclizumab (5 doses of 1 mg/kg every 2 weeks starting on day 0) and etanercept (50 mg intravenously 1 hour pretransplantation). Maintenance immunosuppression consisted of sirolimus (target whole blood trough levels, 5-15 ng/mL) and reduced-dose tacrolimus (target whole blood trough levels, 3-6 ng/mL). At 1 month posttransplantation, tacrolimus was gradually replaced with mycophenolate mofetil (750-1000 mg, twice daily) and tacrolimus was either discontinued or reduced to a target trough level of less than 3 ng/mL. Unfortunately, these results may not be widely applicable, since the study used only ideal, high BMI pancreas donors and low BMI, insulin sensitive recipients.

Other immunosuppressive regimens may prove equally or more efficacious with reduced side-effect profiles compared to the current standard. Recent data from other transplant trials suggest that the combination of TAC/Mycophenolate Mofetil (MMF) results in equal prevention of acute rejection with less impairment of renal function when compared to SRL/TAC.[80] This finding has also been reflected in islet transplantation.[81,82] Furthermore, in solid organ transplants the SRL/TAC combination has been associated with higher withdrawal rates due to toxicity.[83,84] The combination of TAC/MMF has recently been shown to improve long-term renal allograft function compared to SRL/TAC.[85] Limited experience with this combination in islet transplantation seems to suggest that it may be equally as effective with fewer side-effects,[5,67] but no controlled trials have yet been performed.

Tissue sensitization is an important issue in the setting of multiple islet transplants. Panel reactive antibodies increased (>15%) after transplantation in 12% of recipients.[5] It has also been demonstrated that the presence of pretransplant anti-HLA antibodies is associated with loss of C-peptide.[86] The rate of complete graft failure (C-peptide negative) is 5.9 times higher if the pretransplant PRA was >15% in either class I or II. The outcome was particularly poor if the pretransplant HLA antibodies were donor specific (either classes I or II) as the risk of graft failure is 7.3 times higher in this group compare to HLA negative group. This suggests that combination TAC/SRL immunosuppression may not control the alloimmune response in this presensitized population and individuals with a PRA >15% may require more aggressive inductive and maintenance immunosuppression, or perhaps represent a group that may not benefit from islet transplantation. At the University of Alberta, thymoglobulin induction and higher therapeutic target levels of tacrolimus (10 ng/ml if tolerated) are used in recipients with a PRA >15% or alternatively whole pancreas transplantation is considered.

## Clinical Transplantation and Metabolic Outcomes

Standard indications for islet transplantation include adults with type 1 diabetes mellitus for more than five years with recurrent neuroglycopenia, including reduced awareness of hypoglycemic episodes or severe glycemic lability that does not respond to exogenous insulin therapy. Major exclusion criteria include noncorrectable coronary artery disease; a body-mass index of more than 30 kg/m$^2$ (although some studies have been more restrictive); an insulin requirement of more than 1.0 U per kilogram of body weight per day; a glycated hemoglobin level of more than 12%; inadequate renal reserve, which was defined as a serum creatinine level of more than 1.5 mg per deciliter (133 μmol per liter), a creatinine clearance of less than 80 ml per minute per 1.73 m2 of body-surface area, or an albumin level of more than 300 mg per 24-hours (macroalbuminuria); and negative results on serologic analysis for Epstein-Barr virus at the time of assessment.

Despite isolated success with single donor protocols, in most cases insulin independence still requires at least two separate islet infusions from different pancreas donors. At an experienced center such as the University of Alberta, daily insulin requirements are reduced after the first islet infusion by a mean of 52% (range: 12-100%) and glycemic stability is usually greatly improved.[87] However, most recipients do not achieve insulin independence before receiving a second islet infusion for a total transplanted islet mass of at least 10,000 islets/kg. Recipients who demonstrate a reduction in daily insulin dose by >50% after the first islet infusion are more likely to achieve insulin independence after the second islet infusion than those whose daily insulin requirements decline by <50%.[35] The 12-month insulin independence rate for recipients with at least two islet infusions is approximately 70 - 80% by Kaplan-Meier survival analysis.[5,35]

In the Immune Tolerance Network international multicenter trial of the Edmonton Protocol, 36 subjects received a total of 77 islet infusions,

with 11 subjects (31%) receiving one infusion, 9 (25%) receiving two infusions and 16 (44%) receiving three infusions.[6] Of the 36 subjects, one year after the final transplant, 16 (44%) were insulin independent, 10 (28%) had partial graft function (C-peptide positive) and 10 (28%) had complete graft loss. A total of 21 subjects (58%) attained insulin independence with good glycemic control at any point throughout the trial. Of these subjects, 16 (76%) required insulin again at two years; 5 of the 16 subjects who reached the primary end point (31%) remained insulin-independent at two years (Fig. 3A,B). There was a positive correlation between attainment of insulin independence and autoantibody status (Fig. 3C). C-peptide secretion was detectable (0.3 ng per milliliter) in 70% of subjects at two years (Fig. 3D). Importantly, a positive correlation was also observed between the amount of previous experience with islet transplantation of a transplant center and the attainment of insulin independence. At sites where four or more transplants had been performed in the preceding two years, the primary end point (1-year insulin independence) was reached by 12 of 18 subjects (67%) as compared with only 4 of 18 subjects (22%) at sites where fewer than four transplants had been performed.

Insulin independence rates tend to decline with time, so that five years after transplantation, generally less than 10% of recipients remain off insulin.[5] The reason for this decline is not clear, but may involve direct immunosuppressive toxicity, allo or autoimmune rejection, or islet cell apoptosis potentially reflecting the islet's natural life span.[88-91] However, the majority of recipients who resume insulin injections continue to have functioning islet grafts, as evidenced by the persistence of detectable serum C-peptide. Complete loss of C-peptide secretion after transplantation occurs in a minority (~15%) of recipients. The importance of even partial graft function is illustrated by the glycated hemoglobin levels in C-peptide positive versus C-peptide negative transplant recipients. The hemoglobin A1C is normalized in insulin-independent recipients and remains near-normal in islet transplant recipients who remain C-peptide positive, while it is elevated in recipients who lose all C-peptide secretion (Fig. 4).

Other than insulin independence, the most immediate and prolonged benefit of islet transplantation is stabilization of blood glucose, which occurs in almost all recipients, usually after a single islet infusion. Using hypoglycemia and glycemic lability scores,[92] there is a clear reduction in the incidence and severity of hypoglycemia and glycemic lability after transplantation that can lasts several years, although this tends to wane in recipients who lose graft function. Continuous glucose monitoring demonstrates that C-peptide positive recipients who used exogenous insulin continued to benefit from excellent, stable blood glucose control, while eliminating or significantly reducing the incidence of hypoglycemia compared to type 1 diabetes controls (Fig. 5).[93] This finding is important because it suggests that, although insulin independence is an important goal for most islet transplant recipients, it is not necessary in order to benefit from islet transplantation.

Another potential benefit of islet transplantation may be a reduction of long-term secondary complications of diabetes, although this is not yet clear. There is some evidence that IAK transplant recipients show improved endothelial and cardiovascular function compared to kidney transplant recipients.[94,95] Kidney graft survival rates also appear to improve with concomitant islet transplant.[96] Quality of life appears to improve initially after islet transplantation, due primarily to a reduced fear of hypoglycemia, but declines with the loss of insulin independence.[97,98] However, data from controlled studies are not yet available, so a reduction in secondary diabetic complications is not certain and will require more detailed, long-term investigation to be fully known.

## Complications and Side Effects

The most common procedure-related adverse events are abdominal pain and nausea (Table 1). Intraperitoneal hemorrhage has occurred

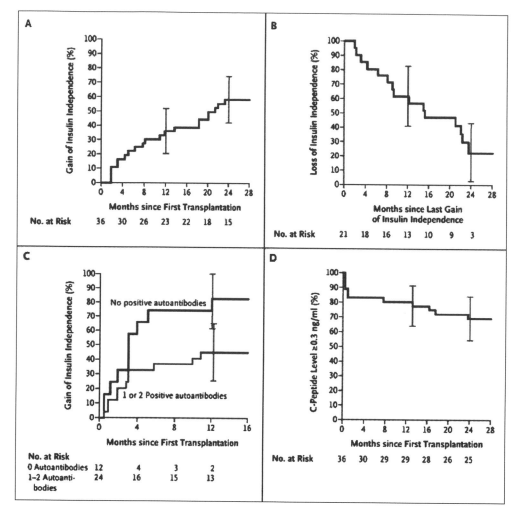

Figure 3. Results of the international trial of the Edmonton Protocol for islet transplantation. Panel A shows the interval between the first transplantation and insulin independence; Panel B shows the subsequent loss of insulin independence among 16 of these 21 subjects during the next 28 months. Panel C shows insulin independence since the last transplantation according to the number of beta cell autoantibodies detected before subjects underwent the last transplantation: 85% for the 12 subjects who had no positive autoantibodies and 46% for the 24 subjects who had one or two positive autoantibodies. Panel D shows the percentage of subjects who had a basal C-peptide level of at least 0.3 ng per milliliter after transplantation. (Adapted with permission from Shapiro et al. New Engl J Med 2006; 355(13):1318-30;[6] ©2006 Massachusetts Medical Society).

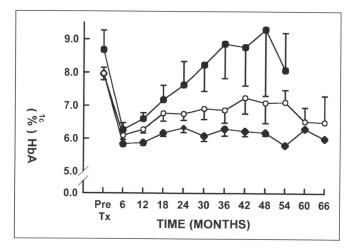

Figure 4. HbA1c (mean ± SE) posttransplantation in subjects who lost all graft function (●), versus those subjects who had some graft function but had to resume insulin (O) and those subjects who remained insulin independent (◆). (Reprinted with permission from: Ryan EA et al. Diabetes, 2005; 54(7):2060-69;[5] ©2005 The American Diabetes Association).

in approximately 10% of recipients,[5,6] however recent improvements in sealing the catheter tract after islet infusion have greatly reduced the risk of bleeding.[99] Portal hypertension can occur acutely during islet infusions, but portal pressures tend to normalize after the acute phase of transplantation.[100] Portal vein thrombosis has occurred in a small number (~4%) of recipients but has generally been limited to branch veins and have resolved with appropriate anticoagulation. Posttransplant elevation of liver enzymes (54%) and catheter-related puncture of the gallbladder (3%) can also occur although these tend to be self-limited.[35]

The most common adverse effect of immunosuppressive therapy is mucosal ulceration involving the tongue and/or buccal mucosa occurring in the vast majority (~90%) of recipients.[101] This is presumed to be a consequence of sirolimus therapy and tends to be dose-dependent. Increased use of lipid-lowering or antihypertensive agents has occurred in about half of all recipients. The rate of cytomegalovirus (CMV) conversion is low (6%) and there has been no overt CMV disease observed.[5,102] Posttransplant anemia is common (80%) and leukopenia is also frequently seen (75%), but severe neutropenia is uncommon.[5,6] Posttransplant weight loss is also often seen. The appearance of intrahepatic periportal steatosis has been observed using magnetic resonance imaging (MRI) in 20-30% of recipients and may occur in a

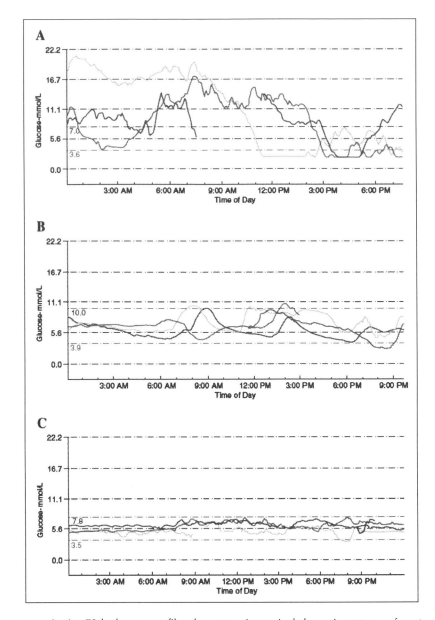

Figure 5. Continuous glucose monitoring 72-h glucose profiles demonstrating typical glycemic patterns of nontransplanted type 1 diabetes subjects (A), islet transplant recipients who had some graft function, but required exogenous insulin (B) and islet transplant recipients who were completely insulin independent (C). (Adapted with permission from Paty et al. Diabetes Technol Ther 2006; 8(2):165-73).

larger proportion.[103,104] These changes are thought to be benign, due to the local effects of insulin on surrounding liver parenchyma. They also appear to be reversible, since complete resolution has been observed in one patient whose graft failed completely. However, the impact of fatty changes in the liver on underlying hepatic glucose metabolism is not yet known.

Immunosuppressive agents can be nephrotoxic and may be partially responsible for an elevation of serum creatinine and a reduction in glomerular filtration rate (GFR) after transplantation.[82] Interestingly, a small number of patients developed proteinuria which resolved upon withdrawal of sirolimus,[81] a finding which has recently been observed in other transplant settings.[105,106] Whether this finding represents a

**Table 1. Adverse events associated with clinical islet transplantation (approximate % incidence)**

| Procedure-Related | % | Immunosuppressive-Related | % |  | % |
|---|---|---|---|---|---|
| Elevated liver enzymes | 54% | Oral ulcers | 90% | New/increased albuminuria | 50% |
| Abdominal pain | 50% | Anemia | 80% | Acne | 39% |
| Nausea/vomiting | 50% | Leukopenia | 75% | Fatigue | 39% |
| Peritoneal hemorrhage | 10% | Diarrhea | 64% | Hepatic Steatosis | 31% |
| Portal vein thrombosis | 4% | Headache | 56% | Ovarian cysts | 20% |
| Gall-bladder puncture | 3% | Neutropenia | 53% | Misc.—tremor, arthralgia, |  |
|  |  | Decreased GFR | 50% | pneumonitis, hematuria, infection |  |

particular risk of combined TAC/SRL immunosuppression is not yet clear. An increase in retinal bleeds has also been seen. This is to be expected in patients who undergo a marked correction in their glycemic control after transplantation.[107-109] The risk of posttransplant infection and malignancy also exist. So far, among University of Alberta islet transplant recipients, there has been one case of severe infection; pulmonary aspergillosis, requiring intravenous antifungal therapy. This patient experienced mild respiratory symptoms and recovered fully after treatment. One recipient had a well-differentiated thyroid neoplasm and another had two basal-cell skin cancers removed. However, there is no way of establishing whether these neoplasms were related to immunosuppressive therapy.

## Future Directions

Many challenges remain in the field of islet transplantation, including the need for more effective, less toxic immunosuppression. Particular attention is being paid to the search for agents that could potentially induce immune tolerance to the islet allograft.[110,111] The rationale for using the islet transplant model for testing immune tolerance is that lack of efficacy is not potentially fatal, as it would be with other transplant models. Novel immunosuppressive agents that may promote tolerance in islet transplantation include; (1) agents that interfere with T-cell activation; (2) T-cell depleting agents; (3) agents that inhibit lymphocyte trafficking and recruitment (Fig. 6). A promising approach to interfering with T-cell activation involves blocking the CD28:B7 and CD40:CD40 ligand costimulatory pathways. Early trials of a humanized anti-CD40L monoclonal antibody (hu5C8, ruplizumab) in pancreatectomized, islet-transplanted non-human primates, resulted in prolonged insulin independence with little toxicity.[112] However, human phase 1 trials of ruplizumab were terminated because of unexpected thromboembolic complications.[113] More recently, the development of a second-generation cytotoxic

T-lymphocyte antigen/immunoglobulin G fusion protein, belatacept (LEA-29Y),[114] has shown even greater efficacy in primate models without evidence of thromboembolism.[115,116] A phase 3 multicenter trial of belatacept for renal transplants is currently underway and a pilot study of this agent in islet transplantation is also planned.

Another promising approach to immune tolerance is by T-cell depletion. One of the most potent agents is the humanized anti-CD52 antibody, alemtuzumab (Campath-1H). While its precise mechanism is not clear, it appears to prevent T-cell activation via CD45 signaling, but does not interfere with T-cell receptor activation.[117] When used as an induction agent, it results in prolonged lymphocyte depletion (months to years).[118] In early randomized controlled trials of kidney transplantation, no significant differences in acute rejection rates were observed between alemtuzumab-treated patients and those using cyclosporine, azathioprine and steroids.[119,120] A pilot study of alemtuzumab induction and sirolimus monotherapy in renal transplants was initially associated with a high rate of humoral rejection.[121] However, three year follow-up data showed good graft (96%) and patient (100%) survival.[122] Furthermore, a relatively large proportion of patients (57%) were able to be maintained on monotherapy. However, because of the higher incidence of early rejection, a brief course of a calcineurin inhibitor was recommended. A small pilot trial in islet transplant recipients suggested that alemtuzumab with sirolimus monotherapy-treated patients had no better graft function than those treated with the standard Edmonton Protocol.[123] However, given the promising results in other transplant settings, studies are ongoing to examine the efficacy of alemtuzumab induction with TAC and MMF maintenance immunosuppression in islet transplantation. Another, potentially useful T-cell depleting agent is the humanized, anti-CD3 monoclonal antibody hOKT3-γ1 (Ala-Ala), because of its demonstrated efficacy in controlling the autoimmune response in non-obese diabetic (NOD) mice.[124,125] A pilot study of this agent in human subjects with early-onset type 1 diabetes resulted

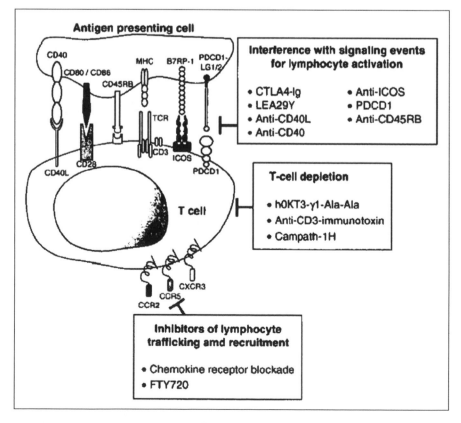

Figure 6. Novel agents with the potential for promoting tolerance in islet transplantation and their points of action. CTLA—cytotoxic T-cell lymphocyte antigen; ICOS—inducible T-cell costimulator; MHC—major histocompatibility complex; PDCD1—programmed cell death 1; PDCD1-LG1/2—PDCD1 ligand 1 and 2; TCR—T-cell receptor. (Adapted with permission from: Nanji SA, Shapiro AM. BioDrugs, 2004; 18(5):315-28;[123]).

in only a small reduction in daily insulin use, but improved metabolic control after one year.[126] In another study, a small group of patients (n = 6) received transplants with cultured islets using anti-CD3 induction and TAC/SRL maintenance therapy. Four of the six subjects became insulin independent after receiving a single islet infusion.[59]

A third approach to inducing immune tolerance is by blocking lymphocyte trafficking and recruitment. A novel agent FTY720, a sphingosine-1-phosphate receptor modulator that interferes with T-cell antigen interaction, has shown early promise. In a small (n = 5) nonhuman primate model of islet transplantation, induction with the anti-IL-2 agent basiliximab followed by maintenance therapy with everolimus (an mTOR inhibitor, related to sirolimus) and FTY720 resulted in all five recipients becoming insulin independent after transplantation after receiving 10,000 islets each.[127] Each of these approaches to inducing immune tolerance in islet transplantation has promise, but more study is needed to determine whether any of them are an improvement on current immunosuppressive regimens.

Currently, the availability of islets from cadaveric donors is extremely limited. If islet transplantation is to be more widely available, new sources of islets must be found or developed. This may be accomplished by improving isolation techniques or the development of alternative sources of transplantable insulin-secreting tissue, such as stem cells[128-132] or animal-derived cells.[133,134] Other proposed means of increasing functional beta cell mass include the use of possible beta cell proliferative agents, such as GLP-1 analogues or other growth factors.[135] In addition, further efforts need to be made to better understand the reasons for the gradual decline in insulin independence after transplantation.

## Conclusion

Islet transplantation, as a means of restoring endogenous insulin secretion in selected individuals with type 1 diabetes, has made significant progress in terms of clinical outcomes over the last several years. Yet, because of the limited availability of donor islets, the risks and side effects associated with the procedure and chronic immunosuppression and the loss of insulin secretion over time, the use of islet transplantation as a therapeutic tool continues to be limited to a small group of individuals with very difficult to control diabetes. A number of challenges remain to be overcome, including improving islet isolation yield, developing newer, less toxic immunosuppressive agents and potentially utilizing alternate sources of islets. If each of these challenges is met, islet transplantation has the potential to be used much more widely for the treatment of type 1 diabetes.

## References

1. Robertson RP. Islet transplantation as a treatment for diabetes—a work in progress. N Engl J Med 2004; 350(7):694-705.
2. Diagnosis and classification of diabetes mellitus. Diabetes Care 2006; 29 Suppl 1:S43-48.
3. The effect of intensive treatment of diabetes on the development and progression of long-term complications in insulin-dependent diabetes mellitus. The Diabetes Control and Complications Trial Research Group. N Engl J Med 1993; 329(14):977-86.
4. Cryer PE. Hypoglycemia is the limiting factor in the management of diabetes. Diabetes Metab Res Rev 1999; 15(1):42-46.
5. Ryan EA, Paty BW, Senior PA et al. Five-year follow-up after clinical islet transplantation. Diabetes 2005; 54(7):2060-69.
6. Shapiro AM, Ricordi C, Hering BJ et al. International trial of the Edmonton protocol for islet transplantation. N Engl J Med 2006; 355(13):1318-30.
7. Lehmann R, Weber M, Berthold P et al. Successful simultaneous islet-kidney transplantation using a steroid-free immunosuppression: two-year follow-up. Am J Transplant 2004; 4(7):1117-23.
8. Toso C, Baertschiger R, Morel P et al. Sequential kidney/islet transplantation: efficacy and safety assessment of a steroid-free immunosuppression protocol. Am J Transplant 2006; 6(5 Pt 1):1049-58.
9. Ballinger WF, Lacy PE. Transplantation of intact pancreatic islets in rats. Surgery 1972; 72(2):175-86.
10. Reckard CR, Barker CF. Transplantation of isolated pancreatic islets across strong and weak histocompatibility barriers. Transplant Proc 1973; 5(1):761-63.
11. Najarian JS, Sutherland DE, Matas AJ et al. Human islet transplantation: a preliminary report. Transplant Proc 1977; 9(1):233-36.
12. Sutherland DE, Matas AJ, Najarian JS. Pancreatic islet cell transplantation. Surg Clin North Am 1978; 58(2):365-82.
13. Hinshaw DB, Jolley WB, Kaiser JE et al. Islet autotransplantation after pancreatectomy for chronic pancreatitis with a new method of islet preparation. Am J Surg 1981; 142(1):118-22.
14. Farney AC, Najarian JS, Nakhleh RE et al. Autotransplantation of dispersed pancreatic islet tissue combined with total or near-total pancreatectomy for treatment of chronic pancreatitis. Surgery 1991; 110(2):427-37; discussion 37-39.
15. Wahoff DC, Papalois BE, Najarian JS et al. Autologous islet transplantation to prevent diabetes after pancreatic resection. Ann Surg 1995; 222(4):562-75; discussion 75-79.
16. White SA, London NJ, Johnson PR et al. The risks of total pancreatectomy and splenic islet autotransplantation. Cell Transplant 2000; 9(1):19-24.
17. Pyzdrowski KL, Kendall DM, Halter JB et al. Preserved insulin secretion and insulin independence in recipients of islet autografts. N Engl J Med 1992; 327(4):220-26.
18. Robertson RP, Lanz KJ, Sutherland DE et al. Prevention of diabetes for up to 13 years by autoislet transplantation after pancreatectomy for chronic pancreatitis. Diabetes 2001; 50(1):47-50.
19. Linetsky E, Bottino R, Lehmann R et al. Improved human islet isolation using a new enzyme blend, liberase. Diabetes 1997; 46(7):1120-23.
20. Ricordi C, Lacy PE, Finke EH et al. Automated method for isolation of human pancreatic islets. Diabetes 1988; 37(4):413-20.
21. Lakey JR, Warnock GL, Shapiro AM et al. Intraductal collagenase delivery into the human pancreas using syringe loading or controlled perfusion. Cell Transplant 1999; 8(3):285-92.
22. Robertson GS, Chadwick DR, Contractor H et al. The optimization of large-scale density gradient isolation of human islets. Acta Diabetol 1993; 30(2):93-98.
23. Noel J, Rabinovitch A, Olson L et al. A method for large-scale, high-yield isolation of canine pancreatic islets of Langerhans. Metabolism 1982; 31(2):184-87.
24. Ricordi C, Finke EH, Lacy PE. A method for the mass isolation of islets from the adult pig pancreas. Diabetes 1986; 35(6):649-53.
25. Rajotte RV, Warnock GL, Evans MG et al. Isolation of viable islets of Langerhans from collagenase-perfused canine and human pancreata. Transplant Proc 1987; 19(1 Pt 2):918-22.
26. Warnock GL, Cattral MS, Rajotte RV. Normoglycemia after implantation of purified islet cells in dogs. Can J Surg 1988; 31(6):421-26.
27. Warnock GL, Cattral MS, Evans MG et al. Mass isolation of pure canine islets. Transplant Proc 1989; 21(2):3371-72.
28. Secchi A, Socci C, Maffi P et al. Islet transplantation in IDDM patients. Diabetologia 1997; 40(2):225-31.
29. Bretzel RG, Brandhorst D, Brandhorst H et al. Improved survival of intraportal pancreatic islet cell allografts in patients with type-1 diabetes mellitus by refined peritransplant management. J Mol Med 1999; 77(1):140-43.
30. International Islet Transplant Registry Newsletter #8. Giessen, Germany: Justus-Liebig University of Giessen; 1999.
31. Shapiro AM, Lakey JR, Ryan EA et al. Islet transplantation in seven patients with type 1 diabetes mellitus using a glucocorticoid-free immunosuppressive regimen. N Engl J Med 2000; 343(4):230-38.
32. Olefsky JM, Kimmerling G. Effects of glucocorticoids on carbohydrate metabolism. Am J Med Sci 1976; 271(2):202-10.
33. Delaunay F, Khan A, Cintra A et al. Pancreatic beta cells are important targets for the diabetogenic effects of glucocorticoids. J Clin Invest 1997; 100(8):2094-98.
34. Ryan EA, Lakey JR, Rajotte RV et al. Clinical outcomes and insulin secretion after islet transplantation with the Edmonton protocol. Diabetes 2001; 50(4):710-19.
35. Ryan EA, Lakey JRT, Paty BW et al. Successful islet transplantation: Continued insulin reserve provides long-term glycemic control. Diabetes 2002; 51(7):2148-57.
36. Ricordi C, Lakey JR, Hering BJ. Challenges toward standardization of islet isolation technology. Transplant Proc 2001; 33(1-2):1709.
37. Lakey JR, Burridge PW, Shapiro AM. Technical aspects of islet preparation and transplantation. Transpl Int 2003; 16(9):613-32.
38. Watt PC, Mullen Y, Benhamou PY et al. Donor factors affecting successful isolation of islets from the human pancreas. Transplant Proc 1994; 26(2):594-95.

39. Benhamou PY, Watt PC, Mullen Y et al. Human islet isolation in 104 consecutive cases. Factors affecting isolation success. Transplantation 1994; 57(12):1804-10.

40. Lakey JR, Warnock GL, Rajotte RV et al. Variables in organ donors that affect the recovery of human islets of Langerhans. Transplantation 1996; 61(7):1047-53.

41. Lakey JR, Rajotte RV, Warnock GL et al. Cold ischemic tolerance of human pancreas: assessment of islet recovery and in vitro function. Transplant Proc 1994; 26(6):3416.

42. Ketchum RJ, Nicolae M, Jahr H et al. Analysis of donor age and cold ischemia time as factors in cadaveric human islet isolation. Transplant Proc 1994; 26(2):596-97.

43. Jassem W, Fuggle SV, Rela M et al. The role of mitochondria in ischemia/reperfusion injury. Transplantation 2002; 73(4):493-99.

44. Obermaier R, Von Dobschuetz E, Benthues A et al. Exogenous and endogenous nitric oxide donors improve post-ischemic tissue oxygenation in early pancreatic ischemia/reperfusion injury in the rat. Eur Surg Res 2004; 36(4):219-25.

45. Hennige AM, Lembert N, Wahl MA et al. Oxidative stress increases potassium efflux from pancreatic islets by depletion of intracellular calcium stores. Free Radic Res 2000; 33(5):507-16.

46. Kuroda Y, Kawamura T, Suzuki Y et al. A new, simple method for cold storage of the pancreas using perfluorochemical. Transplantation 1988; 46(3):457-60.

47. Tsujimura T, Kuroda Y, Kin T et al. Human islet transplantation from pancreases with prolonged cold ischemia using additional preservation by the two-layer (UW solution/perfluorochemical) cold-storage method. Transplantation 2002; 74(12):1687-91.

48. Tsujimura T, Kuroda Y, Avila JG et al. Resuscitation of the ischemically damaged human pancreas by the two-layer method prior to islet isolation. Transplant Proc 2003; 35(7):2461-62.

49. Ricordi C, Fraker C, Szust J et al. Improved human islet isolation outcome from marginal donors following addition of oxygenated perfluorocarbon to the cold-storage solution. Transplantation 2003; 75(9):1524-27.

50. Fujino Y, Kuroda Y, Suzuki Y et al. Preservation of canine pancreas for 96 hours by a modified two-layer (UW solution/perfluorochemical) cold storage method. Transplantation 1991; 51(5):1133-35.

51. Matsumoto S, Kuroda Y, Fujita H et al. Resuscitation of ischemically damaged pancreas by the two-layer (University of Wisconsin solution/perfluorochemical) mild hypothermic storage method. World J Surg 1996; 20(8):1030-34.

52. Kuroda Y, Fujita H, Matsumoto S et al. Protection of canine pancreatic microvascular endothelium against cold ischemic injury during preservation by the two-layer method. Transplantation 1997; 64(7):948-53.

53. Fujino Y, Suzuki Y, Tsujimura T et al. Possible role of heat shock protein 60 in reducing ischemic-reperfusion injury in canine pancreas grafts after preservation by the two-layer method. Pancreas 2001; 23(4):393-98.

54. Salehi P, Hansen MA, Avila JG et al. Human islet isolation outcomes from pancreata preserved with Histidine-Tryptophan Ketoglutarate versus University of Wisconsin solution. Transplantation 2006; 82(7):983-85.

55. Avila JG, Wang Y, Barbaro B et al. Improved Outcomes in Islet Isolation and Transplantation by the Use of a Novel Hemoglobin-based $O_2$ Carrier. Am J Transplant 2006.

56. Gaber AO, Fraga DW, Callicutt CS et al. Improved in vivo pancreatic islet function after prolonged in vitro islet culture. Transplantation 2001; 72(11):1730-36.

57. Gaber AO, Fraga D. Advances in long-term islet culture: the memphis experience. Cell Biochem Biophys 2004; 40(3 Suppl):49-54.

58. Goss JA, Schock AP, Brunicardi FC et al. Achievement of insulin independence in three consecutive type-1 diabetic patients via pancreatic islet transplantation using islets isolated at a remote islet isolation center. Transplantation 2002; 74(12):1761-66.

59. Hering BJ, Kandaswamy R, Harmon JV et al. Transplantation of cultured islets from two-layer preserved pancreases in type 1 diabetes with anti-CD3 antibody. Am J Transplant 2004; 4(3):390-401.

60. Bottino R, Balamurugan AN, Tse H et al. Response of human islets to isolation stress and the effect of antioxidant treatment. Diabetes 2004; 53(10):2559-68.

61. Ichii H, Wang X, Messinger S et al. Improved human islet isolation using nicotinamide. Am J Transplant 2006; 6(9):2060-68.

62. Korsgren O, Nilsson B, Berne C et al. Current status of clinical islet transplantation. Transplantation 2005; 79(10):1289-93.

63. Moberg L, Johansson H, Lukinius A et al. Production of tissue factor by pancreatic islet cells as a trigger of detrimental thrombotic reactions in clinical islet transplantation. Lancet 2002; 360(9350):2039-45.

64. Johansson H, Lukinius A, Moberg L et al. Tissue factor produced by the endocrine cells of the islets of langerhans is associated with a negative outcome of clinical islet transplantation. Diabetes 2005; 54(6):1755-62.

65. Ricordi C, Strom TB. Clinical islet transplantation: advances and immunological challenges. Nat Rev Immunol 2004; 4(4):259-68.

66. Sehgal SN. Rapamune (RAPA, rapamycin, sirolimus): mechanism of action immunosuppressive effect results from blockade of signal transduction and inhibition of cell cycle progression. Clin Biochem 2006; 39(5):484-89.

67. Hering BJ, Wijkstrom M. Sirolimus and islet transplants. Transplant Proc 2003; 35(3 Suppl):187S-90S.

68. Vu MD, Qi S, Xu D et al. Tacrolimus (FK506) and sirolimus (rapamycin) in combination are not antagonistic but produce extended graft survival in cardiac transplantation in the rat. Transplantation 1997; 64(12):1853-56.

69. Chen H, Qi S, Xu D et al. Combined effect of rapamycin and FK 506 in prolongation of small bowel graft survival in the mouse. Transplant Proc 1998; 30(6):2579-81.

70. Qi S, Xu D, Peng J et al. Effect of tacrolimus (FK506) and sirolimus (rapamycin) mono- and combination therapy in prolongation of renal allograft survival in the monkey. Transplantation 2000; 69(7):1275-83.

71. McAlister VC, Gao Z, Peltekian K et al. Sirolimus-tacrolimus combination immunosuppression. Lancet 2000; 355(9201):376-77.

72. McAlister VC, Peltekian KM, Malatjalian DA et al. Orthotopic liver transplantation using low-dose tacrolimus and sirolimus. Liver Transpl 2001; 7(8):701-8.

73. Soulillou JP, Cantarovich D, Le Mauff B et al. Randomized controlled trial of a monoclonal antibody against the interleukin-2 receptor (33B3.1) as compared with rabbit antithymocyte globulin for prophylaxis against rejection of renal allografts. N Engl J Med 1990; 322(17):1175-82.

74. Kirkman RL, Shapiro ME, Carpenter CB et al. A randomized prospective trial of anti-Tac monoclonal antibody in human renal transplantation. Transplantation 1991; 51(1):107-13.

75. Nashan B, Light S, Hardie IR et al. Reduction of acute renal allograft rejection by daclizumab. Daclizumab Double Therapy Study Group. Transplantation 1999; 67(1):110-15.

76. Vincenti F, Nashan B, Light S. Daclizumab: outcome of phase III trials and mechanism of action. Double Therapy and the Triple Therapy Study Groups. Transplant Proc 1998; 30(5):2155-58.

77. Oberholzer J, Toso C, Triponez F et al. Human islet allotransplantation with Basiliximab in type I diabetic patients with end-stage renal failure. Transplant Proc 2002; 34(3):823-25.

78. Froud T, Ricordi C, Baidal DA et al. Islet transplantation in type 1 diabetes mellitus using cultured islets and steroid-free immunosuppression: miami experience. Am J Transplant 2005; 5(8):2037-46.

79. Hering BJ, Kandaswamy R, Ansite JD et al. Single-donor, marginal-dose islet transplantation in patients with type 1 diabetes. JAMA 2005; 293(7):830-35.

80. Kaplan B, Kirk AD. Tacrolimus and sirolimus: when bad things happen to good drugs. Am J Transplant 2006; 6(7):1501-2.

81. Senior PA, Paty BW, Cockfield SM et al. Proteinuria developing after clinical islet transplantation resolves with sirolimus withdrawal and increased tacrolimus dosing. Am J Transplant 2005; 5(9):2318-23.

82. Senior PA, Zeman M, Paty BW et al. Changes in renal function after clinical islet transplant: a four year observational study. Am J Transplant (In press).

83. Mendez R, Gonwa T, Yang HC et al. A prospective, randomized trial of tacrolimus in combination with sirolimus or mycophenolate mofetil in kidney transplantation: results at 1 year. Transplantation 2005; 80(3):303-9.

84. Kobashigawa JA, Miller LW, Russell SD et al. Tacrolimus with mycophenolate mofetil (MMF) or sirolimus vs. cyclosporine with MMF in cardiac transplant patients: 1-year report. Am J Transplant 2006; 6(6):1377-86.

85. Gallon L, Perico N, Dimitrov BD et al. Long-term renal allograft function on a tacrolimus-based, pred-free maintenance immunosuppression comparing sirolimus vs. MMF. Am J Transplant 2006; 6(7):1617-23.

86. Campbell PM, Salam A, Ryan EA et al. Pretransplant HLA antibodies are Associated with Reduced C-peptide secretion post islet cell transplant. Am J Transplant;In press.

87. Ryan EA, Lakey JR, Paty BW et al. Successful islet transplantation: continued insulin reserve provides long term control. Diabetes 2002; 51(7):2148-57.

88. Robertson RP. Pancreatic islet transplantation for diabetes: successes, limitations and challenges for the future. Mol Genet Metab 2001; 74(1-2):200-5.

89. Ricordi C, Inverardi L, Kenyon NS et al. Requirements for success in clinical islet transplantation. Transplantation 2005; 79(10):1298-300.

90. Lakey JR, Mirbolooki M, Shapiro AM. Current status of clinical islet cell transplantation. Methods Mol Biol 2006; 333:47-104.

91. Bonner-Weir S. Beta-cell turnover: its assessment and implications. Diabetes 2001; 50 Suppl 1:S20-24.

92. Ryan EA, Shandro T, Green K et al. Assessment of the severity of hypoglycemia and glycemic lability in type 1 diabetes subjects undergoing islet transplantation. Diabetes 2004; 53(4):955-62.

93. Paty BW, Senior PA, Lakey JR et al. Assessment of glycemic control after islet transplantation using the continuous glucose monitor in insulin-independent versus insulin-requiring type 1 diabetes subjects. Diabetes Technol Ther 2006; 8(2):165-73.

94. Fiorina P, Folli F, Bertuzzi F et al. Long-term beneficial effect of islet transplantation on diabetic macro-/microangiopathy in type 1 diabetic kidney-transplanted patients. Diabetes Care 2003; 26(4):1129-36.

95. Fiorina P, Gremizzi C, Maffi P et al. Islet transplantation is associated with an improvement of cardiovascular function in type 1 diabetic kidney transplant patients. Diabetes Care 2005; 28(6):1358-65.

96. Fiorina P, Folli F, Zerbini G et al. Islet transplantation is associated with improvement of renal function among uremic patients with type I diabetes mellitus and kidney transplants. J Am Soc Nephrol 2003; 14(8):2150-58.

97. Johnson JA, Kotovych M, Ryan EA et al. Reduced fear of hypoglycemia in successful islet transplantation. Diabetes Care 2004; 27(2):624-25.

98. Poggioli R, Faradji RN, Ponte G et al. Quality of life after islet transplantation. Am J Transplant 2006; 6(2):371-78.

99. Villiger P, Ryan EA, Owen R et al. Prevention of bleeding after islet transplantation: lessons learned from a multivariate analysis of 132 cases at a single institution. Am J Transplant 2005; 5(12):2992-98.

100. Casey JJ, Lakey JR, Ryan EA et al. Portal venous pressure changes after sequential clinical islet transplantation. Transplantation 2002; 74(7):913-15.

101. Ryan EA, Paty BW, Senior PA et al. Risks and side effects of islet transplantation. Curr Diab Rep 2004; 4(4):304-9.

102. Barshes NR, Lee TC, Brunicardi FC et al. Lack of cytomegalovirus transmission after pancreatic islet transplantation. Cell Transplant 2004; 13(7-8):833-38.

103. Markmann JF, Rosen M, Siegelman ES et al. Magnetic resonance-defined periportal steatosis following intraportal islet transplantation: a functional footprint of islet graft survival? Diabetes 2003; 52(7):1591-94.

104. Bhargava R, Senior PA, Ackerman TE et al. Prevalence of hepatic steatosis after islet transplantation and its relation to graft function. Diabetes 2004; 53(5):1311-17.

105. Letavernier E, Pe'raldi MN, Pariente A et al. Proteinuria following a switch from calcineurin inhibitors to sirolimus. Transplantation 2005; 80(9):1198-203.

106. Straathof-Galema L, Wetzels JF, Dijkman HB et al. Sirolimus-associated heavy proteinuria in a renal transplant recipient: evidence for a tubular mechanism. Am J Transplant 2006; 6(2):429-33.

107. Bandello F, Vigano C, Secchi A et al. Effect of pancreas transplantation on diabetic retinopathy: a 20-case report. Diabetologia 1991; 34 Suppl 1:S92-94.

108. Progression of retinopathy with intensive versus conventional treatment in the Diabetes Control and Complications Trial. Diabetes Control and Complications Trial Research Group. Ophthalmology 1995; 102(4):647-61.

109. Early worsening of diabetic retinopathy in the Diabetes Control and Complications Trial. Arch Ophthalmol 1998; 116(7):874-86.

110. Truong W, Hancock WW, Anderson CC et al. Coinhibitory T-cell signaling in islet allograft rejection and tolerance. Cell Transplant 2006; 15(2):105-19.

111. Merani S, Truong WW, Hancock W et al Chemokines and their receptors in islet allograft rejection and as targets for tolerance induction. Cell Transplant 2006; 15(4):295-309.

112. Kenyon NS, Chatzipetrou M, Masetti M et al. Long-term survival and function of intrahepatic islet allografts in rhesus monkeys treated with humanized anti-CD154. Proc Natl Acad Sci USA 1999; 96(14):8132-37.

113. Buhler L, Alwayn IP, Appel JZ et al. Anti-CD154 monoclonal antibody and thromboembolism. Transplantation 2001; 71(3):491.

114. Larsen CP, Pearson TC, Adams AB et al. Rational development of LEA29Y (belatacept), a high-affinity variant of CTLA4-Ig with potent immunosuppressive properties. Am J Transplant 2005; 5(3):443-53.

115. Pearson TC, Trambley J, Odom K et al. Anti-CD40 therapy extends renal allograft survival in rhesus macaques. Transplantation 2002; 74(7):933-40.

116. Adams AB, Shirasugi N, Durham MM et al. Calcineurin inhibitor-free CD28 blockade-based protocol protects allogeneic islets in nonhuman primates. Diabetes 2002; 51(2):265-70.

117. Magliocca JF, Knechtle SJ. The evolving role of alemtuzumab (Campath-1H) for immunosuppressive therapy in organ transplantation. Transpl Int 2006; 19(9):705-14.

118. Bloom DD, Hu H, Fechner JH et al. T-lymphocyte alloresponses of Campath-1H-treated kidney transplant patients. Transplantation 2006; 81(1):81-87.

119. Vathsala A, Ona ET, Tan SY et al. Randomized trial of Alemtuzumab for prevention of graft rejection and preservation of renal function after kidney transplantation. Transplantation 2005; 80(6):765-74.

120. Watson CJ, Bradley JA, Friend PJ et al. Alemtuzumab (CAMPATH 1H) induction therapy in cadaveric kidney transplantation—efficacy and safety at five years. Am J Transplant 2005; 5(6):1347-53.

121. Knechtle SJ, Pirsch JD, Fechner JH et al. Campath-1H induction plus rapamycin monotherapy for renal transplantation: results of a pilot study. Am J Transplant 2003; 3(6):722-30.

122. Barth RN, Janus CA, Lillesand CA et al. Outcomes at 3 years of a prospective pilot study of Campath-1H and sirolimus immunosuppression for renal transplantation. Transpl Int 2006; 19(11):885-92.

123. Nanji SA, Shapiro AM. Islet transplantation in patients with diabetes mellitus: choice of immunosuppression. BioDrugs 2004; 18(5):315-28.

124. Chatenoud L, Thervet E, Primo J et al. Anti-CD3 antibody induces long-term remission of overt autoimmunity in nonobese diabetic mice. Proc Natl Acad Sci USA 1994; 91(1):123-27.

125. Chatenoud L, Primo J, Bach JF. CD3 antibody-induced dominant self tolerance in overtly diabetic NOD mice. J Immunol 1997; 158(6):2947-54.

126. Herold KC, Hagopian W, Auger JA et al. Anti-CD3 monoclonal antibody in new-onset type 1 diabetes mellitus. N Engl J Med 2002; 346(22):1692-98.

127. Wijkstrom M, Kenyon NS, Kirchhof N et al. Islet allograft survival in nonhuman primates immunosuppressed with basiliximab, RAD and FTY720. Transplantation 2004; 77(6):827-35.

128. Soria B, Skoudy A, Martin F. From stem cells to beta cells: new strategies in cell therapy of diabetes mellitus. Diabetologia 2001; 44(4):407-15.

129. Bonner-Weir S, Sharma A. Pancreatic stem cells. J Pathol 2002; 197(4):519-26.

130. Lechner A, Habener JF. Stem/progenitor cells derived from adult tissues: potential for the treatment of diabetes mellitus. Am J Physiol Endocrinol Metab 2003; 284(2):E259-66.

131. Hussain MA, Theise ND. Stem-cell therapy for diabetes mellitus. Lancet 2004; 364(9429):203-5.

132. Otonkoski T, Gao R, Lundin K. Stem cells in the treatment of diabetes. Ann Med 2005; 37(7):513-20.

133. Dufrane D, D'Hoore W, Goebbels RM et al. Parameters favouring successful adult pig islet isolations for xenotransplantation in pig-to-primate models. Xenotransplantation 2006; 13(3):204-14.

134. Rood PP, Cooper DK. Islet xenotransplantation: are we really ready for clinical trials? Am J Transplant 2006; 6(6):1269-74.

135. Baggio LL, Drucker DJ. Therapeutic approaches to preserve islet mass in type 2 diabetes. Annu Rev Med 2006; 57:265-81.

# CHAPTER 30

# Metabolic Indicators of Islet Graft Dysfunction

Raquel N. Faradji, Kathy Monroy, Misha Denham, Camillo Ricordi and Rodolfo Alejandro*

## Abstract

Assessing β-cell mass and function is of great importance in the islet transplant setting but it has been challenging. Although achieving insulin independence has been one of the most important end points of islet transplantation (IT), it is critical that it is associated with good glycemic control. Numerous tests have been suggested for the assessment of graft dysfunction. Unfortunately, there are no effective indicators for the early detection of graft dysfunction in IT patients.[1] In this chapter we discuss the different metabolic approaches for the assessment of graft dysfunction, nevertheless it is important to recognize that there is no consensus on which test is best to study β-cell mass and function in patients after IT. The early detection of graft dysfunction could facilitate the identification of underlying causes and guide in the management of transplanted patients, especially if appropriate anti-rejection therapies become available.

## Introduction

Islet transplantation (IT) under steroid free immunosuppression offers a potential treatment for patients with type 1 diabetes complicated with severe hypoglycemia and hypoglycemia unawareness.[2-6] Despite the achievement of a more stable and physiological glycemic control than with exogenous insulin administration, most patients sustain a decline of islet allograft function over time and insulin independence is gradually lost .[2,5,7,8]

The probability of insulin independence after islet transplantation is approximately 75% at 1 year.[2,7,9] Graft dysfunction requiring the reintroduction of exogenous insulin is seen in approximately 60% of the patients by the second year.[2,7,9] Although many factors may play a role in leading to graft dysfunction and failure (e.g., impaired engraftment, toxicity from the immunosuppressant agents, allo-rejection, recurrence of autoimmunity and β-cell exhaustion), most of the time the events resulting in dysfunction of the transplanted islet mass cannot be identified.

Assessing β-cell functional islet mass is of great importance in the IT setting. Achieving insulin independence is one of the most important end points of IT but it must be associated with good glycemic control. Several tests have been suggested for the assessment of graft dysfunction. Unfortunately, there are no effective indicators for the early detection of graft dysfunction in IT patients.[1] In this chapter we discuss the different metabolic approaches for the assessment of graft dysfunction. Nevertheless it is important to recognize that there is no consensus on which test is best to study β-cell mass and function in patients after IT. The early detection of graft dysfunction could facilitate the identification of underlying causes and guide in the management of transplanted patients, especially if appropriate anti-rejection therapies become available.

## Definitions

Insulin independence is defined as C-peptide positive ($>0.3$ ng/ml) recipients that were able to maintain HbA1c $\leq 6.5\%$, fasting capillary blood glucose levels below 7.8 mmol/L (140 mg/dL) and two hour postprandial glucose levels $\leq 10$ mmol/L (180 mg/dl), respectively without requiring exogenous insulin administration.[10]

Graft dysfunction is defined as C-peptide positive ($>0.3$ ng/ml) recipients that presented with fasting capillary glucose $>7.8$ mmol/L and/or two hour postprandial capillary glucose $>10.0$ mmol/L in three or more occasions in 1 week (or if two sequential laboratory A1c values are $>6.5\%$), requiring the reintroduction of exogenous insulin therapy.[10]

## Ninety-Minute Glucose after Mixed Meal Tolerance Test

There are several metabolic assays that can help ascertain the status of the transplanted islets and can help assess the response of the islets over time. β-cell response to meals is evaluated by the Mixed Meal Tolerance Test (MMTT). This test is being widely used to study the response following therapeutic interventions in new onset type 1 diabetes[11] and to monitor the function of transplanted islets.[12,13] Evaluation of the 90-minute glucose (90 min-Glc) after MMTT is considered a good acute measure of β-cell reserve following islet transplantation.[10,12,13] It also measures the homeostatic ability of the individual to handle a secretagogue load. A 90 min-Glc level $\leq 10.0$ mmol/L (180 mg/dl) is considered a good indicator of graft function,[14] therefore it is a helpful test for post islet transplant long term follow up.

## Acute Insulin Release to Glucose ($AIR_{glu}$) and Acute Insulin Release to Arginine ($AIR_{arg}$)

Plasma insulin concentrations after pulse intravenous injection of glucose (IVGTT) show an early rise that peaks within five minutes and subsides within ten minutes ("first-phase" insulin release, $AIR_{glu}$) and declines towards the prestimulation level.[15] This glucose load stimulates insulin release by increasing intracellular ATP, leading to closure of islet potassium channels, influx of calcium and insulin secretion. First-phase insulin release allows for the comparison of serum insulin responses between subjects, or in the same subject over time. A decrease in first-phase insulin secretion in response to intravenous glucose is an early feature of the β-cell dysfunction typically seen in both type 1[15-17] and type 2 diabetes.[18]

β-cell function can also be assessed through the measurement of plasma insulin concentrations after intravenous stimulation with arginine ($AIR_{arg}$).[19] Arginine is a cationic amino acid that directly leads to depolarization of the β-cell, calcium influx and insulin release, even at normal blood glucose concentrations.

*Corresponding Author: Rodolfo Alejandro—Diabetes Research Institute, 1450 NW 10th Avenue (R-134), Miami, FL 33136 USA.
Email: ralejand@med.miami.edu

*Chronic Allograft Failure: Natural History, Pathogenesis, Diagnosis and Management*, edited by Nasimul Ahsan. ©2008 Landes Bioscience.

In certain clinical situations, when it may be desirable to avoid hyperglycemia, such as the period immediately following islet transplantation $AIR_{arg}$ may be preferred to $AIR_{glu}$.[20] Both $AIR_{glu}$ and $AIR_{arg}$ stimulate insulin release through different mechanisms. The different response to these secretagogues is evident in early type 1 and 2 diabetes, where $AIR_{arg}$ is typically preserved after $AIR_{glu}$ is lost.[17,21,22]

The response to arginine can also be increased when the glucose is clamped at an hyperglycemic level ($AIR_{argmax}$); this is known as the glucose potentiation of arginine induced insulin secretion (GPAIS) and has been used as an in vivo metabolic test of insulin secretory reserve in animals and in humans,[18,20,23-27] but it is a much more cumbersome test than the IVGTT or the Arginine Stimulation test.

In pancreas transplantation[20] $AIR_{argmax}$ was found to be a good measure of insulin secretory reserve. $AIR_{glu}$ and $AIR_{arg}$ had a good correlation with $AIR_{argmax}$, allowing for simpler tests to be performed in a regular basis. Both $AIR_{glu}$ and $AIR_{arg}$ decrease with time as β-cell function decreases, although $AIR_{arg}$ is still present when $AIR_{glu}$ disappears. $AIR_{glu}$ was found to have a good relationship with fasting glucose levels.[20]

In IT there have been results indicating that $AIR_{glu}$ correlates better than $AIR_{arg}$ with measures of glycemic control such as fasting plasma glucose and glucose level at 2 h during Oral Glucose Tolerance test (OGTT), proposing $AIR_{glu}$ as a better clinical measure for this population.[13] There is a significant correlation between $AIR_{arg}$ and $AIR_{glu}$ which proposes that either measure can be used for the assessment of β-cell dysfunction and insulin secretory reserve. Furthermore, both $AIR_{glu}$ and $AIR_{arg}$ have been shown to correlate with islet mass transplanted.[13] A markedly impaired first-phase insulin secretory response to intravenous glucose and a blunted secretory response to intravenous arginine have been described in IT patients with time.[13,28] In the IT setting a poor acute response has been documented and attributed to either loss of islet cell mass or decreased β-cell function. $AIR_{glu}$ has shown a statistically significant correlation with 90 min-Glc after a MMTT[13] and was found to be closely related to glycemic control. We have shown that a decrease in the $AIR_{glu}$ occurs before reintroduction of exogenous insulin therapy and may serve as one of the earliest markers in the deterioration of islet allograft function.[28] In general, it appears that $AIR_{glu}$ is a better measure of β-cell function and glycemic control and $AIR_{arg}$ of β-cell mass.

## β-*Score*

Ryan et al[14] recently proposed the use of the β-score as a measure of clinical success after islet transplantation. The β-score is determined from the fasting plasma glucose, A1c, insulin requirements and the determination of stimulated C-peptide levels 90 minutes after MMTT, at any given time point. The score is obtained by assigning 0, 1 or 2 points to each of the components. This proposed scoring system gives 2 points for normal fasting glucose (<5.5 mmol/L, <100 mg/dL), A1c (<6.1%), stimulated C-peptide (≥0.3 nmol/L, ≥0.9 ng/mL) and absence of insulin or oral hypoglycemic agent use. No points are awarded if the fasting glucose is in the diabetic range (>7 mmol/L, >126 mg/dL), the A1c is >7.0%, C-peptide secretion is absent on stimulation (<0.1nmol/L, <0.3 ng/mL), or daily insulin use is more than 0.24 units/kg. One point was given for intermediate values. The score may range from 0 (no graft function) to a score of 8 being interpreted as having good graft function.[14] The β-score correlated well with the 90 min-Glc after MMTT.[14] The β-score provides a simple clinical scoring system that encompasses glycemic control, diabetes therapy and endogenous insulin secretion it gives an integrated measure of β-cell function.[14]

## C-Peptide Glucose Ratio (CP/G)

The relatively low variability, high reproducibility and close relationship of C-peptide measurements in the systemic circulation to endogenously-secreted insulin in the portal system makes this a suitable assay for monitoring β-cell function over time.[22,29,30] A measurement has been proposed to correct for the dependency of plasma C-peptide levels on glycemic values.[10] Correction of C-peptide (ng/ml) for glucose levels was performed using formula:

$$\text{C-peptide/Glucose Ratio} = \frac{\text{C-peptide (ng/ml)} \times 100}{\text{Glucose (mg/dl)}}$$

In this study the CP/G ratio was used to assess graft dysfunction. The CP/G was found to be a superior diagnostic tool to detect graft dysfunction by predicting a high 90 min-Glc (>10 mmol/L, >180 mg/dL) when compared to C-peptide alone (p = 0.01). CP/G and C-peptide, correlated with the total mass of islets infused when assessed at 7 days and 1 and 3 months posttransplant, suggesting that these values are also useful in the assessment of islet engraftment. CP/G correlated with other measures for β-cell function after islet transplantation such as 90 min-Glc after MMTT and the clinical scoring system β-score (Fig. 1)[10]; suggesting that it represents a valuable surrogate marker of islet graft dysfunction that can be easily monitored in transplanted patients. One of the advantages of the CP/G is that it is a simple test that can be estimated at any time point during the posttransplant period, given the ease of its measurement taken from a single fasting blood sample. It can be used as a parameter of insulin secretion on the basal state and can have an immediate application in the assessment of graft dysfunction after islet transplantation. A high

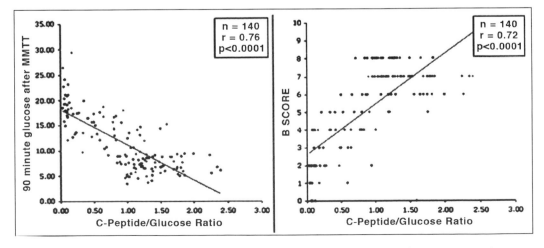

Figure 1. Correlation between C-peptide/Glucose ratio to both 90-minute glucose after MMTT and β-score in 140 tests performed in 22 subjects before and after islet transplantation.

CP/G value provides reassurance that a given patient has adequate islet graft function. A lower value could indicate early graft dysfunction and the requirement of more complicated metabolic and immunologic work up to guide prompt implementation of antirejection and graft rescue therapy.[10] Limitations of the CP/G are that it does not account for the overall patient metabolic state and that it does not correct for the degree of insulin sensitivity or resistance. Similar measures have been proposed by other groups.[31-33]

## Insulin Secretion, Insulin Sensitivity and the Disposition Index

The measurement of insulin sensitivity ($S_I$) is a surrogate marker of the ability of insulin to inhibit glucose production at the level of the liver and enhance glucose utilization; it is inversely proportional to insulin resistance.[34] Insulin resistance is the prominent defect in patients who develop type 2 diabetes but is also a recognized side effect of certain immunosuppressive agents. Calcineurin inhibitors are well known to cause this effect.[35] More recently, sirolimus, which is commonly used in IT, has also demonstrated a deleterious effect on insulin sensitivity.[36] The gold standard for the measurement of insulin sensitivity is the euglycemic hyperinsulinemic glucose clamp[1,37]; however this test is technically difficult to perform, more invasive and may pose greater risk to the patient. Consequently measures that correlate with these results have been sought; these include the homeostasis model assessment (HOMA), calculated from a fasting insulin level,[38] or the insulin sensitivity, insulin secretion and disposition index obtained during the frequently sampled intravenous glucose tolerance (FSIGT)[39-41,42] or the oral glucose tolerance test described by Cobelli et al.[43]

These models have been shown to correlate well with the 'clamp' standard in normal controls, as well as some type 2 diabetes models. Data in IT is lacking. Understanding the effect of insulin sensitivity on insulin secretory dynamics posttransplant is important because insulin resistance imposes an increased demand on β-cell function to maintain the same level of glycemia.[44] For this reason the Clinical Islet Transplantation Consortium decided to include the insulin modified FSIGT as one of the tests to be performed in islet transplant recipients. The acute insulin response to glucose ($AIR_{glu}$), insulin sensitivity ($S_I$) and disposition index (DI) can be obtained from the FSIGT test. The disposition index ($DI = AIR_{glu} \cdot S_I$) provides a composite measure of β-cell function, since it relates the effect of insulin sensitivity ($S_I$) on first-phase insulin secretion ($AIR_{glu}$).

## Continuous Glucose Monitoring System

The Continuous Glucose Monitoring System (CGMS) has become a very helpful tool in the metabolic evaluation of patients with type 1 diabetes treated with IT.[8,45] Furthermore, CGMS has been used to evaluate the metabolic control in IT patients, in patients who achieved insulin independence it provides a display of the glycemic stability obtained with IT as compared to pretransplant conditions.[8,46] CGMS has been proposed as an early indicator of graft dysfunction.[8,47] A recent study at our institution[47] found that there is a significant increase in glucose variability and percentage of time spent in hyperglycemia >7.8 mmol/L (%HGT >7.8,) at 2 to 3 months before graft dysfunction when compared to 5 to 9 months before graft dysfunction. Suggesting that these values obtained from CGMS recordings may be used in IT as early indicators of graft dysfunction. It is important to note that the significant difference seen for %HGT was seen with a threshold for hyperglycemia set at 7.8 mmol/L (140 mg/dL). The same could not be detected by setting the threshold at 10 mmol/L (180 mg/dL) by CGMS (%HGT >10) in these patients. These data also showed that the data obtained from CGMS recordings is a better tool to detect early graft dysfunction when compared to capillary blood glucose measurements at the same threshold of 7.8 mmol/L.[47]

## Immunological and Other Markers to Assess Graft Dysfunction

Dr. Norma Kenyon's group at our Center has previously reported that elevation in Granzyme B (GB) may be associated with graft dysfunction in nonhuman primates[48] and clinical islet transplant recipients.[49] Elevated and sustained cytotoxic lymphocyte gene (CLG) expression correlated with eventual onset of hyperglycemia and reintroduction of exogenous insulin therapy.[49] Since infection can occasionally lead to elevation of CLG expression it is important to identify additional markers of immune activation and islet destruction. There are still no reliable markers of the immune status of islet transplant recipients that could lead to rescue interventions before graft loss. Currently, immunosuppressive drug levels and clinical status guide immunosuppressive drug dosing. Novel tools to detect ATP levels of stimulated CD4+ lymphocytes (Immuknow, Cylex, FDA approved)[50] have been recently proposed to help identifying subjects at risk of infection or graft rejection in solid organ recipients.[51] It has been recently shown that after islet transplantation varying levels of insulin mRNA can be detected in the peripheral blood and that this may correlate with graft dysfunction presumably due to beta cell apoptosis.[52,53] Stimulated CD4+ ATP levels and insulin mRNA levels (by real-time quantitative RT-PCR amplification) may be of assistance in monitoring the islet graft and the immune status of islet transplant recipients and complement the CLG expression panel. β-cell loss has various underlying processes and it is important to recognize and discriminate between immune and non-immune causes of graft dysfunction.

## Conclusions

In order to determine early graft dysfunction a complimentary and global approach should be done between the tests previously discussed; which should be correlated between each other and with other immunological measures. If correlations exist, we could generate useful indicators to initiate appropriate rescue therapies, thereby helping to limit graft loss and thus preventing metabolic deterioration in the long term, so we can ultimately reach the desirable goal of providing insulin independence with sustained metabolic control in islet transplant recipients.

From the metabolic perspective, the simplest tests available to monitor graft dysfunction are the CP/G ratio and the 90 min-Glc after a MMTT. The insulin modified FSIGT will undoubtedly provide significantly more information, since it will assess both first phase insulin release to glucose ($AIR_{glu}$) as well as insulin sensitivity ($S_I$).

In the future, it is possible that a combination of metabolic, imaging and immunological studies will be able to accurately detect early graft dysfunction and therefore guide in to appropriate rescue therapy.

### *Acknowledgements*

Study supported by grants from the National Institutes of Health/NIDDK (R01 DK52802), NCRR (GCRCMO1RR16587; U42RR016603), Juvenile Diabetes Foundation International-JDRFI (4-2004-361), State of Florida and Diabetes Research Institute Foundation.

### References

1. Ferrannini E, Mari A. Beta cell function and its relation to insulin action in humans: a critical appraisal. Diabetologia 2004; 47(5):943-56.
2. Froud T, Ricordi C, Baidal DA et al. Islet transplantation in type 1 diabetes mellitus using cultured islets and steroid-free immunosuppression: Miami experience. Am J Transplant 2005; 5(8):2037-46.
3. Shapiro AM, Lakey JR, Ryan EA et al. Islet transplantation in seven patients with type 1 diabetes mellitus using a glucocorticoid-free immunosuppressive regimen. N Engl J Med 2000; 343(4):230-38.
4. Shapiro AM, Ricordi C, Hering B. Edmonton's islet success has indeed been replicated elsewhere. Lancet 2003; 362(9391):1242.

5. Shapiro AM, Ricordi C, Hering BJ et al. International trial of the Edmonton protocol for islet transplantation. N Engl J Med 2006; 355(13):1318-30.

6. Ricordi C. Islet transplantation: a brave new world. Diabetes 2003; 52(7):1595-603.

7. Ryan EA, Paty BW, Senior PA et al. Five-year follow-up after clinical islet transplantation. Diabete 2005; 54(7):2060-69.

8. Geiger MC, Ferreira JV, Hafiz MM et al. Evaluation of metabolic control using a continuous subcutaneous glucose monitoring system in patients with type 1 diabetes mellitus who achieved insulin independence after islet cell transplantation. Cell Transplant 2005; 14(2-3):77-84.

9. Close N, Alejandro R, Hering B et al cond annual analysis of the collaborative islet transplant registry. Transplant Proc 2007; 39(1):179-82.

10. Faradji RN, Monroy K, Messinger S et al. Simple measures to monitor beta-cell mass and assess islet graft dysfunction. Am J Transplant 2007; 7(2):303-8.

11. Greenbaum CJ, Harrison LC. Guidelines for intervention trials in subjects with newly diagnosed type 1 diabetes. Diabetes 2003; 52(5):1059-65.

12. Alejandro R, Lehmann R, Ricordi C et al. Long-term function (6 years) of islet allografts in type 1 diabetes. Diabetes 1997; 46(12):1983-89.

13. Ryan EA, Lakey JR, Paty BW et al. Successful islet transplantation: continued insulin reserve provides long-term glycemic control. Diabetes 2002; 51(7):2148-57.

14. Ryan EA LJ, Paty BW et al. Beta-Score—An assessment of beta-cell function after islet transplantation. Diabetes Care 2005; (28):343-47.

15. Ferrannini E, Pilo A. Pattern of insulin delivery after intravenous glucose injection in man and its relation to plasma glucose disappearance. J Clin Invest 1979; 64(1):243-54.

16. Brunzell JD, Robertson RP, Lerner RL et al. Relationships between fasting plasma glucose levels and insulin secretion during intravenous glucose tolerance tests. J Clin Endocrinol Metab 1976; 42(2):222-29.

17. Bardet S, Rohmer V, Maugendre D et al. Acute insulin response to intravenous glucose, glucagon and arginine in some subjects at risk for type 1 (insulin-dependent) diabetes mellitus. Diabetologia 1991; 34(9):648-54.

18. Ward WK, Bolgiano DC, McKnight B et al. Diminished B-cell secretory capacity in patients with noninsulin-dependent diabetes mellitus. J Clin Invest 1984; 74(4):1318-28.

19. Blachier F, Mourtada A, Sener A et al. Stimulus-secretion coupling of arginine-induced insulin release. Uptake of metabolized and nonmetabolized cationic amino acids by pancreatic islets. Endocrinology 1989; 124(1):134-41.

20. Robertson RP. Consequences on beta-cell function and reserve after long-term pancreas transplantation. Diabetes 2004; 53(3):633-44.

21. Ganda OP, Srikanta S, Brink SJ et al. Differential sensitivity to beta-cell secretagogues in "early," type I diabetes mellitus. Diabetes 1984; 33(6):516-21.

22. Palmer JP, Benson JW, Walter RM et al. Arginine-stimulated acute phase of insulin and glucagon secretion in diabetic subjects. J Clin Invest 1976; 58(3):565-70.

23. Seaquist E, Robertson R. Effects of hemipancreatectomy on pancreatic alpha and beta-cell function in healthy human donors. J Clin Invest 1992; 89:1761-66.

24. Teuscher A, Seaquist E, Robertson R. Diminished insulin secretory reserve in diabetic pancreas transplant and nondiabetic kidney transplant recipients. Diabetes 1994; 43:593-98.

25. Ward WK, Wallum BJ, Beard JC et al. Reduction of glycemic potentiation. Sensitive indicator of beta-cell loss in partially pancreatectomized dogs. Diabetes 1988; 37(6):723-29.

26. McCulloch DK, Koerker DJ, Kahn SE et al. Correlations of in vivo beta-cell function tests with beta-cell mass and pancreatic insulin content in streptozocin-administered baboons. Diabetes 1991; 40(6):673-79.

27. Larsen MO, Rolin B, Wilken M et al. Measurements of insulin secretory capacity and glucose tolerance to predict pancreatic beta-cell mass in vivo in the nicotinamide/streptozotocin Gottingen minipig, a model of moderate insulin deficiency and diabetes. Diabetes 2003; 52(1):118-23.

28. Baidal D, Froud T, Hafiz M et al. Acute Insulin Release to Intravenous Glucose is the Best Indicator of Islet Allograft Dysfunction. Transplantation 2004; 78(2)(Supplement 1):177.

29. Eaton RP, Allen RC, Schade DS et al. Prehepatic insulin production in man: kinetic analysis using peripheral connecting peptide behavior. J Clin Endocrinol Metab 1980; 51(3):520-28.

30. Van Cauter E, Mestrez F, Sturis J et al. Estimation of insulin secretion rates from C-peptide levels. Comparison of individual and standard kinetic parameters for C-peptide clearance. Diabetes 1992; 41(3):368-77.

31. Matsumoto S, Yamada Y, Okitsu T et al. Simple evaluation of engraftment by secretory unit of islet transplant objects for living donor and cadaveric donor fresh or cultured islet transplantation. Transplant Proc 2005; 37(8):3435-37.

32. Yamada Y, Fukuda K, Fujimoto S et al. SUIT, secretory units of islets in transplantation: An index for therapeutic management of islet transplanted patients and its application to type 2 diabetes. Diabetes Res Clin Pract 2006.

33. Secchi A, Pontiroli AE, Traeger J et al. A method for early detection of graft failure in pancreas transplantation. Transplantation 1983; 35(4):344-48.

34. Kahn SE, Prigeon RL, McCulloch DK et al. Quantification of the relationship between insulin sensitivity and beta-cell function in human subjects. Evidence for a hyperbolic function. Diabetes 1993; 42(11):1663-72.

35. Heisel O, Heisel R, Balshaw R et al. New onset diabetes mellitus in patients receiving calcineurin inhibitors: a systematic review and meta-analysis. Am J Transplant 2004; 4(4):583-95.

36. Teutonico A, Schena PF, Di Paolo S. Glucose metabolism in renal transplant recipients: effect of calcineurin inhibitor withdrawal and conversion to sirolimus. J Am Soc Nephrol 2005; 16(10):3128-35.

37. DeFronzo RA, Tobin JD, Andres R. Glucose clamp technique: a method for quantifying insulin secretion and resistance. Am J Physiol 1979; 237(3):E214-23.

38. Matthews DR, Hosker JP, Rudenski AS et al. Homeostasis model assessment: insulin resistance and beta-cell function from fasting plasma glucose and insulin concentrations in man. Diabetologia 1985; 28(7):412-19.

39. Pacini G, Bergman RN. MINMOD: a computer program to calculate insulin sensitivity and pancreatic responsivity from the frequently sampled intravenous glucose tolerance test. Comput Methods Programs Biomed 1986; 23(2):113-22.

40. Pacini G, Tonolo G, Sambataro M et al. Insulin sensitivity and glucose effectiveness: minimal model analysis of regular and insulin-modified FSIGT. Am J Physiol 1998; 274(4 Pt 1):E592-99.

41. Bergman RN, Prager R, Volund A et al. Equivalence of the insulin sensitivity index in man derived by the minimal model method and the euglycemic glucose clamp. J Clin Invest 1987; 79(3):790-800.

42. Bergman RN, Ader M, Huecking K et al. Accurate assessment of beta-cell function: the hyperbolic correction. Diabetes 2002; 51(Suppl 1):S212-20.

43. Dalla Man C, Caumo A, Basu R et al. Minimal model estimation of glucose absorption and insulin sensitivity from oral test: validation with a tracer method. Am J Physiol Endocrinol Metab 2004; 287(4):E637-43.

44. Rickels MR, Naji A, Teff KL. Insulin sensitivity, glucose effectiveness and free fatty acid dynamics after human islet transplantation for type 1 diabetes. J Clin Endocrinol Metab 2006; 91(6):2138-44.

45. Kessler L, Passemard R, Oberholzer J et al. Reduction of blood glucose variability in type 1 diabetic patients treated by pancreatic islet transplantation. Diabetes Care 2002; 25(12):2256-62.

46. Paty B, Ryan E, Shapiro A et al. Intrahepatic islet transplantation in type 1 diabetic patients does not restore hypoglycemic hormonal counterregulation or symptom recognition after insulin independence. Diabetes 2002; 51:3428-34.

47. Faradji RN, Monroy K, Riefkohl A et al. Continuous glucose monitoring system for early detection of graft dysfunction in allogenic islet transplant recipients. Transplant Proc 2006; 38(10):3274-76.

48. Han D, Xu X, Pastori RL et al. Elevation of cytotoxic lymphocyte gene expression is predictive of islet allograft rejection in nonhuman primates. Diabetes 2002; 51(3):562-66.

49. Han D, Xu X, Baidal D et al. Assessment of Cytotoxic Lymphocyte Gene Expression in the Peripheral Blood of Human Islet Allograft Recipients: Elevation Precedes Clinical Evidence of Rejection. Diabetes 2004; 53(9):2281-90.

50. Zeevi A, Britz JA, Bentlejewski CA et al. Monitoring immune function during tacrolimus tapering in small bowel transplant recipients. Transpl Immunol 2005; 15(1):17-24.

51. Cadillo-Chavez R, de Echegaray S, Santiago-Delpin EA et al. Assessing the risk of infection and rejection in Hispanic renal transplant recipients by means of an adenosine triphosphate release assay. Transplant Proc 2006; 38(3):918-20.

52. Ritz-Laser B, Oberholzer J, Toso C et al. Molecular detection of circulating beta-cells after islet transplantation. Diabetes 2002; 51(3):557-61.

53. Berney T, Mamin A, James Shapiro AM et al. Detection of insulin mRNA in the peripheral blood after human islet transplantion predicts deterioration of metabolic control. Am J Transplant 2006; 6(7):1704-11.

# CHAPTER 31

# Pancreas and Islet Allograft Failure

Patrick G. Dean,* Yogish Kudva and Mark D. Stegall

## Introduction

The treatment of diabetes mellitus is aimed at improving glycemic control. Establishing relatively tight control using exogenous insulin has been shown to reduce the ophthal-mologic, neurologic and renal complications of diabetes mellitus.[1] However, achieving a glycosylated hemoglobin (Hgb A1C) level low enough to reduce secondary complications (a level of approximately 7.0%) is associated with an increased risk of severe, life-threatening hypoglycemia. Both whole organ pancreas transplantation and isolated islet transplantation offer the possibility of even better glycemic control (average Hgb A1C <6.0%) with the addition of counter-regulatory feedback hormones that avoid life-threatening hypoglycemia.

The primary setting in which a whole-organ pancreas transplant is used to treat diabetes is that of the relatively young diabetic patient who requires a kidney transplant for end-stage renal disease secondary to diabetic nephropathy. In this setting, the pancreas transplant is either transplanted as a simultaneous pancreas-kidney transplant (SPK) or a pancreas after kidney transplant (PAK). In addition, some nonuremic diabetic patients choose transplantation to alleviate life-threatening hypoglycemic unawareness or other diabetes-associated complications. In this setting, either solitary pancreas transplant alone (PTA) or islet transplantation may be employed.

Achieving success in pancreas and islet transplantation has been difficult. While measurable progress has been made over the past decade, significant problems persist-both technical and immunologic. This chapter reviews the current status of this rather problematic area of transplantation in an attempt to shed light on what is needed for continued progress.

## Whole Organ Pancreas Transplantation

The first successful pancreas transplant was performed in conjunction with a kidney transplant at the University of Minnesota in 1966.[2] Since this initial experience, the technical aspects of the procurement and transplant operations have been refined and more powerful immunosuppressive regimens have been developed.[3] These advances have significantly improved pancreatic graft survival and have led to more widespread application of pancreas transplantation as treatment for diabetes mellitus. There are currently 2,513 candidates awaiting SPK transplantation and another 1,754 candidates awaiting either PAK or PTA transplants. Thus, a combined 4,267 candidates are awaiting pancreas transplants in the United States. In contrast, only 2,879 and 2,964 candidates are awaiting heart or lung transplantation, respectively.[4] As of 2004, more than 23,000 pancreas transplants had been performed worldwide, according to data from the United Network for Organ Sharing (UNOS) and the International Pancreas Transplant Registry (IPTR).[5] In the US, the number of pancreas transplants has increased over the past decade with a significant increase in PAK and PTA allografts. In 2004, 880 SPK, 419 PAK and 185 PTA transplants were performed in the United States.[6]

In general, the success of pancreas transplantation is lower than that of kidney transplantation with PTA and PAK allografts experiencing lower survival rates compared to SPK transplants or deceased donor kidney transplants. The one-year graft survival (defined as freedom from insulin therapy and including death with function) for those pancreata transplanted in the United States during 2000-2004 was 85% for SPK (n = 3,841), 78% for PAK (n = 1,109) and 76% for PTA (n = 429).[5] By comparison, the one-year solitary deceased donor kidney transplant graft survival was greater than 90% in the same era (see Table 1). Patient survival rates for recipients of all three types of transplants were greater than 95% at one year and 88% at three years.

While the early graft and patient survival rates continue to improve, late graft loss remains common and long-term outcomes are suboptimal. The ten-year graft survival rates (including death with function) for recipients transplanted in 1992-1993 were 46% for SPK, 17% for PAK and 17% for PTA, highlighting a clear distinction between the higher graft survival of

### Table 1. Graft survival

| | Graft Survival | | |
|---|---|---|---|
| | 1 Year (Tx 2002-2003) | 3 Year (Tx 2000-2003) | 5 Year (Tx 1998-2003) |
| SPK-Pancreas | 86% | 79% | 71% |
| SPK-Kidney | 92% | 85% | 77% |
| PAK | 78% | 67% | 57% |
| PTA | 78% | 64% | 57% |
| Islet | 80% | n/a | 8% |
| NonECD DDK | 91% | 80% | 69% |
| ECD DDK | 81% | 67% | 52% |
| LDK | 95% | 88% | 80% |
| Liver | 82% | 73% | 67% |
| Heart | 87% | 79% | 72% |
| Lung | 82% | 63% | 46% |

Abbreviations: SPK = simultaneous pancreas-kidney transplant; PAK = pancreas after kidney transplant; PTA = pancreas transplant alone; ECD = expanded criteria donor; DDK = deceased donor kidney. With the exception of islet transplants, graft survival rates were obtained from the OPTN/SRTR 2005 Annual Report.[5] Data regarding islet transplant recipients were obtained from the most recent published report of the Edmonton group.[41] Graft survival for pancreas and islet transplants was defined as insulin independence.

*Corresponding Author: Patrick G. Dean—Division of Transplantation Surgery, Mayo Clinic College of Medicine, 200 First Street SW, Rochester, MN 55905, USA. Email: dean.patrick2@mayo.edu

*Chronic Allograft Failure: Natural History, Pathogenesis, Diagnosis and Management*, edited by Nasimul Ahsan. ©2008 Landes Bioscience.

SPK vs. either PTA or PAK transplants.[5] By comparison, the estimated ten-year graft survival rates for heart transplants and solitary deceased donor kidney transplants performed during this same era are 55% and 51%, respectively.[7,8] The ten-year patient survival rates for those transplanted in 1992-1993 were 69% for SPK, 40% for PAK and 74% for PTA.[5]

Over the past decade, data regarding some of the major causes of whole-organ pancreas transplant failure have emerged. We will examine each in turn while emphasizing the current gaps in our knowledge.

## Causes of Pancreas Allograft Failure

### Technical Failure

Technical failure accounts for a greater proportion of pancreas graft losses compared to other solid organ transplants.[9] Technical failure includes the following causes of graft loss: thrombosis, hemorrhage, infection, pancreatitis and exocrine leak (see Table 2). In a review of 937 pancreas transplants performed at a single center from 1994-2003, Humar et al reported an overall technical failure rate of 13.1% (123/937). Thrombosis accounted for 52% (n = 64; 6.8% of the overall number of transplants) of the 123 graft losses and pancreatitis led to 20.3% (n = 25) of technical failures. The technical failure rate was highest for SPK transplants (15.3%) and was 12.3% for PAK transplants and 11.4% for PTA transplants. Multivariate analysis showed recipient BMI greater than 30 kg/m$^2$ (relative risk [RR] = 2.42, $P$ = 0.0003), donor BMI >30 kg/m$^2$ (RR = 1.66, $P$ = 0.06), preservation time >24 hours (RR = 1.87, $P$ = 0.04) and donor cause of death other than trauma (RR = 1.58, P = 0.04) to be risk factors for technical failure.[10] Sollinger et al reported a thrombosis rate of only 1% (n = 5) in a review of 500 SPK transplants performed during 1985-1997. In this series, bleeding (n = 4), infection (n = 3) and pancreatitis (n = 3) were also infrequent causes of pancreas graft loss. The authors attributed the low incidence of technical failure to standardized donor management to reduce pancreatic edema, pancreas procurement by a highly trained and experienced team, the use of Viaspan® solution for preservation and the avoidance of a venous extension graft for the portal vein.[11] Data from UNOS and the IPTR suggest that the overall rate of technical failure has decreased with time. In the period including 1988-1989, the technical failure rates were 12% for SPK, 12% for PAK and 24% for PTA transplants. For the 2002-2003 period, the technical failure rates for the 3 transplant types were 6%, 8% and 7%, respectively (P = 0.0001).[5] We recently reviewed our own experience with 194 solitary (121 PAK and 73 PTA) pancreas transplants performed during 1998-2005. In this series, the technical failure within the first 30 days was 11% (n = 21). Of these 21 grafts, 18 were lost to thrombosis, 1 was lost to acute pancreatitis and another was removed due to profound hemorrhage.[12]

Thus, technical failure remains a major cause of graft loss. In general, few clinical studies have objectively examined the causes of thrombosis or the efficacy of any preventative treatment. We believe that this complication merits closer study.

## Immunologic Graft Loss

### Acute Rejection

It is in the area of the prevention of rejection that the most significant progress has been made. Acute cellular rejection rates for pancreas allografts have historically been as high as 70-80%.[13-15] However, immunosuppressive regimens have advanced to the point where many centers currently employ antibody induction therapy. With such regimens, the incidence of acute rejection in the first six months following PTA can be reduced to 8% using rabbit antithymocyte globulin induction, compared to 60% in recipients receiving OKT3 and 50% for those receiving daclizumab.[16] In a recent series comparing the use of antithymocyte globulin to alemtuzumab induction in a corticosteroid avoidance protocol, Kaufman et al report 24-month actual rejection rates of 8.2% and 5.3% ($P$ = 0.70) for patients receiving alemtuzumab and antithymocyte globulin, respectively.[17] In addition, the use of mycophenolate mofetil has led to lower rates of acute pancreas rejection in several trials when compared to azathioprine.[15,18-20] According to the most recent UNOS/IPTR report, graft losses due to immunologic causes in the first year for those transplants performed during 2000-2004 have declined to 10% for PTA, 8% for PAK and 2% for SPK.[5] In an analysis of more than 1,000 pancreas transplants at a single center, Sutherland et al reported an immunologic graft loss rate during the first year of 9% for SPK transplants performed in 1998-2000. The immunologic graft loss rates during the same period were 10% for PAK transplants and 9% for PTA transplants.[21] In the large University of Wisconsin experience, rejection caused 45 of the 102 graft losses reported (9% overall) in 500 SPK transplants. The majority of these grafts were lost to recalcitrant acute cellular rejection.[11] Our analysis of 194 solitary pancreas transplants mentioned above showed that only 4% of PAK allografts and 3% of PTA allografts failed due to acute rejection.[12]

Early studies analyzing the use of portal venous drainage suggested lower acute rejection rates and thus improved graft survival. The mechanism is unclear, but may involve clearance of alloantigen by the liver. However, with improvements in immunosuppression leading to lower acute cellular rejection rates in systemically-drained pancreas transplants, the impetus for portal drainage has decreased and this technique is used in a minority of transplants. The results of several studies comparing portal venous drainage to systemic venous drainage are conflicting regarding the immunologic and graft survival benefits of the former technique.[22-25] Further experimental and clinical investigations will be needed to define the mechanisms of portal tolerance and the benefit to pancreas allografts.

In contrast to kidney transplants, humoral rejection of pancreas transplants has been described only recently.[26] It is probable in such an immunologically stimulatory organ such as a pancreas that humoral rejection is much more prevalent than is commonly realized clinically. It is likely that more detailed studies in this area may improve both early and late outcomes.

### Chronic Rejection

With fewer grafts being lost to technical causes and acute rejection, chronic rejection could emerge as a more common cause of graft loss. The histologic features of this entity have been described and include progressive fibrosis and concentric fibroproliferative endarteritis (see Fig. 1). The fibrosis is accompanied by mononuclear cell infiltration and progressive disappearance of the acinar cells.[27] In addition, recent and organized thrombosis of arteries and veins may also be found in the setting of chronic rejection.[28] However, few studies have described the exact incidence of chronic rejection in pancreas transplantation. Registry data suggest that rejection accounts for 20-30% of graft failures after the first year, but no clear distinction is made between late acute rejection and chronic rejection.[5] In a retrospective analysis of 914 pancreas transplants performed during 1994-2002, Humar et al reported that 70% of grafts

**Table 2. Causes of pancreas allograft failure**

|  | Type of Pancreas Transplant | | |
| --- | --- | --- | --- |
|  | **SPK** | **PAK** | **PTA** |
| Technical | 1-15% | 8-12% | 7-11% |
| Acute Rejection | 2-9% | 4-10% | 3-10% |
| Chronic Rejection | 4-9% | 2% | 3% |
| Death with Function | 19-50% | 12-30% | 11-15% |

Ranges shown in this table summarize the findings of the multiple studies referenced in the text.

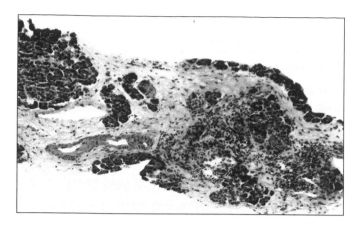

Figure 1. Chronic rejection of a pancreas allograft characterized by atrophy of the acinar parenchyma and mononuclear inflammatory infiltrates (100X magnification).

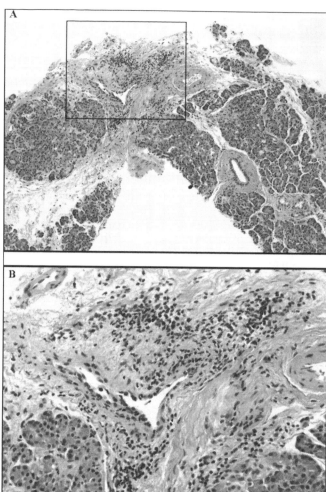

Figure 2. Drachenberg grade II rejection of a pancreas allograft characterized by septal inflammation and mild venous endotheliitis at 100X (A) and 200 X (B) magnifications.

were functioning with a mean follow-up of 39 months. Chronic rejection was the second most common cause of graft loss (after technical failure), accounting for the loss of 8.8% of the grafts (80/914). Multivariate analysis showed the risk factors for chronic rejection to be an isolated pancreas transplant (RR = 3.02, $P$ = 0.002), prior acute rejection episodes (RR = 4.41, $P$ < 0.0001), CMV infection (RR = 0.001, $P$ = 0.001), retransplant (RR = 2.27, $P$ = 0.04) and degree of mismatch at the HLA B locus (RR = 1.68, $P$ = 0.04).[29] Our recent analysis found that only 2% of PAK grafts and 3% PTA grafts were lost to chronic rejection.[12]

In order to prevent graft losses in the future, further efforts to monitor the immunologic status of pancreas allografts are needed. Some centers have advocated the measurement of urinary amylase as a method to monitor the function of bladder-drained allografts.[30] Our own center has employed percutaneous protocol surveillance biopsies at 4 months, 1 year and 5 years after solitary pancreas transplantation to detect subclinical pancreas rejection.[31,32] We recently reviewed all protocol biopsies (n = 150) performed in 84 recipients of 88 pancreas allografts over a 2 year period. The most common finding on these biopsies was Drachenberg grade 0 (n = 113) followed by Drachenberg grade II (minimal) rejection (n = 18; see Fig. 2). Of those patients with grade II findings, 15 subsequently underwent a total of 25 repeat biopsies after receiving no additional immunosuppressive treatment. These biopsies showed that in 13 of the patients, the histologic lesion either improved or failed to progress, while in 2 patients the histologic lesion worsened.[32] Additional studies employing protocol biopsies may provide additional insight into the histologic fate of pancreatic allografts.

No studies currently exist regarding the effects of immunosuppressive medication noncompliance on graft survival. Likewise, the development of alloantibodies, a factor that appears related to late renal allograft loss, has not been studied in pancreas allograft recipients systematically.[33] Lastly, the contribution of recurrent autoimmunity to pancreas allograft failure needs to be further studied.[34,35]

### Death with a Functioning Graft

As the rates of technical failure and immunologic graft loss have decreased, death with a functioning allograft has emerged as a major cause of pancreas graft loss. This trend mirrors that seen in kidney transplantation, where death with function now accounts for up to 43% of graft losses in the first ten years following transplant.[36] Relatively few studies have addressed specifically the issue of death with function in pancreas transplant recipients. However, in the large University of Wisconsin series of SPK transplants, death with a functioning graft accounted for 27 (27%) of 102 graft losses.[11] A recent analysis of 914 pancreas transplants at the University of Minnesota reported death with function as the cause of 19% (51/271) of graft loss.[29] For transplants reported to UNOS/IPTR during the years 2000-2004, death with a functioning

graft represented the cause of graft loss in almost 50% of SPK grafts failing after one year. Death with function accounted for 30% and 15%, respectively, of PAK and PTA graft failures.[5] Of 194 solitary transplants performed in 180 recipients at our institution between 1998-2005, 22 grafts (11%) were lost to death with function.[12]

Any discussion of death following pancreas transplantation must address the survival of transplant recipients relative to comparable patients not receiving a transplant. For SPK transplants, a clear survival benefit has been shown for recipients less than 50 years of age compared to patients with type 1 diabetes mellitus who received a solitary kidney transplant or remained on dialysis.[37] However, a controversy over the survival benefits afforded by solitary pancreas transplants (either PAK or PTA) has recently developed. In fact, a recent analysis by Ventstrom et al reported a survival **disadvantage** for PAK and PTA recipients. The overall relative risk of mortality (compared with patients awaiting the same procedure) over the 4 years of follow-up was 1.57 for PTA and 1.42 for PAK recipients.[38] Stimulated by this report, Gruessner et al performed a similar analysis after excluding patients with multiple listings at different transplant centers, including a more recent patient cohort and extending the follow-up period. This report showed that the mortality at four years for solitary pancreas transplant recipients is not higher than patients on the waiting list.[39] Additional studies with longer follow-up of patients transplanted during the current era may indeed show a survival benefit for solitary pancreas transplantation. However, based upon currently available data, only equivalence in survival can be ascertained.

## *Hyperglycemia with a Functioning Allograft*

Most analyses of pancreas transplantation define graft failure as the resumption of exogenous insulin use. However, this definition may overestimate the true incidence of graft loss, as some patients may have a functioning allograft (defined by high levels of circulating C-peptide) but may still require pharmacologic treatment for hyperglycemia. In an effort to better define this recipient population, we recently analyzed the outcomes of 88 patients undergoing pancreas transplant at our institution during 2001-2002 with a median follow-up of 25 months (range 18-30 months). Of the 88 grafts, 4 were lost to immediate thrombosis, 2 to late thrombosis, and 5 to patient death with function. No graft was lost to acute or chronic rejection. Of the remaining 77 patients, 57 (74%) remained euglycemic, 6 (8%) required pharmacotherapy for less than 1 month and 14 (18%) required pharmacotherapy longer than 1 month. All of those patients requiring pharmacotherapy had detectable circulating C-peptide levels. Factors predicting hyperglycemia included: pretransplant insulin dose, BMI and rejection ($p < 0.0001$, $p = 0.0005$ and $p = 0.0002$, respectively). Hyperglycemia developed in all (6/6) patients requiring >100 U insulin/day pretransplant, in 83% (10/12) requiring >75 U/day and in only 15% (10/65) of those requiring ≤75 U/day ($p < 0.0001$). The median pretransplant hemoglobin A1c for patients developing hyperglycemia was 8.0% (6.7-8.7) compared to 6.6% (5.5-7.4) at 1-year following transplant ($p = 0.0294$).[40] Similar results have been reported by other centers.[41,42] Therefore, the requirement for exogenous insulin may not necessarily imply graft failure, but does represent an imbalance between the insulin requirement of the recipient and the supply produced by the allograft. Diabetes mellitus with detectable C-peptide concentrations after pancreas transplant most likely results from a combination of factors, including a reduction in insulin secretory reserve capacity and increased insulin resistance, inadequate beta cell mass and the diabetogenic effects of tacrolimus and corticosteroids.[43-47] Modifications in patient selection, immunosuppressive regimens and post-transplant care may help decrease the incidence of this condition.

## Islet Transplantation

Islet transplantation offers the possibility of a less-invasive means of establishing glucose control in patients with diabetes mellitus. However, despite a recent increase in enthusiasm for the procedure, long-term results remain relatively poor. One of the largest experiences is that of the University of Alberta, Edmonton where 65 patients have received an islet transplant as of November 1, 2004. Forty-four patients completed the islet transplant as defined by insulin independence, and three further patients received >16,000 islet equivalents (IE)/kg but remained on insulin and were deemed complete. Those who became insulin independent received a total of 799,912 +/- 30,220 IE (11,910 +/- 469 IE/kg). Five subjects became insulin independent after one transplant. Fifty-two patients had two transplants, and 11 patients had three transplants. For these recipients, the median duration of insulin independence was 15 months (interquartile range 6.2-25.5 months).[48]

Using this approach, the 1-year insulin-free success of islet transplantation approaches 80%. This success, matched by a small number of other groups, is much improved over the success rate of previous eras, which hovered around 10%.[49-51] Unfortunately, the late rate of insulin independence has been much lower. Of the few patients who have reached 5-year follow-up, only 8% are insulin independent. However, the majority (approximately 80%) have C-peptide present at 5 years following islet transplant.[48] In contrast, the most-recent 5 year graft survival for PTA in the US was 57%.[5]

The most recent annual report of the Collaborative Islet Transplant Registry (CITR) contains outcomes data for 112 islet alone recipients who have completed at least one follow-up evaluation after their last infusion of islets. Of these 112, 55 (49%) were insulin independent,

39 (35%) were insulin dependent with detectable C-peptide, and 15 (13%) experienced graft failure. The insulin status of three recipients was unknown. At 1-year following the last islet infusion, 57% of the recipients receiving one infusion were insulin independent, 63% receiving two infusions were insulin independent and 47% were insulin independent after three infusions.[52]

The causes of late islet allograft loss are unclear. Recent reports have suggested that the presence of alloantibody either pretransplant or its development following transplant is associated with decreased islet allograft survival.[53,54] These data suggest that late immunologic graft loss may be more common in islet transplant recipients who are maintained on lower levels of immunosuppressive medicines compared to other organ allograft recipients.

## Conclusion

Pancreas transplantation continues to evolve as a treatment for diabetes mellitus. Advances in procurement techniques, the transplant operation and the post-operative management of the recipients have decreased the incidence of technical failure. Improved immunosuppressive regimens have reduced acute rejection rates to the point where this condition rarely causes graft loss. With fewer grafts being lost early, chronic rejection and death with a functioning graft have emerged as major causes of pancreas loss. To decrease the rates of chronic rejection, additional improvements in immunosuppressive regimens and immunologic monitoring will be required. The occurrence of death with function may be reducible by improved recipient selection, risk factor modification, and more thorough post-transplant follow-up.

It appears that the performing a pancreas transplant in selected diabetic patients with end-stage renal disease (either SPK or PAK) is supported by current outcomes. However, the utility of performing either PTA or islet transplantation in nonuremic diabetics is more controversial. We suggest that the major limitations to the wider application of these procedures are not only the risks of the procedures, but also their relative poor long-term success in rendering the recipient insulin-free. Currently, one can argue that PTA maintains an edge over islet transplantation in 5 year success, but that even PTA has an unacceptably high rate of graft loss and is associated with numerous complications, including worsening of native renal function.[55,56] However, the fact that nonuremic diabetics with severe life-threatening hypoglycemia have few options underscores the need for continued innovations to improve the outcomes of transplantation as a treatment for diabetes. Hopefully, more detailed studies of the causes of pancreas and islet allograft failure will continue the recent improvements in this area and will prove transplantation to be a viable option for the treatment of diabetes.

## References

1. DCCT Research Group. The effect of intensive treatment of diabetes on the development and progression of long-term complications in insulin-dependent diabetes mellitus. N Engl J Med 1993; 329(14):977-86.
2. Kelly WD, Lillehei RC, Merkel FK et al. Allotransplantation of the pancreas and duodenum along with the kidney in diabetic nephropathy. Surgery 1967; 61(6):827-37.
3. Odorico JS, Sollinger HW. Technical and immunosuppressive advances in transplantation for insulin-dependent diabetes mellitus. World J Surg 2002; 26(2):194-211.
4. Based upon OPTN data as of August 1, 2006. Available at www.OPTN.org.
5. Gruessner AC, Sutherland DER. Pancreas transplant outcomes for United States (US) and nonUS cases as reported to the United Network for Organ Sharing (UNOS) and the International Pancreas Transplant Registry (IPTR) as of June 2004. Clin Transplant 2005; 19(4):433-55.
6. U.S. Department of Health and Human Services. 2005 Annual Report of the U.S. Organ Procurement and Transplantation Network and the Scientific Registry of Transplant Recipients: Transplant Data 1995-2004. In: Rockville MD, ed. Health Resources and Services Administration, Healthcare Systems Bureau, Division of Transplantation (http://www.hrsa.gov/).

7. Taylor DO, Edwards LB, Boucek MM et al. Registry of the international society for heart and lung transplantation: Twenty-Second Official Adult Heart Transplant Report-2005. J Heart Lung Transplant 2005; 24(8):945-55.
8. Cecka JM. The OPTN/UNOS renal transplant registry. Clin Transpl 2004; 1-16.
9. Reddy KS, Stratta RJ, Shokouh-Amiri MH et al. Surgical complications after pancreas transplantation with portal-enteric drainage. J Am Coll Surg 1999; 189(3):305-13.
10. Humar A, Ramcharan T, Kandaswamy R et al. Technical failures after pancreas transplants: Why grafts fail and the risk factors-a multivariate analysis. Transplantation 2004; 78(8):1188-92.
11. Sollinger HW, Odorico JS, Knechtle SJ et al. Experience with 500 simultaneous pancreas-kidney transplants. Ann Surg 1998; 228(3):284-96.
12. Ramos EJ, Kudva YC, Larson TS et al. Improving long term survival of solitary pancreas transplants. World Transplant Congress 2006. (Abstract.)
13. Rayhill SC, D'Alessandro AM, Odorico JS et al. Simultaneous pancreas-kidney transplantation and living related donor renal transplantation in patients with diabetes: Is there a difference in survival? Ann Surg 2000; 231(3):417-23.
14. Tesi RJ, Henry ML, Elkhammas EA et al. The frequency of rejection episodes after combined kidney-pancreas transplant-the impact on graft survival. Transplantation 1994; 58(4):424-30.
15. Odorico JS, Pirsch JK, Knechtle SJ et al. A study comparing mycophenolate mofetil to azathioprine in simultaneous pancreas-kidney transplantation. Transplantation 1998; 66(12):1751-59.
16. Stegall MD, Kim DY, Prieto M et al. Thymoglobulin induction decreases rejection in solitary pancreas transplantation. Transplantation 2001; 72(10):1671-75.
17. Kaufman DB, Leventhal JR, Gallon LG et al. Alemtuzumab induction and prednisone-free maintenance immunotherapy in simultaneous pancreas-kidney transplantation comparison with rabbit antithymocyte globulin induction - Long-term results. Am J Transplant 2006; 6(2):331-39.
18. Stegall MD, Simon M, Wachs ME et al. Mycophenolate mofetil decreases rejection in simultaneous kidney-pancreas transplantation when combined with tacrolimus or cyclosporine. Transplantation 1997; 64(12):1695-700.
19. Gruessner RWG, Sutherland DER, Drangstveit MB et al. Mycophenolate mofetil in pancreas transplantation. Transplantation 1998; 66(3):318-23.
20. Merion RM, Henry ML, Melzer JS et al. Randomized, prospective trial of mycophenolate mofetil versus azathioprine for prevention of acute renal allograft rejection after simultaneous kidney-pancreas transplantation. Transplantation 2000; 70(1):105-11.
21. Sutherland DER, Gruessner RWG, Dunn DL et al. Lessons learned from more than 1,000 pancreas transplants at a single institution. Ann Surg 2001; 233(4):463-501.
22. Philosophe B, Farney AC, Schweitzer EJ et al. Superiority of portal venous drainage over systemic venous drainage in pancreas transplantation: A retrospective study. Ann Surg 2001; 234(5):689-96.
23. Gaber AO, Shokouh-Amiri MH, Hathaway DK et al. Results of pancreas transplantation with portal venous and enteric drainage. Ann Surg 1995; 221(6):613-24.
24. Stratta RJ, Gaber AO, Hosein Shokouh-Amiri MH et al. A prospective comparison of systemic-bladder versus portal-enteric drainage in vascularized pancreas transplantation. Surgery 2000; 127(2):217-26.
25. Petruzzo P, Da Silva M, Feitosa LC et al. Simultaneous pancreas-kidney transplantation: Portal versus systemic venous drainage of the pancreas allografts. Clin Transplant 2000; 14(4):287-91.
26. Melcher ML, Olson JL, Baxter-Lowe LA et al. Antibody-mediated rejection of a pancreas allograft. Am J Transplant 2006; 6(2):423-28.
27. Papadimitriou JC, Drachenberg CB, Klassen DK et al. Histological grading of chronic pancreas allograft rejection/graft sclerosis. Am J Transplant 2003; 3(5):599-605.
28. Drachenberg CB, Papadimitriou JC, Farney A et al. Pancreas transplantation: The histologic morphology of graft loss. Transplantation 2001; 71(12):1784-91.
29. Humar A, Khwaja K, Ramcharan T et al. Chronic rejection: The next major challenge for pancreas transplant recipients. Transplantation 2003; 76(6):918-23.
30. Sutherland DE, Gruessner R, Gillingham K et al. A single institution's experience with solitary pancreas transplantation: A multivariate analysis of factors leading to improved outcome. Clin Transpl 1991; 141-52.
31. Atwell TD, Gorman B, Larson TS et al. Pancreas transplants: Experience with 232 percutaneous US-guided biopsy procedures in 88 patients. Radiology 2004; 231(3):845-49.
32. Casey ET, Smyrk TC, Burgart LJ et al. Outcome of untreated grade II rejection on solitary pancreas allograft biopsy specimens. Transplantation 2005; 79(12):1717-22.
33. Lee PC, Terasaki PI, Takemoto SK et al. All chronic rejection failures of kidney transplants were preceded by the development of HLA antibodies. Transplantation 2002; 74(8):1192-94.
34. Eisenbarth GS, Stegall M. Islet and pancreatic transplantation—autoimmunity and alloimmunity. N Engl J Med 1996; 335(12):888-90.
35. Sutherland DE, Sibley R, Xu XZ et al. Twin-to-twin pancreas transplantation: Reversal and reenactment of the pathogenesis of type I diabetes. Trans Assoc Am Physicians 1984; 97:80-87.
36. Ojo AO, Hanson JA, Wolfe RA et al. Long-term survival in renal transplant recipients with graft function. Kidney Int 2000; 57(1):307-13.
37. Ojo AO, Meier-Kriesche HU, Hanson JA et al. The impact of simultaneous pancreas-kidney transplantation on long-term patient surival. Transplantation 2001; 71(1):82-90.
38. Venstrom JM, McBride MA, Rother KI et al. Survival after pancreas transplant in patients with diabetes and preserved kidney function. JAMA 2003; 290(21):2817-23.
39. Gruessner RWG, Sutherland DER, Gruessner AC. Mortality assessment for pancreas transplants. Am J Transplant 2004; 4(12):2018-26.
40. Dean PG, Kudva YC, Larson TS et al. Diabetes mellitus following pancreas transplantation. Am J Transplant 2008; 8(1):175-82.
41. Smith JL, Hunsicker LG, Yuh WT et al. Appearance of type II diabetes mellitus in type I diabetic recipients of pancreas allografts. Transplantation 1989; 47(2):304-11.
42. Jones JW, Mizrahi SS, Bentley FR. Type II diabetes after combined kidney and pancreas transplantation for type I diabetes mellitus and end-stage renal disease. Clin Transplant 1996; 10(6 Pt 1):574-75.
43. Smets YF, van der Pijl JW, Frolich M et al. Insulin secretion and sensitivity after simultaneous pancreas-kidney transplantation estimated by continuous infusion of glucose with model assessment. Transplantation 2000; 69(7):1322-27.
44. Christiansen E, Tibell A, Volund A et al. Pancreatic endocrine function in recipients of segmental and whole pancreas transplantation. J Clin Endocrinol Metab 1996; 81(11):3972-79.
45. Ueki M, Yasunami Y, Ina K et al. Diabetogenic effects of FK506 on renal subcapsular islet isografts in rats. Diabetes Res Clin Pract 1993; 20(1):11-19.
46. Paty BW, Harmon JS, Marsh CL et al. Inhibitory effects of immunosuppressive drugs on insulin secretion from HIT-T15 cells and Wistar rat islets. Transplantation 2002; 73(3):353-57.
47. Reynolds RM, Walker BR. Human insulin resistance: The role of glucocorticoids. Diabetes Obes Metab 2003; 5(1):5-12.
48. Ryan EA, Paty BW, Senior PA et al. Five-year follow-up after clinical islet transplantation. Diabetes 2005; 54(7):2060-69.
49. Markmann JF, Deng S, Huang X et al. Insulin independence following isolated islet transplantation and single islet infusions. Ann Surg 2003; 237(6):741-50.
50. Hering BJ, Kandaswamy R, Ansite JD et al. Single-donor, marginal-dose islet transplantation in patients with type 1 diabetes. JAMA 2005; 293(7):830-35.
51. Froud T, Ricordi C, Baidal DA et al. Islet transplantation in type 1 diabetes mellitus using cultured islets and steroid-free immunosuppression: Miami experience. Am J Transplant 2005; 5(8):2037-46.
52. Annual Report of the Collaborative Islet Transplant Registry. In: Rockville MD, ed. The EMMES Corporation, 2005, (Available at ⟨https://web.emmes.com/study/isl/reports/reports.htm⟩).
53. Campbell P, Salam A, Ryan EA et al. Pretransplant HLA antibodies are associated with reduced graft survival after clinical islet transplantation. Am J Transplant 2007; 7(5):1242-48.
54. Campbell P, Al-Saif F, Halpin A et al. Recipients of islet cell transplants are at risk of broad sensitization after failure of the islet transplant. Presented at the 2006 World Transplant Congress, 2006.
55. Mazur MJ, Rea DJ, Griffin MD et al. Decline in native renal function early after bladder-drained pancreas transplantation alone. Transplantation 2004; 77(6):844-49.
56. Gruessner RWG, Sutherland DER, Najarian JS et al. Solitary pancreas transplantation for nonuremic patients with labile insulin-dependent diabetes mellitus. Transplantation 1997; 64(11):1572-77.

# CHAPTER 32

# Chronic Pancreas Allograft Failure

### Elizabeth K. Gross and Rainer W.G. Gruessner*

## Abstract

A pancreas transplant is the only treatment of diabetes mellitus that establishes long-term insulin independence. As of December 31, 2006, about 20,000 pancreas transplants had been performed in the United States, with 1-year graft survival rates of >80% and patient survival rates of >95%. Results have significantly improved over the last decade, primarily thanks to a marked reduction in technical and immunologic failure. The goal of a pancreas transplant is not only to normalize the recipient's glucose metabolism but also to halt or reverse the progression of secondary diabetic complications (e.g., nephropathy, neuropathy, retinopathy) and to improve quality of life. Almost 47% of all pancreas allografts are now functioning at 10 years posttransplant; chronic rejection remains the main cause of late pancreas allograft failure. Herein, we discuss the diagnosis and histopathologic findings of, as well as risk factors for, chronic rejection, including options to decrease its incidence.

## Introduction

Insulin independence in a type 1 diabetic patient was first achieved on December 17, 1966, when William Kelly and Richard Lillehei transplanted a duct-ligated segmental pancreas graft simultaneously with a kidney from a cadaver donor into a 28-year-old uremic woman at the University of Minnesota.[1] With the introduction of cyclosporine (CSA) in the 1980s and tacrolimus in the 1990s, the number of pancreas transplants worldwide has steadily increased (Fig. 1). As of December 31, 2006, a total of 20,014 pancreas transplants in the U.S. and 6,548 pancreas transplants outside the U.S. have been reported to the International Pancreas Transplant Registry (IPTR), which is maintained by the Division of Transplantation at the University of Minnesota.[2]

Pancreas transplants are being performed in three categories: a simultaneous pancreas and kidney transplant (SPK) for uremic diabetic patients; a pancreas after (a previous) kidney transplant (PAK) for posturemic diabetic patients; and a pancreas transplant alone (PTA) for nonuremic patients with brittle diabetes mellitus. Although the SPK category comprises the largest group (about 75%), the number of solitary pancreas transplants (PAK, PTA) has significantly increased over the last decade. The mortality risk not only for SPK but also for PAK and PTA recipients is lower at 1 year posttransplant than the risk of remaining on the waiting list and not undergoing a transplant.[3]

Over the last 40 years, the results of pancreas transplants have dramatically improved, mainly because of two developments: (1) advances in surgical techniques, which have reduced technical failure (TF) rates, and (2) the introduction of new immunosuppressive (IS) agents, in particular tacrolimus, mycophenolate mofetil (MMF), and rapamycin, which have reduced immunologic failure (IF) rates. A recent IPTR analysis for U.S. cases showed patient survival rates of >95% at 1 year posttransplant in each of the three recipient categories and 1-year graft survival rates of 85% for SPK, 78% for PAK, and 77% for PTA recipients.[2]

As a consequence of these improved results, the number of pancreas recipients with 10-year graft function and beyond is steadily increasing (and now represents almost 47% of all pancreas transplants). These recipients enjoy long-term normoglycemic metabolism without the need for multiple daily plasma glucose measurements and insulin injections. They also benefit from the positive effects of long-term pancreas graft function on the secondary complications of diabetes. An increasing body of literature has shown that long-term pancreas graft function not only improves quality of life but also markedly reduces the morbidity and mortality traditionally associated with type 1 diabetes mellitus and its secondary complications. Recipients with >10-year graft function are also a constant reminder that a pancreas transplant, despite numerous improvements in the treatment of diabetes mellitus, currently remains the only treatment that consistently normalizes glucose metabolism long-term (which, except for only a few recipients, has not yet been achieved with islet transplants).[3,4]

Despite these improvements, pancreas grafts continue to be lost at various time points posttransplant (although at a much lower rate now than in the past). The most common reasons for graft loss are rejection (acute or chronic), TF, and death with a functioning graft (DWFG).[2] Although most graft losses occur within the first year posttransplant, the rate of graft failure after the first year is not insignificant, especially in the solitary (PAK, PTA) categories. Chronic pancreas allograft failure is defined as a continuous loss in pancreatic exocrine and/or graft function over time, eventually resulting in the need for insulin resumption.

To further improve the results of pancreas transplants, it is imperative to study the causes of chronic allograft failure. This chapter overviews the most recent IPTR data and offers suggestions for reducing the incidence of chronic graft failure in the future.

## Causes of Chronic Pancreas Allograft Failure

### Early

The most common cause of early graft loss (≤3 months posttransplant) is TF, which refers to graft loss secondary to graft thrombosis, bleeding, infection, leakage, graft pancreatitis, or other causes. About 65% of graft failures within the first 3 months posttransplant are due to TF, as compared with <10% after 1 year posttransplant. In 2002 and 2003, according to IPTR data, the graft failure rate within 3 months posttransplant was 6.4% for SPK, 7.7% for PAK, and 7.4% for PTA recipients.

*Corresponding Author: Rainer W.G. Gruessner—Department of Surgery, University of Arizona, 1501 N. Campbell Avenue, Tuscon, AZ 85724-5066, USA. Email: rgruessner@surgery.arizona.edu

*Chronic Allograft Failure: Natural History, Pathogenesis, Diagnosis and Management*, edited by Nasimul Ahsan. ©2008 Landes Bioscience.

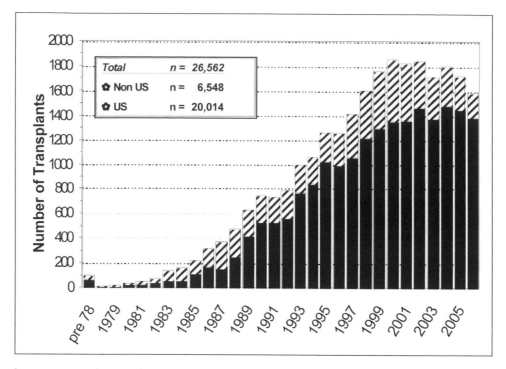

Figure 1. Number of pancreas transplants performed per year in the US and in non-US countries (modified from Gruessner AC, Sutherland DE. Clin Transplant 2005; 19(4):433-55.[2]).

IF as a result of irreversible rejection is the second most common cause of early graft loss (≤3 months posttransplant), accounting for about 12% of all graft losses overall. IF is also the most common cause of graft loss after 3 months posttransplant, accounting for about 50% of all graft losses overall. By recipient category, IF is the most common cause of overall graft loss for PAK and PTA recipients. However, for SPK recipients, TF is the most common cause of graft loss ≤6 months posttransplant and DWFG is the most common thereafter. The most common causes of DWFG are cardiocerebrovascular events and infections; much less common are posttransplant lymphoproliferative disorder (PTLD), trauma, suicide, and graft-versus-host disease (GVHD).

Overall, TF, IF, and DWFG are responsible for over 90% of all graft failures. A small proportion (3% to 5% in each of the three categories) of pancreas recipients never reach insulin independence posttransplant; their outcome is defined as primary nonfunction (Fig. 2).

## Late

TF is an uncommon cause of late graft failure (>1 year posttransplant). When TF occurs after the first year, it is most frequently caused by late leakages of the bladder or of the intestinal anastomosis, graft pancreatitis, and vascular complications (e.g., mycotic pseudoaneurysms,

arterioenteric or arteriocystic fistulas). Timely diagnosis and treatment of late TF decreases the associated morbidity and mortality. The incidence of TF after the first year posttransplant ranges from 4% (SPK) to 10% (PTA). Late thrombosis is usually not a TF, but rather the result of dehydration or rejection (causing decreased intraparenchymal flow or increased vascular resistance).[4]

DWFG as a cause of late graft failure is most commonly noted in SPK recipients, accounting for about 40% of all graft failures after the first posttransplant year, as compared with only 23% in PAK and 14% of PTA recipients. Frequently, DWFG is the result of diabetic micro- and/or macroangiopathy or systemic infections (fungal, viral, bacterial). Close posttransplant monitoring, including regular cardiac reevaluations and avoidance of overimmunosuppression, are key factors in reducing the incidence of DWFG.

IF due to either acute rejection (AR) or chronic rejection (CR) is the most common cause of late graft failure. AR accounts for the majority of cases of IF from 7 to 12 months posttransplant. CR accounts for most cases after the first year posttransplant. AR and CR have also been suspected to be instigating factors in many cases of late graft thrombosis and pancreatitis, because of increased microvascular intragraft resistance.[4]

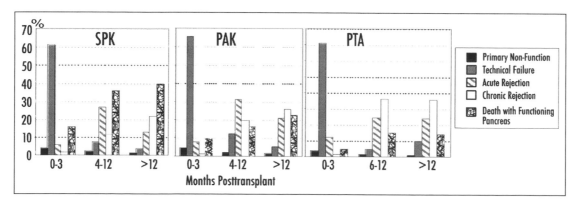

Figure 2. Causes of pancreas allograft loss by time after transplant for primary deceased donor (DD) pancreas transplants, 2000-2004 (modified from Gruessner AC, Sutherland DE. Clin Transplant 2005; 19(4):433-55.[2]).

Reversible AR episodes are a predisposing factor for the development of irreversible graft loss from AR or CR.[5] Most reversible AR episodes occur within the first year posttransplant, but they are not uncommon thereafter in the solitary transplant (PAK, PTA) categories. The mean time interval between transplant and the first reversible AR episode is shortest for PTA and longest for SPK recipients. Likewise, the overall number of reversible AR episodes is highest for PTA and lowest for SPK recipients.[6] Our multivariate analysis at the University of Minnesota of risk factors for AR showed that poor HLA matching at the B locus (in all three recipient categories) and at the A locus (only in the PAK and SPK categories) had a negative impact. In contrast, in all three recipient categories, HLA class II antigens had no impact on the incidence of reversible AR episodes. For recipients ≤45 (vs. >45) years, the incidence of reversible AR episodes was significantly higher, indicating a more vigorous immune response in younger recipients. A previous cytomegalovirus (CMV) infection also increased the incidence of reversible AR episodes for PTA and PAK (but not SPK) recipients. This finding might be the result of underimmunosuppression immediately after successful CMV treatment (Table 1).[4]

Of note, variables that did not affect the incidence of reversible AR episodes included duct drainage type (bladder vs. enteric), donor age (≤45 vs. >45 years), panel-reactive antibody (PRA) levels (>20% vs. ≤20%), and number of transplant (retransplant vs. primary). A difference in reversible AR episodes by gender or race has not been clearly identified. Single-center series have shown an increased risk in blacks, as compared with white and Hispanic populations, while other series have shown no significant difference (14-16).[7-9] Because few risk factors predispose to the occurrence of reversible AR episodes and because many of the aforementioned risk factors cannot be modified (e.g., recipient age), optimizing IS regimens with the primary goal of reducing AR episodes is a prevalent theme in the literature (7-13).[10-16]

CR plays the leading role in late graft loss. Understanding the pathophysiology of and risk factors for CR is critical to improving outcomes. Surprisingly, few reports in the literature address risk factors for CR. Our initial multivariate University of Minnesota analysis identified the following risk factors for CR: a previous reversible AR episode, a solitary pancreas transplant, a CMV infection episode, a retransplant,

and a 1- or 2-antigen mismatch (MM) at the HLA-B locus. Risk factors that did not predispose recipients to CR included a PRA level >20%, recipient age ≤45, a 1- or 2-antigen MM at the HLA-A locus, a 1- or 2-antigen MM at the HLA-DR locus, use of MMF, and use of sirolimus. This initial analysis included all pancreas recipients who underwent a transplant at our institution from June 1994 through December 2002 (n = 914).[5]

A more recent, modified analysis of the University of Minnesota series included only recipients of a technically successful pancreas transplant who had been followed up for at least 3.5 years (n = 659). This analysis identified the following risk factors for CR (Table 2): a reversible AR episode ≤6 months posttransplant (PAK, PTA), a reversible AR episode >6 months posttransplant (SPK, PTA), and a total MM at the HLA-B locus (PTA only). The use of nondepleting antibody induction (PAK), recipient age >45 (PTA), and bladder drainage (PTA) had protective effects. Factors that did not affect CR risk included a PRA level >30%, a total MM at the HLA-A or DR locus, and a CMV infection (Table 2).[4]

According to IPTR analyses, poor HLA matching at the HLA-B locus has consistently been identified as a risk factor for graft loss from irreversible AR and CR, in particular for solitary transplant (PAK, PTA) recipients. Except for a total 4-antigen HLA class I match, or for a total 4-antigen HLA-B MM, we have found no impact of matching on the rates of short- or long-term immunologic graft loss.[2] A number of single-center studies have analyzed the importance of class I and class II HLA matching; most have identified a 1- or 2-antigen class I MM as a risk factor for rejection (1, 6, 18).[17-19] However, other single-center studies have identified a 1- or 2-antigen class II MM as a risk factor. For example, the Ohio State University group identified posttransplant detectable MHC class II-reactive alloantibodies as an independent risk factor for CR in SPK recipients.[20]

According to IPTR data, HLA matching appears to have the greatest impact on the PTA and the least impact on the SPK category. Assessing the impact of HLA matching on the PAK category is more difficult, because so-called "shared mismatches" (i.e., the previous kidney donor's antigens) must also be considered. In contrast to the SPK and PTA recipients, the impact of HLA matching can be studied in PAK recipients for up to 12 antigens. If a PAK recipient has a functioning kidney graft, kidney donor antigens that are not identical with the recipient antigens are regarded as "permissible" antigens.

Aside from HLA matching, other factors also have an impact on the incidence of pancreas graft loss from rejection. As mentioned above, rejection is more common in younger pancreas recipients (age <18 years), who have a more competent immune system, and less common in older recipients (age ≥45 years). Of note, rejection with subsequent graft loss because of noncompliance is negligible in pancreas recipients, as compared with other solid-organ recipients. The reason may be that insulin-dependent diabetics are particularly motivated to avoid recurrent insulin dependency or that they perceive replacement of insulin with IS medications as a favorable tradeoff.

Reducing reversible AR episodes and thereby reducing pancreas graft loss from CR diminishes the need for a retransplant or for steroid and antibody rescue treatment (all of which increase the risk of opportunistic infections and posttransplant malignancies). According to the IPTR, bladder drainage in solitary pancreas transplants continues to provide a graft survival advantage. It allows for early detection of reversible AR episodes via urinary amylase monitoring (hypoamylasuria precedes hyperglycemia by 3 to 7 days). Yet some single-center studies have also suggested a decreased incidence of graft failure from rejection with enteric drainage.[21,22]

Although CR is by far the most common cause of chronic pancreas graft failure, recurrence of autoimmune isletitis has also been reported in a small number of cases as a process that can lead to graft failure. Histologically, this process resembles primary autoimmune isletitis seen

**Table 1. Risk factors for acute pancreas allograft rejection for technically successful transplants, 1994-2002**

|  | PTA (n = 130) | PAK (n = 278) | SPK (n = 222) |
|---|---|---|---|
| HLA-A MMs: 0-1 (vs. 2) | – | + | + |
| HLA-B MMs: 0-1 (vs. 2) | + | + | + |
| HLA-DR MMs: 0-1 (vs. 2) | – | – | – |
| age ≤45 yrs | + + | + + | + + |
| CMV infection | + + | + | |

+ + : p > 0.001 and p ≤ 0.01
+ : p > 0.01 and p ≤ 0.1
– : p > 0.1
MMs: Mismatches
*Variables with no impact in all 3 categories:*
  Donor age ≤45 yrs (vs. >45 yrs)
  Bladder (vs. enteric) drainage
  PRA >20% (vs. ≤20%)
  Pancreas Retransplant
*Variable with no impact in the PAK and SPK categories:*
  Kidney Retransplant

Reproduced with permission from: Gruessner RWG and Sutherland DER. Transplantation of the Pancreas. 2004;[4] with kind permission of Springer Science+Business Media.

**Table 2. Risk factors for chronic pancreas allograft rejection for technically successful transplants, 1994-2003**

|  | SPK | PAK | PTA |
|---|---|---|---|
| Bladder vs. enteric drainage | – | – | ++(0.19) |
| age 10-30 yrs | – | – | – |
| age >45 yrs | +(0.27)* | – | +(0.39) |
| PRA >30% | – | – | – |
| total MM A locus | – | – | – |
| total MM B locus | – | – | +(2.8) |
| total MM DR locus | – | – | – |
| early AR | – | +++(7.6) | ++(2.8) |
| late AR | +(3.0) | – | ++(2.4) |
| CMV infection | – | – | +(2.4) |
| nondepleting antibody | – | +(0.5) | – |

*P Value(Hazard Ratio)
+++: p < 0.001
++ : p > 0.001 and p < 0.01
+ : p > 0.01 and p < 0.1
– : p > 0.1

in patients with preclinical type 1 diabetes mellitus. It is clearly distinct from rejection. Autoimmune recurrence has been noted posttransplant in non- or underimmunosuppressed HLA-identical pancreas recipients. HLA-nonidentical recipients have also been identified, though rarely, as having autoimmune recurrence despite adequate immunosuppression. Work is being done to identify serologic markers of recurrent autoimmunity.

### Impact of IS Therapy

The effect of currently available IS drugs on preventing chronic pancreas graft failure has not been studied in a prospective randomized fashion. It is generally accepted that prevention of reversible AR episodes diminishes the risk of graft loss from CR. For that reason, induction therapy is used with greater frequency in pancreas transplant recipients, as compared with any other solid-organ recipients. In fact, about 90% of solitary pancreas recipients (PAK, PTA) but only about 75% of SPK recipients are placed on induction therapy. According to IPTR data, the use of depleting antibody induction therapy (e.g., ATGAM, Thymoglobulin, Campath) is most prevalent in PAK and PTA recipients (because of their high risk of rejection); the use of depleting vs. nondepleting (e.g., Zenapax, Simulect) antibody induction therapy is nearly equivalent in SPK recipients. Aside from reducing the risk of AR episodes, induction therapy obviates the need for, or minimizes the amount of, calcineurin-inhibitor doses in the immediate posttransplant period. It also protects kidney function (whether native or graft).

In contrast to induction agents, which are usually given over only 5 to 7 days immediately posttransplant, most maintenance IS agents have adverse effects on the recipient's glucose (and/or lipid) metabolism. Tacrolimus, the most frequently used calcineurin inhibitor in pancreas recipients today, has been associated with suppressed insulin production at the transcriptional level, resulting in reduced insulin secretion and glucose intolerance. Although the diabetic effects of tacrolimus are usually not noted in pancreas recipients with full graft function, impaired glucose metabolism as a result of tacrolimus can be noted in recipients with partial or deteriorating graft function. Likewise, cyclosporine (now rarely used by pancreas transplant centers as the calcineurin inhibitor of choice) has been associated with β-cell toxicity, diminished β-cell density, and reduced insulin synthesis and secretion.

Because of the well-documented diabetogenicity of steroids, protocols for steroid withdrawal or avoidance have increasingly been used over the last decade in pancreas recipients.[4,15,23-25]

In contrast to calcineurin inhibitors and steroids, sirolimus has not been linked with impaired glucose metabolism. However, it has been linked with the development of hypercholesterolemia and hypertriglyceridemia. MMF appears to have no adverse effect on glucose or lipid metabolism.

In pancreas recipients with slowly deteriorating graft function, maintenance immunosuppressants that impair glucose metabolism should be dose-reduced or discontinued. In our experience, for pancreas recipients diagnosed with CR and impaired glucose metabolism, the combination of choice is sirolimus (instead of tacrolimus) and MMF. Prospective randomized studies in pancreas recipients diagnosed with chronic allograft rejection are needed to determine which regimen is best for optimizing remaining graft function.

## Laboratory and Clinical Diagnosis

Laboratory diagnosis of chronic pancreas graft failure is nonspecific and depends on the surgical technique used to manage exocrine pancreas secretions.

For bladder-drained transplants, urine amylase has been the most widely used rejection marker. A decrease in urine amylase levels over time, or isolated loss of exocrine function (in the presence of normal plasma glucose levels), has been associated with CR. In our University of Minnesota series, isolated loss of exocrine function (in the presence of partial or full endocrine function) has been noted in about 9% of all pancreas recipients. Of note, some recipients with exocrine, but not endocrine, graft loss have remained insulin-independent for years. But, the probability of being insulin-independent is higher after partial (about 80%) than after total loss (about 30%) of exocrine function, suggesting that the extent of acinar tissue damage reflects the severity of the process leading to endocrine graft failure.[4]

For enteric-drained transplants, laboratory parameters are even less specific (than urine amylase levels) in predicting chronic pancreas graft failure. Serum exocrine markers such as amylase, lipase, and anodal trypsinogen are frequently within the normal range; serum endocrine markers, such as the glucose disappearance rate, are late indicators of deteriorating graft function. Only in enteric-drained SPK recipients can the serum creatinine level be used as a harbinger of rejection. About 95% of all rejection episodes in SPK recipients involve either the kidney graft alone or both the pancreas graft and the kidney graft—isolated pancreas graft rejection is very rare. In most SPK recipients, kidney graft rejection precedes pancreas graft rejection.[6,26]

No clinical symptoms are diagnostic of CR in pancreas recipients. Fever and graft tenderness have been reported in recipients with AR (mostly in the absence of calcineurin inhibitors), but CR is usually symptom-free. The use of imaging studies in diagnosing CR has also been disappointing—such studies can only reveal a usually shrunken pancreas graft in the advanced stages of CR. Thus, diagnosis of CR is basically histopathologic.

Because of the paucity of CR symptoms, graft pancreatectomy as a result of CR is rarely indicated. Fewer than 20% of all irreversibly rejected pancreas grafts require removal (usually in immunosuppressed recipients with a functioning kidney graft). In recipients without a kidney graft, IS can be discontinued after resumption of full-dose insulin therapy—the pancreas graft then shrinks markedly and may not even require removal at the time of a retransplant.[4]

## Histopathology

The gold standard for diagnosing pancreas allograft rejection is via CT- or ultrasound-guided biopsy.

AR, a cell-mediated response, targets acinar and epithelial cells in the pancreas graft. The pathologic findings differ by the severity of AR.[27] In the early stages, AR is characterized by a mixed cellular infiltrate involving connective tissue surrounding the acini. In general, this infiltrate is predominantly composed of lymphocytes, but also includes plasma

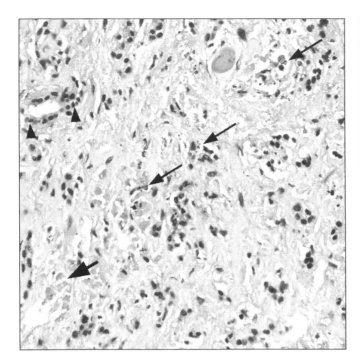

Figure 3. Acute rejection: apoptosis of the acinar cells (small arrows), acinar coagulative necrosis (large arrow), and lymphocytic ductitis (lymphocytes within the ductal epithelium, arrowheads). (Hematoxylin and eosin, original magnification × 400)

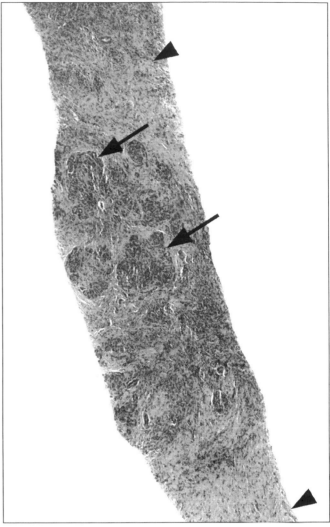

Figure 4. Chronic rejection: acinar atrophy, fibrosis. and chronic inflammation. Arrows mark the residual pancreatic acini, while arrowheads mark areas of fibrosis and chronic inflammation. (Hematoxylin and eosin, original magnification × 40)

cells, neutrophils, and eosinophils. As the severity of AR increases, the inflammatory infiltrate involves acini, and cell necrosis is seen. The inflammatory infiltrate is composed of an increasing percentage of plasma cells and eosinophils (Fig. 3). As rejection progresses, endotheliitis in the vasculature and epithelial damage of ductal cells occur. In severe AR, lymphocytic vasculitis can cause necrosis and fibrosis of the pancreatic parenchyma. Though islets are not directly targeted by inflammatory cells, vascular thrombosis and necrosis involving islets can result. Additionally, hyperglycemia has been described in the setting of AR with histologically normal-appearing islets.[28,29] A grading system has been described for AR that ranges from 0 (normal) to V (severe, with a predominantly lymphocytic or mixed cellular infiltrate, acinar cell necrosis, vasculitis, and ductal endothelial cell injury).[30]

The pathophysiology of chronic pancreas graft rejection is poorly understood. Factors that are suspected to contribute, based on correlation with research conducted in kidney, liver, and heart transplantation, include AR episodes, infections, and, to a smaller extent, toxic effects of pharmaceutical agents.[4] Histologically, CR consists of fibrointimal proliferation (intimal hyperplasia) with luminal narrowing and resultant parenchymal fibrosis[31] (Fig. 4). Previous AR episodes and graft pancreatitis contribute to these changes, but their impact cannot be precisely determined. A grading system has been described for CR that ranges from 0 (normal) to III (more than 60% parenchymal fibrosis, severe arteriopathy, and extensive acinar loss and /or atrophy)[32] (Fig. 4).

Recurrent autoimmune isletitis is histologically identical to primary diabetes. Early in its course, focal lymphocytic infiltration of islets is seen. Other endocrine cells and other tissues (acinar, ductal, and surrounding connective tissue) are spared. As the disease progresses and β–cells become fewer in number, the inflammatory infiltrate diminishes. Late in its course, when β–cells are scarce, the tissue appears normal per light microscopy; islets whose normal composition is 70% to 80% β-cells now contain predominantly α-cells. Immunohistochemical staining for insulin confirms the diagnosis.[33]

## Conclusion

Although the results of pancreas transplants have markedly improved, chronic graft failure, primarily caused by CR, remains an obstacle to further improvement. The CR rate can only be reduced by developing highly potent and less toxic IS agents and protocols that will also minimize infectious and malignant complications. The most common risk factors for CR are a previous AR episode and poor HLA class I matching. Given currently available IS agents, protocols have greatly improved over time, decreasing the incidence of reversible AR episodes within the first year posttransplant from about 80% in the 1980s to about 30% today. Likewise, attention to good HLA matching, particularly in PAK and PTA recipients, has decreased the incidence of CR. If the incidence of CR can be further reduced, the current proportion of pancreas recipients with >10-year graft function, specifically, 47%, can be markedly increased in the future.

### Acknowledgements

The authors wish to acknowledge Angelica C. Gruessner for providing up-to-date IPTR data and Alexander M. Truskinovsky of the University of Minnesota Department of Pathology for providing histopathologic examples of pancreas rejection.

# References

1. Kelly WD, Lillehei RC, Merkel FK et al. Allotransplantation of the pancreas and duodenum along with the kidney in diabetic nephropathy. Surgery 1967; 61(6):827-37.
2. Gruessner AC, Sutherland DE. Pancreas transplant outcomes for United States (US) and non-US cases as reported to the United Network for Organ Sharing (UNOS) and the International Pancreas Transplant Registry (IPTR) as of 2004. Clin Transplant 2005; 19(4):433-55.
3. Gruessner RW, Sutherland DE, Gruessner AC. Survival after pancreas transplantation. JAMA 2005; 293(6):675; author reply 675-76.
4. Gruessner RWG. Immunobiology, diagnosis and treatment of pancreas graft rejection, In: Gruessner RWG and Sutherland DER. Transplantation of the Pancreas. New York: Springer, 2004:364-380.
5. Humar A, Khwaja K, Ramcharan T et al. Chronic rejection: the next major challenge for pancreas transplant recipients. Transplantation 2003; 76(6):918-23.
6. Klassen DK, Weir MR, Schweitzer EJ, Bartlett ST. Isolated pancreas rejection in combined kidney-pancreas transplantation: results of percutaneous pancreas biopsy. Transplant Proc 1995; 27(1):1333-34.
7. Ciancio G, Burke GW, Gomez C et al. Simultaneous pancreas-kidney transplantation in Hispanic recipients with type I diabetes mellitus and end-stage renal disease. Transplant Proc 1997; 29(8):3717.
8. Lo A, Stratta RJ, Egidi MF et al. Outcomes of simultaneous kidney-pancreas transplantation in African-American recipients: a case-control study. Clin Transplant 2000; 14(6):572-79.
9. Danovitch GM, Cohen DJ, Weir MR et al. Current status of kidney and pancreas transplantation in the United States, 1994-2003. Am J Transplant 2005; 5(4 Pt 2):904-15.
10. Humar A, Parr E, Drangstveit MB et al. Steroid withdrawal in pancreas transplant recipients. Clin Transplant 2000; 14(1):75-78.
11. Burke GW 3rd, Ciancio G, Figueiro J et al. Can acute rejection be prevented in SPK transplantation? Transplant Proc 2002; 34(5):1913-14.
12. Burke GW 3rd, Kaufman DB, Millis JM et al. Prospective, randomized trial of the effect of antibody induction in simultaneous pancreas and kidney transplantation: three-year results. Transplantation 2004; 77(8):1269-75.
13. Burke GW 3rd, Ciancio G, Figueiro J et al. Steroid-resistant acute rejection following SPK: importance of maintaining therapeutic dosing in a triple-drug regimen. Transplant Proc 2002; 34(5):1918-19.
14. Reddy KS, Davies D, Ormond D et al. Impact of acute rejection episodes on long-term graft survival following simultaneous kidney-pancreas transplantation. Am J Transplant 2003; 3(4):439-44.
15. Freise CE, Kang SM, Feng S et al. Experience with steroid-free maintenance immunosuppression in simultaneous pancreas-kidney transplantation. Transplant Proc 2004; 36(4):1067-68.
16. Sutherland DE, Gruessner RG, Humar A et al. Pretransplant immunosuppression for pancreas transplants alone in nonuremic diabetic recipients. Transplant Proc 2001; 33(1-2):1656-58.
17. McKenna RM, Takemoto SK, Terasaki PI. Anti-HLA antibodies after solid organ transplantation. Transplantation 2000; 69(3):319-26.
18. Berney T, Malaise J, Morel P et al. Impact of HLA matching on the outcome of simultaneous pancreas-kidney transplantation. Nephrol Dial Transplant 2005; 20 Suppl 2:ii48-53, ii62.
19. Lo A, Stratta RJ, Alloway RR et al. A multicenter analysis of the significance of HLA matching on outcomes after kidney-pancreas transplantation. Transplant Proc 2005; 37(2): 1289-90.
20. Pelletier RP, Hennessy PK, Adams PW et al. Clinical significance of MHC-reactive alloantibodies that develop after kidney or kidney-pancreas transplantation. Am J Transplant 2002; 2(2):134-41.
21. Odorico JS, Becker YT, Groshek M et al. Improved solitary pancreas transplant graft survival in the modern immunosuppressive era. Cell Transplant 2000; 9(6):919-27.
22. Odorico JS, Pirsch JD, Becker YT et al. Results of solitary pancreas transplantation with enteric drainage: is there a benefit from monitoring urinary amylase levels? Transplant Proc 2001; 33(1-2):1700.
23. Heilman RL, Reddy KS, Mazur MJ et al. Acute rejection risk in kidney transplant recipients on steroid-avoidance immunosuppression receiving induction with either antithymocyte globulin or basiliximab. Transplant Proc 2006; 38(5):1307-13.
24. Gruessner RW, Sutherland DE, Parr E et al. A prospective, randomized, open-label study of steroid withdrawal in pancreas transplantation — a preliminary report with 6-month follow-up. Transplant Proc 2001; 33(1-2):1663-64.
25. Cantarovich D, Karam G, Hourmant M et al. Steroid avoidance versus steroid withdrawal after simultaneous pancreas-kidney transplantation. Am J Transplant 2005; 5(6):1332-38.
26. Sutherland DE, Gruessner R, Moudry-Munns K, Gruessner A. Discordant graft loss from rejection of organs from the same donor in simultaneous pancreas-kidney recipients. Transplant Proc 1995; 27(1): 907-8.
27. Nakhleh RE, Sutherland DE. Pancreas rejection. Significance of histopathologic findings with implications for classification of rejection. Am J Surg Pathol 1992; 16(11):1098-107.
28. Sibley RK, Sutherland DE. Pancreas transplantation. An immunohistologic and histopathologic examination of 100 grafts. Am J Pathol 1987; 128(1):151-70.
29. Nakhleh RE, Sutherland DE. Biopsies from pancreas allografts at time of dysfunction: pathologic comparison of allografts which ultimately failed versus those which continued to function. Transplant Proc 1993; 25(1 Pt 2):1194-95.
30. Drachenberg CB, Papadimitriou JC, Klassen DK et al. Evaluation of pancreas transplant needle biopsy: reproducibility and revision of histologic grading system. Transplantation 1997; 63(11):1579-86.
31. Drachenberg CB, Papadimitriou JC, Klassen DK et al. Chronic pancreas allograft rejection: morphologic evidence of progression in needle biopsies and proposal of a grading scheme. Transplant Proc 1999; 31(1-2):614.
32. Papadimitriou JC, Drachenberg CB, Klassen DK et al. Histological grading of chronic pancreas allograft rejection/graft sclerosis. Am J Transplant 2003; 3(5):599-605.
33. Sibley RK, Sutherland DE, Goetz F, Michael AF. Recurrent diabetes mellitus in the pancreas iso- and allograft. A light and electron microscopic and immunohistochemical analysis of four cases. Lab Invest 1985; 53(2):132-44.

# Pathological Aspects of Pancreas Allograft Failure

John C. Papadimitriou* and Cinthia B. Drachenberg

## Abstract

Pancreas allograft failure results from a variety of causes, highly dependent on the time posttransplantation. In the early posttransplantation period pancreas allograft failure is usually related to "technical failures", including thrombosis, infection, pancreatitis, anastomotic leak and bleeding. After the sixth month posttransplantation the majority of graft losses are attributed to chronic allograft rejection. Histological evaluation of the exocrine, endocrine and vascular components allows for an accurate determination of the etiology of pancreas allograft loss. Ancillary studies are in addition necessary for the diagnosis of humoral rejection (i.e., C4d stain), viral infections (i.e., EBV markers) and recurrence of diabetes (i.e., insulin, glucagon).

## Introduction

Improved surgical techniques, availability of potent immunosuppressants, accurate diagnosis of rejection, better treatment of infections and careful selection of donors and recipients have all resulted in excellent one year graft and patient survival after pancreas transplantation.[1-5] Despite all these advances, a proportional improvement in long term allograft survival has not been achieved yet, because similar to other transplanted organs progressive graft sclerosis inexorably develops over time.[6-10]

Schematically, the causes of pancreas graft loss are considered either technical or immunological. The first group includes: thrombosis, infection, pancreatitis, anastomotic leak and bleeding. These complications usually require removal of the organ. The immunological causes include acute and chronic rejection, as well as recurrence of autoimmune isletitis with destruction of insulin producing cells.[1]

Accurate determination of the cause of graft loss requires histological evaluation of the explanted specimen with systematic evaluation of the endocrine and exocrine components and specifically of the arteries.[7] In addition, special stains for immunoglobulins and complement (particularly C4d stain) would allow for the identification of humoral/antibody mediated rejection.[7,11] Furthermore, immunostains for insulin and glucagon are necessary for the identification of beta and alpha cells, respectively when there is a potential diagnosis of recurrent autoimmune disease.[12-14]

Infectious complications involving the graft and the adjacent tissues have heterogeneous clinical and pathological features and should be always considered in the differential diagnosis of graft dysfunction.

## Graft Loss Due to Technical Failure

Technical failure rates have continued to decrease over the years and are in the order of 6-7% for SPK and PAK and 10% for PTA according to the last report of the pancreas transplant registry.[1]

### Idiopathic Graft Thrombosis

Continues to be the leading cause of graft loss due to technical reasons, occurring in 5% of SPK and PAK and 7.8% of PTA.[1,15]

Graft pancreatectomy in idiopathic graft thrombosis (technical failure) is performed early in the posttransplantation period, typically within 48 hr and always within the first 2 weeks.[7]

Upon gross and histological examination the main findings consist of recent thrombosis, occurring within otherwise normal arteries and/or veins and associated ischemic/coagulation necrosis of the parenchyma (Fig. 1A,B). The latter is characterized by the loss of nuclear detail and dissolution of the cellular components. The amount of necrosis varies according to the extent of the thrombosis and the time elapsed until the pancreatectomy. The peripancreatic fat typically shows extensive enzymatic necrosis due to spillage of pancreatic enzymes from the lysed pancreatic cells.

In cases of thrombosis attributed to true "technical failure" the vessels do not show evidence of immune-mediated damage (i.e., transmural or intimal arteritis, fibrinoid necrosis). Accordingly, immunohistochemical stains for immunoglobulins and complement, including C4d, are negative.

Early graft thrombosis has been attributed to the intrinsically low blood flow in the pancreas compared with other solid organs. Perioperative inflammation and edema, as well as microvascular and endothelial damage relating to donor factors and organ preservation, all contribute to further compromise of the blood flow in the early posttransplant period, potentially leading to thrombosis. Accordingly, longer cold ischemia times have been associated with increased incidence of graft thrombosis.[15,17,18]

Grossly evident atherosclerosis in the donor pancreas is a contraindication to transplantation but in occasional grafts lost due to early thrombosis histological evaluation of the arteries demonstrates incipient donor atheromatosis underlying and likely causing the thrombosis (unpublished data). These findings are consistent with the increased rates of technical failure seen with donors older than 50 years as well as with cardio-cerebrovascular disease as the cause of donor death.[1]

### Fungal and Bacterial Infections

These processes affecting the graft and adjacent soft tissues are not uncommon in the early posttransplantation period. In rare occasions these infections are refractory to medical and surgical treatment requiring removal of the graft (Fig. 2A).

Pathological evaluation of pancreas allografts with infectious pancreatitis/peripancreatitis shows mixed inflammation predominantly

*Corresponding Author: Dr. John C. Papadimitriou—Director, Surgical Pathology, University of Maryland Hospital, 22 South Greene St., Baltimore, MD 21201, USA. Email: jpapa001@umaryland.edu

*Chronic Allograft Failure: Natural History, Pathogenesis, Diagnosis and Management*, edited by Nasimul Ahsan. ©2008 Landes Bioscience.

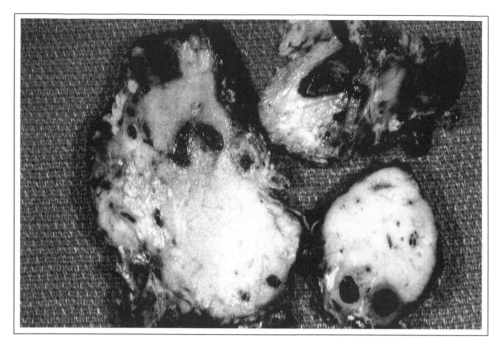

Figure 1A. Gross examination of slices of a pancreas allograft lost at the 2rd day posttransplantation due to idiopathic thrombosis demonstrate occlusion of large arteries and veins by blood clots.

composed of neutrophils and macrophages. Depending on the severity of the infection there are areas of parenchymal liquefactive necrosis and purulent exudates. Typically there is a prominent fibroblastic proliferation not only in the peripancreatic connective tissue, but also extending to fibrous septa in the peripheral, superficial pancreatic parenchyma (Fig. 2B). Accordingly, needle biopsies from grafts with infections pancreatitis/peripancreatitis typically show pronounced septal fibrosis, that should be differentiated from that of chronic rejection.[19] In contrast to the latter the scarring process associated with peripancreatic infections only involves the periphery of the pancreatic parenchyma and long term graft function is not affected if the infection is resolved.

## Non-Infectious (Ischemic) Acute Pancreatitis

Typically seen in the early posttransplantation period has similar features to mild, interstitial pancreatitis occurring in the native pancreas. Histological sections demonstrate neutrophilic infiltrates in septa and acini, edema, interstitial hemorrhage and enzymatic necrosis of the peripancreatic and septal fat (Fig. 3). Idiopathic (non-infectious) pancreatitis is usually secondary to ischemic injury and is associated with an increased risk for subsequent thrombosis.[16]

Histological evaluation of grafts lost due to anastomotic leaks shows necrosis and predominantly neutrophilic inflammation in areas adjacent to suture lines. The duodenal cuff and adjacent segment of anastomotic bowel typically show acute fibrinous or purulent serositis.

Figure 1B. Histological examination of the same specimen as Figure 1A demonstrates that the arterial wall is normal confirming the impression of thrombosis due to "technical failure".

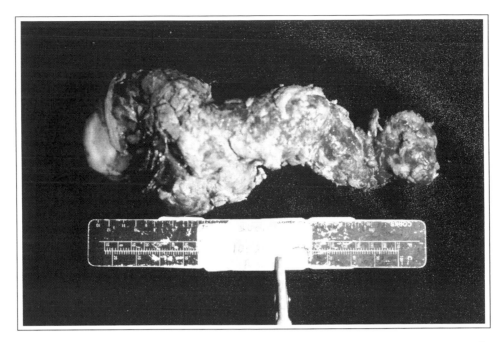

Figure 2A. Macroscopic view of a pancreas graft resected the 2nd week posttransplantation due to severe bacterial peripancreatitis. The organ is covered by purulent exudates.

The parenchyma located away from the leak lacks significant inflammation and the blood vessels are typically normal.

## Graft Loss Due to Immunological Causes

The histological findings in grafts lost for immunological reasons vary, in particular with the time of the graft loss. According to the pancreas transplant registry, loss due to acute rejection occurs more often between the 7th and 12th month, with the frequency decreasing after the first year.[1] In contrast, chronic rejection steadily increases with time after transplantation, representing the most important cause of graft loss after the 1st year.[6,7]

Arterial thrombosis is common in pancreas rejection, being important to remember that acute rejection can be the cause of early graft thrombosis simulating a technical failure.[7] In the case of acute humoral rejection recent thrombosis (platelet and fibrin aggregation) can occur due to endothelial damage and/or activation. Thrombosis may be also secondary to intimal arteritis (endotheliitis) or vasculitis (Fig. 4).

Recent and organized thromboses are also very common in pancreas with transplant arteriopathy (proliferative endarteritis) secondary to chronic rejection.[7]

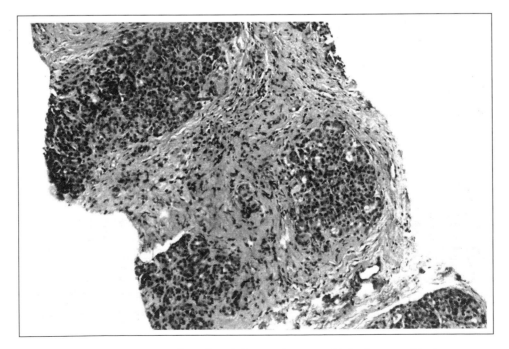

Figure 2B. Intraoperative core biopsy of pancreas allograft with bacterial pancreatitis/peripancreatitis demonstrates active fibroblastic proliferation admixed with inflammatory cells. The acinar parenchyma appears diminished due to the prominent fibrous component; however, the changes in peripancreatitis tend to involve only the periphery of the graft. This is in contrast to the fibrosis in chronic rejection that affects the organ diffusely.

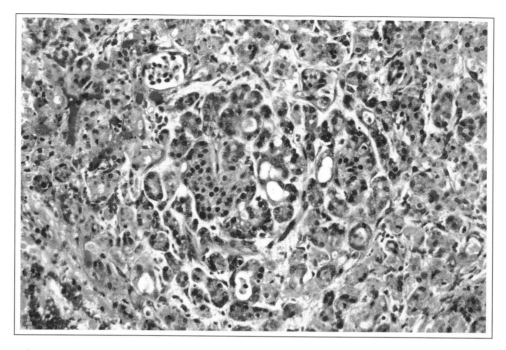

Figure 3. High magnification view of acute, ischemic pancreatitis. The are small areas of acinar necrosis (far left of picture). There is drop-out of acinar cells and scattered clusters of neutrophils.

### Hyperacute Allograft Rejection

Results from the generalized deposition of preformed immuno-globulin and activated complement in the vasculature of the graft with associated endothelial damage and immediate thrombosis. On histological examination there is necrosis of arteries and veins with massive thrombosis and secondary parenchymal necrosis. Immunostains demonstrate immunoglobulin (typically IgG) and complement components in the vascular walls.[7]

### Acute Humoral Rejection

Akin to that seen in renal allografts, can be a cause of early graft failure. Histological evaluation demonstrates absence of inflammation with diffuse staining of capillaries with C4d stain. In a well documented report, graft dysfunction consisting of increase in serum amylase and development of hyperglycemia occurred approximately one month posttransplantation. The allograft biopsy demonstrated diffuse positivity in the interacinar capillaries for the C4d immunostain. Donor specific antibodies were concurrently identified in serum. The patient was treated with rituximab and intravenous immunoglobulin, with resolution of the graft dysfunction.[11]

### Acute Cellular Rejection

Experimental studies of unmodified rejection in pancreatic allografts have shown that the earliest acute changes consist of interstitial infiltrates with prominent lymphocytic infiltration of interlobular veins. In subsequent days there is more generalized lymphocytic inflammation

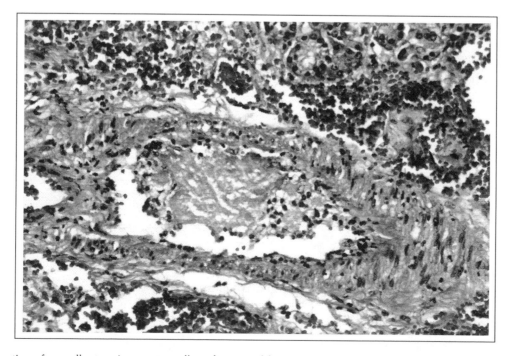

Figure 4. Cross section of a small artery in pancreas allograft resected for acute rejection. Incipient intimal arteritis, is associated with a mural thrombus (asterisks).

with involvement of the acini and damage and drop-out of acinar cells.[19] These findings were confirmed in clinical samples.[20] Based on clinical experience and previously published experimental studies, a histological scheme for the grading of acute allograft rejection was proposed in 1997. In this schema biopsies with no inflammation and biopsies with minimal septal inactive inflammation were considered negative for acute rejection and were graded 0 and I, respectively. More significant histological changes that were considered indicative of acute rejection were classified as follows: active septal inflammation with venous endotheliitis, Grade II; acinar inflammation in addition to septal changes, Grade III; arterial inflammation, Grade IV and changes associated with diffuse parenchymal necrosis, Grade V.[3,21,22] In experimental as well as clinical studies, islets were shown not to represent a primary target of rejection.[3,19] This grading scheme showed good reproducibility, prognostic significance and correlation with other clinical parameters, including response to treatment.[21,22]

In 2005 the process of updating the pancreas allograft grading scheme for acute rejection was started at the Banff Conference for Allograft Pathology (Edmonton, Canada). The updated scheme is expected to be completed in the subsequent 2007 Banff Conference (La Coruña, Spain). Ongoing consensus discussions will attempt to simplify the number and descriptive features of the diagnostic categories by following the principles of the updated 2005 Banff schema for the diagnosis of rejection in the kidney.[23] The new grading scheme will include a humoral rejection category and also guidelines for quantitation of specific histological lesions. More specifically, in recent studies, distinction between pauci- vs. multifocal acinar involvement appears to have far more importance for the long term graft outcome than previously recognized (i.e., the degree of exocrine parenchymal involvement during acute rejection predicts to a large extent the subsequent development of fibrosis).[10,24]

### Chronic Rejection

With the marked improvements achieved in the short-term outcome of pancreas allografts, mostly due to the decrease in the occurrence of acute rejection, late graft loss and chronic rejection overall have become increasingly important issues to address.[6] In contrast to acute rejection that presents with sudden graft dysfunction and can be prevented or successfully treated in the majority of cases, chronic rejection is characterized by a slowly progressive decline in graft function and does not respond well to treatment. Similarly to kidney allografts, in the pancreas chronic rejection is manifested histologically as sclerosis and loss of functional parenchyma. Progressive fibrosis and acinar loss can be observed over a prolonged period of time in serial biopsies.[10]

The pathogenesis of chronic rejection in the pancreas is unclear. It is likely that in the pancreas as in other allografts, the fibrosis represents the end effect of cumulative injury (or injuries) of diverse origins, immunological and non-immunological. It is unclear if in the pancreas similar mechanisms operate for the propagation of tissue damage once a critical amount of parenchymal mass is lost, as is the case in the kidney.[10] The main histological findings in chronic rejection (septal fibrosis, acinar loss) are very similar to those of chronic pancreatitis in native organs. From a pathophysiological point of view, the progressive tissue loss observed in native chronic pancreatitis appears to be also the result of multiple factors, including ductal obstruction by concretions, altered pattern of protein secretion (e.g., lithostathine), oxidative stress,[10,25,26] growth factors over-expression, activation of inflammatory mediators, deregulated immune responses and altered nerve growth with specific neuroimmune interactions.[10] Also, micro-circulatory disturbances have been heavily implicated.[26] Thus, the progress of pancreatic pathology in the native organ is apparently complex, self-perpetuating and with significant similarities to the renal chronic allograft rejection/sclerosis. It is conceivable that in addition to the immune-mediated acinar cell and vascular injury, similar mechanisms to the ones operating in chronic

pancreatitis play some role in the pathogenesis of graft sclerosis. Similar to the changes proposed in the 2005 Banff schema for grading chronic rejection in the kidney, it is recommended that the term 'chronic pancreas graft sclerosis' rather than chronic rejection is used unless there is a clear history of preceding episodes of acute rejection or vascular changes typical of immune mediated injury.[7,23]

The role of vascular injury in pancreas allograft pathology is unequivocal. We have previously demonstrated that vascular thrombosis superimposed on acute (endotheliitis) or chronic (transplant arteriopathy) arterial damage is associated with parenchymal fibrosis/sclerosis[7](Fig. 5). Although arterial changes are typical of chronic pancreatic rejection, grading of pancreas graft sclerosis relies principally on the degree of fibrosis and acinar loss.

Histological grading of chronic pancreas rejection/graft sclerosis in needle biopsies is based on the percentage of septal fibrosis and acinar loss in the core. Septal expansion/fibrosis that in aggregate represents <30% of the biopsy surface is graded I (mild). More important degrees of fibrosis (30-60% and >60%) are graded II (moderate) and III (severe), respectively.[10]

In the differential diagnosis of septal fibrosis potentially leading to a diagnosis of chronic rejection/graft sclerosis, it is important to remember that in peripancreatic abscesses and intra-abdominal fluid collections the superficial pancreatic parenchyma develops striking septal fibrosis. In contrast to chronic rejection that shows disappearance and/or atrophy of the exocrine lobules, in peripancreatic infections/fluid collections the intervening acinar tissue is preserved. This process presents early after the transplantation and has characteristic clinical findings with predominance of the infectious signs and symptoms and in general preservation of the graft function.[19]

### Acute on Chronic Allograft Rejection

In the kidney it has been demonstrated that the risk for chronic rejection increases with the occurrence of interstitial and vascular rejection episodes, particularly the latter.[10,27] Acute rejection episodes occurring late rather than within the first 3 months are also associated with increased risk of chronic rejection in the kidney.[28] In pancreas allografts, late acute rejection (occurring after 12 months posttransplantation) is not unusual and appears to occur more often in solitary pancreas transplants.[8,9,27] Also, repeated episodes of acute rejection, higher grades of acute rejection and late (>1 year) acute rejection are associated with an increased risk of chronic rejection and graft loss[10] (Fig. 6).

### Recurrence of Type I Diabetes Mellitus

Has been well documented and is increasingly recognized as a cause of loss of graft function.[13,14,29] The diagnosis of recurrent disease is made with the combination of sudden or progressive loss of glycemic control associated with selective loss of beta cells in the graft with persistence of the other types of islet cells, particularly alpha cells (Fig. 7A-C). In few cases the active phase of beta cell destruction has been histologically demonstrated; this consists of selective islet mononuclear cell infiltration (isletitis). The inflammation is centered in islets still containing beta cells and isletitis resolves when the beta cells disappear. In addition to the clinical and histologic findings, the diagnosis of recurrent autoimmune disease is aided by the demonstration of islet cell auto-antibodies in serum (i.e., GAD 65 and IA-2). Recurrent autoimmunity operates also in the case of allogeneic islet transplantation.[13,14]

The rarity of recurrent Type I diabetes mellitus in whole pancreas transplantation has been attributed to the inclusion of donor lymphoid tissue with the transplanted pancreas. This would lead to recipient chimerism for a donor T-cell subset potentially associated with some form of advantageous immune modulation.[30]

Inflammation of islets with absence of associated selective loss of beta cells can be occasionally seen in acute allograft rejection. Islet inflammation in these circumstances is proportional to the degree of inflammation

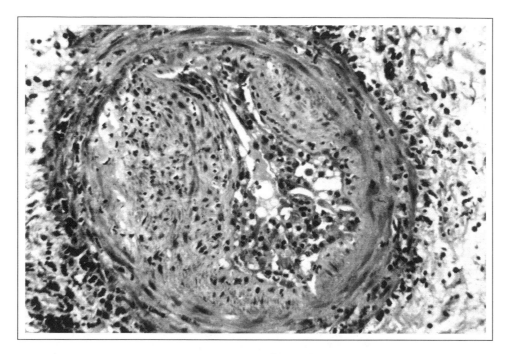

Figure 5. Histological cross section of an artery demonstrates severe luminal narrowing due to acute and chronic rejection. The acute component is represented by the cellular areas (intimal arteritis/endotheliitis) whereas the marked fibrointimal thickening is indicative of chronic vascular rejection.

in the neighboring exocrine parenchyma and usually consists of a mixture of inflammatory cells that may include eosinophils.[3]

Similarly to recurrence of autoimmune disease, reversible injury to the beta cells in calcineurin inhibitor toxicity may present with hyperglycemia. On histological evaluation, however, there is an adequate number of beta cells, albeit these have cytological features of injury (vacuolization) and decreased insulin content upon immunostaining.[31]

### Graft Loss Due to Viral Infectious Complications

Although the incidence of CMV disease in kidney-pancreas transplantations may reach 22% in donor positive-recipient negative

transplantations, the diagnosis of CMV graft pancreatitis is rare.[32-35] In the report of Klassen et al,[35] four patients with CMV pancreatitis were diagnosed on percutaneous needle biopsies 18 weeks to 44 months after transplantation. In all 4 patients prolonged treatment with ganciclovir achieved clinical and histologic resolution of the infection. The clinical presentation of CMV graft pancreatitis is indistinguishable from acute rejection (i.e., increase in serum amylase and lipase).

The histological diagnosis of CMV pancreatitis requires the demonstration of the typical viral cytopathic changes in the infected cells (usually endothelial cells and less frequently acinar cells). The infected cells are very large and contain an intranuclear viral inclusion often

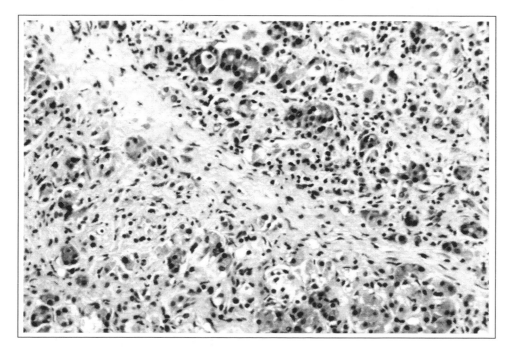

Figure 6. High magnification image of the exocrine/acinar component showing active mononuclear inflammation in a background of increase collagenization (fibrosis) and loss of the acinar cellular elements. This is an example of acute on chronic cellular rejection.

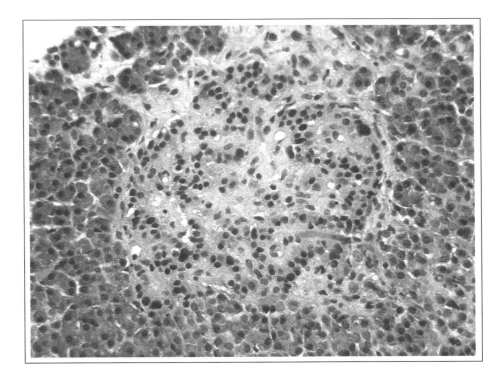

Figure 7A. Pancreas biopsy from patient with recurrence of diabetes mellitus. H&E stained endocrine islet shows increase in eosinophilic extracellular matrix between the acinar cells.

surrounded by a halo ("owls eye"). The cytoplasm contains red granular viral inclusions. The viral changes are associated with various degrees of mononuclear inflammation that very with the severity of the process; necrosis can be also found. Because both CMV infection and acute cellular rejection are characterized by mononuclear infiltrates (i.e., lymphocytes, macrophages), these processes can be confused on needle biopsies if the viral cytopathic changes are not identified.

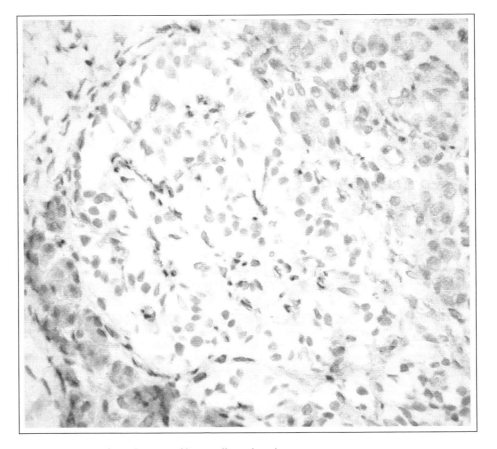

Figure 7B. Stain for insulin shows complete absence of beta cells in the islet.

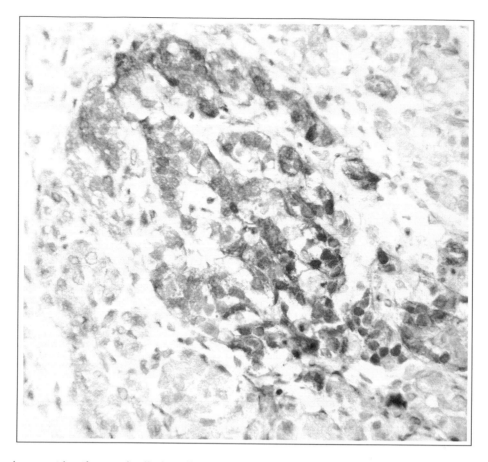

Figure 7C. Stain for glucagon identifies nearly all islet cells as alpha cells.

## EBV-Related Posttransplant Lymphoproliferative Disorder (PTLD)

May involve the pancreas allograft and require explantation. In a previous study, allograft pancreatectomies for PTLD were performed in the first year posttransplantation at a mean of 5 months.[7,36]

Histological examination of grafts of PTLD demonstrates confluent, predominantly mononuclear infiltrates with a characteristic nodular and expansive growth pattern that replaces and displaces the pancreatic parenchyma and adjacent soft tissues. In EBV-PTLD the proliferation is predominantly composed of B-cells with cytologic morphology ranging from mixed polymorphous, to frankly malignant lymphoma. Special studies are necessary to confirm the nature of the process. Specifically, immunostains for EBV-latent membrane protein (LMP-1) and demonstration of EBV-encoded RNAs (EBER) by in-situ hybridization are used for confirmation. PTLD is randomly distributed leaving intervening areas of uninvolved parenchyma. Differently from acute rejection the lymphoid infiltrates in PTLD do not appear to specifically target acini, ducts or blood vessels. Guidelines for the diagnosis of PTLD in needle biopsies have been provided previously.[36]

## References

1. Gruessner AC, DE. Sutherland, Pancreas transplant outcomes for United States (US) and non US cases as reported to the United Network for Organ Sharing (UNOS) and the International Pancreas Transplant Registry (IPTR) as of 2004. Clin Transplant 2005; 19(4):433-55.
2. Kuo PC et al. Solitary pancreas allografts. The role of percutaneous biopsy and standardized histologic grading of rejection. Arch Surg 1997; 132(1):52-57.
3. Drachenberg CB et al. Evaluation of pancreas transplant needle biopsy: reproducibility and revision of histologic grading system. Transplantation 1997; 63(11):1579-86.
4. Bartlett ST et al. Equivalent success of simultaneous pancreas kidney and solitary pancreas transplantation. A prospective trial of tacrolimus immunosuppression with percutaneous biopsy. Ann Surg 1996; 224(4):440-9; discussion 449-52.
5. Klassen DK et al. Pancreas allograft biopsy: safety of percutaneous biopsy-results of a large experience. Transplantation 2002; 73(4):553-55.
6. Humar A et al. Chronic rejection: the next major challenge for pancreas transplant recipients. Transplantation 2003; 76(6):918-23.
7. Drachenberg CB et al. Pancreas transplantation: the histologic morphology of graft loss and clinical correlations. Transplantation 2001; 71(12):1784-91.
8. Stratta RJ. Graft failure after solitary pancreas transplantation. Transplant Proc 1998; 30(2):289.
9. Stratta RJ. Patterns of graft loss following simultaneous kidney-pancreas transplantation. Transplant Proc 1998; 30(2):288.
10. Papadimitriou JC, Drachenberg CB, Klassen DK et al. Histological grading of chronic pancreas allograft rejection/graft sclerosis. Am J Transplant 2003; 3:599-605.
11. Melcher ML et al. Antibody-mediated rejection of a pancreas allograft. Am J Transplant 2006; 6(2):423-28.
12. Sibley RK et al. Recurrent diabetes mellitus in the pancreas iso- and allograft. A light and electron microscopic and immunohistochemical analysis of four cases. Lab Invest 1985; 53(2):132-44.
13. Thivolet C et al. Serological markers of recurrent beta cell destruction in diabetic patients undergoing pancreatic transplantation. Transplantation 2000; 69(1):99-103.
14. Troppmann C et al. Vascular graft thrombosis after pancreatic transplantation: univariate and multivariate operative and non-operative risk factor analysis. J Am Coll Surg 1996; 182(4):285-316.
15. Obermaier R et al. Ischemia/reperfusion-induced pancreatitis in rats: a new model of complete normothermic in situ ischemia of a pancreatic tail-segment. Clin Exp Med 2001; 1(1):51-59.
16. Tyden G et al. Recurrence of autoimmune diabetes mellitus in recipients of cadaveric pancreatic grafts. N Engl J Med 1996; 335(12):860-63.
17. Grewal HP et al. Risk factors for postimplantation pancreatitis and pancreatic thrombosis in pancreas transplant recipients. Transplantation 1993; 56(3):609-12.

18. Jennings WC, Smith J, Corry RJ. Thrombosis in human pancreatic transplantation associated with elevated cyclosporine levels and possible protection by antihypertensive agents. J Okla State Med Assoc 1990; 83(6):255-57.
19. Drachenberg CB, Papadimitriou JC. The inflamed pancreas transplant: histological differential diagnosis. Semin Diagn Pathol 2004; 21(4):255-9.
20. Steiniger B, Klempnauer J. Distinct histologic patterns of acute, prolonged and chronic rejection in vascularized rat pancreas allografts. Am J Pathol 1986; 124(2):253-62.
21. Nakhleh RE, Sutherland DE. Pancreas rejection. Significance of histopathologic findings with implications for classification of rejection. Am J Surg Pathol 1992; 16(11):1098-107.
22. Papadimitriou JC et al. Histologic grading of acute allograft rejection in pancreas needle biopsy: correlation to serum enzymes, glycemia and response to immunosuppressive treatment. Transplantation 1998; 66(12):1741-45.
23. Papadimitriou JC et al. Effectiveness of immunosuppressive treatment for recurrent or refractory pancreas allograft rejection: correlation with histologic grade. Transplant Proc 1998; 30(8):3945.
24. Papadimitriou JC, Drachenberg CB, Klassen DK et al. Diffuse acinar inflammation is the most important histological predictor of chronic rejection in pancreas allografts. Transplantation 2006; 82(1 Suppl 2):223.
25. Pitchumoni CS, Bordalo O. Evaluation of hypotheses on pathogenesis of alcoholic pancreatitis. Am J Gastroenterol 1996; 91(4):637-47.
26. Schilling MK et al. Microcirculation in chronic alcoholic pancreatitis: a laser Doppler flow study. Pancreas 1999; 19(1):21-25.
27. Stratta RJ. Late acute rejection after pancreas transplantation. Transplant Proc 1998; 30(2):646.
28. Basadonna GP et al. Early versus late acute renal allograft rejection: impact on chronic rejection. Transplantation 1993; 55(5):993-95.
29. Solez K, Colvin R, Racusen L et al. Banff '05 Meeting Report: Differential diagnosis of chronic allograft injury and elimination of Chronic Allograft Nephropathy ("CAN") Am J Transplant 2007; In press.
30. Papadimitriou JC, Drachenberg CB. Role of histopathology evaluation in pancreas transplantation. Current Opinion in Organ Transplantation 2002; 7:185-90.
31. Bartlett ST et al. Prevention of autoimmune islet allograft destruction by engraftment of donor T-cells. Transplantation 1997; 63(2):299-303.
32. Drachenberg CB, Klassen DK, Weir MR et al. Islet cell damage associated with tacrolimus and cyclosporine: Morphological features in pancreas allograft biopsies and clinical correlation. Transplantation 1999; 68:396.
33. Lo A et al. Patterns of cytomegalovirus infection in simultaneous kidney-pancreas transplant recipients receiving tacrolimus, mycophenolate mofetil and prednisone with ganciclovir prophylaxis. Transpl Infect Dis 2001; 3(1):8-15.
34. Humar A et al. Cytomegalovirus disease recurrence after ganciclovir treatment in kidney and kidney-pancreas transplant recipients. Transplantation 1999; 67(1):94-7.
35. Keay S. CMV infection and disease in kidney and pancreas transplant recipients. Transpl Infect Dis 1999; 1 Suppl 1:19-24.
36. Klassen DK et al. CMV allograft pancreatitis: diagnosis, treatment and histological features. Transplantation 2000; 69(9):1968-71.
37. Drachenberg CB et al. Epstein-Barr virus-related posttransplantation lymphoproliferative disorder involving pancreas allografts: histological differential diagnosis from acute allograft rejection. Hum Pathol 1998; 29(6):569-77.

# CHAPTER 34

# The Graft:
## Emerging Viruses in Transplantation

Deepali Kumar and Atul Humar*

## Abstract

Emerging infections have become increasingly recognized as causes of morbidity, mortality, graft dysfunction, graft failure and donor-transmitted infections. Specifically, a number of emerging viral pathogens have had significant adverse effects in transplant patients. These viruses may occur as a result of endogenous reactivation of latent viruses, a new exposure from the community, or as a result of donor transmission. This chapter will focus on several illustrative examples of emerging infections including respiratory viruses, adenovirus, West Nile virus and others.

## Introduction

While the term emerging infections can encompass a wide variety of bacterial, fungal and parasitic infections, it is primarily new viral infections that have been highlighted in recently published literature.[1] This chapter will focus on recent and emerging viral infections and their effect on graft function and clinical outcomes in transplant patients. Although there is no standard classification for emerging viral infections in the transplant patient, it is useful to categorize these pathogens in into one of three groups as shown in Figure 1. The first category includes viral infections that transplant patients have likely always had but whose clinical significance is only now being understood. These include human herpesvirus-6, human herpesvirus-7 infection (HHV6 and HHV7), adenovirus infection and parvovirus infection.[1,2] The second category includes infections that have previously been uncommon in transplant patients but are now increasing in importance either due to changing epidemiological factors or changing immunosuppressive regimens. Examples include West Nile virus and BK virus infection respectively (the latter is discussed in a separate chapter). Finally the third group of emerging infections is truly novel pathogens to which transplant patients may be uniquely predisposed. This unique predisposition is due to exogenous immunosuppression or specific transplant factors such as donor related transmission. Potential examples include avian influenza/pandemic influenza and pathogens that may be observed with xenotransplantation (e.g., endogenous retroviruses).[1]

Viral (as opposed to nonviral) emerging infections are a particular problem for transplant recipients for a number of reasons. First, viral pathogens commonly have the ability to establish latency. Therefore such viruses may unknowingly be present either in the recipient or in the donor organ. Current antiviral therapies are generally ineffective against latent viruses, meaning that elimination of latent viruses is usually not possible. Second, detection of viral pathogens may be difficult especially if no specific tests exist or when dealing with unknown or poorly characterized viruses. In transplant patients in particular serological tests are often unreliable. Finally, viral infections are often associated with

indirect sequelae on the allograft. Specifically latent or lytic viral infection may in some settings promote cytokine dysregulation, abnormal antigen recognition and alloreactive T-cell infiltration within the graft leading to damage and dysfunction or potentially triggering either acute or chronic rejection.[3-5] Some examples of emerging viral infections and their implications for transplant patients are discussed in the following chapter. This is not a comprehensive list, but is meant to illustrate the various examples and potential sequelae of such infection in transplant recipients. Respiratory viruses, HHV6 and 7, adenovirus, West Nile virus and, rabies virus will be discussed in more detail with a focus on both direct and indirect effects of viral infection.

## Respiratory Viruses

### Background

Community acquired viral respiratory tract infections (RTI) are common causes of acute respiratory illness in the general community and have been increasingly recognized as common pathogens after solid organ transplantation.[5-11] In this patient population, infection with these pathogens can occasionally result in severe pulmonary disease with significant morbidity and mortality. Traditionally, the most common community respiratory viruses include influenza A and B, parainfluenza serotypes 1, 2 and 3, respiratory syncytial virus (RSV) and adenovirus.[8-10] However, with the use of molecular tests, pathogens such as coronaviruses, rhinoviruses, metapneumovirus and bocavirus are now increasingly recognized as pathogens that may cause both upper and lower tract disease and may have serious sequelae particularly in transplant patients.[13-15] Lung transplant recipients are at particular risk for these infections due to numerous factors. These include potent immunosuppression regimens, abnormal immune response within the allograft, decreased cough due to denervation of the transplanted lung, abnormal lymphatic drainage, impaired mucociliary clearance and direct exposure of the allograft to the environment.[12]

### Direct Effects

Influenza viruses are among the most common and important seasonal respiratory pathogens and severe disease has been described in both solid-organ transplants (SOT) and hematopoietic stem cell transplants (HSCT).[16,17] Studies suggest that between 8-17% of transplant recipients with influenza may progress to lower respiratory tract infection.[16-19] In addition, vaccination strategies are likely not as effective due to attenuated responses in immunosuppressed persons. Reported vaccine response rates in transplant patients vary from 15-86%.[20-23] Since the risk of adverse events is low, the American Society of Transplantation recommends annual influenza vaccination for all transplant recipients.[24]

*Corresponding Author: Atul Humar, Director, Transplant Infectious Diseases, Associate Professor Department of Medicine, University of Alberta, Room 4106 RTF, 8308-114 Street, Edmonton, Alberta, Canada T6G 2E1. Email: ahumar@ualberta.ca

*Chronic Allograft Failure: Natural History, Pathogenesis, Diagnosis and Management*, edited by Nasimul Ahsan. ©2008 Landes Bioscience.

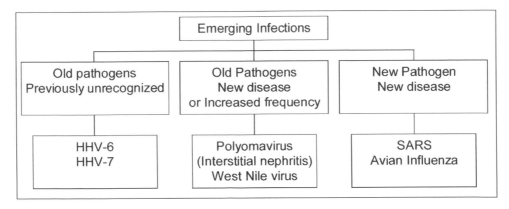

Figure 1. Proposed classification scheme for emerging viral infections in transplantation. Adapted from: Kumar D, Humar A. Curr Opin Infect Dis 2005; 18(4):337-41;[1] with permission from Lippincott Williams & Wilkins.

A link between influenza vaccination and acute rejection was noted in a small study of heart transplant recipients,[25] but this has not been substantiated by other studies.[21,23,26-28] RSV and parainfluenza virus are also a significant cause of lower tract disease in HSCT and organ transplant recipients, particularly lung transplant recipients.[29-31] In this group, such infections may have a high morbidity and mortality.

Molecular methods such as reverse transcriptase-PCR are increasingly being used to identify viral etiologies of RTI in transplant recipients as the diagnostic sensitivity of the commonly used direct fluorescent antibody test and culture based assays are limited. Kumar et al[14] carried out a prospective study of respiratory virus infections in 100 lung transplant recipients. Fifty patients with symptoms consistent with an upper respiratory tract infection were enrolled and matched to fifty asymptomatic lung transplant recipients. DFA, culture and PCR testing of nasopharyngeal swabs and bronchoalveolar lavage samples were performed. The viruses discovered are outlined in Table 1. Based on this study,[14] and confirmed by other studies in which molecular testing is used,[32-34] it is apparent that viruses such as coronavirus, rhinovirus and metapneumovirus are frequently etiologies of viral RTIs in transplant patients and may account for both direct sequelae (including viral pneumonitis) and indirect sequelae such as graft rejection and bacterial or fungal superinfection (see below). We are only beginning to understand the potential clinical impact of these additional viruses in transplant recipients, but severe disease may occur in subsets of patients.

For example, in one study in HSCT recipients, rhinovirus infection prior to engraftment was associated with fatal interstitial pneumonitis in one-third of patients.[35]

### Indirect Consequences of Viral RTIs

Indirect consequences of viral RTI's include triggering acute and chronic rejection and bacterial and fungal superinfection. Indirect effects have been best studied in lung transplant recipients, in who there is accumulating evidence that viral RTIs may trigger bronchiolitis obliterans syndrome (BOS) in a subset of patients.[36-38] In the previously described prospective study of viral RTIs in lung transplant recipients, the 90 day follow-up showed that acute rejection episodes occurred significantly more often in those with respiratory viruses than the matched controls[14] (see Fig. 2). In addition, a decrease in FEV-1 of greater than or equal to 20% was seen in 18% (9/50) of patients with symptomatic respiratory tract infections. Of these nine patients, 6 were classified as having bronchiolitis obliterans syndrome at one year.[14]

Retrospective studies have shown similar findings. Palmer et al[29] reviewed records of 122 adult lung recipients and found 10 episodes of viral RTI with RSV, parainfluenza, influenza or adenovirus. These were identified in patients with pneumonia who underwent bronchoscopy. Death occurred in 2 patients and subsequent BOS in 4 patients. Wendt et al[30] reviewed 19 cases of lower respiratory paramyxovirus infections in 84 lung transplant recipients. The majority received treatment

**Table 1.** *Etiology of community acquired viral respiratory tract infections (RTI) and clinical outcomes in 50 lung transplant recipients and matched controls*

| Characteristic | RTI Patients (n = 50) | Non-RTI Patients (n = 50) | p-Value |
|---|---|---|---|
| Viral etiology | 33 (66%) | 4 (8%) | <0.001 |
|   Rhinovirus 9 | 4 | | |
|   RSV   6 | | | |
|   Parainfluenza | 4 | | |
|   Influenza A | 5 | | |
|   Metapneumovirus | 1 | | |
|   Coronavirus* | 8 | | |
|   Influenza B | 0 | | |
|   Adenovirus | 0 | | |
|   Enterovirus | 0 | | |
| Acute rejection | 8 (16%) | 0 | 0.006 |
| FEV-1 decline (>20%) | 9 (18%) | 0 | 0.003 |
| Percent change in FEV-1 (mean change at 3 months ± SD) | −4.6% | +1.1% | 0.03 |
| Bacterial or fungal superinfection | 3 (6%) | 1(2%) | NS |
| CMV reactivation | 3 (6%) | 3 (6%) | NS |

*Only specimens negative for other viruses were tested for coronaviruses. Adapted from: Kumar D et al. Am J Transplant 2005; 5(8):2031-6;[14] with permission from Blackwell Publishing.

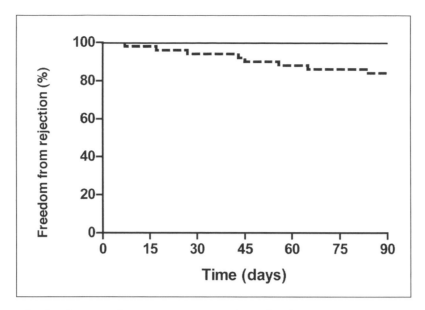

Figure 2. Kaplan-Meier curve for development of acute rejection in a cohort of 50 lung transplant recipients with viral respiratory tract infection (dotted line) compared to 50 matched controls without viral RTI (solid line); p = 0.006 by log-rank statistic. Adapted from: Kumar D et al. Am J Transplant 2005; 5(8):2031-6;[14] with permission from Blackwell Publishing.

with aerosolized ribavirin and only 1 fatality was seen. However 6/18 surviving patients showed a subsequent decline in FEV-1. Holt et al[39] assessed routine surveillance bronchoscopies and bronchoscopies done in patients with pneumonia in 140 adult lung transplant recipients. They found respiratory viruses in 22/140 patients (15.2%) some who had severe symptoms while others were asymptomatic. Rejection and development of permanent impairment of respiratory function was reported in several patients following infection. Several of these studies describe patients with concomitant or subsequent fungal (*Aspergillus* sp.) and bacterial (most commonly *Pseudomonas*) infections. Garantziotis et al[40] described three cases of influenza pneumonia in lung transplant patients. Despite treatment a significant decline in pulmonary function occurred in all 3 patients along with histological evidence of acute rejection or BOS. While the evidence for indirect effects in nonlung transplant recipients is less compelling, RTIs have also been associated with kidney rejection in some studies. For example, in a study of 360 episodes of acute kidney rejection, a correlation was found between previous influenza or adenovirus infection and acute rejection.[41] Similar

associations between rejection and infection with RSV and parainfluenza have been reported in kidney recipients.[42,43]

### Emerging Respiratory Viruses

Two recent emerging infections illustrate several points about the impact a new contagious respiratory virus infection can have on transplant patients and transplant programs in general. These are discussed in more detail below.

### Severe Acute Respiratory Syndrome (SARS)

A classic example of a new pathogen causing a new disease is that of the SARS associated coronavirus (CoV) (see Fig. 3). SARS-CoV was first detected in the Guangdong Province of China in November 2002 and caused atypical pneumonia.[44,45] The illness rapidly spread to a number of different countries across the world. At the end of the outbreak, the total number of cases reported to the World Health Organization was 8,096 and included 774 deaths. In Toronto, Canada, at least two organ transplant recipients (liver and lung transplant) developed SARS

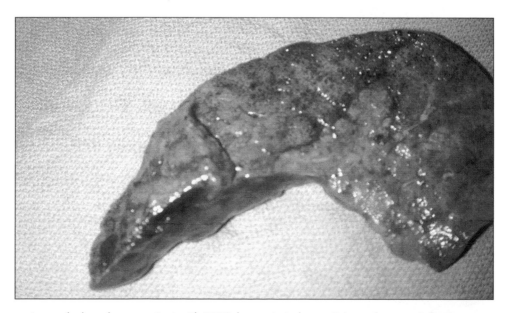

Figure 3. Autopsy specimen of a lung from a patient with SARS demonstrated necrotizing pulmonary infection.

**Table 2. *Hypothetical potential impact of a new contagious viral infection upon transplant patients***

| Potential Consequences of Emerging Infections in Transplant Patients |
| --- |
| Once exposed, more likely to get symptomatic disease ("sentinel chicken") |
| Given contact with Health care setting high risk for exposure |
| Disease more likely to be rapidly progressive and lethal |
| Higher viral burden, higher shedding—Increased infectivity—"Super-spreaders" |
| Potential for donor transmission |

and died.[46,47] Each case was associated with a significant number of secondary infections among family and health care workers. In addition, SARS CoV viral loads in autopsy specimens from the lung transplant recipient were in general several thousand-fold greater than nontransplant patients that had died of SARS.[47,48] Therefore, transplant patients may shed larger amounts of virus than nontransplant recipients and may thereby have the potential to spread the disease to a larger number of secondary contacts (a term dubbed "super-shedders" during the SARS outbreak). In addition the mortality of such a virus may be greater in transplant patients than the general population. Another potential issue identified during the SARS outbreak was the possibility of unknowingly transmitting the SARS CoV from an organ donor to a recipient. Since SARS CoV could be isolated from organs other than lungs, the potential for viral transmission in a donor incubating SARS would conceivably be possible not only for lung transplants but for other organ transplants as well.[1,47,48] In Toronto, the multi-organ transplant program was closed during much of the SARS outbreak. Implementation of clinical donor screening tools and strict isolation precautions allowed the program to resume activity during the midst of the outbreak. Although there has been no recurrence of SARS, this outbreak provided an important lesson for preparedness of the transplant program in the face of newly emerging viruses.[1,49,50] Table 2 summarizes the potential consequences of any contagious emerging viruses in transplantation.

### *Avian Influenza*

Outbreaks of H5N1 influenza in poultry and other birds have spread across the globe at an unprecedented rate. While human transmission has generally involved close contact with infected birds, there is concern the virus may mutate to one that is capable of efficient human-to-human transmission.[51,52] If that occurs, this may herald the onset of the next great influenza pandemic. Currently, great efforts and resources are being dedicated to pandemic preparedness and research in the biology of avian influenza. Murray et al[53] have estimated that 62 million persons would be killed if a pandemic were to occur, based on death rates from the 1918 pandemic. The majority of deaths would be in the developing world. What will be the consequences to transplant patients and transplantation? No cases of H5N1 to date have occurred in transplant patients. However, logical inferences can be made by examining outbreaks of other viruses as well as by looking at the impact of other influenza viruses on transplant patients[50,54] and are similar to those listed in Table 2. These include: (1) in the event of an influenza pandemic, the transplant patient is more likely to develop disease following exposure to the virus; (2) there may also be more opportunities for exposure given that the patient may have frequent contact with the health care setting; (3) disease is likely to be more severe with a higher mortality; (4) viral replication may be greater leading to greater infectivity and shedding of virus. Studies that have evaluated shedding of influenza virus from nasopharyngeal washings in HSCT patients infected with influenza have demonstrated prolonged shedding in this patient population.[16]

In the event of a pandemic, there are also significant implications for the transplant program.[54] There will be concern regarding transmission of pandemic influenza from donors to recipients. In infection, H5N1 influenza virus can be isolated not only from the respiratory tract but is also present in the gastrointestinal tract and other organs.[55,56] Therefore, it is possible that virus is transmissible by a lung transplant as well as other organ transplants. However, more importantly, during a pandemic, resources will be allocated where needed most. Mechanical ventilators and ICU beds will likely be occupied by patients with H5N1 influenza. Therefore, transplant centers may not have sufficient infrastructure capacity (including ventilator beds) to continue performing transplants. The CDC predicts high rates of worker absenteeism may contribute to a breakdown of the normal infrastructure. In the midst of a significant influenza pandemic, it is unlikely that transplant programs will continue to function effectively, leading to further mortality on the transplant waiting list.[54]

## Human Herpesvirus 6 and 7

### *Background*

Both human herpesvirus-6 (HHV-6) and human herpesvirus-7 (HHV-7) are ubiquitous herpesviruses that infect humans early in life.[57,58] A substantial body of literature now exists to suggest that these viruses are potentially pathogenic and clinically important in transplant recipients.[57-69] These viruses are both enveloped double-stranded DNA viruses and along with cytomegalovirus (CMV), make up the β-herpesvirus subgroup of herpesviruses. Both viruses share significant sequence homology to CMV, which is one of the most common and important pathogens in organ transplant recipients.[58]

HHV-6 has been shown to be the cause of roseola (exanthem subitum), a febrile illness of early childhood. Infection in childhood is extremely common and seroprevalence studies indicate that over 90% of healthy adults are seropositive for HHV-6.[70] Like other herpesviruses, HHV-6 can persist in a latent form after the primary infection. The primary target cell for HHV-6 is the CD4+ T-cell (via the CD46 receptor), although the virus appears to have a broad cellular host range. The sites of latency appear to include monocytes and macrophages, oropharyngeal and salivary epithelial cells and possibly other cell types.[60] Based on DNA sequences, there are two distinct variants of HHV-6 designated HHV-6A and HHV-6B. Most infections in transplant recipients appear to be due to variant B. HHV-7 also infects CD4+ T-cells via the CD4 molecule.[58] Again, primary infection occurs in childhood and may be asymptomatic or result in febrile syndromes and roseola. This virus also establishes life-long persistence and latency.

### *Direct Effects of Viral Infection*

HHV-6 and 7 are increasingly becoming recognized as important pathogens following solid-organ and bone marrow transplantation. Reactivation of these viruses appears to be a common event after transplant and has been associated with direct and indirect clinical effects.[57-69]

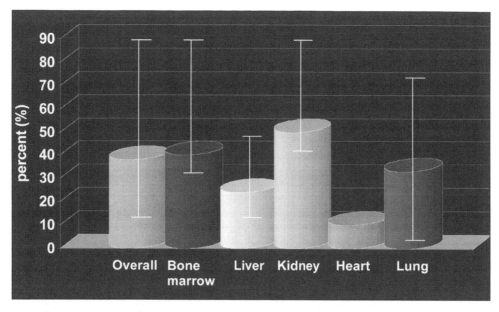

Figure 4. Reported incidence of HHV-6 infection in various transplant types. Error bars demonstrate wide range of reported reactivation rates. Generally the incidence of symptomatic disease is much lower then asymptomatic reactivation.

HHV-6 infection can be defined as evidence for reactivation of latent virus regardless of symptoms. Figure 4 illustrates the relative incidence of HHV-6 infection in different transplant types. The reported incidence varies widely but averages around 30% (range less than 10% to greater than 80%).[61-69] The wide reported range of infections is likely due to different populations studied who may use different immunosuppression protocols and due to the use of different laboratory tests for the diagnosis of infection. HHV-6 reactivation appears to occur quite early after transplantation, usually in the first 2-4 weeks. This is in contrast to CMV infection which usually occurs after the first month of transplantation. In SOT patients, HHV-6 infection has been associated with nonspecific febrile illnesses often accompanied by leukopenia and thrombocytopenia.[62] In fact, CMV, HHV-6 and HHV-7 are often recognized concomitantly suggesting that viral syndromes may often be due to a combination of these viruses.[71] HHV-6 infection has also been reported to cause hepatitis; pneumonitis, encephalitis and colitis in organ transplant recipients.[62,67,72] Directly attributable symptoms due to HHV-7 alone have not been as well recognized as they have for HHV-6. However, HHV-7 viremia appears to be very common in some settings. For example, in a study of 263 livers, heart, kidney and pancreas transplant recipients, HHV-7 viremia was found in 64.3 % of patients.[73] Data from this study suggested that antiviral prophylaxis with ganciclovir or valganciclovir may have decreased the rate of HHV-6 viremia, but had no effect on HHV-7 viremia.[73]

### Indirect Effects of Viral Infection

Apart from there direct effects, a substantial body of literature now exists that suggests that these viruses likely have significant indirect effects exerted through immunomodulatory and immunosuppressive effects[64-66,68,74,75] (see Fig. 5). The indirect consequences of viral replication likely outweigh direct clinical effects.[68] These indirect effects may occur through a number of mechanisms. For example, in vitro data demonstrate that infection of T-cells by HHV-6 results in immune suppression characterized by a down-regulation of IL-2 mRNA and protein synthesis accompanied by a significant reduction in mitogen-driven proliferative responses resulting in a cell mediated immune defect.[76] A general immunosuppressive effect induced by HHV-6 may predispose to further opportunistic infection, or alternatively, HHV-6 may have more specific interactions with other viruses such as CMV. For example, HHV-6 has been demonstrated to modulate cytokine expression in infected mononuclear cells, including potent induction of TNF-α

production.[77] In turn, TNF-α has been shown to stimulate the CMV immediate early gene enhancer/promoter region in a dose-dependent manner resulting in CMV reactivation.[78]

In addition to in vitro data, there is now substantial clinical data from well-designed cohort studies that support the evidence for indirect effects of viral replication. Several studies have found that HHV-6 and -7 infection are risk factors for the development of CMV disease or co-infect patients with CMV disease.[65,66,68,69] Dockrell et al[66] demonstrated that HHV-6 seroconversion was an independent risk factor for the development of symptomatic CMV disease following liver transplantation with a relative risk of 4.0 (CI 2.3-8.0). These findings have been confirmed in studies using more sensitive and specific molecular diagnostic tests.[68] In a study of 200 liver transplant recipients, HHV-6 viremia was an independent risk factor for the development of CMV disease and opportunistic infection.[68] HHV-6 has also been associated with the development of fungal infections.[66,69] Rogers et al[69] evaluated 80 liver transplant recipients by serial cultures for HHV-6 and found infection to be an independent predictor of fungal infections and mortality. Some studies (although results are conflicting) have also associated HHV-6 and HHV-7 with allograft rejection and graft dysfunction. In a study of BAL samples from 87 lung transplant recipients, identification of HHV-6 by PCR was independently associated with the development of BOS.[79] Although these studies may be confounded by use of different diagnostic tests, sample size considerations and difficulty in determining temporal relationships between viral infection and outcomes, most of the literature does support indirect effects of viral replication.

To summarize the existing literature, it appears that reactivation of HHV-6 and HHV-7 are common after transplantation and may result in both direct and indirect effects. Direct effects are most commonly described for HHV-6. Indirect effects appear to be more common including immunomodulatory sequelae and effects on graft function. These need to be studied in greater detail.

## Adenovirus

### Virology

Adenoviruses are ubiquitous viruses that commonly cause respiratory, gastrointestinal and other illnesses. There are over 50 serotypes that can cause human disease with the majority of infections occurring in childhood.[80,81] Adenoviruses are non-enveloped, double stranded DNA

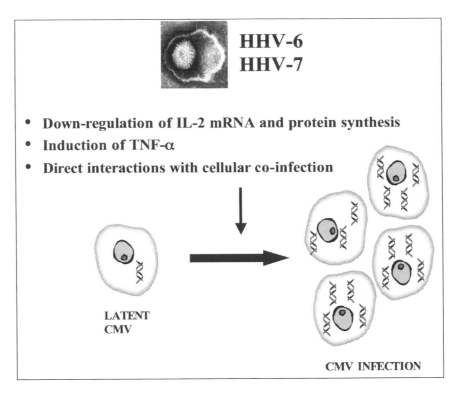

Figure 5. Indirect effects of Human herpesvirus-6 (HHV-6) and Human herpesvirus-7 (HHV-7) infections may outweigh direct consequence of viral replication. Shown is the potential effect of HHV-6 and 7 infection on cytomegalovirus (CMV) reactivation.

viruses whose genome encodes 30 or more structural and nonstructural proteins. It is believed that latent infection is established in lymphoid tissue following early infection during childhood and may persist for the lifetime of the host.[81]

### Direct Effects of Adenovirus

In immunocompetent patients, adenovirus infections are usually self-limited and relatively benign. However, in transplant recipients, infection may be fatal. Multiple different types of serious adenovirus infections have been described in both SOT and HSCT recipients including hepatitis, enteritis, pneumonitis and disseminated disease.[82-88] Severe necrotizing allograft involvement is commonly reported in case series, for example, hemorrhagic pyelonephritis in kidney transplant recipients and necrotizing pneumonia in lung transplant recipients.[84,87] A high mortality has been described in some series although reported mortality rates range from 0-100%. Infections are more common in pediatric recipients, likely due to lack of immunity to the most common adenovirus serotypes.[81,82] Active infection can also likely occur through donor transmission and from reactivation of endogenous latent virus. The relative contributions of these different modes of infection in posttransplant disease are largely unknown. When interpreting studies evaluating the incidence and clinical consequences of adenovirus infection several factors must be taken into account. These include issues in study design (retrospective vs. prospective), diagnostic methodology (molecular vs. culture or other methods), the type of population evaluated (pediatric transplant patients vs. adult) and type of adenovirus infection (viremia vs. allograft infection). These factors account in part for the wide range of reported incidences and clinical effects of adenovirus infection in organ transplant patients.

In a recent review of 55 pediatric solid organ and HSCT patients with adenovirus infection at a single institute, the prevalence rates were calculated at 1 per 16 SOT performed.[89] Gastrointestinal tract disease was the most common and infections occurred early posttransplant (median 1.6 months). Mortality was 14.6% but was exclusively seen in HSCT recipients.[89] In another review of 484 pediatric liver transplant

recipients, 49 (10%) had adenovirus infection.[86] The liver, lung or gastrointestinal tract were the most common sites involved. Mortality was significant and occurred in 9/49 patients (18.4%).[86] Serious infections are also well described in adults. In a study of 191 adult liver transplant recipients, 11 (5.8%) had a positive culture for adenovirus from a variety of sites, of which 7 patients had symptomatic disease with 2 deaths.[83] Lung transplant recipients are at risk for severe necrotizing pneumonitis. In a review of 308 lung transplant recipients, 4 were identified with adenovirus pneumonia. All occurred early posttransplant and all were fatal.[87]

These reports are in contrast to a recently published prospective study of 263 adult liver, kidney, heart and pancreas transplant recipients in which PCR testing of plasma samples for adenovirus was performed at regular intervals posttransplant.[90] Viremia was detected in 19 patients (7.2%) at some point n the first year posttransplant. At the time of viremia, 11 of 19 (58%) patients had no symptoms, while the remaining patients had gastrointestinal, respiratory, or nonspecific symptoms, all relatively mild. In contrast to previous reports, all patients recovered spontaneously without any specific treatment.[90] This suggests that viremia may be a relatively common occurrence when specifically looked for using active molecular surveillance. While cumulatively, most studies suggest a high rate of invasive disease and a significant mortality, important difference may exist in different transplant populations and depending on how data are accrued in various epidemiological studies.

### Indirect Effects of Adenovirus

Given that adenovirus infections (either due to reactivation of latent virus, or acquisition of new infection) may be very common in some settings and often infect the allograft, they may have indirect effects such triggering either acute or chronic allograft rejection. This has emerged as a common theme with viruses that are able to infect the allograft. In a study of pediatric heart transplant recipients, Shirali et al[91] performed PCR studies on 553 consecutive biopsy samples from 149 transplant recipients. A positive PCR for any virus was obtained in 48 samples

(8.7 percent) from 34 patients (23 percent). A number of viruses were found including CMV, enterovirus and others, but adenovirus was the most common virus, being detected in 30 samples. Viral detection and specifically adenovirus, was associated with an adverse cardiac event within three months after the positive biopsy, including acute rejection, coronary vasculopathy and graft loss.[91] In a study evaluating 16 lung and heart-lung recipients, adenovirus was detected in the transplanted lung in 8 patients and was associated with graft failure and histological obliterative bronchiolitis.[92] In contrast, in the study outline previously in which molecular surveillance for viremia was carried out in 263 transplant patients, detection of adenovirus in the blood had no impact on graft rejection or function.[90] It is possible that viral infection involving primarily the allograft may behave differently from self-limited viremic episodes in terms of inducing indirect sequelae.

## West Nile Virus

### Virology and Epidemiology

West Nile virus (WNV) WNV belongs to the genus Flavivirus.[93,94] It is a single stranded positive-sense RNA virus that is transmitted to humans via infected mosquitoes. WNV is maintained in an enzootic cycle involving primarily *Culex sp.* mosquitoes and birds. Birds may develop transient high-titer viremia that allows transmission of the virus to feeding mosquitoes. "Spill over" infections of humans and other mammals may then occur from mosquito vectors. After the bite of an infected mosquito, initial viral replication likely occurs in the skin and regional lymph nodes. A primary viremia seeds the reticuloendothelial system and is followed by a secondary viremia that in some persons may result in seeding of the central nervous system.[93,94] WNV first appeared in North America during an outbreak in New York in 1999 that resulted in several cases of meningoencephalitis.[95] Since then the virus has spread rapidly across the United States and into Canada causing large epidemics of meningoencephalitis. Carefully designed population-based seroprevalence studies have found that in immunocompetent persons about, 80% of infections are asymptomatic while 20% have a self-limited febrile illness that may be accompanied by a maculopapular rash. Based on these studies, the estimated risk of severe neuroinvasive disease, which includes meningitis, encephalitis and acute flaccid paralysis is approximately 1 in 140 infections or less than 1%.[96,97]

### WNV and Transplantation

The ability of the virus to infect the CNS likely relates to a combination of specific viral virulence factors and the host immune response. Both humoral and cell mediated immune responses appear to be important for the control of WNV.[98,99] In mouse models of WNV infection, passive antibody has shown to be protective and IgM production appears to be critical in terminating viremia and preventing CNS invasion.[100,101] In addition, CD8 + T-cells have been shown to traffic into the brain after neuroinvasion. Mice lacking CD8 + T-cells have a significantly higher titer of virus in the CNS.[102] Immunosuppressed animals have prolonged viremia, more neuroinvasive disease, more extensive/severe pathology and a higher fatality rate.[103] Clinical and epidemiologic studies of WNV further support the hypothesis that more severe disease occurs in transplant recipients when compared to immunocompetent individuals.

Transplant recipients may acquire WNV by three primary mechanisms: (1) transfusion transmitted WNV (2) organ donor transmitted infection and (3) community acquired (mosquito) exposure to WNV. Of these three mechanisms, community acquired WNV is the most common route of infection although all three have been described in transplant patients as outlined below.

Pealer et al[104] described 23 cases of transfusion transmitted WNV (TTWNV) in 2003 occurring in the United States. Ten of the 23 patients in the series were immunocompromised (stem cell recipient,

organ transplant, cancer patient) and the reported overall mortality was 29%. This study provided evidence that TTWNV does occur and immunocompromised patients in particular may be at high risk for severe neuroinvasive complications of WNV infection.[104] It was estimated that the risk of TTWNV was approximately 1:30000 blood units or higher in areas with epidemic WNV activity. Based on these studies and a number of other case reports of TTWNV, blood product screening for WNV using a nucleic acid test (NAT) was instituted in Canada and the United States. The initiation of screening (usually using minipools of 6-16 samples) removed thousands of units of potentially infective blood products over the ensuing years and was very successful in reducing the number of cases of TTWNV.[105]

The second mechanism by which transplant recipients may acquire WNV is by organ donor transmission. So far, two separate instances of donor organ transmission of WNV to transplant recipients has also been documented since the emergence of the WNV epidemic in North America.[106,107] In the first instance, a patient that had acquired WNV through transfusion donated a liver, heart and both kidneys to four separate recipients. Three of four recipients developed WNV meningoencephalitis and one developed WNV febrile illness.[106] The time to development of symptoms ranged from 7 to 17 days posttransplant and one of the recipients died due to WNV. In a second instance a donor from New York city with a traumatic head injury donated a liver, one lung and both kidneys to four separate recipients.[107] Two of the 4 recipients (liver and lung) developed severe neuroinvasive WNV disease. Subsequent investigation demonstrated recent community acquired WNV infection in the donor (IgM positive; however PCR testing was negative).

Cases of donor transmitted WNV have meant that transplant programs in areas with WNV activity have had to consider the issue of clinical and laboratory screening of donors. Clinical screening of potential donors is generally simple and an important method of decreasing the chances of organ transmitted WNV.[1] For example, donors with encephalitis, meningitis, or flaccid paralysis in which WNV is a known or potential etiology should generally not be considered. Laboratory screening for WNV in donors is a controversial issue. Several commercially available nucleic acid tests for WNV, which could be potentially applied to screening donors, are available. However, a number of issues need to be considered prior to a particular program instituting a routine WNV NAT testing protocol. These include test availability, approved vs. unapproved tests, the turnaround time for testing and the logistics involved in rapid testing of deceased donors. Also, a medical decision analysis of donor testing reached the conclusion that WNV testing would result in a loss of 452.4 life years due to false positive test results.[108]

Currently no clear guidelines exist regarding NAT testing of organ donors. The Organ Procurement and Transplant Network (OPTN) with the Health Resources and Services Administration (HRSA) in the United States suggests that NAT testing could be considered in areas with significant WNV activity (www.optn.org). As more information is gained about the utility of NAT testing this recommendation may be reassessed. Some centers have instituted NAT testing of donors with rapid turnaround time. If NAT testing is done on a potential donor and is reactive, consideration should be given to deferring the donor unless the recipient has a life-threatening illness where no other life-saving therapies exist. Living donors should avoid activities that place them at risk for WNV infection in the two weeks prior to transplant (e.g., such as camping). In living donors, WNV NAT testing can be done as close to the time of transplant as possible. If the test is reactive, the donation should be deferred for a period of time. Overall, screening of donors for WNV remains controversial. In general, testing would be most reliable in areas with a high prevalence of WNV. The authors would therefore suggest that donor screening be done in areas with significant WNV activity.

Although reports of transfusion and donor transmission of virus are very dramatic, the number of cases is very small compared to those that acquire WNV through the community. In fact, the majority of case reports of WNV infection in transplant recipients are due to mosquito exposure. Several case series suggest a high incidence of neurologic illness with significant morbidity and mortality[109-111] (see Figs. 6, 7). Seroprevalence studies in the general population show the risk of neuroinvasive disease in patients infected with WNV is approximately 1:140 cases or < 1%.[96] We carried out a seroprevalence study in our population of organ transplant recipients shortly after the first occurrence of a WNV outbreak in Canada.[112] The study measured IgG and IgM to WNV in over 800 transplant patients. The rate of a positive IgM antibody (indicating recent infection) was 0.25% (95% CI 0.03-0.88%). Based on this seroprevalence, in conjunction with hospital surveillance data, the risk of neurologic disease was calculated to be 40% (95% CI 16%—80%).[112] Therefore in general, a transplant patient has a much higher risk of neuroinvasive disease if infected with WNV compared to the general population (40% vs. less than 1%). Based on this data, appropriate education of transplant patients is critical during periods of high seasonal WNV activity. Simple measures such as the use of insect repellant and full-sleeves clothing can lead to important reductions in the risk of community-acquired WNV in this patient population. In transplant recipients who do acquire WNV, immunosuppressive therapy should be promptly decreased or discontinued. Remaining therapy is largely supportive or involves use of investigational drugs such as immune globulin preparations or interferon therapy.[94]

To summarize, the emergence of WNV in North America highlights the unique susceptibility of the transplant population. It also demonstrated the number of different routes by which a transplant patient may be infected by a new and emerging infection. A multi-faceted prevention program is the best approach for dealing with such infections. For example, potential measures to prevent WNV during periods of high local WNV activity include (1) clinical and NAT screening of organ donors, (2) NAT testing of blood products and (3) education of transplant patients so that specific precautions are followed during WNV season including the use of enhanced personal protection measures such as insect repellant.

## Unusual Viruses and Allograft Transmission

Latent viral infections are known to be transmitted with the allograft. Common examples of this include herpesviruses such as CMV and EBV. In addition, West Nile Virus donor transmission has been discussed above. A particular concern that has been raised recently, however, is donor transmission of rare viral agents such as rabies and lymphocytic choriomeningitis virus (LCMV).

The rabies virus is classified under the genus Lyssavirus and is present in animal reservoirs including bats, dogs and foxes. Transmission to humans generally occurs from the bite of a rabid animal.[113] Cases in the United States are rare but occur sporadically. More recently, the primary source of human rabies in North America has been transmission by silver-haired bats or Mexican free-tailed bats. The bat variant of rabies appears to be less neurotropic and may also affect fibroblasts. This is in contrast to countries in which rabies is endemic such as India, where dog bites account for up to 97% of human cases.[114] The number of annual human cases in India is estimated to be 20,565 or 2 per 100,000.[114]

Recently, in two separate instances, the transmission of rabies from an infected organ donor to multiple recipients has been documented.[115,116] In 2004, four recipients of a common donor (liver, two kidneys and an arterial segment graft for use in a liver transplant) developed encephalitis within 30 days of transplant and died an average of 13 days after onset of symptoms.[115] Autopsy studies from all patients showed the classic Negri bodies throughout the central nervous system and histopathologic staining was positive for rabies virus antigens. In this case, the donor's toxicology screen was positive for cocaine and marijuana and the cause of death was attributed to progressive subarachnoid hemorrhage resulting in brain death. During the investigation, it was discovered that the donor had reported being bitten by a bat. In a separate incidence that occurred in 2005 in Germany, rabies-associated encephalitis developed in three of six organ transplant recipients of a common donor.[116] The donor had reportedly traveled to India and had contact with stray dogs.

Lymphocytic choriomeningitis virus (LCMV) is an arenavirus maintained in a rodent reservoir.[117] Chronic viral infection occurs in rodents that may then excrete virus in the urine. Humans may

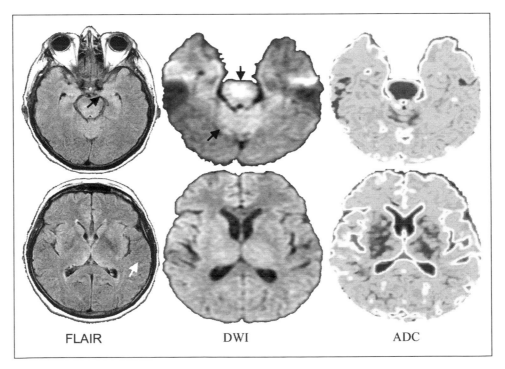

FLAIR                    DWI                    ADC

Figure 6. MRI scan of the brain of a liver transplant patient with West Nile virus meningoencephalitis demonstrating multiple areas of abnormal signal intensity.

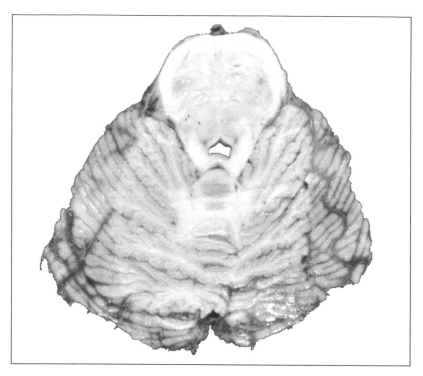

Figure 7. Autopsy specimen of brain tissue from a liver transplant patient with West Nile virus meningoencephalitis. Numerous areas of necrosis were evident (white arrow).

acquire this virus by aerosolized rodent urine, direct contact with rodents or rodent bites. After an incubation period of 5 to 10 days, there is commonly a febrile illness with subsequent rash. A subset of these patients may go on to develop an aseptic meningitis or encephalitis.[117] There is no specific therapy although ribavirin has shown some benefit for other related arenavirus infections. Fischer et al[118] reported transmission of LCMV in two separate incidences involving two donors. The first cluster, transplanted in 2003, consisted of four organ recipients (liver, two kidneys, lung) of a common donor. The second cluster, transplanted in 2005, was also comprised of four organ recipients (liver, two kidneys and lung) from a common donor. Seven of the eight patients died and autopsy studies confirmed LCMV by immunohistochemical staining of tissue for viral antigens. The donor for the 2005 cluster donor had contact with a pet hamster that had an LCMV strain identical to the organ recipients. However, no source of LCMV was found for the 2003 cluster. Despite sensitive nucleic acid detection testing LCMV was not found in the donor.[118]

These cases highlight the importance of transmission of unusual viruses with organ transplantation. Although well publicized it should be emphasized that such cases are a rarity. However, they do highlight the importance of a thorough clinical screening of potential organ donors. Caution should be exercised before accepting organs from donors in whom the cause of death is unknown. This is especially important in instances where donors have meningitis or encephalitis of unknown etiology. While many factors, including the urgency of need in a potentially very ill recipient comes into play, it is the authors' opinion that organs from donors in the latter category should not be used. Since potential donor screening tests of sufficient sensitivity or specificity do not exist for most of these rare pathogens, this is generally not an option. However even if such tests were available, the rarity of such cases would make them difficult to justify in terms of cost or practicality. In addition, for rare diseases, false positive results may lead to unnecessary exclusion of donors whose organs would be perfectly suitable.

## Summary

Emerging infections encompass a wide array of potential pathogens that may affect transplant patients. Of recent interest has been an increasing number of novel or unusual viral pathogens in this patient population. From existing data, it is apparent that these viral infections may have both direct and indirect effects. Direct effects may lead to substantial morbidity and mortality. For some viruses, indirect effects on graft function, acute and chronic rejection, may be even more important than the direct effects of viral infection. In addition in recent years we have witnessed dramatic examples of donor transmission of unusual viral pathogens to multiple recipients often with devastating results. What does the future hold in terms of emerging infections? Further research into the epidemiology and clinical consequences of the different pathogens summarized in this chapter will lead to a greater understanding of both direct and indirect effects. In addition donor-screening protocols will undoubtedly continue to evolve over time to accommodate the potential for new transmission threats. Finally and almost certainly, new, currently unrecognized pathogens will crop up. Again the effects and impact on transplant patients are likely to be unique from the general population.

## References

1. Kumar D, Humar A. Emerging viral infections in transplant recipients. Curr Opin Infect Dis 2005; 18(4):337-41.
2. Boeckh M, Erard V, Zerr D et al. Emerging viral infections after hematopoietic cell transplantation. Pediatr Transplant 2005; 9 Suppl 7:48-54.
3. Cainelli F, Vento S. Infections and solid organ transplant rejection: a cause-and-effect relationship? Lancet Infect Dis 2002; 2(9):539-49.
4. Rubin RH. The direct and indirect effects of infection in liver transplantation: pathogenesis, impact and clinical management. Curr Clin Top Infect Dis 2002; 22:125-54.
5. Ison MG, Hayden FG. Viral infections in immunocompromised patients: what's new with respiratory viruses? Curr Opin Infect Dis 2002; 15(4):355-67.
6. American Society of Transplantation. Guidelines for the Prevention and Management of Infectious Complications of Solid Organ Transplantation: community-acquired respiratory viruses. Am J Transplant 2004; 4:105-9.
7. Barton TD, Blumberg EA. Viral pneumonias other than cytomegalovirus in transplant recipients. Clin Chest Med 2005; 26(4):707-20.

8.  Hodges TN, Torres FP, Marqueson J et al. Community acquired respiratory viruses in lung transplant patients: incidence and outcomes. J Heart Lung Transplant 2001; 20:169-70.

9.  Billings JL, Hertz MI, Wendt CH. Community respiratory virus infections following lung transplantation. Transpl Infect Dis 2001; 3:138-48.

10. Vilchez R, McCurry K, Dauber J et al. Influenza and parainfluenza respiratory viral infection requiring admission in adult lung transplant recipients. Transplantation 2002; 73:1075-78.

11. Monto AS. Epidemiology of viral respiratory infections. Am J Med 2002; 112:4S-12S.

12. Arcasoy SM, Kotloff RM. Lung transplantation. N Engl J Med 1999; 340:1081-91.

13. Gerna G, Vitulo P, Rovida F et al. Impact of human metapneumovirus and human cytomegalovirus versus other respiratory viruses on the lower respiratory tract infections of lung transplant recipients. J Med Virol 2006; 78(3):408-16.

14. Kumar D, Erdman D, Keshavjee S et al. Clinical impact of community-acquired respiratory viruses on bronchiolitis obliterans after lung transplant. Am J Transplant 2005; 5(8):2031-36.

15. Garbino J, Gerbase MW, Wunderli W et al. Lower respiratory viral illnesses: improved diagnosis by molecular methods and clinical impact. Am J Respir Crit Care Med 2004; 170:1197-203.

16. Nichols WG, Guthrie KA, Corey L et al. Influenza infections after hematopoietic stem cell transplantation: risk factors, mortality and the effect of antiviral therapy. Clin Infect Dis 2004; 39(9):1300-6.

17. Vilchez RA, Fung J and Kusne S. The pathogenesis and management of influenza virus infection in organ transplant recipients. Transplant Infect Dis 2002; 4:177-82.

18. Vilchez RA, McCurry K, Dauber J et al. Influenza virus infection in adult solid organ transplant recipients. Am J Transplant 2002: 2:287-91.

19. Ljungman P, Andersson J, Aschan J et al. Influenza A in immunocompromised patients. Clin Infect Dis 1993; 17:244-47.

20. Blumberg EA, Albano C, Pruett T et al. The immunogenicity of influenza virus vaccine in solid organ transplant recipients. Clin Infect Dis 1996; 22:295-302.

21. Duchini A, Hendry RM, Nyberg LM et al. Immune response to influenza vaccine in adult liver transplant recipients. Liver Transpl 2001; 7(4):311-13.

22. Mazzone PJ, Mossad SB, Mawhorter SD et al. The humoral immune response to influenza vaccination in lung transplant patients. Eur Respir J 2001; 18(6):971-76.

23. Soesman NM, Rimmelzwaan GF, Nieuwkoop NJ et al. Efficacy of influenza vaccination in adult liver transplant recipients. J Med Virol 2000; 61(1):85-93.

24. Guidelines for vaccination of solid organ transplant candidates and recipients. Am J Transplant 2004; 4 Suppl 10:160-63.

25. Blumberg EA, Fitzpatrick J, Stutman PC et al. Safety of influenza vaccine in heart transplant recipients. J Heart Lung Transplant 1998; 17(11):1075-80.

26. Lawal A, Basler C, Branch A et al. Influenza vaccination in orthotopic liver transplant recipients: absence of post administration ALT elevation. Am J Transplant 2004; 4(11):1805-9.

27. Kimball P, Verbeke S, Flattery M et al. Influenza vaccination does not promote cellular or humoral activation among heart transplant recipients. Transplantation 2000; 69(11):2449-51.

28. White-Williams C, Brown R, Kirklin J et al. A. Improving clinical practice: should we give influenza vaccinations to heart transplant patients? J Heart Lung Transplant 2006; 25(3):320-23.

29. Palmer SM Jr, Henshaw NG, Howell DN et al. Community respiratory viral infection in adult lung transplant recipients. Chest 1998; 113:944-50.

30. Wendt CH, Fox JM, Hertz MI. Paramyxovirus infection in lung transplant recipients. J Heart Lung Transplant 1995; 14:479-85.

31. Whimbey E, Champlin RE, Couch RB et al. Community respiratory virus infections among hospitalized adult bone marrow transplant recipients. Clin Infect Dis 1996; 22:778-82.

32. Larcher C, Geltner C, Fischer H et al. Human metapneumovirus infection in lung transplant recipients: clinical presentation and epidemiology. J Heart Lung Transplant 2005; 24(11):1891-901.

33. Khalifah AP, Hachem RR, Chakinala MM et al. Respiratory viral infections are a distinct risk for bronchiolitis obliterans syndrome and death. Am J Respir Crit Care Med 2004; 170:181-87.

34. Garbino J, Gerbase MW, Wunderli W et al. Respiratory viruses and severe lower respiratory tract complications in hospitalized patients. Chest 2004; 125:1033-39.

35. Ghosh S, Champlin R, Couch R et al. Rhinovirus infections in myelosuppressed adult blood and marrow transplant recipients. Clin Infect Dis 1999; 29(3):528-32.

36. Husain S, Singh N. Bronchiolitis obliterans and lung transplantation: evidence for an infectious etiology. Semin Respir Infect 2002; 17:310-14.

37. Vilchez RA, Dauber J, Kusne S. Infectious etiology of bronchiolitis obliterans: the respiratory viruses connection—myth or reality? Am J Transplant 2003; 3:245-49.

38. Sharples LD, McNeil K, Stewart S et al. Risk factors for bronchiolitis obliterans: a systematic review of recent publications. J Heart Lung Transplant 2002; 21:271-81.

39. Holt ND, Gould FK, Taylor CE et al. Incidence and significance of noncytomegalovirus viral respiratory infection after adult lung transplantation. J Heart Lung Transplant 1997; 16:416-19.

40. Garantziotis S, Howell DN, McAdams HP et al. Influenza pneumonia in lung transplant recipients: clinical features and association with bronchiolitis obliterans syndrome. Chest 2001; 119:1277-80.

41. Gabriel R, Selwyn S, Brown D et al. Virus infections and acute renal transplant rejection. Nephron 1976; 16:282-86.

42. DeFabritus AM, Riggio RR, David DS et al. Parainfluenza type 3 in a transplant unit. JAMA 1979; 241(4):384-86.

43. Miller RB and Chavers BM. Respiratory syncytial virus infections in pediatric renal transplant recipients. Pediatr Nephrol 10(2).

44. Poutanen SM, Low DE. Severe acute respiratory syndrome: an update. Curr Opin Infect Dis 2004; 17(4):287-94.

45. Peiris JS, Yuen KY, Osterhaus AD et al. The severe acute respiratory syndrome. N Engl J Med 2003; 18;349(25):2431-41.

46. Kumar D, Tellier R, Draker R et al. Severe Acute Respiratory Syndrome (SARS) in a liver transplant recipient and guidelines for donor SARS screening. Am J Transplant 2003; 3(8):977-81.

47. Kumar D, Farcas G, Uy K et al. Severe acute respiratory syndrome (SARS) in transplantation: clinical and virologic findings and implementation of a SARS screening tool. Am J Transplant 2004; suppl 8:425, Abstract #975.

48. Farcas GA, Poutanen SM, Mazzulli T et al. Fatal severe acute respiratory syndrome is associated with multiorgan involvement by coronavirus. J Infect Dis 2005; 15;191(2):193-97. 27.

49. Smith JM, McDonald RA. Emerging viral infections in transplantation. Pediatr Transplant 2006; 10(7):838-43.

50. Allen U, Hebert D, Churchill C et al. The potential impact of SARS on organ transplantation: exercise caution. Pediatr Transplant 2003; 7(5):345-47.

51. Kaye D, Pringle, CR. Avian influenza viruses and their implication for human health. Clin Infect Dis 2005; 40:108.

52. The Writing Committee of the World Health Organization (WHO) Consultation on Human Influenza A/H5. Avian influenza A (H5N1) infection in humans. N Engl J Med 2005; 353:1374-85.

53. Murray CJ, Lopez AD, Chin B et al. Estimation of potential global pandemic influenza mortality on the basis of vital registry data from the 1918-20 pandemic: a quantitative analysis. Lancet 2007; 368(9554):2211-18.

54. Kumar D and Humar A. Pandemic influenza and its implications for transplantation. Am J Transplant 2006; 6(7):1512-17.

55. de Jong, MD, Bach et al. Fatal avian influenza A (H5N1) in a child presenting with diarrhea followed by coma. N Engl J Med 2005; 352:686.

56. Rimmelzwaan GF, van Riel D, Baars M et al. Influenza A virus (H5N1) infection in cats causes systemic disease with potential novel routes of virus spread within and between hosts. Am J Pathol 2006; 168(1):176-83.

57. Singh N. Infections with human herpesvirus 6, 7 and 8 after hemopoietic stem cell or solid organ transplant. In Transplant Infections; Second Edition. Ed. Bowden, Ljungman, Paya. Lippincott Williams and Wilkins; 2004; pp. 367-74.

58. Dockrell DH, Paya CV. Human herpesvirus-6 and -7 in transplantation. Rev Med Virol 2000; 11:23-36.

59. Razonable RR, Paya CV. The impact of human herpesvirus-6 and -7 infection on the outcome of liver transplantation. Liver Transpl 2002; 8(8):651-58.

60. Singh N, Carrigan DR. Human herpesvirus-6 in transplantation: an emerging pathogen. Ann Intern Med 1996; 124:1065-71.

61. Singh N, Carrigan DR, Gayowski T et al. Human herpesvirus-6 infection in liver transplant recipients: documentation of pathogenicity. Transplantation 1997; 64:674.

62. Chang FY, Singh N, Gayowski T et al. Fever in liver transplant recipients: changing spectrum of etiologic agents. Clinical Infect Dis 1998; 26:59.

63. Mendez JC, Dockrell DH, Espy MJ et al. Human beta-herpesvirus interactions in solid organ transplant recipients. J Infect Dis 2001; 183(2):179-184.

64. Dockrell DH, Prada J, Jones MF et al. Seroconversion to human herpesvirus 6 following liver transplantation is a marker of cytomegalovirus disease. J Infect Dis 1997; 176(5):1135-40.

65. Desjardins JA, Gibbons L, Cho E et al. Human herpesvirus-6 reactivation is associated with cytomegalovirus infection and syndromes in kidney transplant recipients at risk for primary cytomegalovirus infection. J Infect Dis 1998; 178:1783-86.

66. Dockrell DH, Mendez JC, Jones M et al. Human herpesvirus 6 seronegativity before transplantation predicts the occurrence of fungal infection in liver transplant recipients. Transplantation 1999; 67:399-403.

67. Singh N, Paterson DL. Encephalitis caused by human herpesvirus-6 in transplant recipients. Transplantation 2000; 69:2474-79.

68. Humar A, Kumar D, Caliendo AM et al. Clinical impact of human herpesvirus 6 infection after liver transplantation. Transplantation 2002; 73:599-604.

69. Rogers J, Rohal S, Carrigan DR et al. Human herpesvirus-6 in liver transplant recipients: role in pathogenesis of fungal infections, neurologic complications and outcome. Transplantation 2000; 69:2566-73.

70. Okuno T, Takahashi K, Balachandra K et al. Seroepidemiology of human herpesvirus 6 infection in normal children and adults. J Clin Microbiol 1989; 27:651-53.

71. Razonable RR, Rivero A, Brown RA et al. Detection of simultaneous beta-herpesvirus infections in clinical syndromes due to defined cytomegalovirus infection.Clin Transplant. 2003; 17(2):114-20.

72. Delbridge MS, Karim MS, Shrestha BM, McKane W. Colitis in a renal transplant patient with human herpesvirus-6 infection. Transpl Infect Dis 2006; 8(4):226-28.

73. Razonable RR, Brown RA, Humar A et al. PV16000 Study Group. Herpesvirus infections in solid organ transplant patients at high risk of primary cytomegalovirus disease. J Infect Dis 2005; 192(8):1331-39.

74. Humar A, Kumar D, Raboud J et al. Interactions between cytomegalovirus, human herpesvirus-6 and the recurrence of hepatitis C after liver transplantation. Am J Transplant 2002; 2(5):461-66.

75. Singh N, Husain S, Carrigan DR et al. Impact of human herpesvirus-6 on the frequency and severity of recurrent hepatitis C virus hepatitis in liver transplant recipients. Clin Transplant 2002; 16(2):92-96.

76. Flammand L, Gosselin J, Stefanescu I et al. Immunosuppressive effect of human herpesvirus 6 on T-cell functions. Blood 1995; 85:1263-71.

77. Flammand L, Gosselin J, D'Addario M et al. Human herpesvirus 6 induces interleukin-1 beta and tumor necrosis factor alpha, but not interleukin-6, in peripheral blood mononuclear cell cultures. J Virology 1991; 65:5105-10.

78. Stein J, Volk HD, Liebenthal C et al. Tumor necrosis factor alpha stimulates the activity of the human cytomegalovirus immediate-early enhancer/promoter in immature monocytic cells. J Gen Virol 1993; 74:2333-38.

79. Neurohr C, Huppmann P, Leuchte H et al. Munich Lung Transplant Group. Human herpesvirus 6 in bronchalveolar lavage fluid after lung transplantation: a risk factor for bronchiolitis obliterans syndrome? Am J Transplant 2005; 5(12):2982-91

80. Hoffman JA. Adenoviral disease in pediatric solid organ transplant recipients. Pediatr Transplant 2006; 10(1):17-25.

81. Kojaoghlanian T, Flomenberg P, Horwitz MS. The impact of adenovirus infection on the immunocompromised host. Rev Med Virol 2003; 13:155-71.

82. Green M, Ljungman P, Michaels MG. Adenovirus, parvovirus B19, papilloma virus and polyomaviruses after hemopoietic stem cell or solid organ transplantation. In: Bowden RA, Ljungman P, Paya C, eds. Transplant Infections, 2nd Edition. Philadelphia: Lippincott, Williams and Wilkins, 2003:399-411.

83. McGrath D, Falagas ME, Freeman R et al. Adenovirus infection in adult orthotopic liver transplant recipients: incidence and clinical significance. J Infect Dis 1998; 177:459-62.

84. Kim SS, Hicks J, Goldstein SL. Adenovirus pyelonephritis in a pediatric renal transplant patient. Pediatr Nephrol 2003; 18(5):457-61.

85. Saad RS, Demetris AJ, Lee RG et al. Adenovirus hepatitis in the adult allograft liver. Transplantation 1997; 64:1483-85.

86. Michaels M, Green M, Wald ER et al. Adenovirus infection in pediatric orthotopic liver transplant recipients. J Infect Dis 1992; 165:170-74.

87. Ohori NP, Michaels MG, Jaffe R et al. Adenovirus pneumonia in lung transplant recipients. Hum Pathol 1995; 26:1073-79.

88. Wreghitt TG, Gray JJ, Ward KN et al. Disseminated adenovirus infection after liver transplantation and its possible treatment with ganciclovir. J Infect 1989; 19:88-89.

89. de Mezerville MH, Tellier R, Richardson S et al. Adenoviral infections in pediatric transplant recipients: a hospital-based study. Pediatr Infect Dis J 2006; 25(9):815-18.

90. Humar A, Kumar D, Mazzulli T et al. PV16000 Study Group. A surveillance study of adenovirus infection in adult solid organ transplant recipients. Am J Transplant 2005; 5(10):2555-59.

91. Shirali GS, Ni J, Chinnock RE et al. Association of viral genome with graft loss in children after cardiac transplantation. N Engl J Med 2001; 344:1498-503.

92. Bridges ND, Spray TL, Collins MH et al. Adenovirus infection in the lung results in graft failure after lung transplantation. J Thorac Cardiovasc Surg 1998; 116:617-23.

93. Petersen LR, Marfin AA. West Nile virus: A primer for the clinician. Ann Intern Med 2002; 137:173-79.

94. Campbell GL, Marfin AA, Lanciotti RS et al. West Nile virus. Lancet Infect Dis 2002; 2:519-29.

95. Nash D, Mostashari F, Fine A et al. The outbreak of West Nile virus infection in the New York City area in 1999. N Engl J Med 2001; 344:1807-14.

96. Mostashari F, Bunning ML, Kitsutani PT et al. Epidemic West Nile encephalitis, New York, 1999: Results of a household-based seroepidemiological survey. Lancet 2001; 358:261-64.

97. Tsai TF, Popovici F, Cernescu C et al. West Nile encephalitis epidemic in southeastern Romania. Lancet 1998; 352:767-71.

98. Gea-Banacloche J, Johnson RT, Bagic A et al. West Nile virus: Pathogenesis and therapeutic options. Ann Intern Med 2004; 140:545-53.

99. Camenga DL, Nathanson N, Cole GA. Cyclophosphamide-potentiated West Nile viral encephalitis: Relative influence of cellular and humoral factors. J Infect Dis 1974; 130:634-41.

100. Diamond MS, Shrestha B, Marri A et al. B cells and antibody play critical roles in the immediate defense of disseminated infection by West Nile encephalitis virus. J Virol 2003; 77:2578-86.

101. Diamond MS, Sitati EM, Friend LD et al. A critical role for induced IgM in the protection against West Nile virus infection. J Exp Med 2003; 198(12):1853-62. Epub 2003.

102. Shrestha B, Diamond MS. Role of CD8+T cells in control of West Nile virus infection. J Virol 2004; 78(15):8312-21.

103. Mateo R, Xiao SY, Guzman H et al. Effects of immunosuppression on West Nile virus infection in hamsters. Am J Trop Med Hyg 2006; 75(2):356-62.

104. Pealer LN, Marfin AA, Petersen LR. 1 Transmission of West Nile Virus through Blood Transfusion in the United States in 2002. N Engl J Med 2003; 349:1236-45.

105. Stramer SL, Fang CT, Foster GA et al. West Nile virus among blood donors in the United States, 2003 and 2004. N Engl J Med 2005; 353(5):451-59.

106. Iwamoto M, Jernigan DB, Guasch A et al. Transmission of West Nile virus from an organ donor to four transplant recipients. N Engl J Med 2003; 348:2196-203.

107. Centers for Disease Control and Prevention (CDC). West Nile virus infections in organ transplant recipients—New York and Pennsylvania. MMWR Morb Mortal Wkly Rep 2005; 54(40):1021-23.

108. Kiberd BA, Forward K. Screening for West Nile virus in organ transplantation: a medical decision analysis. Am J Transplant 2004; 4(8):1296-301.

109. Kumar D, Prasad GV, Zaltzman J et al. Community-acquired West Nile virus infection in solid organ transplant recipients. Transplantation 2004; 77:399-402.

110. Kleinschmidt-DeMasters BK, Marder BA, Levi ME et al. Naturally acquired West Nile virus encephalomyelitis in transplant recipients: clinical, laboratory, diagnostic and neuropathological features. Arch Neurol 2004; 61(8):1210-20.

111. Ravindra KV, Freifeld AG, Kalil AC et al. West Nile virus-associated encephalitis in recipients of renal and pancreas transplants: case series and literature review. Clin Infect Dis 2004; 38(9):1257-60.

112. Kumar D, Drebot MA, Wong SJ et al. A seroprevalence study of West Nile virus infection in solid organ transplant recipients. Am J Transplant 2004; 4(11):1883-88.

113. Jackson AC. Rabies: new insights into pathogenesis and treatment. Curr Opin Neurol 2006; 19(3):267-70.

114. Sudarshan MK, Madhusudana SN, Mahendra BJ et al. Assessing the burden of human rabies in India: results of a national multi-center epidemiological survey. Int J Infect Dis 2006.

115. Srinivasan A, Burton EC, Kuehnert MJ et al. Transmission of rabies virus from an organ donor to four transplant recipients. N Engl J Med 2005; 352(11):1103-11.

116. Hellenbrand W, Meyer C, Rasch G et al. Cases of rabies in Germany following organ transplantation. Euro Surveill 2005; 10(2):E050224.6.

117. Peters CJ. Lymphocytic choriomeningitis virus– an old enemy up to new tricks. N Engl J Med 2006; 354(21):2208-11.

118. Fischer SA, Graham MB, Kuehnert MJ et al. Transmission of lymphocytic choriomeningitis virus by organ transplantation. N Engl J Med 2006; 354(21):2208-11.

# CHAPTER 35

# Cytomegalovirus and Allograft Failure after Solid Organ Transplantation

Hugo Bonatti, Walter C. Hellinger* and Raymund R. Razonable

## Abstract

Allograft rejection and infections are the two major complications of solid organ transplantation. These clinical entities are intimately interrelated, with one predisposing to the other, in a bidirectional relationship. Allograft rejection and its treatment predisposes to various infections, most commonly with cytomegalovirus (CMV). In turn, the occurrence of CMV infection and disease increases the risk of developing acute rejection and chronic allograft dysfunction. In this chapter, we review the clinical and experimental data that support the association between CMV infection and chronic allograft loss after solid organ transplantation. The pathogenetic mechanisms that have been proposed, including immunologic cross-reactivity, the upregulation of immune molecules and the generation of adhesion molecules and inflammatory mediators are evaluated. The implications of the association between CMV and long-term allograft survival are discussed.

## Introduction

Cytomegalovirus (CMV), a ubiquitous β-herpes virus that infects the majority of humans, is the single most common pathogen that negatively influences the outcome of solid organ transplantation (SOT).[1,2] Primary CMV infection, which manifests as benign febrile mononucleosis-like illness in immunocompetent individuals, occurs early in life.[2] By adulthood, up to 80% have serologic evidence of prior CMV infection. Following primary infection, CMV remains latent and often reactivates during immunocompromised states such as after SOT.[2]

The natural history of CMV after SOT is characterized by its reactivation (from an endogenous latent reservoir or in the transplanted allograft) and rapid replication to cause clinical syndromes.[2] CMV disease occurs in 7-32% of SOT recipients, with the risk being lowest in kidney and highest in lung recipients.[2] CMV-seronegative individuals who receive solid organ allograft transplant from CMV-seropositive donors (CMV D+/R−) are at highest risk.[2] Additionally, a wide variety of factors, including acute rejection, intense immunosuppression and viral co-infections heighten the risk of CMV.[2] The incidence of CMV disease has been reduced by antiviral prophylaxis and preemptive therapy.[2-4] However, antiviral prophylaxis has only delayed the onset of CMV so that it occurs most commonly among CMV D+/R− individuals during the third to the sixth month after SOT.[5]

The clinical manifestations of posttransplant CMV disease is categorized into direct and indirect effects.[2] The direct effects of CMV include a febrile illness accompanied by some degree of myelosuppression (CMV syndrome). Tissue-invasive CMV disease may also occur and manifest as allograft dysfunction, including hepatitis, myocarditis, nephritis, pancreatitis, pneumonitis among others. The indirect effects of CMV include an increased predisposition to opportunistic infections such as Epstein Barr virus-associated posttransplant lymphoproliferative disease, fungal and bacterial superinfections, infections with hepatitis C and other herpes viruses. Acute and chronic allograft rejection are the other major indirect effects of CMV.[2]

As will be emphasized in this chapter, the role of CMV in acute rejection, chronic graft dysfunction and allograft survival is suggested by epidemiological, clinical and experimental models. In numerous studies, recipients of kidney, pancreas, liver, heart and lung allografts who developed CMV disease are at higher risk for acute rejection and chronic allograft dysfunction compared to those who did not develop CMV disease. Other studies, however, have contradicted these findings, emphasizing the complex nature of the relationship between CMV and the allograft status. The complexity of this association is likely a reflection of its bidirectional nature. While several studies have shown that CMV is associated with an increased risk of concurrent[6] and subsequent[7] acute rejection, the reverse is also observed, wherein individuals with acute rejection are at higher risk of CMV disease, which is likely a result of a generally more immunosuppressed state.[5,8] Hence, the question of "which comes first—rejection or infection?" is often asked.

The complex relationship between CMV and acute rejection has confounded the analysis of the impact of CMV on long-term allograft function and survival. In several studies, CMV infection was demonstrated as a risk factor for chronic allograft dysfunction in cardiac,[9] lung,[10] liver,[11] pancreas[12] and kidney[13] transplant recipients.[14] On the other hand, as acute rejection often precedes chronic allograft dysfunction, determining whether CMV is a contributor to chronic allograft dysfunction has been a challenge. Currently, chronic allograft dysfunction is considered as the single major cause of long-term allograft failure among SOT recipients.[12,15-18] The inter-related clinical variables of acute rejection and CMV are major risk factors for chronic allograft dysfunction and distinguishing the role of CMV in long-term allograft failure after SOT will be the goal of this chapter.

*Corresponding Author: Walter C. Hellinger—Division of Infectious Diseases, Mayo Clinic, Jacksonville, Florida, USA.
Email: hellinger.walter@mayo.edu

*Chronic Allograft Failure: Natural History, Pathogenesis, Diagnosis and Management*, edited by Nasimul Ahsan. ©2008 Landes Bioscience.

## Pathogenesis

The mechanisms underlying the association between CMV, acute rejection and chronic allograft dysfunction is not well-defined. However, there are at least three possible mechanisms that have been proposed, as follows:

1. CMV may directly invade the parenchyma of transplanted allografts and cause allograft dysfunction. In many cases, this represents the reactivation of donor-derived latent CMV. For example, CMV hepatitis may be observed more commonly in liver transplant recipients, while CMV nephritis may be seen more commonly in kidney transplant recipients. In a study of kidney transplant recipients, a relationship was observed between CMV viremia, allograft dysfunction and CMV-associated glomerular injury.[19] Hence, CMV may cause excessive loss of allografts by inducing specific glomerular lesions,[19] which if left untreated, could result in allograft failure and loss.

2. CMV may infect endothelial cells and the resulting endotheliitis could lead to vasculopathy and thrombosis. The association between CMV and hepatic artery thrombosis has been observed in liver transplant recipients,[20] and this could lead to allograft failure. Heart and kidney transplant recipients who developed CMV had elevated anti-endothelial cell antibodies.[21] This humoral response to endothelial cell antigens suggests endothelial cell injury and implies a risk for vascular rejection, chronic allograft dysfunction and decreased allograft survival.[21]

3. CMV can trigger acute rejection and chronic allograft dysfunction. The rat models of kidney,[22,23] lung,[24] cardiac[25] and liver[26] transplantation support the role of CMV in allograft rejection. However, the mechanisms underlying the association between CMV infection and allograft failure remains unclear. Table 1 lists the proposed mechanism underlying the association between CMV and allograft rejection. One of the proposed mechanisms is the ability of CMV to induce concurrent alloantigen stimulation that generates cytotoxic T lymphocyte responses, which then trigger acute rejection.[27] Data derived from biopsies obtained from CMV-infected transplant recipients indicate that CMV upregulates the expression of adhesion molecules,[24,28,29] thereby facilitating the inflammatory process and the expression of major histocompatibility complex (MHC) antigens on the surface of allograft cells through mediators such as interferon.[30] Additionally, CMV induces smooth muscle proliferation and intimal thickening,[31] which are hallmarks of transplant atherosclerosis—the most common cause of long-term allograft failure after cardiac transplantation.

*Expression of adhesion molecules.* The characteristic feature of acute rejection is the infiltration of the allograft parenchyma with inflammatory lymphocytes. This is mediated by the expression of adhesion molecules such as E-selectin, intercellular adhesion molecule-1 (ICAM-1) and vascular cell adhesion molecule-1 (VCAM-1). Notably, these molecules are upregulated during CMV infection.[23,32] Accordingly, it is believed that, through the upregulation of these molecules, CMV facilitates rejection.[29] In a rat model, CMV caused a very prolonged expression of VCAM-1 and ICAM-1 in vascular endothelium—a finding that was correlated with enhanced histological changes of chronic allograft rejection.[33]

*Upregulation of MHC class antigens.* The administration of immunosuppressive agents such as steroids after SOT is believed to suppress expression of MHC antigens. During allograft rejection, MHC antigens reappear, possibly due to γ-interferon released by activated T-cells. It is proposed that CMV enhances graft rejection by its ability to induce expression of MHC class antigens.[29] There is demonstrable expression of MHC class II antigens on hepatocyte membranes and bile ducts in liver recipients with severe allograft rejection.[24] In the rat model, CMV upregulated MHC class II antigen expression on the surface of heart endothelial cells.[34,35] MHC class II antigens were displayed on most tubular and virtually all endothelial cells during CMV disease.[30] Since recognition of nonself MHC antigens is the major determinant of graft rejection, the upregulation of these molecules by γ-interferon released during CMV infection could contribute to graft rejection.[30] Moreover, antiviral treatment with ganciclovir has been shown to reduce MHC class II expression in a dose-dependent manner.[36] In contrast, the data on MHC class I expression during CMV infection is contradictory. In separate studies, CMV resulted in upregulation[37] or downregulation[32] of MHC class I antigens.

*Immunologic cross-reactivity.* In a murine model, CMV accelerated the development of cardiac rejection.[38] This observation was hypothesized to result from the cross-reactivity between murine CMV epitopes and alloantigens.[38] A sequence homology between the conserved region of HLA-DRβ chain with a 5-amino acid peptide from the CMV immediate early-2 (IE-2) antigen has been suggested as one possible explanation for increased graft rejection during CMV infection.[39]

*Smooth muscle proliferation, endothelial damage and intimal thickening.* One major pathologic finding in transplant allografts undergoing chronic allograft dysfunction is transplant vascular sclerosis (atherosclerosis). The pathogenesis of transplant atherosclerosis appears to involve adhesion molecules and MHC class antigens and hence, CMV, by upregulating the expression of these proteins, is believed to contribute to this pathology. Additionally, CMV infects

**Table 1. Suggested mechanisms by which CMV induces allograft rejection**

| Suggested Mechanisms | Comments |
| --- | --- |
| Expression of adhesion molecules | CMV facilitates allograft rejection by its ability to induce the expression of E-selectin, intercellular adhesion molecule-1 and vascular cell adhesion molecule-1. |
| Upregulation of major histocompatibility complex antigens | The interferon-γ induced expression of MHC Class II molecules could facilitate rejection of the transplanted allograft. |
| Immunologic cross-reactivity | The cross-reactivity between CMV and alloantigens could result in allograft rejection during CMV infection. |
| Smooth muscle proliferation | CMV induces doubling of smooth muscle cell proliferation and contributes to transplant vascular sclerosis. |
| Endothelial damage and intimal thickening | The endothelial injury during CMV infection could lead to transplant vascular sclerosis. CMV has also been shown to induce the secretion of endothelial growth factors. |
| Interstitial fibrosis | CMV infection is associated with high collagen content to DNA ratio. There is increased expression of both types I and II collagen, transforming growth factor β1 and connective tissue growth factor. |

endothelial cells and the endothelial injury could contribute to vascular sclerosis.[21] Elevated levels of anti-endothelial cell antibodies, which is a marker for endothelial injury, have been observed in cardiac and kidney recipients with CMV infection.[21] In the rat model of kidney transplantation, CMV infection resulted in endothelial cell swelling and intimal proliferation in the allograft vasculature.[23] CMV has also been suggested to result in the production of endothelial cell growth factors.

CMV infection results in the induction of molecules associated with increased cell proliferation. In a rat model, CMV infection was associated with doubling of smooth muscle cell proliferation. CMV also significantly induced the transcription of platelet-derived growth factor BB, transforming growth factor-β1 and basic fibroblast growth factor in rats.[36] The rat CMV-encoded chemokine receptor r33 has also been suggested in the development of transplant vascular sclerosis, chronic allograft dysfunction and vascular smooth muscle cell migration.[31] Compared to uninfected controls, allografts from rat CMV-infected subjects had upregulation of chemokine expression and the timing of increased expression paralleled the rat CMV-accelerated neointimal formation. Likewise, in the rat models of small bowel and heart transplantation, CMV accelerated the time to chronic allograft dysfunction by increasing the severity of transplant vascular sclerosis in the small bowel and heart allografts.[40]

Collectively, these experimental data suggest that CMV induces chronic allograft dysfunction through its transplant vascular sclerosis characterized by intimal thickening and smooth muscle proliferation. The role of CMV in this pathology is further supported by the observation that ganciclovir treatment reduced smooth muscle proliferation in a dose-dependent manner.[36]

*Interstitial fibrosis and other histologic changes.* Another major pathologic feature of chronic kidney allograft dysfunction is interstitial fibrosis. In the rat model, CMV-infected rats had accelerated graft fibrosis, characterized by higher collagen content to DNA ratio, when compared to the uninfected rats.[41] An increased expression of type III collagen mRNA was observed in CMV-infected allografts.[41] CMV increases expression of both types I and III collagens and the accumulation of myofibroblasts and CMV enhances total collagen synthesis resulting in interstitial fibrosis characteristic of chronic allograft dysfunction.[42] Furthermore, CMV accelerates chronic allograft dysfunction through its ability to induce the expression of transforming growth factor β1, platelet-derived growth factor-AA and connective tissue growth factor.[43]

The rat CMV model demonstrated enhanced and accelerated tubular epithelial apoptosis that accompanied histological changes of chronic allograft damage.[44] The rat CMV model also demonstrated the increased influx of CD4+ cells and macrophages and an increase in glomerular sclerosis.[45] The rat model further demonstrated that CMV significantly enhanced chronic kidney allograft rejection, with increase in inflammatory infiltrates, glomerular mesanginal matrix and capillary membrane thickening and a more extensive tubular epithelial atrophy.[23]

## CMV and Acute Allograft Rejection

A wealth of clinical studies has suggested that CMV could trigger acute rejection in various SOT settings. In a retrospective study of 301 heart recipients, there was a significantly higher rate of acute rejection during the first 3 months among the 91 patients with CMV compared to the 210 patients without CMV infection.[46] In a retrospective study of 32 lung recipients, CMV infection preceded or occurred concomitantly with acute rejection.[47] In another retrospective study of 325 liver recipients, CMV was significantly associated with a higher number of acute rejection, particularly those with gB1 genotype.[48]

The clinical and experimental models of kidney transplantation provide the most convincing evidence regarding the association between CMV and acute rejection.[49,50] In the rat model,[23] CMV infection was

associated with increased frequency and a more prolonged duration of acute kidney rejection.[23] Clinically, CMV-infected kidney recipients have a high frequency of acute rejection.[30] Some studies even found a correlation between CMV, rejection and graft-related pathology, regardless of symptomatology, while others have noted this correlation only in subjects with symptomatic CMV disease.[51] In a prospective study of 242 kidney recipients, CMV disease was found to be a statistically significant risk factor for acute rejection.[7] In another study involving 477 kidney recipients, even CMV infection alone without disease manifestation was a significant independent predictor of acute rejection. CMV disease was also significantly and independently associated with biopsy-proven tubulo-interstitial rejection.[49] In one of the largest studies conducted in a cohort of 3365 kidney recipients, the presence of CMV was independently associated with acute rejection, which in turn was the major factor that influenced the occurrence of late allograft failure.[52]

Interestingly, even the mere predisposition for CMV, in the absence of active CMV infection and disease, has been significantly correlated with acute rejection. Pretransplant CMV seropositivity was found to be an independent predictor of acute rejection.[53] In a study of 333 kidney recipients, CMV D+/R− individuals had higher rate of acute rejection compared to nonmismatched individuals. In a multiple logistic regression analysis, CMV D+/R− was an independent predictor of acute rejection.[50]

A major observation that lends support to the role of CMV in acute rejection is the impact of antiviral prophylaxis. In a large randomized placebo-controlled clinical trial, the use of valacyclovir as anti-CMV prophylaxis reduced not only the incidence of CMV disease, but also the incidence of biopsy-proven acute rejection by 50% in CMV D+/R− kidney recipients.[54] In another study, kidney recipients who received CMV-immune globulin prophylaxis had lower incidence not only of CMV infection but also of allograft rejection.[55] Moreover, among kidney recipients with histologically-proven acute rejection that was not responsive to conventional anti-rejection therapy, the subgroup of patients with concomitant CMV infection had improvement in allograft function after the institution of anti-CMV therapy.[56]

While all these data support the role of CMV and the clinical utility of anti-CMV drugs in the treatment of some cases of CMV-associated acute rejection, it is also important to emphasize that the data on the effect of antiviral prophylaxis on the incidence of allograft rejection is far from conclusive. In two recent meta-analyses, the use of preemptive anti-CMV therapy[57] or anti-CMV prophylaxis[58] was not significantly associated with a reduction in acute rejection after SOT. In contrast, another meta-analysis observed that anti-CMV prophylaxis and preemptive therapy were significantly associated with reduction in acute rejection after SOT.[59]

## CMV and Chronic Allograft Loss

There have been remarkable improvements in the management of acute rejection across the various SOT populations. This achievement, together with improvements in the prevention and treatment of various infectious syndromes, has allowed the transplant patients to survive longer. As will be discussed in the subsequent sections, chronic allograft dysfunction, sometimes specifically termed as chronic rejection, has become a major entity that negatively impacts the outcome of SOT. As will be evident, the manifestations of chronic allograft dysfunction vary across the different SOT groups and thus will be discussed separately according to transplant type (Table 2).

### *Kidney Transplantation*

The success of kidney transplantation has continually improved and currently, depending on the age group, the one-year graft survival ranges from 86-95% and the five-year graft survival ranges from 58-84% (UNOS data, May 2006). After technically-successful transplantation,

**Table 2. Summary of associations between CMV and allograft rejection in various solid organ transplants**

| Type of Transplant | Features Suggestive of Chronic Rejection | Main Target | Rejection as Cause of Graft Loss | Association with CMV Mismatch | Association with CMV Infection and Disease | Potential Benefits of Antiviral Drugs |
|---|---|---|---|---|---|---|
| Kidney | Glomerular sclerosis, atherosclerosis, interstitial fibrosis | Glomeruli, vascular structures | Common | Established | Established | Suggestive |
| Pancreas | Interstitial fibrosis, acinar loss, thrombosis | Parenchyma, endothelium | Common | Established | Established | Unknown |
| Liver | Vanishing bile duct syndrome | Biliary tree | Rare | Established | Established | Suggestive |
| Small bowel | villous blunting, mucosal inflammation, cryptitis, endothelitis | Vascular structures, mucosa | Rare | Not established | Unknown | Unknown |
| Lung | Bronchiolitis obliterans syndrome | Bronchioles | Common | Established | Established | Highly suggestive |
| Heart | Vasculopathy | Endothelium | Common | Established | Established | Highly suggestive |
| Composite tissue | Not defined | Not defined | Not defined | Unknown | Unknown | Unknown |

allograft failure is mainly due to recurrence of underlying kidney disease or chronic allograft dysfunction. Chronic kidney rejection, which is characterized by interstitial fibrosis, glomerulosclerosis, tubular atrophy and arterial narrowing, is the major reason for the allograft loss after the first year after transplant. Although the pathogenic mechanisms of chronic allograft dysfunction and rejection remains poorly understood, it is suggested that this may result from the production of cytokines and growth factors by different cell types, leading to structural changes in the allograft parenchyma and proliferative remodeling of allograft vasculature. Clinically, this is manifested by a gradual deterioration of graft function. In most studies, acute rejection is the single most important factor that influences the occurrence of chronic allograft dysfunction. In many of these studies, the role of CMV on chronic allograft dysfunction has also been suggested. The association between CMV and the inflammatory responses in the allograft may persist, thereby producing continuous immunological injury that could lead to chronic rejection.

In experimental models, CMV accelerated the progression of chronic graft rejection. As described above, this could be mediated by a combination of mechanisms that include cross-reactivity between alloantigens and CMV-encoded proteins[39,60] and the upregulation of adhesion receptors in endothelial and tubular epithelial cells.[28,35] In the rat model,[23] CMV was associated with accelerated chronic allograft damage index and the degree of serum creatinine elevation was higher.[23]

In a study of 259 kidney recipients, CMV disease was independently associated with chronic allograft nephropathy.[61] Histological and immunohistochemical analyses of biopsy specimens obtained from 50 kidney recipients with allograft rejection demonstrated that the occurrence of CMV was significantly associated with a higher percentage of allograft loss.[62] Another study demonstrated a lower allograft survival among kidney recipients with tissue-invasive CMV disease.[63] Kidney recipients with CMV had significantly increased amount of arteriosclerotic changes in the small arterioles.[64] CMV infection has also been associated with vascular rejection.[65]

Kidney recipients who developed primary CMV infection were more likely to develop chronic allograft dysfunction.[66] In a large study of 3399 cadaveric kidney recipients, a CMV-positive donor was associated with diminished survival.[67] In another large study of kidney recipients, patient and graft survival among 12,239 CMV-seronegative patients was better than 17,636 CMV-seropositive patients.[68] Finally, there is indirect evidence that preemptive antiviral therapy has improved long-term allograft survival. In a retrospective analysis of 169 kidney recipients, patients who received preemptive therapy with ganciclovir had survival rate that was similar to those who did not develop CMV infection, suggesting that preemptive treatment of CMV with ganciclovir may have prevented CMV-induced renal injury and chronic allograft loss.[13]

## Pancreas and Islet Cell Transplanation

The short-term survival of pancreas allografts has improved with advances in immunosuppression. However, the long-term outcome remains modest with one-year survival that ranges from 58% to 82% and the five-year allograft survival that ranges from 40% to 58% (UNOS, accessed May 2005). Chronic rejection is the major cause of allograft failure after a technically-successful pancreas transplantation.

In a single-center retrospective analysis of 914 pancreas recipients (including simultaneous pancreas-kidney, pancreas after kidney and pancreas transplant alone), the majority (70%) continue to maintain adequate function during the follow-up period of 39 months.[63] In this population, the most common causes of allograft loss were technical failure (12.9%) and chronic rejection (8.8%).[63] The most significant risk factor for allograft loss due to chronic allograft dysfunction was a previous episode of acute rejection, with more than 90% of recipients

who lost pancreas allograft due to chronic rejection had a previously treated episode of acute rejection.[63] The episode of acute rejection was also significantly associated with CMV infection.[63] Hence, it is not surprising to observe that CMV was found to be a significant factor for allograft failure due to chronic allograft dysfunction. In another analysis, the occurrence of tissue-invasive CMV disease was observed to correlate with decreased survival of allografts after kidney-pancreas transplantation.[63] In this regard, CMV can also invade the pancreas allograft to cause pancreatitis, which if undiagnosed can lead to allograft failure.[69] The other risk factors for allograft failure due to chronic allograft dysfunction were retransplantation, isolated pancreas transplantation and antigen mismatches at the B loci.[63]

The impact of CMV on the allograft survival after pancreatic islet cell transplantation has not been assessed but initial data suggests that CMV may not be a common infectious complication in this population and that transmission of CMV from a CMV-seropositive donor to a CMV-seronegative recipient occurs rarely, if at all.[70,71]

## Liver Transplantation

The cumulative one- and five-year proportions of graft survival after liver transplantation, over the course of the past 10 years, are 82% and 67%, respectively (UNOS, May 2005). Allograft dysfunction resulting in death or retransplantation beyond one year after transplantation has been an infrequent event throughout the history of liver transplantation, and it has become increasingly so in recent years. The causes of chronic graft loss have also changed. During the 1980s, the 7.5% rate of graft loss after the first year was due predominantly to chronic rejection, opportunistic infection and lymphoma.[72] In the 1990s, the rate of chronic graft loss has fallen to under 5% and the most common causes of late graft dysfunction are recurrent viral hepatitis, predominantly due to hepatitis C and arterial or biliary insufficiency attributed largely, although not exclusively, to technical (surgical) complications.[73,74]

Chronic allograft rejection is usually identified within the first year after transplant.[75] Preceding episodes of acute rejection remain the predominant risk factor for the subsequent appearance of chronic allograft rejection.[18,22,75] With improvements in immunosuppressive therapies, chronic allograft rejection has declined in frequency.[73,76] It is now a complication of less than 5% of liver transplants[75] and a cause of less than one quarter of deaths beyond the first year after transplantation.[73,74]

The development of CMV disease within one year of transplant was shown to increase the risk for graft loss beyond one year after transplantation in two studies of patients transplanted prior to the regular use of CMV prophylaxis and preemptive therapy.[48,77] In one study,[48] the occurrence of CMV infection alone increased the risk for chronic allograft loss. Other studies of patients transplanted prior to the widespread use of CMV prophylaxis and preemptive therapy have identified an association between the development of CMV disease within one year of transplant and an increased risk for death beyond the first year.[78,79]

The causes of graft loss and of death beyond one year after liver transplantation were not reported in the studies associating these outcomes with prior CMV disease. CMV has been found to be a risk factor for chronic rejection in several[80-83] though not all[84] studies. CMV infection was also found to increase the risk for acute rejection in another study[48] which, as indicated above, has been found to be a major risk factor for the subsequent appearance of chronic allograft rejection. Also, studies have identified the high risk CMV D+/R− group[20,85] and the development of CMV infection[79,86,87] as risk factors for the development of hepatic artery thrombosis. One investigation, which failed to identify an association between the high risk CMV D+/R− group and hepatic artery thrombosis, included patients who may have received ganciclovir for prophylaxis, preemptive therapy or treatment.[88]

CMV disease after transplant has also been identified as a risk factor for hepatitis C related cirrhosis and allograft failure after liver transplantation for hepatitis C.[89-91] In one study, CMV disease was associated with increased severity of recurrent hepatitis C after transplant.[92] Even CMV infection alone, without the clinical manifestations of CMV disease, was significantly associated with hepatitis C related mortality after transplantation.[90] In contrast to these studies, CMV infection was not found to be associated with an increased risk for allograft failure after liver transplantation for hepatitis C in a single study of patients who received preemptive ganciclovir treatment.[93]

As reviewed above, graft dysfunction as a cause for graft loss or patient death beyond one year after transplantation is an increasingly infrequent event. Chronic rejection is an increasingly infrequent cause of late hepatic allograft loss. Hepatic arterial insufficiency and recurrent hepatitis C have become important causes of late allograft loss. CMV has been identified as a risk factor for chronic rejection, hepatic artery thrombosis and recurrent hepatitis C of increased severity after liver transplantation. Has prophylaxis or early (preemptive) treatment of CMV after liver transplantation reduced the frequency of chronic rejection, hepatic artery thrombosis or severe recurrent hepatitis C, and more to the point, the frequency of late allograft loss?

The studies which failed to identify an association between the high risk CMV D+/R− group and hepatic artery thrombosis,[88] and an association between CMV and recurrent hepatitis C[93] included patients who were managed during the era of effective anti-CMV therapy. A decreased frequency of acute cellular rejection[94] and a decreased frequency of hepatic artery thrombosi[95] have been associated with the introduction of effective anti-CMV agents for prophylaxis and preemptive therapy. Finally, improved graft survival beyond one year after transplantation has been associated with the introduction of effective anti-CMV preemptive therapy.[79]

As these observations await confirmation, it seems likely that CMV infection during the first year after liver transplantation contributes to the infrequent chronic loss of hepatic allografts by contributing to the development of chronic rejection, hepatic artery thrombosis and severe recurrent hepatitis C. The declining frequencies of chronic allograft loss, chronic rejection and hepatic artery thrombosis may be related to improvements of anti-CMV prophylaxis and preemptive therapy, in addition to the improvements in immunosuppressive therapies and surgical technique. Whether improvements in the prevention and/or early treatment of CMV will reduce the frequency of severe hepatitis C after liver transplantation remains to be seen.

## Heart Transplantion

The one-year cardiac allograft survival ranges from 80% to 90% and the five-year allograft survival ranges from 65% to 72% (UNOS, accessed May 2005). Chronic allograft dysfunction, characterized by allograft vasculopathy, is the major contributing factor for this outcome. The poor tolerance of cardiac allograft to cold ischemia time, the vulnerability to acute rejection and allograft from an older donor have been implicated as factors that contribute to chronic allograft loss.[96,97] The presence of viral genome in the myocardium has also been shown to be predictive of adverse clinical events, including coronary vasculopathy and allograft loss.[98] In this regard, CMV has long been suspected to contribute to the pathogenesis of concentric intimal hyperplasia or graft vasculopathy.[99] This is further supported by the observation that the largest pool of CMV seems to be present in the vascular endothelium.[100] Indeed, the data from cardiac transplantation provides some of the most convincing evidence that CMV is associated with long-term allograft dysfunction.[101] Several studies in this setting have collectively established the contribution of CMV infection to the accelerated development of cardiac allograft vasculopathy.[28,46] Histological examination of cardiac biopsy specimens has demonstrated that cardiac allograft recipients with CMV infection have significantly higher rates of intimal thickening that non-infected controls, especially during the first two years after transplantation.[28] Patients with CMV-positive allografts have

shorter survival and a shorter time to vasculopathy, particularly if the patient is CMV naïve at time of transplant.[102-105] Moreover, patients with CMV infection had higher rates of death and had increased allograft atherosclerosis.[46]

Animal models have shown that CMV-seropositive cardiac allografts develop vasculopathy compared to CMV-seronegative organs.[40,106] It is believed that CMV modulates certain HLA alleles especially in endothelial cells.[107] With the upregulation of HLA class I alleles and the secretion of IL-2 and other cytokines, CMV is postulated to enhance the immunogenicity of allografts, which could then translate into short- and long-term effects.[108] In addition, acute CMV infection induces endothelitis within the cardiac allograft, which could contribute to enhanced allograft arteriosclerosis,[109] and which could result in the generation of anti-endothelial cell antibodies.[21] Consequently, the endothelial function appears to be impaired during CMV infection,[110] and effect that could be abrogated by ganciclovir prophylaxis with or without the use of CMV-hyperimmunoglobulin.[111-113]

## Lung Transplantation

Recent improvements in the management of lung transplantation have resulted in improved early posttransplant survival. However, lung transplantation remains associated with high rates of long-term allograft loss.[114] While the one-year allograft survival ranges from 75% to 100%, the five-year allograft survival remains modest ranging from 30% to 50% (UNOS, accessed May 2005). Chronic lung allograft loss is commonly due to bronchiolitis obliterans syndrome (BOS),[115,116] which is histopathologically characterized by inflammation of small bronchioli and progressive narrowing of the lumen and ultimately terminal damage of the dependent alveolar areas and permanent loss of viable lung tissue.[116]

Originally thought be a purely immunologic process, there is now a different understanding of the pathogenesis of BOS. Several factors such as the gastroesophageal reflux with aspiration of gastric contents have been implicated in some cases.[117]

Various infectious agents such as respiratory syncitial virus and *Chlamydia pneumoniae* have also been associated with BOS, but CMV is the most commonly implicated pathogen that is believed to play a major role.[118] Synergy among these pathogens has also been suggested. This is crucial since pulmonary infections are among the most common complications of lung transplantation. These infections harm the allograft and together with reperfusion injury and rejection,[119-121] could lead to chronic allograft dysfunction. It is likely that the interplay among these factors occur since significant reperfusion injury has also been associated with a higher probability of CMV infection and acute rejection.

The role of CMV in BOS is indirectly suggested by the observation that CMV-seropositive allografts have shorter survival and a shorter time to develop BOS, especially if the recipient is CMV-seronegative prior to lung transplantation.[10,122,123] The development of cytotoxic immune response is crucial in the control of CMV.[124] In a study of 99 lung recipients, the lack of CMV-specific immune response is associated with repeated exacerbations of CMV pneumonitis that led to chronic worsening of allograft function.[122]

Animal models have shown that prophylaxis with either ganciclovir or hyperimmune globulin totally prevented RCMV infection-enhanced tracheal occlusion in a rat lung transplant model.[125] This approach was recently observed in the clinical setting when patients who were receiving combined ganciclovir and anti-CMV hyperimmunoglobulin had improved outcome and a reduction in the risk of bronchiolitis obliterans after lung transplantation.[126] Nevertheless, it is suggested that antiviral prophylaxis will only have a long-term effect on survival and the development of recurrent CMV infection/disease if the immune system can build up an adequate T-cell response.[127]

## *Intestinal Transplantation*

Intestinal transplantation is undertaken for the treatment for a variety of inherited or acquired diseases leading to malabsorbation. Significant improvements in the management of posttransplant complications have resulted in better outcomes, in particular, when considering viral infection.[128] The one- and five-year allograft survival after intestinal transplantation ranges from 60-85% and 40-55% (UNOS, accessed May 2005). While many factors contribute to allograft loss, the role of CMV and other herpes viruses has been suggested.[129] CMV has been implicated in the short- and long-term outcome of intestinal transplantation.[130] CMV exhibits direct and indirect effects on the allograft and the host-immune system. CMV has been associated with reduced patient and allograft survival, increased risk for subsequent bacterial and fungal infections, acute and chronic rejection, along with other complications.[130] It is well known that recurrent episodes of rejection and CMV graft enteritis lead to progressive dysfunction and eventually to destruction of the mucosa and other structures; however, there is no clear histopathological equivalent for chronic dysfunction thus far established other than vascular degeneration and some mucosal degeneration.[131,132] Nonetheless, CMV prophylaxis with ganciclovir and CMV hyperimmuneglobulin improved survival rates and reduced the incidence of CMV infection and disease.[133] However, as the numbers of intestinal transplants are small, there are currently no well-defined data to support the effect of CMV on survival or long-term adverse effects after intestinal transplantation.[134] In the rat model, vasculopathy was observed in intestinal allograft recipients with CMV infection.[40] However, it has to be emphasized that, thus far, a link between CMV and chronic rejection of intestinal allografts in humans has not been established.

## *Composite Tissue Allograft Transplantation*

Composite tissue allograft transplantation represents one of the novel approaches in transplantation medicine. Long-term allograft function after CTA transplantation is a crucial endpoint since short-term function may be difficult to assess as nerve structures must regrow into the allograft and be represented within the central nervous system.[135] Thus far, no definitive evidence of long-term allograft dysfunction is observed after CTA. Accordingly, the role of CMV infection on allograft survival has not been well-studied. Preliminary data however suggest that the overall complication rate was higher when using CMV-seropositive allograft, especially following CMV-mismatch transplantation.[136] Theoretically, after CTA, chronic degeneration of the vascularity and the nerve structures may occur after CMV infection; however, this has not been established thus far.

## Conclusions

With the continuing improvement in the immediate posttransplant care of patients, the survival of the transplanted allograft and of patients has been prolonged. Consequently, chronic rejection has become the major cause of intermediate- and long-term allograft loss and patient mortality. The etiology of chronic allograft dysfunction is multifactorial and several risk factors have been described, such as acute rejection, ischemia/reperfusion injury and viral infections, particularly CMV. In this chapter, several lines of clinical and experimental evidence were provided to support the observation that CMV has a crucial role in the pathogenesis of acute rejection and chronic allograft dysfunction after SOT. This chapter further emphasized that the interrelationship between CMV infection, allograft rejection and allograft survival is highly complex and incompletely understood. Hence, the fundamental question: "do infections trigger allograft rejection or does rejection activate CMV infection?" remains unanswered. Currently, the evidence emanating from studies conducted in kidney recipients argue for the role of CMV on allograft survival. The associations between CMV and cardiac transplant vasculopathy and of CMV and accelerated course of recurrent hepatitis C infection after liver transplantation are intriguing observations of potential clinical significance. Indeed, if these associations are proven, the aggressive prevention of CMV infection and disease could serve as a novel avenue for improving the long-term survival of the allograft and the transplant recipient.

## References

1. Razonable RR, Paya CV. Herpesvirus infections in transplant recipients: current challenges in the clinical management of cytomegalovirus and Epstein-Barr virus infections. Herpes 2003; 10(3):60-65.
2. Paya C, Razonable RR. Cytomegalovirus infection after solid organ transplantation. In: Bowden R, Ljungman P, Paya C, eds. Transplant Infections. Vol 1. Philadelphia, PA: Lippincott Williams and Wilkins 2003; 298-325.
3. Razonable RR, Paya CV. Valganciclovir for the prevention and treatment of cytomegalovirus disease in immunocompromised hosts. Expert Rev Anti Infect Ther 2004; 2(1):27-41.
4. Razonable RR, van Cruijsen H, Brown RA et al. Dynamics of cytomegalovirus replication during preemptive therapy with oral ganciclovir. J Infect Dis 2003; 187(11):1801-8.
5. Razonable RR, Rivero A, Rodriguez A et al. Allograft rejection predicts the occurrence of late-onset cytomegalovirus (CMV) disease among CMV-mismatched solid organ transplant patients receiving prophylaxis with oral ganciclovir. J Infect Dis 2001; 184(11):1461-64.
6. Dittmer R, Harfmann P, Busch R et al. CMV infection and vascular rejection in renal transplant patients. Transplant Proc 1989; 21(4):3600-1.
7. Pouteil-Noble C, Ecochard R, Landrivon G et al. Cytomegalovirus infection—an etiological factor for rejection? A prospective study in 242 renal transplant patients. Transplantation 1993; 55(4):851-57.
8. Kashyap R, Shapiro R, Jordan M et al. The clinical significance of cytomegaloviral inclusions in the allograft kidney. Transplantation 1999; 67(1):98-103.
9. Everett JP, Hershberger RE, Norman DJ et al. Prolonged cytomegalovirus infection with viremia is associated with development of cardiac allograft vasculopathy. J Heart Lung Transplant 1992; 11(3 Pt 2):S133-37.
10. Keenan RJ, Lega ME, Dummer JS et al. Cytomegalovirus serologic status and postoperative infection correlated with risk of developing chronic rejection after pulmonary transplantation. Transplantation 1991; 51(2):433-38.
11. Arnold JC, Portmann BC, O'Grady JG et al. Cytomegalovirus infection persists in the liver graft in the vanishing bile duct syndrome. Hepatology 1992; 16(2):285-92.
12. Humar A, Khwaja K, Ramcharan T et al. Chronic rejection: the next major challenge for pancreas transplant recipients. Transplantation 2003; 76(6):918-23.
13. Akposso K, Rondeau E, Haymann JP et al. Long-term prognosis of renal transplantation after preemptive treatment of cytomegalovirus infection. Transplantation 1997; 63(7):974-76.
14. Li F, Yin M, Van Dam JG, Grauls G et al. Cytomegalovirus infection enhances the neointima formation in rat aortic allografts: effect of major histocompatibility complex class I and class II antigen differences. Transplantation 1998; 65(10):1298-304.
15. Almond PS, Matas A, Gillingham K et al. Risk factors for chronic rejection in renal allograft recipients. Transplantation 1993; 55(4):752-756; discussion 756-57.
16. Boehler A, Estenne M. Obliterative bronchiolitis after lung transplantation. Curr Opin Pulm Med 2000; 6(2):133-39.
17. Cai J, Terasaki PI. Heart transplantation in the United States 2004. Clin Transpl 2004; 331-44.
18. Hayry P, Isoniemi H, Yilmaz S et al. Chronic allograft rejection. Immunol Rev 1993; 134:33-81.
19. Richardson WP, Colvin RB, Cheeseman SH et al. Glomerulopathy associated with cytomegalovirus viremia in renal allografts. N Engl J Med 1981; 305(2):57-63.
20. Madalosso C, de Souza NF, Jr. et al. Cytomegalovirus and its association with hepatic artery thrombosis after liver transplantation. Transplantation 1998; 66(3):294-97.
21. Toyoda M, Galfayan K, Galera OA et al. Cytomegalovirus infection induces anti-endothelial cell antibodies in cardiac and renal allograft recipients. Transpl Immunol 1997; 5(2):104-11.
22. Lautenschlager I, Hockerstedt K, Jalanko H et al. Persistent cytomegalovirus in liver allografts with chronic rejection. Hepatology 1997; 25(1):190-94.
23. Yilmaz S, Koskinen PK, Kallio E et al. Cytomegalovirus infection-enhanced chronic kidney allograft rejection is linked with intercellular adhesion molecule-1 expression. Kidney Int 1996; 50(2):526-37.

24. Steinhoff G, You XM, Steinmuller C et al. Induction of endothelial adhesion molecules by rat cytomegalovirus in allogeneic lung transplantation in the rat. Scand J Infect Dis Suppl 1995; 99:58-60.

25. Lemstrom K, Sihvola R, Bruggeman C et al. Cytomegalovirus infection-enhanced cardiac allograft vasculopathy is abolished by DHPG prophylaxis in the rat. Circulation 1997; 95(12):2614-16.

26. Martelius T, Krogerus L, Hockerstedt K et al. Cytomegalovirus infection is associated with increased inflammation and severe bile duct damage in rat liver allografts. Hepatology 1998; 27(4):996-1002.

27. Vazquez MA. Chronic rejection of renal transplants: new clinical insights. Am J Med Sci 2000; 320(1):43-58.

28. Koskinen PK. The association of the induction of vascular cell adhesion molecule-1 with cytomegalovirus antigenemia in human heart allografts. Transplantation 1993; 56(5):1103-8.

29. Lautenschlager IT, Hockerstedt KA. ICAM-1 induction on hepatocytes as a marker for immune activation of acute liver allograft rejection. Transplantation 1993; 56(6):1495-99.

30. von Willebrand E, Pettersson E, Ahonen J et al. CMV infection, class II antigen expression and human kidney allograft rejection. Transplantation 1986; 42(4):364-67.

31. Streblow DN, Kreklywich CN, Smith P et al. Rat cytomegalovirus-accelerated transplant vascular sclerosis is reduced with mutation of the chemokine-receptor R33. Am J Transplant 2005; 5(3):436-42.

32. Sedmak DD, Knight DA, Vook NC et al. Divergent patterns of ELAM-1, ICAM-1 and VCAM-1 expression on cytomegalovirus-infected endothelial cells. Transplantation 1994; 58(12):1379-85.

33. Kloover JS, Soots AP, Krogerus LA et al. Rat cytomegalovirus infection in kidney allograft recipients is associated with increased expression of intracellular adhesion molecule-1 vascular adhesion molecule-1 and their ligands leukocyte function antigen-1 and very late antigen-4 in the graft. Transplantation 2000; 69(12):2641-47.

34. Ustinov JA, Loginov RJ, Bruggeman CA et al. Cytomegalovirus induces class II expression in rat heart endothelial cells. J Heart Lung Transplant 1993; 12(4):644-51.

35. Ustinov J, Bruggeman C, Hayry P et al. Cytomegalovirus-induced class II expression in rat kidney. Transplant Proc 1994; 26(3):1729.

36. Lemstrom KB, Bruning JH, Bruggeman CA et al. Cytomegalovirus infection-enhanced allograft arteriosclerosis is prevented by DHPG prophylaxis in the rat. Circulation 1994; 90(4):1969-78.

37. van Dorp WT, Jonges E, Bruggeman CA et al. Direct induction of MHC class I, but not class II, expression on endothelial cells by cytomegalovirus infection. Transplantation 1989; 48(3):469-72.

38. Carlquist JF, Shelby J, Shao YL et al. Accelerated rejection of murine cardiac allografts by murine cytomegalovirus-infected recipients. Lack of haplotype specificity. J Clin Invest 1993; 91(6):2602-8.

39. Fujinami RS, Nelson JA, Walker L et al. Sequence homology and immunologic cross-reactivity of human cytomegalovirus with HLA-DR beta chain: a means for graft rejection and immunosuppression. J Virol 1988; 62(1):100-5.

40. Orloff SL, Streblow DN, Soderberg-Naucler C et al. Elimination of donor-specific alloreactivity prevents cytomegalovirus-accelerated chronic rejection in rat small bowel and heart transplants. Transplantation 2002; 73(5):679-88.

41. Inkinen K, Soots A, Krogerus L et al. CMV increases collagen synthesis in chronic rejection in rat renal allograft. Transplant Proc 1999; 31(1-2):1361.

42. Inkinen K, Soots A, Krogerus L et al. Cytomegalovirus increases collagen synthesis in chronic rejection in the rat. Nephrol Dial Transplant 2002; 17(5):772-79.

43. Inkinen K, Soots A, Krogerus L et al. Cytomegalovirus enhance expression of growth factors during the development of chronic allograft nephropathy in rats. Transpl Int 2005; 18(6):743-49.

44. Krogerus L, Soots A, Loginov R et al. CMV accelerates tubular apoptosis in a model of chronic renal allograft rejection. Transplant Proc 2001; 33(1-2):254.

45. van Dam JG, Li F, Yin M et al. Effects of cytomegalovirus infection and prolonged cold ischemia on chronic rejection of rat renal allografts. Transpl Int 2000; 13(1):54-63.

46. Grattan MT, Moreno-Cabral CE, Starnes VA et al. Cytomegalovirus infection is associated with cardiac allograft rejection and atherosclerosis. Jama 1989; 261(24):3561-66.

47. Keller CA, Cagle PT, Brown RW et al. Bronchiolitis obliterans in recipients of single, double and heart-lung transplantation. Chest 1995; 107(4):973-80.

48. Rosen HR, Corless CL, Rabkin J et al. Association of cytomegalovirus genotype with graft rejection after liver transplantation. Transplantation 1998; 66(12):1627-31.

49. Sageda S, Nordal KP, Hartmann A et al. The impact of cytomegalovirus infection and disease on rejection episodes in renal allograft recipients. Am J Transplant 2002; 2(9):850-56.

50. McLaughlin K, Wu C, Fick G et al. Cytomegalovirus seromismatching increases the risk of acute renal allograft rejection. Transplantation 2002; 74(6):813-16.

51. Fietze E, Prosch S, Reinke P et al. Cytomegalovirus infection in transplant recipients. The role of tumor necrosis factor. Transplantation 1994; 58(6):675-80.

52. Pallardo Mateu LM, Sancho Calabuig A, Capdevila Plaza L et al. Acute rejection and late renal transplant failure: risk factors and prognosis. Nephrol Dial Transplant 2004; 19 Suppl 3:iii38-42.

53. Chen JH, Mao YY, He Q et al. The impact of pretransplant cytomegalovirus infection on acute renal allograft rejection. Transplant Proc 2005; 37(10):4203-7.

54. Lowance D, Neumayer HH, Legendre CM et al. Valacyclovir for the prevention of cytomegalovirus disease after renal transplantation. International Valacyclovir Cytomegalovirus Prophylaxis Transplantation Study Group. N Engl J Med 1999; 340(19):1462-70.

55. Borchers AT, Perez R, Kaysen G et al. Role of cytomegalovirus infection in allograft rejection: a review of possible mechanisms. Transpl Immunol 1999; 7(2):75-82.

56. Reinke P, Fietze E, Ode-Hakim S et al. Late-acute renal allograft rejection and symptomless cytomegalovirus infection. Lancet 1994; 344(8939-8940):1737-38.

57. Strippoli GF, Hodson EM, Jones C et al. Preemptive treatment for cytomegalovirus viremia to prevent cytomegalovirus disease in solid organ transplant recipients. Transplantation 2006; 81(2):139-45.

58. Hodson EM, Jones CA, Webster AC et al. Antiviral medications to prevent cytomegalovirus disease and early death in recipients of solid-organ transplants: a systematic review of randomised controlled trials. Lancet 2005; 365(9477):2105-15.

59. Kalil AC, Levitsky J, Lyden E et al. Meta-analysis: the efficacy of strategies to prevent organ disease by cytomegalovirus in solid organ transplant recipients. Ann Intern Med 2005; 143(12):870-80.

60. Beck S, Barrell BG. Human cytomegalovirus encodes a glycoprotein homologous to MHC class-I antigens. Nature 1988; 331(6153):269-72.

61. Sola R, Diaz JM, Guirado L et al. Significance of cytomegalovirus infection in renal transplantation. Transplant Proc 2003; 35(5):1753-55.

62. Kooijmans-Coutinho MF, Hermans J, Schrama E et al. Interstitial rejection, vascular rejection and diffuse thrombosis of renal allografts. Predisposing factors, histology, immunohistochemistry and relation to outcome. Transplantation 1996; 61(9):1338-44.

63. Mayoral JL, Brayman KL, Gillingham KJ et al. Influence of pretransplant cytomegalovirus serology and treatment of rejection upon disease after kidney and kidney plus pancreas transplantation. Transplant Proc 1992; 24(3):929-31.

64. Helantera I, Koskinen P, Tornroth T et al. The impact of cytomegalovirus infections and acute rejection episodes on the development of vascular changes in 6-month protocol biopsy specimens of cadaveric kidney allograft recipients. Transplantation. 2003; 75(11):1858-64.

65. Tenschert W, Dittmer R, Harfmann P et al. Vascular rejection of renal allografts is linked to CMV IgG, positive organ donor. Transplant Proc 1991; 23(5):2641-42.

66. Besse T, Malaise J, De Meyer M et al. Renal allograft outcome from cytomegalovirus seronegative donor into cytomegalovirus seronegative recipient: poor prognosis after seroconversion. Transplant Proc 2000; 32(2):408-10.

67. Gerstenkorn C, Robertson H, Bell A et al. CMV infection as a contributory factor for renal allograft injury and loss. Transplant Proc 2001; 33(4):2461-62.

68. Fitzgerald JT, Gallay B, Taranto SE et al. Pretransplant recipient cytomegalovirus seropositivity and hemodialysis are associated with decreased renal allograft and patient survival. Transplantation 2004; 77(9):1405-11.

69. Fernandez-Cruz L, Sabater L, Gilabert R et al. Native and graft pancreatitis following combined pancreas-renal transplantation. Br J Surg 1993; 80(11):1429-32.

70. Hafiz MM, Poggioli R, Caulfield A et al. Cytomegalovirus prevalence and transmission after islet allograft transplant in patients with type 1 diabetes mellitus. Am J Transplant 2004; 4(10):1697-702.

71. Barshes NR, Lee TC, Brunicardi FC et al. Lack of cytomegalovirus transmission after pancreatic islet transplantation. Cell Transplant 2004; 13(7-8):833-38.

72. Asfar S, Metrakos P, Fryer J et al. An analysis of late deaths after liver transplantation. Transplantation 1996; 61(9):1377-81.

73. Jain A, Reyes J, Kashyap R et al. Long-term survival after liver transplantation in 4,000 consecutive patients at a single center. Ann Surg 2000; 232(4):490-500.

74. Rabkin JM, de La Melena V, Orloff SL et al. Late mortality after orthotopic liver transplantation. Am J Surg 2001; 181(5):475-79.
75. Jain A, Demetris AJ, Kashyap R et al. Does tacrolimus offer virtual freedom from chronic rejection after primary liver transplantation? Risk and prognostic factors in 1,048 liver transplantations with a mean follow-up of 6 years. Liver Transpl 2001; 7(7):623-30.
76. Ludwig J, Hashimoto E, Porayko MK et al. Failed allografts and causes of death after orthotopic liver transplantation from 1985 to 1995: decreasing prevalence of irreversible hepatic allograft rejection. Liver Transpl Surg 1996; 2(3):185-91.
77. de Otero J, Gavalda J, Murio E et al. Cytomegalovirus disease as a risk factor for graft loss and death after orthotopic liver transplantation. Clin Infect Dis 1998; 26(4):865-70.
78. Falagas ME, Paya C, Ruthazer R et al. Significance of cytomegalovirus for long-term survival after orthotopic liver transplantation: a prospective derivation and validation cohort analysis. Transplantation 1998; 66(8):1020-28.
79. Seehofer D, Rayes N, Neumann UP et al. Changing impact of cytomegalovirus in liver transplantation—a single centre experience of more than 1000 transplantations without ganciclovir prophylaxis. Transpl Int 2005; 18(8):941-48.
80. O'Grady JG, Alexander GJ, Sutherland S et al. Cytomegalovirus infection and donor/recipient HLA antigens: interdependent cofactors in pathogenesis of vanishing bile-duct syndrome after liver transplantation. Lancet 1988; 2(8606):302-5.
81. Manez R, Kusne S, Green M et al. Incidence and risk factors associated with the development of cytomegalovirus disease after intestinal transplantation. Transplantation 1995; 59(7):1010-14.
82. Evans PC, Soin A, Wreghitt TG et al. An association between cytomegalovirus infection and chronic rejection after liver transplantation. Transplantation 2000; 69(1):30-35.
83. Candinas D, Gunson BK, Nightingale P et al. Sex mismatch as a risk factor for chronic rejection of liver allografts. Lancet 1995; 346(8983):1117-21.
84. Paya CV, Wiesner RH, Hermans PE et al. Lack of association between cytomegalovirus infection, HLA matching and the vanishing bile duct syndrome after liver transplantation. Hepatology 1992; 16(1):66-70.
85. Oh CK, Pelletier SJ, Sawyer RG et al. Uni- and multi-variate analysis of risk factors for early and late hepatic artery thrombosis after liver transplantation. Transplantation 2001; 71(6):767-72.
86. Gunsar F, Rolando N, Pastacaldi S et al. Late hepatic artery thrombosis after orthotopic liver transplantation. Liver Transpl 2003; 9(6):605-11.
87. Seehofer D, Rayes N, Tullius SG et al. CMV hepatitis after liver transplantation: incidence, clinical course and long-term follow-up. Liver Transpl 2002; 8(12):1138-46.
88. Shi LW, Stewart GJ, Verran D et al. Cytomegalovirus serology status and early hepatic artery thrombosis following adult liver transplantation. Transplant Proc 2003; 35(1):421-22.
89. Burak KW, Kremers WK, Batts KP et al. Impact of cytomegalovirus infection, year of transplantation and donor age on outcomes after liver transplantation for hepatitis C. Liver Transpl 2002; 8(4):362-69.
90. Razonable RR, Burak KW, van Cruijsen H et al. The pathogenesis of hepatitis C virus is influenced by cytomegalovirus. Clin Infect Dis 2002; 35(8):974-81.
91. Rosen HR, Chou S, Corless CL et al. Cytomegalovirus viremia: risk factor for allograft cirrhosis after liver transplantation for hepatitis C. Transplantation 1997; 64(5):721-26.
92. Humar A, Kumar D, Raboud J et al. Interactions between cytomegalovirus, human herpesvirus-6 and the recurrence of hepatitis C after liver transplantation. Am J Transplant 2002; 2(5):461-66.
93. Teixeira R, Pastacaldi S, Davies S et al. The influence of cytomegalovirus viraemia on the outcome of recurrent hepatitis C after liver transplantation. Transplantation 2000; 70(10):1454-58.
94. Slifkin M, Ruthazer R, Freeman R et al. Impact of cytomegalovirus prophylaxis on rejection following orthotopic liver transplantation. Liver Transpl 2005; 11(12):1597-602.
95. Hellinger WC, Bonatti H, Machicao VI et al. Effect of antiviral chemoprophylaxis on adverse clinical outcomes associated with cytomegalovirus after liver transplantation. Mayo Clin Proc 2006; 81(8):1029-33.
96. Stoica SC, Cafferty F, Pauriah M et al. The cumulative effect of acute rejection on development of cardiac allograft vasculopathy. J Heart Lung Transplant 2006; 25(4):420-25.
97. Tuzcu EM, Hobbs RE, Rincon G et al. Occult and frequent transmission of atherosclerotic coronary disease with cardiac transplantation. Insights from intravascular ultrasound. Circulation 1995; 91(6):1706-13.
98. Shirali GS, Ni J, Chinnock RE et al. Association of viral genome with graft loss in children after cardiac transplantation. N Engl J Med 2001; 344(20):1498-503.
99. Mehra MR, Uber PA. TOR inhibitors and cardiac allograft vasculopathy: is inhibition of intimal thickening an adequate surrogate of benefit? J Heart Lung Transplant 2003; 22(5):501-4.
100. Westphal M, Lautenschlager I, Backhaus C et al. Cytomegalovirus and proliferative signals in the vascular wall of CABG patients. Thorac Cardiovasc Surg 2006; 54(4):219-26.
101. Dhaliwal A, Thohan V. Cardiac allograft vasculopathy: the Achilles' heel of long-term survival after cardiac transplantation. Curr Atheroscler Rep 2006; 8(2):119-30.
102. Fateh-Moghadam S, Bocksch W, Wessely R et al. Cytomegalovirus infection status predicts progression of heart-transplant vasculopathy. Transplantation 2003; 76(10):1470-74.
103. Koskinen P, Lemstrom K, Mattila S et al. Cytomegalovirus infection associated accelerated heart allograft arteriosclerosis may impair the late function of the graft. Clin Transplant 1996; 10(6 Pt 1):487-93.
104. Weis M, Kledal TN, Lin KY et al. Cytomegalovirus infection impairs the nitric oxide synthase pathway: role of asymmetric dimethylarginine in transplant arteriosclerosis. Circulation 2004; 109(4):500-5.
105. Valantine HA. The role of viruses in cardiac allograft vasculopathy. Am J Transplant 2004; 4(2):169-77.
106. Zeng H, Waldman WJ, Yin DP et al. Mechanistic study of malononitrileamide FK778 in cardiac transplantation and CMV infection in rats. Transplantation 2005; 79(1):17-22.
107. Hosenpud JD, Chou SW, Wagner CR. Cytomegalovirus-induced regulation of major histocompatibility complex class I antigen expression in human aortic smooth muscle cells. Transplantation 1991; 52(5):896-903.
108. Koskinen PK, Krogerus LA, Nieminen MS et al. Cytomegalovirus infection-associated generalized immune activation in heart allograft recipients: a study of cellular events in peripheral blood and endomyocardial biopsy specimens. Transpl Int 1994; 7(3):163-71.
109. Koskinen P, Lemstrom K, Bruggeman C et al. Acute cytomegalovirus infection induces a subendothelial inflammation (endothelialitis) in the allograft vascular wall. A possible linkage with enhanced allograft arteriosclerosis. Am J Pathol 1994; 144(1):41-50.
110. Petrakopoulou P, Kubrich M, Pehlivanli S et al. Cytomegalovirus infection in heart transplant recipients is associated with impaired endothelial function. Circulation 2004; 110(11 Suppl 1):II207-12.
111. Luckraz H, Charman SC, Wreghitt T et al. Does cytomegalovirus status influence acute and chronic rejection in heart transplantation during the ganciclovir prophylaxis era? J Heart Lung Transplant 2003; 22(9):1023-27.
112. Antretter H, Hofer D, Hangler H et al. [Is it possible to reduce CMV-infections after heart transplantation with a three-month antiviral prophylaxis? 7 years experience with ganciclovir]. Wien Klin Wochenschr 2004; 116(15-16):542-51.
113. Bonaros NE, Kocher A, Dunkler D et al. Comparison of combined prophylaxis of cytomegalovirus hyperimmune globulin plus ganciclovir versus cytomegalovirus hyperimmune globulin alone in high-risk heart transplant recipients. Transplantation 2004; 77(6):890-97.
114. Trulock EP, Edwards LB, Taylor DO et al. Registry of the International Society for Heart and Lung Transplantation: twenty-second official adult lung and heart-lung transplant report—2005. J Heart Lung Transplant 2005; 24(8):956-67.
115. Snell GI, Shiraishi T, Griffiths A et al. Outcomes from paired single-lung transplants from the same donor. J Heart Lung Transplant 2000; 19(11):1056-62.
116. Kroshus TJ, Kshettry VR, Savik K et al. 3rd. Risk factors for the development of bronchiolitis obliterans syndrome after lung transplantation. J Thorac Cardiovasc Surg 1997; 114(2):195-202.
117. Hartwig MG, Appel JZ, Davis RD. Antireflux surgery in the setting of lung transplantation: strategies for treating gastroesophageal reflux disease in a high-risk population. Thorac Surg Clin 2005; 15(3):417-27.
118. Kotsimbos TC, Snell GI, Levvey B et al. Chlamydia pneumoniae serology in donors and recipients and the risk of bronchiolitis obliterans syndrome after lung transplantation. Transplantation 2005; 79(3):269-75.
119. Shreeniwas R, Schulman LL, Berkmen YM et al. Opportunistic bronchopulmonary infections after lung transplantation: clinical and radiographic findings. Radiology 1996; 200(2):349-56.
120. Wallwork J. Risk factors for chronic rejection in heart and lungs—why do hearts and lungs rot? Clin Transplant 1994; 8(3 Pt 2):341-44.
121. Jaramillo A, Fernandez FG, Kuo EY et al. Immune mechanisms in the pathogenesis of bronchiolitis obliterans syndrome after lung transplantation. Pediatr Transplant 2005; 9(1):84-93.

122. Bonatti H, Tabarelli W, Ruttmann E et al. Impact of cytomegalovirus match on survival after cardiac and lung transplantation. Am Surg 2004; 70(8):710-14.

123. Duncan AJ, Dummer JS, Paradis IL et al. Cytomegalovirus infection and survival in lung transplant recipients. J Heart Lung Transplant 1991; 10(5 Pt 1):638-44; discussion 645-46.

124. Zeevi A, Morel P, Spichty K, et al. Clinical significance of CMV-specific T helper responses in lung transplant recipients. Hum Immunol 1998; 59(12):768-75.

125. Tikkanen JM, Kallio EA, Bruggeman CA et al. Prevention of cytomegalovirus infection-enhanced experimental obliterative bronchiolitis by antiviral prophylaxis or immunosuppression in rat tracheal allografts. Am J Respir Crit Care Med 2001; 164(4):672-79.

126. Ruttmann E, Geltner C, Bucher B et al. Combined CMV prophylaxis improves outcome and reduces the risk for bronchiolitis obliterans syndrome (BOS) after lung transplantation. Transplantation 2006; 81(10):1415-20.

127. Westall G, Kotsimbos T, Brooks A. CMV-specific CD8 T-cell dynamics in the blood and the lung allograft reflect viral reactivation following lung transplantation. Am J Transplant 2006; 6(3):577-84.

128. Bueno J, Green M, Kocoshis S et al. Cytomegalovirus infection after intestinal transplantation in children. Clin Infect Dis 1997; 25(5):1078-83.

129. Pascher A, Klupp J, Schulz RJ et al. CMV, EBV, HHV6 and HHV7 infections after intestinal transplantation without specific antiviral prophylaxis. Transplant Proc 2004; 36(2):381-82.

130. Todo S, Tzakis A, Reyes J et al. Small intestinal transplantation in humans with or without the colon. Transplantation 1994; 57(6):840-48.

131. Parizhskaya M, Redondo C, Demetris A et al. Chronic rejection of small bowel grafts: pediatric and adult study of risk factors and morphologic progression. Pediatr Dev Pathol 2003; 6(3):240-50.

132. Meyer D, Gasser M, Heemann U et al. Investigating chronic rejection processes after experimental liver/small bowel transplantation. Transplant Proc 2002; 34(6):2261-62.

133. Tzakis AG. Cytomegalovirus prophylaxis with ganciclovir and cytomegalovirus immune globulin in liver and intestinal transplantation. Transpl Infect Dis 2001; 3 Suppl 2:35-39.

134. Furukawa H, Kusne S, Abu-Elmagd K et al. Effect of CMV serology on outcome after clinical intestinal transplantation. Transplant Proc 1996; 28(5):2780-81.

135. Lanzetta M, Petruzzo P, Margreiter R et al. The International Registry on Hand and Composite Tissue Transplantation.

136. Schneeberger S, Lucchina S, Lanzetta M et al. Cytomegalovirus-related complications in human hand transplantation. Transplantation 2005; 80(4):441-47.

# Hepatitis C Virus Infection as a Risk Factor for Graft Loss after Renal Transplantation

Jose M. Morales* and B. Dominguez-Gil

## Abstract

Liver disease is an important complication after renal transplantation and Hepatitis C virus (HCV) infection is the most frequent cause of liver disease. Clinical course is irrelevant in the short-term, excepting rare cases of fibrosing cholestatic hepatitis. However, in the long run, HCV infection can lead to important liver complications such as cirrhosis, hepatocarcinoma and death. Because Interferon is contraindicated in renal transplant patients, the best way is to treat these patients in dialysis before transplantation. Interferon monotherapy is recommended because Ribavirin induces haemolytic anaemia. Most of the patients with sustained virological response remain HCV RNA negative after transplantation. HCV positive renal transplant patients have a higher risk for developing proteinuria, chronic rejection, infections and posttransplant diabetes. Long-term patient and graft survival rates are lower in HCV-positive patients than in HCV-negative graft recipients. Mortality is higher, mainly as a result of liver disease and infections. Notably, HCV infection is an independent risk factor for graft loss and the presence of posttransplant diabetes and HCV-related glomerulonephritis can contribute to graft failure. Despite this, transplantation is the best option for the HCV-positive patient with end-stage renal disease. Finally, several measures after transplantation to minimize the consequences of HCV infection should be recommended. Adjustment of immunosuppression and careful follow-up in the outpatient clinic for early detection of proteinuria, renal insufficiency, infection, diabetes or worsening of liver disease are mandatory.

## Introduction

Liver disease is one of the leading causes of death in long-term survivors after renal transplantation.[1,2] Hepatitis C virus (HCV) infection is currently the main cause of chronic liver disease in renal transplant patients.[1-3] The special relationship between HCV infection and kidney transplantation is initiated in the also special relationship between this infection and dialysis therapy. Patients in hemodialysis have represented a high risk group for HCV infection. This derives from the fact that the hemodialysis technique has usually implied a risk for a direct or indirect contact with contaminated blood, but also due to the frequent need for blood transfusions among these patients in the past. Fortunately, the frequency of HCV infection among dialysis patients has progressively decreased due to the generalization of serologic screening of blood donors, the use of erythropoietin and the application of universal precaution measures in hemodialysis units.[3]

Therefore, it is easy to understand that HCV infection in kidney transplant patients usually starts with the presence of the infection while the patient is in the waiting list, as a consequence of the acquisition of the infection during dialysis therapy.[1] As the frequency of HCV infection has decreased in dialysis patients, the frequency has also decreased among patients in the waiting list and therefore also among kidney transplanted patients. In fact, our collaborative study performed in Spain which retrospectively analysed the frequency of anti-HCV antibodies in kidney transplanted patients during the years 1990, 1994 and 1998, described a progressive and significant decline in the prevalence of anti-HCV antibodies, with values of 29.5%, 19% and 10%, respectively.[4]

A second phenomenon to be underlined when talking about HCV infection and kidney transplantation is the fact that HCV infection may be transmitted through transplantation itself.[5-7] Nevertheless, the routine serologic screening of donors before transplantation today makes peritransplant transmission of HCV infection anecdotic when using HCV negative donors. Finally, we have increasing information on the impact that kidney transplantation and the chronic use of immunosuppressive therapy have on the natural history of HCV infection and also on the impact that HCV infection itself has on the outcome of kidney transplantation.

In this chapter we are going to review first the impact of kidney transplantation on HCV infection and liver disease, second the most important complications of this infection, as glomerular diseases, diabetes mellitus and the influence on rejection and infection and third the impact on patient and particularly graft survival. Finally we will suggest several recommendations to minimize/prevent the consequences of HCV infection in renal transplant patients.

## Impact of Kidney Transplantation on HCV Infection and Liver Disease

### Epidemiology, Prevalence and Diagnostic Tests of HCV Infection after Kidney Transplantation

Most HCV-positive graft recipients acquired the infection on dialysis, as we commented before, but alternative possibilities include preoperative transfusions and organ transplantation.[1-3]

The prevalence of anti-HCV antibodies by ELISA2/3 varies between 10% and 49% depending on the center, country, race, geographic origin of the recipient, mode of dialysis therapy (hemodialysis versus peritoneal dialysis), time on dialysis, number of blood transfusions, retransplantation, presence of anti-hepatitis B core antigen and history of intravenous drug abuse.[1-3]

Most ELISA2/3 positive patients have detectable HCV RNA in the serum. This viremic state persists in almost all transplanted patients.

*Corresponding Author: José M. Morales—Renal Transplant Unit Nephrology Department, Hospital 12 de Octubre, Avenida de Cordoba s/n 28041-Madrid, Spain. Email: jmorales@h12o.es

*Chronic Allograft Failure: Natural History, Pathogenesis, Diagnosis and Management*, edited by Nasimul Ahsan. ©2008 Landes Bioscience.

The viral load increases 1.8-30.3 times with respect to viral titers before kidney transplantation in HCV RNA positive patients, suggesting that immunosuppressive therapy may facilitate viral replication.[6] In fact, it has been reported that Mycophenolate mofetil[8] and antithymocyte globulin[1] increase HCV viremia. However, it has been recently reported that Cyclosporin inhibits the replication of HCV in cultured hepatocytes.[9] HCV-RNA titers do not differ between patients with or without posttransplant liver disease[1] and are not clearly related to liver disease progression.[10]

## Clinical Course of HCV Infection after Kidney Transplantation

With the exception of patients developing fibrosing cholestatic hepatitis,[11] HCV infection after kidney transplantation has a benign course, although bioquemical and histological abnormalities appear in the long term.[5,7,12] Nevertheless, 20 to 51% of HCV positive kidney transplant patients may have normal transaminase levels, though this does not necesarily mean a normal histology.[7] Any case, the state of the healthy carrier, with normal transaminase levels, a positive RNA and a normal histology, has been also described in up to 10% of anti-HCV positive kidney transplant patients.[13]

The risk of developing a chronic liver disease after kidney transplantation seems to depend on several factors, as the duration and the severity of HCV infection before transplantation, the histopathology, the coinfection by Hepatitis B virus (HBV), the time after transplantation and the kind of immunosupression.[1-3] It must be outlined the fact that the use of anti-lymphocyte preparations seems to be related with a higher risk of developing a liver disease.[1]

## Pathology of HCV Infection after Kidney Transplantation

Liver biopsies in selected patients with chronic elevation of ALT have documented severe liver disease, e.g., chronic active hepatitis or cirrhosis in up to 20% of HCV-positive transplant recipients.[7] The prevalence was less when biopsies were performed in all HCV-positive patients, regardless of ALT levels. Glicklich and Kapoian[14] reported in 164 liver biopsies performed soon after transplantation in HCV-positive patients that chronic hepatitis was common (81%), but cirrhosis infrequent (7%).

When describing the histological evolution of HCV infection after kidney transplantation, the literature is scarce and few series have described such evolution on the basis of protocol liver biopsies, independently of transaminase values. In one study, a minimun of three sequential liver biopsies per patient were performed in 51 HCV RNA positive kidney transplant patients. Time after kidney transplantation was over 6 years and none of the patients had previously received interferon therapy. Three evolutive histological patterns were described: the degree of fibrosis remained stable in 20 patients, fibrosis increased in 21 patients and fibrosis progressively improved in the remaining 10 cases. In this study, baseline liver fibrosis and a high diversity of HVR 1 region in HCV genome behaved as independent risk factors for the regression of fibrosis. The authors concluded that HCV infection does not negatively influence liver histology in the long term in 50% of kidney transplant patients.[15]

With regards to the comparison of the histological evolution of liver disease in HCV kidney transplant patients versus immunecompetent patients, contradictory results have been published. Zylberberg et al[16] described a faster progression of the liver histological activity and fibrosis in kidney transplant patients than in no immunecompromised patients infected by HCV. On the contrary, Alric et al observed a slow progression of liver fibrosis in HCV infected kidney transplant patients, which was also inferior to that observed in infected patients with a normal renal function.[10] These authors suggested that the use of different immunosuppressive protocols by the two groups may explain these differeces.

Finally, a recent study compared the liver histology of 38 HCV infected kidney transplant patients with that of a matched cohort of 38 HCV infected patients with end stage renal disease.[17] This group observed a higher proporticon of cases with septal fibrosis, confluent necrosis and steatosis in transplanted versus non transplanted patients, suggesting that kidney transplantation might modify the natural history of HCV infection in patients with end stage renal disease.

## Treatment of HCV Infection after Kidney Transplantation

The problem of anti-HCV therapy is that Interferon (IFN) increases the risk of allograft dysfunction and therefore its use in kidney transplant patients is contraindicated, with the exception of patients with fibrosing cholestatic hepatitis.[3,18,19] Also, Ribavirin and Amantadine monotherapies have no impact on HCV viremia.[18]

Therefore, the best strategy is to treat HCV infection in patients on dialysis before transplantation.[2,3,18-25] Taking into account the available literature, we suggest a therapeutic schema on the management of HCV infection in dialysis patients on the waiting list for renal transplantation, which is represented in Figure 1.

In dialysis patients, until more information with pegylated α-IFN is available, the recommended therapy is α-IFN monotherapy.[18-25] Because ribavirin induces haemolytic anaemia, its use in patients with end stage renal disease is contraindicated.[18] The treatment would consist on α-IFN, 3 million units three times a week for 48 weeks, independently of HCV genotypes. This treatment obtains a sustained virological response in around 40% of treated patients.[18] Interestingly, an early response to α-IFN, defined as a negative HCV RNA at three months is a good predictor for a long-term sustained virological response.[18] During this treatment, these patients should be out of the waiting list for transplantation. Notably, most of these HCV-positive patients with long-term sustained virological response remain HCV RNA negative after transplantation[25] and HCV-related complications, as glomerulonephritis, significantly decrease.[22] However, It is important to note that HCV-positive patients without a response to α-IFN, as well as HCV-positive patients who refuse to be treated with α-IFN should also be transplanted, because their expectancy of life is better than HCV-positive patients who remain in dialysis.[6,27,28]

# Complications of HCV Infection after Renal Transplantation

The most important complications induced by HCV infection after kidney transplantation are glomerulonephritis and posttransplant diabetes mellitus (PTDM).[2] HCV infection can also influence acute rejection and clearly increases the incidence of infections.[3] In this way, we could anticipate that these complications may constitute a direct or indirect explanation for the diminished graft and patient survival observed among HCV infected kidney transplant patients.

## Renal Disease Induced by HCV Infection

HCV infection has been related to different extrahepatic problems, including hematologic, dermatologic, autoimmune and nephrologic disorders.[2] Glomerular lesions have been described in native and transplanted kidneys of HCV infected patients.[2,3,29-36] The most frequent gomerular lesions observed in HCV positive kidney transplant patients are crioglobulinemic or no crioglobulinemic membranoproliferative glomerulonephritis (MPGN)[33,34] and membranous glomerulonephritis (MGN).[35] Transplant glomerulopathy,[2] anticardiolipine related thrombotic microangiopathy[37] and fibrilar glomerulonephritis[38] have also been described. In fact, HCV infection is considered as a predictor for the developement of proteinuria in kidney transplantation.[39] The most frequent clinical picture is MPGN, usually recurring after a second or a third kidney transplant and the second in frequency is MGN. Notably, HCV-associated renal disease does not correlate with the severity of liver disease.[35]

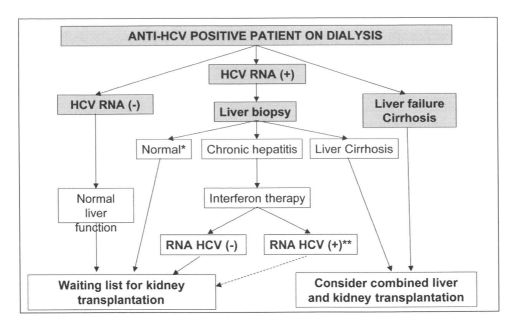

Figure 1. Anti-HCV positive patient on dialysis. *Interferon therapy may be considered to prevent liver disease progression and decrease HCV related morbidity (PTDM, posttransplant glomerulonephritis) after transplantation. **Interferon non responders patients and those refusing to be treated should be included in the waiting list, because HCV positive kidney transplant patiens exhibit a better survival outcome when compared to HCV positive patients remaining in dialysis

The pathogenesis of MPGN and MGN seems to be based on the deposition of immune complexes containing viral RNA in the glomerulus, paradoxically happening in immune compromised patients.[2] Viral antigens have been detected by immunohistochemistry[40] and by in situ hybridation.[41] It has been also reported that laser capture microdissection is a useful method for measuring HCV RNA genomic sequences and HCV core protein in kidney structures such as glomeruli and tubules in patients with HCV-related glomerulonephritis.[42] Wornle et al recently found that Toll-like receptor 3 (TLR3) mRNA expression was clearly elevated in mesangial cells in HCV-related glomerulonephritis and was associated with enhanced proinflammatory cytokines. They hypothesized that immune complexes containing viral RNA activate mesangial TLR3 during HCV infection inducing to chemokine/cytokine release and effecting proliferation and apoptosis. This finding suggests a novel role of TLR3 in HCV-related glomerulonephritis that could establish a link between viral infections and glomerulonephritis.[43]

Because IFN is contraindicated in kidney transplant patients, there is no specific therapy for the treatment of HCV related glomerular lesions in kidney transplant patients. In our unit, in case of an abrupt clinical picture with nephrotic syndrome, with or without renal insufficiency, steroid boluses are applied. In case of developing nephrotic proteinuria with preserved renal function, antiproteinuric agents, mainly angiotensin converting enzyme inhibition(ACEI) and/or angiotensin receptor antagonist II (ARA II) are used.[44] The importance of the treatment with Rituximab in postransplat crioglobulinemic MPGN remains to be determined.[45] Finally, it has been recently observed that treatment with IFN before transplantation may decrease the incidence of posttransplant HCV related glomerulonephritis.[22]

## Postransplant Diabetes Mellitus

HCV infection seems to be related to a higher incidence of diabetes mellitus.[2] This observation applies to non transplanted HCV positive patients, to liver transplant patients whose cause of end stage liver disease has been HCV infection itself and finally to HCV infected kidney transplant patients.[2] In a recent meta-analysis of ten observational studies in kidney transplantation, Fabrizi et al observed that HCV infection increased the incidence of PTDM with an OR of 3.97.[46]

The administration of tacrolimus in HCV infected kidney transplant patients has been described to increase even more the incidence of diabetes mellitus.[47,48] This posttransplant complication may be one of the factors explaining a decreased patient and graft survival among HCV positive kidney transplant patients, since PTDM has been described as an independent risk factor for graft loss and patient death.

The mechanisms explaining the relationship between HCV infection and diabetes mellitus are likely to involve insulin resistance caused by inhibitory actions of the virus on insulin regulatory pathways in the liver.[48] Besides, regarding HCV infection and tacrolimus use, a pharmacokinetic explanation may be considered. Tacrolimus pharmacokinetics is afected by HCV infection; in fact, HCV replication seems to slow tacrolimus metabolism.[49] Therefore, the use of a similar initial dose of tacrolimus in HCV positive and negative kidney transplant patients does probably induce an overexposure to tacrolimus in the formers, at least in the first days after transplantation. This overexposure could explain a higher frequency of PTDM in HCV positive kidney transplant patients treated with tacrolimus. The practical attitude would be to avoid the use of tacrolimus in HCV positive kidney transplant patients, to use lower initial doses of this calcineurin inhibitor or to apply steroid sparing immunosuppressive strategies.

## Acute Rejection and Infection

It is still a controversial issue whether HCV infection decreases or on the contrary enhances the risk of acute rejection. We described that HCV induces a state of immunedeficiency, based on a reduction in the rate of naive T helper lymphocytes and an alteration in the proliferative responses to mitogens of T lymphocytes.[50] These alterations should relate to a decreased incidence of acute rejection, which in fact has been described by some groups.[1-3]

On the contrary, other groups found a similar[4,39] or even a higher incidence of acute rejection in HCV positive patients.[51] Of note, the same conditions related to HCV infection are the ones also determining a higher risk for acute rejection. This means, a longer history of renal disease, previous transplants and blood transfusions are all risk factors for HCV infection in kidney transplant patients. Besides, these situations determine a higher immunological risk. This could be the reason why, despite a virus related immunedeficiency, HCV positive kidney transplant patients may exhibit a higher than expected risk of acute rejection in some series. As an example, Forman et al observed a higher incidente of antibody mediated acute rejection in HCV positive patients in a series of 354 kidney transplant patients.[51] When adjusted for a panel reactive antibody (PRA) >20%, the Cox regression analysis did not identify HCV infection as an independent risk factor for the development of antibody mediated acute rejection.

Because of the previous statements, one may wonder if immunological damage of the graft in HCV positive patients, related to the coexistence of immunological high risk may be the reason why death censored graft survival has been described to be lower in HCV positive kidney transplant patients. As previously said, there are conflicting data regarding the incidence of acute rejection among HCV positive versus HCV negative kidney transplant patients. However, information is lacking with regards to the incidence of subclinical acute rejection in these patients. In a recent published Spanish series of 435 kidney transplant patients with a protocol biopsy performed in the first 6 months after transplantation, it was described that subclinical acute rejection with chronic allograft nephropahy and hepatitis C infection were both independent risk factors for graft loss.[52] Therefore, we could conclude that the immunological damage of the graft does not explain on its own the higher incidence of graft loss in HCV positive kidney transplant patients.

The immune compromised status of the HCV infected kidney transplant patient is also evident in some of the series which identify infection as one of the main causes of death in HCV positive kidney transplant patients in comparison with HCV negative.[53-55] In fact, in the past it was reported that patients with non-A non-B hepatitis had "a marked increase of life-threatening extrahepatic complications".[5] This finding has been demonstrated in HCV-positive patients who had more frequent postoperative infections and potentially fatal infections of the central nervous system, lungs and blood stream (such as cytomegalovirus, tuberculosis, sepsis). Infections are enhanced by heavy immunosuppression.[9,56,57] Therefore aggressive immunosupresion should be avoided if possible in HCV positive patients. Unfortunately, the frequent condition of high immunological risk of the HCV positive patient makes difficult to find the ideal immunosuppresive strategy for these patients.

## The Impact of HCV Infection in Graft and Patient Survival after Kidney Transplantation

It has been documented that the survival of HCV-positive patients after renal transplantation is significantly better than that of matched patients who remain in the waiting list.[1,27,28] Therefore, renal transplantation is the best therapy for patients with HCV infection and end-stage renal disease.[1,3,27,28] However, HCV positive patients after renal transplantation have a lower patient and graft survival compared with HCV negative patients.[4,53-55,58-60]

### *Influence of HCV Infection on Patient Survival*

Several studies demonstrated that patients with HCV infection after renal transplantation exhibited a similar survival in the short-term than non-infected renal transplant patients.[61-65] However in the long-term the situation is different in most of the series, showing that HCV patients have a significant lower survival than HCV negative patients.[4,53-55,58-60] The experience of Hospital Necker is very illustrative: HCV did not adversely affect 5-yr survival[64] but, after a longer follow-up, survival was clearly lower[53] As an example, Mathurin et al published in a case control study that 216 HCV-positive recipients matched with 216 control subjects that ten-year patient survival was significantly lower in HCV positive patients: 65.5% versus 85.3% p < 0.001.[54] In the multivariate analysis, HCV, biopsy-proven cirrhosis, age and year of transplantation were independent risk factors of 10 yr survival in renal transplant patients. Twenty-one percent of deaths were due to liver disease in HCV-positive patients. Interestingly, in patients with liver biopsies, patient survival in those with cirrhosis was not different

at 5 yr, but definitively lower at 10 yr compared to patients with little fibrosis (85% versus 77% and 26% versus 62%, p < 0.05 respectively). In a multicenter study conducted in Spain including 488 HCV-positive from a total population of 3365 patients transplanted from 1990-1998, we also found that HCV infection was an independent risk factor for patient death and the proportion of deaths due to liver disease was higher than negative HCV patients (13.85% versus 0.6% p = 0.03).[4] In spite of this, 10 yr patient survival was 77.5% versus 84.5% in HCV positive versus HCV negative patients, including only functioning grafts after one year,. This survival Fig is acceptable, taking into account that these HCV positive are high risk patients. In fact, the mortality rate progressively increased; it was 10% at 10 yr and 20% at 20 yr.[6]

In a meta-analysis of observational studies recently reported including 6345 patients, the increased mortality in the HCV positive population (relative risk 1.79 with a 95% confidence interval of 1.57 to 2.03) has been corroborated.[60] It was at least partially related to an increase in liver-related death and the frequency of cirrhosis and hepatocellular carcinoma as causes of death was significantly higher in HCV positive compared with HCV negative patients in 6 of the 8 studies (from USA, France, Spain, Taiwan, Germany and Sweden) included in the analysis.[55] Cardiovascular and infectious diseases were also important causes of death in these HCV positive patients included in this systematic review. Extrahepatic complications of HCV infection, such as PTDM, can also contribute to mortality. In fact, HCV infection and PTDM separately considered are independent risk factors for patient death. In this way, a study from USRDS registry showed that the presence of PTDM was associated with lower patient survival in HCV positive patients.[28] Even in HCV patients in the waiting list for transplantation, diabetes mellitus is more prevalent and is a major factor for mortality.[66] A retrospective study also found that HCV infection in kidney-pancreas transplant patients significantly increased the risk of patient death.[67]

A role for heavy immunosuppression, probably increasing HCV viral replication after renal transplantation, is also suggested by the observation that quadruple therapy with monoclonal or polyclonal antibodies was associated with more frequent instances of liver disease,[1] although a recent communication from USA Registry showed that antibody induction is associated with improved survival figures in HCV positive patients.[68] Although comparative studies using different immunosuppressive protocols are not available, current information regarding survival figures, incidence of acute rejection and infectious complications suggest that immunosuppression should be adjusted depending on liver histology.[3] In this way, all conventional current immunosuppressive drugs can be used in HCV positive patients. However, it should be noted again that the risk of PTDM is higher in HCV positive patients treated with Tacrolimus.[47]

### *Influence of HCV Infection on Graft Survival*

Similar to patient survival, HCV infection did not influence graft survival in the short-term.[61-65] However, in most of the recent studies with longer term follow-up patients with HCV infection exhibited lower graft survival compared with HCV negative patients.[4,53-55,58-60,69] In the Spanish study previously mentioned, as an example, 10 yr death censored graft survival was 69% in HCV-positive patients, significantly lower compared with HCV-negative patients 79% (p < 0.0001).[4] Although the lower graft survival may reflect lower patient survival in some reports,[3] in several but not in all series HCV infection was an independent risk factor for graft loss.[54-62] In fact, in the meta-analysis above mentioned including 6345 patients, the presence of anti-HCV positive antibodies was an independent and significant factor for graft failure (relative risk 1.56, 95% CI 1.35-1.80 p = 0.019) in four of the eight studies included in the analysis. It is important to note that these four significant studies included the majority of the patients used for this meta-analysis, this means 4, 613 (73%) patients.[60]

Two problems associated with HCV infection can contribute to a decreased censoring-death graft survival after renal transplantation, such as the development of proteinuria due to chronic allograft nephropathy and/or HCV-associated glomerulonephritis and PTDM.

As we commented before, HCV infection is an independent risk factor for proteinuria after transplantation. Hestin et al reported that persistent proteinuria developed more frequently in HCV positive than in HCV negative patients.[39] At 1 year, the probability of proteinuria was 19.5% in the HCV-positive and 7.5% in HCV-negative patients. At 3 years, proteinuria was 22.9% versus 10.7% and at five years was 45.1% and 13.1%, respectively. However, there was no difference in graft and patient survival at five years between the two groups. The histopathological study in patients with proteinuria (unfortunately few patients) demonstrated a non significantly higher frequency of de novo glomerular lesions in HCV positive patients. Notably chronic rejection was the most frequent lesion in both groups, showing that nearly half of HCV positive patients had transplant glomerulopathy. Although other experiences found a similar incidence of transplant glomerulopathy among HCV-positive versus HCV-negative patients,[22] other authors suggested a possible association between HCV infection and transplant glomerulopathy.[71,72] In fact, Cosio et al found that compared with a group of 105 patients without transplant glomerulopathy, the prevalence of HCV antibodies was significantly higher in patients with chronic transplant glomerulopathy (1.9% versus 33% respectively, p = 0.0004). These authors postulated that HCV, directly or indirectly, or by inducing the release of cytokines, such as interferon, may produce endothelial cell lesion leading to transplant glomerulopathy.[71] Mahmoud et al also found a higher incidence of proteinuria and chronic rejection in a series of HCV positive with HCV RNA in the serum[62] and Bruchfeld et al published that HCV infection was more important for graft loss than time on renal-replacement therapy.[69]

Our experience is Spain is very illustrative: we studied the characteristics of 3365 renal transplants during the 1990s and risk factors associated with death-censored graft failure. Despite worsening of surrogate parameters of renal quality (significantly increase of donor age) and poorer HLA matching (significantly increase of HLA mismatches), graft survival improved during this decade. Acute rejection decreased (from 39% to 25% p < 0.0001) and the prevalence of HCV infection decreased

---

**Table 1. Preventive measures to minimize the consequences of HCV infection after renal transplantation**

**Measures While Patient is on Waiting List**

- Avoid blood transfusions
- Isolate HCV-positive patients
- Treat with pegylated or alpha Interferon if chronic hepatitis is documented by biopsy and HCV-RNA is detected
- Treatment with Pegylated or alpha Interferon in HCV RNA positive patients without chronic hepatitis to decrease posttransplant morbidity- associated with HCV infection (glomerulonephritis, diabetes) could be discussed

**Measures in the Perioperative Period of Renal Transplantation**

- Avoid blood transfusions
- Do no transplant kidneys from HCV-positive donors into HCV-negative recipients and in patients HCV RNA negative after Interferon therapy

**Measures after Renal Transplantation**

  "HCV-positive patients must be followed closely to detect deterioration of liver function, infectious diseases, proteinuria or posttransplant diabetes"

**Immunosuppression**

- Use non-aggressive immunosuppressive protocol: no routine use of ATG, ALG or OKT3, except in immunologically high-risk patients
- Immunosuppression in the maintenance phase depending on the severity of liver disease
- All immunosuppressive drugs can be used in HCV-positive patients

**Liver Disease**

- At each visit, liver enzymes, bilirubin and prothrombin time should be measures; HCV serology and HCV RNA should be tested at least twice a year.
- Liver biopsy should be considered in patients with abnormal liver function (diagnosis, prognosis, modulaton of immunosuppression, possible treatment)
- Extremely important: If HCV-positive patients present severe cholestasis, liver biopsy should be performed immediately. If fibrosing cholestatic hepatitis is present, Interferon therapy should be considered on a case-by-case basis.
- In patients with cirrhosis, ultrasonography and alpha-fetoprotein levels should de monitored frequently to detect early liver carcinoma
- In patients with terminal hepatic failure, liver transplantation should be considered
- Avoid potentially hepatotoxic drugs and alcohol

**Infections**

- Be on the alert for severe and opportunistic infections
- In the case of fever, start effective antibiotic treatment early

**HCV-Related Glomerulonephritis**

- Be on the alert for proteinuria and/or microhematuria
- In patients with proteinuria/microscopic hematuria a graft biopsy should be performed
- In patients with HCV-related glomerulonephritis antiproteinuric drugs should be started

**HCV-Related Posttransplant Diabetes Mellitus**

- Be on the alert for hyperglucemia
- Patients with hyperglucemia should be referred to a diabetologist
- To use an immunosuppressive regimen to optimize antirejection efficacy minimizing the risk of posttransplant diabetes

from 29% to 10% p < 0.0001).[70] These two major-time dependent modifications may be related to this improvement: a reduction in the prevalence of acute rejection and a dramatic reduction in the prevalence of HCV positive patients. Both significantly contributed to counterbalance the detrimental effect of decreasing renal allograft quality and worsening of some recipient-dependent factors. The presence of HCV infection was an independent risk factor for graft loss and the follow-up data revealed a steady increase of serum creatinine between 3 months and 1 year, an increase of proteinuria between 3 months and 1 year and the 1 year proteinuria was also higher in patients with HCV infection.[4] As described in other studies, chronic rejection was the most frequent cause of graft loss.[61] Therefore, HCV infection seems to be associated with greater rates of proteinuria, chronic rejection and graft loss. In fact, as we commented before, in a study with protocol biopsies at six months subclinical acute rejection with chronic allograft nephropathy and HCV infection were both independent risk factor for graft loss.[52] Notably, in a recent study pretransplant INF therapy in 50 HCV positive patients significantly decreased the incidence of chronic allograft nephropathy; in other words absence of IFN therapy before transplantation was a significant risk factor for chronic allograft nephropathy.[73]

The presence of posttransplant cryoglobulinemic or noncryoglobulinemic MPGN and MGN associated to HCV infection contribute to the development of graft failure.[33-35] Concerning type I MPGN, it has been reported an accelerated loss of the graft.[33,34] In MGN, the clinical course and the development of renal failure seems to be similar in patients with and without HCV infection.[32] The exact role of these glomerular lesions on graft loss is unknown because there are no prospective data focusing in this problem and the policy of graft biopsies is not uniform. However, retrospective data suggested that the prevalence of these glomerular lesions associated to HCV infection is less than 5%.[30] An early diagnosis and therapy with steroids and/or ACEI/ARB may be beneficial. Results with IFN before transplantation to maintain HCV RNA negative after transplantation and to decrease the incidence of posttransplant glomerulonephritis.are encouraging.[22]

In renal transplant patients, the complications of PTDM on morbidity, mortality and graft survival are well established.[28,74,75] As we commented before, there is a clear evidence that HCV infection increased the incidence of PTDM.[46] Also, many studies have demonstrated separately that HCV infection and PTDM are independent risk factors, not only for patient death but for death censored graft loss.[28,73] Therefore, it seems logical to think that both entities together can contribute to decrease graft survival after renal transplantation. Concerning tacrolimus therapy in these HCV-positive patients it is important to remark that although Tacrolimus increases PTDM,[47] it also has a protective effect on death censored graft survival and patient survival.[73] Early detection and therapy of PTDM together with the use of an immunosuppressive regimen to minimize the risk of hyperglycaemia and to optimize antirejection efficacy in HCV-positive patients could be important to decrease graft loss. Notably, pretransplant IFN therapy in HCV positive patients may also decrease the incidence of PTDM.[76]

## Conclusions

Hepatitis C virus infection is the most frequent cause of liver disease after renal transplantation. Although in the short-term clinical course is irrelevant, HCV infection in the long-term can lead to severe liver complications such as cirrhosis, hepatocarcinoma and death. Because Interferon is contraindicated after renal transplantation, treatment before transplantation with interferon monotherapy is recommended. Most of the patients with sustained virological response remain HCV RNA negative after transplantation and HCV-related posttransplant morbidity decreases. However HCV positive patients without response or that refused Interferon therapy should also be transplanted because survival after dialysis is better than dialysis. HCV renal transplant patients have a higher risk for developing proteinuria, chronic

rejection, infection and PTDM. Long-term patient and graft survival rates are lower in HCV-positive compared with HCV negative patients. Mortality is higher due to liver disease and infections. Remarkably, HCV infection is an independent risk factor for graft loss and seems to be associated with greater rates of proteinuria, chronic rejection and graft loss. The presence of PTDM and HCV-related glomerulonephritis can contribute to graft failure. Despite this, renal transplantation is the best option for the HCV-positive patient with end-stage renal disease. To minimize the consequences of HCV infection after renal transplantation several measures should be recommended (see Table 1): adjustment of immunosuppressive therapy and careful follow-up for early detection of proteinuria, infection, diabetes or worsening of liver disease are mandatory.

## References

1. Pereira BJ, Levey AS. Hepatitis C virus infection in dialysis and renal transplantation. Kidney Int 1997; 51:981-99.
2. Roth D. Hepatitis C virus: the nephrologists s view. Am J Kidney Dis 1995; 25:3-16.
3. Morales JM, Campistol JM. Transplantation in the patient with hepatitis C. J Am Soc Nephrol 2000; 11:1343-53.
4. Morales JM, Dominguez-Gil B, Sanz-Guajardo D et al. The influence of hepatitis B and hepatitis C virus infection in the recipient on late renal allograft failure. Nephrol Dial Transplant 2004; 19 (supply 3):72-76.
5. PereIra BJ, Levey A. Hepatitis C in organ transplantation: its significance and influence on transplantation policies. Curr Opin Nephrol Hypertens 1993; 2:912-22.
6. Pereira BJ. Natov SN, Bouthot BA et al. Effects of hepatitis C infection and renal transplantation on survival in end-stage renak disease. The New England Organ Bank Hepatitis C Sudy Group. Kidney Int 1998; 53:1374-80.
7. Vosnides GG. Hepatitis C in renal transplantation. Kidney Int 1997; 52:843-61.
8. Rostaing L, Izopet J, Sandres K et al. Changes in hepatitis C virus RNA viremia concentrations in long-term renal transplant patients after introduction of mycophenolate mofetil. Transplantation 2000; 69:991-95.
9. Watashi K, Hijikata M, Hosaka M et al. Cyclosporin A suppresses replication of hepatitis C virus genome in cultured hepatocytes. Hepatology 2003; 38:1282-86.
10. Alric L, Di Marino V, Selves J et al. Long-term impact of renal transplantation on liver fibrosis during hepatitis C virus infection. Gastroenterology 2002; 123(5):1494-99.
11. Muñoz de Bustillo E, Ibarrola C, Colina F et al. Fibrosing cholestatic hepatitis in hepatitis C virus-infected transplant recipients. J Am Soc Nephrol 1998; 9:1109-13.
12. Morales JM, Campistol JM, Andres A et al. Hepatitis C virus and renal transplantation. Curr Opin Nephrol Hypertens 1998; 7:177-83.
13. Haem J, Berthoux P, Mosnier JF et al. Clear evidence of the existence of healthy carriers of hepatitis C virus among renal transplant recipients. Transplantation 1996; 62:699-700.
14. Glicklich D, Kapoian T. Should the hepatitis C positive end stage renal disease patient be transplanted? Seminar Dial 1996; 9:5-8.
15. Kamar N, Rostaing L, Selves J et al. Natural history of hepatitis C virus-related liver fibrosis after renal transplantation. Am J Transplant 2005; 5(7):1704-12.
16. Zylberberg H, Nalpas B, Carnot F et al. Severe evolution of chronic hepatitis C in renal transplantation: a case control study. Nephrol Dial Transplant 2002; 17:129-33.
17. Perez RM, Ferreira AS, Medina-Pestana JO et al. Is hepatitis C more aggressive in renal transplant patients than in patients with end-stage renal disease? J Clin Gastroenterol 2006; 40:444-48.
18. Kamar, N, Ribes D, Izopet J et al. Treatment of hepatitis C virus infection (HCV) alter renal transplantation: implications for HCV-positive diálisis awaiting a kidney transplant. Transplantation 2006; 82:853-56.
19. Fabrizi F, Lunghi G, Dixit V et al. Meta-analysis: antiviral therapy of hepatitis c virus-related liver disease in renal transplant patients. Aliment Pharmacol Ther 2006; 24:1413-22.
20. Barril G, Gonzalez-Parra E, Alcazar R et al. Guia sobre enfermedades viricas en hemodiálisis. Nefrología 2004; 24(supl II):43-66.
21. European Best Practice Guidelines for Renal Transplantation. Section I: Evaluation, selection and preparation of the potential recipient. Nephrol Dial Transplant 2000; 15(supply 7):3-38.
22. Cruzado JM, Casanovas-Taltabull T, Torras J et al. Pretransplant Interferon prevents hepatitis C virus-associated glomerulonephritis in renal allografts by HCV-RNA clearance. Am J Transplant 2003; 3:357-60.
23. Campistol JM, Esforzado N, Morales JM. Hepatitis C virus-positive patients on the waiting list for renal transplantation. Semin Nephrol 2002; 22(4):361-64.

24. Rostaing L, Chatelut E, Payen JL et al. Pharmacokinetics of alfaIFN-2b in chronic hepatitis C virus patients undergoing chronic haemodialysis or with normal renal function:Clinical implications. J Am Soc Nephrol 1998; 9:2344-48.

25. Kamar N, Toupamce O, Buchler M et al. Evidence that clearance of hepatitis C virus RNA alter alpha interferon therapy in dialysis patients is sustained alter renal transplantation. J Am Soc Nephrol 2003; 14:2092-98.

26. Fabrizi F, Dulai G, Dixit V et al. meta-analysis: interferon for the treatment of chronic hepatitis in dialysis patients. Aliment Pharmacol Ther 2003; 18:1071-81.

27. Knoll GA, Tankersley MR, Lee JY et al. The impact of renal transplantation on survival in hepatitis C positive end-stage renal disease patients. Am J Kidney Dis 1997; 29:606-14.

28. Abbott KC, Lentine KL, Bucci JR et al. the impact of transplantation with deceased donor hepatitis positive kidneys on survival in wait-listed long-term dialysis patients. Am J Transplant 2004; 4(12):2032-37.

29. Johnson RJ, Wilson R, Yamabe H et al. Renal manifestations of hepatitis C virus infection. Kidney Int 1994; 46:1255-63.

30. Morales JM, Morales E, Andres Aet al. Glomerulonephritis associated with hepatitis C virus infection. Curr Opin Nephrol Hypertens 1999; 8:205-11.

31. D Amico G. Renal involvement in hepatitis C virus infection: cryoglobulinemic glomerulonephritis. Kidney Int 1998; 54:650-71.

32. Morales JM, Campistol JM, Andres A et al. Glomerular diseases in patients with hepatitis C virus infection after renal transplantation. Curr Opin Nephrol Hypertens 1997; 6:511-15.

33. Cruzado JM, Gil-Vernet S, Ercilla G et al. Hepatitis C virus-associated membranoproliferative glomerulonephritis in renal allografts. J Am Soc Nephrol 1996; 7:2469-75.

34. Roth D, Cirocco R, Zucker K et al. De novo membranoproliferative glomerulonephritis in hepatitis C virus-infected renal allografts recipients. Transplantation 1995; 59:1676-82.

35. Morales JM, Pascual J, Campistol JM et al. Membranous glomerulonephritis associated with hepatitis C virus infection in renal transplant patients. Transplantation 1997; 63:1634-39.

36. Cruzado JM, Carrera M, Torras J et al. Hepatitis C virus infection and de novo glomerular lesions in renal allografts. Am J Transplant 2001; 1(2):171-78.

37. Baid S, Pascual M, William Jr WW et al. Renal thrombotic microangiopathy associated with anticardiolipin antibodies in hepatitis C positive allograft recipients. J Am Soc Nephrol 1999; 10:146-53.

38. Markowitz GS, Cheng JT, Colvin RB et al. Hepatitis C viral infection is associated with fibrillary glomerulonephritis and immunotactoid glomerulopathy. J Am Soc Nephrol 1998; 9:2244-52.

39. Hestin D, Guillemin F, Castin N et al. Pretransplant hepatitis C virus infection: A predictor of proteinuria after renal transplantation. Transplantation 1998; 65:741-44.

40. Kasuno K, Ono T, Matsumori A et al. Hepatitis c virus associated tubulo-interstitial injury. Am J Kidney Dis 2003; 41:767-75.

41. Hoch B, Jucknevicius I, Liapis H. Glomerular injury associated with hepatitis C infection. A correlation with blood and tissue HCV PCR. Semin Diag Pathol 2002; 19:175-87.

42. Sansonno D, Lauletta G, Montrone M et al. Hepatitis C virus RNA and core protein in kidney glomerular and tubular structures isolated with laser microdissection. Clin Exper Immunology 2005; 140:498-506.

43. Wornle M, Schmid H, Banas et al. Novel role of toll-like receptor 3 in hepatitis C-associated glomerulonephritis. Am J Pathol 2006; 168:370-85.

44. Gonzalez E, Esforzado N, Usera G et al. Long-term results of hepatitis C virus-associated glomerular lesions alter renal transplantation. Am J Transplant 2006 (abstract).

45. Basse G, Ribes D, Kamar N et al. Rituximab therpay for mixed cryoglobulinemia in seven renal transplant patients. Transplant Proc 2006; 38:2308-10.

46. Fabrizi F, Martin P, Dixit V et al. Posttransplant diabetes mellitus and HCV seropositive status after renal transplantation: meta-analysis of clinical studies. Am J Transplant 2005; 5:2433-40.

47. Bloom RD, Rao V, Weng F et al. Association of hepatitis C with posttransplant diabetes en renal transplant patients on tacrolimus. J Am Soc Nephrol 2002; 13:1374-80.

48. Bloom RD, Lake JR. Emerging issues in hepatitis C virus positive liver and kidney transplant recipients. Am J Transplant 2006; 6(10):2232-37.

49. Manzanares C, Moreno M, Castellanos F et al. Influence of hepatitis C virus infection on FK 506 blood levels in renal transplant patients. Transplant Proc 1998; 30:1264-65.

50. Corell A, Morales JM, Mandroño A et al. Immunosuppression induced by hepatitis C virus infection reduces acute renal transplant rejection. Lancet 1995; 346:1497-98.

51. Forman JP, Tolkoff-Rubin N, Pascual M et al. Hepatitis C, acute humoral rejection and renal allograft survival. J Am Soc Nephrol 2004; 15:3249-55.

52. Moreso F, Ibernon M, Goma M et al. Subclinical rejection associated with chronic allograft nephropathy in protocol biopsies as a risk factor for late graft loss. Am J Transplant 2006; 6:747-52.

53. Legendre C, Garrigue V, Le Bihan C et al. Harmful long-term impact of hepatitis C virus infection in kidney transplant recipients. Transplantation 1998; 65:667-70.

54. Mathurin P, Mouquet C, Poynard T et al. Impact of hepatitis B and C virus on kidney transplantation outcome. Hepatology 1999; 29:257-63.

55. Gentil MA, Rocha Jl, Rodriguez-Algarra G et al. Impaired kidney transplant survival in patients with antibodies to hepatitis C virus. Nephrol Dial Transplant 1999; 14:2455-59.

56. Rao KV, MaJ. Chronic viral hepatitis enhances the risk of infection but not acute rejection in renal transplant recipients. Transplantation 1996; 62:1765-69.

57. Mitwalli AH, Alam A, Al-Wakeel J et al. Effect of chronic viral hepatitis on graft survival in Saudi renal transplant patients. Nephron Clin Pact 2006; 102:c72-80.

58. Breitenfeldt MK, Rasenak J, Berthold H et al. Impact of hepatitis B and C on graft loss and mortality of patients after renal transplantation. Clin Transplant 2003; 21:300-6.

59. Aroldi A, Lampertico P, Montagnino G et al. Natural history of hepatitis B and C in renal allograft recipients. Transplantation 2005; (15)79:1132-36.

60. Fabrizi F, Martin P, Dixit V et al. Hepatitis C virus antibody status and survival after renal transplantation: meta-analysis of observational studies. Am J Transplant 2005; 5(6):1452-61.

61. Meier-Kriesche HU, Ojo AO, Hanson J et al. Hepatitis C antibody status and outcomes in renal transplant recipients. Transplantation 2001; 72:241-45.

62. Mahmoud IM, Elhabashi AF, Elsawy E et al. the impact of hepatitis C virus viremia on renal graft and patient survival: a 9-year prospective study. Am J Kid Dis 2004; 43:131-39.

63. Roth D, Zucker K, Cirocco R et al. The impact of hepatitis C virus infection on renal allograft recipients. Kidney Int 1994; 45:238-44.

64. Pol S, Legendre C, Saltiel C et al. Hepatitis C virus in kidney recipients, epidemiology and impact of renal transplantation. J Hepatol 1992; 15:202-6.

65. Ponz E, Campistol JM, Bruguera M et al. Hepatitis C virus infection among kidney transplant recipients. Kidney Int 1991; 40:748-51.

66. Bloom RD, Sayer G, Fa K et al. Outcome of hepatitis C virus-infected kidney transplant candidates who remain on the waiting list. Am J Transplant 2005; 5:139-44.

67. Stehman-Breen CO, Psaty BM, Emerson S et al. Association of hepatitis C virus infection with mortality and graft survival in kidney-pancreas transplant recipients. Transplantation 1997; 64:281-86.

68. Luan FL, Schaubel DE, Jia XY. Impact of immunosuppressive regimen on patient survival among kidney transplant recipients by hepatitis C status. Am J Transplant 2006 (abstract).

69. Bruchfeld A, Wilczek H, Elinder CG. Hepatitis C infection, time in renal-replacement therapy and outcome after kidney transplantation. Transplantation 2004; 78:745-50.

70. Seron D, Arias A, Campistol JM et al. Late renal allograft failure between 1990 and 1998 in Spain: a changing scenario. Transplantation 2003; 76:1588-94.

71. Cosio FG, Roche Z, Agrawal A et al. Prevalence of hepatitis C virus in patients with idiopathic glomerulopathies in native and transplanted kidneys. Am J Kid Dis 1996; 28:752-58.

72. Gallay BJ, Alpers CE, Davis Cl et al. Glomerulonephritis in renal allografts associated with hepatitis C infection: A possible relationship with transplant glomerulopathy in two cases. Am J Kidney Dis 1995; 26:662-67.

73. Mahmoud IM, Sobh MA, El-Habashi AF et al. Interferon therpay in hemodialysis patients with chronic hepatitis C: study of tolerance, efficacy and posttransplantation course. Nephron Clin Pract 2005; 100:c133-39.

74. Kasiske BJL, Snyder JJ, Gilbertson D et al. Diabetes mellitus and kidney transplantation in the United States. Am J transplant 2003; 3:178-85.

75. Davidson J, Wilkinson A, Dantal J et al. New-onset diabetes after renal transplantation: 2003 International consensus guidelines. Proceedings of an international expert panel meeting. Transplantation 2003; 75(10 Suppl):SS3-24.

76. Gursoy M, Guvener N, Koksal R et al. Impact of HCV infection on development of posttransplant diabetes mellitas in renal allograft recipients. Transplant Proc 2000; 32:561-562.

# CHAPTER 37

# Polyomavirus Type BK-Associated Nephropathy and Renal Allograft Graft Loss:
## Natural History, Patho-Physiology, Diagnosis and Management

Nasimul Ahsan*

## Abstract

In recent years, polyomavirus type BK-associated nephropathy (PVAN) has emerged as an important cause of renal allograft dysfunction and graft loss. It is estimated to affect up to 10% of renal transplant recipients, with allograft failure rate as high as 80%. After primary infection in immunocompetent individuals, the virus remains latent in the urinary tract and may become reactivated under conditions of immunodeficiency from organ transplantations. Conflicting information has been reported on risk factors of PVAN in transplant recipients and the pathogenesis of tissue damage in PVAN has become a subject of considerable interest. There are also recent reports suggesting that polyomavirus BK (BKV) creates a transcriptional microenvironment that promotes chronic allograft fibrosis. Quantitation of BKV deoxy ribonucleic acid (DNA) in the blood and urine, identification of virus laden "decoy cells" in urine, and histopathologic demonstration of viral inclusions in renal tubules are the applicable diagnostic methods. Early diagnosis and appropriate interventions of PVAN before irreversible damage to the allograft kidney can result in marked improvement of outcome. While various antiviral agents have been tried to treat PVAN, current management aims at modification and/or improvement in the host's immune status. In this chapter, the natural history, pathogenesis and management of chronic renal allograft failure due to PVAN will be discussed.

## Virology

The polyomaviruses and papillomaviruses were previously considered subfamilies of the family *Papovaviridae*, which derived its name from three of its members: rabbit papilloma virus, mouse polyomavirus, and simian vacuolating agent or simian virus 40 (SV 40). Recently, the International Committee on Taxonomy of Viruses has recognized polyomaviruses and papillomaviruses as independent virus families. Viruses of these two families have different biological characteristics and are unrelated both immunologically and genetically. Thirteen members of the family *Polyomaviridae* have been identified, which include two human pathogens, JC virus (JCV) and BKV.[1] BKV and JCV, and macaque polyomavirus, SV 40 are very similar biologically, and each virus has over 70% nucleotide sequence similarity with the other two. Using a data set of 33 whole genomes of polyomaviruses, Crandall et al[2] showed that a robust estimate of the evolutionary relationships exist among the polyomaviruses. BK and JCV were first isolated in 1971 from immunocompromised patients. Padgett et al[3] isolated and partially characterized JCV from the brain of a patient (with the initials J.C.) with Hodgkin's lymphoma who died of progressive multifocal leukoencephalopathy. Gardner et al[4] isolated BKV from the urine of a Sudanese renal transplant patient (with the initials B.K.) who developed ureteral stenosis and was shedding inclusion-bearing epithelial cells in his urine.

Polyomaviruses have the following properties: small size of the virion (diameter 40-45 nm), naked icosahedral capsid, superhelical double-strand circular DNA genome of molecular weight $3.2 \times 10^6$, shared nucleotide sequences with other polyomaviruses, and nuclear site of multiplication. The nonenveloped virion has icosahedral symmetry and 72 pentameric capsomers. The virion is made up of protein (88%) and a single copy of a circular double-stranded DNA molecule (12%), which has about 5,300 base pairs.[1,5] The viral genome is functionally divided into (i) a noncoding control region (NCCR) (0.4 kb), (ii) an early coding region (2.4 kb), which codes for tumors antigens: large T (T-ag), middle T (in mouse and hamster viruses), and small T (t-ag), and (iii) a late coding region (2.3 kb), which codes for viral capsid proteins VP1, VP2, and VP3 and agnoprotein. The NCCR is located between the early and late regions and contains the T-ag binding sites. It contains (i) the origin of DNA replication (*ori*) and (ii) the regulatory regions for early and late transcription control sequences during viral life cycle (Fig. 1).[6] The sequence blocks in NCCR are arbitrarily referred to by the alphabetical designations P, Q, R, and S. These blocks serve as regulatory regions, or enhancer elements, and are believed to contain several transcription factor binding sites, which putatively modulate viral transcription. Naturally occurring BKV strain in the kidney and urine usually have an archetypal regulatory region.[6-8]

The early and late coding regions are transcribed from different strands of the DNA molecule. The direction of early and late transcription is divergent, with opposite strands participating in these processes, starting from the origin of replication. T-ag is a multifunctional protein with distinct ability to bind host cell regulatory proteins. T-ag controls both viral DNA replication, and gene transcription, and interferes with host cell transcription factors.[9,10] During replication, viral DNA associates with host cell histones to form mini viral chromosomes, which are structurally indistinguishable from host cell chromatin.[11,12] Each pentamer of the viral icosahedron consists of five VP1 molecules and one molecule of VP2 or VP3. VP1 [molecular mass (mm) 39,600] is the major capsid protein (accounts for more than 70% of the virion

*Nasimul Ahsan—Professor of Medicine, Mayo Clinic - College of Medicine 4205 Belfort Road, Suite 1100, Jacksonville, Florida 32216-5876, USA. Email: ahsan.nasimul@mayo.edu

*Chronic Allograft Failure: Natural History, Pathogenesis, Diagnosis and Management*, edited by Nasimul Ahsan. ©2008 Landes Bioscience.

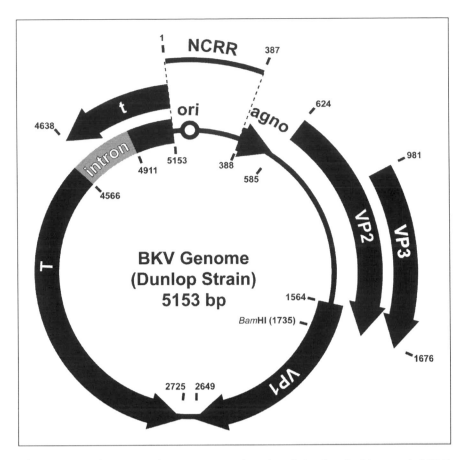

Figure 1. Genomic map of BK virus Dunlop strain. The BKV genome is a closed circular, double-stranded DNA molecule ~5kb in size. The coding regions for the early genes, large and small T antigens (T and t) are transcribed in the counter clockwise direction and the late genes, Agnoprotein (*agno*) and VP1-3, are transcribed in the clockwise direction. The noncoding regulatory region (NCRR) is ~387bp and includes the origin of replication (*ori*). Genomes of BKV contains a unique *Bam*H1 site located in VP1 that is useful for whole genome cloning. Reprinted from: Cubitt CL, Stone GL. Graft 2000; 5:S28-S35, with permission of SAGE Publications, Inc..

protein mass) and it mediates viral attachment to the receptors on susceptible cells. It contains epitopes for neutralization, hemagglutination inhibition (HAI), and other virus-specific and shared immunologic determinants. VP2 (mm 37,300) and VP3 (mm 25,700) are minor capsid proteins.[13,14] Agnoprotein differs from all other early and late proteins in that it localizes primarily in the cytoplasmic and perinuclear regions of the infected cell. Unlike viral capsid proteins, it is not detectable in the virion and its intracellular distribution has led to the suggestion that agnoprotein may promote release of virion from the cell.[15] Agnoprotein also play a role in the stability of microtubules and preservation of the infected cell via interaction with tubulin.[16]

### Isolation and Propagation

BKV can be propagated in human epithelial cells and fibroblasts. For isolation of BKV, human embryonic kidney (HEK) cells, diploid lung fibroblasts, and urothelial cells are suitable.[17] During the course of infection cytopathic effects typical of polyomavirus infections (rounding of cells containing cytoplasmic vacuoles) and formation of BKV plaques on HEK monolayers may take several weeks, whereas BKV-T antigen may be detected in infected cultures in 1 or 2 days.[18] Because BKV agglutinates human red blood cells of O blood type, hemagglutination can be used as a laboratory assay for quantifying virus. Both polyclonal and monoclonal antibodies to the viral T or capsid proteins are used in immunocytochemistry assays to follow the stages of BKV infection.[18-20]

### Life Cycle

Depending on the host cell, polyomaviruses cause either permissive or nonpermissive infections. All polyomaviruses multiply in the nucleus and during permissive infection the viruses cause characteristic, often pathognomonic nuclear changes and result in cell death. Urothelial cells infected with BKV may result, in renal tubulo-interstitial changes, ureteral obstruction and tubular injury. BKV also undergo nonpermissive infection when only the viral T-ag and t-ag are made, resulting in cell transformation in tissue culture of rodent cells. In case of BKV, transformed cells exhibit BKV-T antigen and contain multiple copies of BKV-DNA. BKV-DNA is integrated into the host cell genome in rodent cells, but in human cells, it may remain as free unintegrated copies.[21,22] The transformed cells can induce tumors in the appropriate animal hosts. BKV can also induce clastogenic events in infected cells, resulting in chromosomal damage, translocations, and unstable multicentric chromosomes leading to further DNA damage and ultimately cell transformation and cell death.[23]

### Pathogenesis

The pathogenesis of a polyomavirus infection involves the following sequence of events: (i) entry of virus into the body, (ii) multiplication at the entry site, (iii) viremia with transport of virus to the target organs, and (iv) multiplication in the target organs. VP1 interacts with specific receptors present on susceptible cells, mediates virion entry into the cell by endocytosis; virus is then transported to the nucleus, where it is uncoated.[24] BKV enters into the host cell via α (2-3)-linked sialic acids receptor.[25,26] After multiplication in the nucleus, virus reaches the target organs by the hematogenous route. The viral determinants that affect host range and tissue specificity of BKV are located in the enhancer/promoter elements in the regulatory regions[27] and the early regions of these viruses.[28,29] With respect to BKV, the route of infection is not known; while BKV is seldom recovered from the respiratory tract, the rapid acquisition of antibodies in the first few years of life is consistent with virus transmission by the respiratory route.[30] Although BKV-IgM in cord blood and BKV-DNA in fetal and placental tissues have been reported, there is controversy about the role of transplacental transmission of BKV.[31,32] Other potential sources of BKV infection are blood products and renal allografts.[33,34] BKV has also been identified in other organs including heart, spleen, lung, colon, and liver. Primary infection may be accompanied by transient viruria and in the immunocompetent hosts, BKV persists indefinitely as latent infections. BKV also persists in the kidney and B-lymphocytes for an indefinite period of time.[35]

## Clinical Features

In healthy children, primary infection with BKV is rarely associated clinical disease. In a prospective study, 11 out of 66 children with respiratory illness demonstrated BKV seroconversion; seven of these children had mild respiratory disease and four were asymptomatic. BKV was isolated from the urine of one of the children showing seroconversion. Unintegrated BKV DNA was identified in the tonsillar tissue of five of 12 children with recurrent respiratory disease.[36] In immunocompromised children, primary BKV infection may cause cystitis or nephritis. Primary BKV infection may also present with encephalitis. Following primary infection viruses persist indefinitely as "latent" infections of the kidney.

BKV can be reactivated after many years, usually by states of acquired (cell-mediated) immunosuppression: pregnancy, acquired immune deficiency states, neoplasm, systemic lupus erythematosus, nephrotic syndrome, and bone marrow and organ transplantation.[1,37] In immunocompromised patients BKV reactivation has been found to be associated with nephropathy, ureteritis, ureteral stenosis, hemorrhagic cystitis, autoimmune conditions, and various neoplastic processes[1] (Fig. 2). BKV related vasculopathy, a new tropism, has been described recently, in which a fatal case of disseminated BKV infection in a renal transplant recipient was associated with BKV multiplication in endothelial cells.[38] There are reports PVAN affecting native kidneys in recipients of nonrenal solid organ transplant.[39-42] Genomic sequences of BKV have also been reported to be associated with various auto-immune conditions and neoplastic disorders.[9,43-45]

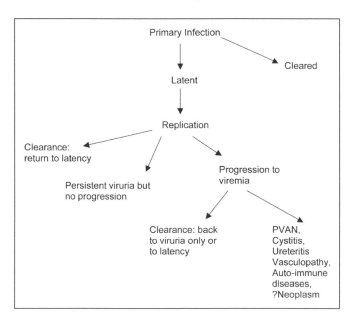

Figure 2. Logical outcomes of polyomavirus infection: primary infection to persistent viremia/viruria, viral clearance and PVAN. Reprinted with permission from: Agha I, Brennan DC. BK virus and immunosuppressive agents. In: Ahsan N, ed. Polyomaviruses and Human Diseases. New York/Austin: Springer Science/Landes Bioscience, 2006:174-84.[95]

## Renal Transplantation—Polyomavirus Associate Nephropathy

### Animal Model

Lee et al[46] reported an animal model and showed that polyomavirus preferentially replicates in the allogenic kidney grafts, accelerating graft failure mimicking PVAN seen in human renal transplant recipients. To compare the impact of native and allograft kidneys in individual mice, heterotropic kidney transplantation was preformed without removal of the native kidneys. Animals were then infected with molecularly cloned and plaque purified mouse polyomavirus wild-type strain A2. Using this model, the investigators showed that acute mouse polyomavirus infection accentuates an anti-donor T cell response and creates massive viral replication in the allograft leading to acceleration of graft rejection. Involvement of a heightened anti-donor T-cell response, in addition to direct virus mediated cytolysis of the kidney was thought to be the potential cause of graft injury. However, there are several caveats in this model: (i) murine polyoma virus, which is a distinct virus (from human polyomavirus), (ii) experiments were conducted without immunosuppression (IS), and (iii) mice were acutely infected, rather than reactivation of virus (as seen in human disease).

In a kidney transplant model, Gorder et al[47] described a new polyomavirus (cynomolgus polyoma virus—CPV) from renal tubules of cynomolgus monkeys (*Macaca fascicularis*) treated with cyclosporine and azathioprine. This virus has 84% DNA sequence homology to SV40. Most of the animals infected with polyomavirus developed lethargy, anorexia and had acute renal failure due to polyomavirus induced interstitial nephritis in the renal graft. In addition, several grafts had extensive rupture and destruction of collecting ducts and demonstrated endarteritis and focal hemorrhage indicative of active cellular rejection. None of the animals with detectable virus in the allograft had infections of the native kidney. In the renal graft, the peak frequency of infection was from day 21-48 after transplant and during this study, no particular association of polyomavirus with any of the immunosuppressive agents was evident.

### Human

Infections in renal transplant recipients have been studied by several investigators[34,38,48-55] and have been frequently reviewed.[1,24,56-71] In a multi-center serologic study of nearly 500 renal allograft recipients and donors in the United States, Andrews et al[34] reported that seropositive donor demonstrated increased risk of primary and reactivated infections with BKV infections. Virus shedding in urine of renal transplant recipients has been monitored by a variety of techniques, including urinary cytology, immunoassay, electron microscopy, virus isolation, enzyme linked immunoassays, nucleic acid hybridization, and polymerase chain reaction (PCR).[72-90] In prospective studies, 25-44% of renal transplant patients excrete virus in their urine in the post-transplant period. The duration of excretion ranges widely, from transient viruria to excretion over several weeks or several months. In a single center prospective longitudinal study (1-year follow-up) involving 104 consecutive renal transplant patients, Bressollette-Bodin and colleagues reported that DNAuria and DNAemia occurred in 57% and 29% of patients, respectively.[91] The kidney of a seropositive donor may initiate infections in the recipient. Infections may be either reactivations or primary infections affecting up to 5% of renal allograft recipients in about 40 weeks (range 6-150) post-transplantation. Persistent BKV infections have been associated with irreversible graft loss in more than 50% of the cases over 12-240 weeks of follow-up.[69] Patients often present with an asymptomatic rise in serum creatinine, necessitating further investigation. Unfortunately, in most cases, renal dysfunction is representative of a late stage of PVAN, occurring as a result of BKV-associated renal allograft injury. Confirmation of the diagnosis requires biopsy of the allograft, additionally, the use of advanced techniques (vide infra) are needed to distinguish tubular injury due to the virus from concurrent acute rejection. The infections also appear to be responsible for some of the cases of ureteral obstruction.[92,93]

## Table 1. Presumed risk factors for polyomavirus-associated nephropathy

| Risk Factor | Increased With |
|---|---|
| **Immunosuppression** | |
| • Triple combinations | Tacrolimus, Mycophenolate mofetil, Prednisone |
| • Drug levels | Tacrolimus trough levels >8 ng/mL |
| • Drug dosing | Mycophenolate mofetil >2 g/day Prednisone |
| • Anti-rejection treatment | Anti-lymphocyte globulin Methylprednisolone pulses |
| **Patient determinants** | |
| • Age | >50 years |
| • Gender | Male > female |
| • Race | White > other |
| • BKV seropositivity | Antibody titer < 1:40 |
| • BKV-specific T-cells | Interferon-γ production Polymorphism linked to low interferon-γ expression |
| • Comorbidity | Diabetes mellitus |
| **Cytomegalovirus infection** | |
| • Organ determinants | |
| • Immunologic injury | HLA-mismatches Prior acute rejections |
| • Drug toxicity | Tacrolimus |
| **Viral determinants** | |
| • Noncoding control region | Rearrangements ? Point mutations in GM-CSF response element ? |
| • Viral capsid protein-1 | Mutations in serotype domain ? |

Reprinted with permisssion from: Hirsch HH, Drachenberg CB, Steiger J, Ramos E. Polyomavirus-associated nephropathy in renal transplantation: critical issues of screening and management. In: Ahsan N, ed. Polyomaviruses and Human Diseases. New York/Austin: Springer Science/Landes Bioscience, 2006:160-73.[56]

## Risk Factors

Potential risk factors of PVAN can be categorized as donor-related, recipient-related, transplant-virus related, virus-related, or immunosuppressant related (Table 1).[56] Most of these risk factors are unavoidable, unmodifiable or their risk contribution has not been consistently shown from study to study, perhaps because the type and intensity of immunosuppressive agents may override any individual or combination of risk factors.

There are two competing hypotheses regarding the source of BKV infection: donor related and recipient related. In the former, the transmission may occur directly through the transplanted organ in recipients who have never been exposed to BKV. High concordance rate (80%) of BKV reactivation among recipient pairs has been found by matching "molecular finger prints" among recipient pairs with BKV reactivation. Donor antibody titer correlated inversely to the onset of post-transplant viruria (p = 0.001) and directly proportional to duration of viruria (p = 0.014) and peak urine viral titers (p = 0.005).[94] Since BKV resides in the renal epithelium, its reactivation in recipient may be due to defective immune surveillance in immunosuppressed recipients. Occurrence of PVAN has also been noted in recipients with pretransplant bilateral native nephrectomies and felt to be from reactivation of virus residing in the ureteral and urinary bladder mucosa.

## Immunosuppresion (IS) and PVAN

Several clinical studies have examined the roles of net IS, as well as individual agents in PVAN. Risk factors include treatment of rejection episodes and increasing viral replication under potent immunosuppressive drugs such as tacrolimus, sirolimus, or mycophenolate mofetil. High dose of corticosteroid therapy has been associated with an increased risk for active PVAN. However, induction therapy with antilymphocytes does not appear to increase the risk of PVAN.[53,56,59-64,95,96] In a single center study, Hirsch and colleagues[48] reported that induction therapy with antilymphocyte preparations (antithymocyte globulin and monoclonal anti-CD25 antibodies) was not significantly associated with decoy cell shedding or PVAN, but corticosteroid pulse antirejection therapy was related to these conditions (p <0.05). Multivariate logistic-regression models showed that the number of corticosteroid pulses was linked to BKV replication (p = 0.01; relative risk [RR] = 1.21; 95% CI, 1.08-1.360 and PVAN (p = 0.02; RR = 1.38; 95% CI, 1.04-1.68). Similarly in a study, Brennan et al[96] reported that induction therapy with rabbit antilymphocyte globulin was not associated with an increased incidence of BK viruria, viremia, or sustained viruria or viremia compared with those who did not receive rabbit antilymphocyte globulin.

### BKV Specific Immunity: Humoral and Cellular Response

Studies on BKV induced immune responses consists mainly of morphological presentation, indirect evidence, and pathophysiological concepts, derived from indirect evidence and experimental tumor models. The role of BKV specific antibody titers was investigated in the setting of PVAN. Using HIA, Hirsch et al[48] reported that 77% of adult kidney transplant recipients were seropositive before transplantation; however the seropositive status was not protective to BKV replication and development of PVAN. On the other hand, BKV seropositive donors were shown to increase the risk for subsequent BKV infection.[34,94] In patients with BKV replication and PVAN, BKV specific immunoglobulin G (IgG) antibody levels were found to increase in patient with declining viremia or past PVAN.[97]

There is now evidence that failure to mount cellular immune response is linked to BKV replication and development of PVAN. Using ELISpot assays, Comoli et al[98] found that in comparison to healthy seropositive individual, BKV-specific interferon-γ (INF-γ) secreting lymphocytes were detectable at a much lower mean frequencies in BKV-seropositive pediatric kidney transplant recipients, whereas no BKV specific INF-γ secreting cells were detectable in patients with PVAN. The frequencies of INF-γ secreting lymphocytes were found to increase as BKV loads in plasma and urine declined with reduction in overall IS. The human leukocyte antigen (HLA) restricted T cells may also play important role in the form of immune reconstitution disease when HLA matching is optimal and may explain the paradoxical worse graft survival in renal allograft with PVAN and better HLA matches.[99] In a mismatched allogenic environment, HLA unrestricted T cells may cause immune mediated damage and BKV directed killing.[100,101] Further studies will be required to gain insight into BKV specific cellular immunity.

### Chronic Allograft Fibrosis

Mannon et al[102] undertook a transcriptional evaluation of kidney allograft biopsies from recipients' with PVAN, acute rejection (AR), and stable allograft function. These investigators showed that renal allograft with PVAN transcribed pro-inflammatory genes equal in character and higher in magnitude to that seen in AR. Importantly in the context of chronic renal allograft loss, BKV infection creates a transcriptional microenvironment that promotes graft fibrosis. In both PVAN and AR groups, Banff histological scores immune staining of inflammatory infiltrates were similar, however, compared to AR groups, transcriptional profiles of PVAN was significantly higher ( PVAN vs. AR): CD8 65.9 ± 18.8 vs. 30.9 ± 2.0, INF-γ 5.1 ± 17.0 vs. 14.0 ± 7.3, perforin 153.8 ± 50.4 vs. 12.1 ± 7.3, and CXCR3 49.9 ± 12.8 vs. 15.6 ± 3.8 (p <0.001). Moreover, in PVAN, the levels of transcripts of genes related to matrix protein collagen I, IV and fibronectin were significantly elevated compared to AR (p = 0.03, 0.04 and <0.01, respectively). Other transcriptional molecules, associated with graft fibrosis including matrix collagens, tumor growth factor-β (TGF-β) [9.4 ± 4.5 vs. 3.2 ± 1.7, p = 0.001], matrix metalloproteins (MMP)-2 [4.6 ± 2.5 vs. 0.3 ± 0.1-fold; p<0.001] and MMP-9 [87.8 ± 65.1- fold vs. 11.8 ± 6.5-fold; p<0.002] were also significantly elevated. Additionally, several markers (α-smooth muscle antigen and S100A4) of epithelial mesenchymal transformation, a process that mediates late graft deterioration following transplantation were found to be higher in PVAN than AR. Taken together these findings for the first time providing new insights into intra-renal inflammatory processes that promotes graft loss.

## Prevention and Control

BKV infections are extremely common and are essentially harmless except when the host is immunologically impaired. In the United States, antibodies to BKV are acquired by 50% of the children by the age of 3-4 years. The antibody prevalence to BKV reaches nearly 100% by the age of 10-11 years and then declines to around 70-80% in the older age groups. Primary infections with BKV in healthy children are rarely associated with illness. BKV seroconversion is associated with mild respiratory illness. Reactivations are brought about not only by significant IS, as in renal transplant recipients and human immunodeficiency virus infected individuals, but also by more subtle factors, such as pregnancy, diabetes, and old age. There have been no attempts to devise strategies for the prevention and control of these infections.[1,35,37]

## Diagnosis

### *Markers of PVAN—Screening and Adjunctive Tests*

Different screening assays for polyomavirus replication have been investigated, such as urine cytology (decoy ells), quantitative assay of urinary BKV DNA or VP1 mRNA load, or electron microscopy for polyomavirus particles. Data from a five-center study comparing a range of BK virions and DNA load indicate that the quantitative real-time PCR results are equivalent within 1 log10. Most quantitative viral load assays use an arbitrary detection limit of 1,000 DNA copies per ml, hence reporting viral load <1,000 copies per ml as "below detection" or "negative". Micturation intervals and fluctuations of urine content may contribute inter-assay variations. Urine cytology and VP1 mRNA assays are susceptible to preanalytic hazards such as the type and duration of processing and shipment. Polyomavirus DNA load may vary depending on whether or not supernatants, cell pellets, or resuspended urine are used for DNA preparation. Highly sensitive qualitative assays for polyomavirus DNA, such as nested PCR applied to cell pellets, may substantially decrease specificity and positive predictive value of PVAN.[57] The results from two prospective studies indicate that BKV replication in the urine examined by either urine cytology (decoy cell) or DNA PCR precedes BKV viremia by a median of 4 weeks, and histologically documented BKV nephropathy by a median of 12 weeks. At three months, the 80th and 50th percentile of all viruric and viremic patients was reached, respectively, whereas PVAN was observed in 0/23 and 1/10 (10%) patients. Between 2 to 5 years post-transplantation, less than 5 percent of all cases occurred.[48,96]

It is recommended that positive screening results be confirmed within four weeks, and/or followed by adjunct quantitative diagnostic assays for higher positive predictive value such as quantitation of BKV DNA load in plasma or urine or VP1 mRNA load in urine. For these assays, threshold values yielding a specificity of >93% for PVAN have been proposed (plasma BKV load >10, 000 copies/ml; VP1 mRNA load >6.5 x 10^5 copies /ng total RNA). Although further validation of threshold values in plasma and urine is needed, persisting polyomavirus loads above these thresholds for >3 weeks is highly suggestive of PVAN (presumptive) and evaluation with allograft biopsy is recommended. To monitor polyomavirus course of disease, quantitative polyomavirus testing in plasma and urine should be performed every 2-4 weeks. (Fig. 3).[48,56,57,95]

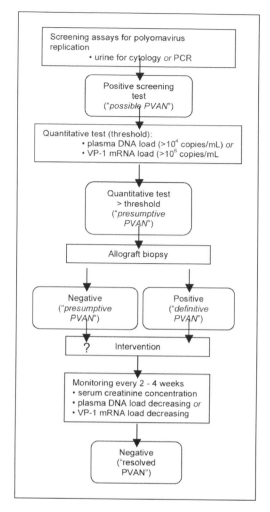

Figure 3. Polyomavirus replication: Recommended screening and intervention. Reprinted with permission from: Hirsch HH, Drachenberg CB, Steiger J, Ramos E. Polyomavirus-associated nephropathy in renal transplantation: critical issues of screening and management. In: Ahsan N, ed. Polyomaviruses and Human Diseases. New York/Austin: Springer Science/Landes Bioscience, 2006:160-73.[56]

## PVAN—Pathologic Diagnosis

In Chapter 38, Nickeleit and colleagues have discussed the pathologic findings of PVAN at great detail. Cytomorphology of urinary epithelial cells is helpful as an indication of polyomavirus excretion in urine.[1,55,61,68,74,75,85] Virus-infected epithelial cells are enlarged "decoy cells", and their nuclei contain a single, homogeneous, large, pale basophilic inclusion that may occupy the entire nuclear area. Polyomavirus-infected urinary cells should not be mistaken for cells infected with cytomegalovirus, which are generally smaller, basophilic or eosinophilic, and surrounded by a halo and contains intracytoplasmic inclusions. The rough-textured nuclear chromatin of a malignant cell differs greatly from the structureless inclusion in a polyomavirus-infected cell. The cytological findings by themselves are not definitive because they cannot distinguish between BKV and JCV infections, and virus excretion in urine may occur without marked cytological abnormalities.

Histological presentation BKV nephropathy has been described recently.[68,103,104] Cytopathic changes in renal tubules reflecting viral multiplication consist of enlarged nuclei with smudgy chromatin, intranuclear inclusions, rounding and detachment. These have been classified into: (a) stage A: focal medullary arrays involvement of tubular cells, (b) stage B: extensive renal involvement with multifocal or diffuse cytopathic alterations, necrosis, profound inflammatory response, and early fibrosis, and (c) stage C: characterized by interstitial fibrosis, scarring, and calcification.

*Table 2. Management and outcomes in polyomavirus-associated nephropathy*

| Source | PVN Cases | Anti-Rejection Therapy | Baseline Immunosuppression | | | Graft Loss Due to PVN |
|---|---|---|---|---|---|---|
| | | | Decrease | Increase | No Change | |
| Hussain et al[107] | 14 | 11 | 14 | — | — | 9/14 (64%) |
| Randhawa et al[59] | 22 | 12 | 16 | — | 6 | 10/22 (45%) |
| Ahuja et al[108] | 10 | 9 | 10 | — | — | 7/10 (70%) |
| Mengel et al[109] | 7 | — | 4 | 1 | 2 | 5/7 (71%) |
| Nickeleit et al[89,90,110,111] | 11 | n/a | n/a | n/a | n/a | 5/11 (45%) |
| Howell et al[123] | 7 | 4 | 6 | 1 | — | 1/7 (14%) |
| Huralt de Ligny et al[124] | 10 | 9 | 8 | 2 | — | 2/10 (20%) |
| Rahamimov et al[125] | 7 | 7 | 6 | — | 1 | 1/7 (14%) |
| Li et al[124] | 6 | 3 | 6 | — | — | 1/6 (17%) |
| Ramos et al[112] | 67 | — | 52 | — | 15 | 11/67 (16%) |
| Buehrig et al[113] | 18 | 3 | 18 | — | — | 7/18 (39%) |
| Trofe et al[114] | 13 | — | 10 | — | 3 | 7/13 (54%) |
| Trofe et al[121] | 10 | — | 10 | — | — | 1/10 (10%) |
| Barri et al[127] | 8 | 1 | 7 | — | 1 | 1/8 (13%) |
| Ginevri et al[122] | 5 | — | 5 | — | — | 1/5 (20%) |
| Hirsch et al[48] | 5 | 4 | 5 | — | — | 0/5 (0%) |
| Totals | 220 | 63 (29%) | 177 (80%) | 4 (2%) | 28 (13%) | 69/220 (31%) |

PVN, polyomavirus-associated nephropathy; n/a, data not available. Reprinted with permission from: Roskopf J, Trofe J, Stratta RJ, Ahsan N. Pharmacologic options for the management of human polyomavirus. In: Ahsan N, ed. Polyomaviruses and Human Diseases. New York/Austin: Springer Science/Landes Bioscience, 2006:228-54.[116]

## PVAN, Concurrent Rejection and Clinical Outcome of Allograft

In the presence of PVAN, the diagnosis of acute rejection may be difficult. There is consensus that endarteritis, fibrinoid vascular necrosis, and glomerulitis (Banff II and III) as well as C4d deposits along peritubular capillaries should be regarded as evidence of concurrent rejection. The diagnosis of tubulointerstitial rejection (Banff I) is challenging since the inflammatory infiltrates may indicate a response to antigens of the virus and/or virus induced tubular cell necrosis. The demonstration of lymphocytic infiltrates, marked tubulitis and tubular HLA-DR expression in areas lacking polyomavirus replication may support the diagnosis of Banff I rejection.[103-106]

A better understanding of how to manage PVAN is slowly evolving. Historically, it was difficult to exclude concurrent acute rejection on biopsy. Therefore, many patients were given intensified immunosuppressive regimens to treat acute rejection. Hussain et al[107] described their experience in patients diagnosed with PVAN. The authors compared seven patients who received OKT3 or equine antithymocyte globulin (Atgam®) at the time of diagnosis, to seven patients who did not. All 7 (100%) patients in the OKT3/Atgam group suffered grafts loss; only 2 (29%) grafts were lost in the other group. Randhawa and colleagues[55] reported their initial experience with 12 patients diagnosed with PVAN and were given empiric steroids. Only 1/12 (8%) patients cleared the virus, and 3/12 (25%) had a partial response. The other eight nonresponders were managed with a reduction in baseline IS, with graft failure occurring in 8/12 (67%) patients. Another study[108] reported 70% graft loss, when 9/10 (90%) patients with PVAN were given steroids at diagnosis and one patient also received Atgam followed by OKT3. Mengel et al[109] prospectively evaluated biopsies of seven patients with PVAN. Acute rejection was found in 6/7 biopsies. IS was initially reduced four patients and overall, graft loss due to PVAN was 71% (5/7). In separate studies, Nickeleit and colleagues reported that 45% (5/11) patients diagnosed with PVAN and concurrent acute rejection developed graft loss when an antirejection therapy was administered.[89,90,110,111] Other authors have reported better outcomes associated with the initial use of antirejection therapy in the setting of PVAN (Table 2).

Several authors have reported their clinical experience with decreasing IS when PVAN is initially diagnosed. Ramos et al,[112] reported their experience in 67 patients with diagnosis of PVAN. Majority of the patients were treated with calcineurin inhibitor, mycophenolate mofetil, and prednisone based immunotherapy. An initial decrease in IS was performed in 52 patients, while no specific IS intervention was made in 15 patients. Eight patients (15.3%) developed acute rejection and six (11.5%) became negative for PVAN. The overall rate of graft loss was 11/67 (16.4%). In comparison to a case-matched polyomavirus-negative control group, patients with PVAN had significantly reduce graft survival (p = 0.0004). Buehrig and colleagues[113] reported their experience with 18 patients diagnosed with PVAN; all patients were ultimately managed with a decrease in maintenance IS. Satisfactory results occurred in 11 patients (61%), while the remaining seven (39%) had poor outcomes (graft loss, increased severity of PVAN on repeat biopsy, or serum creatinine >3 mg/dl at 6 months after diagnosis). Trofe et al[114] reported a graft loss in 7/13 (54%) of patients diagnosed with PVAN; ten patients were managed with a decrease in maintenance IS. On the other hand, in a prospective study involving [78] renal transplant recipients, Hirsch et al[48] reported an overall graft survival rate in the group was 100% in 5 patients who were diagnosed with PVAN.

### Treatment

Human are natural hosts for polyomaviruses. Recently, Zaragoza et al,[115] performed experiments to test the suitability of squirrel monkeys (*Saimiri sciureus*) as an experimental model for BKV and SV40 infection. Four squirrel monkeys received intravenous inoculation with BKV Gardner strain—*pre* and *post*-cyclophosphamide therapy. Viral assessment was determined by quantitative real-time PCR amplification of viral DNA from blood, urine and tissue. The authors found that squirrel monkeys were susceptible to infection with BKV and concluded that the squirrel monkey would be a suitable model for animal studies for experimental BKV infection and may facilitate studies of viral entry, pathogenesis, and anti-viral therapy. The treatment of PVAN is unsatisfactory, since no uniformly effective anti-viral agents are available at this time. Current management strategies aim at the judicious lowering, switching and discontinuation of the dosage of the immunosuppressive therapy to allow clearance of BKV (Table 3).[1,56,57,95,116-128]

**Table 3. Modification of maintenance immunosuppression in renal transplant patients with PVAN**

| Strategy | Intervention | Comment |
|---|---|---|
| **Switching** | | |
| TAC → CYC | Trough levels 100-150 ng/mL | • Consider in patients on TAC-MMF-PRE combinations due to additional MMF lowering effect of CYC |
| MMF → AZA | Dosing ≤100 mg/day | • Consider in patients on TAC-MMF-PRE with prior rejections |
| TAC → SIR | Trough levels <6 ng/mL | • Consider in patients with calcineurin inhibitor toxicity |
| MMF → LEF | Trough levels >30 ng/mL | • Consider in patients with concurrent cytomegalovirus infection |
| **Decreasing** | | |
| TAC | Trough levels <6 ng/mL | • Consider in patients with limited early disease or "presumptive PVAN" |
| MMF | Dosing <1 g/day | • Consider in patients with limited early disease or "presumptive PVAN" |
| CYC | Trough levels 100 ng/mL | • Consider in patients with limited early disease or "presumptive PVAN" |
| **Stopping** | | |
| TAC | •Continue dual therapy | • Consider in patients with presumptive PVAN, with stable graft function |
| or | TAC-PRE | during the second year posttransplant without prior acute rejection episodes |
| MMF | CYC-PRE | |
| or | MMF-PRE | |
| CYC | SIR-PRE | |

Reprinted with permission from: Hirsch HH, Drachenberg CB, Steiger J, Ramos E. Polyomavirus-associated nephropathy in renal transplantation: critical issues of screening and management. In: Ahsan N, ed. Polyomaviruses and Human Diseases. New York/Austin: Springer Science/Landes Bioscience, 2006:160-73.[56]

Funk et al[49] in a retrospective analysis of BKV plasma load in renal transplant recipients undergoing allograft nephrectomy (n = 3) or changes in immunosuppressive regimens (n = 12). The authors reported that after nephrectomy, BKV clearance was fast (viral half life [$t_{1/2}$], 1-2h) or moderately fast ($t_{1/2}$, 20-38h) and after changing immunosuppressive regimens was $t_{1/2}$, 6h-17 days. One large study of patients with prospective monitoring of urine and blood, and preemptive withdrawal of the anti-metabolite upon development of viremia, showed that this strategy resulted in clearance of viremia and viruria, and appeared to prevent progression to PVAN without increasing the risk of AR.[96] Another study prospective showed that viremia and viruria could resolve or decrease over time with standard reductions in IS, without preemptive withdrawal of any component of the immunosuppressive regimen.[127] In patients with progressive graft dysfunction not responding to this maneuver, antiviral treatment has been employed with variable results.[116,129] Retinoic acid, DNA gyrase inhibitors, Cidofovir and 5'-brome2'-deoxyuridine inhibit polyoma virus replication in vitro. In clinical trials, Cidofovir was effective when administered at 20% of the dosage recommended for treatment of CMV.[130-134] In clinical trial 5'-brome2'-deoxyuridine has been ineffective but vidarabine has reportedly produced dramatic remission in a patients with post-bone marrow transplant hemorrhagic cystitis.[135] There are several recent reports of successful reversal of renal allograft function in patients with PVAN, the largest of which was published by University of Pittsburgh Transplant Program.[132] Cidofovir was administered at doses 0.2-1 mg/kg every 1-4 weeks intervals, renal function improved in 31% (5/16) and stabilized in another 31% (5/16) patients. Intravenous immunoglobulin (IVIG) has been shown to contain antibodies against BKV and clinical reduction in BK viremia had been demonstrated when IVIG was administered at a dose of 2-5 g/kg divided over 2-5 days.[136] Interferon has some activity against BKV in vitro but has no effect on BK viruria in renal transplant patients.[137] Early results with two malonoitrilamide compounds, FK-778 and FK-779 showed that these agents demonstrated anti-polyomavirus (simian and murine) activity and were able to decrease free virus production.[138] Recently several centers have reported that treatment with leflunomide, the parent compound of malonoitrilamide resulted in improved graft

function in patients with biopsy proven PVAN.[139-142] The efficacy of these strategies is unclear, because reduction of IS has been used prior to institution of all anti-viral therapies. Also in vitro study has shown that the EC50 for either leflunomide or cidofovir is higher than what may be achieved clinically with conventional dosing.[134] In case of coexisting acute rejection, a two-step procedure of immediate antirejection may be followed by reducing the maintenance IS might be indicated.[56,57,95,116] A proposed algorithm for the pharmacologic management of patients with PVAN is given in Figure 4.[116] In some situations, allograft and native nephrectomies were carried out in order to remove the source of infection and then successful retransplantation was performed.[143,144] Ramos et al[145] reviewed data from 10 patients who lost their renal allograft due to PVAN and later on underwent retransplantation. Of these 10 patients, 7 had transplant nephrectomy. The authors showed that neither nephrectomy of the infected graft before retransplantation nor the choice of immunosuppressive agent influenced the risk of PVAN recurrence. The minimum time to retransplantation after kidney failure was 1 month; however, it remains unclear what the optimum time before retransplantation should be. More recently, there are reports of successful retransplantation during active BK viremia with or without transplant nephrectomy.[146,147] It has been recommended however that allowing enough time to mount an antiviral immune response prior to retransplantation seems appropriate.[148]

## Conclusion

PVAN as a result of BKV infection has emerged as a serious complication of kidney transplantation. Affecting up to 10% of all kidney transplant patients, toady BKV is the most common viral infections in these patient population. According to several clinical studies, graft loss due to PVAN ranged from 10% to >80% in kidney transplant recipients. The best approaches to thwarting the progression of BKV infection to PVAN are early diagnosis and intervention. Identification of potential risk factors PVAN may allow for the prevention or preemption PVAN. Reduction in overall of IS and antiviral therapies may prevent progression of allograft dysfunction. Retransplantation remains a viable option for those patients who experience graft loss due to end-stage PVAN.

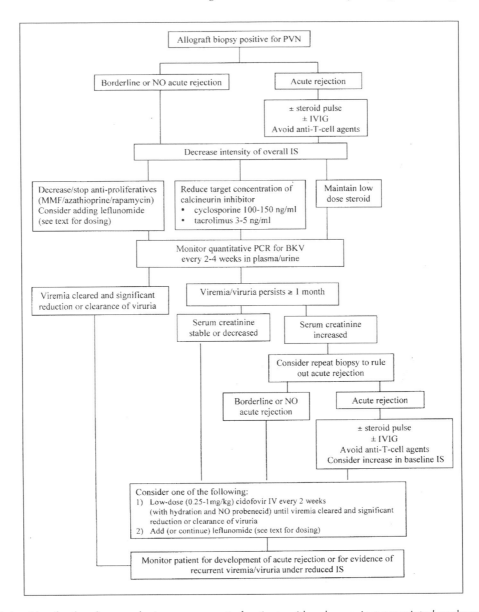

Figure 4. Proposed algorithm for the pharmacologic management of patients with polyomavirus-associated nephropathy. Reprinted with permission from: Roskopf J, Trofe J, Stratta RJ, Ahsan N. Pharmacologic options for the management of human polyomavirus. In: Ahsan N, ed. Polyomaviruses and Human Diseases. New York/Austin: Springer Science/Landes Bioscience, 2006:228-54.[116]

## References

1. Ahsan N, Shah KV. Polyomaviruses and Human Diseases. In: Ahsan N, ed. Polyomaviruses and Human Diseases. New York/Austin: Springer Science+Business Media/Landes Bioscience, 2006:1-18.
2. Crandall KA, Perez-Losada M, Christensen RG et al. Phylogenomics and molecular evolution of polyomaviruses. In: Ahsan N, ed. Polyomaviruses and Human Diseases. New York/Austin: Springer Science+Business Media/Landes Bioscience, 2006:46-59.
3. Padgett BL, Walker DL, Zu Rhein GM et al. Cultivation of papova-like virus from human brain with progressive multifocal leucoencephalopathy. Lancet 1971; i:1257-60.
4. Gardner SD, Field AM, Coleman DV et al. New human papovavirus (B.K.) isolated from urine after renal transplantation. Lancet 1971; i:1253-57.
5. Frisque RJ, Hofstetter C, Tyagarajan SK. Transforming activities of JC virus early protein. In: Ahsan N, ed. Polyomaviruses and Human Diseases. New York/Austin: Springer Science+Business Media/Landes Bioscience, 2006:288-309.
6. Cubitt C. Molecular genetics of the BK virus. In: Ahsan N, ed. Polyomaviruses and Human Diseases. New York/Austin: Springer Science+Business Media/Landes Bioscience, 2006:85-95.
7. Moens U, Johansen T, Johansen JI et al. Noncoding control region of naturally occurring BK virus variants: Sequence comparison and functional analysis. Virus Genes 1995; 10:261-75.
8. Kristoffersen AK, Johnsen JI, Seternes OM et al. The human polyomavirus BK T antigen induces genes expression in human cytomegalovirus. Virus Res 1997; 52:61-71.
9. Rekvig OP, Bendiksen S, Moens U. Immunity and autoimmunity induced by polyomaviruses: Clinical, experimental and theoretical aspects. In: Ahsan N, ed. Polyomaviruses and Human Diseases. New York/Austin: Springer Science+Business Media/Landes Bioscience, 2006:117-47.
10. Moens U, Seternes OM, Johansen B et al. Mechanism of transcriptional regulation of cellular genes SV40 larger T- and small t-antigens. Virus Genes 1997; 15:135-54.
11. McGhee JD, Felsenfeld G. Nucleosome structure. Annu Rev Biochem 1980; 49:1115-56.
12. Kornberg RD, Lorch Y. Twenty-five years of nucleosome, fundamental particle of the eukaryotic chromosome. Cell 1999; 98:285-94.
13. Viscidi RP, Clayman B. Serological cross reactivity between polyomavirus capsid. In: Ahsan N, ed. Polyomaviruses and Human Diseases. New York/Austin: Springer Science+Business Media/Landes Bioscience, 2006:73-84.
14. Rayment I, Baker TS, Casper DL et al. Polyoma virus capsid structure at 22.5A resolution. Nature 1982; 295:13-31.
15. Resnick J, Shenk T. Simian virus 40 agnoprotein facilitates normal nuclear location of the major capsid polypeptide and cell-to-cell spread of virus. J Virol 1986; 60:1098-106.

16. Endo S, Okada Y, Nishihara H et al. JC virus agnoprotein co localizes with tubulin. J Neuro Virol 2003; 9:10-14.
17. Beckmann A, Shah K. Propagation and primary isolation of JCV and BKV in urinary epithelial cell cultures. In: Sever JL, Madden DL, eds. Polyomaviruses and human neurological diseases. New York: Alan R Liss, 1983:3-14.
18. Marshall WF, Telenti A, Proper J et al. Rapid detection of polyomavirus BK by a shell vial cell culture assay. J Clin Microbiol 1990; 28:1613-15.
19. Knowles WA, Gibson PE, Gardner SD. Serological typing scheme for BK-like isolates of human polyomavirus. J Med Virol 1989; 28:118-23.
20. Marshall J, Smith AE, Cheng SH. Monoclonal antibody specific for BK virus large T antigens. Oncogene 1991; 6:1673-76.
21. Pater MM, Pater A, Di Mayorca G et al. BK virus-transformed inbred hamster brain cells: Status of viral DNA in subclones. Mol Cell Biol 1982; 2:837-44.
22. Takemoto KK, Linke H, Miyarnura T et al. Persistent BK papovavirus infection of transformed human fetal brain cells. J Virol 1979; 29:1177-85.
23. Thiele M, Grabowski G. Mutagenic activity of BKV and JCV in human and other mammalian cells. Arch Virol 1990; 113:221-33.
24. Randhawa P, Vats A, Shapiro R. The pathobiology of polyomavirus infection in man. In: Ahsan N, ed. Polyomaviruses and Human Diseases. New York/Austin: Springer Science+Business Media/Landes Bioscience, 2006:148-59.
25. Eash S, Manley K, Gsparovie M et al. The human Polyomaviruses. Cell Mol Life Sci 2006; 63:865-76.
26. Ashok A, Atwood WJ. Virus receptors and tropism. In: Ahsan N, ed. Polyomaviruses and Human Diseases. New York/Austin: Springer Science+Business Media/Landes Bioscience, 2006:60-72.
27. Kenney S, Natarajan V, Strike D et al. JC virus enhancer-promoter active in human brain cells. Science 1984; 226:1337-39.
28. Corallini A, Pagnani M, Caputo A et al. Cooperation in oncogenesis between BK virus early region gene and the activated human c-Harvey ras oncogene. J Gen Virol 1988; 69:2671-79.
29. Knepper JE, di Mayorca G. Cloning and characterization of BK virus- related DNA sequences from normal and neoplastic human tissues. J Med Virol 1987; 21:289-99.
30. Sundsfjord A, Spein AR, Lucht E et al. Detection of human polyomavirus BK DNA in nasopharyngeal aspirates from children with respiratory infections but not in saliva from immunodeficient and immunocompetent patients. J Clin Microbiol 1994; 32:1390-94.
31. Taguchi F, Nagaki D, Saito M et al. Transplacental transmission of BK virus in human. Jpn J Micrbiol 1975; 19:395-98.
32. Shah K, Daniel R, Madden D et al. Serological investigation of BK papovavirus infection pregnant women and their offspring. Infect Immun 1980; 30:29-35.
33. Arthur RR, DagostinS, Shah KV. Detection of BK virus and JC virus in urine and brain tissue by polymerase chain reaction. J Clin Microbiol 1989; 27:1174-79.
34. Andrews CA, Shah KV, Daniel RW et al. A serological investigation of BK virus and JC virus infections in recipients of renal allografts. J Infect Dis 1988; 158:176-81.
35. Doerries K. Human polyomavirus JC and BK persistent infection. In: Ahsan N, ed. Polyomaviruses and Human Diseases. New York/Austin: Springer Science+Business Media/Landes Bioscience, 2006:102-16.
36. Goudsmit J, Wertheim-van Dillen P, van Strein A et al. The role of BK virus in acute respiratory tract disease and the presence of BKV DNA in tonsils. J Med Virol 1982; 10:91-99.
37. Knowles W. Discovery and epidemiology of the human polyomaviruses BK virus (BK) and JC virus (JCV). In: Ahsan N, ed. Polyomaviruses and Human Diseases. New York/Austin: Springer Science+Business Media/Landes Bioscience, 2006:19-45.
38. Petrogiannis-Haliotis T, Sakoulas G, Kirby J et al. BK-related polyomavirus vasculopathy in a renal-transplant recipient. N Engl J Med. 2001; 345:1250-55.
39. Pavlakis M, Haririan A, Klassen DK. BK virus infection after nonrenal transplantation. In: Ahsan N, ed. Polyomaviruses and Human Diseases. New York/Austin: Springer Science+Business Media/Landes Bioscience, 2006:185-89.
40. Schwartz A, Mengel M, Haller H et al. Polyoma virus nephropathy in native kidneys after lung transplantation. Am J Transplantation 2005; 5:2582-85.
41. Barton TD, Blumberg EA, Doyle A et al. A prospective cross-sectional study of BK virus infection in nonrenal solid organ transplant recipients with chromic renal dysfunction. Transpl Infect Dis 2006; 8:102-7.
42. Barber CEH, Hewlett TJC, Geldenhuys L et al. BK virus nephropathy in a heart transplant recipient: Case report and review of literature. Transpl Infect Dis 2006; 8:113-21.

43. Lee W, Langhoff E. Polyomavirus in human cancer development. In: Ahsan N, ed. Polyomaviruses and Human Diseases. New York/Austin: Springer Science+Business Media/Landes Bioscience, 2006:310-18.
44. Barbanti-Boradano G, Sabbioni S, Martini et al. BK virus, JC virus and simian virus 40 infection in humans, and association with human tumors. In: Ahsan N, ed. Polyomaviruses and Human Diseases. New York/Austin: Springer Science+Business Media/Landes Bioscience, 2006:319-41.
45. Rollison DEM. Epidemiologic studies of polyomaviruses and cancer; previous findings, methodologic challenges and future directions. In: Ahsan N, ed. Polyomaviruses and Human Diseases. New York/Austin: Springer Science+Business Media/Landes Bioscience, 2006:342-55.
46. Lee ED, Kemball CC, Wang J et al. A mouse model for polyomavirus-associated nephropathy of kidney transplants. Am J Transplantation 2006; 6:913-22.
47. Gorder MA, Pelle PD, Henson JW et al. A new member of the polyoma virus family causes interstitial nephritis, ureteritis, and enteritis in immunocompromised cynomolgus monkeys. Am J Pathol 1999; 154:1273-84.
48. Hirsch HH, Knowles W, Dickenmann M et al. Prospective study of polyomavirus type BK replication and nephropathy in renal-transplant recipients. N Engl J Med 2002; 347:488-96.
49. Funk G, Steiger J, Hirsch HH. Rapid dynamics of polyomavirus type BK in renal transplant recipient. J Infect Dis 2006; 193:80-87.
50. Munoz P, Fogeda M, Bouza E et al. Prevalence of BK virus replication among recipients of solid organ transplants. Clin Infect Dis 2005; 41:1720-25.
51. Ramos E, Drachenberg CB, Portocarrero M et al. BK virus nephropathy diagnosis and treatment: Experience at the University of Maryland Renal Transplant Program. Clin Transpl 2002; 143-53.
52. Schmitz M, Brause M, Hetzel G et al. Infection with polyomavirus type BK after renal transplantation. Clin Nephrol 2003; 60:125-29.
53. Ginevri F, De Santis R, Comoli P et al. Polyomavirus BK infection in pediatric kidney-allograft recipients: A single-center analysis of incidence, risk factors, and novel therapeutic approaches. Transplantation 2003; 75:1266-70.
54. Lin PL, Vats AN, Green M. BK virus infection in renal transplant recipients. Pediatr Transplant 2001; 5:398-405.
55. Randhawa PS, Finkelstein S, Scantlebury V et al. Human polyoma virus-associated interstitial nephritis in the allograft kidney. Transplantation 1999; 67:103-9.
56. Hirsch HH, Drachenberg CB, Steiger J et al. Polyomavirus-associated nephropathy in renal transplantation: Critical issues of screening and management. In: Ahsan N, ed. Polyomaviruses and Human Diseases. New York/Austin: Springer Science+Business Media/Landes Bioscience, 2006:160-73.
57. Hirsch HH, Brennan C, Drachenberg CB et al. Polyomavirus-associated nephropathy in renal transplantation: Interdisciplinary analyses and recommendations. Transplantation 2005; 79:1277-86.
58. Randhawa P, Brennan DC. BK virus infection in transplant recipients: An overview and update. Am J Transplantation 2006; 6:2000-5.
59. Hariharan S. BK virus nephritis after renal transplantation. Kidney Int 2006; 69:655-62.
60. Hirsch HH. BK virus: Opportunity makes a pathogen. Clin Infect Dis 2005; 41:354-60.
61. Nickeleit V, Singh HK, Mihatsch MJ. Polyomavirus nephropathy: Morphology, pathophysiology, and clinical management. Curr Opin Nephrol Hypertens. 2003; 12:599-605.
62. Hirsch HH, Steiger J. Polyomavirus BK. Lancet Infect Dis 2003; 3:611-23.
63. Kazory A, Ducloux D. Renal transplantation and polyomavirus infection: Recent clinical facts and controversies. Transpl Infect Dis 2003; 5:65-71.
64. Hirsch HH, Steiger J. Polyomavirus BK. Lancet Infect Dis 2003; 3:611-23.
65. Fishman JA. BK virus nephropathy—Polyomavirus adding insult to injury. N Engl J Med 2002; 347:527-30.
66. Hirsch HH. Polyomavirus BK nephropathy: A (re-)emerging complication in renal transplantation. Am J Transplant 2002; 2:25-30.
67. Randhawa P, Vats A, Shapiro R et al. BK virus: Discovery, epidemiology, and biology. Graft 2002; 5:S19-27.
68. Nickeleit V, Steiger J, Mihatsch M. BK virus infection after kidney transplantation. Graft 2002; 5:S46-57.
69. Mylonakis E, Goes N, Rubin RH et al. BK virus in solid organ transplant recipients: An emerging syndrome. Transplantation 2001; 72:1587-92.
70. Randhawa PS, Demetris AJ. Nephropathy due to polyomavirus type BK. N Engl J Med 2000; 342:1361-63.
71. Binet I, Nickeleit V, Hirsch HH. Polyomavirus infections in transplant recipients. Curr Opin Organ transplant 2000; 5:210-16.

72. Lundstig A, Dilner J. Serological diagnosis of human polyomavirus infection. In: Ahsan N, ed. Polyomaviruses and Human Diseases. New York/Austin: Springer Science+Business Media/Landes Bioscience, 2006:96-101.

73. Vats A, Randhawa P, Shapiro R. Diagnosis and treatment of BK virus-associated transplant nephropathy. In: Ahsan N, ed. Polyomaviruses and Human Diseases. New York/Austin: Springer Science+Business Media/Landes Bioscience, 2006:213-27.

74. Purighalla R, Shapiro R, McCauley J et al. BK virus infection in a kidney allograft diagnosed by needle biopsy. Am J Kidney Dis 1995; 26:671-73.

75. Singh HK, Bubendorf L, Mihatsch MJ et al. Urine cytology findings of polyomavirus infections. In: Ahsan N, ed. Polyomaviruses and Human Diseases. New York/Austin: Springer Science+Business Media/Landes Bioscience, 2006:201-12.

76. Jin L. Molecular methods for identification and genotyping of BK virus. Methods Mol Biol 2001; 165:33-48.

77. Ding R, Medeiros M, Dadhania D et al. Noninvasive diagnosis of BK virus nephritis by measurement of messenger RNA for BK virus VP1 in urine. Transplantation 2002; 74:987-94.

78. Randhawa P, Zygmunt D, Shapiro R et al. Viral regulatory region sequence variations in kidney tissue obtained from patients with BK virus nephropathy. Kidney Int 2003; 64:743-47.

79. Randhawa PS, Vats A, Zygmunt D et al. Quantitation of viral DNA in renal allograft tissue from patients with BK virus nephropathy. Transplantation 2002; 74:485-88.

80. Randhawa PS, Khaleel-Ur-Rehman K, Swalsky PA et al. DNA sequencing of viral capsid protein VP-1 region in patients with BK virus interstitial nephritis. Transplantation 2002; 73:1090-94.

81. Leung AY, Chan M, Tang SC et al. Real-time quantitative analysis of polyoma BK viremia and viruria in renal allograft recipients. J Virol Methods 2002; 103:51-56.

82. Bergallo M, Merlino C, Bollero C et al. Human polyoma virus BK DNA detection by nested PCR in renal transplant recipients. New Microbiol 2002; 25:331-34.

83. Baksh FK, Finkelstein SD, Swalsky PA et al. Molecular genotyping of BK and JC viruses in human polyomavirus-associated interstitial nephritis after renal transplantation. Am J Kidney Dis 2001; 38:354-65.

84. Boldorini R, Zorini EO, Fortunato M et al. Molecular characterization and sequence analysis of polyomavirus BKV-strain in a renal-allograft recipient. Hum Pathol 2001; 32:656-59.

85. Fogazzi GB, Cantu M, Saglimbeni L. 'Decoy cells' in the urine due to polyomavirus BK infection : Easily seen by phase-contrast microscopy. Nephrol Dial Transplant 2001 ;16:1496-98.

86. Randhawa P, Baksh F, Aoki N et al. JC virus infection in allograft kidneys: Analysis by polymerase chain reaction and immunohistochemistry. Transplantation 2001; 71:1300-3.

87. Limaye AP, Jerome KR, Kuhr CS et al. Quantitation of BK virus load in serum for the diagnosis of BK virus-associated nephropathy in renal transplant recipients. J Infect Dis 2001; 183:1669-72.

88. Chen CH, Wen MC, Wang M et al. A regulatory region rearranged BK virus is associated with tubulointerstitial nephritis in a rejected renal allograft. J Med Virol 2001; 64:82-88.

89. Nickeleit V, Hirsch HH, Zeiler M et al. BK-virus nephropathy in renal transplants-tubular necrosis, MHC-class II expression and rejection in a puzzling game. Nephrol Dial Transplant 2000; 15:324-32.

90. Nickeleit V, Klimkait T, Binet IF et al. Testing for polyomavirus type BK DNA in plasma to identify renal-allograft recipients with viral nephropathy. N Engl J Med 2000; 342:1309-15.

91. Bressollette-Bodin C, Coste-Burel M, Hoirmant M et al. A prospective longitudinal study of BK virus infection in 104 renal transplant recipients. Am J Transplant 2005; 5:1926-33.

92. Coleman D, Mackenzie S, Gardner S et al. Human polyomavirus (BK) infection and ureteral stenosis in renal allograft recipients. J Clin Pathol 1978; 31:338-47.

93. Constantinescu A, Ahsan N, Lim JW. Polyomavirus allograft nephropathy - parenchymal and extra-parenchymal manifestations. Graft 2002; 5: S98-103.

94. Bohl DL, Storch GA, Ryschkewitsch C et al. Donor origin of BK virus in renal transplantation and role of HLA C7 in susceptibility to sustained BK viremia. Am J Transplant 2005; 5:2213-21.

95. Agha I, Brennan DC. BK virus and immunosuppressive agents. In: Ahsan N, ed. Polyomaviruses and Human Diseases. New York/Austin: Springer Science+Business Media/Landes Bioscience, 2006:174-84.

96. Brennan DC, Agha I, Bohl DL et al. Incidence of BK with tacrolimus versus cyclosporine and impact of preemptive immunosuppression reduction. Am J Transplant 2005; 5:582-94.

97. Hariharan S, Cohen EP, Vasudev B et al. BK virus-specific antibodies and BKV DNA in renal transplant recipient with BKV nephritis. Am J Transplant 2005; 5:2719-24.

98. Comoli P, Azzi A, MaccarioR et al. Polyomavirus BK-specific immunity after kidney transplantation. Transplantation 2004:78:1229-32.

99. Drachenberg CB, Papadimetriou JC, Mann D et al. Negative impact of human leukocyte antigen matching in the outcome of polyomavirus nephropathy. Transplantation 2005; 80:276-78.

100. Comoli P, Basso S, Azzi A et al. Dendritic cells pulsed with polyomavirus BK antigen induce ex vivo polyoma BK virus-specific cytotoxic T-cell lines in seropositive healthy individuals and renal transplant recipients. J Am Soc Nephrol 2003; 14:3197-204.

101. Awadalla Y, Randhawa P, Ruppert K et al. HLA mismatching increases the risk of BK virus nephropathy in renal transplant recipients. Am J Transplant 2004; 4:1691-96.

102. Mannon RB, Hofmann SC, Kampen RL et al. Molecular evaluation of BK polyomavirus nephropathy. Am J Transplant 2005; 5:2883-93.

103. Nickeleit V, Singh HK, Mihatsch MJ. Latent and productive polyomavirus infections of renal allografts; morphological, clinical and pathophysiological aspects. In: Ahsan N, ed. Polyomaviruses and Human Diseases. New York/Austin: Springer Science+Business Media/Landes Bioscience, 2006:190-200.

104. Drachenberg CB, Hirsch HH, Ramos E et al. Polyomavirus disease in renal transplantation: Review of pathological findings and diagnostic methods. Human Pathol 2005; 36:1245-55.

105. Nickeleit V, Zeiler M, Gudat F et al. Detection of complement degradation product c4d in renal allografts: diagnostic and therapeutic implications. J Am Soc Nephrol 2002; 13:242-51.

106. McGilvary ID, Lajoie G, Humar A et al. Polyomavirus infection and acute vascular rejection in a kidney allograft: Coincidence or mimicry? Am J Transplant 2003; 3:501-4.

107. Hussain S, Bresnahan BA, Cohen EP et al. Rapid kidney allograft failure in patients with polyoma virus nephritis with prior treatment with anti-lymphocyte agents. Clin Transplant 2002; 16:43-47.

108. Ahuja M, Cohen EP, Dayer AM et al. Polyoma virus infection after renal transplantation. Use of immunostaining as a guide to diagnosis. Transplantation 2001; 71:896-99.

109. Mengel M, Marwedel M, Radermacher J et al. Incidence of polyomavirus-nephropathy in renal allografts: Influence of modern immunosuppressive drugs. Nephrol Dial Transplant 2003; 18:1190-96.

110. Binet I, Nickeleit V, Hirsch HH et al. Polyomavirus disease under new immunosuppressive drugs: A cause of renal graft dysfunction and graft loss. Transplantation 1999; 67:918-22.

111. Nickeleit V, Hirsch HH, Binet I et al. Polyomavirus infection of renal allograft recipients: From latent infection to manifest disease. J Am Soc Nephrol 1999; 10:1080-89.

112. Ramos E, Drachenberg CB, Papadimitriou JC et al. Clinical course of polyoma virus nephropathy in 67 renal transplant patients. J Am Soc Nephrol 2002; 13:2145-51.

113. Buehrig CK, Lager DJ, Stegall MD et al. Influence of surveillance renal allograft biopsy on diagnosis and prognosis of polyomavirus-associated nephropathy. Kidney Int 2003; 64:665-73.

114. Trofe J, Gaber LW, Stratta RJ et al. Polyomavirus in kidney and kidney-pancreas transplant recipients. Transpl Infect Dis 2003; 5:21-28.

115. Zaragoza C, Li R, Fhle GA et al. Squirrel monkeys support replication of BK virus more efficiently than simian virus 40; an animal model for human BK virus infection. J of Virol 2005; 79:1320-26.

116. Roskopf J, Trofe J, Stratta RJ et al. Pharmacologic options for the management of human polyomavirus. In: Ahsan N, ed. Polyomaviruses and Human Diseases. New York/Austin: Springer Science+Business Media/Landes Bioscience, 2006:228-54.

117. Trofe J, Hirsch HH, Ramos E. polyomavirus-associated nephropathy: Update of clinical management in kidney transplant patients. Transpl Infect Dis 2006; 8:76-85.

118. Lipshutz GS, Flechner SM, Govani MV et al. BK nephropathy in kidney transplant recipients treated with a calcineurin inhibitor free immunosuppression regimen. Am J Transplant 2004; 4:2132-34.

119. Hirsch HH, Mohaupt M, Klimkait T. Prospective monitoring of BK virus load after discontinuing sirolimus treatment in a renal transplant patient with BK virus nephropathy. J Infect Dis 2001; 184:1494-5, (author reply 5-6).

120. Gineveri Hirsch HH, Mohaupt M, Klimkait T. Prospective monitoring of BK virus load after discontinuing sirolimus treatment in a renal transplant patient with BK virus nephropathy. J Infect Dis 2001; 184:1494-5.

121. Trofe J, Cavallo T, First MR et al. Polyomavirus in kidney and kidney-pancreas transplantation: A defined protocol for immunosuppression reduction and histologic monitoring. Transplant Proc 2002; 34:1788-89.

122. Ginevri F, de Santis R, Comoli P et al. Polyomavirus BK infection in pediatric kidney-allograft recipients: A single-center analysis of incidence, risk factors, and novel therapeutic approaches. Transplantation 2003; 875:1266-77.
123. Howell DN, Smith SR, Butterly DW et al. Diagnosis and management of K polyomavirus interstitial nephritis I renal transplant recipients. Transplantation 1999; 68:1279-88.
124. Hurault de Ligny B, Etienne I, Francois A et al. Polyomavirus induced acute tubulo-interstitial nephritis in renal transplant recipients. Transplant Proc 2003; 32:2760-61.
125. Rahamimov R, Lustiga S, Tovar A et al. BK polyoma virus nephropathy in kidney transplant recipient: The role of new immunosuppressive agents. Transplant Proc 2003; 35:605.
126. Li RM, Mannon RB, Kleiner D et al. BK virus and SV-40 coinfection in polyomavirus nephropathy. Transplantation 2002; 74:1497-504.
127. Bari YM, Ahmad I, Ketel BL et al. Polyoma viral infection in renal transplantation: The role of immunosuppressive therapy. Clin Transplan 2001; 15:240-46.
128. Celik B, Shapiro R, vats A et al. Polyomavirus allograft nephropathy: sequential assessment of histologic viral load, tubulitis, and graft function following changes in immunosuppression. Am J Transplant 2003; 3:1378-82.
129. Fishman JA. BK nephropathy: What is the role of antiviral therapy? Am J Transplant 2003; 3:99-100.
130. Vats A, Randhawa PS, Shapiro R. Diagnosis and treatment of BK virus-associated transplant nephropathy. In: Ahsan N, ed. Polyomaviruses and Human Diseases. New York/Austin: Springer Science+Business Media/Landes Bioscience, 2006:213-27.
131. Scantlebury V, Shapiro R, Randhawa P et al. Cidofovir: A method of treatment for BK virus-associated transplant nephropathy. Graft 2002; 5:S82-87.
132. Vats A, Shapiro R, Singh Randhawa P et al. Quantitative viral load monitoring and cidofovir therapy for the management of BK virus-associated nephropathy in children and adults. Transplantation 2003; 75:105-12.
133. Kadambi PV, Josephson MA, Williams J et al. Treatment of refractory BK virus-associated nephropathy with cidofovir. Am J Transplant 2003; 3:186-91.
134. Bjorang O, Tveitan H, Midtvedt K et al. Treatment of polyomavirus infection with cidofovir in a renal-transplant recipient. Nephrol Dial Transplant 2002; 17:2023-25.
135. Held TK, Biel SS, Nitsche A et al. Treatment of BK virus-associated hemorrhagic cystitis and simultaneous CMV reactivation with cidofovir. Bone Marrow Transplant 2000; 26:347-50.
136. Sener A, House AA, Jevnikar AM et al. Intravenous immunoglobulin as a treatment of BK virus associated nephropathy: One year follow-up of renal allograft recipients. Transplantation 2006; 81:117-20.
137. Cheesman SH, Black PH, Rubin RH et al. Interferon and BK Papovavirus—Clinical and laboratory studies. J Infect Dis 1980; 141:157-61.
138. Snoeck R, Andrei G, Lilja HS et al. Activity of malonoitrilamide compounds against murine and simian polyomavirus. 5th International conference on New Trends in Clinical and Experimental Immunosuppression. Switzerland: 2002, (Abstract).
139. Josephson MA, Gillen D, Javaid B et al. Treatment of renal allograft polyoma BK virus infection with leflunomide. Transplantation 2006; 81:704-10.
140. Foster PF, Wright F, McLean D et al. Leflunomide administration as an adjunct in treatment of BK-polyoma viral disease in kidney allografts. Am J Transplant 2003; 3:421, (abstract).
141. Josephson MA, Javaid B, Kadambi PV et al. Leflunomide in solid organ transplantation and polyoma virus infection. In: Ahsan N, ed. Polyomaviruses and Human Diseases. New York/Austin: Springer Science+Business Media/Landes Bioscience, 2006:255-65.
142. Farasati NA, Shapiro R, Vats A et al. Effects of leflunomide and cidofovir on replication of BK-virus in an in vitro culture system. Transplantation 2005; 79:116-18.
143. Poduval RD, Meehan SM, Woodle ES et al. Successful retransplantation after renal allograft loss to polyoma virus interstitial nephritis. Transplantation 2002; 73:1166-69.
144. Boucek P, Voska L, Saudek F. Successful retransplantation after renal allograft loss to polyoma virus interstitial nephritis. Transplantation 2002; 74:1478.
145. Ramos E, Vincenti F, Lu WX et al. Retransplantation in patients with graft loss caused by polyoma virus nephropathy. Transplantation 2004; 77:131-33.
146. Womer KL, Meier-Kriesche HU, Bucci CM et al. Preemptive retransplantation for BK virus nephropathy: Successful outcome despite active viremia. Am J Transplant 2006; 6:209-13.
147. Ginevri F, Pastorino N, De Santis R et al. Retransplantation after kidney graft loss due to polyoma BK virus nephropathy: Successful outcome without original allograft nephrectomy. Am J Kidney Dis 2003; 42:821-25.
148. Hirsch HH, Ramos E. Retransplantation after polyomavirus-associated nephropathy: Just do it. Am J Transplant 2006; 6:7-9.

# Polyomavirus Allograft Nephropathy:
## Clinico-Pathological Correlations

Volker Nickeleit* and Harsharan K. Singh

## Abstract

Polyomavirus nephropathy (PVN) is primarily caused by a productive intra-renal BK virus infection. It is often an iatrogenic complication due to long term over immunosuppression and frequently leads to chronic kidney dysfunction and failure. Post renal transplantation, PVN has emerged as a major problem affecting up to 10% of all kidney grafts, most commonly within the first 6-18 months post surgery. Recent advances in our understanding of the development and progression of PVN have resulted in the definition of clinically significant disease stages: A (early), B (florid) and C (sclerosing). This chapter provides a detailed description of polyomavirus induced acute and chronic renal injury. Morphologic changes are correlated with the clinical presentation and graft outcome. Key biologic aspects that are pertinent to patient management are highlighted. New 'morphology based' strategies, i.e., urine electron microscopy and improved decoy cell analyses, to reliably diagnose PVN non-invasively are discussed and incorporated into an updated diagnostic algorithm for patient screening and monitoring.

## Introduction

Polyomaviruses are of no clinical significance in the immune competent individual. Under immunocompromised conditions, however, the BK- and JC-polyomavirus strains can enter into "organ-destructive" replicative cycles and cause disease. Clinically significant infections/disease caused by BK-viruses are primarily seen in the urinary tract, i.e., the bladder and kidneys, whereas JC-viruses mainly cause neurologic disorders, i.e., progressive multifocal leukoencephalopathy. Early reports that had linked productive infections with BK viruses to the development of ureteral necrosis and stenosis[1] could not be confirmed in a recent series which demonstrated BK viruses in only 8% (2/25) of histologically analyzed necrotic ureters.[2]

Polyomavirus allograft nephropathy (PVN) post kidney transplantation was first described nearly 3 decades ago as a single case report by the pathologist Mackenzie.[3] It was, however, not until the introduction of potent third generation immunosuppressive drugs, mainly high dose tacrolimus and mycophenolate mofetil, into clinical management that transplant centers world-wide experienced "an outbreak" of PVN as a mostly iatrogenic complication.[4-10] Currently, PVN has a prevalence of 1% to 10% and is associated with a graft failure rate of 50% in some centers. It exceeds productive CMV graft infections by approximately 100 times.[9,11-15] Likely, the incidence of PVN will decrease in the future under altered and optimized immunosuppression. A recent report from the Mayo Clinic found a highly significant reduction of the incidence of PVN from 10.5% to 2.5% following the clinical introduction of

"low dose" maintenance tacrolimus immunosuppression.[9] At present, however, PVN still constitutes a major clinical challenge. Specific and potent antiviral drugs to treat productive polyomavirus infections are not available. Consequently, much emphasis is placed on patient screening and a diagnosis of PVN in an early disease stage that often responds favorably to our limited therapeutic options mainly consisting of a reduction of the maintenance immunosuppression.

Most of our current disease knowledge has been gained by studying patients; thus far, reliable animal models mirroring PVN in humans have not been developed.[16] Many aspects of PVN and infections with polyomaviruses including risk factors and treatment options have been discussed in Chapter 37 in this book. In the following paragraphs, we will primarily focus on morphologic changes and biologic concepts that help to understand and manage PVN. We will place special emphasis on recently developed 'morphologic' strategies to accurately diagnose PVN non-invasively that will be critically compared to conventional PCR techniques. In the future, chronic graft failure due to PVN should be the exception rather than the rule.

## Background

Infections with polyomaviruses and PVN are characterized by several key features (see Table 1):

A. PVN is caused by the re-activation of latent intra-renal/intra-graft, i.e., donor derived, polyomaviruses under long-lasting and intense immunosuppression, such as high dose tacrolimus or mycophenolate-mofetil. High latent polyomavirus loads are detected in approximately 5% of all normal/donor kidneys. It is presumably this sub-group of organs that is at increased risk for the development of PVN post transplantation. Latent infections with polyomaviruses do not cause any functional or morphologic changes; they can only be detected by molecular techniques. Of note: circulating antibodies against BK virus, which are found at different titer levels in 66%-90% of healthy adults,[17-19] do not accurately reflect latent intra-renal BK virus loads. In a pilot analysis of 30 deceased nontransplant patients without BKN we compared plasma anti-BK virus antibody titers (by the hemaglutinine-inhibition assay, HAI) to the corresponding latent intra-renal BK virus load levels. Sero-negativity (HAI titers below 1:128) was found in 1/30 patients (3%); this patient showed a corresponding high, latent intra-renal BK virus load. Normal antibody titer ranges (HAI: 1:128-1:1024) were observed in 26/30 patients (87%), of which 16 were without latent intra-renal BK-virus infections.

*Corresponding Author: V. Nickeleit—Associate Professor of Pathology, Director: Nephropathology Laboratory, Department of Pathology, The University of North Carolina at Chapel Hill, Chapel Hill, NC 27599-7525, USA. Email: volker_nickeleit@med.unc.edu

*Chronic Allograft Failure: Natural History, Pathogenesis, Diagnosis and Management*, edited by Nasimul Ahsan. ©2008 Landes Bioscience.

### Table 1. Polyomavirus infections (modified from ref. 13)

| | |
|---|---|
| **Primary Infection** | Initial infection of host organism with polyomaviruses including viremic spread to permissive tissues; insignificant clinical symptoms |
| **Latent Infection** | Dormant asymptomatic infections of permissive cells (e.g., renal tubular, transitional cells) following the primary infection; virus detection only with molecular techniques |
| **Serologic Evidence of an Infection** | Varying antibody titer levels found in nearly all healthy children and 60%-90% of asymptomatic adults; no correlation with latent viral load levels; IgG and IgM titer levels increase during productive infections |
| **Viral Activation** | Evidence of polyomavirus replication: (1) decoy cells or free virions in the urine; (2) viral detection by PCR in the urine, serum or cerebrospinal fluid. Might be seen as a transient, asymptomatic event; viral activation is always part of viral disease |
| **Viral Disease (PVN, PML, hemorrhagic cystitis)** | Histologic evidence of viral activation in organs (cytopathic signs and/or positive immunohistochemistry or in-situ hybridization signals) and associated virally induced tissue injury*, often associated with clinical symptoms |

PVN: polyomavirus allograft nephropathy; PML: progressive multifocal leukoencephalopathy. * PVN stage A shows only minimal acute tubular injury and often no renal dysfunction

### Table 2. PVN: Mode of kidney injury and dysfunction

| Virally induced acute tubular injury (ATN) | |
|---|---|
| Focal | no/mild renal dysfunction |
| Diffuse | pronounced renal dysfunction |
| Early ATN | reversible changes (possible restitutio ad integrum) |
| Chronic ATN | irreversible changes (fibrosis, tubular atrophy) |

The remaining 3/30 patients (10%) demonstrated high plasma anti-BK virus antibody titers (HAI ≥1:2048); in one patient without a latent intra-renal BK virus infection (personal observation,[20]). Thus, it seems that dormant intra-renal BK-virus load levels cannot be extrapolated from plasma antibody titers. Consequently, kidneys from sero-negative donors may contain latent BK-viruses that can predispose the recipient for the development of PVN post grafting (also see Table 1 in the chapter by Ahsan).

B. Slight changes in the immune status can lead to transient, mostly asymptomatic and self-limiting activation of latent polyomaviruses,[21] especially in the urothelium, which harbors latent BK virus infections in 43% of individuals (personal observation). Such activation is characterized by the detection of free viral particles in the urine (by electron microscopy or PCR techniques) and viral inclusion bearing cells, so-called "decoy cells" in urine cytology specimens (see below). Signs of transient, asymptomatic viral activation can occasionally also be detected in serum samples by real time PCR. Such polyomavirus (re)activation is commonly not accompanied by PVN.[4,22,23] PVN, on the other hand, is always associated and typically preceded by the activation of polyomaviruses.[6,19,22,24-26]

C. PVN is typically diagnosed 10-12 months post transplantation with only anecdotal cases reported as early as 6 days and as late as 6 years post grafting.[14,26] Depending on the extent of virally induced tubular injury (Table 2), patients clinically present with varying degrees of allograft dysfunction. Serum creatinine levels vary from normal (PVN stage A) to markedly increased (PVN stages B and C).[26-30] Not surprisingly, PVN has been diagnosed in surveillance protocol biopsies,[11,28,30] and

thus, clinical evidence of graft dysfunction is only ill suited to optimally time a diagnostic graft biopsy in order to establish an early diagnosis.

D. PVN is best diagnosed histologically in a graft biopsy, which additionally provides prognostically relevant information on the disease stage, concomitant rejection and potential other abnormalities, such as calcineurin inhibitor induced toxicity. Since PVN affects the kidney in a focal fashion, adequate samples are needed to guarantee an optimal diagnostic yield, i.e., two biopsy cores obtained with a 15 or 16 gauge needle. The diagnosis may be missed in 25%-37% of cases if only one small core of renal cortex is sampled.[13,31] A new strategy for accurately diagnosing PVN non-invasively is urine electron microscopy and the detection of three-dimensional, cast-like polyomavirus clusters[32] (see below).

E. PVN is nearly always caused by a productive infection with the BK-virus strain. Only a minority of cases (approximately one third) show activation of polyoma-BK- and JC-viruses simultaneously with, as yet, undetermined biological significance.[33,34] Polyomavirus nephropathies that are only induced by a productive JC- or SV-40 virus infection are exceptionally rare.[35,36] PVN is hardly ever seen in association with a concurrent second viral infection.[15,37] Morphologic changes induced by productive BK-, JC-, or SV-40 polyomavirus infections are identical; ancillary techniques such as immunohistochemistry, in-situ hybridization, or PCR are required for the exact identification of viral strains.

F. PVN is typically limited to the transplant and "the worst case scenario" is graft failure. Systemic signs of an infection including involvement of the native kidneys[38] are generally not seen (with only exceptional case reports of systemic viral disease, a potential case of BK virus associated hemophagocytic syndrome and PVN associated hemorrhagic cystitis[39-43]). Thus, patients are usually not at risk for generalized disease or lethal viral spread.

## PVN: Morphologic Findings

PVN is defined as an intraparenchymal productive polyomavirus infection (BK-virus >> JC-virus) of the kidney with light microscopic (i.e., intra-nuclear viral inclusion bodies), or immunohistochemical or in-situ hybridization evidence of viral replication accompanied by varying degrees of parenchymal damage ranging from minimal to marked (Tables 1-3).

### Gross Pathology

Transplant nephrectomy specimens from patients with PVN show uncharacteristic changes closely resembling fibrotic lesions seen in other chronic diseases leading to graft failure. In PVN the removed allografts are generally slightly decreased in size, firm, with an ill-defined corticomedullary junction and thinned, sclerosed cortex. The renal surface is either smooth or granular. Infarction and large scar formation are not features of PVN.

*Table 3. Polyomavirus allograft nephropathy: Disease stages (modified from refs. 13,47)*

| | |
|---|---|
| **Stage A\*** (early changes) | Viral activation in cortex and/or medulla with intra-nuclear inclusion bodies and/or positive immunohistochemical or in-situ hybridization signals. |
| | No or minimal tubular epithelial cell necrosis/lysis. |
| | No or minimal denudation of tubular basement membranes. |
| | No or minimal interstitial inflammation in foci with viral activation. |
| | No or minimal tubular atrophy and interstitial fibrosis (≤10%). |
| **Stage B\*** (florid changes) | Marked viral activation in cortex and/or medulla. |
| | Marked virally induced tubular epithelial cell lysis and associated denudation of tubular basement membranes. |
| | Interstitial inflammation\*\* (mild to marked). |
| | Interstitial fibrosis and tubular atrophy (minimal to moderate <50%) |
| |     Stage B1—≤25% of specimen involved |
| |     Stage B2—>25% and <50% of specimen involved |
| |     Stage B3—≥50% of specimen involved (and <50% fibrosis) |
| **Stage C\*** (advanced sclerosing changes) | Viral activation in cortex and medulla. |
| | Interstitial fibrosis and tubular atrophy ≥50% of sample. |
| | Tubular epithelial cell lysis and basement membrane denudation (minimal to marked). |
| | Interstitial inflammation (minimal to marked). |
| **Burnt-Out Stage\*\*\*** | No signs of viral replication. |
| | Varying degrees of interstitial fibrosis and tubular atrophy. |

\*Additional signs of BK virus activation are always present: (1) Decoy cells in the urine; (2) BK virus DNA or RNA in the serum and urine; and (3) Free virions and cast-like viral clusters in the urine. \*\* Interstitial inflammation and tubulitis can in some cases mark concurent tubulo-interstitial cellular rejection; rejection induced changes are not part of PVN staging. \*\*\* The burnt-out stage of PVN only shows uncharacteristic changes that cannot be distinguished from nonspecific graft fibrosis; the diagnosis can only be suspected based on the patient's history.

## Light Microscopy

Histologic signs caused by a productive polyomavirus infection of the kidney are characteristically found in epithelial cells lining collecting ducts, tubules, Bowman's capsule (parietal epithelial cells) and the renal pelvis (transitional cells). Viral replication and the assembly of daughter virions result in the formation of intra-nuclear inclusion bodies, cell injury and cell lysis.[5-7,24,26] Most important for the clinical presentation and course of PVN are the tubular changes as severe, virally induced tubular injury and the denudation of tubular basement membranes can ultimately lead to atrophy and fibrosis, i.e., PVN disease stage C (Tables 2 and 3). Despite marked epithelial damage, however, the tubular basement membranes usually remain intact. During early stages of disease (i.e., PVN stages A and B1) they can serve as the structural skeleton for regeneration and restitution once the viral replication ceases. Parietal epithelial cells lining Bowman's capsule can also show signs of viral replication and on occasion form "pseudo-crescents".[24,44,45] Viral inclusion bodies are also found in the transitional cell layer lining the renal pelvis, the (graft) ureter and potentially even the recipient's urinary bladder.[24,41] Polyomavirus replication in the urothelium, however, is not a defining histologic hallmark characterizing PVN since it can also be seen in patients without viral nephropathy as a transient and asymptomatic sign of viral activation[46] -or- in the setting of hemorrhagic cystitis (without PVN) following bone marrow transplantation (Table 1). Glomerular capillary tufts, blood vessels, inflammatory and mesenchymal cells are generally nonpermissive for the replication of BK-/polyomaviruses.

## Staging

Based on the degree of tubulo-interstitial damage, PVN is histologically classified into three disease stages: early "A", florid "B" and late sclerosing "C" (Table 3).[13,24,29,31,47-50] Pertinent to staging are the severity of virally induced epithelial cell injury, interstitial fibrosis and tubular atrophy rather than the extent of interstitial inflammation. Progression from disease stage A to B or C can occur within few months.[12,49]

PVN stage A represents the earliest disease phase with only very focal 'nonlytic' viral replication and inconspicuous denudation of tubular basement membranes; it is often limited to the renal medulla and likely represents re-activated foci of latent BK virus infections (Fig. 1). The interstitium is normal or only shows minimal inflammation and fibrosis. Intra-nuclear viral inclusion bodies are generally found (albeit sometimes few in number). In rare, presumably very early phases of a productive polyomavirus infection, characteristic intra-nuclear viral inclusion bodies may be absent; in these cases, the diagnosis of PVN is solely based on immunohistochemical or in-situ hybridization staining signals. The detection of one intra-nuclear viral inclusion body or alternatively a crisp staining signal in one nucleus (with or without an inclusion body) suffices for proving 'active' viral replication and rendering the diagnosis of PVN stage A.

Florid PVN (stage B) is characterized by marked virally induced tubular injury and epithelial cell lysis, conspicuous denudation of tubular basement membranes and interstitial edema with a mixed, mild to marked inflammatory cell infiltrate (B- and T-lymphocytes, plasma cells and histiocytes) (Fig. 2). Polymorphonuclear leukocytes can be prominent adjacent to severely injured tubules, presumably due to urinary back-leak into the interstial compartment. On occasion, tubulo-centric granulomas rich in histiocytes form in and around virally injured tubules (Fig. 3).

PVN stage C, the sclerosing phase, shows marked interstitial fibrosis and tubular atrophy involving, per definition more than 50% of the tissue sample. Stage C is associated with varying degrees of inflammation and viral replication (Fig. 4).

The burnt-out stage of PVN following the resolution of viral replication demonstrates nonspecific "chronic" changes: varying degrees of interstitial fibrosis and tubular atrophy (ranging from minimal to marked) and occasionally mild lymphocytic inflammation in areas of parenchymal scarring. Any signs of polyomavirus replication by light microscopy and immunohistochemistry/in-situ hybridization are, per definition, no longer detectable. The PVN burnt-out stage closely mimicks so-called "chronic allograft nephropathy"; it can only be suspected based on the patient's history.

The staging of PVN is robust and it carries clinical significance (see below). Additional attempts may be made (and recorded for each disease stage) to semi-quantitatively assess the extent of viral replication/infection in the tubular compartment from cy0 (no replication), cy1 (<10% infected tubules with viral replication in the biopsy sample), cy2 (10%-≤25% infected tubules), cy3 (26%-≤50% infected tubules) and

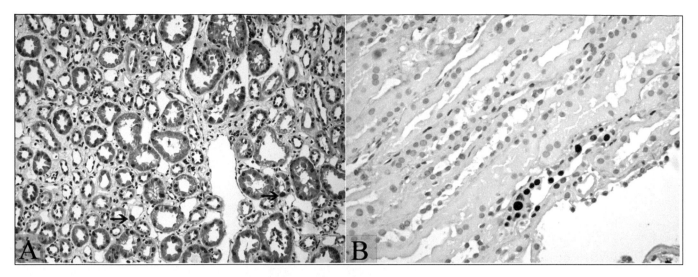

Figure 1. A,B) PVN disease stage A: only rare intra-nuclear polyomavirus inclusion bodies are noted in some tubules (arrows in A) located at the cortico-medullary junction. There is no evidence of significant tubular injury; the interstitial compartment is normal without inflammation. B) Illustrates crisp, focal, intra-nuclear staining in tubular epithelial cells with an antibody directed against the SV-40 T-antigen. Note that only one nephron is affected by polyomavirus replication. Formalin fixed and paraffin embedded tissue sections, 20× original magnification, (A) hematoxylin and eosin stained section; (B) immunohistochemistry.

cy4 (>50% of tubules with viral replication;[48]). Most commonly, PVN stage A presents with cy1, PVN stage B with cy 2, 3, or 4 and PVN stage C with cy 1, 2, or 3. Of note: the 'cy' scoring does not determine the PVN disease stage. Definitive rules to adequately assess and compare "cy" scores have so far not been set, i.e., light microscopy (detection of intra-nuclear inclusion bodies) versus immunohistochemistry (detection of the large T-antigen associated with viral replication) or in-situ hybridization (detection of amplified viral DNA/mature daughter virions). Thus, depending on the approach taken, "cy" scores may vary. In our experience, the presence and extent of a "productive polyomavirus infection/PVN" are best documented by the immunohistochemical detection of the "large T-antigen" expression (see below).

### Inflammation

PVN stages A-C can be associated with varying degrees of interstitial inflammation (Figs. 1-4) that is only poorly understood:[6,12,51]

Scenario #1: The inflammatory response may potentially be beneficial and indicate cellular containment of viral replication by BK virus specific cytotoxic T-lymphocytes[52-54] and the release of gamma interferon.[55] Immunologic reconstitution under decreased immunosuppression during attempts to treat PVN may possibly provide the right window of opportunity for such a cellular anti-viral immune response. However, data to support this concept are currently scant.[56] In one small series the stimulation of BK virus specific CD8+ T-cells was not beneficial but rather detrimental and associated with graft failure.[54]

Scenario #2: In cases of severe, virally induced acute tubular injury, edema and inflammation including the attraction of polymorphonuclear leukocytes may represent a "reactive" phenomenon that is caused by urine back-leak through denuded basement membranes into the interstitial compartment. Similar changes can be seen in native kidneys with severe nephrotoxic forms of acute tubular injury (ATN). Such a "reactive" inflammatory response driven by tubular damage is seen in disease stages B or C.

Scenario #3: Viral replication in markedly damaged tubules can induce the formation of expansile, intra- and peritubular, nonnecrotizing histiocytic granulomas (Fig. 3). More commonly, however, it triggers a diffuse mononuclear (lymphocytic/histiocytic) and plasma cell rich interstitial inflammatory cell infiltrate, edema and tubulitis (Fig. 2). Such inflammatory responses including granuloma formation might be classified as a "type IV hypersensitivity reaction"; they can be seen in PVN disease stages B and C.

Scenario #4: Inflammation rich in lymphocytes can occasionally reflect a cellular (and antibody) mediated immune reaction driven by the release of virions and characterized by the formation of immune complex type deposits in tubular basement membranes (Fig. 5). In such cases, immunofluorescence microscopy and electron microscopy typically show granular, immunoglobulin deposits in tubular basement membranes reminiscent of 'anti-TBM disease/primary tubulointerstitial nephritis with immune complexes' seen in native kidneys.[45,57] This immune mediated inflammation might be classified as a "type III hypersensitivity reaction"; it is reported in PVN disease stages B or C.

Scenario #5: CD8+ lymphocytic allo-responses and acute rejection episodes with tubulitis, transplant endarteritis, or glomerulitis, occasionally C4d positive, have been reported in cases of PVN.[6,26,58-60] It seems that acute allograft rejection in the setting of PVN is an independent, second disease process that is not primarily triggered by viral replication (see chapter by Ahsan).[6,19,26,60] Acute rejection can be seen in PVN stages A-C. It is most easily diagnosed in stage A that should, per definition, lack a significant inflammatory cell infiltrate.[13,47,60]

Except for the potentially transient "immune reconstitution associated" inflammatory reaction (hypothetical scenario #1), all other persistent inflammatory responses are detrimental since they can result in progressive fibrosis and chronic graft failure.[6,24,26,51]

The interpretation of "inflammation" in PVN is challenging, in particular, the diagnosis of acute allograft rejection. Careful histologic analysis and the detection of scattered polymorphonuclear leukocytes adjacent to severely injured and denuded tubules (scenario#2), granulomas (scenario#3), granular immune deposits along tubular basement membranes (scenario#4), or transplant endarteritis, glomerulitis, tubular MHC-class II expression or capillary C4d deposits (scenario#5) may help to guide the diagnostic decision making process. Clinically and prognostically most important is the diagnosis of concurrent allograft rejection, since it should alter the therapeutic strategies.[26,60] Since cases of PVN with marked tubulitis generally fare poorly,[31] unrecognized tubulo-interstitial cellular rejection (Banff type I) may potentially contribute to PVN induced graft demise more frequently than commonly suspected.[60,61] The immunohistochemical phenotyping of the inflammatory cells in PVN has shown plasma cell (CD138) as well as B (CD20) or T-cell (CD3) dominant infiltrates with currently undetermined pathophysiological significance. It is not diagnostically helpful for distinguishing different forms of "viral nephritis" or to diagnose concurrent acute cellular allograft rejection.[51,62,63]

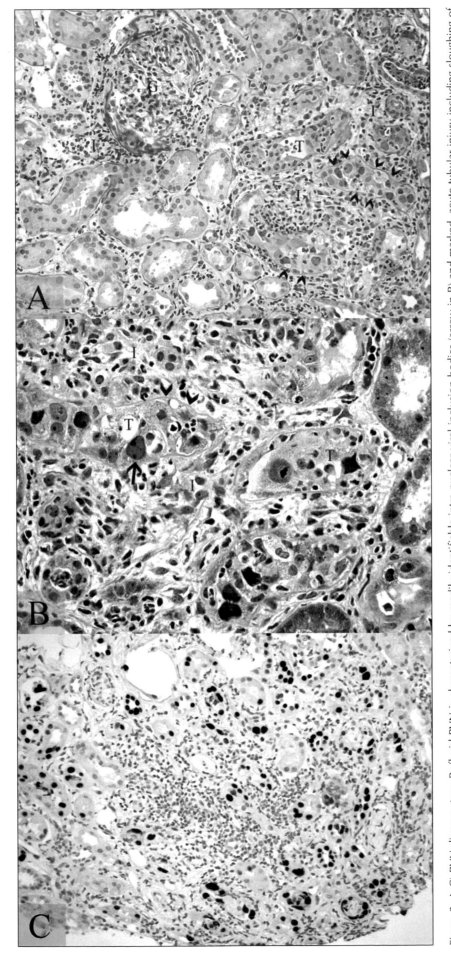

Figure 2. A-C) PVN disease stage B: florid PVN is characterized by readily identifiable intra-nuclear viral inclusion bodies (arrow in B) and marked, acute tubular injury including sloughing of epithelial cells, the denudation of tubular basement membranes (arrow heads in A and B), and focal tubulitis (T = tubules). The interstitium (I) shows edema and a marked, mixed, mononuclear inflammatory cell infiltrate with scattered plasma cells; interstitial fibrosis is inconspicuous. Glomeruli (G) are normal. C) illustrates crisp, diffuse, intra-nuclear staining in tubular epithelial cells with an antibody directed against the SV-40 T-antigen. Formalin fixed and paraffin embedded tissue sections; (A, C) 20× and (B) 40× original magnification; (A) Periodic Acid Schiff (PAS) incubation, (B) trichrome stain, (C) immunohistochemistry.

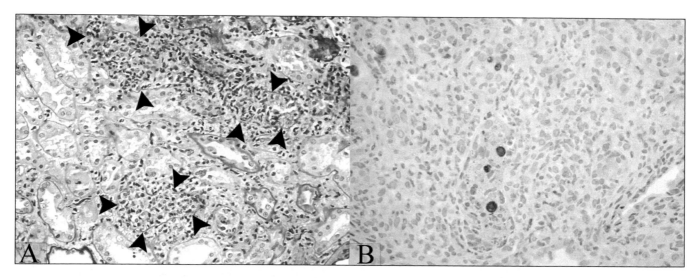

Figure 3. A,B) PVN with a granulomatous type of inflammation. Tubules are distended (arrow heads in A) by a marked, intra-tubular inflammatory reaction rich in histiocytes; affected tubules demonstrate epithelial cells with viral replication (nuclear staining for SV-40 T-antigen illustrated in B). Granulomatous inflammation is also noted in the interstitium surrounding injured tubules (B). Note: the SV-40 T-antigen expression is not detected in inflammatory cell elements (B), and this case did not show any immune complex type deposits along the tubular basement membranes. Formalin fixed and paraffin embedded tissue sections, 20× original magnification; (A) Periodic Acid Schiff (PAS) incubation, (B) immunohistochemistry to detect the SV-40 T-antigen.

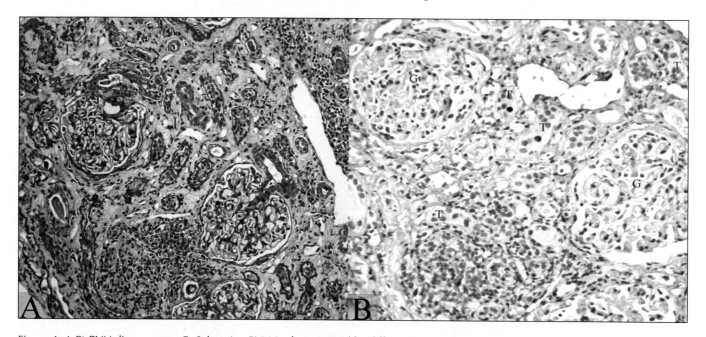

Figure 4. A,B) PVN disease stage C. Sclerosing PVN is characterized by diffuse interstitial fibrosis (I = interstitium) and tubular atrophy involving ≥50% of the biopsy sample as well as varying degrees of inflammation; glomeruli (G) are unremarkable. Note the markedly atrophic tubules in (B) with thickened basement membranes; these 'chronic' changes are irreversible. B) Illustrates very focal intra-nuclear staining in rare tubular epithelial cells with an antibody directed against the SV-40 T-antigen (T = tubules). Formalin fixed and paraffin embedded tissue sections, 20 × original magnification; (A) Periodic Acid Schiff (PAS) incubation, (B) immunohistochemistry. Figure 4 depicts a repeat biopsy obtained from the same patient three months subsequent to the initial histologic diagnosis of PVN stage A illustrated in Figure 1.

## Ancillary Techniques

In tissue specimens, polyomavirus replication is readily detected with commercially available antibodies directed against the SV-40 T-antigen (large T-antigen) that cross-react with all polyomaviruses pathogenic in humans (i.e., BK-, JC- and SV-40). They reliably work in frozen and paraffin sections and give a crisp, purely intra-nuclear staining signal (Figs. 1B, 2C, 3B, 4B). The expression of the T-antigen marks the initial phase of polyomavirus replication. It can precede the formation of intra-nuclear viral inclusion bodies and crisp staining signals may be seen in normal nuclei/tissue[6,24,26,29] as the earliest morphologic sign of viral replication and disease (Tables 1 and 3).

Polyomaviruses can also be identified by in-situ hybridization or with antibodies directed against viral capsid proteins. These latter techniques mostly detect "mature" virions and, therefore, not only mark intra-nuclear but additionally also dense intratubular "cast-like" viral aggregates (Fig. 6). Strain specific antibodies directed against BK-, JC-, or SV-40 viruses are also available, although most of them are primarily used by the research community; their use for diagnostic clinical purposes does not carry any advantage over antibodies to SV-40 T-antigen since PVN is nearly always driven by the replication of BK-viruses. Of note: depending on the technique used, staining results may vary greatly; suboptimal staining protocols can result in

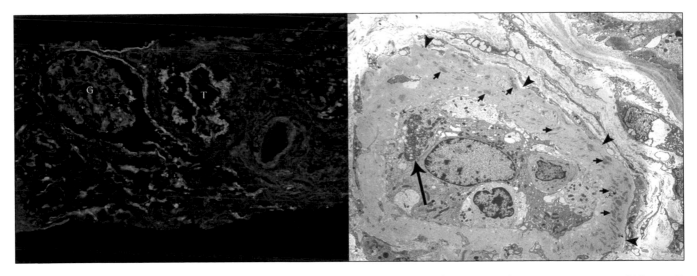

Figure 5. A,B) PVN with a "type III hypersensitivity reaction" that is reminiscent of "anti-TBM-disease" seen in native kidneys. A) Immunofluorescence microscopy with an antibody directed against IgG shows abundant granular immune deposits along a thickened tubular basement membrane (T = tubule). Deposits are also noted in Bowman's capsule (G = glomerulus). B) Electron microscopy reveals a thickened tubular basement membrane (arrow heads) with multiple electron dense immune complex type deposits (short arrows). These deposits do not contain viral capsid proteins based on immunohistochemical staining results.[57] The long arrow in (B) shows polyomavirus aggregates in a tubular epithelial cell. Fresh frozen tissue sample in (A), 20× original magnification, immunofluorescence microscopy to detect IgG deposits. Glutaraldehyde fixed biopsy core (B), 4000 original magnification, electron microscopy.

an artificially low number of "positive signals" rather than a marked decrease of the staining intensity in individual nuclei. Antibodies directed against the BK-virus T-antigen (clone BK.T-1) cross react with unrelated nuclear proteins (Ku86) and result in nonspecific staining signals.[64] Epithelial cells with signs of polyomavirus replication express p53 tumor suppressor proteins[43] and the proliferation marker KI-67 as signs of altered (viral) DNA assembly.

PCR techniques may also be utilized to demonstrate viral DNA or RNA in tissue samples and to confirm the diagnosis of PVN.[65-67] However, PCR results must be interpreted with great caution. Only very strong amplification signals of viral DNA (greater than 10 viral copies per cell equivalent), in the setting of histologically or immunohistochemically demonstrable virally induced cytopathic changes, can be used to confirm the diagnosis of PVN, to identify the viral strain and to distinguish clinically significant productive from clinically insignificant latent polyomavirus infections (see above).[65,66,68-72] The detection of viral RNA in renal biopsy cores by PCR clearly indicates viral replication/PVN. RNA extraction and amplification methods, however, are challenging techniques, susceptible to error and do not provide information which exceeds the results obtained with standard immunohistochemistry (such as the detection of the SV-40 T-antigen).[66,67]

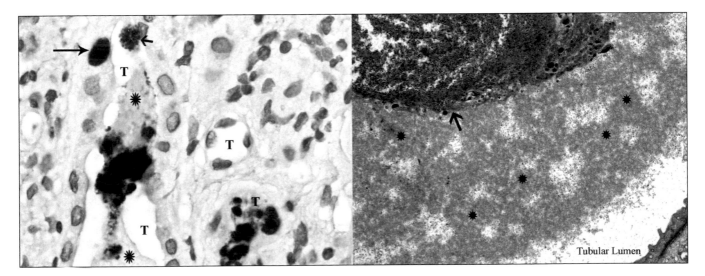

Figure 6. A) In-situ hybridization shows polyomavirus DNA, i.e., presumably mostly "mature" daughter virions, not only in tubular epithelial cell nuclei (brown staining; long arrow) but also within tubular lumens (T = tubules) where the viruses form dense aggregates in close association with proteinaceous casts/Tamm-Horsfall protein (asterisks). Note the necrotic epithelial cell releasing densely arranged clusters of mature virions into the tubular lumen (short arrow). B) Electron microscopy illustrates the release of daughter virions into the tubular lumen. Subsequent to intra-nuclear viral replication polyomaviruses are shed from lysed host cells (arrow) into tubular lumens containing abundant Tamm-Horsfall protein (asterisks). It is believed that three dimensional, cast-like viral aggregates ('Haufen') detected in voided urine samples from patients with PVN form in virally injured tubules and mark intra-renal disease (compare with Fig. 8B). Formalin fixed and paraffin embedded tissue section in (A), 100× (oil) original magnification, in-situ hybridization employing a protocol that allows for the detection of all polyomavirus strains. Glutaraldehyde fixed biopsy core (B), 12500× original magnification, electron microscopy.

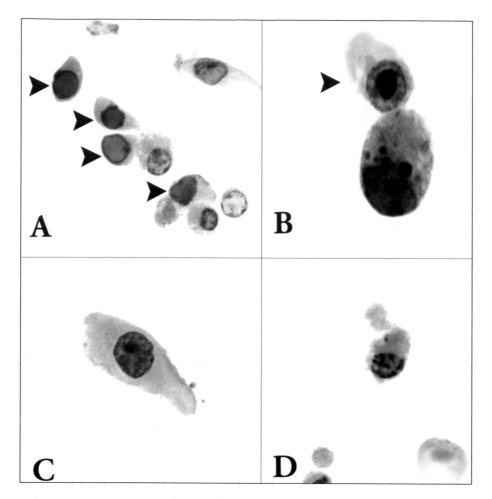

Figure 7. A-D) Urine cytology. The activation and replication of polyomaviruses can easily be monitored by searching for viral inclusion bearing epithelial cells, i.e., decoy cells, in routine urine cytology specimens (decoy cell phenotypes 1-4 illustrated in A-D). The diverse intra-nuclear polyomavirus inclusion bodies are caused by differences in the nuclear architecture with varying combinations of crystalloid viral arrays, evenly dispersed individual daughter virions, and chromatin.[47] If high numbers of uncommon decoy cell phenotypes 2-4 (illustrated in B-D) are noted, i.e., >25% of the total decoy cell population, then PVN can be predicted with >75% probability.[89] Thinprep® specimen of a voided urine sample, Papanicolaou stain, 60x (oil) original magnification.

In PVN, standard immunofluorescence microscopy with a common panel of antibodies directed against IgG, IgA, IgM, kappa and lambda light chains, fibrinogen and complement factor C3 generally does not show any diagnostic staining pattern in tubules, glomeruli, or blood vessels. In some patients, PVN is associated with a type III hypersensitivity reaction, resembling "anti-TBM disease" in native kidneys. These latter cases are characterized by granular, immune deposits along thickened tubular basement membranes that are easily discernible by immunofluorescence microscopy (using various antibodies directed against immunoglobulins and complement factors) or by electron microscopy (Fig. 5). Tubular immune deposits do not seem to contain viral capsid proteins.[57]

PVN and virally induced tubular injury are not associated with marked and diffuse upregulation of MHC-class II (i.e., HLA-DR) in tubular epithelial cells or with the deposition of the complement degradation product C4d along peritubular capillaries.[6,12,73] Both, the detection of C4d and tubular HLA-DR expression can help to establish a diagnosis of (concurrent) allograft rejection.[60]

By electron microscopy, polyomaviruses present as viral particles of 30 to 50 nm in diameter that occasionally form crystalloid aggregates (Fig. 6B).[22,24] Polyomaviruses are ultrastructurally identified by size and their icosahedral capsid structure (Fig. 8); polyomavirus strains cannot be distinguished. Virions are primarily found in the nucleus and rarely in the cytoplasm. Following viral replication, intra-nuclear viral assembly and the subsequent lysis of the host cell,

crystalloid arrays of mature daughter virions are released into tubular lumens rich in Tamm-Horsfall protein (Fig. 6). Tamm-Horsfall protein likely serves as an "adhesive" that facilitates the formation of densely packed, cast-like viral clusters (Fig. 8). These viral clusters are flushed out of injured tubules and excreted with the urine. They can be detected in voided urine samples as pathognomonic markers for PVN (see below).

## Course of PVN

PVN contributes directly to chronic graft failure and loss.[6,74] Unfortunately, treatment options are currently limited since highly specific anti-polyomavirus drugs are not available[75] (also see chapter by Ahsan). Outcome of PVN is largely determined by three factors: (i) the disease stage at time of the initial diagnosis, (ii) the duration of PVN with persistent, virally driven, pro-fibrotic tubulo-interstitial injury (Table 2), and (iii) the extent of pre-existing graft damage, e.g., hypertension induced arterionephrosclerosis, structural calcineurin-inhibitor toxicity etc. Allograft rejection can contribute to graft demise.[6,12,26] The ultimate therapeutic goal is to diagnose PVN early (stage A) and to rapidly stop viral replication, thereby limiting tubular injury and preventing disease progression to irreversible scarring.[11,28,30] Once PVN has resolved and the infection has resumed a latent state, signs of viral activation in the tissue, plasma and urine cease.[6,25,31,59] The sclerosing PVN stage C with greater than 50% interstitial scarring is typically associated with severe allograft dysfunction and rapid loss.[12,24,27,29,31,49]

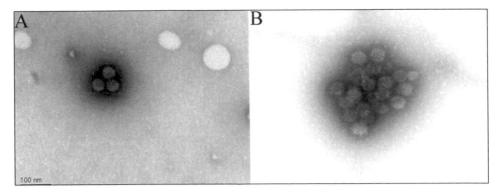

Figure 8. A,B) Voided urine sample. A) polyomavirus activation, no PVN: free viral particles of approximately 45 nanometers in diameter are discernible, consistent with the shedding of polyomaviruses. The detection of free virions is a sign of polyomavirus activation in the urothelium; it does not accurately mark PVN.[79] B) PVN: the detection of three dimensional, cast-like, polyomavirus aggregates ('Haufen') in the urine is diagnostic for intra-renal disease. The detection of 'Haufen' has a positive and negative predictive value for PVN of >95%.[32] Negative staining electron microscopy on voided urine samples, uranyl acetate, 125000× original magnification, electron microscopy.

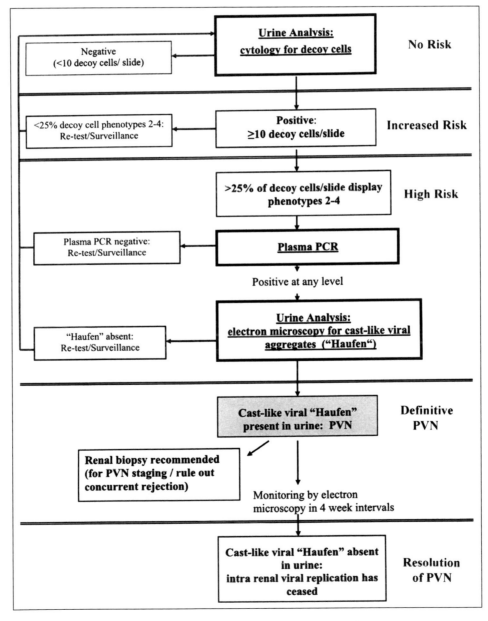

Figure 9. Revised schematic algorithm of a patient screening and monitoring protocol for diagnosing PVN non-invasively. Urine electron microscopy and the detection of cast-like polyomavirus aggregates, so-called "Haufen", are important new strategies to optimize the diagnosis of PVN. Steps one, i.e., decoy cell phenotyping and three, i.e., urine electronmicroscopy and the search for "Haufen", can be performed on a single voided urine sample.

The early PVN disease stages, namely A and B1, can best respond to therapeutic intervention with favorable long-term outcome.[6,11,19,26,28-31,59,76] Drachenberg and colleagues diagnosed PVN stage A on average 8.7 months and stages B and C 15.9 months post transplantation.[76] Under optimal conditions, viral clearance can be achieved within 3 weeks (personal observation in a patient presenting with PVN stage B1) and can result in complete functional restitution.[59] Usually, however, viral resolution that has been observed in up to 53%-78% of patients[12,30,76] is a rather slow process taking many months. Donor kidneys of cadaveric origin, high viral loads at time of diagnosis, i.e., plasma BK levels of $>1.5 \times 10^5$ copies/ml, and >5% of tubules with signs of polyomavirus infection have been identified as potential conditions protracting viral clearance.[30] In diagnostic repeat biopsies, disease progression was observed in 37%-75% of patients initially presenting with PVN stage A and in 13%-60% with stage B.[12,49] In some cases, PVN can progress rapidly. One of our patients advanced from disease stage A to C in three months (Figs. 1 and 4); a similar case was reported by Gaber and colleagues.[12] If, ultimately, PVN resolves (after disease progression), kidney biopsies show increased interstitial fibrosis and tubular atrophy, i.e., the PVN-burnt out stage (Table 3).[6,12] Depending on the extent of irreversible chronic injury, renal allograft function at time of viral resolution can vary from "back to baseline" (best case scenario in the setting of no or mild fibrosis), to chronic renal dysfunction (moderate fibrosis), or graft failure (worst case scenario in cases with marked fibrosis and atrophy).[6,12,30,76]

Although, in recent years, treatment options for PVN have remained limited, disease outcome has nevertheless improved. This is largely due to a high level of clinical suspicion and the implementation of patient screening procedures facilitating an early diagnosis.[11,19,25,26,48,76,77]

## Diagnostic Strategies: Patient Screening and Monitoring

The clinical risk assessment for PVN is based on the observation that all patients with disease show signs of polyomavirus activation in the urine and plasma that often precede viral nephropathy by weeks to months. Over the last years, the systematic introduction of patient screening and monitoring protocols into clinical practice has resulted in significant improvements in the detection and outcome of PVN; the basic principles of these clinical surveillance protocols have been extensively described and reviewed elsewhere (see chapter by Ahsan).[6,13,19,22,25-27,48,78-81] All currently employed screening strategies, i.e., decoy cell quantification by urine cytology (Fig. 7), PCR assays to detect viral DNA or RNA in the urine and plasma, or urine electron microscopy to quantify the shedding of free viral particles (Fig. 8A), have an excellent negative predictive value of 95%-100% to exclude PVN, i.e., no signs of viral activation = no PVN. Far less affirmative, however, are positive screening results ranging in their predictive values from 27% (quantification of decoy cells[6]) to approximately 80% (quantitative plasma PCR assays to detect BK virus DNA with a defined threshold level of $1 \times 10^4$ copies/ml[19]). 'False' positive screening tests for PVN are mainly due to polyomavirus activation in extra renal sites of viral latency, in particular in the urothelium, and can include high levels of BK viremia and/or viruria.[13,46,79,82-86] Significant BK viremia in the absence of PVN is seen in patients with polyomavirus associated inflammation of the urothelium following bone marrow transplantation[83] and occasionally also in nonkidney solid organ transplant recipients.[87] In order to better identify patients at high risk for disease, BK virus cut-off load levels for viruria (PCR readings $>1 \times 10^7$ BK copies/ml urine) and viremia (PCR readings $>1 \times 10^4$ BK copies/ml plasma) have been proposed.[48,78] These "threshold levels", however, have not been confirmed in large multi-center trials, vary from laboratory to laboratory due to nonstandardized PCR assays, are only significant at time of the initial diagnosis and not during persistent PVN to detect viral clearance, and do not reliably

distinguish between nondiseased viral activators and patients with PVN.[13,30,79,83-85] Thus, the predictive diagnostic power of PCR readings in the plasma and urine is often weaker than expected. Of note: a multicenter comparative trial to quantify BK virus load levels by PCR in defined test sample sets demonstrated inter-laboratory differences of the BK DNA load readings of up to 2 logs.[88] At present, clinical strategies for risk assessment, diagnosis and monitoring of PVN can be regarded as helpful, but not ideal.

In the following paragraphs we will briefly discuss two new methods to identify patients with PVN more accurately (Figs. 7-9): a) urine cytology and the quantification of different decoy cell phenotypes and b) urine electron microscopy and the detection of cast-like polyomavirus aggregates. Both studies can conveniently be performed non-invasively in a voided urine sample and can render a definitive diagnosis of PVN based on a single collection. Pending further validation, these methods hold great promise to provide powerful diagnostic information.

A. Urine cytology (Fig. 7): identification and quantification of different decoy cell phenotypes.

So-called "decoy cells" are intra-nuclear, polyomavirus inclusion bearing epithelial cells found in voided urine specimens. Decoy cells are a morphologic sign of polyomavirus activation. They are commonly thought to be primarily of transitional/urothelial origin and can easily be detected and counted in standard Papanicolaou stained ThinPrep® cytology preparations.[22,82] It was already noted during the early days of the PVN epidemic that the shedding of large numbers of decoy cells was associated with an increased risk for the development of PVN post renal transplantation.[4-6] The quantitative assessment of decoy cells is fast, easy, economical and readily available in every major transplant center affiliated with a pathology division. It is commonly used as the first screening method to identify patients at (increased) risk for PVN with an overall positive predictive value of 25%-30% (using a cut-off level of ≥10 decoy cells per ThinPrep® slide to call a sample "decoy positive").[6,13,19,22,25,26,48,79] The predictive yield can—in our experience—be easily increased by assessing not only the total number of decoy cells in a preparation but, additionally, also the different decoy cell variants (Fig. 7). If the common ground-glass decoy cell variant (phenotype 1, Fig. 7A) accounts for <75% of the total decoy cell count in a ThinPrep® specimen and the uncommon clumped variants (phenotypes 2-4, Fig. 7B-D) total >25%, then a patient is at high risk for PVN (positive predictive value >75%,[89] Fig. 9). The increased shedding of clumped decoy cell variants is seen in all PVN disease stages (A-C) and persists during ongoing viral nephropathy.[89] In our experience, the uncommon, clumped decoy cell variants are not typically seen in the urine of patients with polyomavirus associated lower urinary tract infections following bone marrow transplantation, i.e., hemorrhagic cystitis. It, therefore, is appealing to assume that decoy cell phenotypes 2-4 primarily originate from intra-renal/intra-tubular sites where polyomavirus replication often leads to the formation of different intra-nuclear inclusion bodies in tubular cells (see refs. 6, 26, 47). Thus, the easy and rapid identification and quantification of different decoy cell phenotypes may help to mark PVN more accurately, thereby rendering urine cytology a highly attractive and powerful screening tool. Future prospective studies are required to further validate the clinical significance of "decoy cell phenotyping"; until then, plasma PCR tests should additionally be used to adequately assess a patient's risk for PVN (Fig. 9).

B. Urine electron microscopy (Fig. 8): identification of cast-like polyomavirus aggregates, so-called "Haufen".

In renal allograft recipients with PVN, urine is drained through infected nephrons with virally induced tubular injury. As illustrated above (Fig. 6) post replication daughter virions are released, often in the form of crystalloid viral arrays, from necrotic epithelial cells into tubular lumens. The specific intra tubular microenvironment allows for the formation of three-dimensional, cast-like polyomavirus aggregates that are flushed out of the infected kidneys. Consequently, viral clusters, named "Haufen" (German for cluster or stack), can be detected in voided urine samples from patients with PVN by conventional electron microscopy employing negative staining techniques (Fig. 8B). In our experience, "Haufen" are highly sensitive and specific for kidney disease. Their appearance in and disappearance from the urine closely follows the histologic course of PVN in all disease stages (positive and negative predictive values: >95%[32]). A protracted prodromal phase with shedding of "Haufen" prior to the development of PVN does not occur. We could never find "Haufen" in any patient with asymptomatic polyomavirus activation including high levels of viruria (and viremia) or in patients suffering from lower urinary tract polyomavirus infections post bone marrow transplantation (n = 50 control patients). Thus, the qualitative assessment of cast-like polyomavirus clusters in the urine by electron microscopy can be regarded as the first non-invasive test to accurately diagnose PVN (Fig. 9).

Negative staining electron microscopy as a diagnostic tool to facilitate patient management requires "thinking outside the box" for most transplant physicians. Until now, only few investigators have used this technique on urine samples in the setting of polyomavirus infections,[8,79,90,91] and the assessment of "Haufen" to diagnose PVN is a very recent development not yet in common practice. However, negative staining electron microscopy is a very well established and commonly used technique to search for viral pathogens in body fluids. It has a rapid turn around time of approximately 3 hours post sample collection and can be performed in any pathology division with a diagnostic electron microscopy unit at low costs.[79] Negative staining electron microscopy simply has to be added to the list of required tests. We advocate its use and the specific search for cast-like polyomavirus aggregates, "Haufen", in voided urine samples in all patients at high risk in order to rule PVN in or out (Fig. 9).

## Conclusion

Polyomavirus nephropathy (PVN) is a challenging and serious complication mainly affecting renal allografts. Recent advances in risk assessment, the development of novel non-invasive diagnostic techniques, and disease staging will result in improvements of patient management and graft survival. Future research efforts should be directed towards the development of potent and specific anti-polyomavirus drugs and the better definition of risk factors promoting PVN. In addition, improved multi-center cooperations are required in order to validate and standardize clinical tests more efficiently, such as quantitative PCR assays, decoy cell analyses including cell phenotyping, and urine electron microscopy as a non-invasive method to diagnose PVN. We also have to gain better insight into the pathobiological significance of T-cell rich inflammatory cell infiltrates in florid cases of PVN as well as the potential role of JC-virus co-activation. How often does allograft rejection in PVN remain unrecognized, how aggressive should it be treated, and does JC-virus coreplication alter the presentation and course of PVN?

## References
1. Coleman DV, Mackenzie EF, Gardner SD et al. Human polyomavirus (BK) infection and ureteric stenosis in renal allograft recipients. J Clin Pathol 1978; 31:338-47.
2. Karam G, Maillet F, Parant S et al. Ureteral necrosis after kidney transplantation: risk factors and impact on graft and patient survival. Transplantation 2004; 78:725-29.
3. Mackenzie EF, Poulding JM, Harrison PR et al. Human polyoma virus (HPV)—a significant pathogen in renal transplantation. Proc Eur Dial Transplant Assoc 1978; 15:352-60.
4. Binet I, Nickeleit V, Hirsch HH et al. Polyomavirus disease under new immunosuppressive drugs: a cause of renal graft dysfunction and graft loss. Transplantation 1999; 67:918-22.
5. Drachenberg CB, Beskow CO, Cangro CB et al. Human polyoma virus in renal allograft biopsies: morphological findings and correlation with urine cytology. Hum Pathol 1999; 30:970-77.
6. Nickeleit V, Hirsch HH, Zeiler M et al. BK-virus nephropathy in renal transplants-tubular necrosis, MHC-class II expression and rejection in a puzzling game. Nephrol Dial Transplant 2000; 15:324-32.
7. Randhawa PS, Finkelstein S, Scantlebury V et al. Human polyoma virus-associated interstitial nephritis in the allograft kidney. Transplantation 1999; 67:103-9.
8. Howell DN, Smith SR, Butterly DW et al. Diagnosis and management of BK polyomavirus interstitial nephritis in renal transplant recipients. Transplantation 1999; 68:1279-88.
9. Cosio FG, Amer H, Grande JP et al. Comparison of low versus high tacrolimus levels in kidney transplantation: assessment of efficacy by protocol biopsies. Transplantation 2007; 83:411-16.
10. Mengel M, Marwedel M, Radermacher J et al. Incidence of polyomavirus-nephropathy in renal allografts: influence of modern immunosuppressive drugs. Nephrol Dial Transplant 2003; 18:1190-96.
11. Ramos E, Drachenberg C, Hirsch HH et al. BK polyomavirus allograft nephropathy (BKPVN): eight-fold decrease in graft loss with prospective screening and protocol biopsy. Am J Transplant (supplement) WTC 2006 congress abstracts 2006:121.
12. Gaber LW, Egidi MF, Stratta RJ et al. Clinical utility of histological features of polyomavirus allograft nephropathy. Transplantation 2006; 82:196-204.
13. Nickeleit V, Mihatsch MJ. Polyomavirus nephropathy in native kidneys and renal allografts: an update on an escalating threat. Transpl Int 2006; 19:960-73.
14. Sachdeva MS, Nada R, Jha V et al. The high incidence of BK polyoma virus infection among renal transplant recipients in India. Transplantation 2004; 77:429-31.
15. Nada R, Sachdeva MU, Sud K et al. Co-infection by cytomegalovirus and BK polyoma virus in renal allograft, mimicking acute rejection. Nephrol Dial Transplant 2005; 20:994-96.
16. Nickeleit V. Animal models of polyomavirus nephropathy: hope and reality. Am J Transplant 2006; 6:1507-9.
17. Knowles WA, Pipkin P, Andrews N et al. Population-based study of antibody to the human polyomaviruses BKV and JCV and the simian polyomavirus SV40. J Med Virol 2003; 71:115-23.
18. Bohl DL, Storch GA, Ryschkewitsch C et al. Donor Origin of BK Virus in Renal Transplantation and Role of HLA C7 in Susceptibility to Sustained BK Viremia. Am J Transplant 2005; 5:2213-21.
19. Hirsch HH, Knowles W, Dickenmann M et al. Prospective study of polyomavirus type BK replication and nephropathy in renal-transplant recipients. N Engl J Med 2002; 347:488-96.
20. Nickeleit V, Gordon J, Thompson D et al. Antibody titers and latent polyoma-BK-virus (BKV) loads in the general population: potential donor risk assessment for the development of BK-virus nephropathy (BKN) post transplantation. J Am Soc Nephrol (abstracts issue) 2004; 15:524A.
21. Polo C, Perez JL, Mielnichuck A et al. Prevalence and patterns of polyomavirus urinary excretion in immunocompetent adults and children. Clin Microbiol Infect 2004; 10:640-44.
22. Singh HK, Bubendorf L, Mihatsch MJ et al. Urine cytology findings of polyomavirus infections. In: Ahsan N, ed. Polyomaviruses and Human Diseases. New York/Austin: Springer Science+Business Media/Landes Bioscience, 2006:201-12.
23. Tani EM, Montenegro MR, Viana de Camargo JL. Urinary bladder washings in autopsies. A cytopathologic study of 63 cases. Acta Cytol 1983; 27:128-32.
24. Nickeleit V, Hirsch HH, Binet IF et al. Polyomavirus infection of renal allograft recipients: from latent infection to manifest disease. J Am Soc Nephrol 1999; 10:1080-89.
25. Nickeleit V, Klimkait T, Binet IF et al. Testing for polyomavirus type BK DNA in plasma to identify renal allograft recipients with viral nephropathy. N Engl J Med 2000; 342:1309-15.
26. Nickeleit V, Steiger J, Mihatsch MJ. BK Virus Infection after Kidney Transplantation. Graft 2002; 5(suppl):S46-57.
27. Nickeleit V, Singh HK, Mihatsch MJ. Polyomavirus nephropathy: morphology, pathophysiology and clinical management. Curr Opin Nephrol Hypertens 2003; 12:599-605.

28. Buehrig CK, Lager DJ, Stegall MD et al. Influence of surveillance renal allograft biopsy on diagnosis and prognosis of polyomavirus-associated nephropathy. Kidney Int 2003; 64:665-73.

29. Nickeleit V, Mihatsch MJ. Polyomavirus nephropathy: pathogenesis, morphological and clinical aspects. Verh Dtsch Ges Pathol 2004; 88:69-84.

30. Wadei HM, Rule AD, Lewin M et al. Kidney transplant function and histological clearance of virus following diagnosis of polyomavirus-associated nephropathy (PVAN). Am J Transplant 2006; 6:1025-32.

31. Drachenberg CB, Papadimitriou JC, Hirsch HH et al. Histological patterns of polyomavirus nephropathy: correlation with graft outcome and viral load. Am J Transplant 2004; 4:2082-92.

32. Singh HK, Madden V, Detwiler R et al. Detection of three dimensional polyomavirus clusters ("Haufen") by negative staining urine electron microscopy: a sensitive and specific marker of polyomavirus-BK nephropathy. J Am Soc Nephrol (abstracts issue) 2006; 17:760A-61A.

33. Baksh FK, Finkelstein SD, Swalsky PA et al. Molecular genotyping of BK and JC viruses in human polyomavirus-associated interstitial nephritis after renal transplantation. Am J Kidney Dis 2001; 38:354-65.

34. Trofe J, Cavallo T, First MR et al. Polyomavirus in kidney and kidney-pancreas transplantation: a defined protocol for immunosuppression reduction and histologic monitoring. Transplant Proc 2002; 34:1788-89.

35. Kazory A, Ducloux D, Chalopin JM et al. The first case of JC virus allograft nephropathy. Transplantation 2003; 76:1653-55.

36. Milstone A, Vilchez RA, Geiger X et al. Polyomavirus simian virus 40 infection associated with nephropathy in a lung-transplant recipient. Transplantation 2004; 77:1019-24.

37. Bruno B, Zager RA, Boeckh MJ et al. Adenovirus nephritis in hematopoietic stem-cell transplantation. Transplantation 2004; 77:1049-57.

38. Funk GA, Steiger J, Hirsch HH. Rapid dynamics of polyomavirus type BK in renal transplant recipients. J Infect Dis 2006; 193:80-87.

39. Petrogiannis-Haliotis T, Sakoulas G, Kirby J et al. BK-related polyomavirus vasculopathy in a renal-transplant recipient. N Engl J Med 2001; 345:1250-55.

40. Sandler ES, Aquino VM, Goss-Shohet E et al. BK papova virus pneumonia following hematopoietic stem cell transplantation. Bone Marrow Transplant 1997; 20:163-65.

41. Singh D, Kiberd B, Gupta R et al. Polyoma virus-induced hemorrhagic cystitis in renal transplantation patient with polyoma virus nephropathy. Urology 2006; 67:423 e411-23.

42. Esposito L, Hirsch H, Basse G et al. BK virus-related hemophagocytic syndrome in a renal transplant patient. Transplantation 2007; 83:365.

43. Weinreb DB, Desman GT, Burstein DE et al. Expression of p53 in virally infected tubular cells in renal transplant patients with polyomavirus nephropathy. Hum Pathol 2006; 37:684-88.

44. Celik B, Randhawa PS. Glomerular changes in BK virus nephropathy. Hum Pathol 2004; 35:367-70.

45. Nair R, Katz DA, Thomas CP. Diffuse glomerular crescents and peritubular immune deposits in a transplant kidney. Am J Kidney Dis 2006; 48:174-78.

46. Herawi M, Parwani AV, Chan T et al. Polyoma virus-associated cellular changes in the urine and bladder biopsy samples: a cytohistologic correlation. Am J Surg Pathol 2006; 30:345-50.

47. Colvin RB, Nickeleit V. Renal transplant pathology. In: Jennette JC, Olson JL, Schwartz MM, Silva FG, eds. Pathology of the Kidney (vol 2). 6 Edition. Philadelphia, Baltimore, New York, London: Lippincott Williams & Wilkins, 2007:1347-1490.

48. Hirsch HH, Brennan DC, Drachenberg CB et al. Polyomavirus-associated nephropathy in renal transplantation: interdisciplinary analyses and recommendations. Transplantation 2005; 79:1277-86.

49. Drachenberg RC, Drachenberg CB, Papadimitriou JC et al. Morphological spectrum of polyoma virus disease in renal allografts: diagnostic accuracy of urine cytology. Am J Transplant 2001; 1:373-81.

50. van Gorder MA, Della Pelle P, Henson JW et al. Cynomolgus polyoma virus infection: a new member of the polyoma virus family causes interstitial nephritis, ureteritis and enteritis in immunosuppressed cynomolgus monkeys. Am J Pathol 1999; 154:1273-84.

51. Mannon RB, Hoffmann SC, Kampen RL et al. Molecular evaluation of BK polyomavirus nephropathy. Am J Transplant 2005; 5:2883-93.

52. Chen Y, Trofe J, Gordon J et al. Interplay of cellular and humoral immune responses against BK virus in kidney transplant recipients with polyomavirus nephropathy. J Virol 2006; 80:3495-505.

53. Krymskaya L, Sharma MC, Martinez J et al. Cross-reactivity of T-lymphocytes recognizing a human cytotoxic T-lymphocyte epitope within BK and JC virus VP1 polypeptides. J Virol 2005; 79:11170-78.

54. Hammer MH, Brestrich G, Andree H et al. HLA type-independent method to monitor polyoma BK virus-specific CD4 and CD8 T-cell immunity. Am J Transplant 2006; 6:625-31.

55. Abend JR, Low JA, Imperiale MJ. Inhibitory effect of gamma interferon on BK virus gene expression and replication. J Virol 2007; 81:272-79.

56. Comoli P, Azzi A, Maccario R et al. Polyomavirus BK-specific immunity after kidney transplantation. Transplantation 2004; 78:1229-32.

57. Bracamonte E, Leca N, Smith KD et al. Tubular basement membrane immune deposits in association with BK polyomavirus nephropathy. Am J Transplant 2007 (in press).

58. Han Lee ED, Kemball CC, Wang J et al. A mouse model for polyomavirus-associated nephropathy of kidney transplants. Am J Transplant 2006; 6:913-22.

59. Mayr M, Nickeleit V, Hirsch HH et al. Polyomavirus BK nephropathy in a kidney transplant recipient: critical issues of diagnosis and management. Am J Kidney Dis 2001; 38:E13.

60. Nickeleit V, Mihatsch MJ. Polyomavirus allograft nephropathy and concurrent acute rejection: a diagnostic and therapeutic challenge. Am J Transplant 2004; 4:838-39.

61. Celik B, Shapiro R, Vats A et al. Polyomavirus allograft nephropathy: sequential assessment of histologic viral load, tubulitis and graft function following changes in immunosuppression. Am J Transplant 2003; 3:1378-82.

62. Jeong HJ, Hong SW, Sung SH et al. Polyomavirus nephropathy in renal transplantation: a clinicopathological study. Transpl Int 2003; 16:671-75.

63. Ahuja M, Cohen EP, Dayer AM et al. Polyoma virus infection after renal transplantation. Use of immunostaining as a guide to diagnosis. Transplantation 2001; 71:896-99.

64. Zambrano A, Villarreal LP. A monoclonal antibody specific for BK virus large T-antigen (clone BK.T-1) also binds the human Ku autoantigen. Oncogene 2002; 21:5725-32.

65. Randhawa PS, Vats A, Zygmunt D et al. Quantitation of viral DNA in renal allograft tissue from patients with BK virus nephropathy. Transplantation 2002; 74:485-88.

66. Schmid H, Burg M, Kretzler M et al. BK virus associated nephropathy in native kidneys of a heart allograft recipient. Am J Transplant 2005; 5:1562-68.

67. Schmid H, Nitschko H, Gerth J et al. Polyomavirus DNA and RNA detection in renal allograft biopsies: results from a European multicenter study. Transplantation 2005; 80:600-4.

68. Nickeleit V, Singh HK, Gilliland MGF et al. Latent Polyomavirus Type BK Loads in Native Kidneys Analyzed by TaqMan PCR: What Can Be Learned To Better Understand BK Virus Nephropathy? J Am Soc Nephrol (abstracts issue) 2003; 14:424A.

69. Boldorini R, Veggiani C, Barco D et al. Kidney and urinary tract polyomavirus infection and distribution: molecular biology investigation of 10 consecutive autopsies. Arch Pathol Lab Med 2005; 129:69-73.

70. Randhawa P, Shapiro R, Vats A. Quantitation of DNA of polyomaviruses BK and JC in human kidneys. J Infect Dis 2005; 192:504-9.

71. Chesters PM, Heritage J, McCance DJ. Persistence of DNA sequences of BK virus and JC virus in normal human tissues and in diseased tissues. J Infect Dis 1983; 147:676-84.

72. Limaye AP, Smith KD, Cook L et al. Polyomavirus nephropathy in native kidneys of nonrenal transplant recipients. Am J Transplant 2005; 5:614-20.

73. Nickeleit V, Zeiler M, Gudat F et al. Detection of the complement degradation product C4d in renal allografts: diagnostic and therapeutic implications. J Am Soc Nephrol 2002; 13:242-51.

74. Ramos E, Drachenberg CB, Papadimitriou JC et al. Clinical course of polyoma virus nephropathy in 67 renal transplant patients. J Am Soc Nephrol 2002; 13:2145-51.

75. Rinaldo CH, Hirsch HH. Antivirals for the treatment of polyomavirus BK replication. Expert Rev Anti Infect Ther 2007; 5:105-15.

76. Drachenberg CB, Papadimitriou JC, Wali R et al. Improved outcome of polyoma virus allograft nephropathy with early biopsy. Transplant Proc 2004; 36:758-59.

77. Brennan DC, Agha I, Bohl DL et al. Incidence of BK with tacrolimus versus cyclosporine and impact of preemptive immunosuppression reduction. Am J Transplant 2005; 5:582-94.

78. Hirsch HH, Drachenberg CB, Steiger J et al. Polyomavirus associated nephropathy in renal transplantation: critical issues of screening and management. In: Ahsan N, ed. Polyomaviruses and Human Diseases. New York/Austin: Springer Science+Business Media/Landes Bioscience, 2006:160-73.

79. Singh HK, Madden V, Shen YJ et al. Negative-staining electron microscopy of the urine for the detection of polyomavirus infections. Ultrastruct Pathol 2006; 30:329-38.

80. Ding R, Medeiros M, Dadhania D et al. Noninvasive diagnosis of BK virus nephritis by measurement of messenger RNA for BK virus VP1 in urine. Transplantation 2002; 74:987-94.

81. Nickeleit V, Steiger J, Mihatsch MJ. Re: Noninvasive diagnosis of BK virus nephritis by measurement of messenger RNA for BK virus VP1. Transplantation 2003; 75:2160-61 (letter).

82. Koss LG. Chapter 22: The lower urinary tract in the absence of cancer. In: Koss LG, Melamed MR, eds. Koss' Diagnostic Cytology and Its Histopathologic Basis (vol 1). 5 Edition. Philadelphia, Baltimore, New York: Lippincott Wiliams and Wilkins, 2006:738-76.

83. Erard V, Kim HW, Corey L et al. BK DNA viral load in plasma: evidence for an association with hemorrhagic cystitis in allogeneic hematopoietic cell transplant recipients. Blood 2005; 106:1130-32.

84. Erard V, Storer B, Corey L et al. BK virus infection in hematopoietic stem cell transplant recipients: frequency, risk factors and association with postengraftment hemorrhagic cystitis. Clin Infect Dis 2004; 39:1861-65.

85. Leung AY, Suen CK, Lie AK et al. Quantification of polyoma BK viruria in hemorrhagic cystitis complicating bone marrow transplantation. Blood 2001; 98:1971-78.

86. Bressollette-Bodin C, Coste-Burel M, Hourmant M et al. A prospective longitudinal study of BK virus infection in 104 renal transplant recipients. Am J Transplant 2005; 5:1926-33.

87. Razonable RR, Brown RA, Humar A et al. A longitudinal molecular surveillance study of human polyomavirus viremia in heart, kidney, liver and pancreas transplant patients. J Infect Dis 2005; 192:1349-54.

88. Gordon J, Brennan D, Limaye AP et al. Multicenter validation of polyomavirus BK quantification for screening and monitoring of renal transplant recipients. Am J Transplant 2005; (suppl. 11, 5):381A-82A.

89. Singh HK, Shen YJ, Detwiler R et al. Polyomavirus inclusion bearing decoy cell phenotypes: role in the management of patients with polyomavirus-BK nephropathy (BKN). J Am Soc Nephrol (abstracts issue) 2006; 17:761A.

90. Tong CY, Hilton R, MacMahon EM et al. Monitoring the progress of BK virus associated nephropathy in renal transplant recipients. Nephrol Dial Transplant 2004; 19:2598-605.

91. Alexander RT, Langlois V, Tellier R et al. The prevalence of BK viremia and urinary viral shedding in a pediatric renal transplant population: a single-center retrospective analysis. Pediatr Transplant 2006; 10:586-92.

# Pharmacotherapeutic Options in Solid Organ Transplantation

Jennifer Trofe,* Anikphe Imoagene-Oyedeji and Roy D. Bloom

## Abstract

Over the past decade, advances in immunosuppressive therapies have resulted in lower rates of acute rejection and consequently, significant improvements in patient and graft survival after solid organ transplantation. Increasingly successful outcomes have focused attention on the complications of long-term immunosuppression. Immunosuppressive regimens must be balanced to not only minimize the patient's risk for rejection, but also to avoid the risk of adverse effects such as infection, metabolic complications and noncompliance. Innovative immunosuppressive strategies such as steroid or calcineurin inhibitor avoidance or withdrawal and agents with novel mechanisms of action are currently being evaluated in an effort to mitigate complications of immunotherapies. This chapter will briefly review the immune response to the allograft and then focus on a review of the efficacies and toxicities of current and future immunosuppressive therapies for prevention and treatment of allograft rejection after solid organ transplantation.

## Introduction

The past two decades have seen significant advancements in transplantation. medical and surgical improvements in organ transplantation. Introduction of more potent immunosuppressants in the 1990s have fueled the investigation and implementation of new and unique regimens to prevent rejection. The use of these contemporary agents has been accompanied by a dramatic decline in acute rejection rates across all solid organ transplants. However, improved immunosuppressive efficacy has now served to focus transplant clinicians on potential toxicities as well as the financial implications of individual therapies.

In previous eras, rates of acute rejection and patient and graft survival at 6 or 12 months posttransplant were accepted endpoints for immunosuppressive trials. The uniformly improved outcomes observed with the modern immunosuppressive arsenal, have made it very difficult to compare individual therapies or regimens using these traditional metrics. Most modern immunosuppressive trials now utilize composites of all these endpoints to better evaluate efficacy. Moreover, few trials are conducted for longer than three years and extrapolation of these relatively short-term results may not be applicable in the long-term.

In light of the foregoing, it is apparent that the focus of immunosuppressive therapy must shift to the long term management of "the disease of immunosuppression". In addition to preventing rejection, these therapies must be tailored to reduce adverse effects (nephrotoxicity, cardiovascular and metabolic complications, infections and malignancies), address adherence and financial issues and improve quality of life to significantly impact long term patient and graft survival.

This chapter provides an overview of the currently available immunosuppressive agents that are used in solid organ transplantation, commencing with a brief review of the allo-immune response and the history of immunosuppression. It is followed by an outline of clinically relevant aspects of current immunosuppressive agents, including mechanisms of action, pharmacokinetics, adverse effects, drug dosing and therapeutic drug monitoring, as well as their use in therapy in organ transplantation. Finally, we discuss factors involved in customizing immunosuppressive regimens to individual patients and possible future research directions for these therapies.

## Overview of the Immune Response to Solid Organ Allografts

Both adaptive and innate immune responses are involved in allo-immunity. The adaptive immune system is composed of humoral (B-lymphocytes) and cell-mediated immunity (T-lymphocytes).The innate immune system comprises macrophages, dendritic cells, natural killer (NK) cells and noncellular components such as complement regulatory proteins that interact with the adaptive immune system to influence immune responses. Thymic-derived T-cells play a central role in cell-mediated immunity and bone marrow-derived B-cells are responsible for humoral immunity. The adaptive immune system generates specific responses against foreign antigens by the clonal expansion of antigen specific lymphocytes.

In the humoral immune response, B-lymphocytes recognize specific "nonself antigen" via the B-cell receptor complex (BCR). When activated, they secrete antibodies that bind specifically with foreign antigen and activate other effectors of the immune response. Cell mediated responses involve recognition of "nonself antigen" via the T-cell receptor (TCR) and generally exert effects locally through both cell-to-cell contact and secretion of soluble mediators (e.g., cytokines, chemokines). Cell-mediated immunity is important for direct cellular lysis, graft versus host disease, solid organ transplant rejection and delayed-type hypersensitivity reactions. Foreign antigen is presented to T-and B-cells by professional antigen presenting cells (APCs) such as dendritic cell, macrophages, B-cells. Appropriate costimulatory signaling between accessory molecules (CD28, CD80/86, ICOS, PD-1, CD40, CD40L) on these pairs of cells is usually required for activation of T-and B-lymphocytes. Other nonmarrow derived cells such as specialized stromal cells and endothelial cells often secrete cytokines that provide an appropriate microenvironment for immune cells to function. Stromal and endothelial cells may occasionally become nonprofessional APCs when there is persistence of the inflammatory milieu.

*Corresponding Author: Jennifer Trofe— Department of Pharmacy, Hospital of the University of Pennsylvania, 3400 Spruce Street, Philadelphia, PA 19104, USA. Email: jennifer.trofe@uphs.upenn.edu

*Chronic Allograft Failure: Natural History, Pathogenesis, Diagnosis and Management*, edited by Nasimul Ahsan. ©2008 Landes Bioscience.

The transplanted organ contains numerous antigens that are recognized as foreign by the host's immune system. The resulting allo-immune response can be described in various phases: First an induction phase involves antigen recognition, T-cell and B-cell activation, differentiation and expansion; an effector phase when allograft injury occurs; and lastly a resolution phase during which the immune response to the allograft diminishes but with emergence of residual memory to donor antigens. Anti-donor memory cells are a barrier to transplantation.

During the induction phase allo-immune responses may be direct (where donor APCs present alloantigen to host T-cells or indirect (host dendritic cells that had infiltrated within the graft traffic to secondary organs where they present allo-antigen to T-cells).

The effector phase marks the emergence of effector T-cells from lymphoid organs to infiltrate the allograft and orchestrate an inflammatory response. Primed T-cells produce cytokines that amplify the immune response, mediate cytotoxicity and delayed-type hypersensitivity reactions all of which lead to allograft injury and destruction. Additionally, B-cells produce allo-antibody against donor HLA and nonHLA antigens and proliferate and differentiate into plasma cells. Antibodies generally result in Fc receptor mediated phagocytosis by cells of the innate immune system as well as complement-mediated lysis and fibrin deposition. This all culminates in neutrophil recruitement and increased vascular permeability. Thus within hours to days after transplantation, an immune response to reject the allograft is set in place by the host immune system.

T-cell immune responses are the target of most immunotherapies. An understanding of this process is therefore helpful to understanding the mechanism of action of these agents. T-cell activation requires the delivery of three distinct signals. Signal 1 is delivered through the CD3 complex on the T-cell after engagement of the TCR-MHC/allopeptide complex. The second, costimulatory signal is provided by the interaction of CD80/86 on APCs with CD28 on T-cells (signal 2). Delivery of signals 1 and 2 contribute to T-cell activation through several pathways including: the calcium-calcineurin pathway, the RAS-mitogen-activated protein (MAP) kinase pathway and the nuclear factor-κβ pathway. These pathways trigger the expression of many new molecules such as CD154, interleukin-2, interleukin-15, CD25 and other cytokines. Interleukin -2 activates the "mammalian target of rapamycin" (m-TOR) pathway to provide the trigger for proliferation (signal 3).

The goal of all pharmacotherapeutic interventions in transplantation is to promote long term stable allograft acceptance and disrupt the host alloimmune responses that lead to "allograft rejection." T-cells responding mainly through the direct pathway, but also partly by indirect pathway of antigen recognition mediate acute cellular rejection. The indirect pathway of antigen recognition against nonMHC and MHC antigens may be important for chronic immune rejection. When preformed antibodies against donor antigens are present at the time of transplantation, antibody mediated rejection occurs and often results in hyperacute rejection. However, activation of naïve B-cells, or reactivation of memory B-cells during the first few days to weeks after transplantation, may result in acute humoral rejection, in which vascular inflammation is often seen. The majority of untreated individuals rapidly reject their allografts. For a more detailed discussion on the immunology and pathogenesis of rejection in solid organ transplantation the interested reader is referred to the earlier chapters of this book where these processes are described in further detail.

## History of Immunosuppression

The first successful endeavor into human transplantation occurred in 1954 when a kidney was transplant between identical twins.[1] The recipient experienced immediate graft function and survived for nine years posttransplant until his allograft failed from recurrent glomerulonephritis. As more successful transplants were performed between identical twins, interest arose in developing immunosuppressive therapies so that the benefit of transplantation could be extended beyond perfectly matched donor-recipient pairs.[2] The initial experience with transplants between-non-identical twins were associated with high early rejection rates and were of little overall benefit. Some of the earliest attempts at prevention of rejection consisted of total body irradiation, in combination with corticosteroids. The majority of these attempts were uniformly fatal.[3] Nevertheless, these early beginnings provided the driving force to the pursuit of more effective anti-rejection therapies.

Since the early experience with clinical transplantation, there has been ongoing investigation for the discovery of effective immunosuppressive therapies (Fig. 1). The first clinical trial of an immunosuppressive agent in renal transplant recipients occurred in 1962 with the development of azathioprine.[4] One year allograft survival rates were improved to as high as 50% when combined with corticosteroids.

In 1966, antithymocyte globulins were developed as an adjunctive agent to the double drug regimen of azathioprine and corticosteroids and as a treatment for rejection.[5] Although antithymocyte globulins slightly improve graft survival, their utility was limited by manufacturing issues and adverse events.

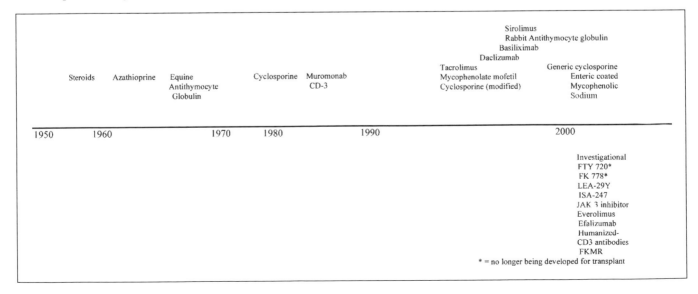

Figure 1. History of immunosuppression.

In 1979, the first large clinical trial with cyclosporine was reported.[6] With the introduction of cyclosporine in the 1980s, one year renal allograft graft survival rates in the 1980s increased to more than 80%.[7] Cyclosporine also greatly contributed to the advancement of successful heart, liver, pancreas and lung transplantation.

Shortly thereafter, muromonab CD-3 antibody (OKT3) became available for the treatment of rejection.[8] Similar to antithymocyte globulin, this antibody preparation was also associated with significant adverse events. One of the most limiting adverse events was the formation of idiotypic antibodies against the product, thereby limiting its efficacy in repeated courses.

Until 1995, only 5 immunosuppressants were approved for use in organ transplantation. Since then, the number of new agents has more than doubled. The introduction of these generally more potent immunosuppressive agents (modified cyclosporine, tacrolimus, mycophenolate mofetil, rabbit antithymocyte globulin, basiliximab, daclizumab and sirolimus) has led to further decreases in acute rejection for all solid organ transplant recipients.[9] At the end of 2004, in the United States, 168,761 recipients were living with functioning solid organ transplants.[10]

Despite improvements in rejection rates and 1 year allograft survival, there has not been proportionate increases in 5 to 10 year long term outcomes.[11] Interestingly, cardiovascular disease, not rejection, is the primary cause of death among all solid organ transplant recipients.[12] Nephrotoxicity from extended exposure to calcineurin inhibitors such as tacrolimus and cyclosporine greatly impacts patient survival and is an important cardiovascular disease risk. A recent analysis of 69,321 patients who received nonrenal transplants in the United States between 1990 and 2000, found the five-year risk of chronic kidney disease (CKD) to range from 7 to 21% depending on the type of organ transplanted. Moreover, CKD in these populations was associated with a four-fold increase in the risk of death.[13]

Current research is focusing on development of immunosuppressive agents and strategies to lessen adverse effects and promote long term-patient and graft survival. The hope is that future research will result in the development of effective immunosuppressive agents and strategies with less adverse effects, which in turn may lead to improved adherence, quality of life and prolonged patient and graft survival.

## Review of Immunosuppressive Agents

Immunosuppression can be achieved by depletion of lymphocytes, diversion of lymphocyte trafficking, or blockade of pathways of lymphocyte activation.[14] Multiple agents are often employed in immunosuppressive regimens to block various signals in the immune cascade.

Immunosuppressive agents can be classified as small-molecule drugs, depleting and nondepleting protein drugs (polyclonal and monoclonal antibody agents), fusion proteins and glucocorticoids.[14] In this section, we shall review currently available immunosuppressive agents that are used in clinical practice.

## Corticosteroids

Corticosteroids remain an important component of maintenance immunosuppression and also play a critical in the treatment of rejection. The main form of corticosteroid used in the United States for prevention and treatment of rejection are the glucocorticoids; intravenous methylprednisolone and oral prednisone, respectively.

### Mechanism of Action

Glucocorticoids freely cross cell membranes and bind to an intracellular nuclear steroid receptor. Once bound, this complex migrates into the cell nucleus and regulates transcription of glucocorticoid receptor genes.[15] Glucocorticoids act by multiple pathways and decreasing the transcription of key cytokines involved in rejection and inflammation, such as, interleukins-1, 2, 4, 6, 10, tumor necrosis factor (TNF)

alpha and interferon (IFN) gamma.[16] Corticosteroids also impair monocyte/macrophage function and decrease the number of circulating CD4+ T-cells.[17] High doses of glucocorticosteroids cause apoptosis of activated T- and B-cells as well[18] and may explain the potent effect of pulse-dose steroids for treatment of rejection.

### Pharmacokinetics

All glucocorticoids are well absorbed and are 65-90% protein bound. Prednisone is metabolized via the liver to its active metabolite prednisolone, yet metabolism may be impaired with hepatic dysfunction. Glucocorticoids are excreted in the urine. The half life elimination is approximately 3 hours but may be up to 5 hours in end stage renal disease.

### Dosing and Therapeutic Monitoring

Currently no assay exists to measure glucocorticoid levels in the clinical setting.[19] Dosing is generally by center-specific protocols. These usually consist of intra-operative intravenous methylprednisolone bolus followed by a taper over a few postoperative days and subsequent conversion to oral prednisone and continued taper to chronic maintenance dose. Depending on the organ transplanted, tapers can vary. In cases of milder rejection, two to three pulse steroid doses of intravenous glucocorticoids are often administered. Although glucocorticoids are hepatically metabolized and renally excreted, dose adjustments rarely need to be made in the setting of hepatic or renal insufficiency in transplant recipients.

### Adverse Effects

The effectiveness of corticosteroids is offset by their multitude of complications. Common adverse events include psychosis, increased risk of infection, impaired wound healing, peptic ulcer disease, weight gain, glucose intolerance, hypertension, hyperlipidemia, glaucoma, cataracts and osteoporosis. These factors can significantly impact on patient risk for cardiovascular morbidity and mortality, compliance, quality of life and ultimately affect long term patient and graft survival.

### Therapeutic Role

Corticosteroids have been used in preventing and treating rejection since 1954. They are commonly used as part of a triple-drug regimen together with a calcineurin inhibitor and an antiproliferative agent. To try to decrease corticosteroid-related complications, research has focused on steroid minimization, withdrawal and avoidance regimens after transplantation.

## Azathioprine

Although this anti-proliferative agent was first utilized to prevent kidney transplant rejection over 40 years ago, it is still occasionally used in current immunosuppressive regimens. Azathioprine is only FDA approved for use in kidney transplantation, though it has been widely incorporated into regimens for other solid organ groups as well.

### Mechanism of Action

The mechanisms of action of azathioprine as an immunosuppressant in organ transplantation, are not entirely understood. Azathioprine is metabolized to 6-mercaptopurine through reduction by glutathione. It is subsequently converted to 6-thiouric acid, 6-methyl-mercaptopurine and 6-thioguanine and incorporated into replicating DNA. The active metabolite, 6-mercaptopurine inhibits formation of thio-inosinic acid which is necessary for de novo purine synthesis. Since lymphocytes lack a salvage pathway for purine synthesis, inhibition of DNA synthesis leads to suppression of T- and B-cell proliferation. Azathioprine may also induce apoptosis of T-cells through CD28 costimulation.[20,21]

### Pharmacokinetics

Azathioprine is the prodrug of 6-mercaptopurine and is approximately 30% protein bound. After metabolism via hepatic xanthine oxidase to 6-mercaptopurine, it is converted to its inactive metabolite by thiopurine S methyltransferase (TPMT). Up to 10% of the population may possess a polymorphism for this enzyme, resulting in lower enzyme activity and increased risk for myelosuppression[17,22] The half life of 6-mercaptopurine is approximately 0.7 to 3 hours. It is renally eliminated and elimination is prolonged with renal failure.

### Adverse Effects

Dose related bone marrow suppression, including leukopenia, megaloblastic anemia and thrombocytopenia, are the most significant adverse events. The dose should be reduced if leukopenia (white blood cell count less than 3000 cells/mm$^3$) occurs. Myelosuppression is usually reversible with dose reduction. Of note, patients on allopurinol (a xanthine oxidase inhibitor) require about a 75% dose reduction in azathioprine since xanthine oxidase is necessary for conversion of 6-mercaptopurine to its inactive metabolites. Dose reductions are also required in severe renal insufficiency. Other adverse effects include gastrointestinal toxicity and infrequently, hepatotoxicity and pancreatitis. Skin cancer has also been associated with long-term use of azathioprine and recently a mechanism of action involving drug induced sensitivity of DNA to ultraviolet A light has been described. [23]

### Dosing and Monitoring

Azathioprine is administered on the day of, or a few days prior to, transplant at a dose of 3-5 mg/kg/day given once daily as either orally or by intravenous route. This dose is reduced to maintenance of 1-3 mg/kg/day. There are no monitoring parameters for azathioprine in routine transplant clinical practice. However, testing for polymorphisms of thiopurine S methyltransferase, as described in the pharmacokinetic section above, may be useful in guiding initial dosing.

### Therapeutic Role

Azathioprine has been used as an adjunctive maintenance therapy to prevent rejection. It is mostly used in combination with a calcineurin inhibitor and corticosteroids and is not used to treat a rejection episode. Although azathioprine's patent expired in 1979, a generic formulation was not FDA approved until 1996.[24] Generic azathioprine has been shown to be safe and effective in transplant recipients when compared to the azathioprine brand product, Imuran.[25]

## Mycophenolate Mofetil (MMF) and Enteric-Coated Mycophenolate Sodium (EC-MPS)

Due to improved efficacy, MMF replaced azathioprine as the adjunctive anti-proliferative immunosuppressant.[26] It is approved for use with cyclosporine and corticosteroids in heart, liver and kidney recipients. Recently, an enteric coated formulation of mycophenolate formulation (EC_MPS) was approved as an adjunct agent for prevention of rejection after kidney transplantation in conjunction with cyclosporine and corticosteroids.

### Mechanism of Action

MMF is metabolized to mycophenolic acid (MPA) which causes noncompetitive, reversible inhibition of inosine monophosphate dehydrogenase (IMPDH). This is the rate-limiting enzyme in the de novo synthesis of guanine nucleotide. Most cells can synthesize guanine nucleotides via the IMPDH pathway or the salvage pathway. Lymphocytes do not possess a salvage pathway and hence blockage of the IMPDH pathway results in decreased T- and B-cell proliferation.

### Pharmacokinetics

MMF is the morpholinoethyl ester of MPA. MMF has 94% bioavailablity and is highly protein bound (97% bound to albumin). It is rapidly converted to the active metabolite MPA in the liver. MPA is then glucuronidated to MPA glucuronide (MPAG), the inactive

metabolite. However, MPAG may undergo enterohepatic recirculation and is reformed and reabsorbed more distally in the gut. This results in a second MPA peak six to twelve hours after the dose. Cyclosporine inhibits biliary excretion of MPAG into the gastrointestinal tract thereby decreasing overall MPA exposure.[27] This effect does not occur with tacrolimus or sirolimus. Inactive MPAG is eliminated through the urine. The half life of MMF is 18 hours and is prolonged in renal insufficiency. Severe hepatic dysfunction can also result in increased MPA levels due to impairment of glucuronidation. MPA exposure immediately posttransplant is approximately one third lower than three to six months posttransplant.[28] Pharmacokinetics of EC-MPS are similar to those of MMF with the exception of a longer time to reach maximal plasma concentrations of MPA and greater absolute bioavailability.[29,30]

### Adverse Events

One of the major complications of MMF therapy is gastrointestinal adverse events (diarrhea, nausea, vomiting)[31] Reductions in MMF dosage may result in an improvement in gastrointestinal adverse events, but must be balanced against the risk of rejection.[32] EC-MPS releases MPA in the small intestine and may reduce upper gastrointestinal tract side effects. Other dose related adverse effects include leukopenia and thrombocytopenia.

### Dosing and Monitoring

MMF is commercially available as tablets, capsules, oral suspension and intravenous formulation. Dosing, efficacy and toxicity are similar regardless of administration mode.[33-35] The recommended MMF dose is 1 to 3 grams/day in two divided doses. EC-MPS is available in 180 mg and 360 mg tablets. EC-MPS at a dose of 720 milligrams twice daily is equivalent in efficacy to 1000 milligrams twice daily of MMF.[36] The MPA area under the curve is higher when MPA therapies are used with tacrolimus or sirolimus and may be lower when it is used in combination with cyclosporine.[37,38] Doses should be lowered for gastrointestinal adverse or hematological adverse effects as well as for severe hepatic or renal insufficiency. Mycophenolate dosage reductions are usually made without therapeutic drug monitoring. Although methods are available to monitor MPA levels, the assays are not widely utilized secondary to availability and complicated methodology.[39] Nevertheless, it may be useful to monitor levels in patients with toxicity, or suspected problems with drug absorption.[40]

### Therapeutic Role

MMF in combination with cyclosporine and corticosteroids is more effective agent than azathioprine for the preventing allograft rejection.[41-44] MMF is also effective when used together with tacrolimus.[45] Current research is focused on evaluating mycophenolate as a kidney-sparing drug in kidney recipients with calcineurin inhibitor mediated nephrotoxicity.[46] Initial trials in kidney recipients comparing EC-MPS to MMF in conjunction with cyclosporine and corticosteroids suggest equivalent efficacy for preventing rejection.[47,48] A nonsignificant trend towards less gastrointestinal adverse effects was noted with EC-MPS, although neither trial was designed to evaluate differences in gastrointestinal tolerability. Current research is investigating whether individualizing mycophenolate dosage to a target 2-hour MPA area under the curve will reduce the risk of rejection and adverse effects.[28,49]

## Sirolimus and Everolimus

In 1999, sirolimus was approved for use with cyclosporine and corticosteroids to prevent kidney allograft rejection[50,51] and its immunosuppressive role has continued to evolve. Everolimus is related to sirolimus but with an improved pharmacokinetic profile and greater bioavailability.[46] It is not approved for use in the United States.

## Mechanism of Action

Sirolimus and everolimus are macrocylic lactones that bind to the intracellular immunophilin known as FK binding protein. This complex binds to the mammalian target of rapamune (mTOR) causing inhibition of signal transduction and leading to cell cycle arrest of T- and B-cells in the G1-S cell cycle phase. TOR inhibitors also inhibit interleukin-2 induced proliferation of T- and B-cells. Finally, sirolimus demonstrates antiproliferative effects on vascular smooth muscle and fibroblast growth factors; resulting in inhibition of neointimal hyperplasia. Unlike tacrolimus which also binds to FK binding protein, sirolimus and everolimus do not inhibit calcineurin inhibitor activity.

## Pharmacokinetics

Sirolimus is associated with large intra and interpatient pharmacokinetic variability. Its absorption is rapid and highly variable. It is highly protein bound (92%) and is extensively metabolized by hepatic CYP3A4 and p-glycoprotein. The half life of sirolimus is extremely long at 62 hours. Time to peak levels is one to three hours. The majority of this drug is excreted through the feces. In contrast, everolimus is associated with greater absorption and has a shorter half life of 16-19 hours compared to sirolimus.[52]

## Adverse Events

Sirolimus is associated with hyperlipidemia, leukopenia, anemia and thrombocytopenia, effects that are generally dose-related.[53] Dose-dependent interstitial pneumonitis can also occur and may resolve over weeks to months.[54] Due to its antiproliferative effects on fibroblasts, wound healing complications such as lymphoceles can occur, especially early posttransplant.[55] Sirolimus is therefore often delayed until at least a month posttransplant. We believe that consideration should also be given to holding sirolimus, if possible, for two to four weeks before other major surgeries and up to four weeks post-operatively. To mitigate the risk of rejection, our own practice is to replace sirolimus during this time with another agent that has a lesser impact on wound healing. Other frequently observed sirolimus related side effects include mouth ulcers and edema.[56] Although initially believed to be less nephrotoxic than calcineurin inhibitors,[57] sirolimus is now well-recognized to cause reversible proteinuria, varying from low-grade to nephritic range, that may abate with discontinuation of the drug. In kidney transplant recipients, sirolimus has been associated with deterioration of chronic allograft nephropathy if proteinuria is present and it may also prolong delayed graft function.[58] Sirolimus may cause hemolytic uremic syndrome[59] that may reverse with discontinuation of therapy. Due to similar mechanism of action, adverse events of everolimus are anticipated to be similar to those of sirolimus.

## Dosing and Monitoring

Sirolimus is dosed orally once daily and is monitored by trough levels. Some clinicians give a loading dose followed by a trough level guided maintenance dose. Because sirolimus and cyclosporine potentiate one another's blood levels, initial recommendations were to take sirolimus four hours after the cyclosporine dose.[60] This is not necessary when used with tacrolimus.[61] Doses do not need modification in renal insufficiency, but should be reduced by 33% for hepatic impairment. Due to its long half life, it only requires weekly monitoring.[62] The optimal target trough level is not yet clearly established, though higher sirolimus levels are associated with more adverse events. Trough levels range from 5 ng/mL to 15 ng/mL depending on the regimen being used. We target lower sirolimus trough levels when the drug is used in combination with calcineurin inhibitors. Since sirolimus is metabolized by CYP3A4, drugs that inhibit or induce this enzyme will affect sirolimus levels and dose adjustments may be necessary.[62,63] Everolimus requires twice daily dosing due to its shorter half life and trough level monitoring will also be required.

## Therapeutic Role

The role of sirolimus in organ transplantation continues to evolve.[64] It has been used in conjunction with calcineurin inhibitors with the goal of sparing long-term nephrotoxicity from this latter class of therapy.[65] This may take the form of calcineurin inhibitor minimization or elimination.[66,67] Because of wound healing issues, the potentiation effect of calcineuin inhibitors, as well as the adverse effect profile described above, sirolimus should be used cautiously early after transplant. The safety of de novo use of sirolimus in liver, heart or lung transplantation is not established. Experience has mainly focused on conversion from calcineurin inhibitors.[64] However, determining the benefit of this strategy is limited by the lack of controlled investigation in this area. It has been proposed that effective sirolimus conversion requires intervention at early stages of renal insufficiency.[65] Sirolimus has also been shown to suppress tumor growth in experimental studies,[68] and may possess theoretical advantages over other immunosuppressive agents with respect to posttransplant neoplasia.[69,70]

Everolimus is currently being evaluated in kidney and heart recipients in combination with cyclosporine. Similar efficacy to mycophenolate mofetil and azathioprine has been demonstrated.[46,71] Everolimus has been shown to reduce the incidence and severity of heart allograft vasculopathy[72] and may also possess anti-tumor activity in vitro.[73]

# Calcineurin Inhibitors (Cyclosporine and Tacrolimus)

The introduction of nonmodified cyclosporine into the clinical arena in the early 1980s revolutionized transplantion by significantly improving patient and graft survival. Twenty years later calcineurin inhibitor-based immunosuppressive protocols, mainly incorporating tacrolimus or modified cyclosporine remain the standard of care in transplantation. Both cyclosporine and tacrolimus are approved for prevention of rejection in kidney, liver and heart transplant recipients.

## Mechanism of Action

Cyclosporine and tacrolimus bind to cytoplasmic immunophilins called cyclophilin and FK binding protein, respectively. This results in production of a complex that inhibits calcium-sensitive phosphatase, calcineurin. T-lymphocytes are extremely sensitive to calcineurin inhibition. Calcineurin inhibition causes inhibition of lymphokine gene transcription, mainly interleukin-2 and interferon-gamma which are necessary for T-cell activation. Although they have similar mechanisms of action, tacrolimus is 10-100 times more potent than cyclosporine in vitro and as a result, in clinical practice tacrolimus is given in doses up to 50-fold lower than those of cyclosporine.[74,75]

## Pharmacokinetics

Bioavailability of nonmodified cyclosporine varies greatly ranging from less than 10% in liver transplant recipients and up to 89% in renal transplant recipients. Modified cyclosporine is less dependent on the presence of food, bile acids or gastrointestinal mobility compared to nonmodified cyclosporine resulting in 30% improved absorption. Overall, oral cyclosporine bioavailability is approximately 30% that of intravenous cyclosporine.

On the other hand, tacrolimus absorption can also be variable and food taken within 15 minutes of the dose decreases the absorption by 27%. Bioavailability varies from 7 to 28%. Both calcineurin inhibitors are highly protein bound (greater than 90%). Unlike cyclosporine, clamping of a t-tube in liver transplant recipients does not alter trough concentrations or area under the curve of tacrolimus. Tacrolimus takes 0.5 to four hours to peak and has a half life of 21-61 hours. Nonmodified cyclosporine peaks at two to six hours with some patients experiencing a second peak at five to six hours postdosing. In contrast, modified cyclosporine peaks at 1.5 to 2 hours postdosing. Half life eliminations of nonmodified and modified cyclosporine are biphasic, with terminal half lives of 19 and 9 hours respectively. Both cyclosporine and tacrolimus are extensively metabolized by CYP3A4 enzymes and are mainly eliminated via the fecal route.

### Adverse Events

Both cyclosporine and tacrolimus are associated with numerous adverse effects, some of which are dose-dependent. Two concerning adverse events with these agents are acute and chronic nephrotoxicity.[76] This is most likely related to renal vasoconstriction involving both afferent and efferent arterioles as well as tubular toxicity. There is some evidence that tacrolimus maybe less nephrotoxic than cyclosporine at clinically used target trough concentrations.[45] Hepatotoxicity presenting as hyperbilirubinemia and elevated transaminases occurs with both agents, as well, particularly when levels and doses are high. Cardiovascular related adverse events of both agents include hypertension, hyperlipidemia and glucose intolerance.[77] Tacrolimus is associated with a lower incidence of hypertension and hyperlipidemia and a higher tendency to diabetogenicity when compared to cyclosporine. Electrolyte abnormalities including hyperkalemia (secondary to renal tubular acidosis), hypomagnesemia (secondary to tubular wasting) and hyperuricemia more likely with cyclosporine thasn tacrolimus also occur with both agents.[78] Other adverse events include neurotoxicity which may manifest as paresthesias, tremors, seizures and encephalopathy.[79] Although initial studies suggested a higher incidence of neurotoxicity with tacrolimus, this may have been a result of high tacrolimus doses and levels rather than the drug itself. Thrombotic microangiopathy has been described with both cyclosporine and tacrolimus and may reverse with discontinuation of the drug.[80] Case reports have shown that switching from one agent to the other is sometimes also a successful treatment option. Other differences in adverse events are that cyclosporine is associated with hirsutism and gingival hyperplasia while tacrolimus can be associated with alopecia, but not gingival hyperplasia.[81]

### Dosing and Monitoring

Both tacrolimus and cyclosporine are available as oral and intravenous formulations. When used intravenously, significant dose reduction (one third of the oral dose) are required due to increased bioavailability compared to oral preparations. Many transplant centers administer this in two to three divided doses or a continuous infusion, titrating upward to therapeutic target ranges. Occasionally, patients may be converted from nonmodified to modified cyclosporine preparations. Although 1:1 milligram conversion from nonmodified to modified cyclosporine is recommended, close monitoring of levels is required due to improved absorption with modified cyclosporine which may necessitate lower dosage requirements to maintain similar target cyclosporine levels.[82,83]

Doses and target levels are dependent on the type of organ transplant, the donor and recipient risk factors for rejection and the recipient's predilection to calcineurin inhibitor associated adverse events. Doses of tacrolimus may range from 0.05 to 0.2 milligrams per kilogram per day, while doses of cyclosporine can range from 2-10 mg/kg/day with both agents usually being administered in two divided doses 12 hours apart. Therapeutic trough ranges can vary from 5-20 ng/mL for tacrolimus and 100-400 ng/mL for cyclosporine respectively.

Trough levels measurements for tacrolimus or cyclosporine can be performed by a variety of techniques. Cyclosporine can be measured by high performance liquid chromatography (HPLC) which measures primarily parent compound and minimizes the contribution of metabolites. Since this technique is labor-intensive and difficult to standardize, trough levels can also be measured by flourescence polarization immunoassay (FPIA) and enzyme-multiplied-immunoassay techniques (EMIT). Both FPIA and EMIT measure parent compound and metabolites. Therefore, HPLC results for cyclosporine are approximately 20-25% lower FPIA or EMIT.[18] This is important to note when transplant recipients utilize laboratories that use different assays.

Tacrolimus level monitoring is performed via enzyme-linked immunosorbent assay (ELISA) or a microparticle enzyme immunoassay (MEIA).[84] Both methods measure parent and metabolite compound. Tacrolimus trough concentrations are thought to have a good correlation with the area under the curve. However, this may not necessarily be the case for cyclosporine preparations. Interest has also arisen in measuring cyclosporine levels at or near their maximum concentration (C2 level drawn 2-hours after a morning dose).[85,86] This newer approach has not been implemented on a wide-scale for a number of reasons including: (i) C2 measurements vary in range dependent on the organ transplanted, the recipient and the method used, (ii) logistical limitations in obtaining well-timed blood samples, (iii) patient compliance especially in situations where individuals were previously having trough levels measured. Since calcineurin inhibitors are metabolized by CYP3A4 enzymes, drugs that inhibit or induce this enzyme will affect tacrolimus and cyclosporine trough levels and dose adjustments may be necessary. The reader is referred to a recent case presentation and comprehensive review on this topic.[87] Dosing adjustments for cyclosporine and tacrolimus are needed in both hepatic and renal impairment.

### Therapeutic Role

Calcineurin inhibitors remain pivotal in preventing rejection. Although some comparative studies demonstrate similarities in efficacy between tacrolimus and cyclosporine, tacrolimus has emerged as the calcineurin inhibitor of choice in the United States based on lower acute rejection rates[45] and a more favorable adverse event profile.[88] Cyclosporine and tacrolimus concentrations are both affected by pharmacokinetic factors such as ethnicity, hepatitis C status, time after transplantation, patient age, donor liver characteristics, hematocrit and albumin concentrations, diurnal rhythm, food administration, corticosteroid dosage, diarrhea and cytochrome P450 enzymes and P-glycoprotein expression.[89] Due to complex pharmacokinetics and narrow therapeutic indexes, as well as patient specific variations in absorption, distribution and elimination, continual therapeutic drug monitoring is necessary for both tacrolimus and cyclosporine based immunosuppressive regimens.[90] Despite their potent anti-rejection effects, nephrotoxic, metabolic and cardiovascular complications of calcineurin inhibitors may negatively impact long term patient and graft survival. Over recent years, research has focused on calcineurin inhibitor withdrawal or avoidance regimens in liver,[91] kidney[92] pancreas transplant,[93] heart[94] and lung transplant.[95] However, due to concerns of increased immunological injury in many of these studies, these practices have not yet gained widespread acceptance. Our own preferred approach has been to rather employ a calcineurin inhibitor minimizing strategy through the use of (i) induction therapy and (ii) triple therapy maintenance regimens combined with an antiproliferative agent such as MMF and low-dose corticosteroids.

### Generic Cyclosporine Formulations

The patent for cyclosporine expired in 1995 resulting in the development of both modified and nonmodified cyclosporine generic products. The availability of generic cyclosporine formulations for use in organ transplant has been wrought with controversy.[96] Recent retrospective single center and registry analyses of cyclosporine-treated kidney recipients found that patients receiving the generic product had an increased six month rejection incidence and lower one year graft survival rates (Opelz G available at http://www.ctransplant.org).[97] Large prospective studies are necessary to further evaluate this phenomenon.

An FDA requirement for approval of a generic product is that it must demonstrate bioequivalence to the brand product when administered at the same dose under the same circumstances.[98] The FDA defines bioequivalence as the absence significant difference in bioavailability of two drugs under similar testing conditions. Pharmacokinetic parameters evaluated to assess bioavailability include: time to achievement of maximum concentration (Tmax), maximum concentration

(Cmax) and area under the time versus concentration curve (AUC). The FDA requires that the 90% confidence interval of the relative mean Cmax and AUC of the generic drug be within 80-125% of the brand product. Many transplant clinicians have argued that this is too great a difference for a drug like cyclosporine, which is associated with a narrow therapeutic index. Moreover, assuming that one generic has 80% bioequivalence and a comparing generic has 125% bioequivalence, this theoretically means that there could be up to a 45% difference in bioequivalence between the two drugs, so that generics should not be used interchangeably. Patients should be educated to notify the transplant center whenever switches in the dispensed brand or generic occur, so that appropriate blood level monitoring can be performed. An American Society of Transplantation report on generic immunosuppressants concluded that although generics (such as cyclosporine) appear appropriate in low risk patients, insufficient data exists to make recommendations on their use in potentially "at risk populations" such as African American or pediatric recipients.[24] Since generic formulations for both modified and nonmodified cyclosporine preparations exist, it is also important to educate patients to know exactly which formulation they are receiving.

### Investigational Calcineurin Inhibitors ISA 247

A cyclosporine analogue (ISA 247) is also being developed. A recent in vitro study has shown this agent to be a more potent immunosuppressant than cyclosporine.[99] ISA 247 is currently in phase II clinical trials in renal transplantation.[100]

### Modified Release Tacrolimus

A modified release form of tacrolimus (FK506E (MR4) is being developed for a once daily dosing regimen to improve patient compliance. Preliminary trial data in liver and kidney recipients suggest similar drug exposure, safety and efficacy to standard tacrolimus and improved inter-subject variability.[101]

## Antibody Preparations

Antibody preparations have played a role in transplantation since the 1960s. Antibody agents, are typically used as induction therapy or to treat acute rejection. Induction therapy is administered perioperatively and appears to reduce the risk for early rejection.[102] It may delay introduction of calcineurin inhibitors in the postoperative period, with theoretical benefit of mitigating nephrotoxicity, although this is not completely established. Antibody preparations also effectively treat acute rejection, either in the setting where high dose steroid pulse therapy has failed to effect a reversal, or as a primary treatment, particularly where histological injury is more severe. Although antibody induction therapy has been linked to the development of posttransplant lymphoproliferative disorder in some retrospective analyses, the data remains inconclusive at this time.[103] Due to an increased risk of infection, appropriate antibiotic immunoprophylaxis should be instituted at the time of therapy.[104]

## Monoclonal Antibodies

### Muromonab-CD3 (OKT-3)

The monoclonal antibody muromonab-CD3 was approved in the 1980s. Over time it has been used as both an induction agent and for treatment of steroid-resistant rejection episodes.

### Mechanism of Action

Muromonab-CD3, is a murine monoclonal IgG2 antibody that binds to the CD3 receptor complex on the T-cell surface. This initially results in transient activation of T-cells and cytokine release and ultimately blocks T-cell proliferation. Within minutes after administration, CD3 positive T-cells become undetectable in the peripheral circulation. The mechanism underlying this "disappearance" is likely multifactorial,

including CD3 receptor modulation, internalization, binding of the antibody to the T-cell receptor on the cell surface, compliment dependent lysis and opsonization. Muromonab CD3 can hence produce non-immunocompetent T-cells incapable of antigen recognition.[105]

### Pharmacokinetics

The half life of muromonab-CD3 is about 18 hours. After three to five days, lymphocytes re-appear in circulation but are immunologically imcompetent in the absence of CD3 expression. T-cell function usually returns to normal within 48 hours of therapy discontinuation,[106] so that adequate maintenance immunosuppressive levels are required at this time to prevent a rebound acute rejection episode from occurring.

### Adverse Events

The cytokines released after muromonab CD3 binds to T-cell receptor, particulary after the first dose, lead to most of the side effects of this therapy.[107] The cytokines released include interleukin 2,6, tumor necrosis factor alpha and interferon gamma.[108] Cytokine syndrome typically occurs one to three hours after the first dose and may last up to 12-16 hours. Premedications with acetaminophen, diphenhydramine and corticosteroids may attenuate the first-dose response. Fever,chills, headache, flu-like symptoms and hypotension are common and aseptic meningitis and arthralgias are well decribed. Nausea and vomiting may also occur. The serum creatinine usually increases after the initial 2 to 3 doses of antibody, then declines. Clear chest X-rays within 24 hours predose as well as diuresis to limit weight gain to fewer than 3% over seven days prior to first dose administration are recommended. This helps to prevent acute respiratory distress and pulmonary edema that can result from release of tumor necrosis factor-alpha and sequestration of neutrophils in the lungs. Fortunately, reactions lessen with repeated doses. Muromonab-CD3 can also lead to formation of human anti-mouse antibodies (HAMA) after a treatment course. These neutralizing antibodies may diminish efficacy of subsequent courses of therapy and should be checked before retreatment is instituted.[109] Concurrent administration of calcineurin inhibitors with muromonab-CD3 may prevent formation of these neutralizing antibodies.

### Dosing and Monitoring

Muromonab CD-3 is usually administered as an intravenous push of 5 milligrams per day. Daily doses of 2.5 mg have been used for induction, while doses up to 10 mg per day have been used for rejection treatment. The duration of therapy varies from 3-14 days according to the indication for therapy and patient's responses. Although some clinicians use muromonab lymphocyte sub-set analyses or serum levels of muromonab-CD3 to guide dosing,[110] there are no prospective trials to base this practice on.

### Therapeutic Role

Due to development of newer, efficacious, less toxic antibody preparations, muromonab-CD3 use is mainly limited to treatment of corticosteroid resistant rejection episodes.

## IL-2 Receptor (Anti-CD25) Daclizumab and Basiliximab

Daclizumab and Basiliximab were developed as induction agents to prevent acute rejection.[111-114] Daclizumab is a humanized monoclonal antibody with only the complimentary region derived from murine complex. In contrast, basiliximab is a chimeric monoclonal antibody with the variable region of the antibody containing murine components.

### Mechanism of Action

Activated T-cells express CD25, the alpha chain of the interleukin-2 receptor. Both basiliximab and daclizumab selectively saturate this receptor, thereby blocking IL-2 mediated T-cell proliferation.

## Pharmacokinetics

Due to differences in composition (chimeric versus humanized), the half life of the chimeric and humanized antibodies vary. The half life of basiliximab is approximately 7 days while that of daclizumab is 20 days.

## Adverse Events

Since most murine portions of these antibodies have been replaced with human amino acid sequences, adverse events are uncommon. Unlike muromonab-CD3, neither agent is associated with cytokine release and sensitization is rare and not considered to be of clinical significance. Repeated administration of either basiliximab or daclizumab are generally well tolerated. However, a few cases of anaphylaxis with repeated courses have been described.[115]

## Dosing and Monitoring

Daclizumab is usually administered as a peripheral intravenous infusion over fifteen minutes at a dose of 1 mg/kg every two weeks posttransplant for a total of five doses. Basiliximab is given via peripheral intravenous infusion over twenty minutes at a dose of 20 mg on post-operative day zero and four only. Although flow cytometric measurements of CD25 cells and urinary excretion of soluble interleukin-2 receptor alpha have been performed in clinical trials,[116-118] these practices have not been widely adapted in the clinical setting.

## Therapeutic Role

IL-2 receptor blockers are used for induction after kidney, heart and liver transplantation where they reduce acute rejection.[111,113,119,120] However, a similar benefit has not been demonstrated in lung transplant patients.[121-123] Moreover, due to their mechanism of action, IL-2 receptor blockers are ineffective to treat rejection. They have little effect on preventing ischemic reperfusion injury or delayed graft function after kidney transplant. Abbreviated two-dose daclizumab regimens have been evaluated in both kidney and liver transplant recipients.[116,124-126] Outcomes when compared to standard dosing of basiliximab have been conflicting and require further study. Interleukin 2 receptor antagonists have been evaluated in steroid-withdrawal regimens in both kidney and liver transplant recipients with and without hepatitis C.[125,127-129] Further research from large multicenter prospective randomized studies is necessary to elucidate the role of interleukin 2 receptor antagonists in these settings.

## Polyclonal Antibodies (Equine and Rabbit Antithymocyte Globulin)

Polyclonal antibodies were the first antibody preparation used in clinical transplantion. Early preparations were plagued with significant batch to batch variations in potency. Two commercially available preparations currently exist in the United States in transplantation; equine antithymocyte globulin and rabbit antithymocyte globulin. Although both agents were developed to treat steroid-resistant and severe acute rejection, they are also used as induction agents, mainly in kidney transplantation. Rabbit antithymocyte globulin has mostly replaced equine antithymocyte globulin, as it has been shown to be a more efficacious antibody preparation in the treatment of kidney rejection.[130]

## Mechanism of Action

Polyclonal antibodies preparations are purified immunoglobulins isolated from serum of animals (rabbits or horses) immunized with human thymocytes. Lymphocyte depletion occurs by complement mediated lysis, removal by the reticuloendothelial system and antibody-mediated cell-dependent cytotoxicity.[131] Lymphocytes that return to circulation after cessation of therapy have blunted proliferative responses.

## Pharmacokinetics

The half life of rabbit antithymocyte globulin is two to three days while the half life of equine antithymocyte globulin is 1.5 to twelve days.

## Adverse Events

Since polyclonal antibodies preparations contain foreign proteins, most of their adverse events are related to their structure. Premedications with acetaminophen, diphenhydramine and corticosteroids are recommended to be given before each dose to help ameliorate these effects. Fever, chills and head-aches are common adverse events with the first few doses. Leukopenia and thrombocytopenia also occur due to the nonspecificity of these antibodies. Reductions in doses are recommended in these settings. Serum sickness may occur, particularly in recipients with previous exposure to these antibodies either through prior treatment or due to environmental exposure to equine or rabbit antibodies.[132] Since polyclonal agents increase the cumulative immunosuppressive load in recipients, anti-infective prophylaxis for viral, fungal and pneumocystis pneumonia infections is necessary.[104]

## Dosing and Monitoring

Rabbit antithymocyte globulin is dosed at 1.5 milligram per kilogram per dose, usually administered as a 4 to 6 hour infusion via a central line. Duration of therapy is dependent on its indication. Induction therapy usually comprises three to five doses, with the first dose frequently being administered intra-operatively; while treatment of steroid-resistant rejection usually consists of seven to ten doses. Recent data has shown that rabbit antithymocyte globulin may also be administered via a peripheral line, with 1,000 units of heparin and 20 milligrams of hydrocortisone added to the infusate to reduce the incidence of thrombosis and local tissue reaction.[133,134] This route of administration can allow dosing to continue in the ambulatory setting.

Equine antithymocyte globulin is dosed at 10-20 milligrams per kilogram per dose. An intradermal sensitivity skin test is recommended prior to administration of equine antithymocyte globulin, but not rabbit antithymocyte globulin. Similar to rabbit antithymocyte globulin, duration of therapy is dependent on its indication. The dosing regimens are generally similar to those for rabbit antithymocye globulin.

Unlike muromonab-CD3, polyclonal antibodies are seldom associated with formation of neutralizing antibodies capable of inhibiting drug effectiveness with repeated administration, due to an insufficient anti-idiotypic response against the multiple idiotypic specificities of the polyclonal anti-T-cell preparations.[135] Monitoring can be performed through flow cytometric measurement of CD2 or CD3 counts. This technique was shown to minimize the amount of rabbit antithymocyte a kidney transplant recipient receives as induction therapy, while still effectively preventing rejection.[136]

## Therapeutic Role

Rabbit antithymocyte globulin has mostly replaced equine antithymocyte globulin.[130,137] Rabbit antithymocyte globulin may also decrease the risk of delayed graft function in kidney recipients when initially administered intra-operatively as opposed to post-operatively[138,139] Use of rabbit antithymocyte globulin has also extended to heart,[140] lung,[141] liver[142] and pancreas transplant[143] as induction agents and as treatment for rejection episodes.

Table 1 summarizes the available immunosuppressive agents that are in clinical use, their mechanisms of action, adverse events and current dosing guidelines.

# General Principles of Immunosuppression in Transplantation

The risk of rejection is greatest during the first six posttransplant months. The current immunotherapeutic approach of most transplant clinicians is to expose recipients to the greatest immunosuppressive load during the peri-transplant period through the use of induction agents. Although the rejection risk diminishes over time, it never completely disappears; consequently, patients generally require some maintenance immunotherapy for the life of the functioning organ. Goals of immunosuppression after organ transplantation are; (i) prevent organ rejection

**Table 1.** *Pharmacotherapeutic options in solid organ transplantation*

| Drug | Mechanism of Action | Adverse Effects | Dosing Guidelines |
|---|---|---|---|
| Corticosteroids | Inhibition of transcription of genes encoding proinflammatory cytokines involved in rejection. Impairs monocyte-macrophage function and causes lymphocyte apoptosis in high doses | Infections, glucose intolerance, hypertension, impaired wound healing, skin fragility, growth retardation, osteoporosis, obesity, suppression of hypothalamic-pituitary-adrenal axis, cataracts, glaucoma, psychosis | Variable rate of taper with aim to minimize dose as early as possible. Biannual bone densitometry and antiresorptive therapy and supplementation with calcium and vitamin D recommended if maintenance dose is greater than 5 mg/day of prednisone |

**CALCINEURIN INHIBITORS (CNIs)**

| Drug | Mechanism of Action | Adverse Effects | Dosing Guidelines |
|---|---|---|---|
| Cyclosporine | Cyclosporine-cyclophilin complex prevents T-cell activation by inhibition of calcineurin. Inhibition causes inhibition of lymphokine gene transcription, mainly interleukin-2 and interferon-gamma which are necessary for T-cell activation | Infections, increased neoplasia, nephrotoxicity, hypertension, hyperlipidemia, glucose intolerance, hyperkalemia, hyperuricemia, gout, hypomagnesemia, CNS toxicities; tremors, paresthesias, seizures, encephalopathy, thrombotic microangiopathy, hirsuitism, gingival hyperlasia, hepatotoxicity | Used in combination therapy with steroids, antimetabolites and/or antiproliferative agents. Rarely as monotherapy Dosed 12-hourly to achieve a variable target therapeutic 12-hr trough level of 100 to 400 ng/mL or C-2 level of 800 – 1500 ng/mL depending on type of allograft and time posttransplant. Levels can be checked 2-3 days after dosage adjustment |
| Tacrolimus | Tacrolimus-FKBP complex inhibits calcineurin, leading to impaired T-cell activation | Infections, increased neoplasia, hyperglycemia, neurotoxicity including tremor and encephalopathy, akinetic mutism, paraesthesias; hepatotoxicity, nephrotoxicity, hyperkalemia, hypomagnesemia, alopecia, hyperlipidemia, diarrhea | Usually given in combination with antimetabolite/steroid or alone, rarely as monotherapy.Target level depend on the use of concomitantly administered drugs and organs transplanted. By immunoassay targe 12-hour trough levels of 5-20 ng/mL. After first 3-6 months posttransplant levels can be dropped to lower ranges, levels can be checked 2-3 days after dosage adjustment |

**MAMMALIAN TARGET OF RAPAMYCIN (m-TOR) INHIBITORS**

| Drug | Mechanism of Action | Adverse Effects | Dosing Guidelines |
|---|---|---|---|
| Sirolimus/ Everolimus | Binds FKBP and drug-FKBPcomplex inhibits m-TOR signal transduction with arrest of cell growth in G1-S cell cycle phase and blocks IL-2 mediated cell proliferation, inhibits EGF and FGF | Infections, anemia, leucopenia, hyperlipidemia, impaired wound healing serous effusions, lymphoceles, thrombocytopenia, diarrhea, interstitial pneumonitis, edema, mouth sores | Generally used with CNIs, dosed once daily to target 24-hr trough level on 4-8 ng/mL. Higher doses may be required when used in calcineurin inhibitor-free regimen. Following dosage adjustment should wait 5-7 days before rechecking level. Most patients require concomitant statin therapy for hyperlipidemia |

**ANTIMETABOLITES**

| Drug | Mechanism of Action | Adverse Effects | Dosing Guidelines |
|---|---|---|---|
| Azathioprine | Competitive inhibitor of purine synthesis | Leukopenia, thrombocytopenia, megaloblastic anemia, hepatotoxicity, pancreatitis, increased malignancies especially of skin that is related to drug-induced sensitivity of DNA to UV radiation | Usually used in combination with CNIs and corticosteroids at a dose of 1-5 mg/kg depending on time. Reduce/discontinue if WBC <4 000, chronic anemia, or in presence of severe infection or new malignancy. Significant dose reduction required for patients on allopurinol or other xanthine oxidase inhibitor |
| Mycophenolic Acid | Noncompetitive reversible inhibitors of inosine monophosphate dehydrogenase; inhibits purine synthesis | Gastrointestinal upsets; nausea, vomiting, diarrhea, dyspepsia; leukopenia, anemia, thrombocytopenia | Usually used in combination with CNIs and corticosteroids; 2 g/day in divided doses with cyclosporine 1-1.5 g/day with tacrolimus. Reduce/discontinue dose if WBC <4 000, for chronic anemia, persistent GI disturbance, or in presence of severe infection or in new malignancy. |

*continued on next page*

*Table 1. Continued*

| Drug | Mechanism of Action | Adverse Effects | Dosing Guidelines |
|---|---|---|---|
| **ANTIBODY PREPARATIONS** | | | |
| Muromonab-CD3 | Specifically binds to CD3 receptor complex on T-cells and modulates function. It causes cytokine release, lympholysis and blockage of proliferation | Cytokine release syndrome: fever, chills, pulmonary edema, hypotension, myalgia, headache, flu-like symptoms, aseptic meningitis, arthralgias, thrombosis, increased infections, increased neoplasia | 2.5-10 mg/day for 3-14 days depending on indication. Premedication with acetaminophen, diphenhydramine and corticosteroids is advised with first dose to minimize adverse events and cytokine release. Diuretics often required to prevent pulmonary edema during OKT3 therapy. |
| Anti-CD25 antibodies Basiliximab Dacluzimab | Specifically target CD25 on activated lyrphocytes and block T-cell proliferation | Adverse events rare. Infections may occur, hypersensitivity may occur and anaphylaxis has been reported with repeated doses. | Basiliximab is given as two 20 mg doses, first on day 0 and (within 2 h prior to transplant) the second on day 4 post-transplant. The recommended dosage of daclizumab dose is 1 mg/kg IV infusion within 24 h before transplantation (day 0), then every 14 days for four doses (protocol may vary between centers) |
| Polyclonal Antithy-mocyte globulin (equine and rabbit) | Lymphocyte depletion through numerous T ce;l markers. It causes cytokine release and blockage of lymphocyte proliferation. | Leukopenia, thrombocytopenia, cytokine release syndrome, fever, nausea, diarrhea, chills, pulmonary edema, hypotension, myalgia, headache, flu-like symptoms, arthralgias, increased infections, increased neoplasia | Rabbit antithymocyte globulin is dosed at 1.5mg/kg/day and equine antithymocyte globulin is dosed at 10-20mg/kg/day. Both are administered for 3-14 days depending on indication. Premedications with acetaminophen, diphenydramine and corticosteroids is advised with first dose to minimize adverse events and cytokine release. An intradermal sensitivity skin test is recommended prior to administration of equine antithymocyte globulin only. |

by promoting stable graft acceptance or tolerance; (ii) to minimize non-immune drug toxicities including kidney dysfunction and cardiovascular disease and; (iii) to avoid the cumulative long term consequences of immunodeficiency including infection and malignancies. To this end, ongoing immunosuppressive dosage refinement is required to safely optimize graft function while mitigating adverse effects. The most commonly used treatment regimen involves quadruple therapy, where antibody induction up front is followed by a combination of three other agents including steroids, calcineurin inhibitors and antiproliferative therapies. Maintenance therapy typically comprises of a calcineurin inhibitor with two other adjunctive agents, usually an antimetabolite such as mycophenolic acid therapies or azathioprine and/or steroids. Occasionally, TOR inhibitors replace the calcineurin inhibitor or the antimetabolite. In the United States, tacrolimus has emerged as the caclineurin inhibitor of choice and mycophenolic acid therapies have largely replaced azathioprine in clinical practice.

Selection and dosing of immunosuppressive agents are dynamic because of the tremendous interpatient variability. This variability includes host factors such as age, gender, ethnicity, prior history of autoimmune disease or allograft rejection, donor factors such as degree of HLA mismatch, cold ischemia time and other general factors like the number and type of allografts present in a particular recipient. Table 2 summarizes some of the known factors that predict alloreactivity and in essence determine the amount of immunosuppression that will be required. Tailoirng an immunosuppressive regimen to a patient requires precise balance reducing rates of graft loss coupled with acceptable long term risks and safety profiles of the agents chosen. Part of the toxicity of immunosuppressive agents is linked to their role in inducing immunodeficiency and defective immunosurveillance in the host. These effects result in increased infections and neoplasia in recipients who are generally at low risk of rejection. Conversely, inadequately immunosuppressed patients are likely to develop allograft rejection, while infectious and cancer-related complications would be unexpected.

The use of lymphocyte depleting antibody induction diminishes the initial immune response to alloantigen and the attendant risk of acute rejection. While daclizumab and basiliximab are not lymphocyte depleting, by virtue of their binding to the IL-2 receptor, they are capable of successfully interrupting the signaling processes in the initial adaptive immune response to the allograft. The effects of some antibodies may be durable, lasting up to three to twelve months after transplantation. For this reason, the exposure to induction therapy has the capacity to greatly reduce the amount of maintenance therapy required for long term immunosuppression. However, the injudicious use of lymphocyte depleting antibody induction therapy culminates in "overimmunosuppression" and predisposes patients to higher rates of opportunistic infections and malignancies than observed in non-induced patients.

### Treatment of Acute Rejection

Another important role of immunosuppressive agents is treatment of acute rejection. Fortunately, because of advancements in immunotherapeutics, acute rejection rates have significantly declined over the past decade. The treatment of rejection requires reinforcement and intensification of the immunosuppressive regimen, usually in the form of higher doses of intravenous corticosteroids, antibody lymphocyte depleting drugs (muromonab-CD3, equine or rabbit antithymocyte globulin), increased doses of calcineurin inhibitors and in some instances, addition of other adjunctive agents. As a consequence of enhanced immunosuppression dosing, adverse drug effects and the cumulative immunosuppressive load increases. This greater immunosuppressive burden can lead to a higher incidence of opportunistic infections and malignancies when compared to recipients who have not experienced rejection.

*Table 2. Predictors of high immune reactivity and rejection*

| Patient Factors | Donor/Allograft Factors |
|---|---|
| Age (young>older) | Cadaveric vs LRT |
| Gender (male>female) | (cadaver organs reject more than living donor organs) |
| Race (AA>whites>Asians) | |
| No of previous transplants | Ischemia time |
| h/o autoimmune disease (e.g., SLE) | (cold ischemia upregulates HLA, the longer the time more the rejection) |
| Previous h/o rejection | |
| HLA mismatch | Delayed graft function (DGF) |
| High PRA | (increased risk of rejection with DGF) |
| Chronic illness, malnutrition- less rejection | Type of allograft |
| | (skin>bowels>islets>lung>pancreas>SPK>kidney>heart>liver) |
| Poor compliance | Multiple Organ transplants |
| | (Livers and hearts protect 2nd organs) |

## Alternative Dosing Strategies and Immune Reactivity Monitoring

The multitude of adverse effects associated with antirejection therapies have offset some of the benefit related to their immunosuppressive efficacy. Notable side effects of the more commonly used maintenance therapies, include to a varying degree, hypertension, hyperlipidemia, new onset diabetes mellitus, nephrotoxicity and anemia, all of which contribute to significant cardiovascular morbidity and mortality, the scourge of successful solid organ transplantation. In prior eras, when alternative treatments were limited and when rejection rates were high, these adverse effects were accepted as the immutable cost of transplant immunosuppression.[144] However, with our current, expanded and improved immunosuppressive armamentarium and lower rates of immunological injury, posttransplant regimens aimed at decreasing long term complications can now be explored. This has fuelled the development of sparing regimens, with a particular emphasis on steroids and calcineurin inhibitors. Various sparing regimens have been tested, including strategies on avoidance, typically defined by complete avoidance of a therapy from time of transplant (i.e., never administered), withdrawal (usually defined by initial use of a therapy with subsequent elimination) and minimization (defined by dose reduction of a therapy to a minimum dose/level without withdrawal). Unfortunately, these terms are often used interchangeably in the literature, limiting meaningful comparison between strategies. Study endpoints typically reported with sparing regimens include clinically proven acute rejection rates, protocol biopsy proven sub-acute rejection and shorter term patient and graft survival.

### Steroid Sparing Regimens

While steroid withdrawal is widely practiced in liver[145] and heart recipients,[146] it is far more controversial in kidney transplantation. Most of the initial steroid withdrawal studies in kidney recipients revealed an increased risk of both acute rejection and allograft loss following steroid elimination during any point of the patient's posttransplant course.[147-149] Studies with more contemporary immunosuppression have consistently shown enhanced outcomes after steroid withdrawal when compared to earlier investigations.[150,151] However, most of the more recent favorable analyses have been limited by either retrospective or uncontrolled study design, use of lower risk patients and/or short follow-up. At this time, most of the ongoing randomized controlled trials have suggested a trend towards a higher rejection rate in patients withdrawn from steroids compared to those maintained on this therapy;[152] moreover, a clear metabolic benefit (in terms of weight gain, new onset diabetes) is yet to be convincingly demonstrated in the groups of patients from whom steroids were eliminated. It remains to be established whether early steroid withdrawal confers any benefit over early steroid minimization (e.g., prednisone tapered to 5 mg daily by 3 to 4 weeks posttransplant).

Until such time, we recommend an early steroid minimization protocol rather than withdrawal. Where transplant clinicians are committed to withdrawal, we believe that caution should be exercised. Optimal candidates that could be considered for corticosteroid sparing therapy include those at low risk of rejection, as well as patients at greatest risk for corticosteroid side effects (children, postmenopausal women, patients with a history of psychological disorder).

### Calcineurin Inhibitor Sparing Regimens

Several recent studies have examined calcineurin inhibitor sparing regimens.[66 67,153] To date, most calcineurin inhibitor withdrawal or avoidance studies have been with cyclosporine treated kidney recipients, under immunosuppressive coverage with either mycophenolate mofetil or mycophenolate mofetil and sirolimus.[92] Cyclosporine withdrawal after the addition of high dose sirolimus to the regimen at three months postkidney transplant has also been reported.[154] Because sirolimus potentiates cyclosporine, it is not surprising that serum creatinine was better and blood pressure was lower in patients withdrawn from cyclosporine compared to the arms that received sirolimus and cyclosporine.

Acute rejection rates have also been slightly increased slightly increased in cohorts withdrawn from cyclosporine.[84] While it is certainly likely that sirolimus is less nephrotoxic as a stand-alone agent than cyclosporine,[64] it remains to be established whether this benefit is still present compared to chronic low dose tacrolimus. Overall, before calcineurin inhibitor elimination strategies can be embraced, randomized controlled trials are needed to shed more light on the safety and efficacy of these regimens.

Among nonkidney transplant recipients, there has been a great desire to eliminate calcineurin inhibitors, with a particular emphasis on sparing chronic nephrotoxicity. This is being increasingly investigated in lung, heart[155] and liver transplant recipients[156] through the addition of sirolimus at some point beyond three months posttransplant. Most published reports have taken the form of retrospective case reports and/or case series. While some of these experiences show promise, larger prospective, preferably randomized controlled trials are needed to shed more light on the optimal management of these patients.

## Monitoring of Immune Reactivity

The greatest concern with immunosuppression minimization is the risk of increasing rejection rates. Efforts have focused on the development of techniques predict rejection risk, monitor surrogates for acute rejection and measuring cumulative immune activity. These methods include assays that measure immune activity (such as ELISPOT and Cylex assay), donor specific antibodies, genomics, proteomics and metabolomics.[157-160] Measurements of cytokines and their components in blood, urine or the allograft itself are also being investigated.[161,162]

To date, none of these practices have been associated with sufficient sensitivity or specificity to be used alone. However, it is hoped that future methods and technologies may lead to the development of rapid, inexpensive and noninvasive approaches to tailor immunosuppression to individual recipient needs.

## Emerging and Adjunctive Therapies

In current clinical practice, many transplant recipients are presensitized to transplantation antigens prior to engraftment. Through desensitization of these patients, a deliberate attempt is made to overcome recognized and well-established immune barriers by removal of preformed antibodies. Adjunctive therapies such as plasmapheresis, intravenous immunoglobulin, antilymphocyte antibody therapy together with most of the contemporary immunosuppressive agents have been used in this setting. Rituximab (Rituxan) is a monoclonal antibody against the B-cell surface antigen CD20, initially approved as an oncology therapy for treating lymphoma. In kidney transplantation, it has been used along with intravenous immunoglobulin as an anti-B-cell therapy for desensitization of high risk transplant recipients and treatment of acute humoral rejection.[163] Alemtuzumab is a humanized anti-CD52 monoclonal antibody that was originally approved for the treatment of relapsed or refractory chronic lymphocytic leukemia (CLL). The utility of Alemtuzumab for desensitization, induction and treatment of rejection has been explored.[164-166] Larger studies with longer follow-up to assess long term outcomes are needed to draw a conclusion on the role of these agents in modern transplantation.

Current immunosuppressive drug development is focused on further attenuating drug specific adverse effects, while promoting long term patient and graft survival. Two agents under investigation are LEA 29Y and CP 690,500. Belatacept is a second-generation CTLA4Ig with increased affinity to CD80 and CD86 that has recently completed phase 2 trials in kidney transplant.[167] It is a biologic fusion protein that prevents signal 2 of the alloimmune response. Because it blocks costimulation, it has been proposed for use as chronic induction therapy or maintenance therapy in allograft recipients. This could potentially become a new paradigm in immunosuppression therapy that eliminates the need for either calcineurin inhibitors or steroids that will ultimately minimize drug toxicities and prolong graft survival.[168]

CP 690, 500 is a JAK3 inhibitor currently in phase 2 trials .[169] JAK3 is highly expressed in lymphoid cells and binds specifically to the gammac receptor which is shared by several interleukin receptors including IL-2, IL-4, IL-7, IL-15 and IL-21. JAK3 inhibitors disrupt signals from the gammac chain and consequently block proliferation and differentiation of lymphoctes.

As our understanding of immune mechanisms underlying stable graft acceptance and regulation of the immune response increases, a potential role for customized cellular therapies for transplant recipients may become a future reality. In the earlier era of transplantation modest success was achieved in promoting graft acceptance with donor specific transfusions although the precise mechanisms remained unclear and the increasing transmission of viral infections by blood products led to the abandonment of this practice. Currently some investigators are exploring the role of allogeneic donor bone marrow transplants at the time of transplant in tolerizing the recipient to the graft.[170]

## Conclusion

In the past, the traditional endpoints for evaluating immunosuppression therapies were focused almost entirely on six and twelve month posttransplant rejection rates. However, commonly observed remarkably low rejection rates together with ever-improving short term outcomes has now forced the transplant community to redefine traditional efficacy metrics, to focus on extending graft and patient survival and to mitigate immunosuppression related complications. With the availability of newer more potent agents, efforts can and

should be made to reduce toxicities associated with long term use of immunosuppressants, such as calcineurin inhibitor associated nephrotoxicity. Emerging strategies include sparing regimens and regimens aimed at achieving tolerance.

It is clear that current and future immunosuppressive therapies must be individually tailored to both the recipient's short and long term needs. Moreover, although the majority of transplant recipients require life-long immunosuppression, instead of simply considering it as an insurmountable obstacle of posttransplant care, it should periodically be re-assessed and diversified to accommodate the patient's ever-changing posttransplant needs. Finally, immunosuppressive regimens should be individualized to account for the synergistic effects of not only short term efficacy and toxicity, but also donor and recipient immunologic factors, recipient pre-existing comorbidities, long term efficacy and toxicity, recipient financial burden, compliance and quality of life.

## References

1. Merrill JP, Murray JE, Harrison JH et al. Successful homotransplantation of the human kidney between identical twins. JAMA 1956; 160:277-82.
2. Sayegh MH, Carpenter CB. Transplantation 50 years later—progress, challenges and promises. N Engl J Med 2004; 351(26):2761-2766.
3. Kuss R, Legrain M, Mathe G et al. Homologous human kidney transplantation. Experience with six patients. Postgrad Med J 1962; 38:528-31.
4. Murray JE, Merrill JP, Harrison JH et al. Prolonged survival of human-kidney homografts by immunosuppressive drug therapy. New Engl J Med 1963; 268:1315-23.
5. Kashiwagi N, Brantigan CO, Brettschneider L et al. Clinical reactions and serologic changes after the administration of heterologous antilymphocyte globulin to human recipients of renal homografts. Ann Intern Med 1968; 68(2):275-86.
6. Calne RY, Rolles K, White DJ et al. Cyclosporin A initially as the only immunosuppressant in 34 recipients of cadaveric organs: 32 kidneys, 2 pancreases and 2 livers. Lancet 1979; 2(8151):1033-36.
7. Kahan BD. Cyclosporine. N Engl J Med 1989; 321(25):1725-38.
8. Cosimi AB, Burton RC, Colvin RB et al. Treatment of acute renal allograft rejection with OKT3 monoclonal antibody. Transplantation 1981; 32(6):535-39.
9. Meier-Kriesche HU, Li S, Gruessner RW et al. Immunosuppression: evolution in practice and trends, 1994-2004. Am J Transplant 2006; 6(5 Pt 2):1111-31.
10. Organ Procurement and Transplant Network. Prevalence of people living with a functioning transplant at end of year 1995 to 2004. http:\\www.optn.on
11. Meier-Kriesche HU, Schold JD, Srinivas TR et al. Lack of improvement in renal allograft survival despite a marked decrease in acute rejection rates over the most recent era. Am J Transplant 2004; 4(3):378-83.
12. Organ Procurement and Transplant Network. Prevalence of people living with a functioning transplant at end of year 1995 to 2004. In: OPTN/SRTR 2005 annual report: summary tables, transplant data 1995 to 2004. Table 1.14. (Accessed November 9, 2006 at http://www.ustransplant.org).
13. Ojo AO, Held PJ, Port FK et al. Chronic renal failure after transplantation of a nonrenal organ. N Engl J Med 2003; 349(10):931-40.
14. Halloran PF. Immunosuppressive drugs for kidney transplantation. N Engl J Med 2004; 351(26):2715-29.
15. Buckbinder L, Robinson RP. The glucocorticoid receptor: molecular mechanism and new therapeutic opportunities. Curr Drug Targets Inflamm Allergy 2002; 1(2):127-36.
16. Hricik DE, Almawi WY, Strom TB. Trends in the use of glucocorticoids in renal transplantation. Transplantation 1994; 57(7):979-89.
17. Taylor AL, Watson CJ, Bradley JA. Immunosuppressive agents in solid organ transplantation: Mechanisms of action and therapeutic efficacy. Crit Rev Oncol Hematol 2005; 56(1):23-46.
18. Zand MS. Immunosuppression and immune monitoring after renal transplantation. Semin Dial 2005; 18(6):511-19.
19. Frey BM, Frey FJ. Clinical pharmacokinetics of prednisone and prednisolone. Clin Pharmacokinet 1990; 19(2):126-46.
20. Tiede I, Fritz G, Strand S et al. CD28-dependent Rac1 activation is the molecular target of azathioprine in primary human CD4+ T-lymphocytes. J Clin Invest 2003; 111(8):1133-45.
21. Maltzman JS, Koretzky GA. Azathioprine: old drug, new actions. J Clin Invest 2003; 111(8):1122-24.

22. Warrington JS, Shaw LM. Pharmacogenetic differences and drug-drug interactions in immunosuppressive therapy. Expert Opin Drug Metab Toxicol 2005; 1(3):487-503.

23. Parrish JA. Immunosuppression, skin cancer and ultraviolet A radiation. N Engl J Med 2005; 353(25):2712-13.

24. Alloway RR, Isaacs R, Lake K et al. Report of the American Society of Transplantation conference on immunosuppressive drugs and the use of generic immunosuppressants. Am J Transplant 2003; 3(10):1211-15.

25. Haroldson JA, Somerville KT, Carlson S et al. A retrospective assessment of safety, efficacy and pharmacoeconomics of generic azathioprine in heart-transplant recipients. J Heart Lung Transplant 2001; 20(3):372-74.

26. Ciancio G, Miller J, Gonwa TA. Review of major clinical trials with mycophenolate mofetil in renal transplantation. Transplantation 2005; 80(2 Suppl):S191-200.

27. Zucker K, Rosen A, Tsaroucha A et al. Unexpected augmentation of mycophenolic acid pharmacokinetics in renal transplant patients receiving tacrolimus and mycophenolate mofetil in combination therapy and analogous in vitro findings. Transpl Immunol 1997; 5(3):225-32.

28. Shaw LM, Nawrocki A, Korecka M et al. Using established immunosuppressant therapy effectively: lessons from the measurement of mycophenolic acid plasma concentrations. Ther Drug Monit 2004; 26(4):347-51.

29. Tedesco-Silva H, Bastien MC, Choi L et al. Mycophenolic acid metabolite profile in renal transplant patients receiving enteric-coated mycophenolate sodium or mycophenolate mofetil. Transplant Proc 2005; 37(2):852-55.

30. Budde K, Glander P, Diekmann F et al. Review of the immunosuppressant enteric-coated mycophenolate sodium. Expert Opin Pharmacother 2004; 5(6):1333-45.

31. Behrend M. Adverse gastrointestinal effects of mycophenolate mofetil: aetiology, incidence and management. Drug Saf 2001; 24(9):645-63.

32. Sollinger HW. Mycophenolates in transplantation. Clin Transplant 2004; 18(5):485-92.

33. Pescovitz MD, Conti D, Dunn J et al. Intravenous mycophenolate mofetil: safety, tolerability and pharmacokinetics. Clin Transplant 2000; 14(3):179-88.

34. Bunchman T, Navarro M, Broyer M et al. The use of mycophenolate mofetil suspension in pediatric renal allograft recipients. Pediatr Nephrol 2001; 16(12):978-84.

35. Armstrong VW, Tenderich G, Shipkova M et al. Pharmacokinetics and bioavailability of mycophenolic acid after intravenous administration and oral administration of mycophenolate mofetil to heart transplant recipients. Ther Drug Monit 2005; 27(3):315-21.

36. Arns W, Breuer S, Choudhury S et al. Enteric-coated mycophenolate sodium delivers bioequivalent MPA exposure compared with mycophenolate mofetil. Clin Transplant 2004(Online early):1-8.

37. Hubner GI, Eismann R, Sziegoleit W. Drug interaction between mycophenolate mofetil and tacrolimus detectable within therapeutic mycophenolic acid monitoring in renal transplant patients. Ther Drug Monit 1999; 21(5):536-39.

38. van Gelder T, Hilbrands LB, Vanrenterghem Y et al. A randomized double-blind, multicenter plasma concentration controlled study of the safety and efficacy of oral mycophenolate mofetil for the prevention of acute rejection after kidney transplantation. Transplantation 1999; 68(2):261-66.

39. van Gelder T, Le Meur Y, Shaw LM et al. Therapeutic drug monitoring of mycophenolate mofetil in transplantation. Ther Drug Monit 2006; 28(2):145-54.

40. Shaw LM, Korecka M, Venkataramanan R et al. Mycophenolic acid pharmacodynamics and pharmacokinetics provide a basis for rational monitoring strategies. Am J Transplant 2003; 3(5):534-42.

41. Eisen HJ, Kobashigawa J, Keogh A et al. Three-year results of a randomized, double-blind, controlled trial of mycophenolate mofetil versus azathioprine in cardiac transplant recipients. J Heart Lung Transplant 2005; 24(5):517-25.

42. A blinded, randomized clinical trial of mycophenolate mofetil for the prevention of acute rejection in cadaveric renal transplantation. The Tricontinental Mycophenolate Mofetil Renal Transplantation Study Group. Transplantation 1996; 61(7):1029-37.

43. Placebo-controlled study of mycophenolate mofetil combined with cyclosporin and corticosteroids for prevention of acute rejection. European Mycophenolate Mofetil Cooperative Study Group. Lancet 1995; 345(8961):1321-25.

44. Wiesner R, Rabkin J, Klintmalm G et al. A randomized double-blind comparative study of mycophenolate mofetil and azathioprine in combination with cyclosporine and corticosteroids in primary liver transplant recipients. Liver Transpl 2001; 7(5):442-50.

45. Gonwa T, Johnson C, Ahsan N et al. Randomized trial of tacrolimus + mycophenolate mofetil or azathioprine versus cyclosporine + mycophenolate mofetil after cadaveric kidney transplantation: results at three years. Transplantation 2003; 75(12):2048-53.

46. Patel JK, Kobashigawa JA. Everolimus: an immunosuppressive agent in transplantation. Expert Opin Pharmacother 2006; 7(10):1347-55.

47. Budde K, Curtis J, Knoll G et al. Enteric-coated mycophenolate sodium can be safely administered in maintenance renal transplant patients: results of a 1-year study. Am J Transplant 2004; 4(2):237-43.

48. Salvadori M, Holzer H, de Mattos A et al. Enteric-coated mycophenolate sodium is therapeutically equivalent to mycophenolate mofetil in de novo renal transplant patients. Am J Transplant 2004; 4(2):231-36.

49. Pawinski T, Hale M, Korecka M et al. Limited sampling strategy for the estimation of mycophenolic acid area under the curve in adult renal transplant patients treated with concomitant tacrolimus. Clin Chem 2002; 48(9):1497-504.

50. Kahan BD. Efficacy of sirolimus compared with azathioprine for reduction of acute renal allograft rejection: a randomised multicentre study. The Rapamune US Study Group. Lancet 2000; 356(9225):194-202.

51. MacDonald AS. A worldwide, phase III, randomized, controlled, safety and efficacy study of a sirolimus/cyclosporine regimen for prevention of acute rejection in recipients of primary mismatched renal allografts. Transplantation 2001; 71(2):271-80.

52. Neumayer HH. Introducing everolimus (Certican) in organ transplantation: an overview of preclinical and early clinical developments. Transplantation 2005; 79(9 Suppl):S72-75.

53. Hong JC, Kahan BD. Sirolimus-induced thrombocytopenia and leukopenia in renal transplant recipients: risk factors, incidence, progression and management. Transplantation 2000; 69(10):2085-90.

54. Morelon E, Stern M, Israel-Biet D et al. Characteristics of sirolimus-associated interstitial pneumonitis in renal transplant patients. Transplantation 2001; 72(5):787-90.

55. Goel M, Flechner SM, Zhou L et al. The influence of various maintenance immunosuppressive drugs on lymphocele formation and treatment after kidney transplantation. J Urol 2004; 171(5):1788-92.

56. Mahe E, Morelon E, Lechaton S et al. Cutaneous adverse events in renal transplant recipients receiving sirolimus-based therapy. Transplantation 2005; 79(4):476-82.

57. Morales JM, Wramner L, Kreis H et al. Sirolimus does not exhibit nephrotoxicity compared to cyclosporine in renal transplant recipients. Am J Transplant 2002; 2(5):436-42.

58. Simon JF, Swanson SJ, Agodoa LY et al. Induction sirolimus and delayed graft function after deceased donor kidney transplantation in the United States. Am J Nephrol 2004; 24(4):393-401.

59. Reynolds JC, Agodoa LY, Yuan CM et al. Thrombotic microangiopathy after renal transplantation in the United States. Am J Kidney Dis 2003; 42(5):1058-68.

60. Kaplan B, Meier-Kriesche HU, Napoli KL et al. The effects of relative timing of sirolimus and cyclosporine microemulsion formulation coadministration on the pharmacokinetics of each agent. Clin Pharmacol Ther 1998; 63(1):48-53.

61. McAlister VC, Mahalati K, Peltekian KM et al. A clinical pharmacokinetic study of tacrolimus and sirolimus combination immunosuppression comparing simultaneous to separated administration. Ther Drug Monit 2002; 24(3):346-50.

62. Mahalati K, Kahan BD. Clinical pharmacokinetics of sirolimus. Clin Pharmacokinet 2001; 40(8):573-85.

63. Zimmerman JJ. Exposure-response relationships and drug interactions of sirolimus. Aaps J 2004; 6(4):e28.

64. Lee VW, Chapman JR. Sirolimus: its role in nephrology. Nephrology (Carlton) 2005; 10(6):606-14.

65. Early sirolimus conversion is superior to late sirolimus conversion in reversing renal dysfunction in liver transplant recipients. Transplantation 2006; 82(1 Suppl 2):232.

66. Flechner SM, Kurian SM, Solez K et al. De novo kidney transplantation without use of calcineurin inhibitors preserves renal structure and function at two years. Am J Transplant 2004; 4(11):1776-85.

67. Alexander JW, Goodman HR, Cardi M et al. Simultaneous corticosteroid avoidance and calcineurin inhibitor minimization in renal transplantation. Transpl Int 2006; 19(4):295-302.

68. Majewski M, Korecka M, Kossev P et al. The immunosuppressive macrolide RAD inhibits growth of human Epstein-Barr virus-transformed B-lymphocytes in vitro and in vivo: A potential approach to prevention and treatment of posttransplant lymphoproliferative disorders. Proc Natl Acad Sci USA 2000; 97(8):4285-90.

69. Mathew T, Kreis H, Friend P. Two-year incidence of malignancy in sirolimus-treated renal transplant recipients: results from five multicenter studies. Clin Transplant 2004; 18(4):446-49.

70. Stallone G, Schena A, Infante B et al. Sirolimus for Kaposi's sarcoma in renal-transplant recipients. N Engl J Med 2005; 352(13):1317-23.

71. Lorber MI, Mulgaonkar S, Butt KM et al. Everolimus versus mycophenolate mofetil in the prevention of rejection in de novo renal transplant recipients: a 3-year randomized, multicenter, phase III study. Transplantation 2005; 80(2):244-52.

72. Eisen H. Long-term cardiovascular risk in transplantation—insights from the use of everolimus in heart transplantation. Nephrol Dial Transplant 2006; 21 Suppl 3:iii9-13.

73. Majewski M, Korecka M, Joergensen J et al. Immunosuppressive TOR kinase inhibitor everolimus (RAD) suppresses growth of cells derived from posttransplant lymphoproliferative disorder at allograft-protecting doses. Transplantation 2003; 75(10):1710-17.

74. Takeuchi H, Okuyama K, Konno O et al. Optimal dose and target trough level in cyclosporine and tacrolimus conversion in renal transplantation as evaluated by lymphocyte drug sensitivity and pharmacokinetic parameters. Transplant Proc 2005; 37(4):1745-47.

75. Karamperis N, Povlsen JV, Hojskov C et al. Comparison of the pharmacokinetics of tacrolimus and cyclosporine at equivalent molecular doses. Transplant Proc 2003; 35(4):1314-18.

76. Liptak P, Ivanyi B. Primer: Histopathology of calcineurin-inhibitor toxicity in renal allografts. Nat Clin Pract Nephrol 2006; 2(7):398-404.

77. Boots JM, Christiaans MH, van Hooff JP. Effect of immunosuppressive agents on long-term survival of renal transplant recipients: focus on the cardiovascular risk. Drugs 2004; 64(18):2047-73.

78. Vanrenterghem YF. Which calcineurin inhibitor is preferred in renal transplantation: tacrolimus or cyclosporine? Curr Opin Nephrol Hypertens 1999; 8(6):669-74.

79. Bechstein WO. Neurotoxicity of calcineurin inhibitors: impact and clinical management. Transpl Int 2000; 13(5):313-26.

80. Chiurchiu C, Ruggenenti P, Remuzzi G. Thrombotic microangiopathy in renal transplantation. Ann Transplant 2002; 7(1):28-33.

81. Tanabe K. Calcineurin inhibitors in renal transplantation: what is the best option? Drugs 2003; 63(15):1535-48.

82. Bartucci MR, Bayer L, Brooks BK et al. Conversion from Sandimmune to Neoral in organ transplant recipients. J Transpl Coord 1998; 8(4):227-233; quiz 234-25.

83. Curtis JJ, Lynn M, Jones PA. Neoral conversion from Sandimmune in maintenance renal transplant patients: an individualized approach. J Am Soc Nephrol 1998; 9(7):1293-300.

84. Hardinger KL, Koch MJ, Brennan DC. Current and future immunosuppressive strategies in renal transplantation. Pharmacotherapy 2004; 24(9):1159-76.

85. Citterio F. Evolution of the therapeutic drug monitoring of cyclosporine. Transplant Proc 2004; 36(2 Suppl):420S-25S.

86. Kahan BD. Therapeutic drug monitoring of cyclosporine: 20 years of progress. Transplant Proc 2004; 36(2 Suppl):378S-91S.

87. Formea CM, Evans CG, Karlix JL. Altered cytochrome p450 metabolism of calcineurin inhibitors: case report and review of the literature. Pharmacotherapy 2005; 25(7):1021-29.

88. Gaston RS. Current and evolving immunosuppressive regimens in kidney transplantation. Am J Kidney Dis 2006; 47(4 Suppl 2):S3-21.

89. Staatz CE, Tett SE. Clinical pharmacokinetics and pharmacodynamics of tacrolimus in solid organ transplantation. Clin Pharmacokinet 2004; 43(10):623-53.

90. Shaw LM, Holt DW, Keown P et al. Current opinions on therapeutic drug monitoring of immunosuppressive drugs. Clin Ther 1999; 21(10):1632-52.

91. Wilkinson A, Pham PT. Kidney dysfunction in the recipients of liver transplants. Liver Transpl 2005; 11(11 Suppl 2):S47-51.

92. Bestard O, Cruzado JM, Grinyo JM. Calcineurin-inhibitor-sparing immunosuppressive protocols. Transplant Proc 2005; 37(9):3729-32.

93. Egidi FM. Management of hyperglycaemia after pancreas transplantation: are new immunosuppressants the answer? Drugs 2005; 65(2):153-66.

94. Meiser B, Reichart B, Adamidis I et al. First experience with de novo calcineurin-inhibitor-free immunosuppression following cardiac transplantation. Am J Transplant 2005; 5(4 Pt 1):827-31.

95. Groetzner J, Wittwer T, Kaczmarek I et al. Conversion to sirolimus and mycophenolate can attenuate the progression of bronchiolitis obliterans syndrome and improves renal function after lung transplantation. Transplantation 2006; 81(3):355-60.

96. Ponticelli C. Generic cyclosporine: a word of caution. J Nephrol 2004; 17 Suppl 8:S20-24.

97. Taber DJ, Baillie GM, Ashcraft EE et al. Does bioequivalence between modified cyclosporine formulations translate into equal outcomes? Transplantation 2005; 80(11):1633-35.

98. Federal Food, Drug, Cosmetic Act 21 USC, 355 (j) & (B).

99. Birsan T, Dambrin C, Freitag DG et al. The novel calcineurin inhibitor ISA247: a more potent immunosuppressant than cyclosporine in vitro. Transpl Int 2005; 17(12):767-71.

100. Dumont FJ. ISAtx-247 (Isotechnika/Roche). Curr Opin Investig Drugs 2004; 5(5):542-50.

101. Wente MN, Sauer P, Mehrabi A et al. Review of the clinical experience with a modified release form of tacrolimus [FK506E (MR4)] in transplantation. Clin Transplant 2006; 20 Suppl 17:80-84.

102. Szczech LA, Berlin JA, Aradhye S et al. Effect of anti-lymphocyte induction therapy on renal allograft survival: a meta-analysis. J Am Soc Nephrol 1997; 8(11):1771-77.

103. Dharnidharka VR. Posttransplant lymphoproliferative disease: association with induction therapy? Drugs 2006; 66(4):429-38.

104. Fishman JA, Rubin RH. Infection in organ-transplant recipients. N Engl J Med 1998; 338(24):1741-51.

105. Norman DJ. Mechanisms of action and overview of OKT3. Ther Drug Monit 1995; 17(6):615-20.

106. Chatenoud L. Immunologic monitoring during OKT3 therapy. Clin Transplant 1993; 7(4 Pt 2):422-30.

107. Sgro C. Side-effects of a monoclonal antibody, muromonab CD3/orthoclone OKT3: bibliographic review. Toxicology 1995; 105(1):23-29.

108. DeVault GA, Jr. Kohan DE, Nelson EW et al. The effects of oral pentoxifylline on the cytokine release syndrome during inductive OKT3. Transplantation 1994; 57(4):532-40.

109. Thistlethwaite JR, Jr. Stuart JK, Mayes JT et al. Complications and monitoring of OKT3 therapy. Am J Kidney Dis 1988; 11(2):112-19.

110. Schroeder TJ, First MR, Hurtubise PE et al. Immunologic monitoring with Orthoclone OKT3 therapy. J Heart Transplant 1989; 8(5):371-80.

111. Vincenti F, Kirkman R, Light S et al. Interleukin-2-receptor blockade with daclizumab to prevent acute rejection in renal transplantation. Daclizumab Triple Therapy Study Group. N Engl J Med 1998; 338(3):161-65.

112. Nashan B, Moore R, Amlot P et al. Randomised trial of basiliximab versus placebo for control of acute cellular rejection in renal allograft recipients. CHIB 201 International Study Group. Lancet 1997; 350(9086):1193-98.

113. Kahan BD, Rajagopalan PR, Hall M. Reduction of the occurrence of acute cellular rejection among renal allograft recipients treated with basiliximab, a chimeric anti-interleukin-2-receptor monoclonal antibody. United States Simulect Renal Study Group. Transplantation 1999; 67(2):276-84.

114. Nashan B, Light S, Hardie IR et al. Reduction of acute renal allograft rejection by daclizumab. Daclizumab Double Therapy Study Group. Transplantation 1999; 67(1):110-15.

115. Baudouin V, Crusiaux A, Haddad E et al. Anaphylactic shock caused by immunoglobulin E sensitization after retreatment with the chimeric anti-interleukin-2 receptor monoclonal antibody basiliximab. Transplantation 2003; 76(3):459-63.

116. Lin M, Ming A, Zhao M. Two-dose basiliximab compared with two-dose daclizumab in renal transplantation: a clinical study. Clin Transplant 2006; 20(3):325-29.

117. ter Meulen CG, Jacobs CW, Wetzels JF et al. The fractional excretion of soluble interleukin-2 receptor-alpha is an excellent predictor of the interleukin-2 receptor-alpha status after treatment with daclizumab. Transplantation 2004; 77(2):281-86.

118. Kovarik JM, Kahan BD, Rajagopalan PR et al. Population pharmacokinetics and exposure-response relationships for basiliximab in kidney transplantation. The US Simulect Renal Transplant Study Group. Transplantation 1999; 68(9):1288-94.

119. Marino IR, Doria C, Scott VL et al. Efficacy and safety of basiliximab with a tacrolimus-based regimen in liver transplant recipients. Transplantation 2004; 78(6):886-91.

120. Hershberger RE, Starling RC, Eisen HJ et al. Daclizumab to prevent rejection after cardiac transplantation. N Engl J Med 2005; 352(26):2705-13.

121. Brock MV, Borja MC, Ferber L et al. Induction therapy in lung transplantation: a prospective, controlled clinical trial comparing OKT3, anti-thymocyte globulin and daclizumab. J Heart Lung Transplant 2001; 20(12):1282-90.

122. Burton CM, Andersen CB, Jensen AS et al. The incidence of acute cellular rejection after lung transplantation: a comparative study of anti-thymocyte globulin and daclizumab. J Heart Lung Transplant 2006; 25(6):638-47.

123. Hachem RR, Chakinala MM, Yusen RD et al. A comparison of basiliximab and anti-thymocyte globulin as induction agents after lung transplantation. J Heart Lung Transplant 2005; 24(9):1320-26.

124. Pham K, Kraft K, Thielke J et al. Limited-dose Daclizumab versus Basiliximab: a comparison of cost and efficacy in preventing acute rejection. Transplant Proc 2005; 37(2):899-902.

125. Washburn WK, Teperman LW, Heffron TG et al. A novel three-dose regimen of daclizumab in liver transplant recipients with hepatitis C: a pharmacokinetic and pharmacodynamic study. Liver Transpl 2006; 12(4):585-91.

126. Sellers MT, McGuire BM, Haustein SV et al. Two-dose daclizumab induction therapy in 209 liver transplants: a single-center analysis. Transplantation 2004; 78(8):1212-17.

127. Filipponi F, Callea F, Salizzoni M et al. Double-blind comparison of hepatitis C histological recurrence Rate in HCV+ Liver transplant recipients given basiliximab + steroids or basiliximab + placebo, in addition to cyclosporine and azathioprine. Transplantation 2004; 78(10):1488-95.

128. Vincenti F. Interleukin-2 receptor antagonists and aggressive steroid minimization strategies for kidney transplant patients. Transpl Int 2004; 17(8):395-401.

129. Nelson DR, Soldevila-Pico C, Reed A et al. Anti-interleukin-2 receptor therapy in combination with mycophenolate mofetil is associated with more severe hepatitis C recurrence after liver transplantation. Liver Transpl 2001; 7(12):1064-70.

130. Gaber AO, First MR, Tesi RJ et al. Results of the double-blind, randomized, multicenter, phase III clinical trial of Thymoglobulin versus Atgam in the treatment of acute graft rejection episodes after renal transplantation. Transplantation 1998; 66(1):29-37.

131. Norman DJ. Antilymphocyte antibodies in the treatment of allograft rejection: targets, mechanisms of action, monitoring and efficacy. Semin Nephrol 1992; 12(4):315-24.

132. Tanriover B, Chuang P, Fishbach B et al. Polyclonal antibody-induced serum sickness in renal transplant recipients: treatment with therapeutic plasma exchange. Transplantation 2005; 80(2):279-81.

133. Wiland AM, Fink JC, Philosophe B et al. Peripheral administration of thymoglobulin for induction therapy in pancreas transplantation. Transplant Proc 2001; 33(1-2):1910.

134. Marvin MR, Droogan C, Sawinski D et al. Administration of rabbit antithymocyte globulin (thymoglobulin) in ambulatory renal-transplant patients. Transplantation 2003; 75(4):488-89.

135. Regan J, Campbell K, van Smith L et al. Characterization of anti-thymoglobulin, anti-Atgam and anti-OKT3 IgG antibodies in human serum with an 11-min ELISA. Transpl Immunol 1997; 5(1):49-56.

136. Peddi VR, Bryant M, Roy-Chaudhury P et al. Safety, efficacy and cost analysis of thymoglobulin induction therapy with intermittent dosing based on CD3+ lymphocyte counts in kidney and kidney-pancreas transplant recipients. Transplantation 2002; 73(9):1514-18.

137. Hardinger KL, Schnitzler MA, Miller B et al. Five-year follow up of thymoglobulin versus ATGAM induction in adult renal transplantation. Transplantation 2004; 78(1):136-41.

138. Hardinger KL, Schnitzler MA, Koch MJ et al. Thymoglobulin induction is safe and effective in live-donor renal transplantation: a single center experience. Transplantation 2006; 81(9):1285-89.

139. Goggins WC, Pascual MA, Powelson JA et al. A prospective, randomized, clinical trial of intraoperative versus postoperative Thymoglobulin in adult cadaveric renal transplant recipients. Transplantation 2003; 76(5):798-802.

140. Uber WE, Uber LA, VanBakel AB et al. CD3 monitoring and thymoglobulin therapy in cardiac transplantation: clinical outcomes and pharmacoeconomic implications. Transplant Proc 2004; 36(10):3245-49.

141. Krasinskas AM, Kreisel D, Acker MA et al. CD3 monitoring of antithymocyte globulin therapy in thoracic organ transplantation. Transplantation 2002; 73(8):1339-41.

142. Bogetti D, Sankary HN, Jarzembowski TM et al. Thymoglobulin induction protects liver allografts from ischemia/reperfusion injury. Clin Transplant 2005; 19(4):507-11.

143. Trofe J, Stratta RJ, Egidi MF et al. Thymoglobulin for induction or rejection therapy in pancreas allograft recipients: a single centre experience. Clin Transplant 2002; 16 Suppl 7:34-44.

144. Lo A, Alloway RR. Strategies to reduce toxicities and improve outcomes in renal transplant recipients. Pharmacotherapy 2002; 22(3):316-28.

145. Everson GT, Trouillot T, Wachs M et al. Early steroid withdrawal in liver transplantation is safe and beneficial. Liver Transpl Surg 1999; 5 (4 Suppl 1):S48-57.

146. Keogh A, Macdonald P, Harvison A et al. Initial steroid-free versus steroid-based maintenance therapy and steroid withdrawal after heart transplantation: two views of the steroid question. J Heart Lung Transplant. 1992; 11(2 Pt 2):421-27.

147. Offermann G, Schwarz A, Krause PH. Long-term effects of steroid withdrawal in kidney transplantation. Transpl Int 1993; 6(5):290-92.

148. Ahsan N, Hricik D, Matas A et al. Prednisone withdrawal in kidney transplant recipients on cyclosporine and mycophenolate mofetil—a prospective randomized study. Steroid Withdrawal Study Group. Transplantation 1999; 68(12):1865-74.

149. Sinclair NR. Low-dose steroid therapy in cyclosporine-treated renal transplant recipients with well-functioning grafts. The Canadian Multicentre Transplant Study Group. Cmaj 1992; 147(5):645-57.

150. Woodle ES, Vincenti F, Lorber MI et al. A multicenter pilot study of early (4-day) steroid cessation in renal transplant recipients under simulect, tacrolimus and sirolimus. Am J Transplant 2005; 5(1):157-66.

151. Hricik DE, Knauss TC, Bodziak KA et al. Withdrawal of steroid therapy in African American kidney transplant recipients receiving sirolimus and tacrolimus. Transplantation 2003; 76(6):938-42.

152. Tan JY, Zhao N, Wu TX et al. Steroid withdrawal increases risk of acute rejection but reduces infection: a meta-analysis of 1681 cases in renal transplantation. Transplant Proc 2006; 38(7):2054-56.

153. Vincenti F, Ramos E, Brattstrom C et al. Multicenter trial exploring calcineurin inhibitors avoidance in renal transplantation. Transplantation 2001; 71(9):1282-87.

154. Oberbauer R, Segoloni G, Campistol JM et al. Early cyclosporine withdrawal from a sirolimus-based regimen results in better renal allograft survival and renal function at 48 months after transplantation. Transpl Int 2005; 18(1):22-28.

155. Bloom RD, Doyle AM. Kidney disease after heart and lung transplantation. Am J Transplant 2006; 6(4):671-79.

156. Mehrabi A, Fonouni H, Kashfi A et al. The role and value of sirolimus administration in kidney and liver transplantation. Clin Transplant 2006; 20 Suppl 17:30-43.

157. Wishart DS. Metabolomics: the principles and potential applications to transplantation. Am J Transplant 2005; 5(12):2814-20.

158. Kowalski RJ, Post DR, Mannon RB et al. Assessing relative risks of infection and rejection: a meta-analysis using an immune function assay. Transplantation 2006; 82(5):663-68.

159. Gibney EM, Cagle LR, Freed B et al. Detection of donor-specific antibodies using HLA-coated microspheres: another tool for kidney transplant risk stratification. Nephrol Dial Transplant 2006; 21(9):2625-29.

160. Hricik DE, Heeger PS. Minimization of immunosuppression in kidney transplantation. The need for immune monitoring. Transplantation 2001; 72(8 Suppl):S32-35.

161. Muthukumar T, Dadhania D, Ding R et al. Messenger RNA for FOXP3 in the urine of renal-allograft recipients. N Engl J Med 2005; 353(22):2342-51.

162. Li B, Hartono C, Ding R et al. Noninvasive diagnosis of renal-allograft rejection by measurement of messenger RNA for perforin and granzyme B in urine. N Engl J Med 2001; 344(13):947-54.

163. Becker YT, Samaniego-Picota M, Sollinger HW. The emerging role of rituximab in organ transplantation. Transpl Int 2006; 19(8):621-28.

164. Csapo Z, Benavides-Viveros C, Podder H et al. Campath-1H as rescue therapy for the treatment of acute rejection in kidney transplant patients. Transplant Proc 2005; 37(5):2032-36.

165. Magliocca JF, Knechtle SJ. The evolving role of alemtuzumab (Campath-1H) for immunosuppressive therapy in organ transplantation. Transpl Int 2006; 19(9):705-14.

166. Morris PJ, Russell NK. Alemtuzumab (Campath-1H): a systematic review in organ transplantation. Transplantation 2006; 81(10):1361-67.

167. Vincenti F, Larsen C, Durrbach A et al. Costimulation blockade with belatacept in renal transplantation. N Engl J Med 2005; 353(8):770-81.

168. Vincenti F. New clinical trials in kidney transplantation. Chronic biologicals: a new era of immunosuppression? Paper presented at: World Transplant Congress; 2006; Boston, Massachusetts.

169. Tedesco Silva H, Jr. Pinheiro Machado P, Rosso Felipe C et al. Immunotherapy for De Novo renal transplantation: what's in the pipeline? Drugs 2006; 66(13):1665-84.

170. Tan HP, Smaldone MC, Shapiro R. Immunosuppressive preconditioning or induction regimens : evidence to date. Drugs 2006; 66(12):1535-45.

# INDEX